Elements of Astrochemistry: Chemical History of Matter in the Universe

Elements of Astrochemistry: Chemical History of Matter in the Universe

Edited by:

Maria Emilova Velinova

ARCLER PRESS

www.arclerpress.com

Elements of Astrochemistry: Chemical History of Matter in the Universe

Maria Emilova Velinova

Arcler Press
224 Shoreacres Road
Burlington, ON L7L 2H2
Canada
www.arclerpress.com
Email: orders@arclereducation.com

ISBN: 978-1-77407-771-9 (Hardcover)

DECLARATION

ABOUT THE EDITOR

Maria Velinova is Ph.D. holder in Quantum chemistry at the University of Sofia since April 2012. Her major research experience is in the field of Computational Chemistry, especially in statistical mechanics methods applied to different sorts of biomolecules. Member of the Laboratory of Quantum and Computational Chemistry at the University of Sofia.

TABLE OF CONTENTS

Section-5: Laboratory-Based Astrochemistry

LIST OF CONTRIBUTORS

D. Shulyak
Max-Planck Institut für Sonnensystemforschung, Justus-von-Liebig-Weg 3, 37077 Göttingen, Germany

M. Rengel
Max-Planck Institut für Sonnensystemforschung, Justus-von-Liebig-Weg 3, 37077 Göttingen, Germany

A. Reiners
Institute for Astrophysics, Georg-August University, Friedrich-Hund-Platz 1, 37077 Göttingen, Germany

U. Seemann
Institute for Astrophysics, Georg-August University, Friedrich-Hund-Platz 1, 37077 Göttingen, Germany

F. Yan
Institute for Astrophysics, Georg-August University, Friedrich-Hund-Platz 1, 37077 Göttingen, Germany

B. M. Giuliano
Max-Planck-Institut für extraterrestrische Physik, Gießenbachstrasse 1, 85748 Garching, Germany

A. A. Gavdush
Prokhorov General Physics Institute of the Russian Academy of Sciences, 119991 Moscow, Russia
Bauman Moscow State Technical University, 105005 Moscow, Russia

B. Müller
Max-Planck-Institut für extraterrestrische Physik, Gießenbachstrasse 1, 85748 Garching, Germany

K. I. Zaytsev
Prokhorov General Physics Institute of the Russian Academy of Sciences, 119991 Moscow, Russia
Bauman Moscow State Technical University, 105005 Moscow, Russia

T. Grassi
Universitäts-Sternwarte München, Scheinerstr. 1, 81679 München, Germany
Excellence Cluster Origin and Structure of the Universe, Boltzmannstr. 2, 85748 Garching bei München, Germany

A. V. Ivlev
Max-Planck-Institut für extraterrestrische Physik, Gießenbachstrasse 1, 85748 Garching, Germany

M. E. Palumbo
INAF – Osservatorio Astrofisico di Catania, Via Santa Sofia 78, 95123 Catania, Italy

G. A. Baratta
INAF – Osservatorio Astrofisico di Catania, Via Santa Sofia 78, 95123 Catania, Italy

C. Scirè
INAF – Osservatorio Astrofisico di Catania, Via Santa Sofia 78, 95123 Catania, Italy

G. A. Komandin
Prokhorov General Physics Institute of the Russian Academy of Sciences, 119991 Moscow, Russia

S. O. Yurchenko
Bauman Moscow State Technical University, 105005 Moscow, Russia

P. Caselli
Max-Planck-Institut für extraterrestrische Physik, Gießenbachstrasse 1, 85748 Garching, Germany

Paolo Ciarcelluti
Web Institute of Physics, Via Fortore 3, 65015 Montesilvano, Italy

Masahiro Kawasaki
Institute for Cosmic Ray Research, The University of Tokyo, Kashiwa 277-8582, Japan
Kavli IPMU (WPI), UTIAS, The University of Tokyo, Kashiwa 277-8583, Japan

Kazunori Kohri
Theory Center, IPNS, KEK, Tsukuba 305-0801, Japan
The Graduate University of Advanced Studies, Tsukuba 305-0801, Japan
Rudolf Peierls Centre for Theoretical Physics, The University of Oxford,1 Keble Road, Oxford OX1 3NP, UK

Takeo Moroi
Department of Physics, The University of Tokyo, Tokyo 113-0033, Japan
Kavli IPMU (WPI), UTIAS, The University of Tokyo, Kashiwa 277-8583, Japan

Yoshitaro Takaesu
Department of Physics, The University of Tokyo, Tokyo 113-0033, Japan
Research Institute for Interdisciplinary Science, Okayama University,Okayama 700-8530, Japan

Naoki Yoshida
Kavli Institute for the Physics and Mathematics of the Universe, University of Tokyo, Kashiwa, Chiba 277-8583, Japan
Department of Physics, University of Tokyo, Bunkyo, Tokyo 113-0033, Japan

Takashi Hosokawa
Jet Propulsion Laboratory, California Institute of Technology, Pasadena, CA 91198, USA

Kazuyuki Omukai
Department of Physics, Kyoto University, Kyoto 606-8502, Japan

Maria Bergemann
Max Planck Institute for Astronomy, 69117 Heidelberg, Germany

Andrew J. Gallagher
Max Planck Institute for Astronomy, 69117 Heidelberg, Germany

Philipp Eitner
Max Planck Institute for Astronomy, 69117 Heidelberg, Germany
Ruprecht-Karls-Universität, Grabengasse 1, 69117 Heidelberg, Germany

Manuel Bautista
Department of Physics, Western Michigan University, Kalamazoo, Michigan 49008, USA

Remo Collet
Stellar Astrophysics Centre, Department of Physics and Astronomy, Aarhus University, 8000 Aarhus C, Denmark

Svetlana A. Yakovleva
Department of Theoretical Physics and Astronomy, Herzen University, St. Petersburg 191186, Russia

Anja Mayriedl
Montessori-Schule Dachau, Geschwister-Scholl-Str. 2, 85221 Dachau, Germany

Bertrand Plez
LUPM, UMR 5299, Université de Montpellier, CNRS, 34095 Montpellier, France

Mats Carlsson
Rosseland Centre for Solar Physics, University of Oslo, PO Box 1029 Blindern, 0315 Oslo, Norway
Institute of Theoretical Astrophysics, University of Oslo, PO Box 1029 Blindern, 0315 Oslo, Norway

Jorrit Leenaarts
Institute for Solar Physics, Department of Astronomy, Stockholm University, AlbaNova University Centre, 106 91 Stockholm, Sweden

Andrey K. Belyaev
Department of Theoretical Physics and Astronomy, Herzen University, St. Petersburg 191186, Russia

Camilla Hansen
Max Planck Institute for Astronomy, 69117 Heidelberg, Germany

Valentine Wakelam
Laboratoire d'astrophysique de Bordeaux, Univ. Bordeaux, CNRS, B18N, allée Geoffroy Saint-Hilaire, Pessac, 33615, France

Emeric Bron
Instituto de Ciencias de Materiales de Madrid (CSIC), Madrid, 28049, Spain
LERMA, Obs. de Paris, PSL Research University, CNRS, Sorbonne Universités, UPMC Univ. Paris 06, ENS, F-75005, France

Stephanie Cazaux
Faculty of Aerospace Engineering, Delft University of Technology, Delft, Netherlands
Leiden Observatory, Leiden University, P.O. Box 9513, Leiden, NL 2300 RA, Netherlands

Francois Dulieu
LERMA, Université de Cergy Pontoise, Sorbonne Universités, UPMC Univ. Paris 6, PSL Research University, Observatoire de Paris, UMR 8112 CNRS, 5 mail Gay Lussac 95000 Cergy Pontoise, France

Cécile Gry
CNRS, LAM, Laboratoire d'Astrophysique de Marseille, Aix Marseille Univ, Marseille, France

Pierre Guillard
Sorbonne Universités, UPMC Univ. Paris 6 & CNRS, UMR 7095, Institut d'Astrophysique de Paris, 98 bis bd Arago, Paris, 75014, France

Emilie Habart
Institut d'Astrophysique Spatiale, Univ. Paris-Sud & CNRS, Univ. Paris-Saclay - IAS, bâtiment 121, univ Paris-Sud, Orsay, 91405, France

Liv Hornekær
Dept. Physics and Astronomy, Aarhus University, Ny Munkegade 120, Aarhus C, 8000, Denmark

Sabine Morisset
Institut des Sciences Moléculaires d'Orsay, ISMO, CNRS, Université Paris-Sud, Université Paris Saclay, Orsay, F-91405, France

Gunnar Nyman
Department of Chemistry and Molecular Biology, University of Gothenburg, Gothenburg, SE 412 96, Sweden

Valerio Pirronello
Dipartimento di Fisica e Astronomia, Universitá di Catania, Via S. Sofia 64, Catania, 95123 Sicily, Italy

Stephen D. Price
Chemistry Department, University College London, 20 Gordon Street, London WC1H 0AJ, UK

Valeska Valdivia
Laboratoire AIM, Paris-Saclay, CEA/IRFU/DAp - CNRS, Université Paris Diderot, Gif-sur-Yvette Cedex, 91191, France

Gianfranco Vidali
201 Physics Bldg., Syracuse University, Syracuse, NY, 13244, USA

Naoki Watanabe
Institute of Low Temperature Science, Hokkaido University, Sapporo, Hokkaido, 060-0819, Japan

J. Chantzos
Center for Astrochemical Studies, Max-Planck-Institut für extraterrestrische Physik, Gießenbachstraße 1, 85748 Garching, Germany

V. M. Rivilla
INAF – Osservatorio Astrofisico di Arcetri, Largo Enrico Fermi 5, 50125 Firenze, Italy

A. Vasyunin
Ural Federal University, Ekaterinburg, Russia
Visiting Leading Researcher, Engineering Research Institute "Ventspils International Radio Astronomy Centre" of Ventspils University of Applied Sciences, Inženieru 101, Ventspils 3601, Latvia

E. Redaelli
Center for Astrochemical Studies, Max-Planck-Institut für extraterrestrische Physik, Gießenbachstraße 1, 85748 Garching, Germany

L. Bizzocchi
Center for Astrochemical Studies, Max-Planck-Institut für extraterrestrische Physik, Gießenbachstraße 1, 85748 Garching, Germany

F. Fontani
INAF – Osservatorio Astrofisico di Arcetri, Largo Enrico Fermi 5, 50125 Firenze, Italy

P. Caselli
Center for Astrochemical Studies, Max-Planck-Institut für extraterrestrische Physik, Gießenbachstraße 1, 85748 Garching, Germany

H.M. Cuppen
Institute for Molecules and Materials, Radboud University Nijmegen, Heyendaalseweg 135, 6525 AJ Nijmegen, The Netherlands

C. Walsh
Leiden Observatory, Leiden University, PO Box 9513, 2300 RA Leiden, The Netherlands
School of Physics and Astronomy, University of Leeds, Leeds LS2 9JT, UK

T. Lamberts
Institute for Molecules and Materials, Radboud University Nijmegen, Heyendaalseweg 135, 6525 AJ Nijmegen, The Netherlands
Computational Chemistry Group, Institute for Theoretical Chemistry, University of Stuttgart, Pfaffenwaldring 55, 70569 Stuttgart, Germany

D. Semenov
Max Planck Institute for Astronomy, Königstuhl 17, 69117 Heidelberg, Germany

R.T. Garrod
Depts. of Astronomy & Chemistry, University of Virginia, McCormick Road, PO Box 400319, Charlottesville, VA 22904, USA

E.M. Penteado
Institute for Molecules and Materials, Radboud University Nijmegen, Heyendaalseweg 135, 6525 AJ Nijmegen, The Netherlands

S. Ioppolo
Department of Physical Sciences, The Open University, Walton Hall, Milton Keynes MK7 6AA, UK

M. Minissale
Aix-Marseille Université, CNRS, PIIM, Marseille, France

T. Nguyen
LERMA, Université de Cergy Pontoise, Sorbonne Université, PSL Research University, Observatoire de Paris, UMR 8112 CNRS, 5 mail Gay Lussac, 95000 Cergy Pontoise, France

F. Dulieu
LERMA, Université de Cergy Pontoise, Sorbonne Université, PSL Research University, Observatoire de Paris, UMR 8112 CNRS, 5 mail Gay Lussac, 95000 Cergy Pontoise, France

M. Minissale
Aix-Marseille Université, CNRS, PIIM, Marseille, France

T. Nguyen
LERMA, Université de Cergy Pontoise, Sorbonne Université, PSL Research University, Observatoire de Paris, UMR 8112 CNRS, 5 mail Gay Lussac, 95000 Cergy Pontoise, France

F. Dulieu
LERMA, Université de Cergy Pontoise, Sorbonne Université, PSL Research University, Observatoire de Paris, UMR 8112 CNRS, 5 mail Gay Lussac, 95000 Cergy Pontoise, France

Poorna Srinivasan
Institute of Meteoritics, University of New Mexico, Albuquerque, NM 87131, USA.
Department of Earth and Planetary Sciences, University of New Mexico, Albuquerque, NM 87131, USA.

Daniel R. Dunlap
Center for Meteorite Studies, School of Earth and Space Exploration, Arizona State University, Tempe, AZ 85287, USA.

Carl B. Agee
Institute of Meteoritics, University of New Mexico, Albuquerque, NM 87131, USA.
Department of Earth and Planetary Sciences, University of New Mexico, Albuquerque, NM 87131, USA.

Meenakshi Wadhwa
Center for Meteorite Studies, School of Earth and Space Exploration, Arizona State University, Tempe, AZ 85287, USA.

Daniel Coleff
Jacobs Technology, NASA Johnson Space Center, Mail Code XI3, 2101 NASA Parkway, Houston, TX 77058, USA.

Karen Ziegler
Institute of Meteoritics, University of New Mexico, Albuquerque, NM 87131, USA.
Department of Earth and Planetary Sciences, University of New Mexico, Albuquerque, NM 87131, USA.

Ryan Zeigler
NASA Johnson Space Center, Mail Code XI2, 2101 NASA Parkway, Houston, TX 77058, USA.

Francis M. McCubbin
NASA Johnson Space Center, Mail Code XI2, 2101 NASA Parkway, Houston, TX 77058, USA.

LIST OF ABBREVIATIONS

ALMA	Atacama Large Millimeter/submillimeter Array
ASW	Amorphous Solid Water
BSE	bulk silicate Earth
BWO	Backward-wave oscillator
CAS	Center for Astrochemical Studies
CDM	cold dark matter
CMB	cosmic microwave background
CNSA	China National Space Administration
ESO	European Southern Observatory
FTIR	Fourier transform infrared
HJ	hot Jupiters
IGRINS	Immersion GRating INfrared Spectrograph
ISO	Infrared Space Observatory
LH	Langmuir-Hinshelwood
LROC	Lunar Reconnaissance Orbiter Camera
LSS	large scale structure
MCMC	Markov chain Monte Carlo
NAC	Narrow Angle Camera
NLTE	non-local thermodynamic equilibrium
NOEMA	Northern Extended Millimeter Array
OPR	ortho-para ratio
PAH	polycyclic aromatic hydrocarbon
PDRs	Photo-dissociation regions
QMS	quadrupole mass spectrometer
TPD	temperature-programmed desorption

PREFACE

The research field of Astrochemistry occupies to understand the formation, destruction, and survival of molecules in astrophysical environments, where it takes place the synthesis of the molecules via several kinds of gas-phase reactions, and reactions on dust-grain surfaces. In particular, one of the main problems of Astrochemistry is unraveling the chemical and physical properties of molecular clouds, in which the star and planet formation process occurs. This book describes the origin and evolution of atomic matter in the Universe, considering early Universe, interstellar regions, and the solar system.

Elements of Astrochemistry: Chemical History of Matter in the Universe book begins Section 1 which focuses on the study of the Universe by spectroscopic observations from space and ground. In particular, it discusses remote sensing of exoplanetary atmospheres with ground-based high-resolution near-infrared spectroscopy, broadband spectroscopy of astrophysical ice analogs, and X-IFU synthetic observations of galaxy clusters.

Section 2 deals with the chemical history of the very early universe to the formation of first atoms. In detail, it examines a detailed study of the primordial nucleosynthesis in the presence of mirror dark matter, and of the effects of long-lived massive particles, which decayed during that epoch. Moreover, it discusses the observational constraints on the possibility of heavy element production beyond 7Li in the early universe.

Section 3 treats of the creation of the higher elements in the heart of the stars, by starting to discuss the first generation of stars, black holes, and galaxies. Moreover, it presents a non-local thermodynamic equilibrium model for the formation of Manganese in late-type stars and a study of solar cycle correlation of coronal element abundances in Sun-as-a-star observations. Lastly, it examines the causes of the abundance of live 244Pu in deep-sea reservoirs on Earth.

Section 4 reviews the interstellar chemistry from the viewpoints of theory, experiments, models and observations, ranging the formation of molecular hydrogen, silicates, large polycyclic aromatic hydrocarbons and fullerenes, and isocyanogen. Moreover, it examines the interstellar oxygen depletion problem and the nucleobase synthesis in interstellar ices.

Section 5 provides some examples of laboratory-based astrochemistry, i.e., the experimental study of the penetration of oxygen and deuterium atoms into porous water ice, and the measurement of total electron ionization cross-sections for molecules.

Finally, the last Section 6 focuses on the evolutionary history of the moon and the inner solar system, and their Silica-rich volcanism.

SECTION 1: THE MEASUREMENT OF THE UNIVERSE BY SPECTROSCOPY

1

Remote Sensing of Exoplanetary Atmospheres with Ground-Based High-Resolution Near-Infrared Spectroscopy

D. Shulyak[1], M. Rengel[1], A. Reiners[2], U. Seemann[2], and F. Yan[2]

[1] Max-Planck Institut für Sonnensystemforschung, Justus-von-Liebig-Weg 3, 37077 Göttingen, Germany

[2] Institute for Astrophysics, Georg-August University, Friedrich-Hund-Platz 1, 37077 Göttingen, Germany

ABSTRACT

Context. Thanks to the advances in modern instrumentation we have learned about many exoplanets that span a wide range of masses and composition. Studying their atmospheres provides insight into planetary origin, evolution, dynamics, and habitability. Present and future observing facilities will address these important topics in great detail by using more precise observations, high-resolution spectroscopy, and improved analysis methods.

Citation: D. Shulyak, M. Rengel, A. Reiners, U. Seemann and F. Yan "Remote sensing of exoplanetary atmospheres with ground-based high-resolution near-infrared spectroscopy" A&A, 629 (2019) A109 DOI: https://doi.org/10.1051/0004-6361/201935691

 Open Access funding provided by Max Planck Society.

Aims. We investigate the feasibility of retrieving the vertical temperature distribution and molecular number densities from expected exoplanet spectra in the near-infrared. We use the test case of the CRIRES+ instrument at the Very Large Telescope which will operate in the near-infrared between 1 and 5 μm and resolving powers of R = 100 000 and R = 50 000. We also determine the optimal wavelength coverage and observational strategies for increasing accuracy in the retrievals.

Methods. We used the optimal estimation approach to retrieve the atmospheric parameters from the simulated emission observations of the hot Jupiter HD 189733b. The radiative transfer forward model is calculated using a public version of the τ-REx software package. Results. Our simulations show that we can retrieve accurate temperature distribution in a very wide range of atmospheric pressures between 1 bar and 10−6 bar depending on the chosen spectral region. Retrieving molecular mixing ratios is very challenging, but a simultaneous observations in two separate infrared regions around 1.6 and 2.3 μm helps to obtain accurate estimates; the exoplanetary spectra must be of relatively high signal-to-noise ratio S/N > 10, while the temperature can already be derived accurately with the lowest value that we considered in this study (S/N = 5).

Conclusions. The results of our study suggest that high-resolution near-infrared spectroscopy is a powerful tool for studying exoplanet atmospheres because numerous lines of different molecules can be analyzed simultaneously. Instruments similar to CRIRES+ will provide data for detailed retrieval and will provide new important constraints on the atmospheric chemistry and physics.

Keywords. planets and satellites: atmospheres – planets and satellites: individual: HD 189733 b – radiative transfer – methods: numerical

INTRODUCTION

Studying exoplanetary atmospheres is a key to understanding physicochemical processes, origin and evolution paths, and habitability conditions of these distant worlds. This is usually done with the help of spectroscopic and/or (spectro-)photometric observations from space and from the ground. However, obtaining radiation from an exoplanetary atmosphere at an accuracy sufficient for its detailed investigation is a difficult task due to the large distance to even nearby exoplanets, which results in a very weak photon flux reaching the top of the Earth's atmosphere. Photometric observations are the more efficient than spectroscopic observations in terms of the time needed to achieve a required signal-to-noise ratio (S/N) for the exoplanetary signal. By using photometric observations obtained in a wide wavelength range (which may cover dozens of microns in the infrared domain) it is possible to constrain the temperature stratification and mixing ratios of most abundant molecules in the atmospheres of exoplanets. So far, the atmospheres of hot Jupiters (HJ) have been a subject of intense investigations (e.g., Brogi & Line 2019; Sing et al. 2016; Waldmann et al. 2015a; Schwarz et al. 2015; Line et al. 2013; Lee et al. 2012). These gas giants have masses between 0.5 and 13 MJ and short rotation periods P < 10 days (Wang et al. 2015) so that their atmospheres are hot due to a close distance to their parent stars (Winn et al. 2010). This results in strong irradiation from their atmospheres that is much more easy to detect than the atmospheres of similar exoplanets that are at larger distances from their stars. To date, there are about a dozen HJs that have been investigated using photometric observations from different space missions (e.g., Sing et al. 2016). Spectroscopic techniques, on the other hand, are capable of resolving individual spectral lines of atmospheric molecules, but require much longer observing times and are usually limited to a narrow wavelength range compared to photometry. Nevertheless, modern high-resolution spectroscopy has been successfully used to detect the presence of many molecules in atmospheres of different types of exoplanets via cross-correlation techniques (e.g., Snellen et al. 2010; de Kok et al. 2014; Brogi et al. 2014, 2016; Kreidberg et al. 2015). Unfortunately, extracting profiles of individual molecular lines from available spectroscopic observations remains very challenging due to the high noise level, telluric removal problems, and purely instrumental effects. Recently, Brogi & Line (2019) showed that the cross-correlation technique can be used to asses the temperature structure in the atmospheres of HJs via the differential analysis of weak and strong features in the retrieved exoplanet spectra without the need to resolve profiles of individual lines (see, e.g., Schwarz et al. 2015).

However, only very limited information can be retrieved and it is still not possible to obtain robust results, for example due to the quality of the observed spectrum. This is why most past and present studies utilize spectroscopy to detect the presence of molecules, but they do not attempt to study atmospheric temperature stratification and accurate mixing ratios. The advantage of spectroscopy in providing accurate estimates of molecular mixing ratios is currently limited by the capabilities of the available instruments. In particular, the quality of the observed spectrum is often not good enough to attempt studying atmospheric temperature stratification and accurate mixing ratios. However, with the advent of future instrumentation it will be possible to study the accurate shapes of atmospheric lines on diverse exoplanetary atmospheres. This will open a way to constrain atmospheric chemistry, global circulation (winds), and signatures of trace gases produced by possible biological activity. Motivated by recent progresses in exoplanetary science and advances in instrumentation, in this work we investigate the potential of applying very high-resolution spectroscopy to the study of HJ atmospheres. We focus our research on the CRyogenic high-resolution IR Echelle Spectrometer (CRIRES+) on the Very Large Telescope scheduled for the end of 2019 at the European Southern Observatory (ESO) (Dorn et al. 2014; Follert et al. 2014). CRIRES+ is an upgrade of the wellknown old CRIRES; it has improved efficiency thanks to new detectors, polarimetric capability, and an order of magnitude larger wavelength coverage that can be achieved in singlesetting observation thanks to a new cross-disperser. In particular, this improvement is essential for exoplanetary research because it opens a way to observe many molecular features and to study atmospheric physics in great detail. Similar to the previous version of the instrument, CRIRES+ will operate between 0.9 and 5.3 μm with a highest achievable resolving power of $\lambda/\Delta\lambda = R = 100\ 0001$. Our main goal was to test the near-infrared wavelength domain at very high spectral resolution for the retrieval of atmospheric temperature profiles and mixing ratios of molecular species using CRIRES+ as a test case. The key difference between our research and similar projects is that our aim was to retrieve assumption-free temperature profiles and that we used different near-infrared spectral regions that contain numerous lines of atmospheric molecules expected to be observed with CRIRES+. We simulated observations at different wavelength regions and their combinations, spectral resolutions, and S/N to test our ability to derive accurate temperatures and chemical compositions of HJ atmospheres. We made predictions to identify observational strategies and potential targets to study exoplanet atmospheres with high-resolution spectroscopy. Finally, we note that our results can also be applied to any other present or future instrument with capabilities similar to CRIRES+.

METHODS

Estimating the state of the atmosphere

In our retrieval approach we used an algorithm based on optimal estimation (OE; Rodgers 1976). The main idea of OE is to derive a maximum a posteriori information about parameters that define the state of the system under investigation. The OE approach finds an optimal solution that minimizes the difference between the model and observations under an additional constraint, which is our knowledge of a priori parameter values and their errors. Following Rodgers (2000) we solve the OE problem for the state vector of the system

$$x_{i+1} = x_i + \left[(1+\gamma)S_a^{-1} + K_i^T S_y^{-1} K_i\right]^{-1} \times \left[K_i^T S_y^{-1}(y - y_i) + S_a^{-1}(x_a - x_i)\right],$$

where x_{i+1} and xi are respectively the new and old solutions for the state vector at iteration i + 1 and i, x_a and Sa are the a priori vector and its covariance matrix, y and S_y are the measurements and measurement errors, and y_i and K_i are respectively the model prediction for the state vector xi and the corresponding Jacobian matrix calculated at x_i . An adjustable parameter (γ) is used to fine-tune the balance between the measurement and the a priori constraint (Rodgers 2000; Irwin et al. 2008). We searched for the solution that optimizes the total cost function φ that can be written in the form

$$\phi = (y - y_i)^T S_y^{-1}(y - y_i) + (x_i - x_a)^T S_a^{-1}(x_i - x_a), \quad (2)$$

where the first term is the usual χ^2 computed as a weighted difference between the model and observations, while the second term accounts for the deviation of the state vector from its a priori assumption. After the solution has converged, the final errors on the state vector are

$$S_x = \left(S_a^{-1} + K_i^T S_y^{-1} K_i\right)^{-1}. \quad (3)$$

It is seen that the obtained solution depends on how accurately we know our a priori. By setting large errors on the a priori, the OE eventually turns to a regular nonlinear χ^2 - minimization problem. In the case of solar system planets, the a priori values of physical parameters such as temperature stratification and surface pressures can be known from in situ measurements, if available. This helps to substantially narrow down the parameter range and the application of OE provides robust solutions with realistic error estimates (Hartogh et al. 2010; Irwin et al. 2008; Rengel et al. 2008). In the case of extrasolar planets, little to nothing is usually known about their physical and chemical structures. Nevertheless, even in these cases it is possible to build initial constraints upon some simplistic analytical models and to estimate realistic error bars a priori. Because of its relatively fast computational performance compared to other methods, such as the Markov chain Monte Carlo (MCMC) algorithm, the OE method was successfully used to study the atmospheres of exoplanets (see, e.g., Lee et al. 2012, 2014; Line et al. 2012, 2013; Barstow et al. 2013, 2014, 2017). Like every retrieval technique, OE has its own advantages and weaknesses. An obvious limitation is the use of a priori information, which penalizes the solution toward the one that provides the best balance between the observed and a priori uncertainties. Because the inverse problem we are dealing with is ill posed, there can often be a variety of solutions that provide the same good fit but very different values of fit parameters. This happens if, for example, some of the parameters are strongly correlated and/or the data quality is poor. In these cases using the a priori constraints helps to keep retrievals from finding the solution with the lowest χ^2 but not expected from the physical point of view (i.e., strong fluctuations in temperature profile, molecular mixing ratios that are too far from the predictions of chemical models, among others.). Unfortunately, in the case of exoplanets the use of inappropriate a priori information can lead retrievals to fall into local minima rather than finding the true solutions. In the OE method this difficulty is easy to overcome by choosing large enough errors on (poorly known) a priori parameters whose influence on the final solution is thus substantially reduced. Usually, a family of retrievals with different sets of initial guesses and a priori errors are used to ensure that the found solution is robust (see, e.g., Lee et al. 2012). Another limitation of the OE method is the assumption that the distribution of the a posteriori parameters are Gaussian. This is usually the case for data that is of high spectral resolution and S/N, which is satisfied in most of our retrievals presented below. When the data is of low quality the usual workaround is again to perform retrievals with multiple initial guesses to ensure that the global minimum is found. The relative comparison of different retrieval methods can be found in Line et al. (2013). In our work we do not favor any of the retrieval methods specifically. The choice of the retrieval approach normally depends on the goals of the particular investigation, quality of the available data, number of free parameters, available computing resources, among others. Here we chose the OE method because (1) it has long been used for the analysis of high-resolution spectroscopic data of planets in our solar system, which means that the expertise can be easily shared between research fields; (2) future instruments will eventually provide us with high-resolution data for the brightest exoplanets, and this data will be of better quality compared to what we have now, thus allowing the OE method to use all its advantages; and (3) we carried out assumption-free retrievals of temperature distribution, which means that the local value of temperature in each atmospheric layer is a free parameter in our approach; because the atmosphere is split into several tens of layers, this results in too many free parameters to be treated with purely Bayesian approaches.

Forward model

In order to use the OE method, an appropriate radiative transfer forward model is needed that computes the outgoing radiation from the visible planetary surface for a given set of free parameters. We did this by employing the Tau Retrieval for Exoplanets (τ-REx) software package (Waldmann et al. 2015b,a). The τ-REx forward model uses up-to-date molecular cross sections based on line lists provided by the EXOMOL2 project (Tennyson & Yurchenko 2012) and HITEMP (Rothman et al. 2010). These cross sections are pre-computed on a grid of temperature and pressure pairs and are stored in binary opacity tables that are available for a number of spectral resolutions. The continuum opacity includes Rayleigh scattering on molecules and collisionaly induced absorption due to H2-H2 and H2-He either after Abel et al. (2011, 2012) or Borysow et al. (2001); Borysow (2002) and Borysow & Frommhold (1989), respectively. The original line opacity tables contain high-resolution cross sections for each molecular species and for a set of temperature and pressure pairs. Thus, we optimized the public version of τ-REx3 to compute spectra in wide wavelength intervals and with very high spectral resolution R = 100 000 expected for CRIRES+ via the more efficient usage of opacity tables inside the code. Finally, we adapted our retrieval code for multiprocessing which is essential for the analysis of altitude dependent quantities (e.g., T, mixing ratios, winds) when the number of free parameters can be very high. In order to keep solutions stable and to ensure that the retrieved altitude dependent quantities (e.g., temperature) are smooth, it is usually assumed that these quantities are correlated with a characteristic correlation length l_{corr} expressed in pressure scale heights, for example. Then, the off-diagonal elements of the a priori co-variance matrix can be written as

$$S_{ij} = \sqrt{S_{ii}S_{jj}}\exp\left(-\frac{|\ln p_i/p_j|}{l_{corr}}\right), \tag{4}$$

where p_i and p_j are the pressures at atmospheric layers i and j, respectively, and l_{corr} is the number of pressure scale heights within which the correlation between layers drops by a factor of e. We found that a value of l_{corr} = 1.5

provides a necessary amount of smoothing without losing the physical information that we retrieve, and thus we adopted this value in all our retrievals (the same value of l_{corr} was also used in Irwin et al. 2008)

If no correlation between atmospheric layers is used then the inverse of co-variance matrices can be analytically computed. To the contrary, when the correlation length is non-zero the Sa matrix contains off-diagonal elements whose values can differ by many orders of magnitude which leads rounding-off errors in the numerical matrix inversion. In order to reduce numerical problems we normalize the state vector co-variance matrix Sa by its diagonal elements and re-normalize obtained solution in Eq. (2) accordingly. In our retrievals we make no assumptions about the altitude dependence of temperature profile. However, we assumed constant mixing ratios of molecular species. We do this to be consistent with the original work by Lee et al. (2012) (see next section). This also decreases the number of free parameters in our model and significantly reduces calculation time.

The final set of free parameters in our retrieval model includes 50 temperature values corresponding to different altitude levels, four molecular species (H_2O, CO, CO_2, CH_4), and a continuum scaling factor for each wavelength region that we investigated. The scaling factors account for any possible normalization inaccuracies between observed and predicted spectra. This adjustment will be needed because after the reduction each CRIRES+ spectra will be normalized to an arbitrary chosen level (e.g., pseudo continuum). In this work we make no attempts to simulate these very complicated reduction steps that should be a subject of a separate investigation. Instead, our simulations represent an ideal case where we do not consider all instrument systematics and data reduction inaccuracies. Thus, we normalized the observed spectrum to some arbitrary value (mean flux level in our case) and did the same with the fluxes computed with our forward model. The additional scaling factor is then applied to our forward model to account for any remaining mismatch between the observed and modeled spectra. Clearly, these scaling factors should be very close to unity in the case of accurate retrieval because the observed and predicted spectra were computed with the same code. Still, these additional scaling factors will be needed in retrievals performed on real data and we include them in our model.

In this work we concentrate only on emission spectroscopy. We do this because emission spectra are formed in a very wide range of atmospheric temperatures and pressures, while transmission spectroscopy (performed when a planet transits in front of the host star) senses mainly regions of high altitudes. In addition, the duration of transits of all potential targets are shorter than off-transit times when the day side of the planet is visible. This makes it easier to observe emission spectra with a required noise level.

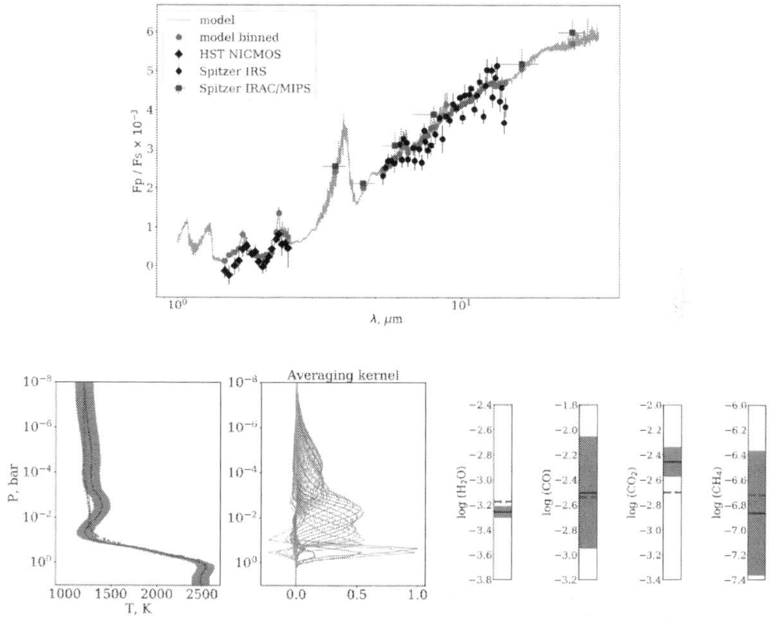

Figure. 1. Top panel: observed day side fluxes of HD 189733b and our best fit model predictions (see legend in the plot). Bottom panels: retrieved T-P profile along with corresponding averaging kernels and mixing ratios of molecular species. We used the T-P profile of Lee et al. (2012) a priori (shown as red crosses). Our best fit model is shown with a solid blue line and the shaded area represents 1σ error bars. The assumed a priori are shown as a green dashed line (which coincides with the red crosses in this particular case). The same color-coding is used for the four side plots of mixing ratios (on the right).

VALIDATION OF THE METHOD

Before addressing the main goal of our work, we validated our method against some representative examples. To do this we chose a test case of a well-studied exoplanet, HD 189733b. This HJ transits a young main sequence star of spectral type K0 at a distance of about 19.3 pc (Bouchy et al. 2005). The atmospheric structure of HD 189733b was studied by different groups and with different methods in the past (see, e.g., Lee et al.

2012; Waldmann et al. 2015a; Brogi & Line 2019). Figure 1 summarizes the retrieved temperature and mixing ratio of H_2O, CO, CO_2, and CH_4 in the atmosphere of HD 189733b using photometric observations obtained with different space missions (see figure legend). We used results of Lee et al. (2012) as our a priori and assumed 100 K uncertainties for our initial temperature profile. We note that here and throughout the paper our a priori is the same as an initial guess in our iterative retrieval approach, although in a more general case these two can differ. The analysis of corresponding averaging kernels for the temperature distribution (second plot in the bottom panel of Fig. 1) shows that the available observations probe temperature structures in a wide range of atmospheric depths between 10^{-6} and 1 bar. The averaging kernels describe a response of the retrieval to a small perturbation in temperature at each atmospheric depth (Rodgers 2000), and thus characterize the contribution of each depth to the final result. The zero values for averaging kernels correspond to the case when no information can be retrieved from corresponding depths and they are nonzero otherwise. It is seen that averaging kernels peak at appropriate atmospheric depths between 10^{-6} and 1 bar, thus indicating that our retrievals are robust at these altitudes; they approach zero outside that range. Figure 1 shows that we retrieved parameters in agreement with original work by Lee et al. (2012). We also confirm the high content of water in the atmosphere of HD 198733 b, and the general shape of the temperature distribution. Finally, the concentrations of CO_2 and CH_4 are badly constrained and we could not measure their content with any initial guess assumed, again in agreement with Lee et al. (2012).

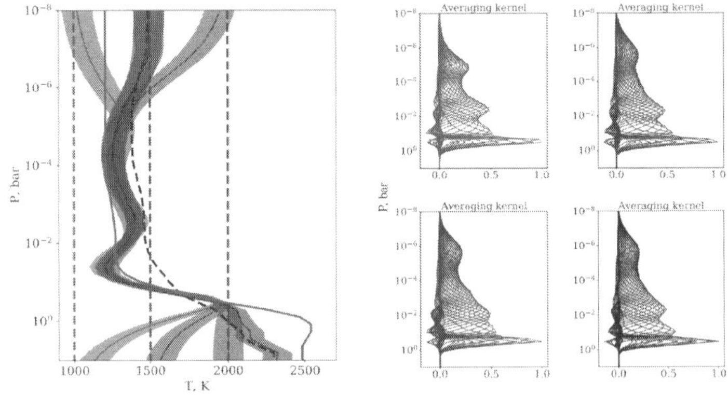

Figure. 2. Left panel: retrieved temperature distribution assuming four different initial guesses: three homogeneous at 1000, 1500, and 2000 K (vertical dashed lines), and the one from Waldmann et al. (2015a) (blue dashed line). The retrieved best fit profiles are shown with solid lines along with corresponding error bars as shaded areas. The profile from Lee et al. (2012) is shown as a solid red line. Right panels: averaging kernels for the four retrieved temperature distributions (from left to right and from top to bottom, color-coded accordingly).

To check the robustness of our results we performed four retrievals with different initial temperature profiles to investigate the sensitivity of our solutions to the initial guess. The result are presented in Fig. 2 where we show retrieved temperature profiles assuming four different initial guesses. In all cases we constrain very similar temperatures which confirms the robustness of our retrieval approach. In the lower and upper atmosphere our solution remains unconstrained because the observations do not sense these altitudes, as can be seen from the right panel of Fig. 2 which shows corresponding averaging kernels for each of the retrieved temperature distribution

Finally, we note that our temperature distribution and that published in Lee et al. (2012) are noticeably different from the temperature found by Waldmann et al. (2015a) (blue dashed line in Fig. 2). The largest deviation is found in the region of the temperature drop around 0.1 bar. This is likely because of the different opacity tables used in the old version of τ-REx (I. Waldmann, priv. comm.). We conclude that our retrieval method is accurate and robust, and that it can be used for atmospheric studies

RESULTS

Compared to its old version, the essential new capability of CRIRES+ is the increased wavelength coverage when observing with a single setting. In one shot, the instrument is able to cover a region corresponding to about one spectral band (i.e., ten times more than the old CRIRES). This gives us the possibility to study many spectral lines simultaneously with our retrieval method and to increase the amount of information that we can learn about exo-atmospheres. In what follows we study several cases of simulated observations at different spectral bands, their combinations, spectral resolving powers, and S/N values. We chose to use five different spectral regions that are free from strong atmospheric telluric absorption and that contain molecular bands of main molecules found in atmospheres of HJs: H2O, CO, CO_2, and CH_4. These regions are: 1.50–1.70, 2.10–2.28, 2.28–2.38, 3.80–4.00, and 4.80–5.00 μm. Each of these regions can be observed with single CRIRES+ setting; however, some gaps between spectral orders within each setting are expected. For each of these regions we computed the spectra of HD 189733b assuming spectral resolutions of R = 100 000 and R = 50 000 (corresponding to slit widths of 4000 and 1000, respectively) and four S/N values of 5, 10, 25, and 50, respectively. The S/N values are given per spectral dispersion element (1 pixel), which would be obtained in real observations after integrating the spectrum profile along the spatial direction on the CCD. We note that the S/N of the planetary spectrum is obtained from the S/N of the stellar spectrum by taking into account a characteristic planet-to-star flux ratio S /N(planet) = F_p/F_s ·S /N(star). The values of S/N(star) can be directly computed using exposure time calculators, for example, as we discuss below.

In order to obtain accurate results, we performed retrievals with different initial guesses and their uncertainties. We used three starting guesses for molecular volume mixing ratios (VMRs) of ±0.5 and −1.0 dex from the true solution. For each of the initial guess we tried uncertainties of 0.5 and 1.0 dex, respectively. By choosing such large uncertainties on the VMRs we avoid the problem of falling into local minima, as was discussed in the previous section. We found that we always converge to nearly the same solution whether 0.5 or 1.0 dex error bars were assumed. However, this was not always the case, especially for data of low S/N, and we discuss these cases below. For the temperature retrievals we tried different uncertainties between 100 and 500 K and found that the value of 200 K allows us to retrieve a temperature structure that is smooth and very close to the true one. Choosing too high temperature errors causes strong nonphysical fluctuations in the finally retrieved temperature distribution and we do not discuss this solution further.

Single spectral region retrievals

In Fig. 3 we show the results of retrievals from five spectral regions and S /N = 10 (the results for other S/N values are provided in Figs. A.1–A.3). The data was simulated using temperature structure and mixing ratios of HD 189733 b as derived in Lee et al. (2012) (red dashed lines in Fig. 3). In all cases we start from an isothermal temperature distribution with T = 1500 K. In this way we can better study the ability of the EO method to recover the true temperature distribution at different altitudes. The initial values for mixing ratios were taken to be −0.5 dex from the true ones with the uncertainty of 1 dex. It is seen that the spectra of HJs in the wavelength range covered by CRIRES+ is sensitive to a wide range of atmospheric pressures between 1 and 10−4 bar. At these altitudes we recover temperatures very close to the true solution. However, everything that is outside this altitude range remains unconstrained due to the lack of information provided by the observed spectra.

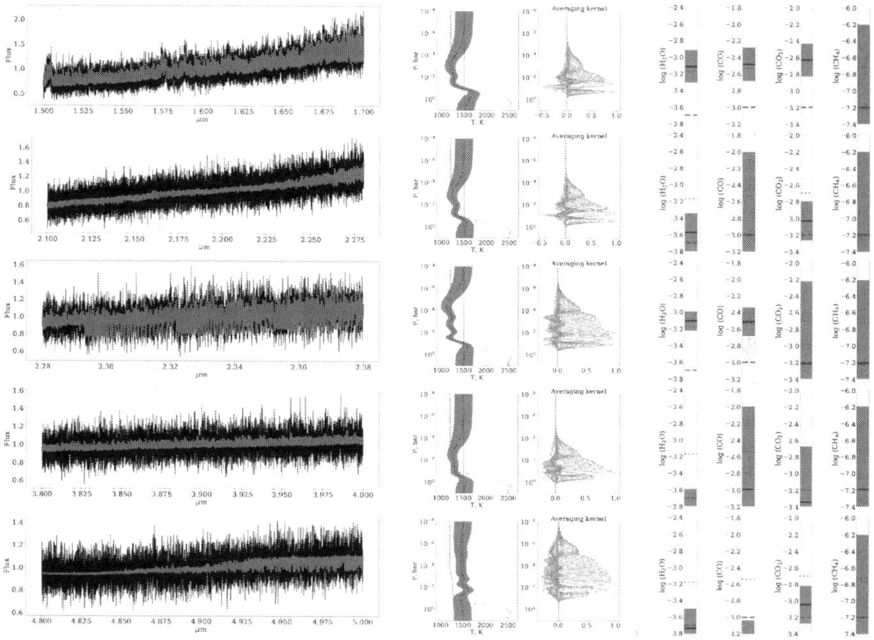

Figure. 3. Retrieved temperature and mixing ratios from five spectral regions with R = 100 000 and S /N = 10. In each panel, the first plot compares the best fit predicted spectrum (solid red line) with the simulated observations (black). The second plot is the temperature distribution as a function of atmospheric pressure (solid blue line) and with error bars shown as shaded areas. The red crosses and green dashed line are the true solution and initial guess, respectively. The third plot shows averaging kernels derived for the temperature distribution. Here averaging kernels for different atmospheric depths are randomly color-coded for clarity. The next four plots are the values of the retrieved mixing ratios of four molecular species (color-coded as in the first plot). The mixing ratios were assumed to be constant with atmospheric depth, and we show their values on the y-axis

The typical temperature errors in the pressure range between 1 and 10−4 bar is 120 K for S /N = 5, 100 K for S /N = 10, 90 K for S /N = 25, and 80 K for S /N = 50. All spectral windows except 4.80 − 5.00 μm appear to be relevant to study temperature stratification in atmospheres of HJs, but only two of them (1.50–1.70) and (2.28–2.38) μm, allow us to constrain

VMRs of at least some molecules. As was mentioned in Lee et al. (2012), the degeneracy between fit parameters may lead to incorrectly derived mixing ratios. We find that this is also the case for high-resolution low S/N data. As an example calculation, our Fig. 4 shows retrievals of H2O and CO from the (2.28–2.38) μm region using three different initial guesses for the molecular VMRs. For S /N = 5 the retrievals converge to nearly the same solution if the initial values of VMRs are 0.5 and 1.0 dex below the true one. However, when the initial guess is 0.5 dex higher than the true value then the algorithm prefers to increase the local temperature in the atmosphere keeping the VMRs at their high values.

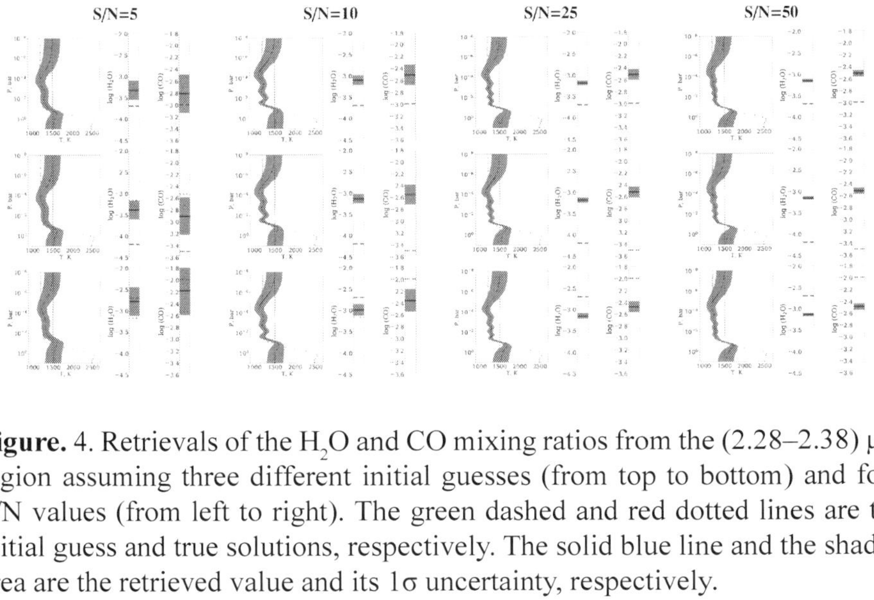

Figure. 4. Retrievals of the H_2O and CO mixing ratios from the (2.28–2.38) μm region assuming three different initial guesses (from top to bottom) and four S/N values (from left to right). The green dashed and red dotted lines are the initial guess and true solutions, respectively. The solid blue line and the shaded area are the retrieved value and its 1σ uncertainty, respectively.

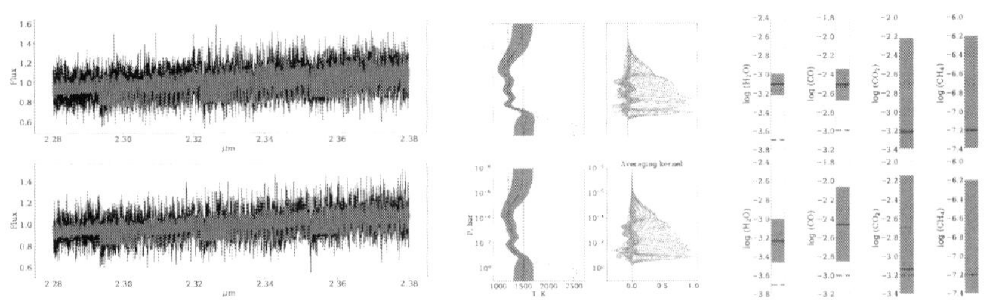

Figure. 5. Comparison of retrieved temperature and mixing ratios of four molecules assuming S /N = 10 and two spectral resolutions of R = 100 000 and R = 50 000 for the top and bottom panels, respectively. The color-coding is the same as in Fig. 3.

The increase in local temperature makes lines of molecules weaker, thus compensating for the high initial VMR. The molecular VMRs and temperature are thus degenerated as all solutions result in the same final normalized cost function ($\varphi = 0.73$). As the S/N of the data increases, the solutions for temperature and VMRs both tend to converge to the true values, while for S/N 6 10 the final values may deviate from the true ones (but are still within the 3σ error bars) depending on the initial guess.

Finally, degrading spectral resolution by a factor of two from R = 100 000 to R = 50 000 does not significantly affect the results of our temperature retrievals, while the accurate retrievals of molecular VMRs become problematic because the obtained uncertainties are large. An example is shown in Fig. 5 for the region 2.28–2.38 μm and S /N = 10. With the degrading of spectral resolution, the amount of information that we can learn from the observed spectra mostly depends on the number of spectral features resolved. Decreasing the resolving power weakens the depth of spectral lines, and thus directly affects retrieval of mixing ratios. At the same time, the profiles of strong molecular lines are still satisfactorily resolved, which provides enough information for the temperature retrieval. It is thus possible to rebin the obtained high-resolution spectrum to a smaller resolution and boost the S/N without greatly affecting our ability to retrieve temperature stratification. Alternatively, our Fig. 5 reflects the case of two observations with different slit widths, but after reaching the same S/N. This would require less telescope time for a wider slit width to reach the same S/N, which could be an affordable choice to reduce the integration time for a fixed S/N.

Retrievals from multiple spectral regions

As a next step we studied the scenario in which the retrievals are done from a combination of different spectral regions. In particular, we investigated combinations of the (2.28–2.38) μm region with the others because, as noted in the previous section, this region is often used for the detection of molecular lines in atmospheres of HJs, and it also provides rich information about the altitude structure of their atmospheres. Figure 6 compares the results of retrievals considering S /N = 10 (the results for other S/N values are provided in Figs. A.4–A.6). As expected, we find that a combination of (1.50–1.70)+(2.28–2.38) μm regions is one of the best in terms of retrieved information.

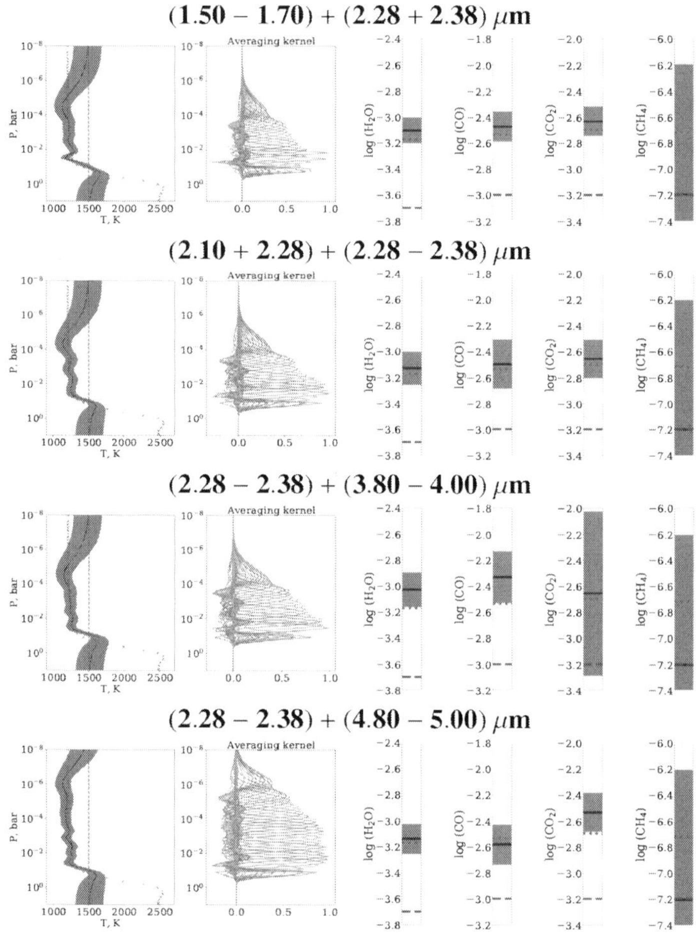

Figure. 6. Retrieved temperature and mixing ratios of four molecular species from a combination of different spectral regions with a reference region 2.28–2.38 μm. The retrievals are shown for the case of S /N = 10. The color-coding is the same as in Fig. 3.

When analyzed separately both these regions already constrain accurate temperatures, and combining them helps to improve our retrievals even further. On the other hand, the combination of (2.28–2.38)+(4.80–5.00) μm gives us the possibility to probe even higher altitudes up to P = 10−6 bar, but the concentration of CO is then derived less accurately compared to the previous case with S /N = 10, and noticeably worse with S /N = 5 (Fig. A.4). In general, retrievals from two different spectral regions help to constrain molecular number densities better. For the same case of (1.50–1.70)+(2.28–

2.38) μm we can now pin down the mixing ratio of CO and CO2 by a factor of about two compared to the case when either of these regions is used (compare the VMR uncertainties in the top panel of Figs. 3 and 6). The typical temperature errors in the pressure range between 1 and 10−4 bar is 100 K for S /N = 5, 90 K for S /N = 10, 80 K for S /N = 25, and 70 K for S /N = 50. We conclude that retrievals from two different spectral regions should be preferred in order to constrain molecular number densities as accurately as possible compared to single-band retrievals, but such observations would require two times longer integration times for the same noise levels.

Finally, simultaneous retrievals from three and four spectral windows (with a maximum wavelength coverage of about 1 μm) did not significantly improve our results for the temperature and mixing ratios compared to the retrievals from two spectra ranges. Thus, the only benefit of observing at many different spectral regions is to measure molecular number densities of as many molecules as possible. However, enough lines of at least three molecules that we considered (H2O, CO, CO2) can be captured already with two cross-disperser settings of CRIRES+. Thus, including additional spectral windows is not justified for the case of atmospheric retrievals.

SUGGESTED OBSERVATION STRATEGIES AND POTENTIAL TARGETS

Our analysis shows that with a single setting, CRIRES+ covers a spectral range containing enough features for exoplanet atmospheric studies. However, not all the spectral regions that we considered in this study are equally good for the retrieval of temperature and molecular concentrations. In the case of a single observation setting, accurately determining the temperature and mixing ratios depends only on the S/N of the spectrum of an exoplanetary atmosphere. Our simulations showed that under the same observing conditions the retrievals of accurate mixing ratios will generally require higher S/N compared to the case when only temperature stratification is investigated. For instance, with S /N = 5 it is possible to derive accurate temperatures in the line forming region of an HJ atmosphere (P > 1 bar), while realistic values of the H2O mixing ratio, for example, can only be obtained with two times higher S/N (see the results for the (2.28–2.38) μm region in Figs. 3 and A.1). In order to retrieve accurate temperature stratification and mixing ratios of H2O, CO, and CO2, the simultaneous observations at two spectral windows (1.50–1.70)+(2.28–2.38) μm and S /N = 10 should be used. As a next choice the (2.10–2.28)+(2.28–2.38) μm or (2.28–

2.38)+(4.80–5.00) μm regions can also be used for robust retrievals. We note that the M band contains many more telluric lines and the performance of the old CRIRES in that region was noticeably worse compared to the performance in K band. When only S /N = 5 could be reached, the choice of a single observation setting is not obvious, and there are two options. On the one hand, using (1.50–1.70) μm region could provide VMRs of H2O, CO, and CO2, but accurate temperatures could only be retrieved up to 10−3 bar. On the other hand, the (2.28–2.38) μm region would still provide accurate measurements of H2O and temperatures up to 10−4 bar, but less accurate CO and no constraints for CO2 (see Fig. A.1). Furthermore, the planet-to-star flux contrast in the (1.50–1.70) μm wavelengths is expected to be lower in HJs compared to the (2.28–2.38) μm region, thus requiring longer integration times for the same S/N. Observing in two spectral regions with S /N = 5 in each of them, we still favor the combination of (1.50–1.70)+ (2.28–2.38) μm. However, for the same integration time, we suggest observing only the (2.28–2.38) μm region obtaining higher S /N = √ 2 × 5 = 7, which allows more precise measurements of H2O and CO. In order to provide estimates for the exposure times needed to study atmospheres of HJs we used predictions of our models and the ESO Exposure Time Calculator (ETC) for CRIRES4 . Because there is no ETC for CRIRES+ available yet, we note that the listed Exposure times are upper limits due to the expected improved performance of the new instrument compared to the old one. From the list of known HJs we selected several of the best targets that can be observed with high-resolution groundbased spectroscopy similar to the capabilities of CRIRES+.

Table 1. Observability of favorable Hot Jupiters for atmospheric retrieval studies with ground-based high resolution spectroscopy.

Name	d (a.u.)	R_s ($R_\odot$)	R_p (R_J)	T_s (K)	T_p (K)	V^m star (mag)	K^m star (mag)	$\langle F_p/F_s \rangle$	t_{exp} (h) ($R = 100\,k/50\,k$)	Visible from ESO? (Yes/No)
51 Peg b	0.052	1.27	1.90	5800	1370	5.46	3.91	1e-3	6/5	Yes
HD 46375 b	0.041	1.00	1.02	5200	1229	7.84	6.00	4e-4	71/61	Yes
HD 73256 b	0.037	1.22	1.00 [(a)]	5636	1545	8.06	6.26	4e-4	91/79	Yes
HD 75289 b	0.046	1.25	1.03	6120	1527	6.36	5.06	4e-4	29/26	Yes
HD 143105 b	0.038	1.23	1.00 [(a)]	6380	1739	6.75	5.52	5e-4	27/23	No
HD 179949 b	0.045	1.19	1.05	6260	1541	6.24	4.94	4e-4	26/23	Yes
HD 187123 b	0.043	1.17	1.00 [(a)]	5714	1433	7.83	6.34	4e-4	98/85	No (too low)
HD 189733 b	0.031	0.81	1.14	4880	1185	7.65	5.54	7e-4	78/68	Yes
HD 209458 b	0.048	1.20	1.38	6100	1467	7.63	6.31	7e-4	95/83	Yes
HD 212301 b	0.036	1.25	1.07	6000	1692	7.76		6e-4	37/32	No (too low)
HD 217107 b	0.073	1.08	1.00 [(a)]	5666	1043	6.16	4.54	2e-4	67/58	No
KELT-20 b	0.054	1.56	1.74	8981	2307	7.58	7.42	1e-3	43/37	No (too low)
τ Boo b	0.046	1.33	1.06	6300	1624	4.49	3.36	4e-4	10/9	Yes
υ And b	0.059	1.63	1.00 [(a)]	6200	1560	4.10	2.86	2e-4	14/12	No (too low)
WASP-18 b	0.021	1.23	1.17	6400	2374	9.30	8.13	2e-3	22/19	Yes
WASP-33 b	0.026	1.44	1.60	7400	2661	8.14	7.47	2e-3	13/11	No (too low)
WASP-76 b	0.033	1.73	1.83	6250	2166	9.52	8.24	2e-3	25/21	Yes

Notes. For each target the table lists its name, distance to the parent star, radii of the star and the planet, stellar effective temperature, planetary equilibrium temperature, stellar magnitudes in V and K bands, mean planet-to-star flux ratio hFp/Fsi calculated between 1 and 5 μm, and exposure time needed to reach an S /N = 5 for the finally obtained exoplanet spectrum using single CRIRES+ setting. The exposure times are given separately for spectral resolutions R = 100 k and R = 50 k, respectively. The last column contains a flag for the observability of the star at the ESO Paranal observatory (Chile). The two primary targets for CRIRES+ are marked with bold. (a)We assumed $R_p = 1.0\ R_J$ due to the lack of information.

We additionally restricted the list of objects by including only those that require less than 100 h of integration time. We list these targets in Table 1. The integration times were computed using stellar and planetary parameters listed in the current version of the Extrasolar Planets Encyclopaedia5 and The Exoplanet Orbit Database6 . The stellar magnitudes in the V and K bands were extracted from the SIMBAD database7 . The planetary equilibrium temperatures were computed assuming atmospheric albedo a = 0.03 expected for HJs (Sudarsky et al. 2000). We calculated planet-to-star flux ratios as an average between 1 and 5 μm assuming a blackbody approximation for stellar and planetary flux. This values will be larger in the red part of the spectrum (typically by a factor of two), and thus the corresponding exposure times that we list in the table will be shorter for the same S/N at the red end of CRIRES+ spectra. However, the performance of old CRIRES detectors in the M band was noticeably lower compared to the K band so that the estimation of realistic exposure times for CRIRES+ at different spectral bands is currently not possible. This is why we provide exposure times for the mean flux contrast and use one of the most efficient instrument settings at the echelle order No. 25 (λeff = 2.267 μm). We note that the S/N for the planetary spectrum is defined by the planet-to-star flux ratio S /N(planet) = F_p/F_s ·S /N(star).

Due to the demand for very high-quality exoplanetary spectra the number of potential targets is very small. Among them, it will be possible to observe only two HJs from the ESO Paranal observatory and with affordable integration times. These exoplanets are 51 Peg b and τ Boo b. For 51 Peg b, a little more than one full night (e.g., 12 h) will be needed to reach S /N = 5 using two instrument settings or S /N = 7 in one setting. This will be sufficient for accurate atmospheric inversions. For our preferable configuration of S /N = 10 and two settings a 48 h integration time will be needed. In case of τ Boo b two full nights will be needed to reach S /N = 5 in two settings, or eight nights for S /N = 10 and two settings. The choice of the corresponding instrument settings will depend on the performance of the CRIRES+ at different infrared bands. Our simulations favors the (1.50–1.70)+(2.28–2.38) μm region because the VMR of three molecules can be accurately constrained, as can the temperature distribution, until P < 10−4 bar. If one wants to look at higher altitudes then the (2.28–2.38)+(4.80–5.00) μm region should be used, but the uncertainties on CO will be slightly larger. In the case of a single setting we suggest the (2.28–2.38) μm region; however, the (1.50–1.70) μm region can also be an option if one wants to constrain the VMRs of as many molecules as possible. Should the sensitivity of CRIRES+ be considerably improved, then the integration times listed in Table 1 would be smaller, thus favoring a detailed analysis of the atmosphere of 51 Peg b with two setting observations.

DISCUSSION

In our analysis of high-resolution simulated observations we used the OE method for the retrieval of atmospheric parameters, such as temperature and molecular mixing ratios. We did not attempt to utilize other methods (e.g., MCMC) primarily because the number of free parameters in our model is high. In this particular case the OE method has the advantage of being relatively fast but still robust. It should be noted that retrieving a detailed structure of planetary atmospheres relies primarily on the data quality and the presence of lines that probe different atmospheric altitudes. In cases when the temperature distribution is expected to be very smooth, or when the quality of the spectrum is poor so that the altitude structure could not be satisfactorily resolved, it is possible to parameterize altitude-dependent quantities with simple functions containing only few free parameters. For example, we could have parameterized the temperature distribution using a commonly used profile after, Parmentier & Guillot (2014), among others, which requires only four free parameters instead of retrieving a temperature value at each atmospheric layer. This would allow us to use a standard Bayesian parameter search. However, in this work we decided to carry out assumption-free retrievals. Hence, our application of the OE method should not be considered as the only suggested approach to the atmospheric retrievals at high spectral resolution, but rather as one possibility among others, with the ability to retrieve accurate parameters with a proper application

Near-future instruments will deliver data of much better quality and wider spectral range coverage compared to what is available now. That means that at high spectral resolution the shapes of spectroscopic lines originating from the planetary atmospheres could be studied directly. This was not possible to achieve before, and the high-resolution spectroscopy of exoplanets was pushed forward mainly by the detection of molecular species via the cross-correlation technique. This was proven to be very efficient in disentangling the stellar and planetary fluxes from the noisy observations that are additionally contaminated by telluric absorption. Recently, Brogi & Line (2019) made another step forward and suggested the attractive approach of using cross-correlation technique for the retrieval of atmospheric properties. This is done by the mapping of the cross-correlation coefficients to log-likelihood. After this mapping was performed, the standard Bayesian methods could be used to derive the best fit parameters and their errors. The choice of the function that maps cross-correlation coefficients to the log-likelihood space is arbitrary, but should satisfy some basic criteria to make retrievals unambiguous (see Brogi & Line 2019, for more details). The new mapping proposed by Brogi & Line (2019) has certain benefits. For instance, it can be applied to already existing data when the planetary lines are drawn in the noise-dominated spectra. Mapping of the cross-correlation coefficient and the OE method both rely on the presence of hundreds or even thousands of atmospheric lines, although the former method works with the residual spectra, while our implementation of the OE method works with spectra in arbitrary units and the corresponding flux scaling is a free parameter(s) of the model that could be optimized simultaneously with other (atmospheric) parameters. In our particular case the application of the new cross-correlation mapping was not strongly justified because the number of free parameters in our model is too high, making Bayesian postprocessing very time consuming. In addition, we note that with the sufficiently high S/N values of the planetary spectra that are expected to be obtained with future instruments, the profiles of individual planetary lines could be satisfactorily resolved and directly analyzed. This will likely make the use of any crosscorrelation mapping unnecessary favoring other approaches. We note that cross-correlation itself will still be used as a powerful tool for disentangling stellar and planetary spectra. In this work we analyzed five different spectral regions between 1 and 5 μm. In these regions the telluric absorption is expected to be relatively weak compared to the rest of the spectrum. In addition, these regions contain strong bands of molecular species that we investigated (H_2O, CO, CO_2, CH_4).

Another region where it is possible to study planetary atmospheres that we did not consider in this study is the one around 3.5 μm, which is dominated by numerous lines of CH4 and H2O. We carried out several additional retrievals using the 3.50–3.75 μm spectral region, but retrieved much less information compared to the other four regions. We conclude that although this region contains numerous molecular lines that can be easily detected from the usual cross-correlation analysis as shown in de Kok et al. (2014), among others, all these lines are strongly blended with each other and are not favorable for retrieving accurate molecular number densities. In this work we did not account for the instrumental effects such as flat-fielding, order merging, or wavelength calibration. Some of these effects may influence the shape of the observed spectrum (which is a common problem for every echelle spectrograph) and therefore have direct impact on the retrieved parameters. Unfortunately, at this moment we cannot quantify these effects for CRIRES+ because many parameters of the instrument are still under investigation and the actual values will only be known once the instrument has been tested against real observations. Obtaining the best quality planetary spectrum requires a very accurate removal of instrument effects corresponding to the ratio between stellar and planetary continuum fluxes, which is about $\approx 10^{-3}$ for known HJs. Obviously, this accuracy will be very hard to achieve. On the other hand, the global shape of individual molecular bands will likely be preserved. In this case only arbitrary flux normalization to each spectral region under investigation will be needed (as we did in this study) in order to carry out accurate retrievals. Such a normalization ideally keeps the ratio of the planetary (pseudo)continuum-to-line depth unchanged even if the information regarding the true stellar continuum was lost during the data processing stage. For instance, in the wavelength domain covered by CRIRES+, the stellar continuum is very smooth. Therefore, instead of using simple continuum scaling, we also attempted to fit low-order polynomials to the simulated observations in each spectral region and did the same to the predicted spectra at each iteration in our retrievals. By doing this exercise we found that the choice of normalization function does not affect the final result very much. The temperature and mixing ratios varied from retrieval to retrieval, but always within the error bars. This exercise is very simplistic and cannot reflect all the details of the data reduction process planned for CRIRES+ and should not be considered as an approximation for reality. Nevertheless, it tells us that information about the true stellar continuum is not likely to be a major obstacle toward atmospheric retrievals as long as the relative shape of spectral lines is preserved. We note that in our approach the pseudo-continuum could be fit using low-order polynomials to be found during the retrieval process. This leaves the S/N of the planetary spectrum as the main limitation for atmospheric retrievals. Thus, accurate retrievals can still be performed provided that the order merging within a single CRIRES+ setting is accurately done by the pipeline of the instrument. We plan to improve our simulations by considering realistic instrumental effects in our next work. Even if all the instrumental effects are properly taken into account, the telluric absorption represents the next obvious problem. Telluric lines are present in all near-infrared bands covered by CRIRES+. Even in the five carefully chosen regions that we investigated in this paper, many of the telluric lines can be strong depending on local weather conditions. Normally, moderately strong telluric lines can be successfully removed (e.g., Yan & Henning 2018; Brogi et al. 2016), while spectral regions containing the strongest telluric lines are excluded from the analysis.

However, in cases when telluric lines need to be removed, the accuracy of this procedure should again be very high, about $\approx 10^{-3}$ or better. More importantly, there might be numerous very weak telluric lines, with intensities of just a fraction of a percent relative to the stellar continuum, that are unaccounted for in existing telluric absorption models and that could potentially distort the shape of planetary lines. We note that this is of minor importance if only a detection of exoplanetary lines using cross-correlation technique is needed. In that case, the amplitude of the cross-correlation function depends on matching the position of exoplanet lines against a chosen template spectrum. Thus, even if the depth of exoplanetary lines is affected by unremoved telluric absorption, it will not hide the detection of the molecular species. However, it could still be an issue when making retrievals from those distorted planetary lines. The overall impact of these very weak telluric lines should be tested with real observations. One obvious advantage of CRIRES+ is the ability to observe many molecular lines simultaneously, which will surely help to reduce the impact of telluric absorption in atmospheric retrievals

In our simulated observations we used mixing ratios of four molecules (H2O, CO, CO2, CH4) as derived in Lee et al. (2012). Our retrievals showed that we can derive number densities of H2O, CO, and CO2 by using different spectral regions. At the same time, we could not constrain a mixing ratio of CH4 because the assumed mixing ratio is very low ($1.9 \times 10-7$) and the lines of CH4 are too weak compared to the lines of other three molecules. This is in agreement with the predictions of equilibrium chemical calculations (e.g., Gandhi & Madhusudhan 2017; Malik et al. 2017; Mollière et al. 2015). We note that CH4 has a very rich spectrum in the near-infrared with many bands present in the wavelength range of CRIRES+. In atmospheres of HJs with equilibrium temperatures below 1000 K or with C/O > 1, the lines of CH4 should become strong enough for the purpose of abundance analysis. As a test case we simulated an additional set of observations with the increased methane mixing ratio of CH4 = 5×10^{-4} and performed retrievals from the (3.80–4.00) μm region and from the combined (2.28–2.38)+(3.80–4.00) μm region. In both cases we were able to obtain definite constraints on the methane mixing ratio, with slightly more accurate values when derived from the combined (2.28–2.38)+(3.80–4.00) μm region, just as expected. Thus, using CRIRES+ it will also be possible to constrain methane concentration in planets with cooler atmospheric temperatures than those considered here.

SUMMARY

High-resolution spectroscopy in the near-infrared is a powerful tool for the study of absorption and emission lines to estimate the state of extra-solar atmospheres. Until now, it was mainly used for the detection of different molecular species via transit spectroscopy. In this work we made another step forward and investigated the potential of high-resolution spectroscopy to study molecular number densities and temperature stratification in exoplanet atmospheres by applying the optimal estimation method to the simulated observations at different near-infrared bands. We used a modified τ-REx forward model to simulate emission (i.e., out-of-transit) spectra for different molecular mixing ratios, spectral resolutions, and S/N values. The results of our investigation are summarized below.

- The spectra of HJs in the wavelength range of CRIRES+ is sensitive to a very wide range of atmospheric pressures between 1 and 10^{-6} bar.
- Avoiding regions with strong telluric contamination, the best region for deriving mixing ratios of H_2O, CO and CO_2 is around 1.6 μm, and the best region for studying temperature stratification and mixing ratios of H_2O and CO is around 2.3 μm. The mixing ratio of CH4 is impossible to constrain accurately in any wavelength range and at the S/N that we considered, due to the very low number density that we assumed in our synthetic

observations. It can be retrieved though, for example in cooler exoplanets that may have a CH_4 mixing ratio of $\approx 10^{-4}$–10^{-3}

- Retrieving mixing ratios simultaneously from two separate spectral regions helps to obtain accurate results for H_2O, CO, and CO_2. In this regard the combination of the 1.6 and 2.3 μm regions looks promising at any S/N, and the combination of 2.3 μm with any other region except 3.9 μm when S/N > 10. However, the retrieval of accurate number densities always requires higher S/N compared to the S/N required for the temperature retrievals. The latter can be accurately retrieved even with the lowest S/N = 5 that we assumed in our simulations.
- In the case of single setting retrievals, observing at the 2.3 μm region provides accurate VMRs of H_2O and CO only if S /N >10. When S /N > 25 the 1.6 μm region provides very accurate VMRs of H_2O, CO, and CO_2. However, to achieve such high S/N values would exceed typical allocated telescope observational times. The best strategy for obtaining reliable mixing ratios of as many molecules as possible is thus to observe at two different infrared bands, at 2.3 μm and at any of the others except the one at 3.9 μm.
- Degrading spectral resolution results in decreasing the sensitivity of our retrievals of molecular concentrations, but the temperature distribution can still be accurately derived. Rebinning the spectra to a lower resolution could help to increase S/N and to obtain estimates on the molecular number densities and atmospheric temperature structure in cases when the S/N of the original data is not sufficient (e.g., S /N < 5)
- With CRIRES+ it will be possible to carry out detailed atmospheric retrievals in two HJs known so far. Planethunting missions such as TESS8 , PLATO9 , and CHEOPS10 are expected to significantly increase the number of exoplanets accessible for current and future ground-based spectroscopic studies

Our study suggests that even though the spectra of exoplanets will likely be obtained with very high noise levels, the atmospheric retrievals can still benefit from numerous molecular lines observed thanks to the wide wavelength coverage of CRIRES+. Moreover, additional important constraints can be provided by utilizing low-resolution data (Brogi et al.

2017). This data is already available for a number of HJs and could be used to break the degeneracy between the retrieved parameters. Thus, in our future work we plan to make another step forward and create a suite for the simultaneous retrieval of atmospheric parameters from low- and high-resolution data.

REFERENCES

1. Abel, M., Frommhold, L., Li, X., & Hunt, K. L. C. 2011, 66th International Symposium On Molecular Spectroscopy, EFC07
2. Abel, M., Frommhold, L., Li, X., & Hunt, K. L. C. 2012, J. Chem. Phys., 136, 044319
3. Barstow, J. K., Aigrain, S., Irwin, P. G. J., Fletcher, L. N., & Lee, J. M. 2013, MNRAS, 434, 2616
4. Barstow, J. K., Aigrain, S., Irwin, P. G. J., et al. 2014, ApJ, 786, 154
5. Barstow, J. K., Aigrain, S., Irwin, P. G. J., & Sing, D. K. 2017, ApJ, 834, 50
6. Borysow, A. 2002, A&A, 390, 779
7. Borysow, A., & Frommhold, L. 1989, ApJ, 341, 549
8. Borysow, A., Jorgensen, U. G., & Fu, Y. 2001, J. Quant. Spectr. Rad. Transf., 68, 235
9. Bouchy, F., Udry, S., Mayor, M., et al. 2005, A&A, 444, L15
10. Brogi, M., & Line, M. R. 2019, AJ, 157, 114
11. Brogi, M., de Kok, R. J., Birkby, J. L., Schwarz, H., & Snellen, I. A. G. 2014, A&A, 565, A124
12. Brogi, M., de Kok, R. J., Albrecht, S., et al. 2016, ApJ, 817, 106
13. Brogi, M., Line, M., Bean, J., Désert, J.-M., & Schwarz, H. 2017, ApJ, 839, L2
14. de Kok, R. J., Birkby, J., Brogi, M., et al. 2014, A&A, 561, A150
15. Dorn, R. J., Anglada-Escude, G., Baade, D., et al. 2014, The Messenger, 156, 7
16. Follert, R., Dorn, R. J., Oliva, E., et al. 2014, Proc. SPIE, 9147, 914719
17. Gandhi, S., & Madhusudhan, N. 2017, MNRAS, 472, 2334
18. Hartogh, P., Jarchow, C., Lellouch, E., et al. 2010, A&A, 521, L49
19. Irwin, P. G. J., Teanby, N. A., de Kok, R., et al. 2008, J. Quant. Spectr. Rad. Transf., 109, 1136
20. Kreidberg, L., Line, M. R., Bean, J. L., et al. 2015, ApJ, 814, 66
21. Lee, J.-M., Fletcher, L. N., & Irwin, P. G. J. 2012, MNRAS, 420, 170
22. Lee, J.-M., Irwin, P. G. J., Fletcher, L. N., Heng, K., & Barstow, J. K. 2014, ApJ, 789, 14
23. Line, M. R., Zhang, X., Vasisht, G., et al. 2012, ApJ, 749, 93

24. Line, M. R., Wolf, A. S., Zhang, X., et al. 2013, ApJ, 775, 137
25. Malik, M., Grosheintz, L., Mendonça, J. M., et al. 2017, AJ, 153, 56
26. Mollière, P., van Boekel, R., Dullemond, C., Henning, T., & Mordasini, C. 2015, ApJ, 813, 47
27. Parmentier, V., & Guillot, T. 2014, A&A, 562, A133
28. Rengel, M., Hartogh, P., & Jarchow, C. 2008, Planet. Space Sci., 56, 1368
29. Rodgers, C. D. 1976, Rev. Geophys. Space Phys., 14, 609
30. Rodgers, C. D. 2000, Inverse Methods for Atmospheric Sounding: Theory and Practice (Singapore: World Scientific Publishing Co)
31. Rothman, L. S., Gordon, I. E., Barber, R. J., et al. 2010, J. Quant. Spectr. Rad. Transf., 111, 2139
32. Schwarz, H., Brogi, M., de Kok, R., Birkby, J., & Snellen, I. 2015, A&A, 576, A111
33. Sing, D. K., Fortney, J. J., Nikolov, N., et al. 2016, Nature, 529, 59
34. Snellen, I. A. G., de Kok, R. J., de Mooij, E. J. W., & Albrecht, S. 2010, Nature, 465, 1049
35. Sudarsky, D., Burrows, A., & Pinto, P. 2000, ApJ, 538, 885
36. Tennyson, J., & Yurchenko, S. N. 2012, MNRAS, 425, 21
37. Waldmann, I. P., Rocchetto, M., Tinetti, G., et al. 2015a, ApJ, 813, 13
38. Waldmann, I. P., Tinetti, G., Rocchetto, M., et al. 2015b, ApJ, 802, 107
39. Wang, J., Fischer, D. A., Horch, E. P., & Huang, X. 2015, ApJ, 799, 229
40. Winn, J. N., Fabrycky, D., Albrecht, S., & Johnson, J. A. 2010, ApJ, 718, L145
41. Yan, F., & Henning, T. 2018, Nat. Astron., 2, 714

2

Broadband Spectroscopy of Astrophysical ICE Analogues

B. M. Giuliano[1], A. A. Gavdush[2,3], B. Müller[1], K. I. Zaytsev[2,3], T. Grassi[4,5], A. V. Ivlev[1], M. E. Palumbo[6], G. A. Baratta[6], C. Scirè[6], G. A. Komandin[2], S. O. Yurchenko[3] and P. Caselli[1]

[1] Max-Planck-Institut für extraterrestrische Physik, Gießenbachstrasse 1, 85748 Garching, Germany
[2] Prokhorov General Physics Institute of the Russian Academy of Sciences, 119991 Moscow, Russia
[3] Bauman Moscow State Technical University, 105005 Moscow, Russia
[4] Universitäts-Sternwarte München, Scheinerstr. 1, 81679 München, Germany
[5] Excellence Cluster Origin and Structure of the Universe, Boltzmannstr. 2, 85748 Garching bei München, Germany
[6] INAF – Osservatorio Astrofisico di Catania, Via Santa Sofia 78, 95123 Catania, Italy

ABSTRACT

Context. Reliable, directly measured optical properties of astrophysical ice analogues in the infrared and terahertz (THz) range are missing from the literature. These parameters are of great importance to model the

Citation: B. M. Giuliano, A. A. Gavdush, B. Müller, K. I. Zaytsev, T. Grassi, A. V. Ivlev, M. E. Palumbo, G. A. Baratta, C. Scirè, G. A. Komandin, S. O. Yurchenko and P. Caselli "Broadband spectroscopy of astrophysical ice analogues - I. Direct measurement of the complex refractive index of CO ice using terahertz time-domain spectroscopy" A&A, 629 (2019) A112 DOI: https://doi.org/10.1051/0004-6361/201935619

 Open Access funding provided by Max Planck Society.

dust continuum radiative transfer in dense and cold regions, where thick ice mantles are present, and are necessary for the interpretation of future observations planned in the far-infrared region.

Aims. Coherent THz radiation allows for direct measurement of the complex dielectric function (refractive index) of astrophysically relevant ice species in the THz range.

Methods. We recorded the time-domain waveforms and the frequency-domain spectra of reference samples of CO ice, deposited at a temperature of 28.5 K and annealed to 33 K at different thicknesses. We developed a new algorithm to reconstruct the real and imaginary parts of the refractive index from the time-domain THz data.

Results. The complex refractive index in the wavelength range 1 mm–150 μm (0.3–2.0 THz) was determined for the studied ice samples, and this index was compared with available data found in the literature.

Conclusions. The developed algorithm of reconstructing the real and imaginary parts of the refractive index from the time-domain THz data enables us, for the first time, to determine the optical properties of astrophysical ice analogues without using the Kramers–Kronig relations. The obtained data provide a benchmark to interpret the observational data from current ground-based facilities as well as future space telescope missions, and we used these data to estimate the opacities of the dust grains in presence of CO ice mantles.

INTRODUCTION

One of the main problems in unraveling the chemical and physical properties of molecular clouds, in which the star and planet formation process takes place, is to estimate correctly the amount of gas contained. The difficulties in the direct observation of molecular hydrogen constrain the possibility to calculate the total mass of a cloud. An easy alternative could be to use carbon monoxide as a tracer of molecular gas, but in dense and cold regions of the interstellar medium and protoplanetary discs, CO is not a good tracer of gas mass because CO molecules preferentially reside on dust grains, forming thick icy mantles (e.g. Dutrey et al. 1998; Caselli et al. 1999). Alternatively, the dust continuum emission is the best available tool to compute a molecular cloud mass, if dust opacities are known.

The advent of Atacama Large Millimeter/submillimeter Array (ALMA) and Northern Extended Millimeter Array (NOEMA) facilities offers the possibility to observe the dust continuum emission in the millimetre and submillimetre part of the electromagnetic spectrum with very high angular resolution and sensitivity. However, to model the dust continuum emission properly it is necessary to have information about its grain size distribution and its chemical composition, since the dust opacity depends directly on these parameters. If we take into account that the dust grains can be covered by ice mantles at the centre of prestellar cores or in protoplanetary disc mid-planes, we also need to investigate how the presence of ices is changing the dust opacities.

Unfortunately, no experimental data are available for these cases, and the interpretation of the dust continuum emission measurements relies on calculated opacity values, such as those tabulated in Ossenkopf & Henning (1994). The goal of our study is to provide laboratory data on the optical properties of CO ice and utilize these data to calculate the opacities of dust grains covered by CO ice mantles. We compare the opacity values obtained by our study to those available in the literature.

The presently available set of data focusses mainly on the determination of the optical constants in the visible and mid-infrared (MIR) range (Hudgins et al. 1993; Ehrenfreund et al. 1997; Baratta & Palumbo 1998; Loeffler et al. 2005; Dartois 2006; Palumbo et al. 2006; Warren & Brandt 2008; Mastrapa et al. 2009).

Far-infrared (FIR) studies on spectral properties of molecular solids, without deriving optical constants, started early, to deepen the understanding of the infrared-active lattice vibrations of simple species. Anderson & Leroi (1966) studied frequencies of CO and N_2 in the range 40–100 cm^{-1}, and Ron & Schnepp (1967) complemented the available information with CO, N_2 and CO_2 intensity studies in the same frequency range. In 1994 Moore and Hudson published a comprehensive study of FIR spectra of cosmic type ices, including ice mixtures. These data included the analysis of amorphous and crystalline phases of the pure molecular ices, and the authors discussed the implications of the results on the identification based on astronomical observations. An estimation of the band strengths in the FIR region for pure ices and ice mixtures relevant for astrophysical environments can be found in Giuliano et al. (2014, 2016).

Recently, the investigation of the terahertz (THz) spectroscopic properties of ice mantles analogues has gained considerable interest (Allodi et al. 2014; Ioppolo et al. 2014; McGuire et al. 2016). This technique allows direct measurement of the intermolecular vibrations in the ice samples related to the lattice structure, which can be connected to their large-scale structural changes and finally to their thermal history. On the contrary, spectroscopic features measured in the MIR frequency range are indicative of the intramolecular vibrations of the sample, which can provide a wealth of information on the molecular identification and chemical reactivity. A comparison of our THz experimental data with that observed in the MIR range could help us to reveal intra- and inter-molecular vibrations. However, this study is beyond the scope of this paper and will be addressed in our future investigations.

Nowadays, numerous spectroscopic methods are extensively used for dielectric measurements at THz frequencies (Lee 2009); these include the following: Fourier transform infrared (FTIR) spectroscopy (Griffiths & de Haseth 1986), Backward-wave oscillator (BWO) spectroscopy (Komandin et al. 2013), spectroscopy based on photomixing (Preu et al. 2011) or parametric conversion (Kawase et al. 1996; Kiessling et al. 2013), and, finally, THz time-domain spectroscopy (THz-TDS; Auston 1975; Van Exter et al. 1989). These methods exploit either continuous-wave or broadband sources, operate in different spectral ranges, and are characterized with different sensitivity and performance. Among these methods, THz-TDS seems to be the most appropriate for studying laboratory analogues of circumstellar and interstellar ices. In contrast to other approaches, THz-TDS yields detection of both amplitude and phase of sub-picosecond THz pulses in a wide spectral range as a result of a single measurement; thus, the reconstruction of the dielectric response of a sample might be performed without using the Kramers–Kronig relations (Martin 1967) and involving additional physical assumptions. Furthermore, THz-TDS yields analysis of separate wavelets forming the time-domain response of a sample; thus, it is a powerful method for the characterization of multilayered samples. Thereby, we selected THz-TDS as a spectroscopic technique for our experiments.

We aim at the extension of the laboratory data in the FIR/THz region, and we show how the employment of the THz-TDS is able to provide direct measurement of the real and imaginary part of the refractive index of the ice sample. The experimental and theoretical methods employed are explained in Sect. 2, the results obtained and how these data are relevant for astrophysical application are presented in sections Sects. 3.3 and 4, respectively; the conclusions are illustrated in Sect. 5.

EXPERIMENTAL AND THEORETICAL METHODS

For this series of experiments a dedicated set-up has been designed and developed in the laboratories of the Center for Astrochemical Studies (CAS) located at the Max Planck Institute for Extraterrestrial Physics in Garching (Germany). The set-up is composed of a closed-cycle He cryocooler coupled to a THz time-domain spectrometer. The cryocooler vacuum chamber is small enough to be hosted in the sample compartment of the THz spectrometer, and it is mounted on a motor controlled translational stage, which ensures the tuning of the cryostat position with respect to the THz beam. The details of the main components of the apparatus and the ice growing procedure are given in the following subsections.

Cryogenic Set-Up

The high power cryocooler was purchased from Advanced Research Systems. The model chosen is designed to handle high heat loads thereby ensuring a fast cooling. This instrument is equipped with a special interface capable of reducing the vibration transmitted from the cold head to the sample holder at the nanometer level. This requirement is important in case of spectroscopic measurements in the THz frequency region, where the induced vibration of the sample can cause the increase of the noise level of the recorded spectra.

The cryostat is placed in a 15 cm diameter vacuum chamber, equipped with four ports for optical access and for the gas inlet. The optical arrangement is designed to work in the transmission configuration. The optical windows chosen for the measurements at the desired frequency range are made of high-resistivity float-zone silicon (HRFZ-Si), purchased by Tydex. This material features a high refractive index of n_{Si} = 3.415, negligible dispersion, and impressive transparency in the desired frequency range. The same material was chosen as a substrate for ice growing. In order to suppress the Fabry-Perot resonances in the THz spectra, caused by multiple reflection of the THz pulse within the windows, the thickness of the three Si windows must be different from each other. For this purpose we chose to use the 1 mm thick Si window as substrate for the ice deposition, and we used two windows of 2 mm and 3 mm thickness as optical access to the THz beam.

We performed measurements with a slightly focussed THz beam configuration in order to mitigate vignetting and diffraction of the beam at the metal components of the vacuum chamber. A schematic overview of the chamber arrangement is sketched in Fig. 1. The pumping station is composed of a turbomolecular pump (84 ls^{-1} nitrogen pumping speed) combined with a backing rotary pump (5 $m^3 h^{-1}$ pumping speed), thus providing a base pressure of about 10^{-7} mbar. The minimum temperature measured at the sample holder in normal operation mode is 5 K.

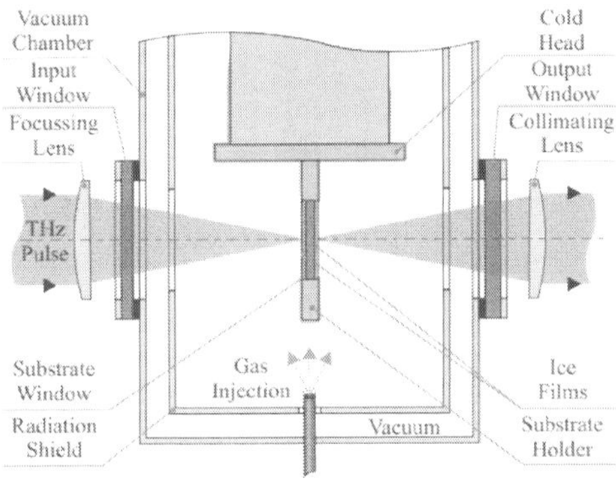

Figure. 1. Sketch of the vacuum chamber of the cryostat coupled to the THz beam at the CAS.

Terahertz Time-Domain Spectrometer

The THz-TDS set-up used for this work was purchased from the company Batop GmbH. The model chosen is the TDS-1008, which has a customized sample compartment in order to allocate the cryostat. This model is based on two photoconductive antennas made of low-temperature-grown gallium arsenide (LT-GaAs), which constitute the emitter and detector of the THz pulse (Lee 2009). The antennas are triggered by a femtosecond laser (TOPTICA, 95 fs, 780 nm) with a pulse repetition rate of 100 MHz and an average input power of 65 mW. Further details on the set-up are provided in Appendix A.

Figure 2a shows the optical path of the laser beam into the optical bench of the spectrometer. Panel b in Fig. 2 shows standard broadband Fourier spectra. The THz pulse registered with the beam path empty is then converted in the blue spectrum spanning the frequency range from 0.05 THz to 3.5 THz with maximal spectral amplitude centred at about 1.0 THz. In TDS, the time-domain THz waveform is converted to the frequency domain using the direct Fourier transform, which yields the frequency-dependent amplitude and phase of the THz wavelet. Since the frequency-domain data are calculated via the direct Fourier transform, the spectral resolution of measurements is determined as $\Delta\nu = 1/\Delta T$, where ΔT is a size of the time-domain apodization filter, chosen to avoid the edge effects (i.e. the Gibbs effect) in the frequency domain. In our experiments we used the 35-ps Tukey apodization filter (see Appendix B), which yields the spectral resolution of about 0.03 THz.

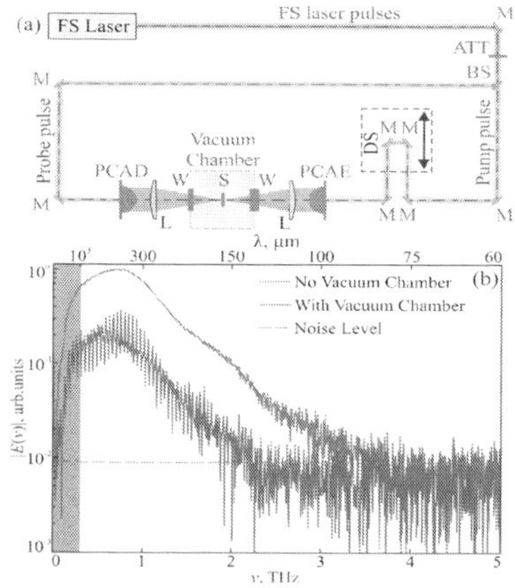

Figure 2. Set-up of THz-TDS for the spectroscopy of ices. *Panel a*: chematic of the set-up, where FS laser stands for the femtosecond laser, M stands for the optical mirrors, ATT stands for the attenuator of the laser beam intensity, BS stands for the optical beams splitter, DS stands for the mechanical double-pass delay stage, PCAE and PCAD stand for the photoconductive antenna-emitter and antenna-detector, respectively, L stands for the TPX lenses, and S and W stand for the HRFZ-Si substrate and windows, respectively. *Panel b*: spectra of THz waveforms $E(\nu)$ transmitted through the empty THz beam path or the THz beam path with the cryostat; the shaded region below $\approx$0.3 THz indicates the spectral range in which distortions from the THz beam diffraction on the aperture of the substrate are expected.

The green spectrum was recorded with the cryostat placed in the sample compartment. It is converted from a waveform which contains both a first THz pulse (i.e. a ballistic one) and a train of satellite pulses originating from the multiple THz wave reflections within the windows. The ballistic THz pulse is delayed in the input/output windows and the substrate of the vacuum chamber. The spectrum of this waveform is slightly suppressed owing to the Fresnel losses and modulated due to the interference of the ballistic pulse and the satellites.

In Fig. 2b, we show the shaded area at lower frequencies, where we expect growing distortions of the experimental data caused by the THz beam diffraction on the aperture of the substrate that is 20 mm in diameter.

Assuming that the THz beam spot formed at the substrate is diffraction-limited, the lateral intensity distribution in the spot is defined by the Bessel function of the first kind (Born & Wolf 1980). The resulting width of the first intensity peak is approximately $(3.8/\pi)(f/D)(c/v)$, where $D = 25$ mm and $f = 67$ mm stand for the diameter and the back focal distance of the focussing lens, respectively. From this model, we deduce the critical frequency of 0.3 THz, below which less than 95% of the beam energy passes through the substrate aperture. Thereby, considering both the spectral sensitivity of our THz-TDS set-up and the diffraction limits, the spectral operation range of our experimental set-up is approximately limited within 0.3–2.0 THz.

The THz-TDS housing is kept under purging with cold nitrogen gas during the entire experiment to mitigate the absorption features due to the presence of atmospheric water in the THz beam path. The residual humidity measured at the sensor was less then 10^{-3}%.

Ice Preparation

The ices are prepared using a standard technique in which the molecular sample in its gaseous form is allowed to enter the vacuum chamber through a stainless steel 6 mm pipe. The gas flux is controlled by a metering valve. Once the gas is expanding inside the vacuum chamber it condensates on the substrate.

For this set of experiments an ice thickness of the order of millimetre is required to fulfil the sensitivity characteristic of the THz-TDS set-up. This value is orders of magnitude higher than the usual ice thickness reached using this deposition technique, which is of the order of μm. To deposit such a thick ice in a reasonable amount of time, we chose fast deposition conditions, in which a considerable amount of gas is introduced in the vacuum chamber. In these conditions it is very difficult to obtain an ice sample which is homogeneous enough to determine its optical properties. To overcome this problem the gas inlet characteristics must be set accordingly.

We decided to remove any directionality from the gas inlet, keeping the pipe end cut at the vacuum chamber wall at a distance of approximately 7 cm from the substrate (see Fig. 1). This configuration creates an ice layer on each side of the substrate. During the deposition, the pressure measured inside the chamber is approximately 10^{-2} mbar. The ice deposition was divided in steps of 4, 5, and 6 min duration, up to a total of 30 min deposition time, in three different experiments. The final temperature was up to 28.5, 31.2, and 33.1 K for each step at 4, 5, and 6 min deposition time, respectively. This increase is due to the condensation of the gas onto the cold surfaces of the cryostat, which is producing a heating rate too fast to be dynamically removed from the cooling system during the deposition. After each step, the THz spectrum has been recorded. This procedure was performed to rule out possible effects on the ice structure due to different deposition temperatures. As interstellar ices can be commonly found at temperatures as low as 10 K, the temperature of the cold substrate has been kept at 14 K, which is the lowest temperature achievable in the set-up in this configuration, because the radiation shield of the sample holder must be removed to ensure that no directionality of the molecular beam is present. Before moving to the next deposition step, the system was allowed to thermalize and spectra recorded after each deposition step have been taken at a temperature of 14 K.

As stated in Urso et al. (2016) the analysis of the Raman and infrared spectra of experiments performed at increasing tempertaures from 17 to 32 K show no profile variation in the band at 2140 cm^{-1}, which could be ascribed to a structural change in the ice morphology.

The waveform recorded in the time domain is compared, as reference for the measurements, with the waveform recorded for the substrate without ice, kept at a reference temperature of 14 K as well. After the deposition was completed, we measured the spectra in different regions of the sample, to ensure that the ice morphology is spatially homogeneous. The results obtained from the spectra measured on a grid of 11 points spaced by 2 mm are in agreement within 10%, indicating a uniform ice formation over the substrate.

DERIVATION OF THE OPTICAL CONSTANTS

In order to determine the optical constants, the ice thickness must be known. The laser interference technique is a well-established method to estimate the thickness of an ice sample deposited on a substrate as a function of the time. The absolute accuracy of this method is approximately within 5%, but the maximum CO ice thickness that we can measure with this technique is limited to 5 μm before the reflected laser signal becomes too weak to be detected, due to scattering losses occurring both in the bulk and on surface of the film. Thus, this technique is well suited for studying thin layers, when the ice thickness in total is below 10 μm (5 μm on each side of the substrate), but is not appropriate for experiments on thick ice samples. In turn, for the thickness estimation of the millimeter-size sample of ice featuring rather low THz absorption, we developed a model to perform an initial estimation of the optical properties and the ice thickness directly from the recorded THz spectra, as described in the following subsections.

Ice parameters modelling

Our model aims to reconstruct the optical properties of ices, which are defined as follows:

$$n(\nu) = n'(\nu) - in''(\nu) \equiv n'(\nu) - i\frac{c}{2\pi\nu}\alpha(\nu), \tag{1}$$

where n' and n'' are the real and imaginary parts of the complex refractive index n, c is the light speed, and α is the amplitude absorption coefficient, which is defined as half of the value of the power absorption coefficient.

Equivalently, we can write

$$n^2(\nu) \equiv \varepsilon(\nu) = \varepsilon'(\nu) - i\varepsilon''(\nu), \tag{2}$$

where ε' and ε'' are the real and imaginary parts of the complex dielectric permittivity ε.

The model describes the THz wave propagation through the substrate with the ice deposited on both surfaces. The reconstruction of the ice parameters proceeds following three main steps.

The first task is modelling the reference and sample waveforms. Because of the focussed arrangement of the THz beams, the electromagnetic wave is assumed to be planar and to interact with the sample interfaces at the normal angle of incidence. This is a common and conventional assumption widely applied in dielectric spectroscopy (Pupeza et al. 2007; Zaytsev et al. 2014). It allows us to describe all the features of the THz pulse interaction with the multilayered sample using the Fresnel formulas, which define the THz wave amplitude reflection at (and transmission through) the interface between the media and the Bouguer-Lambert-Beer law, which defines the absorption and phase delay of the THz wave in a bulk medium. Further details on these assumptions are given in Appendix B.

Figure 3 represents the THz wave propagation through the three layers structure: the first ice film, the HRFZ-Si substrate, and the second ice film, where the symbols 0–3 and N correspond to different components of the plane wave passing through the multilayered structure. As shown in Fig. 3, for the sample spectrum, we took into account the contribution of the ballistic THz pulse (1) and the satellite pulses (2 and 3), caused by the multiple THz wave reflection in the ice films. The mathematical description of the wave propagation can be found in Appendix B.

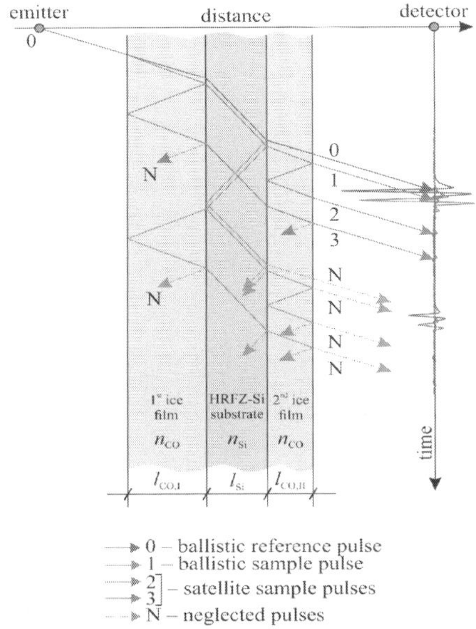

Figure 3. Time–distance diagram illustrating the THz wave propagation through the HRFZ-Si substrate with ice films deposited on its surfaces. Lines 0–3 illustrate the ballistic pulse and satellite pulses transmitted in the direction of the antenna-detector; N stands for unaccounted satellites with larger time delays. The solid lines represent the pulses used for the analysis, while dotted lines correspond to neglected pulses.

The second step consists in estimating the initial thickness $l_{CO,I}$, $l_{CO,II}$ and the initial complex refractive index n_{init} of the two ice layers as shown in Fig. 4. The thickness estimation can be derived from the time delay δt_{01} between the ballistic pulse of the reference and sample waveforms (0 and 1 in Fig. 4), the first satellite pulse (2) and the ballistic pulse (1) of the sample waveform δt_{12}, and the second satellite pulse (3) and the ballistic pulse (1) of the sample waveform δt_{13}, in Fig. 4a. Since the HRFZ-Si has a very high refractive index, we consider that the refractive index of ice n satisfies the inequality $n_0 = 1.0 < n < n_{Si} = 3.415$, where n_0 is the refractive index in vacuum.

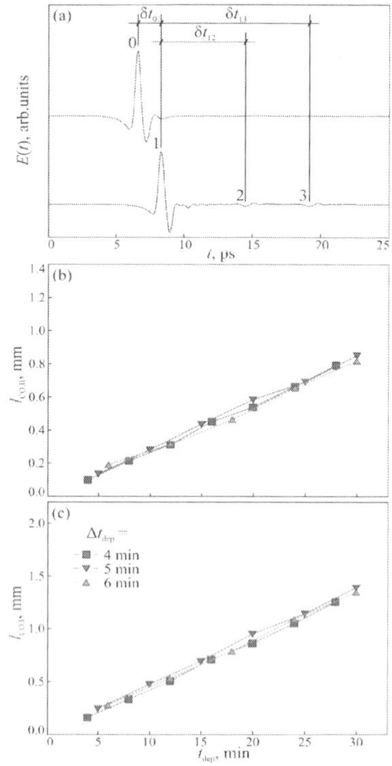

Figure 4. Calculation of the thicknesses $l_{CO,I}$, $l_{CO,II}$ and the complex refractive index n of the CO ice films (see Fig. 3). *Panel a*: time delays between the ballistic THz pulses of the reference (0) and sample (1) waveforms, δt_{01}; the first satellite pulse (2) and the ballistic pulse (1) of the sample waveform, δt_{12}; and the second satellite pulse (3) and the ballistic pulse (1) of the sample waveform, δt_{13} (the pulses are delineated as in Fig. 3). *Panels b and c*: estimates for the thicknesses of the two ice films as a function of the total deposition time t_{dep} for the different deposition intervals of Δt_{dep} = 4, 5, and 6 min; the first assumption for the real part of the refractive index of ice is between $n'_{init} = 1.230$ and 1.255±0.035; for the imaginary part we first set $n''_{init} = 0$.

Then, by neglecting imaginary parts in the complex refractive indexes of media, the first assumptions for the real part of the complex refractive index of ice films and both their thicknesses $l_{CO,I}$ and $l_{CO,II}$ are described as follows:

$$n'_{init} = \frac{\delta t_{12} + \delta t_{13}}{\delta t_{12} + \delta t_{13} - 2\delta t_{01}}, \tag{3}$$

$$l_{CO,I} = \frac{c\,\delta t_{13}}{2n'}, \quad l_{CO,II} = \frac{c\,\delta t_{12}}{2n'}. \tag{4}$$

Equation (3) is obtained from a mathematical model of sample and reference waveform. Further details on the derivation can be found in Appendix B. From Eq. (4) it is possible to obtain information only on the thicknesses of ice films, while the identification of the specific layer (I or II) is not allowed. We observe a linear increase of the ice thickness with the total deposition time. A first assumption for the real part of the complex refractive index of CO ice is in the range $n'_{\text{init}} = 1.230$ to 1.255±0.035 for all the considered deposition intervals; here, the error accounts for an accuracy of the THz pulse peak position estimation. The first assumption for the imaginary part of the complex refractive index of ice has been done considering $\alpha_{\text{init}} = 0$; thus, $n''_{\text{init}} = 0$.

Finally, from the first estimation of the ice thickness and complex refractive index, it is possible to reconstruct the THz dielectric response of ice. The reconstruction procedure is reported in Appendix B, while the results are summarized in Fig. 4. Panels b and c show the growth of the two CO ice layers in time (t), considering different deposition steps of $\Delta t = 4$, 5, and 6 min.

We can compare the results obtained from the THz spectral data to the thickness calculation performed with the well-established laser interference techniques described in Sect. 3.2 to validate the calculation of the ice thickness using this model. The good agreement between the two methodologies confirm the validity of the present analysis.

Laser Interference Technique

In the adopted experimental configuration, a He–Ne laser beam ($\lambda = 632.8$ nm) is directed towards the sample and reflected at near normal incidence both by the vacuum-sample and sample-substrate interfaces. The reflected beam is detected by an external diode detector. It is possible to follow the accretion of the ice film by looking at the interference curve (intensity vs. time) of the reflected laser beam. Further details on the laser interference technique can be found in Urso et al. (2016)[1].

The results obtained with the two techniques have been compared. The data on the accretion of the ice versus time obtained from the analysis of the interference curve, measured at the early stage of the deposition process, are in good agreement with the data obtained from the THz spectra. In addition, using the laser interference technique, we obtain for the CO ice a refractive index $n_{CO} = 1.27$, that is close to the value obtained using the THz technique.

The good agreement between the two methodologies confirm the validity of the present analysis.

Reconstruction of the THz Response

The recorded THz waveforms and their Fourier spectra are presented in Fig. 5 for the optical substrate (used as a reference) and CO ice samples at increasing thicknesses. In panel a the waveform $E(t)$ is shown for the reference (green) and five subsequent deposition steps (black to light red) of approximately 0.45 mm total thickness for each step. The thickness reached after the total deposition time is approximately 2.3 mm, split in two ice layers of $\approx$0.85 mm and $\approx$1.45 mm on top of each side of the substrate. A small source of inhomogeneity can be ascribed to the position of the pipe connected to the pumping system, which is located in a lateral position of the vacuum chamber with respect to the cold substrate.

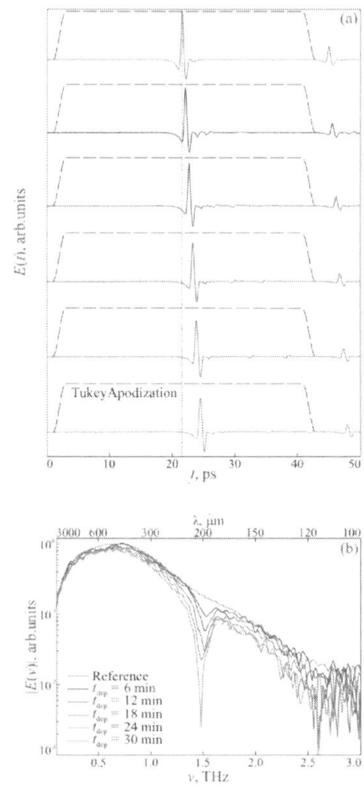

Figure 5. Evolution of the THz pulse and its spectra during the CO ice deposition. *Panel a*: reference waveform $E(t)$ transmitted through the cryostat with the

empty substrate (green), and sample waveforms (black to light red) transmitted through the substrate with the CO ice deposited on its surfaces. *Panel b*: Fourier spectra $|E(\nu)|$ of the reference and sample THz waveforms calculated with the use of Tukey apodization. The waveforms in (*a*) and spectra in (*b*) correspond to different values of the total deposition time t_{dep} (indicated); the deposition intervals are $\Delta t_{dep} = 6$ min.

In the THz spectrum of CO ice, we observe a Lorenz-like resonant peak centred near 1.5 THz (50 cm^{-1}) and a second blurred feature close to 2.5 THz (83 cm^{-1}), masked by the sharp bands produced by the atmospheric water contamination in the spectrometer sample compartment. The highest frequency accessible in our set-up is presently limited by the strong absorption features of residual water and carbon dioxide in the spectrometer case. We plan to change the current set-up to an evacuated case to get rid of the contamination from the residual atmosphere, and we expect to extend the accessible frequency range up to 4 THz.

The estimated deposition rate for these deposition conditions is ≈0.05 mm min^{-1} for layer I and ≈0.03 mm min^{-1} for layer II. These values are in a reasonable agreement with the results obtained employing the laser technique, with which the deposition rate is calculated to be 0.02 mm min^{-1}. This agreement validates our hypothesis that the ice structure of thick ices does not differ significantly from the structure of thin ices, growing homogeneously over time during the deposition. The calculated optical properties are independent from the total thickness of the ice sample, allowing us to relate the laboratory data to the astrophysical ice conditions.

Figure 6 shows the determination of CO ice parameters, which are the refractive index a, the amplitude absorption coefficient b, and both the real c and imaginary d parts of the complex dielectric permittivity; see Eqs. (1) and (2).

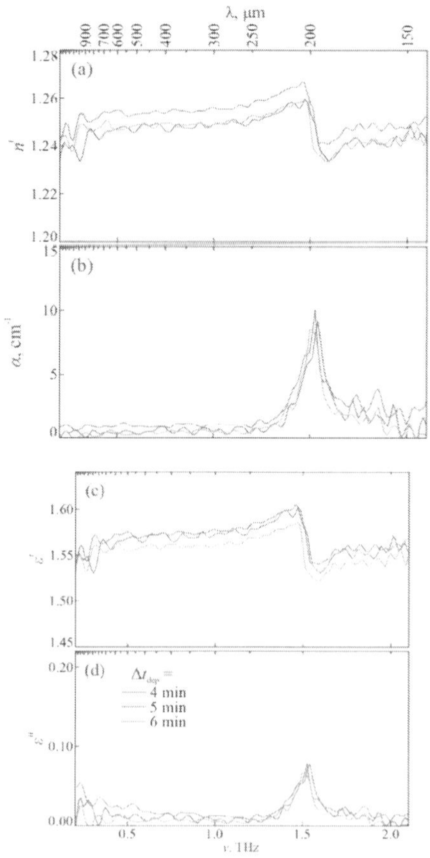

Figure 6. Optical properties of CO ice. (*a*) Real part of the refractive index, (*b*) amplitude absorption coefficient, and (*c*) real and (*d*) imaginary parts of the dielectric permittivity (see Eqs. (1) and (2)). For all deposition intervals, the dielectric curves demonstrate the existence of a Lorenz-like absorption peak, centred near 1.5 THz and featuring similar bandwidth. Distortions of the results seen at frequencies below 0.3 THz (such as an oscillatory character of n and ε' and an increase of ε'' with decreasing frequency) are due to diffraction effects (see Sect. 2.2).

DISCUSSION

A benefit of the direct reconstruction of the optical properties of the ices, provided by THz-TDS, is the detection of the frequency-dependent amplitude and phase of the waveform in a broad frequency range as a result of a single measurement. These data eliminate the need to use the Kramers–

Kronig relations for the reconstruction of the optical properties, excluding additional distortion of the experimental data by edge effects, which frequently appears as a result of the Hilbert integral transformation. This is of particular importance when dealing with broadband spectral kernels which are usually present even when operating at low temperature.

We could also compare the refractive index of CO ice at THz frequencies with that previously calculated in the MIR range, from Hudgins et al. (1993), Ehrenfreund et al. (1997), and Baratta & Palumbo (1998) for a CO ice deposited at 10 K. However, a direct comparison between refrative index values found by different authors is not straightforward. As discussed by Loeffler et al. (2005) and Baratta & Palumbo (2017), the density and in turn the refractive index of an ice sample could strongly depend on the experimental conditions such as temperature, growth angle, and deposition rate. We plan to test the effect of the change in the deposition conditions on the ice structure and check if this change will affect the optical constants. These results will be published in a forthcoming paper.

We compare the spectroscopic signature of the CO ice in our experiments with previous studies available in the literature. Data on FIR spectra of solid CO were reported by Anderson & Leroi (1966) and Ron & Schnepp (1967). These studies investigated the absorption spectra of amorphous CO, deposited at 10 K on a crystalline quartz substrate between 250 and 30 cm^{-1}. Two bands are visible at 50 and 83 cm^{-1} (1.5 and 2.5 THz). The spectral features observed in our experiments are in excellent agreement with these data, even though the 2.5 THz feature is masked by atmospheric water bands in our set-up. The THz-TDS technique has a low sensitivity, requiring very thick ice layers to be detectable. Using our fast deposition rate the MIR vibrational bands of CO were strongly saturated within the first 30 s of deposition. We did some preliminary tests on the ice growing using the vibrational bands in the NIR range and we were able to follow the ice growing up to approximately 160 μm thickness. Also in this case the ice growing is constant and linear in time during deposition, but the error associated with the thickness estimate using the NIR band is large (20–30%).

It is not surprising then that in the data reported by Ioppolo et al. (2014), on THz and MIR spectroscopy of interstellar ice analogues, the CO ice absorption band in the THz region was not observed. The ice thickness in these experiments was estimated in all cases to be less than 10 μm. In our case, the minimum thickness required to be able to observe the 1.5 THz feature is estimated to be of the order of hundreds μm.

The data obtained are then employed to calculate the dust opacity for a given grain size distribution, as reported for example in Ossenkopf & Henning (1994); see Appendix C for additional details on the method we developed to reproduce their results. Following their approach, we report in Fig. 7 the calculated dust opacities assuming different ice coatings and different experimental data for the optical constants of ices. Data from Ossenkopf & Henning (1994) are labelled OH94 and reported for bare grains (green dotted line), thin (blue dotted), and thick ice mantles (orange dotted). The labels V = 0, 0.5, and 4.5 indicate the volume ratio of the core of refractory material to the ice mantle (see Appendix C). Conversely to the present work, Ossenkopf & Henning (1994) assumed a $H_2O:CH_3OH:CO:NH_3$ = 100:10:1:1 mixture ice mantle composition, i.e. water-based, with a minor amount of methanol, carbon monoxide, and ammonia. As described in detail in Appendix C, following the same procedure as in Ossenkopf & Henning (1994), we extended the real and the imaginary parts of the refractive index from Hudgins et al. (1993) to longer wavelengths and we included spherical carbonaceous impurities.

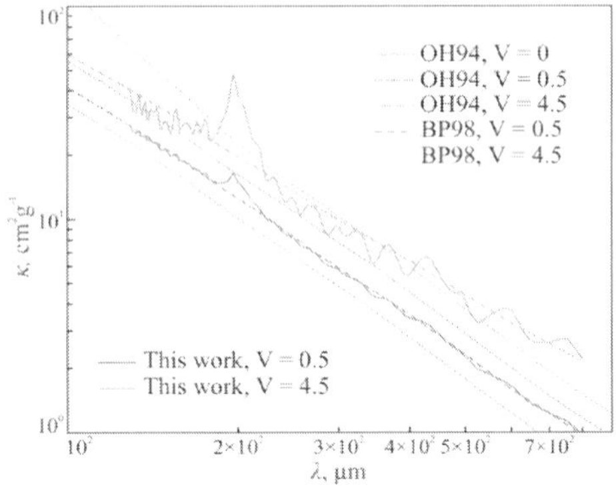

Figure 7. Calculated and reference opacities of astrophysical dust with CO ice and ice mixtures as a function of the wavelength. The dotted lines labelled with OH94 refer to bare grains and ice mixtures by Ossenkopf & Henning (1994), the dashed lines with BP98 to CO ice by Baratta & Palumbo (1998), and the solid lines to the CO data by the present work. The value V indicates the volume ratio of refractory core to ice mantle, for which we follow Ossenkopf & Henning (1994), where V = 0 (black) is the bare grain, V = 0.5 (blue) thin ice, and V = 4.5 (orange) thick ice. See text for additional details.

The dielectric functions found by our experiments refer to pure CO ice, which explains the differences in the opacities shown in Fig. 7 calculated in the 100–1000 μm range (solid lines). To have a more relevant comparison, we calculated the opacity using the same grain distribution and refractory materials as in Ossenkopf & Henning (1994). But in this work we extrapolated the optical values of CO ice coating from Baratta & Palumbo (1998, labelled BP98), where the refractive index of CO ice deposited at 12 K is calculated from the spectrum recorded in the 4400–400 cm^{-1} (2.27–25 μm) infrared spectral range. The opacities of Baratta & Palumbo (Fig. 7, dashed lines) show an agreement with our results, except for the contribution from the presence of a CO ice absorbing feature at approximately 200 μm, which is absent in the extrapolated data, as expected.

When we compare the refractive index, we note that the real part by BP98 ($n' = 1.28$) is reasonably close to our data ($n' = 1.24$), while for the imaginary part the discrepancy is very large, since n'' is completely determined by the absorption feature at 200 microns that is out of the range investigated by Baratta & Palumbo (1998) (see Fig. 6). It is worth mentioning that Anderson & Leroi (1966) and Ron & Schnepp (1967) reported a CO absorption feature at 2.5 THz, which is not visible in our spectra because it is masked by the atmospheric water contamination. This feature should further decrease the actual value of n' at higher frequencies (above 2.5 THz). If we could measure the 2.5 THz feature and extend the calculation of the optical properties to this value, we would probably obtain a lower value of n', slightly increasing the discrepancy with the data by Baratta & Palumbo (1998). The reconstruction of the THz dielectric response of ices without the use of the Kramers–Kronig relations, which is provided by THz-TDS, can provide an independent methodology to determine the optical properties of ice samples and validate the previous studies.

Since the real and imaginary parts of the dielectric constant are employed to compute the opacity, we can infer from the calculated opacity curve that the imaginary part, which shows the biggest difference from the data presented by Baratta & Palumbo (1998), does not play a major role in the determination of the opacity. This conclusion might be different for other absolute values or in a different spectral range.

Values of opacities for some selected wavelengths are also given in Table 1. The motivation for the choice of CO ice arises from the need of investigating the ice mantle properties for sources in which drastic CO depletion is expected, such as prestellar cores or protoplanetary discs mid-planes (e.g. Caselli et al. 1999; Pontoppidan et al. 2003). In these cases, an ice mantle rich in CO can be formed and influence the optical properties of the dust grains. Thus, it is interesting to compare how the opacities change when ice mantles with diverse chemical composition are present. The common molecular species in astrophysical ices, such as H_2O, CO_2, NH_3 and possibly N_2, present absorption features in the FIR range. Therefore, the study of the influence of the spectroscopic features on the opacity of the ice mantles is important and we plan to extend this study to other pure ices and ice mixtures, which could be representative of the different molecular ice compositions in various astrophysical environments.

Table 1. Opacities calculated at selected wavelengths and parameters of the fitting function $\kappa = \kappa_0(\lambda/\lambda_0)^\beta$, where $\lambda_0 = 1\ \mu m$, for the two models with different volume ratios (cf. solid lines in Fig. 7).

$\lambda/\mu m$	$\kappa_{V=0.5}$	$\kappa_{V=4.5}$
200	14.94	38.12
250	8.16	12.58
350	4.28	6.28
500	2.27	4.13
800	0.97	2.22
$\kappa_0/10^5$	2.28	1.28
β	−1.85	−1.65

CONCLUSIONS

In this work we have presented a study on the optical properties of solid CO at temperature and pressure conditions significant for astrophysical applications. While previous data in the MIR frequency range are available in the literature, to the best of our knowledge, this study is the first to provide the complex refractive index and complex dielectric permittivity of CO ice in the THz range.

We have shown that the ability of THz-TDS to measure both the amplitude and phase information about the transmitted pulse provides direct reconstruction of the complex dielectric function of ices without the use of the Kramers–Kronig relations. The THz spectral features of ices can have a large bandwidth, such as the CO absorption line at 1.5 THz. In this case, an implementation of the Kramers–Kronig relations (i.e. the Gilbert transform), relying only on the power transmission/reflection spectrum, could lead to edge effects and resulting distortion of the dielectric response. Such distortions are of particular importance when a spectral feature of the sample is located near the border of the spectral operation range. Our results justify that THz-TDS set-up is an appropriate instrument for accurate measurements of dielectric properties of ices at THz frequencies.

These results have been used to calculate the opacities of the dust grains covered by a CO ice layer. Discrepancies with currently available opacities suggest that measurements such as those presented in this work are needed to provide a better interpretation of dust continuum emission, including dust and gas mass estimates. In addition, they will provide further insight into the radiative transfer processes based on ice analogues optical and physical properties.

[1] A publicly-available interface can be found at http://www.oact.inaf.it/spess/ to calculate the refractive index of the ice sample and derive the theoretical interference curve from the amplitude of the experimental curve. The sample thickness is obtained by comparing the two curves and using a procedure described in a document available at this web page.

[2] https://bitbucket.org/tgrassi/compute_qabs, commit: 8c0812f

[3] This number is chosen manually to select the $^{-1}$ decaying part of the data after the last available resonance.

[4] http://scatterlib.wikidot.com/mie

[5] Adapted here from the FORTRAN version at http://scatterlib.wikidot.com/coated-spheres

ACKNOWLEDGMENTS

The authors acknowledge the anonymous referee for providing very useful comments which significantly improved the quality of the manuscript. Also, the authors acknowledge Mr. Christian Deysenroth for the very valuable contribution in the designing and development of the experimental set-up. We would like to thank Volker Ossenkopf for fruitful discussion to offering insight into the dust modelling methodology. The work of Arsenii A. Gavdush on alignment of the THz-TDS set-up and digital processing of the THz waveforms was supported by the Russian Foundation for Basic Research (RFBR), Project #18-32-00816. The work of Gennady A. Komandin in solving the inverse problem of THz time-domain spectroscopy was supported by the Russian Science Foundation (RSF), Project #18-12-00328. Tommaso Grassi acknowledges the support by the DFG clusterof excellence "Origin and Structure of the Universe" (http://www.universe-cluster.de/). This work was funded by the Deutsche Forschungsgemeinschaft (DFG, German Research Foundation) – Ref no. FOR 2634/1 ER685/11-1. This work has been partially supported by the project PRIN-INAF 2016, The Cradle of Life – GENESIS-SKA (General Conditions in Early Planetary Systems for the rise of life with SKA).

APPENDIX A: THZ-TDS OPTICS

In this section we provide a detailed description of the configuration of the laser beam inside the optical compartment of the set-up, already shown in Fig. 2. The power of the laser beam is attenuated and successively divided into equivalent channels. Thus, the antenna-emitter is pumped and the antenna-detector is probed with an equal average power of about 20 mW. The optical delay between the pump and probe beams is varied using a double-pass linear mechanical delay stage from Zaber with the travel range of 101.6 mm and the positioning accuracy of $< 3\,\mu$m. The THz radiation undergoes 10 kHz electrical modulation to detect synchronously the THz amplitude using a lock-in detection principle.

The THz beam emitted by the photoconductive antenna is collimated by an integrated HRFZ-Si hemispherical lens and then focussed on a substrate window using a polymethylpentene (TPX) lens with a focal length of 67 mm and diameter of 25 mm. After passing through the cryostat vacuum chamber, the beam is collimated by an equal TPX lens in the direction of the antenna-detector. Finally, the THz beam is focussed onto the photoconductive gap of the antenna-detector by an equal integrated HRFZ-Si hemispherical lens. In our measurements, during the waveform detection, we used a time-domain stride of 50 fs, which allows for satisfying the Whittaker–Nyquist–Kotelnikov–Shannon sampling theorem (Nyquist 1928), at time-domain window size of 100 ps and an averaging time of 0.1 s at each time-domain step with no waveform averaging.

The signal measured at the antenna-detector is recorded in the time domain $E(t)$ and converted into its Fourier-spectrum $E(\nu)$ via

$$E(\nu) = \frac{1}{\sqrt{2\pi}} \int_{-\infty}^{+\infty} E(t)\, \mathrm{e}^{-i2\pi\nu t} \mathrm{d}t, \qquad \text{(A.1)}$$

where t and ν stand for time and frequency.

APPENDIX B: RECONSTRUCTION OF THE TERAHERTZ DIELECTRIC PERMITTIVITY

Equations (3) and (4) are obtained as follows. We defined the complex amplitude of electromagnetic wave $E_0 = |E_0|\exp(i\varphi_0)$, which interacts either with a bare substrate (which we use as reference) or with two ice layers (when detecting the sample waveform). The pulses complex amplitudes depend on the initial complex amplitude of electromagnetic wave E_0 and on ices and reference layers, which define the THz-wave reflection and transmission at the interfaces (*R*-operators and *T*-operators, respectively) and its absorption and phase delays in a bulk material (*P*-operators).

Then, we define the amplitudes of the ballistic reference pulse as follows:

$$\begin{aligned} E_{\mathrm{R,b}} &= E_0 P_0 (l_0 - l_{\mathrm{Si}})\, T_{0,\mathrm{Si}}\, P_{\mathrm{Si}} (l_{\mathrm{Si}})\, T_{\mathrm{Si},0} \\ &= \left|E_{\mathrm{R,b}}\right| \exp (i\varphi_{\mathrm{R,b}}), \end{aligned} \tag{B.1}$$

where l is the thickness of the medium, and the symbols 0 and Si indicate the vacuum and HRFZ-Si, respectively,

the ballistic sample pulse as follows:

$$\begin{aligned} E_{\mathrm{S,b}} &= E_0 P_0 (l_0 - l_{\mathrm{Si}} - l_{\mathrm{CO,I}} - l_{\mathrm{CO,II}})\, T_{0,\mathrm{CO}} P_{\mathrm{CO}} (l_{\mathrm{CO,I}}) \\ &\times T_{\mathrm{CO,Si}} P_{\mathrm{Si}} (l_{\mathrm{Si}})\, T_{\mathrm{Si,CO}} P_{\mathrm{CO}} (l_{\mathrm{CO,II}})\, T_{\mathrm{CO},0} \\ &= \left|E_{\mathrm{S,b}}\right| \exp (i\varphi_{\mathrm{S,b}}), \end{aligned} \tag{B.2}$$

where $l_{\mathrm{CO,I}}$ and $l_{\mathrm{CO,II}}$ are defined as in Sect. 3.1,

and the two satellite sample pulses as follows:

$$\begin{aligned} E_{\mathrm{S,1s}} &= E_0 P_0 (l_0 - l_{\mathrm{Si}} - l_{\mathrm{CO,I}} - l_{\mathrm{CO,II}})\, T_{0,\mathrm{CO}} P_{\mathrm{CO}} (l_{\mathrm{CO,I}}) \\ &\times T_{\mathrm{CO,Si}} P_{\mathrm{Si}} (l_{\mathrm{Si}})\, T_{\mathrm{Si,CO}} P_{\mathrm{CO}} (l_{\mathrm{CO,II}})\, T_{\mathrm{CO},0} \\ &\times R_{\mathrm{CO,Si}} P_{\mathrm{CO}}^2 (l_{\mathrm{CO,I}})\, R_{\mathrm{CO},0} \\ &= \left|E_{\mathrm{S,1s}}\right| \exp (i\varphi_{\mathrm{S,1s}}), \\ E_{\mathrm{S,2s}} &= E_0 P_0 (l_0 - l_{\mathrm{Si}} - l_{\mathrm{CO,I}} - l_{\mathrm{CO,II}})\, T_{0,\mathrm{CO}} P_{\mathrm{CO}} (l_{\mathrm{CO,I}}) \\ &\times T_{\mathrm{CO,Si}} P_{\mathrm{Si}} (l_{\mathrm{Si}})\, T_{\mathrm{Si,CO}} P_{\mathrm{CO}} (l_{\mathrm{CO,II}})\, T_{\mathrm{CO},0} \\ &\times R_{\mathrm{CO},0} P_{\mathrm{CO}}^2 (l_{\mathrm{CO,II}})\, R_{\mathrm{CO,Si}} \\ &= \left|E_{\mathrm{S,2s}}\right| \exp (i\varphi_{\mathrm{S,2s}}), \end{aligned} \tag{B.3}$$

which are clearly observed in reference and sample TDS waveforms in Fig. 4a.

Then, by neglecting the phase changes during the reflection at the interfaces of absorbing media as well as the distortion of optical pulses due to dispersion of material parameters, we calculated the phases of these pulses as

$$
\begin{aligned}
\varphi_{R,b} &= \frac{2\pi\nu}{c}\left(n_0\left(l_0 - l_{Si}\right) + n_{Si}l_{Si}\right) \\
&= -\frac{2\pi\nu}{c}\left(l_0 - l_{Si}\left(n_{Si} - 1\right)\right), \\
\varphi_{S,b} &= \frac{2\pi\nu}{c}\left(n_0\left(l_0 - l_{Si} - l_{CO,I} - l_{CO,II}\right) + n_{Si}l_{Si}\right. \\
&\quad \left. + n_{CO}\left(l_{CO,I} + l_{CO,II}\right)\right) \\
&= -\frac{2\pi\nu}{c}\left(l_0 - l_{Si}\left(n_{Si} - 1\right) - \left(l_{CO,I}l_{CO,II}\right)\left(n_{CO} - 1\right)\right), \\
\varphi_{S,1s} &= \frac{2\pi\nu}{c}\left(n_0\left(l_0 - l_{Si} - l_{CO,I} - l_{CO,II}\right) + n_{Si}l_{Si}\right. \\
&\quad \left. + n_{CO}\left(3l_{CO,I} + l_{CO,II}\right)\right) \\
&= -\frac{2\pi\nu}{c}\left(l_0 - l_{Si}\left(n_{Si} - 1\right) - \left(l_{CO,I} + l_{CO,II}\right)\left(n_{CO} - 1\right)\right. \\
&\quad \left. -2l_{CO,I}n_{CO}\right), \\
\varphi_{S,2s} &= \frac{2\pi\nu}{c}\left(n_0\left(l_0 - l_{Si} - l_{CO,I} - l_{CO,II}\right) + n_{Si}l_{Si}\right. \\
&\quad \left. + n_{CO}\left(l_{CO,I} + 3l_{CO,II}\right)\right) \\
&= -\frac{2\pi\nu}{c}\left(l_0 - l_{Si}\left(n_{Si} - 1\right) - \left(l_{CO,I} + l_{CO,II}\right)\left(n_{CO} - 1\right)\right. \\
&\quad \left. -2l_{CO,II}n_{CO}\right),
\end{aligned} \tag{B.4}
$$

where n_0 and n_{Si} are defined as in Sect. 3.1 and n_{CO} is the refractive index of the CO ice. The phase changes are weak since we study rather low-absorbing dielectric materials. The dispersion of HRFZ-Si is very low, while the dispersion in ice is negligible owing to its small thickness.

These phases are used to calculate the time delays between the pulses, which are indicated in Fig. 4a, i.e.

$$
\begin{aligned}
\delta t_{01} &= \frac{\varphi_{S,b} - \varphi_{R,b}}{2\pi\nu} = \frac{n_{CO} - 1}{c}\left(l_{CO,I} + l_{CO,II}\right), \\
\delta t_{12} &= \frac{\varphi_{S,1s} - \varphi_{S,b}}{2\pi\nu} = \frac{2n_{CO}}{c}l_{CO,I}, \\
\delta t_{13} &= \frac{\varphi_{S,2s} - \varphi_{S,b}}{2\pi\nu} = \frac{2n_{CO}}{c}l_{CO,II}.
\end{aligned} \tag{B.5}
$$

Solving this system of equations yields Eqs. (3) and (4).

Figure 5b shows the Fourier spectra $E(\nu)$ of the reference and sample TDS waveforms. In order to filter out the contribution of the satellite THz pulses (caused by the interference in the input and output windows and the substrate) and to improve the analysis of the frequency-domain data, we apply equal apodization procedure (window filtering) to all waveforms,

$$
E_{filt}(t) = E(t)\,H(t - t_0), \tag{B.6}
$$

where $E(t)$ and $E_{\text{filt}}(t)$ stand for the initial and filtered waveforms, $H(t)$ defines the apodization function, and t_0 defines a position of the apodization filter towards the THz waveform, i.e.

$$H(t) = \begin{cases} \frac{1}{2} + \frac{1}{2}\cos\left[\frac{2\pi}{\omega}\left(\frac{t}{\tau} - \frac{\omega}{2}\right)\right], & 0 < \frac{t}{\tau} < \frac{\omega}{2}, \\ 1, & \frac{\omega}{2} < \frac{t}{\tau} < 1 - \frac{\omega}{2}, \\ \frac{1}{2} + \frac{1}{2}\cos\left[\frac{2\pi}{\omega}\left(\frac{t}{\tau} - 1 + \frac{\omega}{2}\right)\right], & 1 - \frac{\omega}{2} < \frac{t}{\tau} < 1. \end{cases} \tag{B.7}$$

The parameter $H(t)$ is a Tukey apodization filter (Tukey et al. 1986) with the width τ, and ω stands for a parameter of the filter smoothness; for $\omega = 0$ the window has a rectangular form and for $\omega = 1$ it is the Hann window Harris (1978). As shown in Fig. 5a, we use the Tukey window centred at the maximum of the reference THz waveform with the smoothness parameter of $\omega = 0.1$ and the width of 40 ps, which yields the frequency-domain resolution of 0.025 THz.

Let us consider the Fresnel formulas (Born & Wolf 1980), defining the THz wave amplitude reflection at (and transmission through) the interface between media m and k as follows:

$$R_{m,k}(\nu) = \frac{n_m(\nu) - n_k(\nu)}{n_m(\nu) + n_k(\nu)},$$
$$T_{m,k}(\nu) = \frac{2n_m(\nu)}{n_m(\nu) + n_k(\nu)}, \tag{B.8}$$

where $R_{m,k}(\nu)$ and $T_{m,k}(\nu)$ stand for coefficients of the complex amplitude reflection and transmission, respectively, while $n_m(\nu)+in_k(\nu)$ is the complex refractive index of the media. The relation between complex amplitudes of the THz wave right after the emitter ($z = 0$), $E_0(\nu)$ and at the position z along the beam axis, $E(\nu,z)$ is given by

$$E(\nu, z) = E_0(\nu)\exp\left(-i\frac{2\pi\nu}{c}n(\nu)z\right). \tag{B.9}$$

If the thicknesses and refractive indexes of all layers are known, Eqs. (B.8) and (B.9) yield description of all peculiarities of the THz pulse interacting with multilayered structures (Zaytsev et al. 2014, 2015; Gavdush et al. 2019).

We derive the equations describing the complex amplitudes of the reference $E_R(\nu)$ and sample $E_S(\nu)$ spectra, assuming only wavelets inside the Tukey apodization. For the reference spectrum, we obtain

$$E_R/E_0 = P_0(l_0 - l_{Si})\, T_{0,Si}\, P_{Si}(l_{Si})\, T_{Si,0}, \tag{B.10}$$

where the indexes 0 and Si correspond to the free space and the HRFZ-Si medium; l_0 and l_{Si} are the total length of the THz beam path and the thickness of the HRFZ-Si substrate, respectively; $T_{0,Si}(\nu)$ and $T_{Si,0}(\nu)$ are the transmission coefficients for the respective interfaces (between the free space and the HRFZ-Si), defined by Eq. (B.8); and $P_0(\nu,z)$ and $P_{Si}(\nu,z)$ are operators describing the THz wave propagation in the free space and the HRFZ-Si, respectively, as given by the exponential factor in Eq. (B.9).

As shown in Fig. 3, for the sample spectrum we take into account the contribution of the ballistic THz pulse (1) and the satellite pulses (2 and 3), caused by the multiple THz wave reflection in the ice films. This yields the following equation:

$$\begin{aligned} E_S/E_0 = {} & P_0(l_0 - l_{Si} - l_{CO,I} - l_{CO,II})\, T_{0,CO} P_{CO}(l_{CO,I}) \\ & \times T_{CO,Si} P_{Si}(l_{Si})\, T_{Si,CO} P_{CO}(l_{CO,II})\, T_{CO,0} \\ & \times \left[1 + R_{CO,Si} P^2_{CO}(l_{CO,I})\, R_{CO,0} \right. \\ & \left. + R_{CO,0} P^2_{CO}(l_{CO,II})\, R_{CO,Si}\right], \end{aligned} \tag{B.11}$$

where the summation terms in the brackets correspond to the wavelets 1, 2, and 3; $l_{CO,I}$ and $l_{CO,II}$ stand for thicknesses of the first and second ice films; $R_{CO,Si}(\nu)$ and $R_{CO,0}(\nu)$ are the transmission coefficients for the respective interfaces, see Eq. (B.8); the definition of the remaining factors (P and T) is similar to that in Eq. (B.10). We point out that the complex amplitudes (E) and all factors (T, R, and P) in Eqs. (B.10) and (B.11) are frequency-dependent. Equations (B.10) and (B.11) form a basis for the reconstruction of the THz dielectric response of ices.

The reconstruction is performed via the following minimization procedure:

$$n(\nu) = \operatorname{argmin}_{n(\nu)} \left[\mathbf{\Phi}(n(\nu), \nu)\right], \tag{B.12}$$

where argmin is an operator which determines the minimum argument of the vector error functional **Φ**. The latter is formed from the complex theoretical T_{Th} and experimental T_{Exp} transfer functions

$$\mathbf{\Phi}(n(\nu), \nu) = \begin{pmatrix} |T_{Th}(n,\nu)| - |T_{Exp}(\nu)| \\ \phi[T_{Th}(n,\nu)] - \phi[T_{Exp}(\nu)] \end{pmatrix}, \tag{B.13}$$

where $|\ldots|$ and $\phi[\ldots]$ are the absolute values and phases of the complex functions, respectively.

We define the theoretical transfer function T_{Th} as the sample spectrum (Eq. (B.11)) normalized by the reference spectrum (Eq. (B.10))

$$T_{Th} = \frac{T_{0,CO}T_{CO,Si}T_{Si,CO}T_{CO,0}}{T_{0,Si}T_{Si,0}} \times P_{CO}(l_{CO,I} + l_{CO,II}) P_0(-l_{CO,I} - l_{CO,II}) \times \left[1 + R_{CO,Si}P_{CO}^2(l_{CO,I}) R_{CO,0} + R_{CO,0}P_{CO}^2(l_{CO,II}) R_{CO,Si}\right]. \quad \text{(B.14)}$$

Considering all reflection and transmission operators in Eq. (B.9), we note that T_{Th} depends only on the refractive index of the HRFZ-Si substrate n_{Si}, which is known a priori and on the parameters of the CO ice to be determined; it excludes the contribution of several factors, such as the unknown complex amplitude of the TDS source $E_0(\nu)$, the unknown total length of the THz beam path l_0, and, finally, the thickness of the HRFZ-Si substrate l_{Si}, which is known, too, but can slightly vary owing to angular deviations of the substrate during the vacuum chamber assembling.

The experimental transfer function T_{Exp} is calculated in a similar manner, relying on the Fourier spectra of the experimental sample E_S and reference E_R waveforms (after applying the Tukey apodization),

$$T_{Exp} = \frac{E_S}{E_R}. \quad \text{(B.15)}$$

We note that all the functions and operators in the theoretical and experimental transfer functions are frequency-dependent, and both transfer functions take into account only the ballistic pulses of the reference and sample waveforms, as well as the first and second ice-related satellite pulses of the sample waveform.

By introducing equal confidence intervals for the refractive index n' and the amplitude absorption coefficient α, as $[n'_{init} - \Delta n', n'_{init} + \Delta n']$ and $[0, \alpha_{max}]$ with $\Delta n'$=0.25 and $\alpha_{max} = 15\ cm^{-1}$, we use the non-linear trust region approach (Coleman & Li 1996) to reconstruct the THz dielectric response of the CO ice in the spectral operation range of 0.3–2.0 THz.

APPENDIX C: OPACITY MODEL BENCHMARK

In order to verify the correctness of the machinery employed to calculate the opacity from the dielectric constants, we reproduce the results found by Ossenkopf & Henning (1994) in their Fig. 5, panels a–c, compact grains case. We describe the methodology employed. The results described can be reproduced by running test_04.py from the publicly available code[2], while Fig. 7 from our paper can be reproduced with test_05.py.

C.1. Dielectric constants

The complex dielectric functions $\varepsilon' - i\varepsilon''$ of the refractory components are taken from Ossenkopf et al. (1992, cool oxygen-rich silicates, their Fig. 10) and from Preibisch et al. (1993, amorphous carbon, their Table 1), while ice is assumed to be a H_2O:CH_3OH:CO:NH_3 = 100:10:1:1 mixture at 10 K from Hudgins et al. (1993, their Table 2A).

For the aims of this work we need to extrapolate the ice data relative to the H_2O:CH_3OH:CO:NH_3 = 100:10:1:1 mixture to longer wavelengths, as done by Ossenkopf & Henning (1994). First we fit the last 45 data points[3] of the imaginary part (i.e. approximately 71−194 μm) with a $f(\lambda) \propto \lambda^{-1}$ function (Ossenkopf, priv. comm.), and we use this to extrapolate ε'' with 200 linearly spaced wavelength points, in the range 71−800 μm. To retrieve the real part of ε at each $\omega = 2\pi c \lambda^{-1}$ point, we apply the Kramers–Kronig (e.g., Bohren & Huffman 1983) relations in the discrete form

$$\varepsilon'(\omega) = 1 + \frac{2}{\pi}\Omega(\omega), \qquad \text{(C.1)}$$

with $\Omega(\omega)$ the discrete integral over the positive frequency ranges using the composite trapezoidal rule with the integrand $\frac{\varepsilon''(\omega_i)\,\omega_i}{\omega_i^2 - \omega^2}$, and excluding ω_i where the denominator of the argument vanishes (i.e. the Cauchy principal value of the corresponding finite integral).

After the extrapolation, the ice dielectric functions are modified by mixing spherical inclusions of amorphous carbon using the Bruggeman effective medium approximation (Bruggeman 1935), with a volume filling fraction of 0.11 and 0.013 for the thin and the thick ice cases, respectively.

C.2. Absorption Coefficients

The absorption efficiency $Q_{abs}(\lambda, a)$ is function of the wavelength and grain size a, and computed with the routine BHMIE.PY for the bare grains[4] and with BHCOAT.PY for the coated grains[5], both from Bohren & Huffman (1983). In Ossenkopf & Henning (1994) there are three cases: bare grains without ice; thin ice, which has a volume ratio of $V = 0.5$; and thick ice, where $V = 4.5$. The radius of the refractory core a and the fraction V determine the radius of the mantle $a_{coat} = a(V+1)^{1/3}$.

C.3. Opacities

With Q_{abs} it is possible to retrieve the opacities averaged on the grain size distribution $\varphi(a)$ as

$$\kappa(\lambda) = \frac{\pi}{C} \int_{a_{min}^{coat}}^{a_{max}^{coat}} \varphi(a)\, a^2 Q(\lambda, a)\, da, \quad \text{(C.2)}$$

where a_{min}^{coat} to a_{max}^{coat} is the range in which the size distribution is valid including coating, and

$$C = \frac{4}{3}\pi\rho_0 \int_{a_{min}}^{a_{max}} \varphi(a)\, a^3 da, \quad \text{(C.3)}$$

where a_{min} to a_{max} is the range where the size distribution is valid, but considering only the refractory material, i.e. silicates or carbonaceous dust.

Ossenkopf & Henning (1994) assume $a_{min} = 5\times10^{-7}$ cm, $a_{max} = 2.5\times10^{-5}$ cm, $\rho_0 = 2.9$ g cm^{-3} (silicates) and $\rho_0 = 2$ g cm^{-3} (amorphous carbon), and $\varphi(a)=a^{-3.5}$.

The total opacity is $\kappa_{tot}(\lambda)=0.678\kappa_{Si}(\lambda)+0.322\kappa_{AC}(\lambda)$, where the two terms are, respectively, the silicates and carbonaceous opacities including ice coating, and the two coefficients are calculated from the volume ratio discussed in Sect. 3.1 of Ossenkopf & Henning (1994). In particular, Ossenkopf & Henning assume a volume ratio of the refractory components $V_{AC}/V_{Si} = 0.69$ (their Sect. 3.1), which can be converted into the corresponding mass ratio $M_{AC}/M_{Si} = 0.4758$ using the relation $M_i = \rho_i V_i$, where ρ_i are the bulk densities of the two refractory components (being in Ossenkopf & Henning the opacity defined per unit mass of the refractory material), so that $\kappa_{tot} = (M_{AC}\kappa_{AC}+M_{Si}\kappa_{Si})/(M_{AC}+M_{Si})$.

REFERENCES

1. Allodi, M. A., Ioppolo, S., Kelley, M. J., McGuire, B. A., & Blake, G. A. 2014, Phys. Chem. Chem. Phys. (Incorporating Faraday Transactions), 16, 3442
2. Anderson, A., & Leroi, G. E. 1966, J. Chem. Phys., 45, 4359
3. Auston, D. H. 1975, Appl. Phys. Lett., 26, 101
4. Baratta, G. A., & Palumbo, M. E. 1998, J. Opt. Soc. Am. A, 15, 3076
5. Baratta, G. A., & Palumbo, M. E. 2017, A&A, 608, A81
6. Bohren, C. F., & Huffman, D. R. 1983, Absorption and Scattering of Light by Small Particles (New York: Wiley)
7. Born, M., & Wolf, E. 1980, Principles of Optics, 6th edn. (UK: Pergamon Press)
8. Bruggeman, D. A. G. 1935, Ann. Phys., 416, 636
9. Caselli, P., Walmsley, C. M., Tafalla, M., Dore, L., & Myers, P. C. 1999, ApJ, 523, L165
10. Coleman, T., & Li, Y. 1996, SIAM J. Optim., 6, 418
11. Dartois, E. 2006, A&A, 445, 959
12. Dutrey, A., Guilloteau, S., Prato, L., et al. 1998, A&A, 338, L63
13. Ehrenfreund, P., Boogert, A. C. A., Gerakines, P. A., Tielens, A. G. G. M., & van Dishoeck, E. F. 1997, A&A, 328, 649
14. Gavdush, A. A., Chernomyrdin, N., Malakhov, K., et al. 2019, J. Biomed. Opt., 24, 027001
15. Giuliano, B. M., Escribano, R. M., Martín-Doménech, R., Dartois, E., & Muñoz Caro, G. M. 2014, A&A, 565, A108
16. Giuliano, B. M., Martín-Doménech, R., Escribano, R. M., Manzano-Santamaría, J., & Muñoz Caro, G. M. 2016, A&A, 592, A81
17. Griffiths, P., & de Haseth, J. 1986, Fourier Transform Infrared Spectroscopy (New York, NY, USA: John Wiley + Sons)
18. Harris, F. J. 1978, Proc. IEEE, 66, 5
19. Hudgins, D. M., Sandford, S. A., Allamandola, L. J., & Tielens, A. G. G. M. 1993, ApJS, 86, 713
20. Ioppolo, S., McGuire, B. A., Allodi, M. A., & Blake, G. A. 2014, Faraday Discuss., 168, 461
21. Kawase, K., Sato, M., Taniuchi, T., & Ito, H. 1996, Applied Physics Letters, 68, 2483

22. Kiessling, J., Breunig, I., Schunemann, P. G., Buse, K., & Vodopyanov, K. L. 2013, New J. Phys., 15, 105014
23. Komandin, G. A., Chuchupal, S. V., Lebedev, S. P., et al. 2013, IEEE Trans. Terahertz Sci. Technol., 3, 440
24. Lee, Y. S. 2009, Principles of Terahertz Science and Technology (New York, USA: Springer)
25. Loeffler, M. J., Baratta, G. A., Palumbo, M. E., Strazzulla, G., & Baragiola, R. A. 2005, A&A, 435, 587
26. Martin, P. C. 1967, Phys. Rev., 161, 143
27. Mastrapa, R. M., Sandford, S. A., Roush, T. L., Cruikshank, D. P., & Dalle Ore, C. M. 2009, ApJ, 701, 1347
28. McGuire, B. A., Ioppolo, S., Allodi, M. A., & Blake, G. A. 2016, Phys. Chem. Chem. Phys., 18, 20199
29. Nyquist, H. 1928, Trans. Am. Inst. Electr. Eng., 47, 617
30. Ossenkopf, V., & Henning, T. 1994, A&A, 291, 943
31. Ossenkopf, V., Henning, T., & Mathis, J. S. 1992, A&A, 261, 567
32. Palumbo, M. E., Baratta, G. A., Collings, M. P., & McCoustra, M. R. S. 2006, Phys. Chem. Chem. Phys. (Incorporating Faraday Transactions), 8, 279
33. Pontoppidan, K. M., Fraser, H. J., Dartois, E., et al. 2003, A&A, 408, 981
34. Preibisch, T., Ossenkopf, V., Yorke, H. W., & Henning, T. 1993, A&A, 279, 577
35. Preu, S., Döhler, G. H., Malzer, S., Wang, L. J., & Gossard, A. C. 2011, J. Appl. Phys., 109, 061301
36. Pupeza, I., Wilk, R., & Koch, M. 2007, Opt. Express, 15, 4335
37. Ron, A., & Schnepp, O. 1967, J. Chem. Phys., 46, 3991
38. Tukey, J., Cleveland, W. S., & Brillinger, D. R. 1986, The Collected Works of John W. Tukey. Volume I: Time Series, 1949–1964 (Wadsworth Statistics/Probability Series), 1st edn. (Wadsworth Advanced Books & Software)
39. Urso, R. G., Scirè, C., Baratta, G. A., Compagnini, G., & Palumbo, M. E. 2016, A&A, 594, A80
40. Van Exter, M., Fattinger, C., & Grischkowsky, D. 1989, Appl. Phys. Lett., 55, 337

41. Warren, S. G., & Brandt, R. E. 2008, J. Geophys. Res. (Atmospheres), 113, D14220
42. Zaytsev, K., Gavdush, A., Karasik, V., et al. 2014, J. Appl. Phys., 115, 193105
43. Zaytsev, K., Gavdush, A., Chernomyrdin, N., & Yurchenko, S. 2015, IEEE Trans. Terahertz Sci. Technol., 5, 817

SECTION 2:
FROM THE VERY EARLY UNIVERSE TO THE FIRST ATOMS

3

Big Bang Nucleosynthesis in Visible and Hidden-Mirror Sectors

Paolo Ciarcelluti[1]

[1]Web Institute of Physics, Via Fortore 3, 65015 Montesilvano, Italy

ABSTRACT

One of the still viable candidates for the dark matter is the so-called mirror matter. Its cosmological and astrophysical implications were widely studied, pointing out the importance to go further with research. In particular, the Big Bang nucleosynthesis provides a strong test for every dark matter candidate, since it is well studied and involves relatively few free parameters. The necessity of accurate studies of primordial nucleosynthesis with mirror matter has then emerged. I present here the results of accurate numerical simulations of the primordial production of both ordinary nuclides and nuclides made of mirror baryons, in presence of a hidden mirror sector with unbroken parity symmetry and with gravitational interactions only. These elements are the building blocks of all the structures forming in the Universe; therefore, their chemical composition is a key ingredient for astrophysics with mirror dark matter. The production of ordinary nuclides

Citation: Paolo Ciarcelluti "Big Bang Nucleosynthesis in Visible and Hidden-Mirror Sectors" Advances in High Energy Physics Volume 2014, Article ID 702343, 7 pages http://dx.doi.org/10.1155/2014/702343

shows differences from the standard model for a ratio of the temperatures between mirror and ordinary sectors $x = T'/T \gtrsim 0.3$, and they present an interesting decrease of the abundance of . For the mirror nuclides, instead, one observes an enhanced production of ^4He, which becomes the dominant element for $x \lesssim 0.5$, and much larger abundances of heavier elements.

INTRODUCTION

The nature of the dark matter of the Universe is still completely unknown, and mirror matter represents one of the possible promising candidates. Its postulation derives theoretically from the necessity to restore the parity symmetry of the physical laws and phenomenologically from the need of describing the physics at any scale, from the whole cosmos to the elementary particles. In its basic theory, mirror matter is formed by baryons with exactly the same properties as our ordinary baryons, but with opposite handedness (right) of weak interactions so that globally the system of all particles together (ordinary and mirror) is parity symmetric. All the particles and the coupling constants are the same; then the physical laws of mirror matter are the same as that of ordinary matter, but the only interaction between the two kinds of particles is gravitational, while the other fundamental interactions act separately in each sector. This is valid also for electromagnetic interactions, meaning that a mirror photon would interact with mirror baryons but not with the ordinary ones, making mirror matter invisible to us and detectable just via their gravitational effects. To this basic model of interactions, it is possible to add other interactions involving mixings between ordinary and mirror particles. The most important of them is, at the present, the kinetic mixing of photons that in the mirror paradigm would be responsible of the positive results of the dark matter direct detection experiments. Extensive reviews on mirror matter at astroand particle physics levels can be found, for example, in [1–4]. The original idea and the first applications of mirror matter are present in [5–8]. One of the key points in the macroscopic mirror theory is that, even if the physical laws are the same as ordinary matter, the initial conditions can be different. This means that the densities of particles and their temperatures can be different, leading to the need of only two free parameters that describe the basic mirror model defined as

$$x \equiv \left(\frac{s'}{s}\right)^{1/3} \approx \left(\frac{T'}{T}\right),$$
$$\beta \equiv \frac{\Omega'_b}{\Omega_b}, \quad (1)$$

where $s(s'), T(T')$, and $\Omega_b(\Omega'_b)$ are, respectively, the ordinary (mirror) entropy density, photon temperature, and cosmological baryon density.

Different cosmological parameters for initial temperatures and densities mean different cosmological evolutions for the two kinds of particles, in particular concerning the key phenomena of Big Bang nucleosynthesis (BBN), recombination, cosmic microwave background (CMB), large scale structure (LSS) formation, and the following evolution at lower scales, as galactic and stellar formation and evolution. The CMB and LSS were well studied [3, 9–13], and a recent work [14] has even shown that mirror matter can fit the observations with the same level of accuracy as generic cold dark matter (CDM). Primordial nucleosynthesis was studied in the past in several works [1, 6, 7, 15–17], obtaining historically the first bound on mirror matter parameters [18]. In fact, if, for example, the temperature of the mirror particles would be the same as that of ordinary ones, the contribution of mirror relativistic species to the Hubble expansion rate at BBN epoch would be equivalent to that of an effective number of extra- (massless) neutrino families $\Delta N_\nu \simeq 6.14$. This would be in conflict with any estimate, even the most conservative one. Then, considering the weak bound $\Delta N_\nu \lesssim 1$ and just applying the approximate relation $\Delta N_\nu \approx 6.14x^4$, one obtains $x \lesssim 0.7$ [15, 18]. The 4th power of x gives a mild dependence on this parameter. In view of the definition 1 of x, this simply means that the mirror particles should have a lower temperature than the ordinary ones in the early Universe.

One of the peculiarities of mirror matter is that not only it influences the ordinary BBN but also it has its own mirror BBN. This is a parallel primordial nucleosynthesis that is influenced by the ordinary baryonic matter, in an analogous way and for the same reason as mirror matter influences ordinary BBN. The big difference is that, while ordinary BBN receives very low influence by the mirror matter, since $\Delta N_\nu \propto x^4$ and $x<1$, the opposite happens for the mirror BBN, since instead for it $\Delta N_\nu \propto x^{-4}$!.

As the mirror particles have a lower temperature than the ordinary ones, the conditions required to start the nucleosynthesis are reached at earlier cosmological times, which mean different conditions, and in particular a larger cosmic expansion rate. This effect was studied in previous works, obtaining as a main result an increased production of mirror helium He' in comparison with the ordinary one. This last effect was studied also in the presence of the photon-mirror photon kinetic mixing, obtaining similar results [19, 20]. The primordial abundance Y' of He' is dependent on the inverse of the parameter x and can reach very high values, up to $Y' = 0.8$-0.9 for low values $x \sim 0.1$. All these studies considered the approximate relation $x \simeq T'/T$ expressed in (1) to be valid. Indeed this relation is valid along most of the history of the Universe, but not at the period of BBN, since at this time the electron-positron annihilations in both ordinary and mirror sectors heat the respective photons at different times, inducing large differences between x and T'/T, up to 30–40%. This effect and the thermodynamics of the early Universe were studied in detail in [21]. Here we use the results of that work for the degrees of freedom and temperatures in the two particle sectors, in order to obtain a more detailed numerical description of the ordinary and mirror primordial nucleosynthesis processes, and more carefully predict the primordial abundances. Some preliminary results of this analysis were previously presented in [1, 16, 17]

An accurate study of BBN in ordinary and mirror particle sectors is important for the following reasons.

- At the present status of knowledge, the standard theory of BBN is essentially dependent on just one parameter, the baryonic asymmetry $\eta = n_b/n_\gamma$. If compared with other cosmological tools, as, for example, CMB, which have much more free parameters, one can easily understand its importance as a key test for any dark matter candidate.
- In standard BBN there is the still open "lithium problem," related to the discrepancy between observations and predictions of the primordial abundance of this nuclide. This suggests the need for new physics beyond the standard model; then it is important to understand if mirror matter can alleviate this problem
- The interpretation of DAMA and other direct detection experiments in terms of mirror matter is dependent on the abundance of mirror helium He and heavier elements (so-called "metals"); then a correct estimate of their abundances is crucial in order to verify this hypothesis [22, 23].

- The mirror BBN furnishes the primordial chemical composition of dark matter, which sets the initial conditions for the formation and evolution of structures at cosmic, galactic, and subgalactic scales. The effect of the enhanced abundance of He on the evolution of mirror stars has already been studied, showing a large effect [24]. It is fundamental to estimate also the abundances of metals, which are responsible for many opacity processes of matter, involved in fragmentation processes forming galaxies and stars.

MODELS

I consider models in which the dark matter is made of mirror matter and there are no interactions between ordinary and mirror particles except for the gravitation that links the two sectors. Each set of particles is assumed to be in thermodynamical equilibrium (as usual in primordial nucleosynthesis studies) independently of the other one. This implies that there are no entropy exchanges between the two sectors then the entropy densities are separately conserved, and the parameter x is constant by definition. Using the definition of the entropy density [25], the parameter x is

$$x \equiv \left(\frac{s'}{s}\right)^{1/3} = \left[\frac{q'(T')}{q(T)}\right]^{1/3} \frac{T'}{T}, \qquad (2)$$

where $q(T)$ and $q'(T')$ are the ordinary and mirror entropic degrees of freedom. The ratio of entropic degrees of freedom in the two sectors determines the mirror temperature as a function of the ordinary one, once the parameter x is fixed. This ratio is not constant at the times of primordial nucleosynthesis, since the e^+-e^- annihilations heat the photons of each sectors independently, and they happen at different times in the early Universe. This and related aspects were studied in detail in [21]. Here we use the same treatment, valid for the ranges of temperatures which we are interested in (below ~10 MeV). It is based on the numerical solution of the equations for the conservations of ordinary and mirror entropies and the absence of entropy exchanges (constancy of x):

$$\frac{(7/8)\,q_e\,(T) + q_\gamma}{(7/8)\,q_\nu}\left(\frac{T}{T_\nu}\right)^3 = \frac{22}{21},$$

$$\frac{(7/8)\,q_e\left(T'\right) + q_\gamma}{(7/8)\,q_\nu}\left(\frac{T'}{T'_\nu}\right)^3 = \frac{22}{21},$$

$$\frac{\left[(7/8)\,q_e\left(T'\right) + q_\gamma\right]T'^3 + (7/8)\,q_\nu T'^3_\nu}{\left[(7/8)\,q_e\,(T) + q_\gamma\right]T^3 + (7/8)\,q_\nu T^3_\nu} = x^3, \qquad (3)$$

where q_i is the entropic degrees of freedom of species$i$, Tv and $T_v^{'}$] are the temperatures of the ordinary and mirror neutrinos.

Since ordinary and mirror particles have the same microphysics, I assume that the neutrino decoupling temperature is the same in each sector, which means, due to the initial difference of temperatures, that the decouplings of ordinary and mirror neutrinos take place at different times in the early Universe. This simplification is justified by the fact that, despite being the mirror sector colder, the final value of the neutrino to photon temperature ratio is the same in both sectors

Together with x, the second free parameter for the mirror BBN is the mirror baryon to photon ratio (or baryonic asymmetry) $\eta' = n'_b/n'_\gamma$, which can be expressed in terms of the ordinary baryon to photon ratio $\eta = n_b/n_\gamma$ and the mirror parameters x and β:

$$\eta' = \beta x^{-3}\eta > \eta, \qquad (4)$$

where the inequality is due to the bounds on x and the expected values for β>1.

Since the nuclear physics is the same for ordinary and mirror matter, it is possible to use and modify a preexisting code for primordial nucleosynthesis that numerically solves the equations governing the production and evolution of nuclides. The choice is the well-tested and fast WagonerKawano code [26, 27], which has enough accuracy for the purposes of this analysis. The numerical code has been doubled to include the mirror sector and modified in order to take into account the evolution of the temperature of the mirror particles and the degrees of freedom of both sectors, according to the aforementioned treatment [21]. For the neutron lifetime we consider the value τ= 885.7 s, while for the final baryon to photon the ratio $\eta = 6.14 \cdot 10^{-10}$. We consider the usual standard number of neutrino families for ordinary matter $N_\nu = 3.04$. Then, the only two free parameters of the code are the mirror ones, x and β. Several models are computed for x ranging from 0.1 to 0.7 and β from 1 to 5, which are the values of cosmological interests.

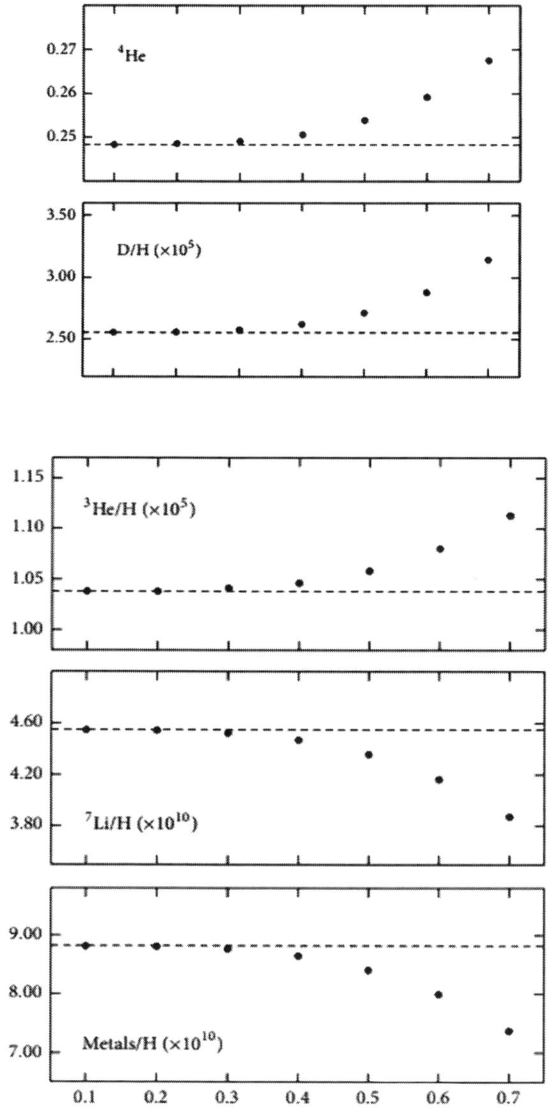

Figure 1: Primordial abundances of ordinary ^{4}He, D, ^{3}He, ^{7}Li, and metals (elements heavier than ^{4}He) for several values of x and compared with the predictions of the standard model (dashed lines).

RESULTS

For each couple of mirror parameters (x, β) the primordial abundances of both ordinary and mirror elements are derived. The ordinary BBN, as expected, is independent on the density of mirror baryons (mirror baryonic asymmetry); then it depends, once fixed the microphysical parameters and the ordinary baryonic asymmetry, on just one parameter, the ratio of entropies x. The mirror BBN, instead, is clearly dependent on both x and the cosmic mirror baryonic density, expressed by the parameter β.

Table 1: Elements produced in the ordinary sector. The last row includes all elements with atomic mass larger than 7.

	Standard	$x = 0.1$	$x = 0.2$	$x = 0.3$	$x = 0.4$	$x = 0.5$	$x = 0.6$	$x = 0.7$
n/H (10^{-16})	1.161	1.161	1.160	1.159	1.510	1.505	1.527	2.044
p	0.7518	0.7518	0.7516	0.7511	0.7495	0.7463	0.7409	0.7326
D/H (10^{-5})	2.554	2.555	2.558	2.575	2.618	2.709	2.874	3.144
T/H (10^{-8})	8.064	8.065	8.076	8.132	8.280	8.588	9.146	10.07
^{3}He/H (10^{-5})	1.038	1.038	1.038	1.041	1.046	1.058	1.080	1.113
^{4}He	0.2483	0.2483	0.2485	0.2491	0.2506	0.2538	0.2592	0.2675
^{6}Li/H (10^{-14})	1.111	1.111	1.113	1.124	1.151	1.210	1.318	1.499
^{7}Li/H (10^{-10})	4.549	4.548	4.543	4.523	4.468	4.356	4.162	3.871
^{7}Be/H (10^{-10})	4.266	4.266	4.260	4.238	4.177	4.051	3.832	3.502
^{8}Li+/H (10^{-15})	1.242	1.242	1.243	1.251	1.269	1.306	1.370	1.464

Table 2: Elements produced in the mirror sector. The last row includes all elements with atomic mass larger than 7

	$x = 0.1$ ($\beta = 5$)	$x = 0.2$ ($\beta = 5$)	$x = 0.3$ ($\beta = 5$)	$x = 0.4$ ($\beta = 5$)	$x = 0.5$ ($\beta = 5$)	$x = 0.6$ ($\beta = 5$)	$x = 0.7$ ($\beta = 5$)
n/H	$5.762 \cdot 10^{-25}$	$2.953 \cdot 10^{-24}$	$2.590 \cdot 10^{-22}$	$2.908 \cdot 10^{-21}$	$1.840 \cdot 10^{-20}$	$6.858 \cdot 10^{-20}$	$1.726 \cdot 10^{-19}$
p	0.1735	0.2840	0.3646	0.4357	0.4966	0.5488	0.5924
D/H	$1.003 \cdot 10^{-12}$	$3.090 \cdot 10^{-10}$	$4.838 \cdot 10^{-9}$	$2.240 \cdot 10^{-8}$	$6.587 \cdot 10^{-8}$	$1.553 \cdot 10^{-7}$	$3.279 \cdot 10^{-7}$
T/H	$9.679 \cdot 10^{-21}$	$4.999 \cdot 10^{-16}$	$1.238 \cdot 10^{-13}$	$2.603 \cdot 10^{-12}$	$2.108 \cdot 10^{-11}$	$1.030 \cdot 10^{-10}$	$3.722 \cdot 10^{-10}$
^{3}He/H	$3.282 \cdot 10^{-6}$	$3.522 \cdot 10^{-06}$	$3.740 \cdot 10^{-6}$	$3.949 \cdot 10^{-6}$	$4.172 \cdot 10^{-6}$	$4.415 \cdot 10^{-6}$	$4.691 \cdot 10^{-6}$
^{4}He	0.8051	0.7233	0.6351	0.5648	0.5035	0.4514	0.4077
^{6}Li/H	$7.478 \cdot 10^{-21}$	$1.241 \cdot 10^{-18}$	$1.309 \cdot 10^{-17}$	$4.460 \cdot 10^{-17}$	$1.016 \cdot 10^{-16}$	$1.923 \cdot 10^{-16}$	$3.361 \cdot 10^{-16}$
^{7}Li/H	$1.996 \cdot 10^{-7}$	$7.162 \cdot 10^{-8}$	$3.720 \cdot 10^{-8}$	$2.289 \cdot 10^{-8}$	$1.535 \cdot 10^{-8}$	$1.086 \cdot 10^{-8}$	$7.962 \cdot 10^{-9}$
^{7}Be/H	$1.995 \cdot 10^{-7}$	$7.159 \cdot 10^{-8}$	$3.675 \cdot 10^{-8}$	$2.236 \cdot 10^{-8}$	$1.489 \cdot 10^{-8}$	$1.041 \cdot 10^{-8}$	$7.885 \cdot 10^{-9}$
^{8}Li+/H	$4.354 \cdot 10^{-9}$	$3.458 \cdot 10^{-10}$	$5.926 \cdot 10^{-11}$	$1.396 \cdot 10^{-11}$	$3.827 \cdot 10^{-12}$	$1.168 \cdot 10^{-12}$	$3.949 \cdot 10^{-13}$
	$x = 0.1$ ($\beta = 1$)	$x = 0.2$ ($\beta = 1$)	$x = 0.3$ ($\beta = 1$)	$x = 0.4$ ($\beta = 1$)	$x = 0.5$ ($\beta = 1$)	$x = 0.6$ ($\beta = 1$)	$x = 0.7$ ($\beta = 1$)
n/H	$8.888 \cdot 10^{-17}$	$1.110 \cdot 10^{-16}$	$1.915 \cdot 10^{-16}$	$1.620 \cdot 10^{-16}$	$2.058 \cdot 10^{-16}$	$1.399 \cdot 10^{-16}$	$2.076 \cdot 10^{-16}$
p	0.1772	0.2831	0.3675	0.4400	0.5028	0.5566	0.6017
D/H	$1.331 \cdot 10^{-6}$	$4.086 \cdot 10^{-6}$	$7.094 \cdot 10^{-6}$	$1.018 \cdot 10^{-5}$	$1.352 \cdot 10^{-5}$	$1.743 \cdot 10^{-5}$	$2.235 \cdot 10^{-5}$
T/H	$3.068 \cdot 10^{-9}$	$1.192 \cdot 10^{-8}$	$2.190 \cdot 10^{-8}$	$3.228 \cdot 10^{-8}$	$4.358 \cdot 10^{-8}$	$5.675 \cdot 10^{-8}$	$7.328 \cdot 10^{-8}$
^{3}He/H	$5.228 \cdot 10^{-6}$	$6.119 \cdot 10^{-6}$	$6.880 \cdot 10^{-6}$	$7.566 \cdot 10^{-6}$	$8.232 \cdot 10^{-6}$	$8.931 \cdot 10^{-6}$	$9.719 \cdot 10^{-6}$
^{4}He	0.8226	0.7168	0.6326	0.5602	0.4974	0.4436	0.3984
^{6}Li/H	$8.638 \cdot 10^{-15}$	$1.422 \cdot 10^{-14}$	$1.660 \cdot 10^{-14}$	$1.747 \cdot 10^{-14}$	$1.790 \cdot 10^{-14}$	$1.845 \cdot 10^{-14}$	$1.951 \cdot 10^{-14}$
^{7}Li/H	$5.712 \cdot 10^{-8}$	$1.867 \cdot 10^{-8}$	$8.930 \cdot 10^{-9}$	$4.953 \cdot 10^{-9}$	$2.948 \cdot 10^{-9}$	$1.811 \cdot 10^{-9}$	$1.120 \cdot 10^{-9}$
^{7}Be/H	$5.711 \cdot 10^{-8}$	$1.863 \cdot 10^{-8}$	$8.878 \cdot 10^{-9}$	$4.896 \cdot 10^{-9}$	$2.891 \cdot 10^{-9}$	$1.755 \cdot 10^{-9}$	$1.064 \cdot 10^{-9}$
^{8}Li+/H	$2.036 \cdot 10^{-10}$	$1.468 \cdot 10^{-11}$	$2.514 \cdot 10^{-12}$	$5.944 \cdot 10^{-13}$	$1.657 \cdot 10^{-13}$	$5.184 \cdot 10^{-14}$	$1.814 \cdot 10^{-14}$

In Table 1 the primordial abundances of elements produced by ordinary nucleosynthesis, for different values of x compared with the standard model of nucleosynthesis (absence of dark matter), are reported. Protons and 4 He are expressed in mass fraction, all the other nuclei in ratios to the proton abundance. In the last row, indicated with 8 Li+, the contributions of the elements with atomic mass larger than 7 are included all together. The evolution of the abundances has been followed until the end of BBN process (at $T \sim 8 \cdot 10^{-4}$ MeV). It is evident that the differences with the standard BBN appear only for $x > 0.1$ (for $x = 0.1$ they are of order 10^{-4} or less), and for $x = 0.3$ they are limited to below 1%, but they increase for increasing values of x. The abundances of most elements (D, T, 3 He, ^{4} He, ^{6} Li, and ^{8} Li+) increase with x, while those of ^{7} Li and ^{7} Be decrease. This predicted decrease for ^{7} Li is an interesting result, since it goes exactly in the direction required to solve the still open "lithium problem." At first sight, the entity of the decrease should not be sufficient to solve this problem of standard BBN, but could certainly alleviate it. A dedicated statistical analysis will help to better evaluate this interesting possibility carried by the mirror matter. The trends of the observable primordial abundances, namely, ^{4} He, D, ^{3} He, ^{7} Li, and metals (the sum of the abundances of all elements heavier than 4 He), are plotted in Figure 1 as functions of x and compared with the standard model. As expected, the trend with x is not linear, since it is (indirectly) dependent on the ordinary degrees of freedom that scale as x^4 . This dependence is the reason of the negligible effects predicted at lower x. The results of the models for mirror nuclei are shown in Table 2, which is the analogous of Table 1. Since in this case the models depend on both the ratio of entropies and the ratio of baryonic densities, they are computed for the same different values of x as for the ordinary BBN, and for two different values of β, chosen at the extremes of the range (1 and 5) in order to maximize the effects of their change.

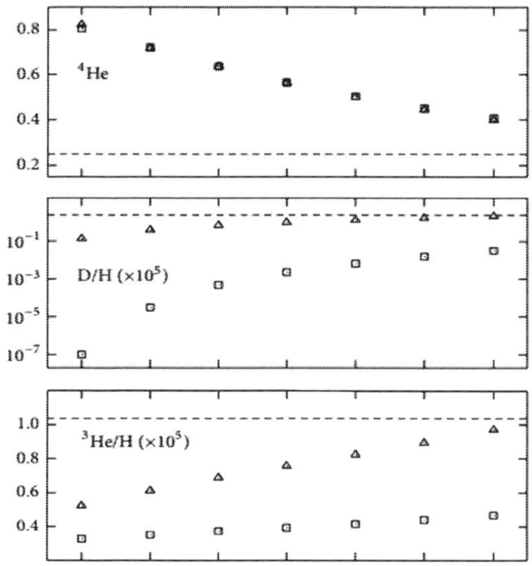

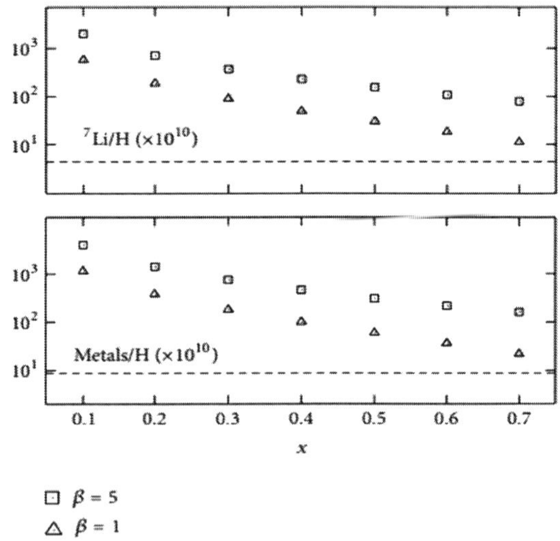

Figure 2: Primordial abundances of mirror ^{4}He, D, ^{3}He, ^{7}Li, and metals (elements heavier than ^{4}He) for several values of x and two different β and compared with the predictions of the standard model (dashed lines).

One immediately sees that the mirror BBN is very different from the ordinary one. This is what in fact one expects; since the contribution of the ordinary particles to the mirror degrees of freedom has a dependence as x^{-4}, then it is significant and becomes higher for lower x. As expected, for higher values of x, the primordial abundances of mirror nuclides become less different from the ordinary ones, since the temperature of the mirror particles becomes higher, and then similar to that of the ordinary ones, in view of the approximate relation $T' \sim xT$. In addition, the same general trend is observed for lower values of β that means baryonic densities similar to the ordinary ones. It is not simple to describe the trends of the mass fractions by changing sectors and parameters, as the final abundances depend on many physical processes acting together, but one can try to summarize some results. Comparing the mirror nuclei with the ordinary ones, one observes the following: much less residual neutrons and considerably less protons that essentially went to build ^{4}He nuclei; much more 4 He (clearly the dependence on x is the opposite as for the protons); several orders of magnitude less D,

T, ^{3}He; much less ^{6}Li for β=5 and similar abundances for β=1; much more ^{7}Li, ^{7}Be, and ^{8}Li+. Considering the trends with x, mirror abundances of n, p, ^{4}He, and ^{8}Li+ have opposite trends than the corresponding ordinary elements. Comparing the predictions obtained for the different values of β, one observes the following: the trends with x are the same for each element; the abundances of ^{4}He are very similar; the abundances of ^{3}He become almost double going from β=5 to β=1; for the lower β there is much more D, T, and ^{6}Li (some orders of magnitude) and much less ^{7}Li, ^{7}Be, and ^{8}Li+ (around one order). Analogously to what was done for the ordinary matter, in Figure 2 I plot the abundances of mirror ^{4}He, D, ^{3}He, ^{7}Li, and metals, as functions of x and for the two values of β.The previously mentioned trends, namely, the growing similarity with the standard model abundances for higher x and lower β, are evident. Differently from the ordinary nuclei, the mirror ones are not directly observable, but their primordial abundances are a key ingredient for studies of the following evolution of the Universe at all scales and for the aforementioned interpretation of the dark matter direct detection experiments. The computed mass fraction for mirror ^{4}He is in qualitative agreement with what predicted by previous analytical studies [1, 15], confirming that it is larger than the ordinary one for every x and becomes the dominant mass contribution for $x \lesssim 0.5$, meaning that dark matter would be dominated by mirror helium. Another important result of the simulations is the prediction of a much larger abundance of metals produced by mirror nucleosynthesis. These elements have a large influence on the opacity of mirror matter, which has an important role in many astrophysical processes, as, for example, the fragmentation of primordial gas during the phase of contraction. In order to complete the analysis, I show in Figure 3 the evolution of the abundances of ordinary and mirror D, ^{3}He, ^{4}He, and metals. The models used have the parameters x = 0.4 and β=5. The evolutions of standard and ordinary abundances are very similar (and for this reason the standard ones are not shown in the figure), while the mirror ones have a similar shape, but different values. In particular, they appear shifted towards earlier times, as a consequence of the smaller temperature of the primordial mirror plasma.

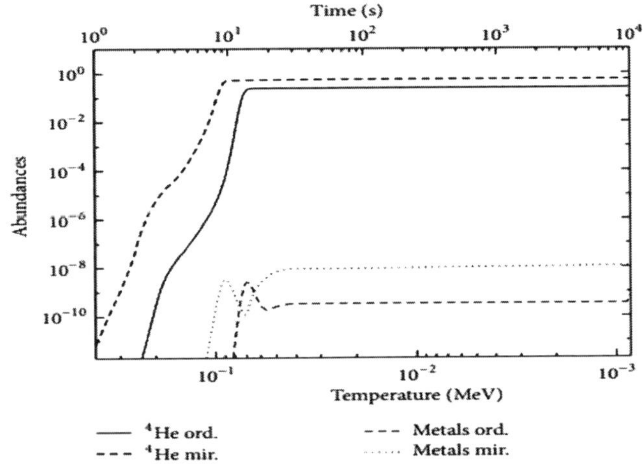

(a)

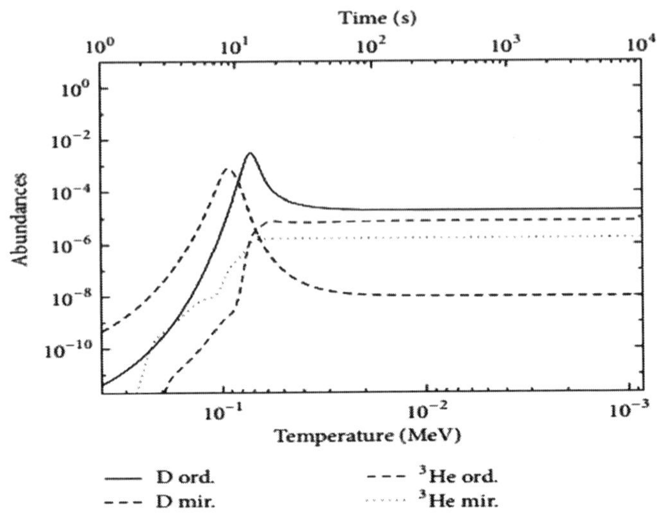

(b)

Figure 3: Time/temperature evolution of nuclides during ordinary and mirror primordial nucleosynthesis. The models have the mirror parameters $x = 0.4$ and $\beta=5$.

CONCLUSIONS

In this paper I present the detailed study of the primordial nucleosynthesis in presence of mirror dark matter in its basic model with only gravitational interactions between ordinary and mirror particles. The BBN is studied for both kinds of matter, using an accurate treatment of the thermodynamics of the early Universe, based on the work done in [21], which considers the changes in the radiation temperatures due to the two e^+-e^- annihilations. The present analysis shows the results of accurate numerical simulations and updates all the previous works. Fixing the cosmological parameters to their standard values, only the two free mirror parameters are considered, namely, the ratio of entropies x and the ratio of baryonic densities β. Both ordinary and mirror nucleosynthesis are followed until their ends, obtaining the evolution and final abundances of primordial elements in every sector. For the ordinary nuclides, they depend only on the parameter x, while for the mirror ones they are dependent also on β. As expected, the upper bound $x < 0.7$ limits the effect of mirror particles on ordinary nucleosynthesis that is negligible for $x = 0.1$ and starts to be around few percent for $x = 0.3$, with a dependence growing with x. An interesting unexpected result is the prediction of a lower abundance of ^{7} Li. This effect could help to alleviate the "lithium problem," but it requires a future dedicated statistical analysis. In the mirror sector, the Big Bang nucleosynthesis produces in a similar way mirror nuclides, but with different abundances as a consequence of its different initial conditions. In particular, as previously analytically predicted, there is an enhanced production of mirror 4 He that becomes the dominant nuclide for $x \lesssim 0.5$ and arrives at more than 80% for the lowest values of x. This effect has a very small dependence on β. In addition, there is a much larger (few orders of magnitude) production of mirror metals (elements heavier that ^{4} He). Even if their abundances are anyway very low, they could have consequences on the opacity of dark matter and on its many related astrophysical phenomena. This work provides the primordial chemical composition of the mirror dark matter, which has to be used in studies of the evolution of the Universe at all scales and in the analyses of the dark matter direct detection experiments.

REFERENCES

1. P. Ciarcelluti, "Cosmology with mirror dark matter," *International Journal of Modern Physics D*, vol. 19, no. 14, pp. 2151–2230, 2010.
2. L. B. Okun, "Mirror particles and mirror matter: 50 years of speculation and search," *Physics-Uspekhi*, vol. 50, no. 4, pp. 380–389, 2007.
3. P. Ciarcelluti, *Cosmology of the mirror universe [Ph.D. thesis]*, 2003.
4. R. Foot, "Experimental implications of mirror matter-type dark matter," *International Journal of Modern Physics A*, vol. 19, no. 23, pp. 3807–3818, 2004.
5. T. D. Lee and C. N. Yang, "Question of parity conservation in weak interactions," *Physical Review*, vol. 104, no. 1, pp. 254–258, 1956.
6. S. Blinnikov and M. Khlopov, "Possible astronomical effects of mirror particles," *Soviet Astronomy*, vol. 27, pp. 371–375, 1983.
7. M. Y. Khlopov, G. Beskin, N. Bochkarev, L. Pustylnik, and S. Pustylnik, "Observational physics of mirror word," *Soviet Astronomy*, vol. 35, p. 21, 1991.
8. R. Foot, H. Lew, and R. R. Volkas, "A model with fundamental improper spacetime symmetries," *Physics Letters B*, vol. 272, no. 1-2, pp. 67–70, 1991.
9. P. Ciarcelluti, "Cosmology with mirror dark matter II: cosmic microwave background and large scale structure," *International Journal of Modern Physics D*, vol. 14, no. 2, pp. 223–256, 2005.
10. P. Ciarcelluti, "Cosmology with mirror dark matter I: linear evolution of perturbations," *International Journal of Modern Physics D*, vol. 14, no. 2, pp. 187–222, 2005.
11. Z. Berezhiani, P. Ciarcelluti, D. Comelli, and F. L. Villante, "Structure formation with mirror dark matter: CMB and LSS," *International Journal of Modern Physics D*, vol. 14, no. 1, pp. 107–120, 2005.
12. P. Ciarcelluti, "Structure formation, CMB and LSS in a mirror dark matter scenario," *Frascati Physics Series*, vol. 555, p. 1, 2004, http://arxiv.org/abs/astro-ph/0409629.
13. A. Y. Ignatiev and R. R. Volkas, "Mirror dark matter and large scale structure," *Physical Review D*, vol. 68, no. 2, Article ID 023518, 11 pages, 2003.
14. P. Ciarcelluti and Q. Wallemacq, "Is dark matter made of mirror matter? Evidence from cosmological data," *Physics Letters B*, vol. 729, pp. 62–66, 2014.

15. Z. Berezhiani, D. Comelli, and F. L. Villante, "The early mirror universe: inflation, baryogenesis, nucleosynthesis and dark matter," *Physics Letters B*, vol. 503, no. 3-4, pp. 362–375, 2001.
16. P. Ciarcelluti, "Astrophysical tests of mirror dark matter," *AIP Conference Proceedings*, vol. 1038, pp. 202–210, 2008.
17. P. Ciarcelluti, "Early universe cosmology with mirror dark matter," *AIP Conference Proceedings*, vol. 1241, p. 351, 2010, http://arxiv.org/abs/0911.3592.
18. Z. G. Berezhiani, A. D. Dolgov, and R. N. Mohapatra, "Asymmetric inflationary reheating and the nature of mirror universe," *Physics Letters B*, vol. 375, no. 1–4, pp. 26–36, 1996. | Zentralblatt MATH | MathSciNet
19. P. Ciarcelluti and R. Foot, "Early universe cosmology in the light of the mirror dark matter interpretation of the DAMA/Libra signal," *Physics Letters B*, vol. 679, no. 3, pp. 278–281, 2009.
20. P. Ciarcelluti and R. Foot, "Primordial He' abundance implied by the mirror dark matter interpretation of the DAMA/Libra signal," *Physics Letters B*, vol. 690, no. 5, pp. 462–465, 2010.
21. P. Ciarcelluti and A. Lepidi, "Thermodynamics of the early universe with mirror dark matter," *Physical Review D*, vol. 78, no. 12, Article ID 123003, 7 pages, 2008.
22. R. Foot, "Mirror dark matter and the new DAMA/LIBRA results: a simple explanation for a beautiful experiment," *Physical Review D*, vol. 78, no. 4, Article ID 043529, 10 pages, 2008.
23. R. Foot, "Mirror dark matter interpretations of the DAMA, CoGeNT and CRESST-II data," *Physical Review D*, vol. 86, no. 2, Article ID 023524, 10 pages, 2012.
24. Z. Berezhiani, P. Ciarcelluti, S. Cassisi, and A. Pietrinferni, "Evolutionary and structural properties of mirror star MACHOs," *Astroparticle Physics*, vol. 24, no. 6, pp. 495–510, 2006.
25. E. W. Kolb and M. S. Turner, *The Early Universe*, Frontiers of Physics 69, Addison-Wesley, New York, NY, USA, 1990.
26. R. V. Wagoner, "Big-Bang nucleosynthesis revisited," *The Astrophysical Journal*, vol. 179, pp. 343–360, 1973.
27. L. Kawano, "Let's go: early universe. 2. Primordial nucleosynthesis: the computer way," FERMILAB-PUB-92-004-A, 1992.

4

Revisiting Big-Bang Nucleosynthesis Constraints on Long-Lived Decaying Particles

Masahiro Kawasaki[a,b], Kazunori Kohri[c,d,e], Takeo Moroi[f,b], and Yoshitaro Takaesu[f,g]

[a]Institute for Cosmic Ray Research, The University of Tokyo, Kashiwa 277-8582, Japan
[b] Kavli IPMU (WPI), UTIAS, The University of Tokyo, Kashiwa 277-8583, Japan
[c] Theory Center, IPNS, KEK, Tsukuba 305-0801, Japan
[d] The Graduate University of Advanced Studies, Tsukuba 305-0801, Japan
[e] Rudolf Peierls Centre for Theoretical Physics, The University of Oxford,1 Keble Road, Oxford OX1 3NP, UK
[f] Department of Physics, The University of Tokyo, Tokyo 113-0033, Japan
[g] Research Institute for Interdisciplinary Science, Okayama University,Okayama 700-8530, Japan

ABSTRACT

We study effects of long-lived massive particles, which decay during the big-bang nucleosynthesis (BBN) epoch, on the primordial abundances of light elements. Compared to the previous studies, (i) the reaction rates of the standard BBN reactions are updated, (ii) the most recent observational data

Citation: Masahiro Kawasaki, Kazunori Kohri, Takeo Moroi, and Yoshitaro Takaesu. "Revisiting big-bang nucleosynthesis constraints on long-lived decaying particles" Phys. Rev. D 97, 023502 https://doi.org/10.1103/PhysRevD.97.023502

of light element abundances and cosmological parameters are used, (iii) the effects of the interconversion of energetic nucleons at the time of inelastic scatterings with background nuclei are considered, and (iv) the effects of the hadronic shower induced by energetic high energy anti-nucleons are included. We compare the theoretical predictions on the primordial abundances of light elements with latest observational constraints, and derive upper bounds on relic abundance of the decaying particle as a function of its lifetime. We also apply our analysis to unstable gravitino, the superpartner of the graviton in supersymmetric theories, and obtain constraints on the reheating temperature after inflation.

INTRODUCTION

The big-bang nucleosynthesis (BBN) is one of the most important predictions of big-bang cosmology. At the cosmic temperature around 0.1 MeV, the typical energy of the cosmic microwave background (CMB) photon becomes sufficiently lower than the binding energies of light elements so that the light elements can be synthesized with avoiding the dissociation due to the scattering with background photons. The cross sections for nuclear reactions governing the BBN are well understood so that a precise theoretical calculation of the primordial light element abundances is possible with the help of numerical calculations. In addition, the primordial abundances of the light elements are well extracted from astrophysical observations. Comparing the theoretical predictions with the observational constraints, a detailed test of the BBN is now possible; currently the predictions of light elements (deuterium D and 4He) in the standard BBN (SBBN) reasonably agree with observations. It has been well recognized that, if there exists new physics beyond the standard model which may induce non-standard BBN reactions, the predictions of the SBBN change.#1 In particular, with a long-lived unstable particle which decays into electromagnetic [13, 14, 15, 16, 17, 18, 19, 20] or hadronic [21, 22, 23, 24, 25, 26, 27, 28, 29, 30, 31, 32] particles, the light element abundances are affected by photodissociation, hadrodissociation, and p $\leftrightarrow$ n conversion processes. In order not to spoil the agreements between the theoretical predictions and observational constraints, upper bounds on the primordial abundances of the unstable particles are obtained. Such a constraint has been intensively studied in the past. Remarkably, the BBN constraints may shed light on beyond-the-standard-model particles to which collider studies cannot impose constraints. One important example is gravitino, which is the superpartner of graviton in supersymmetric (SUSY) models [33]. The gravitino is very weakly interacting, and is produced by the scattering processes of particles in the thermal bath after inflation; the primordial abundance of gravitino is approximately proportional to the reheating temperature after inflation. Consequently, with the reheating temperature being fixed, we may acquire constraints on the gravitino mass assuming that gravitino is unstable. Such a constraint can be converted to the upper bound on the reheating temperature after inflation, which provides important information in studying cosmology based on SUSY models. The purpose of this paper is to revisit the BBN constraints on long-lived particles, which we call X, taking into account recent progresses in theoretical and observational studies of the primordial abundances of light elements. Theoretically, the understandings of the cross sections of the SBBN reactions have been improved, which results in smaller uncertainties in the theoretical calculations of the light element abundances. In addition, the observational constraints on the primordial abundances of light elements have been updated. These affect the BBN constraints on the primordial abundances of long-lived particles. In this paper, we study the BBN constraints on the primordial abundance of long-lived exotic particles, which we parameterize by using the so-called yield variable:

$$Y_X \equiv \left[\frac{n_X}{s}\right]_{t \ll \tau_X}, \tag{1.1}$$

where n_X is the number density of X, s is the entropy density, and the quantity is evaluated at the cosmic time much earlier than the lifetime of X (denoted as τ_X). We take into account theoretical and observational progresses. In particular, compared to previous studies:

- The reaction rates of the SBBN reactions are updated.
- The most recent observational constraints on the primordial abundances of light elements are adopted.
- The calculation of the evolution of the hadronic showers induced by energetic nucleons from the decay is improved. • We include the effect of hadronic shower induced by energetic anti-nucleons from the decay.[#2]

We consider various possible decay modes of long-lived particles and derive upper bounds on their abundances. We also apply our analysis to the study of the effects of unstable gravitino on light element abundances. For such a study, we adopt several patterns of mass spectra of superparticles (i.e., squarks, sleptons, gauginos, and Higgsinos) suggested by viable SUSY models, based on which partial decay rates of gravitino are calculated. The organization of this paper is as follows. In Section 2, we summarize the observational constraints on light element abundances we adopt in our analysis. In Section 3, we explain how the theoretical calculation of light element abundances is performed with taking into account the effects of the decay of long-lived particles. Upper bounds on the primordial abundances of generic decaying particles are given in Section 4. Then, our analysis is applied to the case of unstable gravitino in Section 5. Section 6 is devoted for conclusions and discussion.

OBSERVED ABUNDANCES OF LIGHT ELEMENTS

We first summarize the current observational constraints on the primordial abundances of light elements D, ^{4}He, ^{3}He and ^{7}Li. In the following (A/B) denotes the ratio of number densities of light elements A and B, and the subscript "p" indicates the primordial value.

D

The primordial abundance of D is inferred from D absorption in damped Lyα systems (DLAs). Most recently Cook et al. [35] measured D/H by observing a DLA toward QSO SDSS J1358+6522. Moreover they reanalyzed four previously observed DLAs and from the total five DLA data they obtained the primordial D abundance as

$$(D/H)_p = (2.53 \pm 0.04) \times 10^{-5}. \tag{2.1}$$

The quoted error is much smaller (by a factor of ~ 5) than those obtained in the previous study. The improvement of the D measurement is a main reason why we obtain more stringent BBN constraints than those in the previous work [28] as seen in later sections

^{3}He

The ^{3}He abundance is measured in protosolar objects. As described in the previous work [28] we use the ratio ^{3}He/D as observational constraint instead of using ^{3}He/H. This is because chemical evolution can increase or decreases the ^{3}He abundance and it is difficult to infer the primordial value for ^{3}He. On the other hand the D abundance always decrease in chemical evolution and D is more fragile than 3He. Consequently, the ratio ^{3}He/D increases monotonically with time, which allows us to use the measured ^{3}He/D as an upper bound on the primordial value [36]. From 3He abundances observed in protosolar clouds [37], we adopt

$$(^3\text{He}/\text{D})_p < 0.83 + 0.27. \tag{2.2}$$

^{4}He

The primordial mass fraction of ^{4}He, ^{Y}p, is inferred from measurement of recombination lines of HeII (and HII) emitted from extra-galactic HII regions. Izotov, Thuan and Guseva [38] reported a new determination of Y^p with the use of the infrared as well as visible ^{4}He emission lines in 45 extragalactic HII regions. Their result is

$$Y_p = 0.2551 \pm 0.0022. \tag{2.3}$$

After Ref. [38] was reported, Aver, Olive and Skillman [39] reanalyzed the data of Ref. [38]. They estimated the 4He abundance and its error by using Markov chain Monte Carlo (MC) analysis and obtained

$$Y_p = 0.2449 \pm 0.0040. \tag{2.4}$$

Thus, the two values are inconsistent, and the discrepancy is more than a 2σ level. If we adopt the baryon-photon ratio η determined by Planck, the BBN prediction for Y_p is well consistent with Eq. (2.4) but not with Eq. (2.3). For this reason we adopt the value given in Eq. (2.4) as a constraint on Y_p. We will also show how constraints change if we adopt Eq. (2.3).

^{7}Li (and ^{6}Li)

The primordial abundance of ^{7}Li was determined by the measurement of ^{7}Li in the atmospheres of old metal-poor stars. The observed ^{7}Li abundances in stars with [Fe/H] = −(2.5 − 3) showed almost a constant value (log10(^{7}Li/H) ‘ −9.8) called Spite plateau which was considered as primordial.[#3] However, the plateau value turns out to be smaller than the standard BBN prediction by a factor nearly 3. In fact, Ref. [40] reported the plateau value log10(^{7}Li/H) = −9.801 ± 0.086 while the BBN prediction is log10(^{7}Li/H) = −9.35 ± 0.06 for the central value of η suggested from the CMB data [41] (see, Section 3). This discrepancy is called the lithium problem. Moreover, the recent observation shows much smaller ^{7}Li abundances (log10(^{7}Li/H) < −10) for metal-poor stars with metalicity below [Fe/H] ~ −3 [40]. Thus, the situation of the ^{7}Li observation is now controversial. Since we do not know any mechanism to make ^{7}Li abundances small in such metal poor stars, we do not use ^{7}Li as a constraint in this paper. We do not use 6Li either because ^{6}Li abundance is observed as the ratio to the number density of ^{7}Li

BBN WITH DECAYING PARTICLES

In this section, we explain how we calculate the light-element abundances taking into account the effects of decaying massive particles. Our procedure is based on that developed in Refs. [27, 28] with several modifications which will be explained in the following subsections.

Overview

The BBN constraints strongly depends how X decays. In order to make our discussion simple, we first concentrate on the case where X decays only into a particle and its antiparticle. With the final-state particles being fixed, the subsequent decays of the daughter particles from X decay as well as the hadronization processes of colored particles (if emitted) are studied by using the PYTHIA 8.2 package [42]. We note here that, even if X primarily decays into a pair of electromagnetically interacting particles (like e^+e^- or $\gamma\gamma$ pair), colored particles are also emitted by the final-state radiation; such effects are included in our study. The effects of the long-lived decaying particles depend on their decay modes, which are classified into the following two categories:

Radiative decay

In the decay of the long-lived particles, high-energy photons and electromagneticallycharged particles are emitted.

Hadronic decay

High-energy colored particles (i.e., quarks or gluons) are emitted in the decay.

We first overview them in the following. (A flow-chart of the effects induced by such highenergy decay products is shown in Fig. 1.)

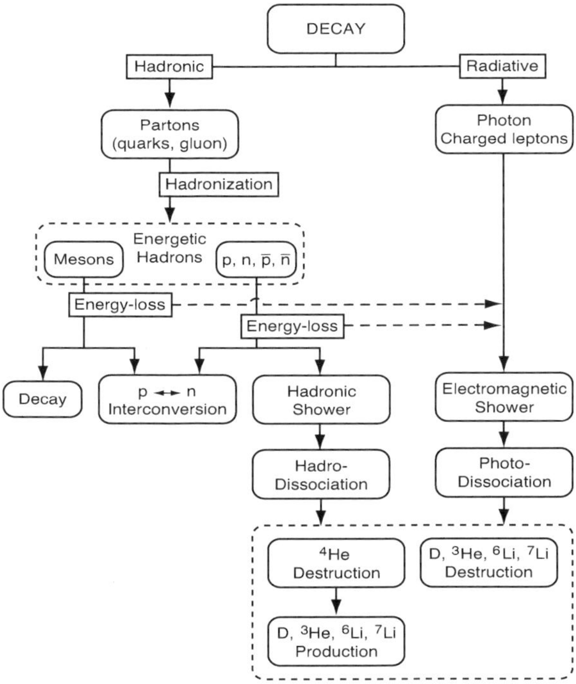

Figure 1: Flow-chart of the hadronic decay of massive particles. New effects induced by "energetic" anti-nucleons (¯n and ¯p) are included in this study

Radiative decay modes

Energy injections by energetic electromagnetic particles induce electromagnetic showers through their scatterings off the background photons and electrons [11, 13, 14, 15, 16, 17, 18, 26, 30, 32]. The electromagnetic particles include γ and $e^{\pm}$ as well as charged hadrons. Energetic photons in the shower can destroy light elements, in particular, D and ^{4}He, produced by the standard BBN (SBBN) reactions. (Such processes are called photodissociations). The photon spectrum in the electromagnetic shower is determined by the total amount of the visible energy injected by the decay and the temperature of the background thermal bath [11, 13, 14]; in particular, at a high precision, the normalization of the photon spectrum is proportional to the total amount of the energy injection. In our numerical calculation, the total visible energy, which is the sum of the energies of photon and charged particles after the hadronization, is calculated, based on which the normalization of the photon spectrum is determined. The photodissociation processes become effective when the threshold energy for the e^+e^- pair creation by scattering of high energy photons and background photons, which is approximately given by $E_{\rm th}^{(\gamma)} \sim m_e^2/22T$ [13], is larger than the threshold energy for the dissociation processes of light elements. For the photodissociation of ^{4}He, this is the case at the cosmic temperature lower than ~ 1 keV (which corresponds to the cosmic time $t \gtrsim 10^6$ sec).

With the photodissociation of ^{4}He, both ^{3}He and D are copiously produced. Thus, for the case where the lifetime of the long-lived particle is longer than ~ 10^6 sec, the D and ^{3}He abundances give stringent constraints on the radiatively-decaying modes. The photodissociation of D becomes effective at higher temperature because of the smallness of its binding energy; the photodissociation of D may be important for long-lived particle with lifetime longer than ~ 104 sec, for which significant destruction of D takes place [15, 17]. Another effect of the electromagnetic shower on BBN is non-thermal production of ^{6}Li. The energetic T and 3He are produced through photodissociation of ^{4}He and they scatter off the background ^{4}He and synthesize ^{6}Li. The productions of ^{7}Li and ^{7}Be due to energetic ^{4}He scattered by energetic photons through inelastic $\gamma + {}^4$He are negligible.

Hadronic decay modes

In hadronic decays, the emitted colored particles fragment into hadrons such as pions, kaons, nucleons (i.e., neutron n and proton p), and anti-nucleons (i.e., anti-neutron $\bar{n}$ and antiproton $\bar{p}$). Hereafter $N = p$ or n ($\bar{N} = \bar{p}$ or $\bar{n}$) denotes the nucleon (anti-nucleon). Energetic hadrons, in particular, the nucleons, induce hadronic shower and hadrodissociation processes. In addition, even after being stopped, some of the hadrons (in particular, charged pions and nucleons) change the neutron-to-proton ratio in the background plasma, which affects the ^{4}He and D abundances. The most important effects of the hadronic decay modes are summarized as follows:

When a massive particle decays into hadrons at $T \gtrsim \mathcal{O}(0.1)$ MeV (i.e., when the lifetime is shorter than $\sim \mathcal{O}(10^2)$ sec), high-energy hadrons are stopped in the thermal plasma [28]. Extra pions and nucleons affect the neutron-to-proton ratio after the neutron freeze-out by interchanging background p and n through the strongly-interacting interconversion processes like destroy background nuclei [22, 23, 24].

$$\pi^+ + n \longrightarrow p + \pi^0, \quad (3.1)$$

$$\pi^- + p \longrightarrow n + \pi^0. \quad (3.2)$$

There also exist similar interconversion processes caused by injected nucleons. The neutron-to-proton ratio n/p increases due to the strongly-interacting conversions, resulting in the increase of the abundances of ^{4}He and D [21, 25].

If the background temperature at the time of the decay is low enough, the emitted hadrons cannot be stopped in the plasma; this is the case for energetic n and $T \lesssim 0.1$ MeV ($T \lesssim 30$ keV) [28]. The energetic nucleons scatter off and

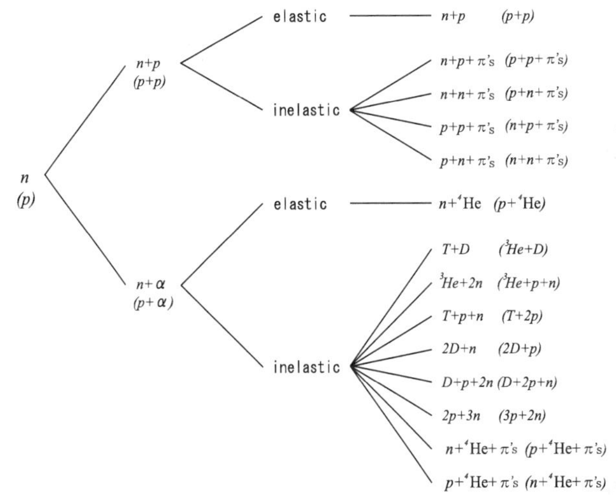

Figure 2: Schematic picture of hadronic shower induced by a high energy projectile n (or p) which scatters off the background proton or the background ^{4}He.

The processes considered in this study are summarized in Figs. 2 and 3. In particular, through the destruction of 4He by the highenergy neutrons, an overproduction of D may occur, which significantly constrains the primordial abundance of the hadronically decaying long-lived particles [27, 28, 31, 34]. In addition, energetic ^{3}He, ^{4}He and T produce ^{7}Li, ^{7}Be and ^{6}Li non-thermally through scattering off the background ^{4}He.

New implementations in the numerical calculation

As mentioned before, our analysis is based on Refs. [27, 28]. In the present analysis, we improve the calculation of the light element abundances. Here, we summarize the new implementations in the numerical calculation added after Refs. [27, 28, 31].

- We update the SBBN reaction rates and their uncertainties.

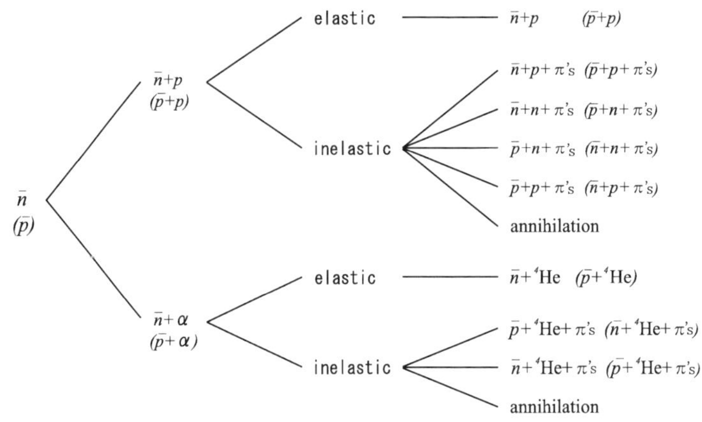

Figure 3: Same as Fig. 2, but for a high energy projectile ¯n (or ¯p).

- We revise the algorithm to calculate the evolution of the hadronic shower induced by the injections of energetic p and n. In particular, we include the p ↔ n interconversion of the energetic nucleons via inelastic scatterings.[#4] In inelastic scattering processes of energetic nucleon (i.e., p, n, ¯p, or ¯n) with background nuclei, it is likely that the projectile nucleon carries away most of the energy while other final state particles are less energetic; the energies of the final-state particles other than the most energetic one are typically $K_T \sim 140$ MeV. The conversion of the projectile nucleon via the inelastic scatterings affects the (effective) energy loss rate of the nucleon as we discuss below.
- We newly include the effects of energetic ¯p and ¯n injections by the decay

SBBN reactions

Compared to the previous studies [28, 31], we update the SBBN reaction rates adopting those given in Ref. [43, 44], which are obtained by fitting relatively-new experimental data. In order to take into account uncertainties in the reaction rates, we perform MC simulations to estimate error propagations to the light element abundances. For given values of the lifetime and the primordial abundance of X, we perform 1, 000 runs of the calculation of the light element abundances with assuming that the reaction rates (as well as other parameters in the calculations) are random Gaussian variables. In each run, the reaction rate Ri for i-th reaction is determined as

$$R_i(T) = \bar{R}_i(T) + \sigma_i(T)\, y_i, \tag{3.3}$$

where $\bar{R}_i$ and σ_i are the central value and the standard deviation of the reaction i, and y_i is the temperature independent Gaussian variable with probability distribution

$$P(y) = \frac{1}{\sqrt{2\pi}} \exp\left(-\frac{1}{2}y^2\right). \tag{3.4}$$

Notice that, as explicitly shown in Eq. (3.3), some of uncertainties of reaction rates depend on the cosmic temperature. Throughout each run, such a reaction rate is determined by adding the temperature-dependent uncertainty multiplied by the temperature-independent random Gaussian variable to the central value. If the upper and lower uncertainties of a reaction rate are asymmetric, we take upward and downward fluctuations with equal probability, assuming that they obey one-sided Gaussian distributions.

In order to check the consistency between the observational constraints summarized in Section 2 and the SBBN predictions, we compare them in Fig. 4; we plot the SBBN values of the light element abundances as functions of the baryon to photon ratio $\eta = n_B/n_\gamma$. From the top to the bottom, we plot (i) the mass fraction of ^{4}He (Y_p), (ii) the deuterium to hydrogen ratio (D/H), (iii) the ^{3}He to hydrogen ratio (^{3}He/H), (iv) the ^{7}Li to hydrogen ratio (^{7}Li/H), and (v) the helium 3 to deuterium ratio (^{3}He/D). The theoretical predictions show 2σ bands due to the uncertainties in experimental data of cross sections and lifetimes of nuclei. The boxes indicate the 2σ observational constraints (see Section 2). The CMB observations provide independent information about the baryon-to-photon ratio; we adopt

$$\eta = (6.11 \pm 0.04) \times 10^{-10} \quad (68\%\ \text{C.L.}), \tag{3.5}$$

which is based on the TT, TE, EE+lowP+BAO analysis of the Planck collaboration, $\Omega_B h^2 = 0.02229^{+0.00029}_{-0.00027}$ (95% C.L.), where Ω_B is the density parameter of baryon and h is the Hubble constant in units of 100 km/sec/Mpc.) [41]. In Fig. 4, we also show the CMB constraint on η at 2σ (the vertical band). We can see that the SBBN predictions for the baryon-to-photon ratio suggested from the CMB observations are in reasonable agreements with observations. In our numerical calculation, we use the value of η given in Eq. (3.5).

The effects of the high energy n and p injected into the thermal bath have been studied in Refs. [22, 23, 24, 27, 28]. Compared to the previous studies, we newly include the interconversion reactions between energetic p and n at the time of the inelastic scatterings with pion emissions (see Fig. 2). Once an energetic p or n is injected into thermal plasma, it can be converted to n or p through inelastic scatterings off the background nuclei, i.e., p or 4He. (Hereafter, the background 4He is denoted as αBG.) As we discuss below, such an

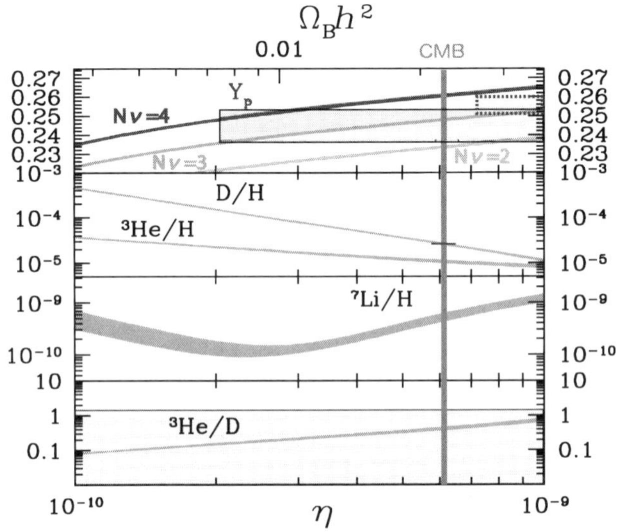

Figure 4: Theoretical predictions of light element abundances as functions of the baryon to photon ratio $\eta = nB/n\gamma$. (The upper horizontal axis shows ΩBh^2, where ΩB is the density parameter of baryon and h is the Hubble constant in units of 100 km/sec/Mpc.) The vertical band shows $\eta = (6.11 \pm 0.08) \times 10^{-10}$, which is the 2σ band of baryon-to-photon ratio suggested by the CMB observations [41]. For Y_p, we plot results for effective number of neutrino species $N_\nu = 2$, 3, and 4 (from the bottom to the top). The boxes indicate the observational values with 2σ uncertainties. (For Y_p, the box surrounded by the solid and dashed lines correspond to Eq. (2.4) and Eq. (2.3), respectively.) interconversion affects the (effective) energy loss rates of nucleons during their propagation in thermal plasma, resulting in a change of the hadrodissociation rates

The effects of the interconversions of energetic p and n are important when the cosmic temperature becomes lower than ~ 0.1 MeV. This is because the energetic nucleons are stopped in the thermal plasma if the background temperature is high enough; as we have mentioned, the neutron is likely to be stopped when T & 0.1 MeV while the proton is stopped when T & 30 keV. Once stopped, the nucleons do not induce hadrodissociation processes.

An energetic neutron does not cause hadronic showers if it is converted to proton, even though it would have a sufficient time to induce hadronic showers without the interconversion. Thus, the inclusion of the energetic p ↔ n conversion makes the BBN constraints milder in particular when $30\ \mathrm{keV} \lesssim T \lesssim 0.1\ \mathrm{MeV}$ (i.e., $10^2\ \mathrm{sec} \lesssim t \lesssim 10^3\ \mathrm{sec}$) The inelastic scatterings occur with the targets of p and αBG. The interconversion rates are estimated as follows:

Because of the lack of the experimental data of a cross section of each mode for the n (projectile)+ p (target) and p (projectile)+ p (target) processes shown in Fig. 2, we assume that all of the cross sections for those 8 modes (n+p, p+p → p+p+π's, p+n+ π's, n+p+π's, n+n+π's) are equal. (In the case of p+p → n+n+π's, at least two pions should be emitted and hence we consider the process only if it is kinematically allowed.) In addition, we neglect the effects of the interconversions induced by the emitted pions from the inelastic scatterings. That is because those pions decay before they scatters off the background particles.

With inelastic n+αBG scatterings, the energetic n can be again converted to p. Because of the lack of experimental data, we assume that, when the inelastic processes with pion emissions are concerned, the cross section for n +4 He scattering is equal to that for p+^{4}He scattering. Since the rates of the inelastic scatterings n+^{4}He → n+^{4}He and p+4He are relatively small [45], effects of the interconversion due to the inelastic n+αBG scatterings are unimportant and does not change the hadronic shower evolution.

We also comment on the non-thermal production processes of ^{6}Li, ^{7}Li, and ^{7}Be. With the inelastic scatterings N + αBG → · · · , the final state may contain energetic T, ^{3}He, and ^{4}He. They may scatter off αBG to produce heavier elements, i.e., ^{6}Li, ^{7}Li, and 7Be [22, 23, 24, 28]. Although we do not use ^{6}Li and ^{7}Li to constrain the primordial abundance of X, these nonthermal production processes are included in our numerical calculation In particular, we include the processes $^4\mathrm{He} + \alpha_{\mathrm{BG}} \to ^6\mathrm{Li} + \cdots,\ ^7\mathrm{Li} + \cdots,\ \mathrm{and}\ ^7\mathrm{Be} + \cdots,$ whose effects were not taken into account in Ref. [28]. For the study of these processes, energy distribution of 4He produced by the inelastic hadronic scatterings is necessary. We determine energy distribution of the final-state ^{4}He using the prescription of Appendix C in Ref. [28]. Notice that, for the non-thermal production processes of ^{6}Li induced by T and ^{3}He, the energy distributions of T and 3He produced by the hadrodissociations of $^\alpha$BG are obtained by fitting experimental data (see Ref. [28]).

Injections of energetic $\bar{n}$ or $\bar{p}$

In the present study, we newly include the effects of the injections of energetic anti-nucleons, $\bar{n}$ or $\bar{p}$. We only consider the scatterings of energetic anti-nucleons off background p and αBG, and the anti-nucleons are treated as sources of hadronic showers.#5 The high-energy nucleons in the hadronic showers induced by $\bar{n}$ and $\bar{p}$ can destroy αBG and further produce copious high-energy daughter nuclei. In Fig. 3, we show a schematic picture for the reactions induced by energetic anti-nucleons. Because of the lack of experimental data for the processes including the anti-nucleons, we adopt several approximations and assumptions on the reactions induced by the energetic anti-nucleon. We note here that, in adopting approximations or assumptions, we require that the constraints become conservative in order not to over-constrain the properties of long-lived particles.

First, let us consider the scattering of energetic anti-nucleons with background p. We use the experimental data of the differential cross sections for the $\bar{p}$-p scatterings by referring Ref. [46] for the total and elastic cross sections and Ref. [47] for the annihilation cross sections. Because of the lack of experimental data, we use the data for the $\bar{p}$-p scatterings to estimate the cross sections for other processes. (i) We assume that the differential cross sections for $\bar{n}$-p scatterings are the same as those of corresponding $\bar{p}$-p scatterings paired in Fig. 3. We expect that this assumption is reasonable because the Coulomb corrections to the cross sections are estimated to be less than a few percent for the energy of the energetic nucleons of our interest [28]. (ii) Through inelastic scatterings off background p, strongly-interacting p ↔ n interconversion reactions occur with emitting pions. Then, there are four possible combinations of final-state anti-nucleon and scattered nucleon; the final-state anti-nucleon may be $\bar{p}$ or $\bar{n}$, while the nucleon may be p or n (see Fig. 3). We assume that the differential cross sections for these processes are equal and use the data for the $\bar{p}$-p scatterings for all of these processes because cross sections for some of the final states are unknown. Here one remark is that the process $\bar{n}$(projectile) +p(target) → $\bar{p}$+n needs at least two pion emission so we take into account it if kinematically allowed.

Next, we discuss scatterings of energetic anti-nucleons off αBG. Unfortunately, experimental data of the cross sections for the scattering processes of anti-nucleon with 4He are insufficient to perform a detailed study. Therefore, we only consider the energy-loss and interconversions of anti-nucleon induced by the inelastic scatterings with αBG. Approximating that the energetic anti-nucleon scatters off individual nucleons in ^{4}He, the cross sections for the scattering processes with ^{4}He are taken to be four times larger than those with background nucleons. Notice that this assumption is reasonable when the energy of $\bar{N}$ is larger than the binding energy of ^{4}He.#6 In addition, for the scatterings induced by the antinucleon, we neglect the subsequent reactions induced by the cascade products from $^{\alpha}$BG; in other words, we neglect the destruction of $^{\alpha}$BG and the recoil energy of $^{\alpha}$BG (if it is not destroyed). These would give the most conservative bounds. For annihilation reactions of anti-nucleon with background p or αBG, we could expect emissions of energetic hadrons, by which light element abundances should be affected. However, because we do not have sufficient data of differential cross sections, we conservatively neglect any effects after the annihilation of anti-nucleon. Thus, in our calculation, only the effect of the pair annihilations of anti-nucleon is to reduce the number of energetic antinucleon.

ξ parameters

In the present study, the effects of the hadrodissociation are parameterized by the functions called ξAi (with Ai being n, D, T, 3He, α, ^{6}Li, ^{7}Li and ^{7}Be) [22, 23, 24, 27, 28]. For Ai 6= α, $_{\xi \text{Ai}}$ is the total number of Ai produced by the hadronic decay of one parent particle X, while ξα is the total number of αBG destroyed. With the properties of the long-lived particle X being given, ξ_{Ai} depend on three quantities: cosmic temperature T, the mass fraction of ^{4}He, and the baryon to photon ratio η.

It is notable that ξ_{Ai} is defined to be a value just after the shower evolutions. Notice that the timescale of the shower evolution is much shorter than those of any nuclear reactions in SBBN. Therefore, the face value of ξAi does not represent a net increase or decrease of the nuclei Ai . For example, ^{6}Li is processed in a SBBN reaction ^{6}Li + p → ^{3}He + ^{4}He, even after the value of ξ^{6}Li is fixed after the evolutions of the hadronic shower.

In the case of our interest, the hadronic showers are triggered by the injections of highenergy nucleons (i.e., N = p and n) and anti-nucleons (i.e., $\bar{N}$ = $\bar{p}$ and $\bar{n}$) which originate from the hadronizations of colored particles emitted by the decay of long-lived particles. ξ_{Ai} are given by the convolutions of two functions as

$$\xi_{A_i}(T, Y_p, \eta) = \sum_{N=n,\bar{n},p,\bar{p}} \int \tilde{\xi}_{A_i,N}(E_N, T, Y_p, \eta) \frac{dn_N}{dE_N} dE_N, \tag{3.6}$$

where (dn_N/dE_N) is the energy spectrum of the injected nucleon N with its energy EN per a single massive particle decay, while $\tilde{\xi}_{Ai,N}$ is the number of the light element Ai produced (or destroyed for Ai = α) by the products of hadronic shower induced by the injection of single N with energy EN . We calculate the functions $\tilde{\xi}_{Ai,N}$ based on the procedure explained in Ref. [28].[#7]

In Figs. 5, 6, 7, 8 and 9, we show $\tilde{\xi}_{n,N}$, $\tilde{\xi}_{D,N}$, $\tilde{\xi}_{T,N}$, $\tilde{\xi}^3_{He,N}$, and $\tilde{\xi}_{\alpha,N}$, respectively, as functions of the kinetic energy EN for N = n, ¯n, p and ¯p, taking T = 4 keV. Notice that, at T = 4 keV, the energetic neutrons with En > O(1) TeV loose their kinetic energy immediately down to ~ 1 TeV [31]. Thus, ˜ξα,N become insensitive to EN for EN & 1 TeV.

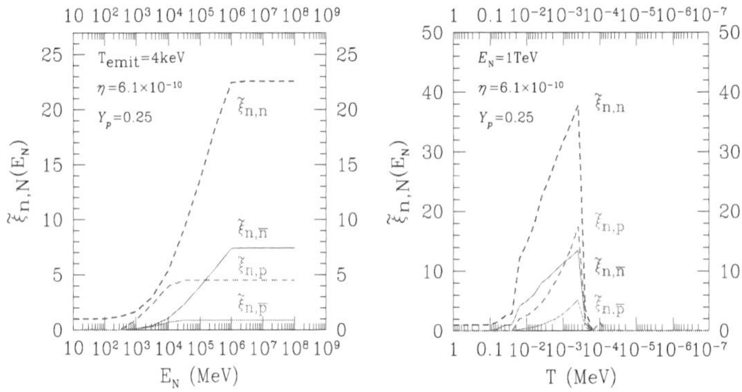

Figure 5: $\tilde{\xi}_{n,N}$ for T = 4 keV as functions of the kinetic energy (left) and those for EN = 1 TeV as functions of the cosmic temperature (right). Here, N = n, ¯n, p and ¯p, and we take $\eta = 6.1 \times 10^{-10}$ and $Y_p = 0.25$.

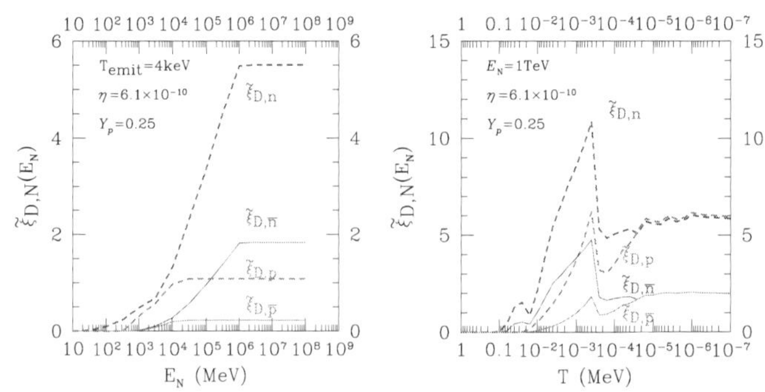

Figure 6: ˜ξD,N for T = 4 keV as functions of the kinetic energy (left) and those for EN = 1 TeV as functions of the cosmic temperature (right). Here, N = n, ¯n, p and ¯p, and we take $\eta = 6.1 \times 10^{-10}$ and Yp = 0.25.

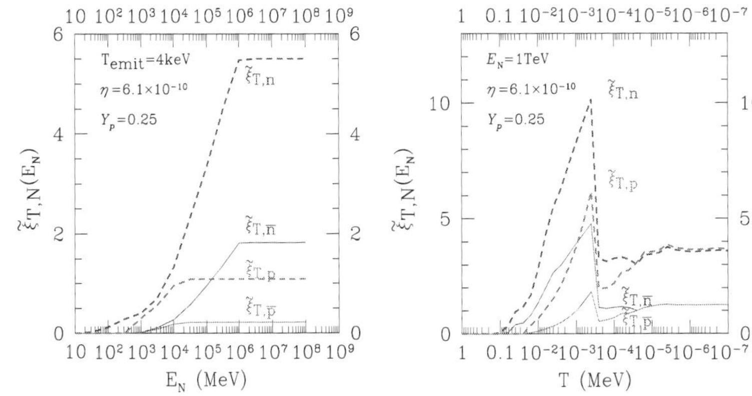

Figure 7: $\tilde{\xi}_{T,N}$ for T = 4 keV as functions of the kinetic energy (left) and those for EN = 1 TeV as functions of the cosmic temperature (right). Here, N = n, $\bar{n}$, p and $\bar{p}$, and we take $\eta = 6.1 \times 10^{-10}$ and Yp = 0.25.

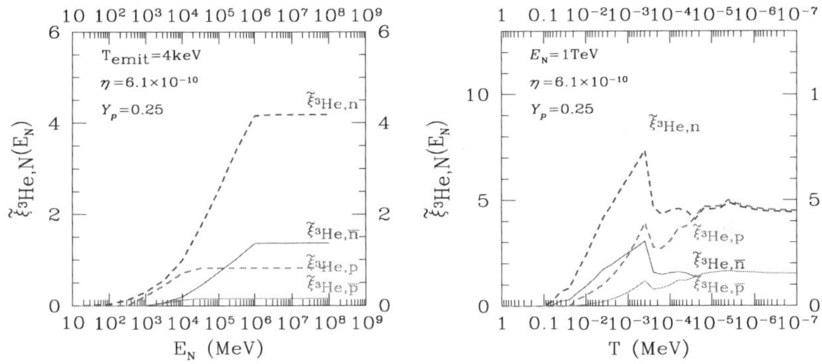

Figure 8: $\tilde{\xi}_{3He,N}$ for T = 4 keV as functions of the kinetic energy (left) and those for EN = 1 TeV as functions of the cosmic temperature (right). Here, N = n, $\bar{n}$, p and $\bar{p}$, and we take $\eta = 6.1 \times 10^{-10}$ and Yp = 0.25.

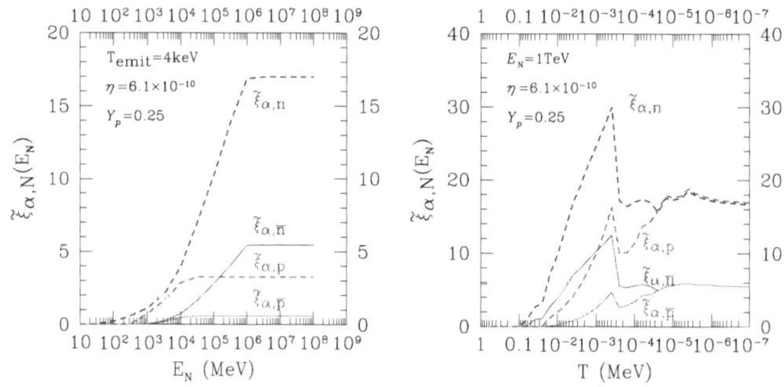

Figure 9: $\tilde{\xi}_{\alpha,N}$ for T = 4 keV as functions of the kinetic energy (left) and those for EN = 1 TeV as functions of the cosmic temperature (right). Here, N = n, $\bar{n}$, p and $\bar{p}$, and we take $\eta = 6.1 \times 10^{-10}$ and $Y_p = 0.25$.

On the same token, emitted high-energy protons with $E_n > O(10)$ GeV loose their kinetic energy down to ~ 10 GeV. The figures also show how the $\tilde{\xi}_{Ai,N}$ parameters depend on the cosmic temperature, taking EN = 1 TeV. As one can see, $\tilde{\xi}_{Ai,N}$ parameters increase as the temperature decreases until T ~ 0.5 keV; this is because the mean free path of energetic nucleons becomes longer as the temperature drops. We can also see sharp drop-offs of the $\tilde{\xi}$-parameters at T ~ 0.5 keV; this comes from the fact that, when the cosmic temperature is lower than ~ 0.5 keV, energetic neutrons decay before scattering off background nuclei and that the neutron contributions to the hadrodissociations become negligible.

From the figures, we find that the injections of $\bar{n}$ and $\bar{p}$ should change the total number of the destroyed αBG by ~ 20 – 30 %. As will be shown later, the constraints become stronger by 10 – 30 % when we include the hadrodissociations by anti-nucleons. On the other hand, by the effects of the interconversions at inelastic scatterings, the constraints become weaker by 50 – 80 %. That is because high-energy protons which can be produced by interconversion from a projectile neutron tend to be stopped more easily than neutrons through electromagnetic interactions inside the thermal plasma. With ξ_{Ai} (T, Y_p, η) being given, the effects of the hadrodissociation are included into the Boltzmann equations which govern the evolutions of light element abundances; for A_i = n, T, D, 3He, we use

$$\frac{dn_{A_i}}{dt} = \left[\frac{dn_{A_i}}{dt}\right]_{\rm SBBN} + \left[\frac{dn_{A_i}}{dt}\right]_{\rm photodis} + n_X \Gamma_X \xi_{A_i}, \tag{3.7}$$

while

$$\frac{dn_{^4\mathrm{He}}}{dt} = \left[\frac{dn_{^4\mathrm{He}}}{dt}\right]_{\mathrm{SBBN}} + \left[\frac{dn_{^4\mathrm{He}}}{dt}\right]_{\mathrm{photodis}} - n_X \Gamma_X \xi_\alpha. \tag{3.8}$$

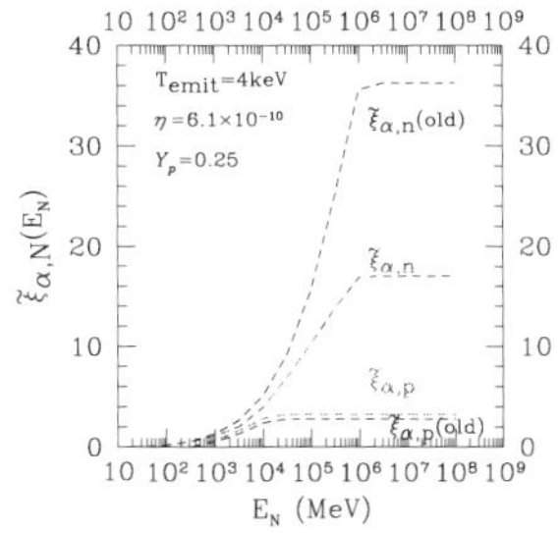

Figure 10: $\tilde{\xi}_{\alpha,N}$ for T = 4 keV as functions of the kinetic energy, based on the current method (red and blue) as well as those based on the old method which is without the interconversions between n and p at inelastic scattering (black). Here, we take the same parameters as those used in the left panel of Fig. 9.

Here, the subscript "SBBN" and "photodis" imply the reaction rates due to the SBBN and photodissociation processes. Notice that the effects of the p ↔ n interconversions of background nucleons are included in the SBBN contributions by properly modifying the number densities of background p and n. For 6Li, 7Li and 7Be, we include non-thermal production processes induced by secondary energetic T, 3He, 4He. Such energetic nuclei are produced by the hadronic scattering processes of background α as well as by the electromagnetic processes with energetic photon. For the non-thermal processes induced by hadrons, the effects are parameterized by the following quantity (with Af = ^{6}Li, ^{7}Li and ^{7}Be) [22, 23, 24, 28]

$$\xi_{A_f} = \sum_{A_i = \mathrm{T}, ^3\mathrm{He}, ^4\mathrm{He}} \int dE_{A_i}^{(\mathrm{in})} f_{A_i}(E_{A_i}^{(\mathrm{in})}) \int^{E_{A_i}^{(\mathrm{in})}} dE_{A_i} \left(\frac{dE_{A_i}}{dt}\right)^{-1} n_\alpha \sigma_{A_i+\alpha_{\mathrm{BG}} \to A_f + \cdots}(E_{A_i}) \beta_{A_i} P_{A_f}, \tag{3.9}$$

where f_{Ai} is the energy distribution of Ai produced by the scattering or hadrodissociation processes of α_{BG}, $(dE_{A_i}/dt)^{-1}$ is the energy-loss rate, $\sigma_{A_i+\alpha_{\mathrm{BG}} \to A_f + \cdots}$ is the production cross section of Af , βAi is the velocity, and PAf is the survival probability of Af after production. (For Af = ^{6}Li, Ai = T, ^{3}He, and 4He contributes, while only Ai = ^{4}He is relevant for Af = ^{7}Li and ^{7}Be.) Then, for Ai =^{6}Li, ^{7}Li and ^{7}Be, the Boltzmann equations are given in the following form:

$$\frac{dn_{A_i}}{dt} = \left[\frac{dn_{A_i}}{dt}\right]_{\rm SBBN} + \left[\frac{dn_{A_i}}{dt}\right]_{\rm photodis} + \left[\frac{dn_{A_i}}{dt}\right]_{\gamma+\alpha_{\rm BG}\to\cdots} + n_X\Gamma_X\xi_{A_i}. \tag{3.10}$$

Here, $[dnA_i /dt]_{\gamma+\alpha BG}\to\cdots$ denotes the effects of the non-thermal production processes initiated by the photodissociation of α_{BG} [16], which exists only for $A_i = {}^6Li$.

CONSTRAINTS ON GENERIC DECAYING PARTICLES

Constraints from each light elements

We first investigate the BBN constraints on generic decaying particles. In Fig. 10 we show the constraints on the decaying particles which primarily decay into e +e − or b ¯b. In the figure, we show the constraints on the combination $m_X Y_X$ (with mX being the mass of X) as functions of the lifetime τX for the final states e +e − and b ¯b. #8

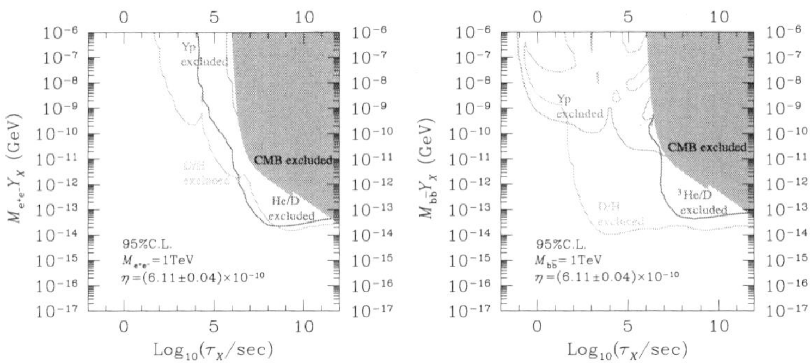

Figure 11: Constraints on mXYX vs. τX plane, assuming that the main decay modes are e +e − (left) and b ¯b (right). The BBN constraints comes from ^4He (green), D (cyan) and $^3He/D$ (red). The orange shaded region is excluded by the CMB spectral distortion.

When the decaying particles mainly decay into e +e −, the hadronic branching ratio is small. In fact, hadrons are produced through the decay process X → e + + e − + q + ¯q, but the branching ratio for a such process is suppressed by ∼ $O((\alpha/4\pi)^2) \sim 10^{-6}$. Thus, most of the constraints are due to radiative decay. At the cosmic time τX ∼ 104 − 106 sec only D is destroyed by energetic photons that are not thermalized by photon-photon processes, so photodissociation of D gives the stringent constraint for τX ~ $10^4 - 10^6$ sec. At t & 106 sec, 4He is also destroyed by photodissociation processes, which leads to non-thermal production of D and 3He. Thus, the stringent constraints are imposed by overproduction of D and ^{3}He.#9 It is seen that the abundance of the X is also constrained for τX . 10^4 sec, which is due to hadrodissociation of ^{4}He. Since the photons with energy larger than O(1) MeV are quickly thermalized at t . 10^4 sec, they cannot destroy light elements. On the other hand, hadrons like proton and neutron can destroy ^{4}He and produce D and ^{3}He non-thermally. The resultant constraint due to hadrons is weak because the hadronic branching ratio is small in this case.

In the case where the X mainly decays into b ¯b (Fig. 10 (right)), the stringent constraints come from hadrodissociation of light elements. The high energy quarks emitted in the decay induce hadronic showers in which ^{4}He nuclei are destroyed by energetic nucleons. The hadrodissociation of ^{4}He leads to overproduction of D, which gives a stringent constraint, in particular, for τX ~ $10^2 - 10^7$ sec The photodissociation of ^{4}He also produces D and gives a stringent constraint for $\tau_X \gtrsim 10^7$ sec where effects of hadrodissociation and photodissociations are roughly comparable. In addition, non-thermal production of ^{3}He by photodissociation gives a significant constraint for $\tau_X \gtrsim 10^7$ sec. For τ_X . $\gtrsim$ 100 sec the constraint coming from ^{4}He overproduction is most stringent. At the early stage of BBN (t ~ 1 – 100 sec) interconversion of protons and neutrons is the most important process which almost determines the final 4He abundance. The strongly interacting conversion increases n/p from its standard value. As a result more ^{4}He is produced, from which we obtain the constraint for τ_X . $\gtrsim$ 100 sec. From the figure one notice that there appears the constraint from D/H for τ_X ~ 0.1 – 100 sec. The constraint around $m_X Y_X \sim 10^{-9}$ GeV comes from overproduction of D due to larger n/p. Since the change of n/p affects the abundance of ^{4}He more significantly, the D/H constraint is weaker. The D/H constraint around $m_X Y_X \sim 10^{-7}$ GeV needs some caution. In this parameter region large n/p already makes the abundance of 4He much larger than the standard value. The D abundance is sensitive to the change of ^{4}He abundance because D is residual after synthesizing ^{4}He. The ^{4}He abundance has O(10)% theoretical uncertainty in this parameter region, which drastically increases the uncertainty of the D abundance. As a result, it makes difficult to obtain a reliable constraint from D. Since this region is ruled out by overproduction of ^{4}He, this is not an obstacle to obtain constraints on the properties of X

In Fig. 10, the constraints from the CMB are also shown. When the decay of X injects electromagnetic energy into the background plasma, the spectrum of the CMB is distorted. Since photon number changing processes, like the double Compton scattering, are not efficient at t $\gtrsim$ 10^7 sec, the resultant CMB spectrum deviates from the Planck distribution if X decay at such an epoch [49, 50]. The COBE set a stringent constraint on the spectrum distortion of the CMB [51], from which we obtain the upper bound on the abundance of X, as shown in Fig. 10. It is seen that the BBN constraints are more stringent than those from the CMB for both decay modes.

In Fig. 10, the constraints from the CMB are also shown. When the decay of X injects electromagnetic energy into the background plasma, the spectrum of the CMB is distorted. Since photon number changing processes, like the double Compton scattering, are not efficient at t $\gtrsim$ 10^7 sec, the resultant CMB spectrum deviates from the Planck distribution if X decay at such an epoch [49, 50]. The COBE set a stringent constraint on the spectrum distortion of the CMB [51], from which we obtain the upper bound on the abundance of X, as shown in Fig. 10. It is seen that the BBN constraints are more stringent than those from the CMB for both decay modes.

CONSTRAINTS ON VARIOUS DECAY MODES

Now, we show constraints for the cases with various main decay modes. We first consider the cases where X decays into colored particles. In Fig. 11 we show the combined constraints on the abundance and lifetime of X whose main decay mode is $u\bar{u}$, $b\bar{b}$, $t\bar{t}$ or gg; in Fig. 11, only the most stringent constraint is shown combining the constraints from D, ^{4}He, and ^{3}He/D. For all decay modes shown in this figure, the decay products are colored particles and produce hadronic showers whose total energy is roughly equal to m_X. Therefore, the resultant constraints are similar. As described above, hadrons produced by the decay affects the BBN for $\tau_X \lesssim 10^7$ sec. As a result, the constraints from overproduction of ^{4}He due to enhance of n/p is most stringent for $\tau_X \lesssim 10^2$ sec while overproduction of D due to hadrodissociation of 4He gives the strongest constraint for $\tau_X \gtrsim 10^2 - 10^7$ sec. We also show how the constraints depend on mX, taking mX = 0.03, 0.1, 1, 10, 100 and 1000 TeV. Since the number of hadrons produced through the hadronization process depends on mX as m^{δ}_X with $\delta \sim 0.3$, the constraints on $m_X Y_X$ from hadrodissociation become weaker as

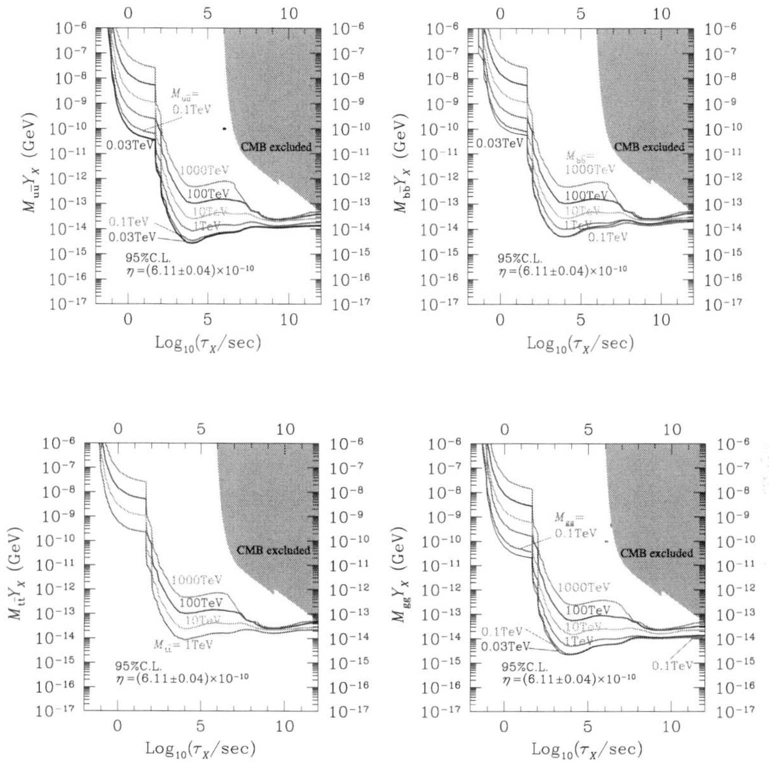

Figure 12: Constraints on $m_X Y_X$ vs. τX plane, assuming that the main decay modes are uu‾ (upper left), b ‾b (upper right), tt‾ (lower left) and gg (lower right). The black, red, green, blue and magenta solid lines denote the BBN constraints for mX = 0.03, 0.1, 1, 10, 100 and 1000 TeV, respectively. The orange shaded regions are excluded by the constraint from the CMB spectral distortion.

mX increases. On the other hand, the most stringent constraints for τX $\gtrsim$ 10^7 sec come from the photodissociation of ^{4}He, which leads to the overproduction of D and ^{3}He. Since the effects of photodissociations are determined by the total energy injection, the constraint only depends on $m_X Y_X$ as confirmed in Fig. 11.

In Fig. 12, we show the combined constraints on X which mainly decays into e +e −, τ +τ −, γγ or W+W−, taking mX = 0.03, 0.1, 1, 10, 100 and 1000 Te_V. For the decay into e +e −, as described in Section 4.1, the constraints are determined by photodissociation effect for τ_X & 10^4 sec (destruction of D for $\tau_X \sim 10^4 - 10^6$ sec and overproduction of 3He for τ_X & 10^6 sec). Because the branching ratio into qq¯ is small, the effect of hadronic decay is significant only for τ_X . 10^4 sec where the photodissociation does not take place. The constraints for the case that X mainly decays into τ +τ − is also shown. Since the branching ratio into qq¯ is as small as the e +e − case, the constraints for τX & 104 sec are almost same as Fig. 12 (upper left); the constraints are slightly weaker because some amount of the energy is carried away by neutrinos. The effect of hadrodissociation is seen for τX . 104 sec, which is similar to the e +e − decay. The constraints from the change of n/p are seen for τX . 102 sec, which is due to the mesons produced by the τ decay. The constraints on the decay mode γγ are shown in Fig. 12 (lower left). For τX & 106 sec, the photodissociation is important and the constraint is similar to that for e +e − decay. However, compared to the e +e − case, the qq¯ production rate at the decay is relatively large ~ O(α/4π) ~ $10^{-4} - 10^{-3}$. Therefore, the constraints from the hadronic processes is more important than that from photodissociation for τX . 10^6 sec. In Fig. 12 (lower right), we show the constraints on the decay into W+W−. Since W bosons further decay into hadrons with branching ratio of about 0.67, effects of the hadronic decay are important and hence the resultant constraints looks similar to those on qq¯ decay modes.

So far we have adopted Eq. (2.4) as the observational constraint of 4He. Now let us see how the constraints change if we adopt the other constraint (2.3). In Fig. 13, the BBN constraint from D, 3He/D and 4He are shown, assuming that the main decay mode is b ¯b. As mentioned in Section 2, the 4He abundance (2.3) estimated by Izotov, Thuan and Guseva is significantly larger than that obtained by Aver, Olive and Skillman given in Eq. (2.4) and not consistent with the standard BBN if we use the baryon-to-photon ratio determined by Planck. Their observation becomes consistent with BBN if the abundance of 4He is increased by the decay of X. When X mainly decays into b ¯b, there appears a region consistent with Eq. (2.3) as well as with other light-element abundances, i.e., τX ~ $10^{-1} - 10^4$ sec and $m_X Y_X \sim 10^{-10} - 10^{-6}$ GeV. This is due to the effect that the decay of X with such lifetime induces the p ↔ n conversion, resulting in the increase of the abundance of 4He. If the 4He abundance (2.3) is confirmed, it may suggest the existence of a long-lived hadronically decaying particle which solves the discrepancy between the standard BBN and Eq. (2.3). On the other hand, if the hadronic branching ratio is much smaller, like the case that the main decay mode is e

+e −, there is no region consistent with Eq. (2.3) together with observational constrains on D and 3He/D.

Before closing this section, we comment on the effects of the decaying particles on abundances of 7Li and 6Li although we do not use them to derive the constraints on the decaying particles. In Fig. 14, abundances of ^{7}Li and ^{6}Li are shown, assuming that the main decay mode is b ¯b or e +e −.

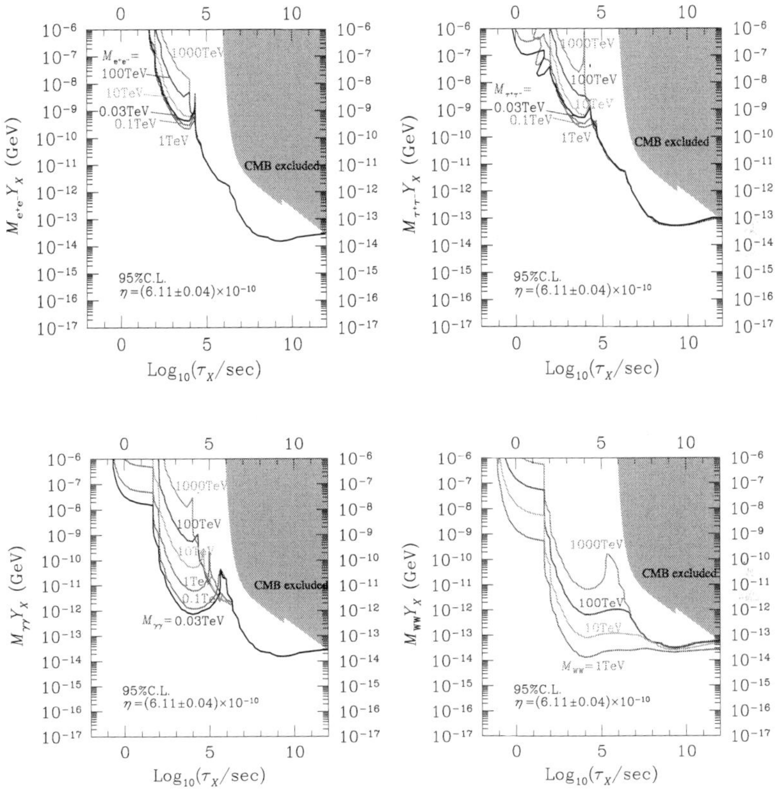

Figure 13: Constraints on $m_X Y_X$ vs. τ_X plane, assuming that the main decay modes are e +e − (upper left), τ +τ − (upper right), γγ (lower left) and W+W− (lower right). The black, red, green, blue and magenta solid lines denote the BBN constraints for m_X = 0.03, 0.1, 1, 10, 100 and 1000 Te_V, respectively. The orange shaded regions are excluded by the constraint from the CMB spectral distortion.

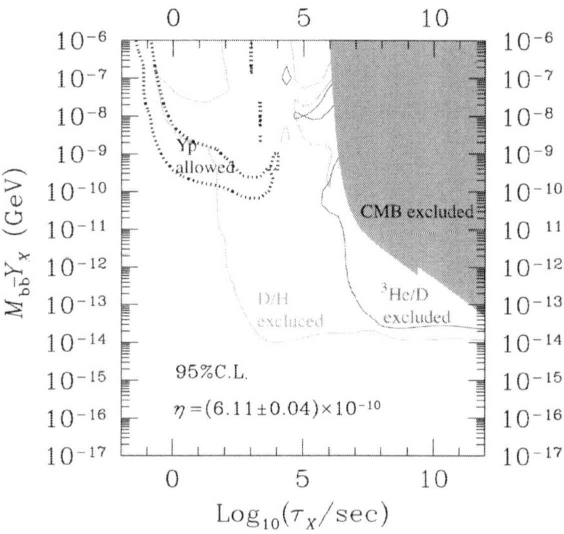

Figure 14: Constraints on $m_X Y_X$ vs. τ_X plane, assuming that m_X = 1 TeV and that the main decay mode is b ¯b. The solid cyan, solid red and dashed lines denote the BBN constraints from D, ^{3}He/D and 4He, respectively. Here, we use Eq. (2.3) as the observational constraint on Y_p. The orange shaded region is excluded by the CMB spectral distortion.

Notably, for $\tau_X \sim 10^2 - 10^3$ sec, primordial ^{7}Li abundance is reduced by non-thermal neutrons; ^{7}Be, which is one of the origins of primordial ^{7}Li, can be converted to ^{7}Li by non-thermal neutrons as ^{7}Be+n $\rightarrow^7$ Li+p, then ^{7}Li is destroyed by the SBBN reaction 7Li +p $\rightarrow$4 He +4 He. Such effects may dominate over the non-thermal production of 7Li due to energetic T, 3He, and 4He from the photodissociations and scattering of ^{4}He, resulting in the net decrease of the 7Li abundance. For the case of b ¯b mode, the constraints from D/H, ^{3}He/D and 4He exclude the parameter region where the ^{7}Li abundance significantly decreases. On the other hand, ^{7}Li abundance can be reduced for the main decay mode of e +e − without conflicting with other constraints, if $\tau_X \sim$ O(100) sec and mXYX ~ 10^{-7} GeV. This might provide a solution to the ^{7}Li problem in the SBBN.

In Fig. 14 it is seen that non-thermal production of ^{6}Li due to the decaying particles is significant for $\tau_X \gtrsim 10^3$ sec. Taking into account the constraints from the other light elements, ^{6}Li/^{7}Li can be as large as 10^{-2} for the b ¯b decay mode, while it can be O(0.1) for the e +e − mode. On the contrary, ^{6}Li is hardly produced in SBBN (i.e., ^{6}Li/ 7Li . 10^{-4}). Thus, if a significant amount of ^{6}Li/^{7}Li is observed in low-metal stars, it would give an evidence for the existence of decaying particles in the early universe. In particular, for the e +e − mode, it is remarkable that ^{6}Li/7Li is significantly large with simultaneously solving the ^{7}Li problem at around $\tau_X \sim 10^3$ sec and $m_X Y_X \sim 10^{-7.5}$ GeV.

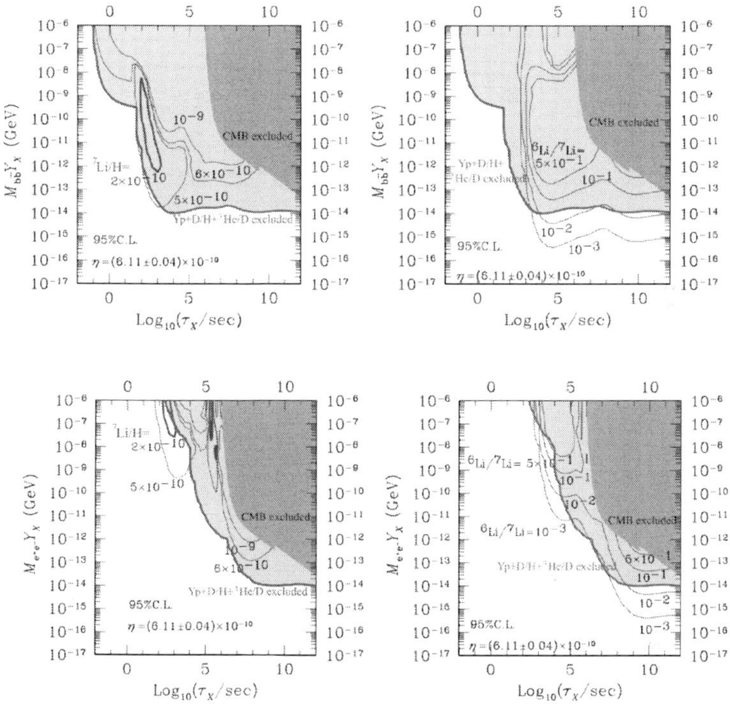

Figure 15: Abundances of ^{7}Li/H (left) and ^{6}Li/7Li (right) for decay particles with mass 1 TeV which decay mainly into b ¯b (upper) and e +e − (lower). The constraints from other light elements and CMB are shown in pink shaded region (surrounded by red solid line) and orange shaded region, respectively

GRAVITINO

One of the important candidates of long-lived particles is the gravitino in SUSY models. Gravitino is the superpartner of the graviton, and interacts very weakly because its interaction is suppressed by inverse powers of the (reduced) Planck scale MPl ‘ 2.4 × 10^{18} GeV. Because of the weakness of its interaction, the lifetime of the gravitino may become so long that its decay products affect the light element abundances (if the gravitino is unstable). Assuming R-parity conservation, the gravitino becomes unstable if it is not the lightest superparticle (LSP). Even when the gravitino is the LSP, the next-to-the lightest superparticle (NLSP) decays into gravitino with very long lifetime, and hence the BBN constraints on the properties of the NLSP are derived, e.g., see Refs. [31, 52]. Here, we pay particular attention to the former case where the gravitino is unstable because, in such a case, we can obtain an upper bound on the reheating temperature after inflation in order not to overproduce the gravitino. Applying the analysis of the non-standard BBN processes discussed in the previous sections, we study the effects of the gravitino decay on the light element abundances and derive the upper bound on the reheating temperature. The primordial abundance of the gravitino is sensitive to the reheating temperature after inflation,[#10] which we define

$$T_{\rm R} \equiv \left(\frac{10}{g_*(T_{\rm R})\pi^2} M_*^2 \Gamma_{\rm inf}^2 \right)^{1/4}, \tag{5.1}$$

with Γ_{inf} being the decay rate of the inflaton, and g∗(TR) being the effective number of the massless degrees of freedom at the time of reheating. In our study, we use the value suggested by the minimal SUSY standard model (MSSM), g∗(TR) = 228.75. The gravitino is produced via the scattering processes of MSSM particles in the thermal bath. The Boltzmann equation for the number density of gravitino (denoted as $n_{3/2}$) is given by

$$\frac{dn_{3/2}}{dt} + 3Hn_{3/2} = C_{3/2},$$

where $C_3/_2$ is the thermally averaged collision term. The most precise calculation of the collision term includes hard thermal loop resummation to avoid infrared singularity [56, 57, 58], and $C_{3/2}$ is parametrized as

$$C_{3/2} = \sum_{i=1}^{3} \frac{3\zeta(3)T^6}{16\pi^3 M_{\rm Pl}^2} \left(1 + \frac{M_i^2(T)}{3m_{3/2}^2} \right) c_i g_i^2 \ln\left(\frac{k_i}{g_i}\right), \tag{5.3}$$

where i = 1, 2, and 3 correspond to the gauge groups $U(1)_Y$, $SU(2)_L$, and $SU(3)_C$, respectively. Here, $M_i(T)$ is the gaugino mass parameter at the renormalization scale Q = T, and g_i is the gauge coupling constant. In addition, ci and k_i are numerical constants. Refs. [56, 57] give (c1, c2, c3) = (11, 27, 72) and (k_1, k_2, k_3) = (1.266, 1.312, 1.271), while the coefficients from Ref. [58] are (c_1, c_2, c_3) = (9.90, 20.77, 43.34) and (k_1, k_2, k_3) = (1.469, 2.071, 3.041).[#11] As one can see, the collision term is typically $C_{3/2}$ ~ T 6/M2

Pl (as far as T is higher than the masses of MSSM particles). Then, the yield variable of the gravitino, which is defined as $Y_{3/2}$ hereafter, is approximately proportional to the reheating temperature for the parameter region of our interest. Consequently, the BBN constraints on the gravitino abundance can be converted to the upper bound on the reheating temperature TR. We numerically solve the Boltzmann equation (5.2) (as well as the evolution equation of the universe based on Einstein equation) to accurately calculate the primordial abundance of gravitino. Assuming the MSSM particle content up to the GUT scale, the gravitino abundance can be well fit by the following formula:

$$Y_{3/2} \simeq y_0 T_{\mathrm{R}}^{(8)} \left[1 + \delta_0^{(1)} \ln T_{\mathrm{R}}^{(8)} + \delta_0^{(2)} \ln^2 T_{\mathrm{R}}^{(8)}\right] + \sum_{i=1}^{3} y_i T_{\mathrm{R}}^{(8)} \left[1 + \delta_i^{(1)} \ln T_{\mathrm{R}}^{(8)} + \delta_i^{(2)} \ln^2 T_{\mathrm{R}}^{(8)}\right] \left(\frac{M_i^{(\mathrm{GUT})}}{m_{3/2}}\right)^2, \quad (5.4)$$

Where $T_{\mathrm{R}}^{(8)} \equiv T_{\mathrm{R}}/10^8 \text{ GeV and } M_i^{(\mathrm{GUT})}$ i are the gaugino masses at the GUT scale which is taken to be 2 × 1016 GeV in our analysis. Numerical constants in the above fitting formula based on Refs. [56, 57] and those for Ref. [58] are summarized in Tables 1 and 2, respectively. The primordial abundance of gravitino based on Refs. [56, 57] is smaller than that based on Ref. [58]. Thus, we perform our numerical analysis with the former set of the coefficients in order to derive conservative constraint. The bound based on Ref. [58] can be obtained by translating the bound on T_R to that on the primordial abundance $Y_{3/2}$ using, for e.g., Eq. (5.4); the bound on T_R based on Ref. [58] is at most 2^{-3} times more stringent than that based on Refs. [56, 57].

Because the interaction of the gravitino is governed by the SUSY, the partial decay rates of the gravitino are determined once the mass spectrum of the MSSM particles are known. For a precise calculation of the the upper bound on the reheating temperature, we fix the mass spectrum of the MSSM particles and calculate the decay widths of the gravitino. We consider several sample points; the mass spectrum of each sample point is summarized in Table 3. The sample point 1, which is based on the so-called CMSSM [60] parametrized by the universal scalar mass m0, unified gaugino mass $M_{1/2}$, universal tri-linear coupling constants A_0 (with respect to the corresponding Yukawa coupling constants), tan β (which is the ratio of the vacuum expectation values of up- and down-type Higgs bosons), and the sign of the SUSY invariant Higgs mass parameter μ. For the sample point 1, we take

$m_0 = 433$ GeV, $M_{1/2} = 970$ GeV, $A_0 = -3020$ GeV, $\tan\beta = 14$, and $\mu > 0$. As one can see, in the sample point 1, all the MSSM particles are lighter than ~ 2.5 TeV, and the discovery of some of the MSSM particles are expected in future LHC experiment. The sample point 2 is also based on the CMSSM, with the underlying parameters of $m_0 = 5000$ GeV,

Table 1: Numerical constants for the formula (5.4), which gives a fitting formula for the primordial gravitino abundance, based on Refs. [56, 57]

	$i = 0$	$i = 1$	$i = 2$	$i = 3$
y_i	2.3×10^{-14}	1.3×10^{-16}	1.6×10^{-15}	1.3×10^{-14}
$\delta_i^{(1)}$	0.015	0.055	0.013	−0.042
$\delta_i^{(2)}$	−0.0005	0.0081	0.0001	0.0006

Table 2: Numerical constants for the formula (5.4) based on Ref. [58].

	$i = 0$	$i = 1$	$i = 2$	$i = 3$
y_i	4.4×10^{-14}	1.5×10^{-16}	2.1×10^{-15}	3.2×10^{-14}
$\delta_i^{(1)}$	−0.014	0.066	0.015	−0.089
$\delta_i^{(2)}$	0.0054	0.0042	0.0002	0.0031

$M_{1/2} = 700$ GeV, $A_0 = -8000$ GeV, $\tan\beta = 10$ and $\mu > 0$. At this point, the μ-parameter is relatively small, and the Higgsino-like neutralino becomes the LSP. In the sample points 3 and 4, we consider the case where the sfermion masses are above 100 TeV while gaugino masses are around the TeV scale.#12 Such a mass spectrum is motivated in the so-called pure gravity mediation model [61, 62, 63], in which scalar masses originate from the direct K¨ahler interaction between the SUSY breaking field and the MSSM chiral multiplet, while the gaugino masses are from the effect of anomaly mediation [64, 65]. Then, the gaugino masses are given in the following form:

$$M_1 = \frac{g_1^2}{16\pi^2}(11F_\Phi + L), \tag{5.5}$$

$$M_2 = \frac{g_2^2}{16\pi^2}(F_\Phi + L), \tag{5.6}$$

$$M_3 = \frac{g_3^2}{16\pi^2}(-3F_\Phi), \tag{5.7}$$

where F_{Φ} is the expectation value of the compensator multiplet,#13 while L parametrize the effect of Higgs-Higgsino loop on the gaugino masses. In a large class of models, L is of order FΦ for the scale below the masses of heavy Higgses and Higgsinos. The gaugino masses for the sample points 3 and 4 are obtained by adopting (F_{Φ}, L) = (131 TeV, 218 TeV), and (82 TeV, 87 TeV), respectively. Notice that, above the mass scale of the heavy Higgs and Higgsino, the gaugino mass parameters are obtained by Eqs. (5.5) – (5.7), with taking L → 0. For the calculation of the primordial abundance of the gravitino, we take account of this effect, and the thermally averaged gravitino production cross sections for the points 3 and 4 are evaluated by taking vanishing L

With fixed mass spectrum of the MSSM particles, we vary the gravitino mass and derive the upper bound on the reheating temperature as a function of the gravitino mass. In our calculation, all the two-body tree-level decay processes of the gravitino are taken into account. We also include three-body decay processes $\tilde{G} \to \tilde{\chi}_0 q\bar{q}$ (with $\tilde{G}$, $\tilde{\chi}_0$ and q denoting gravitino, the lightest neutralino and u, d, s, c or b quark, respectively), if the mass splitting between gravitino and the lightest neutralino is less than the Z-boson mass. This is because those three-body decay processes can have substantial contribution to hadronic emissions from gravitino decays to the lightest neutralino when the two-body decay process $\tilde{G} \to \tilde{\chi}_1^0 Z$ is kinematically forbidden. For the sample points 3 and 4, the lightest chargino $(\tilde{\chi}_1^{\pm})$ has degenerate mass with the lightest neutralino and behaves as the LSP. We therefore include the off-shell $W^{\pm}$ induced three-body decay processes, $\tilde{G} \to \tilde{\chi}_1^{\pm} q\bar{q}'$, as well in our calculation if $m_{3/2} - m_{\tilde{\chi}_1^{\pm}} < m_W$.

The partial decay rates of the gravitino are calculated by using MadGraph5 aMC@NLO v2.1 package [66] with gravitino interactions being implemented via FeynRules v2.3 [67, 68, 69, 70]. Subsequent decay and the hadronization processes of the decay products of gravitino are simulated by using PYTHIA 8.2 package [42]. For given reheating temperature, the primordial abundance of the gravitino is calculated by numerically solving Eq. (5.2). For the thermally averaged gravitino production cross section, we adopt the results of Refs. [56, 57]. Following the procedure discussed in the previous sections, we calculate the light element abundances, taking into account the effects of the decay products of the gravitino. The constraints on the gravitino abundance is then translated to the upper bound on the reheating temperature. The upper bound on T_R for the sample points 1, 2, 3, and 4 are shown in Fig. 15. As one can see, a severe upper bound of $\sim 10^5 - 10^6$ GeV is obtained from the overproduction of D for relatively light gravitino mass (i.e., for $m_{3/2} \lesssim$ a few TeV). This is because, when the gravitino is lighter than a few TeV, the lifetime of the gravitino is longer than ~ 103 sec so that thc constraint from D is significant, as discussed in the previous section. On the contrary, for heavier gravitino mass (i.e., for $m_{3/2} \gtrsim 10$ TeV), for which the lifetime of the gravitino becomes shorter than $\sim 10^2$ sec, the most stringent constraint comes from the overproduction of ^{4}He due to the p ↔ n conversion.

We also comment on the constraints based on the ^{4}He abundance given in Eq. (2.3), which is inconsistent with the SBBN prediction. As mentioned in the previous section, with a hadronically decaying long-lived particle, the primordial ^{4}He abundance may become consistent with Eq. (2.3). In the present case, gravitino has sizable branching ratio for hadronic decay modes. In Fig. 15, we also show a region consistent with the ^{4}He abundance (2.3) estimated by Izotov, Thuan and Guseva. We can see that the allowed region exists for $m_{3/2} \sim O(10)$ TeV and TR ~ $O(10^9)$ GeV.

Notice that the reheating temperature is bounded from above in order not to overclose the universe by the LSP produced from the decay of the gravitino. For the parameter region of our interest, the gravitino decays at the cosmic temperature lower than the freeze-out temperature of the LSP. Thus, the density parameter of the LSP from the decay of the gravitino is evaluated as.

Table 3: The mass spectrum of the MSSM particles as well as the gaugino masses at the GUT scale for the sample points adopted in our analysis. Here, $m_{\tilde{u}_{R,i}}$ and $m_{\tilde{u}_{L,i}}$ ($m_{\tilde{d}_{R,i}}$ and $m_{\tilde{d}_{L,i}}$) are masses of right- and left-handed sups (sdowns) in i-th generation, respectively, while $m_{\tilde{e}_{R,i}}$ and $m_{\tilde{e}_{L,i}}$ ($m_{\tilde{\nu}_{L,i}}$) are masses of right- and left-handed charged sleptons (sneutrinos) in i-th generation, respectively. In addition, $m_{\tilde{t}_1}$, $m_{\tilde{b}_1}$, and $m_{\tilde{\tau}_1}$ ($m_{\tilde{t}_2}$, $m_{\tilde{b}_2}$, and $m_{\tilde{\tau}_2}$) are lighter (heavier) stop, sbottom, and stau masses, respectively, while $m_{\tilde{\nu}_{\tau L}}$ is the tau-sneutrino mass. Furthermore, $m_{\chi_i^0}$, $m_{\chi_i^{\pm}}$, and $m_{\tilde{g}}$ are neutralino, chargino, and gluino masses, respectively, while mh and mA are masses of the lightest Higgs boson and CP-odd Higgs boson, respectively. All the mass parameters are given in units of GeV.

	Point 1	Point 2	Point 3	Point 4
$m_{\tilde{u}_{R,1,2}}$	1907	5242	$\sim 1\times 10^5$	$\sim 1\times 10^5$
$m_{\tilde{d}_{R,1,2}}$	1898	5054	$\sim 1\times 10^5$	$\sim 1\times 10^5$
$m_{\tilde{u}_{L,1,2}}(m_{\tilde{d}_{L,1,2}})$	1980 (1982)	5082 (5083)	$\sim 1\times 10^5$	$\sim 1\times 10^5$
$m_{\tilde{e}_{R,1,2}}$	562	4801	$\sim 1\times 10^5$	$\sim 1\times 10^5$
$m_{\tilde{e}_{L,1,2}}(m_{\tilde{\nu}_{L,1,2}})$	771 (767)	5093 (5092)	$\sim 1\times 10^5$	$\sim 1\times 10^5$
$m_{\tilde{t}_1}$	977	1455	$\sim 1\times 10^5$	$\sim 1\times 10^5$
$m_{\tilde{t}_2}$	1635	3603	$\sim 1\times 10^5$	$\sim 1\times 10^5$
$m_{\tilde{b}_1}$	1608	3607	$\sim 1\times 10^5$	$\sim 1\times 10^5$
$m_{\tilde{b}_2}$	1843	4990	$\sim 1\times 10^5$	$\sim 1\times 10^5$
$m_{\tilde{\tau}_1}$	417	4735	$\sim 1\times 10^5$	$\sim 1\times 10^5$
$m_{\tilde{\tau}_2}$	732	5062	$\sim 1\times 10^5$	$\sim 1\times 10^5$
$m_{\tilde{\nu}_{\tau L}}$	723	5061	$\sim 1\times 10^5$	$\sim 1\times 10^5$
$m_{\chi_1^0}$	417	187	1000	500
$m_{\chi_2^0}$	791	−209	1470	880
$m_{\chi_3^0}$	−1836	321	$\sim 1\times 10^5$	$\sim 1\times 10^5$
$m_{\chi_4^0}$	1838	618	$\sim 1\times 10^5$	$\sim 1\times 10^5$
$m_{\chi_1^{\pm}}$	791	199	1000	500
$m_{\chi_2^{\pm}}$	1839	618	$\sim 1\times 10^5$	$\sim 1\times 10^5$
$m_{\tilde{g}}$	2131	1817	3000	2000
m_h	124	126	125	125
m_A	1906	1000	$\sim 1\times 10^5$	$\sim 1\times 10^5$
$M_{1,2,3}^{(\mathrm{GUT})}$	970	700	(2583, 376, −1137)	(1619, 236, −716)

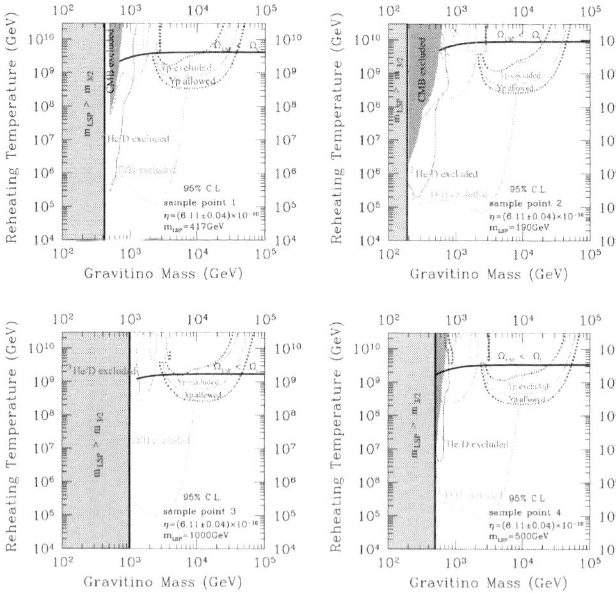

Figure 16: Upper bound on the reheating temperature TR as a function of $m_{3/2}$ at 95% C.L. for models of 1) Point 1 (upper left), 2) Point 2 (upper right), 3) Point 3 (lower left), and 4) Point 4 (lower right), respectively. The regions surrounded by the black-dotted line indicate the region consistent with Eq. (2.3).

$$\Omega_{\rm LSP}^{\rm (decay)} = \frac{m_{\rm LSP} Y_{3/2} s_{\rm now}}{\rho_{\rm crit}},$$

where snow is the entropy density of the present universe, and ρ_{crit} is the critical density. We show the contour of $\Omega_{\rm LSP}^{\rm (decay)} = \Omega_c$ (with $\Omega_c \simeq 0.26$ [41] . being the density parameter of the cold dark matter).#14 We can see that, with the present choice of the MSSM mass spectrum, the upper bound from the overclosure of the universe is $10^9 - 10^{10}$ GeV, which is less stringent when the gravitino mass is smaller than ~ 40 − 50 TeV.

Before closing this section, we comment on the implication of our result on the leptogenesis [71], in which the baryon asymmetry of the universe originates from the lepton asymmetry generated by the decay of right-handed neutrinos. In order to generate enough amount of baryon asymmetry via thermal leptogenesis, the reheating temperature is required to be higher than ~ 10^9 GeV [72, 73]. Thus, for a viable scenario of thermal leptogenesis, scenarios realizing the the gravitino mass of O(10) TeV is suggested, like the pure gravity mediation scenario [61, 62, 63]. Notice that such a scenario works irrespective of the observational constraint on the 4He abundance, Eq. (2.3) or Eq. (2.4).

CONCLUSIONS AND DISCUSSION

We have revisited and updated the BBN constraints on long-lived particles. Compared with the previous analysis we have improved the following points. First, the SBBN reactions and their uncertainties are updated. Second, we have revised the hadronic shower calculation taking into account p $\leftrightarrow$ n conversion in inelastic scatterings of energetic nucleons off the background p or 4He. Third, we have included the effects of the hadronic showers induced by the injections of energetic anti-nucleons (̄p and ̄n). Finally we have used the most recent observational data for the abundances of 4He and D and the cosmological parameters. We have obtained the constraints on the abundance and lifetime of long-lived particles with various decay modes. They are shown in Figs. 11 and 12. The constraints become weaker when we include the p $\leftrightarrow$ n conversion effects in inelastic scatterings because energetic neutrons change into protons and stop without causing hadrodissociations. On the other hand, inclusion of the energetic anti-nucleons makes the constraints more stringent. In addition, the recent precise measurement of the D abundance leads to stronger constrains. Thus, in total, the resultant constraints become more stringent than those obtained in the previous studies.

We have also applied our analysis to unstable gravitino. We have adopted several patterns of mass spectra of superparticles and derived constraints on the reheating temperature after inflation as shown in Fig. 15. The upper bound on the reheating temperature is $\sim 10^5 - 10^6$ GeV for gravitino mass m3/2 less than a several TeV and $\sim 10^9$ GeV for $m_{3/2} \sim O(10)$ TeV. This implies that the gravitino mass should be $\sim O(10)$ TeV for successful thermal leptogenesis

In obtaining the constraints, we have adopted the observed 4He abundance given by Eq. (2.4) which is consistent with SBBN. On the other hand, if we adopt the other estimation (2.3), 4He abundance is inconsistent with SBBN. However, when long-lived particles with large hadronic branch have lifetime $\tau_X \sim 0.1 - 100$ sec and abundance $m_X Y_X \sim 10^{-9}$, Eq. (2.3) becomes consistent with BBN.

In this work, we did not use ^{7}Li in deriving the constraints since the plateau value in ^{7}Li abundances observed in metal-poor stars (which had been considered as a primordial value) is smaller than the SBBN prediction by a factor 2–3 (lithium problem) and furthermore the recent discovery of much smaller ^{7}Li abundances in very metal-poor stars cannot be explained by any known mechanism. However, the effects of the decaying particles on the ^{7}Li and ^{6}Li abundances are estimated in our numerical calculation. Interestingly, if we assume that the plateau value represents the primordial abundance, the decaying particlcs which mainly decays into e +e − can solve the lithium problem for $\tau X \sim 10^2 - 10^3$ sec and $m_X Y_X \sim 10^{-7}$.

REFERENCES

1. D. Lindley, Astrophys. J. 294, 1 (1985).
2. M. Y. Khlopov and A. D. Linde, Phys. Lett. 138B, 265 (1984).
3. F.Balestra et al., Sov. J. Nucl. Phys. 39, 626 (1984).
4. J. R. Ellis, J. E. Kim and D. V. Nanopoulos, Phys. Lett. 145B, 181 (1984).
5. R. Juszkiewicz, J. Silk and A. Stebbins, Phys. Lett. 158B, 463 (1985).
6. J. R. Ellis, D. V. Nanopoulos and S. Sarkar, Nucl. Phys. B 259, 175 (1985).
7. J. Audouze, D. Lindley and J. Silk, Astrophys. J. 293, L53 (1985).
8. D. Lindley, Phys. Lett. B 171, 235 (1986).
9. M. Kawasaki and K. Sato, Phys. Lett. B 189, 23 (1987).
10. R. J. Scherrer and M. S. Turner, Astrophys. J. 331, 19 (1988).
11. J. R. Ellis, G. B. Gelmini, J. L. Lopez, D. V. Nanopoulos and S. Sarkar, Nucl. Phys. B 373, 399 (1992).
12. M. Y. Khlopov, Y. L. Levitan, E. V. Sedelnikov and I. M. Sobol, Phys. Atom. Nucl. 57, 1393 (1994) [Yad. Fiz. 57, 1466 (1994)].
13. M. Kawasaki and T. Moroi, Prog. Theor. Phys. 93, 879 (1995).
14. M. Kawasaki and T. Moroi, Astrophys. J. 452, 506 (1995).
15. E. Holtmann, M. Kawasaki, K. Kohri and T. Moroi, Phys. Rev. D 60, 023506 (1999).
16. K. Jedamzik, Phys. Rev. Lett. 84, 3248 (2000).
17. M. Kawasaki, K. Kohri and T. Moroi, Phys. Rev. D 63, 103502 (2001).
18. R. H. Cyburt, J. R. Ellis, B. D. Fields and K. A. Olive, Phys. Rev. D 67, 103521 (2003).
19. M. Kusakabe et al., Phys. Rev. D 79, 123513 (2009).
20. V. Poulin and P. D. Serpico, Phys. Rev. D 91, no. 10, 103007 (2015).
21. M. H. Reno and D. Seckel, Phys. Rev. D 37, 3441 (1988).
22. S. Dimopoulos, R. Esmailzadeh, L. J. Hall and G. D. Starkman, Astrophys. J. 330, 545 (1988).
23. S. Dimopoulos, R. Esmailzadeh, L. J. Hall and G. D. Starkman, Phys. Rev. Lett. 60, 7 (1988).
24. S. Dimopoulos, R. Esmailzadeh, L. J. Hall and G. D. Starkman, Nucl. Phys. B 311, 699 (1989).

25. K. Kohri, Phys. Rev. D 64, 043515 (2001).
26. K. Jedamzik, Phys. Rev. D 70, 063524 (2004).
27. M. Kawasaki, K. Kohri and T. Moroi, Phys. Lett. B 625, 7 (2005).
28. M. Kawasaki, K. Kohri and T. Moroi, Phys. Rev. D 71, 083502 (2005).
29. K. Kohri, T. Moroi and A. Yotsuyanagi, Phys. Rev. D 73, 123511 (2006).
30. K. Jedamzik, Phys. Rev. D 74, 103509 (2006).
31. M. Kawasaki, K. Kohri, T. Moroi and A. Yotsuyanagi, Phys. Rev. D 78, 065011 (2008).
32. R. H. Cyburt, J. Ellis, B. D. Fields, F. Luo, K. A. Olive and V. C. Spanos, JCAP 0910, 021 (2009).
33. S. Weinberg, Phys. Rev. Lett. 48, 1303 (1982).
34. M. Kawasaki, K. Kohri, T. Moroi and Y. Takaesu, Phys. Lett. B 751, 246 (2015).
35. R. Cooke, M. Pettini, R. A. Jorgenson, M. T. Murphy and C. C. Steidel, Astrophys. J. 781, no. 1, 31 (2014)
36. G. Sigl, K. Jedamzik, D. N. Schramm and V. S. Berezinsky, Phys. Rev. D 52, 6682 (1995).
37. J. Geiss and G. Gloeckler, Space Sience Reviews 106, 3 (2003).
38. Y. I. Izotov, T. X. Thuan and N. G. Guseva, Mon. Not. Roy. Astron. Soc. 445, 778 (2014).
39. E. Aver, K. A. Olive and E. D. Skillman, JCAP 1507, no. 07, 011 (2015).
40. L. Sbordone, P. Bonifacio, E. Caffau, H.-G. Ludwig, N. T. Behara, J. I. G. Hernandez, M. Steffen and R. Cayrel et al., Astron. Astrophys. 522, A26 (2010).
41. P. A. R. Ade et al. [Planck Collaboration], Astron. Astrophys. 594, A13 (2016).
42. T. Sjostrand et al., Comput. Phys. Commun. 191, 159 (2015).
43. P. D. Serpico, S. Esposito, F. Iocco, G. Mangano, G. Miele and O. Pisanti, JCAP 0412, 010 (2004).
44. R. H. Cyburt and B. Davids, Phys. Rev. C 78, 064614 (2008).
45. J. P. Meyer, Astron. Astrophys. Suppl. 7 417 (1972).
46. C. Patrignani et al. [Particle Data Group], Chin. Phys. C 40, no. 10, 100001 (2016).
47. E. Bracci, J. P. Droulez, E. Flaminio, J. D. Hansen, D. R. O. Morrison,

and Douglas Robert Ogston, CERN HERA-73-1 (1973).

48. J. Hisano, M. Kawasaki, K. Kohri, T. Moroi and K. Nakayama, Phys. Rev. D 79, 083522 (2009).
49. W. Hu and J. Silk, Phys. Rev. Lett. 70, 2661 (1993).
50. J. Chluba, Mon. Not. Roy. Astron. Soc. 436, 2232 (2013).
51. D. J. Fixsen, E. S. Cheng, J. M. Gales, J. C. Mather, R. A. Shafer and E. L. Wright, Astrophys. J. 473, 576 (1996).
52. K. Ishiwata, M. Kawasaki, K. Kohri and T. Moroi, Phys. Lett. B 689, 163 (2010).
53. M. Endo, K. Hamaguchi and F. Takahashi, Phys. Rev. Lett. 96, 211301 (2006).
54. S. Nakamura and M. Yamaguchi, Phys. Lett. B 638, 389 (2006).
55. M. Kawasaki, F. Takahashi and T. T. Yanagida, Phys. Lett. B 638, 8 (2006).
56. M. Bolz, A. Brandenburg and W. Buchmuller, Nucl. Phys. B 606, 518 (2001) Erratum: [Nucl. Phys. B 790, 336 (2008)].
57. J. Pradler and F. D. Steffen, Phys. Rev. D 75, 023509 (2007).
58. V. S. Rychkov and A. Strumia, Phys. Rev. D 75, 075011 (2007).
59. J. Ellis, M. A. G. Garcia, D. V. Nanopoulos, K. A. Olive and M. Peloso, JCAP 1603, no. 03, 008 (2016).
60. S. P. Martin, Adv. Ser. Direct. High Energy Phys. 21, 1 (2010) [Adv. Ser. Direct. High Energy Phys. 18, 1 (1998)].
61. M. Ibe, T. Moroi and T. T. Yanagida, Phys. Lett. B 644, 355 (2007).
62. M. Ibe and T. T. Yanagida, Phys. Lett. B 709, 374 (2012).
63. N. Arkani-Hamed, A. Gupta, D. E. Kaplan, N. Weiner and T. Zorawski, arXiv:1212.6971 [hep-ph].
64. L. Randall and R. Sundrum, Nucl. Phys. B 557, 79 (1999).
65. G. F. Giudice, M. A. Luty, H. Murayama and R. Rattazzi, JHEP 9812, 027 (1998).
66. J. Alwall et al., JHEP 1407, 079 (2014).
67. C. Duhr and B. Fuks, Comput. Phys. Commun. 182, 2404 (2011).
68. C. Degrande, C. Duhr, B. Fuks, D. Grellscheid, O. Mattelaer and T. Reiter, Comput. Phys. Commun. 183, 1201 (2012).
69. N. D. Christensen et al., Eur. Phys. J. C 73, no. 10, 2580 (2013).

70. A. Alloul, N. D. Christensen, C. Degrande, C. Duhr and B. Fuks, Comput. Phys. Commun. 185, 2250 (2014).
71. M. Fukugita and T. Yanagida, Phys. Lett. B 174, 45 (1986).
72. G. F. Giudice, A. Notari, M. Raidal, A. Riotto and A. Strumia, Nucl. Phys. B 685, 89 (2004).
73. W. Buchmuller, P. Di Bari and M. Plumacher, Annals Phys. 315, 305 (2005)

SECTION 3:
STARS AND THE CREATION OF THE HIGHER ELEMENTS

5

Formation of the First Stars in the Universe

sNaoki Yoshida[1,2], Takashi Hosokawa[3], and Kazuyuki Omukai[4]

[1]Kavli Institute for the Physics and Mathematics of the Universe, University of Tokyo, Kashiwa, Chiba 277-8583, Japan
[2]Department of Physics, University of Tokyo, Bunkyo, Tokyo 113-0033, Japan
[3]Jet Propulsion Laboratory, California Institute of Technology, Pasadena, CA 91198, USA
[4]Department of Physics, Kyoto University, Kyoto 606-8502, Japan

ABSTRACT

The standard theory of cosmic structure formation posits that the present-day rich structure of the universe developed through gravitational amplification of tiny matter density fluctuations left over from the Big Bang. Recent observations of the cosmic microwave background, large-scale structure, and distant supernovae determined the energy content of the universe and the basic statistics of the initial density field with great accuracy. It

Citation: Naoki Yoshida, Takashi Hosokawa, Kazuyuki Omukai, "Formation of the first stars in the universe", Progress of Theoretical and Experimental Physics, Volume 2012, Issue 1, 2012, 01A305, https://doi.org/10.1093/ptep/pts022

has become possible to make accurate predictions for the formation and nonlinear growth of structure through early to the present epochs. We review recent progress in the theory of structure formation in the early universe. Results from state-of-the-art computer simulations are presented. Finally, we discuss prospects for future observations of the first generation of stars, black holes, and galaxies.

INTRODUCTION: THE DARK AGES

The rich structures in the universe we see today, such as galaxies and galaxy clusters, have developed over a very long time. Astronomical observations utilizing large ground-based telescopes discovered distant galaxies [1–3], quasars [4,5], and gamma-ray bursts [6,7] in the universe when its age was less than one billion years old. We can track the evolution of cosmic structure from the present day all the way back to such an early epoch. We can also observe the state of the universe at an even earlier epoch, about 370 000 years after the Big Bang, as the cosmic microwave background (CMB) radiation. The anisotropies of the CMB provide information on the *initial conditions* for the formation of all the cosmic structures. In between these two epochs lies the remaining frontier of astronomy, when the universe was about a few to several million years old. The epoch is called the cosmic Dark Ages [8].

Shortly after the cosmological recombination epoch, when hydrogen atoms were formed and the CMB photons were last scattered, the CMB shifted to infrared, and then the universe would have appeared completely dark to human eyes. A long time had to pass until the first stars were born, which then illuminated the universe once again and terminated the Dark Ages. The first stars are thought to be the first sources of light, and also the first sources of heavy elements that enabled ordinary stellar populations, planets, and, ultimately, life to emerge [9].

Over the past decades, there have been a number of theoretical studies on the yet-unrevealed era in cosmic history. As early as 1953, Schwarzschild and Spitzer [10] speculated on the existence of massive primordial stars in the early universe, based on observational facts such as the very low metallicity of the oldest stars in the Galaxy, the higher frequency of white dwarfs than expected, and the red excess of distant elliptical galaxies. The study of primordial star formation was stimulated by the discovery of CMB in 1965 [11], which firmly established the Big Bang cosmology. It was then thought that the first generation of stars must have been formed from a pristine gas that consisted of only hydrogen and helium. The detailed physical processes in the primordial gas leading to the first star formation have been studied by a number of authors. [12–16] Although steady progress had been seen in theoretical studies, it was not until the late 1990s when the first stars were considered seriously within the framework of the standard cosmological model. Rapid development of super-computers enabled us to take an *ab initio* approach to perform numerical simulations starting from the early universe to the birth of the first stars.

In this article, we review recent progress in the theory of structure formation in the early universe. Theoretical studies hold promise for revealing the detailed process of primordial star formation for two main reasons: (1) the initial conditions, as determined cosmologically, are well-established, so that statistically equivalent realizations of a standard model universe can be accurately generated, and (2) all the important basic physics such as gravitation, hydrodynamics, and atomic and molecular processes in a hydrogen–helium gas are understood. In principle, therefore, it is possible to make solid predictions for the formation of early structure and of the first stars in an expanding universe. We describe some key physical processes. Computer simulations are often used to tackle the highly nonlinear problems of structure formation. We present the results from large-scale cosmological *N*-body hydrodynamic simulations. We conclude the present article by giving future prospects for observations of the first stars.

HIERARCHICAL STRUCTURE FORMATION AND THE FIRST COSMOLOGICAL OBJECTS

We first describe the generic hierarchical nature of structure formation in the standard cosmological model, which is based on weakly-interacting cold dark matter (CDM). The primordial density fluctuations predicted by popular inflationary universe models have very simple characteristics [22]. The density fluctuations are described by a Gaussian random field, and have a nearly scale-invariant power spectrum $P(k) \propto kn$ for wavenumber k with $n \sim 1$. The perturbation power spectrum is processed through the early evolution of the universe [23]. Effectively, the slope of the power-spectrum changes slowly as a function of length scale, but the final shape is still simple and monotonic in the CDM models. The CDM density fluctuations have progressively larger amplitudes on smaller length scales. Hence structure formation is expected to proceed in a "bottom-up" manner, with smaller objects forming earlier.

To obtain the essence of hierarchical structure formation, it is useful to work with the mass variance. This is defined as the root-mean square of mass density fluctuations within a sphere that contains mass M (see Appendix for definition). Figure 1 shows the variance and the collapse threshold at $z=0,5,20$. At $z=20$, the mass of a collapsing halo that corresponds to a 3-σ fluctuation is just about $10^6 M_\odot$. As shown later in Sect. 3, this is the characteristic mass of the first objects in which the primordial gas can cool and condense by molecular hydrogen cooling.

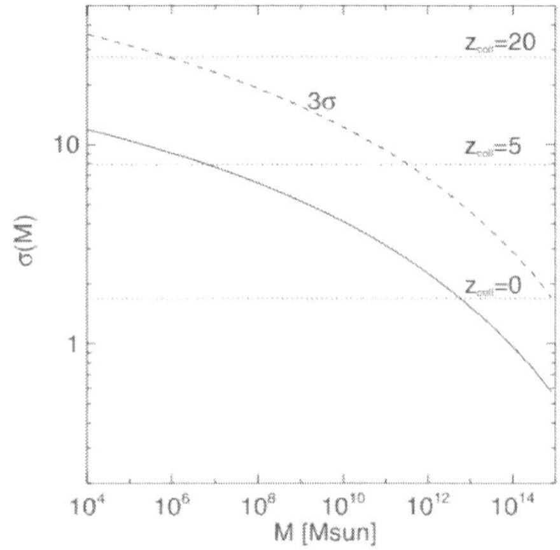

Figure 1. Mass variance and collapse thresholds for a flat CDM model with cosmological constant. The assumed cosmological parameters are: matter density $\Omega_m=0.3$, baryon density $\Omega_b=0.04$, amplitude of fluctuations $\sigma_8=0.9$, and the Hubble constant $H_0=70\,\text{km s}^{-1}\text{ Mpc}^{-1}$. The horizontal dotted lines indicate the threshold for collapse at $z=20,5,0$. We also show the variance for 3-σ fluctuations by a dashed line.

The mass variance is sensitive to the shape of the initial power spectrum. For instance, in warm dark matter models in which the power spectrum has an exponential cut-off at the dark matter particle free-streaming scale, the corresponding mass variance is significantly reduced [24,25]. In such models, early structure formation is effectively delayed, and hence small nonlinear objects form later than in the CDM model. Thus the formation epoch of the first objects and hence the onset of cosmic reionization are directly related to the nature of dark matter and the shape of the primordial density fluctuations [26–28]. We discuss this issue further in Sect. 4.

FORMATION OF THE FIRST COSMOLOGICAL OBJECTS

The basics of the formation of nonlinear dark matter halos are easily understood; because of their hierarchical nature, dark matter halos form in essentially the same fashion regardless of mass and the formation epoch. Halos would form at all mass scales by gravitational instability from nearly scale-free density fluctuations. The first "dark" objects are then well defined, and are indeed halos of a very small mass that is set by the dark matter particles' initial thermal motion [29]. Such small objects remain dark without hosting star(s) in them.

The formation of the first luminous objects involves a variety of physical processes in addition to gravity, and so is much more complicated. However, the established standard cosmological model has enabled us to answer fundamental questions with some confidence, such as *when did the first objects form?*, and *what is the characteristic mass?* Theoretical studies as well as numerical simulations of early structure formation concluded that this process likely began as early as when the age of the universe was less than a million years [8,34].

The initially diffuse cosmic gas falls into the potential well of a dark-matter halo and is heated adiabatically and by weak shocks. For further collapse, condensation, and star formation, the internal energy of the gas must be radiated away efficiently. Specifically, the radiative cooling time must be shorter than the age of the universe at that epoch for the luminous objects to form before being incorporated hierarchically into large objects [30]. Since hydrogen atoms have excitation energies that are too high, radiative cooling proceeds only via a trace amount of molecular hydrogen, which has the first excited state at $\sim 512\,K$.

In the standard CDM model, dense, cold clouds of self-gravitating molecular gas develop in the inner regions of small dark halos and contract into protostellar objects with masses in the range $\sim 100-1000\,M_{\odot}$. Figure 2 shows the projected gas distribution in a cosmological simulation that includes hydrodynamics and primordial gas chemistry [35]. Star-forming gas clouds are found at the knots of filaments, resembling the large-scale structure of the universe, although actually much smaller in mass and size. This manifests the hierarchical nature of structure in the CDM universe.

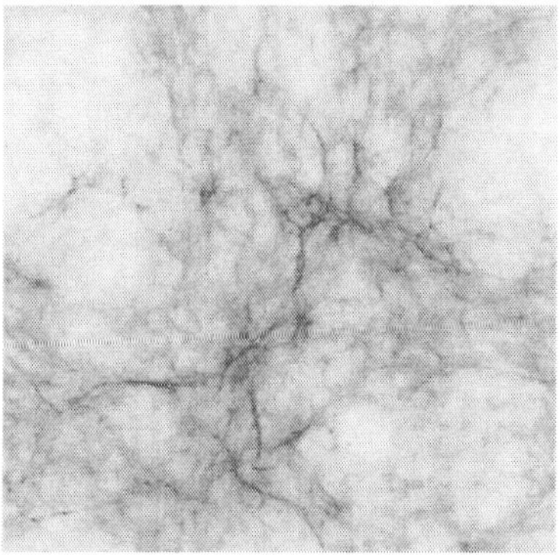

Figure 2. The projected gas distribution at $z=17$ in a cubic volume of $600h^{-1}$ kpc per side. The cooled dense gas clouds appear as bright spots at the intersections of the filamentary structures. From Ref. [35].

In a diffuse cosmic gas of primordial composition, molecular hydrogen (H_2) forms via a sequence of reactions,

$$\mathrm{H} + e^{-} \rightarrow \mathrm{H}^{-} + \gamma, \quad (1)$$

$$\mathrm{H}^{-} + \mathrm{H} \rightarrow \mathrm{H}_2 + e^{-}. \quad (2)$$

In the above set of reactions (*H^- channel*), after the slow first step, the second step follows immediately; the H_2 formation rate is essentially limited by the first step.

H_2 molecules so formed induce the initial cooling and collapse of primordial clouds. The critical temperature for these processes to operate is found to be about 2000 K. This is explained as follows. For simplicity, we consider here H_2 formation in a medium with constant number density n and temperature T, assuming that a virialized halo does not evolve much until significant H_2 formation. [36] The ionization fraction $x=n[H^+]/n$ and the molecular fraction $f=n[H_2]/n$ evolve as

$$\dot{x} = -\alpha_{\mathrm{rec}}\, n\, x^2, \quad (3)$$

where α_{rec} is the radiative recombination coefficient of hydrogen. The recombination proceeds on the timescale

$$t_{\mathrm{rec}} = \frac{1}{\alpha_{\mathrm{rec}} x n}. \quad (4)$$

The solution of Eq. (3) for the initial ionization degree x_0 is given as

$$x = \frac{x_0}{1 + t/t_{\mathrm{rec},0}} \quad (5)$$

where $t_{rec,0} \equiv t_{rec}(x_0)$.

The H_2 fraction is governed by the equation

$$\dot{f} = k_{\rm form}\, n\, (1 - x - 2f)\, x, \qquad (6)$$

where k_{form} is the rate coefficient of the reaction (1).

Using Eq. (5) and approximating $1-x-2f \simeq 1$ for an almost neutral gas, we obtain

$$f(t) = f_0 + \frac{k_{\rm form}}{\alpha_{\rm rec}} \ln\left(1 + \frac{t}{t_{\rm rec,0}}\right). \qquad (7)$$

This solution indicates that, after one recombination time, the H_2 fraction saturates at about

$$f \sim \frac{k_{\rm form}}{\alpha_{\rm rec}} = 4 \times 10^{-5} (T/10^3{\rm K})^{1.52}. \qquad (8)$$

This simple scaling is shown to provide a remarkably good estimate. Figure 3 shows the molecular fraction f against the virial temperature for halos located in a large cosmological simulation. The solid line is an analytical estimate of the H_2 fraction needed to cool the gas within a Hubble time:

$$f_{\rm cool} = \frac{\frac{3}{2}kT}{n \Lambda_{\rm H_2} t_{\rm H}} \qquad (9)$$

where Λ_{H2} is the cooling function of H_2 molecules [37]. In Fig. 3, halos appear to be clearly separated into two populations; those in which the gas has cooled (solid circles), and the others (open circles). The analytic estimate yields a critical temperature of ~2000 K, which indeed agrees very well with the distribution of gas clouds in the f–T plane. There is an important dynamical effect, however. The gas in halos that accrete mass rapidly (primarily by mergers) is unable to cool efficiently owing to gravitational and gas dynamical heating. The effect explains the spread of halos into two populations at $T \sim 2000-5000$ K. Therefore, "minimum collapse mass" models are a poor characterization of primordial gas cooling and gas cloud formation in the hierarchical CDM model. The formation process is significantly affected by the dynamics of gravitational collapse. It is important to take into account the details of halo formation history [35,38].

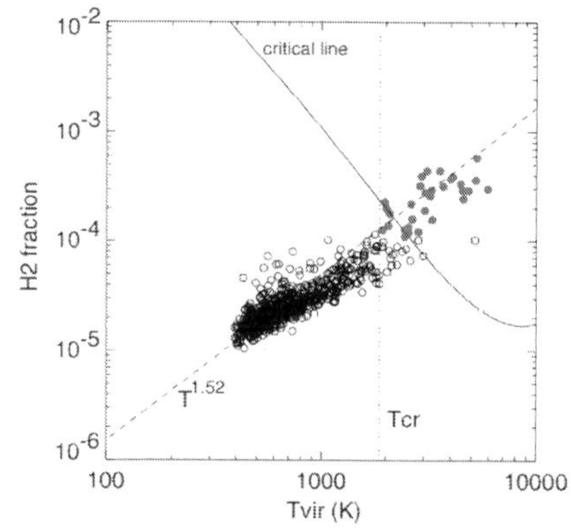

Figure 3. The mass weighted mean H_2 fraction versus virial temperature for halos that host gas clouds (filled circles) and for those that do not (open circles) at $z = 17$ ($t_{age} \sim 300 \times 10^6$ years). The solid curve is the H_2 fraction needed to cool the gas at a given temperature and the dashed line is the asymptotic H_2 fraction (see Eq. (8)). From Ref. [35].

THE ROLE OF DARK MATTER AND DARK ENERGY

The basic formation process of the first objects is described largely by the physics of a primordial gas. Its thermal and chemical evolution specifies a few important mass scales, such as the Jeans mass at the onset of collapse (see Sect. 5). However, when and how primordial gas clouds are formed are critically affected by the particle properties of dark matter, by the shape and the amplitude of the initial density perturbations, and by the overall expansion history of the universe. We here introduce two illustrative examples; a model in which dark matter is assumed to be "warm", and another cosmological model in which dark energy obeys a time-dependent equation of state.

If dark matter is warm, the matter power spectrum has an exponential cut-off at the particle free-streaming scale, and then the corresponding mass variance at small mass scales is significantly reduced [24,25]. The effect is clearly seen in Fig. 4. The gas distribution is much smoother in a model with warm dark matter. For the particular model with a dark matter particle mass of 10 keV, dense gas clouds are formed in filamentary shapes, rather than in blobs embedded in dark matter halos [25,39]. While further evolution of the filamentary gas clouds is uncertain, it is expected that stars are lined up along filaments. Vigorous fragmentation of the filaments, if it occurs, can lead to the formation of multiple low-mass stars.

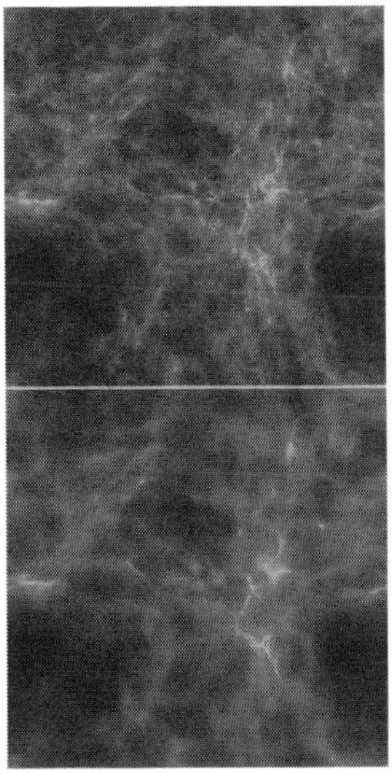

Figure 4. The projected gas distribution at $z=15$ for the standard CDM model (top) and for a WDM model (bottom). We see much smoother matter distribution in the WDM model, in which only a few gas clouds are found. From Ref. [25].

Dark matter particles might affect primordial star formation in a very different way. A popular candidate for dark matter is super-symmetric particles [40], neutralinos for instance. Neutralinos are predicted to have a large cross-section for pair-annihilation. Annihilation products are absorbed in very dense gas clouds, which can counteract molecular cooling [41]. Because primordial gas clouds are formed at the center of dark matter halos, where dark matter density is very large, the annihilation rate and resulting energy input can be significant. While the net effect of dark matter annihilation remains highly uncertain, it would be interesting and even necessary to include the effect if such annihilating dark matter was detected in laboratories.

The nature of dark energy also affects the formation epoch of the first objects [42]. The growth rate of density perturbations is a function of the cosmic expansion parameter, which is determined by the energy content of the universe. In general, the energy density of dark energy can be written as

$$\rho_{\rm DE} \propto \exp\left[\int^{a} -3\,\frac{da'}{a'}(1+w(a'))\right], \qquad (10)$$

where a is the cosmic expansion parameter, and $w(a)$ defines the effective equation of state of dark energy via $P=w\rho$. For the simplest model of dark energy, i.e., Einstein's cosmological constant with $w=-1$, cosmic expansion is accelerated only at late epochs ($z<1$), which is unimportant for early structure formation. However, some dark energy models predict time-dependent equation of state, which effectively shifts the formation epoch to early or later epochs. Figure 5 shows the number of primordial gas clouds as a function of redshift for the standard ΛCDM model and for evolving dark energy models.

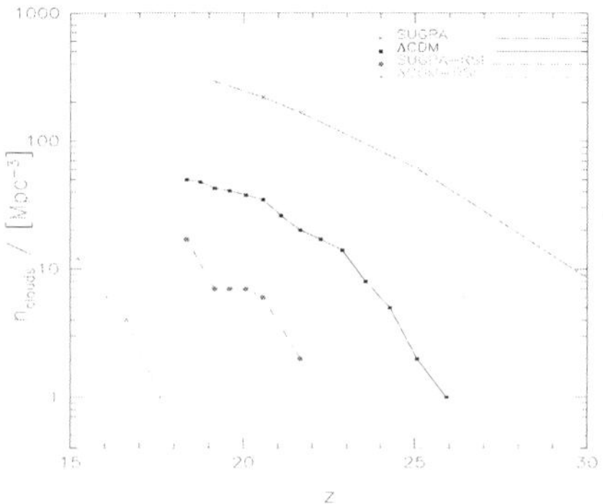

Figure 5. The number of primordial gas clouds at high redshifts for a variety of models; SUGRA (an evolving dark energy), ΛCDM, SUGRA+a running spectral inflation model, and ΛCDM+a running spectral inflation model. From Ref. [42].

Unfortunately, it is extremely difficult to measure the abundance of star-forming gas clouds as a function of time from currently available observations. It is possible to infer how early cosmic reionization began from the large-scale anisotropies of CMB polarization [43], but the CMB polarization measurement does not put tight constraints on the reionization history. We will need to wait for a long time until future radio observations map out the distribution of the intergalactic medium in the early universe by detecting redshifted 21 cm emission from neutral hydrogen [44].

FORMATION OF THE FIRST STARS

We have discussed the condition for star formation in virialized halos in Sect. 3. In this section, we describe more details of the formation process of stars—the first stars.

Direct numerical simulations are a very powerful tool for studying complex gas dynamics in the early cosmological halos. The first simulations of this sort were performed in the last decade [51,52]. These calculations achieved a large dynamic range and implemented primordial gas chemistry, and hence were able to follow the evolution of a primordial gas cloud in detail. Figure 2 shows the projected gas distribution in a cosmological simulation that includes hydrodynamics and primordial gas chemistry [35]. The first objects of $\simeq 10^{5-6} M_\odot$ are found at the knots of filaments, resembling the large-scale structure of the universe, although actually much smaller in mass and size. This manifests the hierarchical nature of structure in the CDM universe. Inside the first object, cold dense cores of self-gravitating gas with mass-scale $\sim 1000 M_\odot$ develop and subsequently contract to form protostars (Fig. 6). The characteristic mass is set by the temperature evolution of the primordial gas, which is shown in Fig. 7. At low densities, the temperature first increases adiabatically because there is no efficient coolant. Then, hydrogen molecules form via the H^- channel with $f \sim 10^{-3}$ at ~ 10 cm^{-3} (see Sect. 3). Cooling by rot-vibrational transitions of H_2 brings the temperature down to a few hundred Kelvin when the density is $\sim 10^4$ cm^{-3}, where the H_2 rotational levels reach local thermodynamic equilibrium. The line cooling efficiency then saturates. Subsequently, the temperature increases but only gradually because of the balance between the saturated H_2 cooling and the ever-increasing compressional heating.

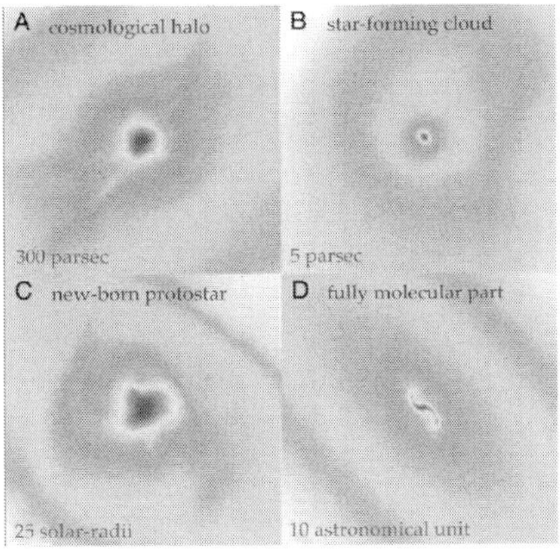

Figure 6. Projected gas distribution around the protostar. The regions shown are, clockwise from top left, the large-scale gas distribution around the cosmological halo (300 pc per side), a self-gravitating, star-forming cloud (5 pc per side), the central part of the fully molecular core (10 astronomical units per side), and the final protostar (25 solar radii per side). We use the density-weighted temperature to color the bottom-left panel, to show clearly the complex structure of the protostar. From Ref. [53].

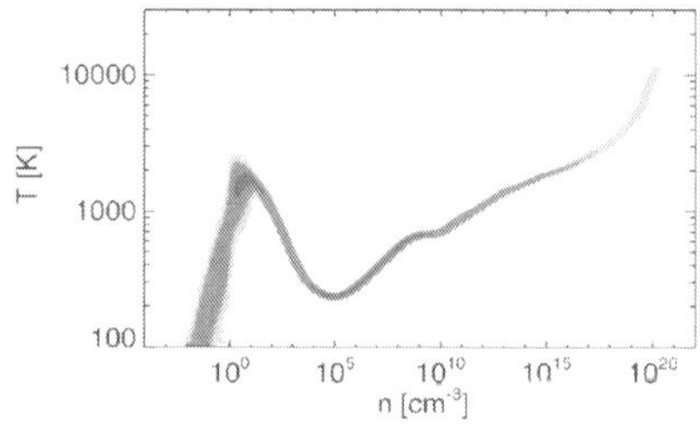

Figure 7. The thermal evolution of a primordial pre-stellar gas cloud. We use the output of a spherical collapse simulation in 3D set-up. The evolutionary track is characterized by the initial adiabatic contraction phase ($n \sim 10^{-2}$–10^1 cm^{-3}), condensation of a dense core by H_2 line cooling ($n \sim 10^1$–10^4 cm^{-3}), the onset of gravitational run-away collapse ($n \sim 10^5$ cm^{-3}), run-away collapse ($n \sim 10^5$–10^{10} cm^{-3}), rapid H_2 formation by three-body reactions ($n \sim 10^{10}$ cm^{-3}), further collapse ($n \sim 10^{10}$–10^{18} cm^{-3}), and the final adiabatic phase ($n > 10^{18}$ cm^{-3}).

During the initial cooling phase, the cloud condenses into filamentary structure. Once the temperature begins to increase, the cloud's deformation stops, yielding a quasi hydrostatic core. The Jeans mass at the temperature inflection point

$$M_J = 1.75 \times 10^3 M_\odot \left(\frac{n}{10^4 \text{ cm}^{-3}}\right)^{-1/2} \left(\frac{T}{200 \text{ K}}\right)^{3/2} \quad (5.1)$$

is thus imprinted as the characteristic mass of the dense molecular cloud.

Further gravitational collapse leading to protostar formation has been studied extensively over the past few decades [15,19]. The fact that H_2 cooling is the main driver of this collapse has been recognized since the late 1960s [12–15]. In the early studies, due to the lack of some important chemical/cooling processes, the temperature was found to increase in low density regimes and thus the predicted protostellar mass was higher than in modern calculations. The first calculation that includes all the important micro-physics, but assumes spherical symmetry, was carried out by Ref. [48]. Recently, a 3D version of this in a cosmological context, an ab initio simulation of the formation of a primordial protostar, has been performed [53].

As seen in Fig. 7, the gas temperature increases by just an order of magnitude while the density increases by over ten orders of magnitude from $n \sim 10^4$ to 10^{18}cm^{-3}. During this phase, although the H_2 molecules are always the dominant coolant, the detailed cooling process changes from H_2 rovibrational line emission ($n \sim 10^4$–10^{13}cm^{-3}), H_2 collision-induced emission (10^{13}–10^{16}cm^{-3}), to chemical cooling by H_2 dissociation (10^{16}–10^{18}cm^{-3}).

The dynamical evolution in this quasi-isothermal phase can be well described by a Larson–Penston-type self-similar solution with modification for the gradual temperature increase, where the central flat density part with the Jeans size is surrounded by the envelope with the power-law ($\propto r^{-2.2}$) density distribution. With some angular momentum present, the central part contracts to an oblate shape due to the centrifugal force. In this specific calculation, however, the rotational motion does not become large enough to prevent the collapse and the rotational velocity remains about half the Keplerian velocity.

When the central density reaches $n \sim 10^{16}$cm^{-3}, the gas becomes optically thick to the H_2 collision-induced absorption, which is the inverse process of collision-induced emission, but the latent heat used for H_2 dissociation works as effective cooling for a while. With H_2 dissociation almost completed, the temperature begins to rise adiabatically at $n \sim 10^{18}$cm^{-3} and the resultant increases of pressure gradient eventually overcome the gravity and halt the collapse at $n \sim 10^{21}$cm^{-3}. This is the moment of birth of a protostar. The protostar has a mass of just 0.01 solar masses, a radius of $\sim 5 \times 10^{11}$ cm, a central particle number density of $\sim 10^{21}$ cm^{-3} and a temperature of about 20 000 Kelvin at its formation. The small mass is expected from the Jeans mass at the final adiabatic phase. At the same time, hydrodynamic shocks are generated at the surface where supersonic gas-infall is suddenly stopped. Although formed as a tiny embryo, the protostar grows rapidly in mass by accretion of ambient matter.

A long-standing question is whether or not a primordial gas cloud experiences fragmentation during its evolution. Although the chemo-thermal instability during the rapid phase of three-body H_2 formation was once considered as a fragmentation mechanism [56,57], later studies [53,55,60] showed it to be too weak to induce fragmentation. However, with large enough angular momentum, the central part of the collapsing core forms a thin disk-like structure, which eventually fragments into binaries or small multiples [58,59,62]. The evolution in the post-collapse phase is also important. After a protostar is formed at the center, a circumstellar disk develops, which then become gravitationally unstable in many cases [63–65]. Although these numerical experiments are still limited to a rather early phase of binary formation and are not yet conclusive about their fates, the likely outcome is the formation of a binary or a small number of multiple systems. Three-dimensional calculations following the entire evolution of a protostellar system are needed for future study.

On the assumption that there is only one stellar seed (protostar) at the center of the parent gas cloud, the subsequent protostellar evolution can be calculated using the standard model of star formation [61,66,67]. For a very large accretion rate characteristic of a primordial gas cloud, $\dot{M} > 10^{-3}\ M_\odot\ \text{year}^{-1}$, a protostar can grow quickly to become a massive star. Figure 8 shows the evolution of protostellar radius and mass for such large accretion rates. The resulting mass when the star reaches the zero-age main sequence is as large as one hundred times that of the sun [55,67].

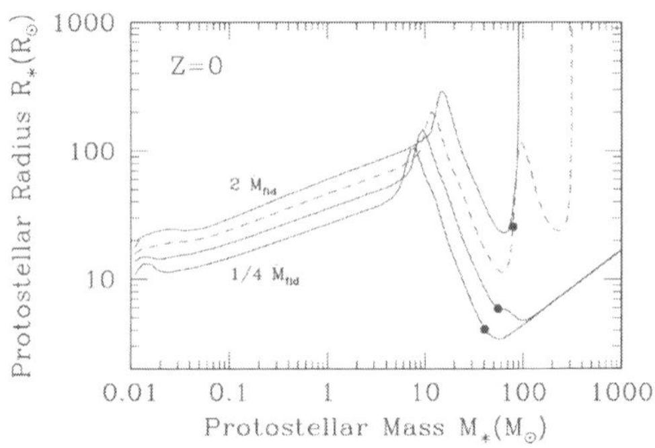

Figure 8. The evolution of the radius and mass of a primordial protostar. The accretion rates assumed are 1/4, 1/2, 1, 2 $\dot{M}_{\text{fid}}$ (from bottom to top) with a fiducial rate of $\dot{M}_{\text{fid}} = 4.4 \times 10^{-3}\ M_\odot\ \text{year}^{-1}$. The solid points indicate the time when hydrogen burning begins. From Ref. [67].

Overall, the lack of vigorous fragmentation, the large gas mass accretion rate, and the lack of a significant source of opacity (such as dust) provide favourable conditions for the formation of massive, even very massive, stars in the early universe [55,68]. One important question remaining is whether or not, and how, gas accretion is stopped. This question is directly related to the final mass of the first stars. A few mechanisms have been suggested that can stop gas accretion and terminate the growth of a protostar [68]. Following the growth of a primordial protostar to the end of its evolution in a 3D simulation will be the next frontier.

THE MASS OF THE FIRST STARS

At the moment of the birth of a tiny embryo star ($\simeq 0.01\ M_\odot$), the star is surrounded by a huge amount of the natal gas cloud with $M \sim 10^3 M_\odot$.

The protostar rapidly grows in mass by gathering these materials by its gravitational pull. The final mass of the star, which predestines the star's evolution, depends on how much gas is accreted onto the star during this stage. The stellar mass could reach several $\times 100 M_{\odot}$ if the mass accretion continued with the typical rate of $\dot{M} \sim 10^{-3} M_{\odot}$ year^{-1}. However, the growth is largely affected by the strength of stellar feedback effect(s) against the accretion flow. If the mass accretion is shut off earlier by some stellar feedback effects, the resulting final stellar mass would be lower. A plausible feedback mechanism is gas heating by stellar radiation, because the stellar luminosity rapidly increases with the stellar mass. In the case of primordial star formation, the accretion envelope has a much lower opacity than those in present-day star formation, because of the lack of dust grains. However, energetic UV photons that ionize the gas could significantly affect the dynamics of the accretion flow [68,70]. The surrounding gas is accreted onto the protostar via a circumstellar disk, and then Hii regions would grow toward the polar directions where the gas density decreases as the protostellar system evolves. The disk would be exposed to stellar UV radiation and then eventually photo-evaporate. The mass accretion could be terminated when the circumstellar disk is completely evaporated. Semi-analytic models predict that this effect terminates the stellar growth when the stellar mass reaches $\sim 100 M_{\odot}$ [68].

Numerical simulations are a powerful and direct method for studying the complex interplay between the accretion flow and stellar radiative feedback. The first study of this sort was presented in Ref. [72], in which a hybrid numerical code is used in order to follow the dynamics of the accretion flow and evolution of an accreting protostar. A 2D axisymmetric radiation hydrodynamic code [73,74], coupled with stellar evolution calculations [67,69,71], is used. Starting with an initial condition based on the cosmological simulations [53,55], we calculated long-term evolution over 0.1×10^6 years after the protostar's birth. Figure 9 presents the simulated evolution of the gas accretion envelope around the protostar. We see that the bipolar Hii region is growing when the stellar mass is $25 M_{\odot}$ (panel b). The gas pressure within the Hii region is much higher than the surroundings owing to its high temperature, which causes dynamical expansion of the Hii region throughout the accretion envelope. Panels (b)–(d) show that the opening angle of the bipolar Hii region increases with time due to dynamical expansion. A blastwave also propagates ahead of the ionization front. When the stellar mass is $42 M_{\odot}$ (panel d), the accretion envelope even outside of the Hii region is shocked and heated up. The stellar UV radiation directly hits the circumstellar disk, which extends to a few × 10^3 AU around the protostar. The disk is photo-evaporating and losing its mass. Figure 10 clearly shows that the stellar growth via mass accretion is limited by the stellar radiative feedback. The mass accretion is completely shut off when the stellar mass is ≃ 43 $M_\odot$, which is the final stellar mass of a very first star.

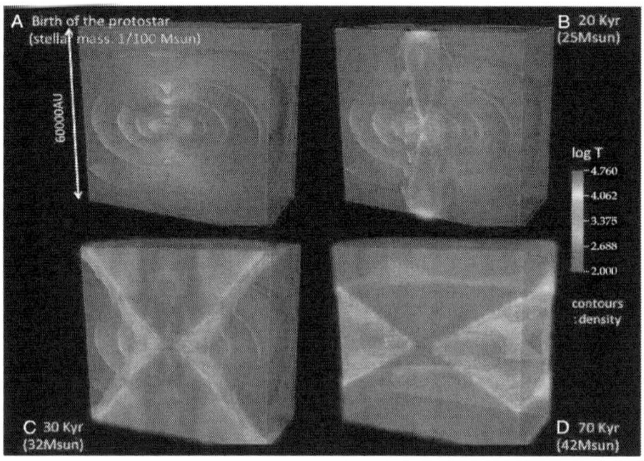

Figure 9. Formation and expansion of an Hii region around a primordial protostar [72]. The four panels (a)–(d) show snapshots at the moments of (a) the birth of an embryo protostar ($t = 0$), (b) $t = 2 \times 10^4$ years, (c) 3×10^4 years, and (d) 7×10^5 years. The protostellar masses for the snapshots are also presented. The colors and contours represent the spatial distributions of the gas temperature and density.

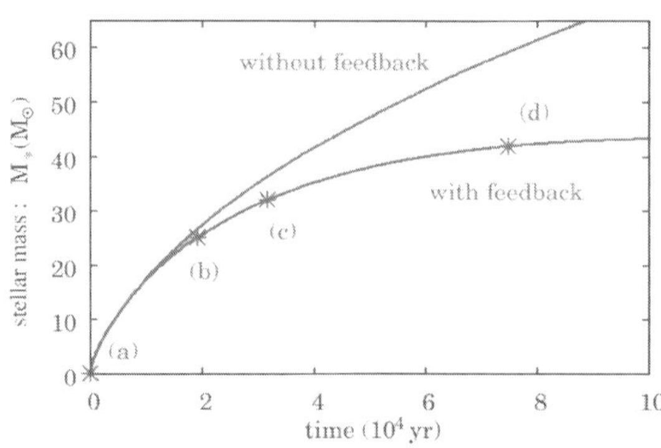

Figure 10. Growth of the stellar mass with time elapsed since the birth of the star. The blue line represents our fiducial case, the evolution of which is presented in Fig. 9. The asterisks on this line mark the moments seen in panels

(a)–(d) in Fig. 9. The red line presents the evolution in a reference case where the stellar UV feedback is turned off while the other settings remain the same.

Our results suggest that a number of the first stars were as massive as the Galactic O-type stars. This challenges the previous theory of early star formation, which posits that the very first stars were extremely massive, exceeding $100 M_{\odot}$. Interestingly, the second-generation primordial stars, which formed from the primordial gas affected by radiative or mechanical feedback from the first stars, would be less massive, several $\times 10 M_{\odot}$ stars. [75,76,110] The lower-mass of the second-generation stars is due to a different gas thermal evolution with additional radiative cooling by H_2 and HD molecules. However, this mode of star formation with efficient HD cooling is suppressed by even weak H_2 photo-dissociating background radiation [110]. If so, the formation process of the later-generation primordial stars would be similar to that of the very first stars. Our protostellar evolution calculations suggest that the typical masses of the primordial stars were always several tens of solar-masses, regardless of their generation.

Such "ordinary" massive stars end their lives as core-collapse supernovae and yield the first heavy elements in the early universe. This explains the fact that no signatures of pair-instability supernovae, which is the fate of very massive stars of $\simeq 150-300 M_{\odot}$ [77,78], have been found in the abundance patterns of the Galactic metal-poor stars [79,80]. Nonetheless, our predicted stellar mass of several $\times$ $10 M_{\odot}$ is still much higher than the typical stellar mass in the present-day universe, $\simeq 0.6 M_{\odot}$. This suggests that there should have been a transition in the star formation mode across cosmic time, probably owing to an increase in the metallicity of star-forming clouds [70,81,82,123].

The effect of magnetic fields is generally thought to be unimportant in primordial star formation because there is no obvious generation mechanism for strong magnetic fields in the early universe. However, recent theoretical studies and direct numerical simulations argue that pre-stellar evolution and protostellar growth can be affected by the existence of magnetic fields [83]. Magnetic fields can be amplified by turbulence-driven dynamo mechanisms [84,85]. Although the dynamical effects of magnetic fields in a primordial gas are probably small [86], currently available numerical simulations still do not show convergence in the final amplitude because the rate of amplification is strongly coupled to the numerical resolution and the strength of turbulence at very small length scales. Magnetic fields might also drive protostellar jets, as in present-day protostellar evolution, so that the net mass growth rate of the protostar can be reduced [87]. These important issues need to be further explored by following the amplification of magnetic fields and magneto-hydrodynamics of a magnetized primordial gas (low-ionization plasma) self-consistently.

FEEDBACK FROM THE FIRST STARS

The emergence of the first generation of stars has important implications for the thermal state and chemical properties of the intergalactic medium in the early universe. At the end of the Dark Ages, the neutral, chemically pristine gas was reionized by ultraviolet photons emitted from the first stars, but also enriched with heavy elements when these stars ended their lives as energetic supernovae. The importance of supernova explosions, for instance, can be easily appreciated by noting that only light elements were produced during the nucleosynthesis phase in the early universe. Chemical elements heavier than lithium are thus thought to be produced exclusively through stellar nucleosynthesis, and they must have been expelled by supernovae to account for various observations of high-redshift systems [88,89].

Feedback from the first stars may have played a crucial role in the evolution of the intergalactic medium and (proto)galaxy formation. A good summary of the feedback processes is found in Ref. [90]. We here review two important effects, and highlight a few unsolved problems.

Radiative Feedback

The first feedback effect we discuss is caused by radiation from the first stars. First stars can cause both negative and positive effects in terms of star-formation efficiency. Far-UV radiation dissociates molecular hydrogen via Lyman–Werner resonances [91–93], while UV photo-ionization heats up the surrounding gas. Photo-ionization also increases the ionization fraction, which in turn promotes H_2 formation. Yet another radiative feedback effect is conceivable; X-rays can promote H_2 production by boosting the free electron fraction in distant regions [94,95]. It is not clear whether overall negative or positive feedback dominates in the early universe.

Three-dimensional calculations [97] show consistently strong negative effects of FUV radiation. Figure 11 shows the distance at which the H_2 dissociation time equals the free-fall time. Hydrogen molecules in gas clouds within a few tens of parsecs are easily destroyed by a nearby massive star. However, gas self-shielding (opacity effects) needs to be taken into account for dense gas clouds. H_2 dissociation becomes ineffective for large column densities of $N_{H2} > 10^{14}\,cm^{-2}$ for an approximately stationary gas [96]. In fact, small halos are *not* optically thin and thus the gas at the center can be self-shielded against FUV radiation [35,97]. Because of the complexities associated with the dynamics, chemistry, and radiative transfer involved in early gas cloud formation, the strength of the radiative feedback still remains uncertain. Recent simulations [98,99] generally suggest that FUV radiation does not completely suppress star formation even for large intensities of $J > 10^{-22}\,erg\,s^{-1}\,Hz\,cm^{-2}$. In contrast with the naive implication of the negative feedback from FUV radiation, star formation can possibly continue in early minihalos. It is intriguing that the recent measurement of CMB polarization does not suggest a very large optical depth to Thomson scattering, perhaps constraining a large contribution to reionization from minihalos [100,101].

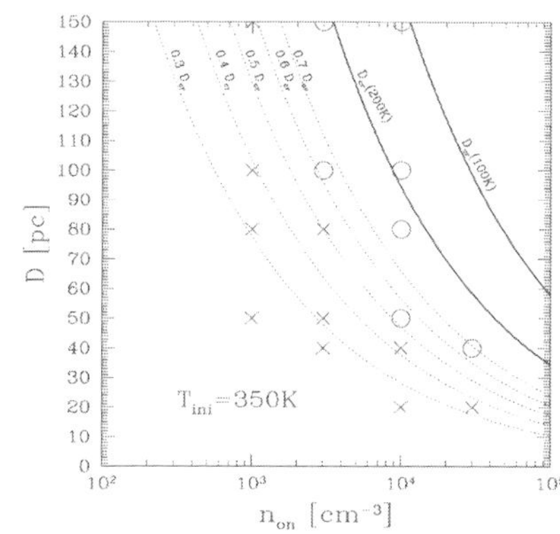

Figure 11. The critical distance from a radiation source at which the cloud can collapse even under photo-dissociating feedback. From Ref. [97].

If the formation of H_2 is strongly suppressed by an FUV background, star formation proceeds in a quite different manner. A primordial gas cloud cools and condenses nearly isothermally by atomic hydrogen cooling. If the gas cloud initially has a small angular momentum, it can collapse to form an intermediate mass black hole via direct collapse [102,103]. Such first black holes might power small quasars. X-ray from early quasars is suggested as a source of a positive feedback effect by increasing the ionization fraction in a primordial gas [95]. However, the net effect is much weaker than one naively expects from simple analytic estimates, unless negative feedback by FUV radiation is absent [104].

Ionizing radiation causes much stronger effects, at least locally. The formation of early Hii regions has been studied by a few groups using radiation hydrodynamics simulations [105–107]. Early Hii regions are different from present-day Hii regions in two aspects. Firstly, the first stars and their parent gas cloud are hosted by a dark matter halo. The gravitational force exerted by dark matter is important in the dynamics of early Hii regions. Secondly, the *initial* gas density profile around the first star is typically steep [51,53,55]. These two conditions make the evolution different from that of present-day local Hii regions.

Figure 12 shows the structure of an early Hii region [109]. The star-forming region is located as a dense molecular gas cloud within a small mass ($\sim 10^6 M_\odot$) dark matter halo. A single massive Population III star with $M_* = 200 M_\odot$ is embedded at the center. The formation of the Hii region is characterized by initial slow expansion of an ionization front (I-front) near the center, followed by rapid propagation of the I-front throughout the outer gas envelope. The transition between the two phases determines a critical condition for complete ionization of the halo. For small mass halos, the transition takes place within a few 10^5 years, and the I-front expands over the halo's virial radius (Fig. 12). The gas in the halo is effectively evacuated by a supersonic shock, with the mean gas density decreasing to $\sim 1\ cm^{-3}$ in a few million years. It takes over tens to a hundred million years for the evacuated gas to be re-incorporated in the halo [109,116]. The most important implication from this result is that star formation in the early universe would be intermittent. Small mass halos cannot sustain continuous star formation.]

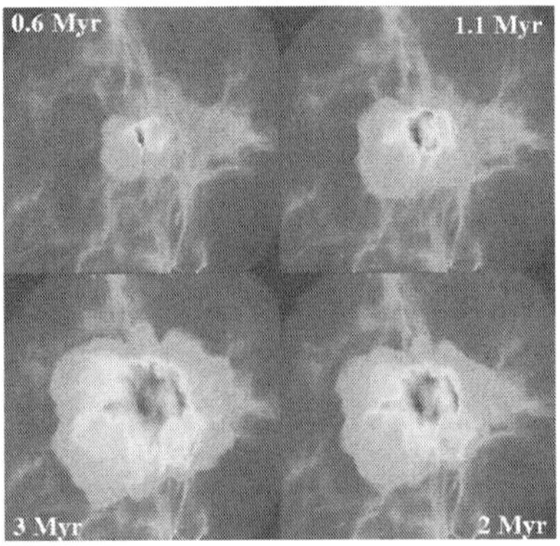

Figure 12. The structure and evolution of an Hii region around a massive Population III star in an early cosmological halo. Each panel has a side length of 7

kilo-parsecs. The Hii region grows, clockwise from the top-left panel. From Ref. [109].

Early gas clouds are expected to be strongly clustered [34,38]. Because even a single massive star affects over a kilo parsec volume, the mutual interactions between nearby star-forming gas clouds may be important. Large-scale cosmological simulations with radiative feedback effects, such as those discussed here, are clearly needed to fully explore the impact of early star formation.

Mechanical Feedback

Massive stars end their lives as supernovae. Such energetic explosions in the early universe are thought to be violently destructive; they expel the ambient gas out of the gravitational potential well of small-mass dark matter halos, causing an almost complete evacuation [112–116]. Since massive stars process a substantial fraction of their mass into heavy elements, SN explosions can cause prompt chemical enrichment, at least locally. They may even provide an efficient mechanism to pollute the surrounding intergalactic medium to an appreciable degree [117,118].

Population III supernova explosions in the early universe were also suggested as a trigger of star formation [119], but modern numerical simulations have shown that the gas expelled by supernovae falls back to the dark halo potential well after about the system's free-fall time [118,120]. The density and density profile around the supernova sites are of particular importance because the efficiency of cooling of supernova remnants is critically determined by the density inside the blastwave. If the halo gas is evacuated by radiative feedback prior to explosion, the supernova blastwave propagates over the halo's virial radius, leading to complete evacuation of the gas even with an input energy of 10^{51} erg. A large fraction of the remnant's thermal energy is lost in 10^5−10^7 year by line cooling, whereas, for even greater explosion energies, the remnant cools mainly via inverse Compton scattering. The situation is clearly different from the local galactic supernova. In the early universe, the inverse Compton process with cosmic background photons acts as an efficient cooling process.

The halo destruction efficiency by a single SN explosion is important for the formation of the first galaxies. A simple criterion, $E_{SN}>E_{bi}$, where E_{bi} is the gravitational binding energy, is often used to determine the destruction efficiency. However, whether or not the halo gas is effectively blown away is determined not only by the host halo mass (which gives an estimate of E_{bi}), but also by a complex interplay of hydrodynamics and radiative processes (Fig. 13). SNRs in dense environments are highly radiative and thus a large fraction of the explosion energy can be quickly radiated away. An immediate implication from this result is that, in order for the processed metals to be transported out of the halo and distributed to the IGM, I-front propagation and pre-evacuation of the gas must precede the supernova explosion. This roughly limits the mass of host halos from which metals can be ejected into the IGM to $< 10^7 M_{\odot}$, i.e., the first generation of stars can be a significant source of early metal-enrichment of the IGM [112,115,118].

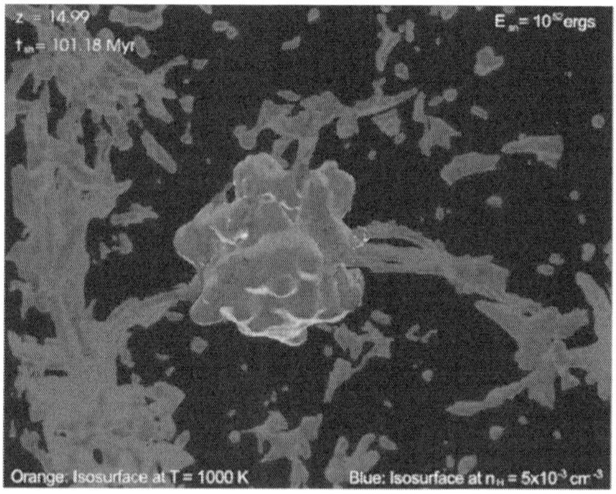

Figure 13. The structure of an early supernova remnant. The shock-front reached a radius of 2 kpc about 100×10^6 years after the explosion. A large explosion energy of 10^{52} erg is assumed for this simulation. From Ref. [125].

Although metal-enrichment and dust production by the first supernovae could result in greatly enhanced gas cooling efficiency, which might possibly change the mode of star formation to that dominated by low-mass stars [121], the onset of these "second-generation" stars may be delayed owing to gas evacuation, particularly in low-mass halos. This again supports the notion that early star formation is likely self-regulating. If the first stars are massive, only one period of star formation is possible for a small halo and its descendants within a Hubble time. The sharp decline in the destruction efficiency at $M_{halo} > 10^7 M_{\odot}$ indicates that the global cosmic star formation activity increases only after a number of large mass ($> 10^{7-8} M_{\odot}$) halos are assembled.

FORMATION OF THE FIRST GALAXIES AND BLACK HOLES

The hierarchical nature of cosmic structure formation (see Sect. 2) naturally predicts that stars or stellar size objects form first, earlier than galaxies form. The first generation of stars set the scene for the subsequent galaxy formation. The characteristic minimum mass of a first galaxy (including dark matter) is perhaps ~ 10^7–$10^8 M_\odot$, in which gas heated up to 10^4–10^5 K by the first star feedback can be retained.

The first galaxies are assembled through a number of large and small mergers, and then turbulence is generated dynamically, which likely changes the star-formation process from a quiescent one (like in minihalos) to a highly complicated but organized one. There have been a few attempts to directly simulate this process in a cosmological context [124,125]. The results generally argue that star formation in a large mass system is still an inefficient process overall. However, a significant difference is that the interstellar medium is likely metal-enriched in the first galaxies. Theoretical calculations [122,123] show that cooling by heavy elements and by dust can make the gas temperature at the onset of run-away collapse substantially lower than for a primordial gas. The lower gas temperature causes two effects; it lowers the Jeans mass ($\propto T^{3/2}/\rho^{1/2}$), and also lowers the mass accretion rate ($\propto c_s^3/G$), thereby providing at least two necessary conditions for low-mass star formation. The combined effects of strong turbulence and metal-enrichment might cause the stellar initial mass function to be close to that in present-day star-forming regions.

In the first galaxies, primordial star formation may proceed in a peculiar manner. Formation of super-massive stars is suggested as a possible outcome in such cases. Super-massive stars could then collapse to massive black holes (BHs), to seed the formation of super-massive BHs in the early universe. It is generally thought that the formation of super-massive stars requires the following two conditions: (i) a star-forming cloud collapses monolithically without fragmentation, and (ii) the accretion rate onto the formed protostar must be high enough ($\gtrsim 0.1\ M_\odot$/year) so that the protostar can indeed grow to be very massive within its lifetime. In terms of physical processes, formation of H_2 molecules must be suppressed in the star-forming cloud in order to keep the gas temperature high. This can be achieved either by strong photo-dissociation [126,127] or by collisional dissociation. [128] Such a cloud collapses nearly isothermally at several thousand Kelvin owing to cooling by atomic hydrogen and by H^- ions. The gas does not go through a rapid cooling phase and thus is expected not to fragment into numerous smaller clumps. [102] The evolution of a protostar with such extremely rapid accretion is rather different from cases with lower accretion rates. [129] The rapidly accreting star inflates to a very large radius, with its effective surface temperature being as low as several thousand K (see Fig. 14). For a very massive star whose luminosity is close to the Eddington value, the mass–radius relation reduces to $R_* \propto M_*^{1/2}$ for a roughly constant surface temperature. Because of the low effective temperature, such a supergiant star does not cause strong radiative feedback effects to halt the gas accretion. This is in stark contrast with the self-regulation mechanism of the growth of the first stars. The outcome is likely the formation of a supermassive star with mass $\gtrsim 10^5 M_\odot$, which eventually collapses directly to a black hole by post-Newtonian instability. Recent numerical simulations show that, through the assembly of the first galaxies, such remnant BHs continue to be fed gases through cold-streaming flows [130].

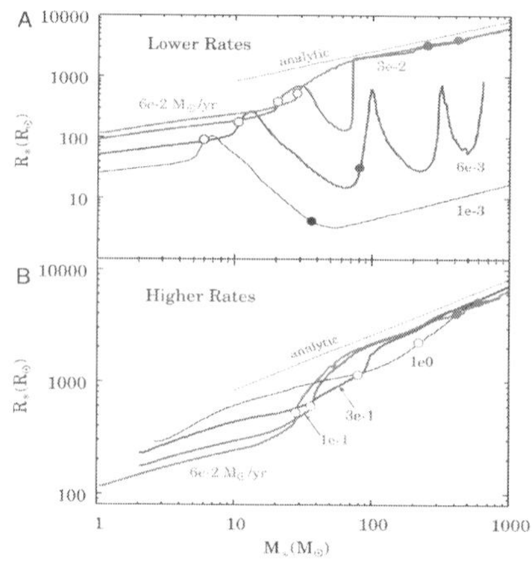

Figure 14. Evolution of the protostellar radius for various accretion rates. *Upper panel:* We compare the evolutionary tracks for $\dot{M} = 10^{-3}\,M_\odot$ year^{-1}, $6\times10^{-3}\,M_\odot$*year*$^{-1}$, $3\times10^{-2}\,M_\odot$year^{-1}, and $6\times10^{-2}\,M_\odot$year^{-1}. The open and filled circles on each curve denote the epochs when $t_{KH}=t_{acc}$ and when the central hydrogen burning begins, respectively. *Lower panel:* same as the upper panel but for even higher accretion rates of $6\times10^{-2}\,M_\odot$year^{-1}, $0.1\,M_\odot$year^{-1}, $0.3\,M_\odot$year^{-1}, and $1\,M_\odot$year^{-1}. In both panels the thin green line represents the mass–radius relation $R \propto M_*^{1/2}$ (see text). From Ref. [129].

Understanding the formation of the first galaxies is very challenging, because of the complexities described above. Nevertheless, it is definitely the subject where theoretical models can be really tested against direct observations in the near future. The first galaxies may be more appropriately called faint protogalaxies, which will be detected by next-generation telescopes. *JWST* will measure the luminosity function of these faint galaxies at $z>7$, which reflects the strength of feedback effects from the first stars [131].

PROSPECTS FOR FUTURE OBSERVATIONS

A number of observational programs are planned to detect the first stars, black holes, and galaxies, both directly and indirectly. We close this review by discussing the prospects for future observations.

The first supernovae and the first galaxies will be the main target of next-generation (near-)infrared telescopes [132,133], and indirect information on the first stars will be obtained from the CMB polarization, the near-infrared background, high-redshift supernovae, gamma-ray bursts, and so-called Galactic archeology.

The seven-year dataset of the *Wilkinson Microwave Anisotropy Probe (WMAP)* yields the CMB optical depth to Thomson scattering, $\simeq 0.088 \pm 0.015$ [134]. This measurement provides an integral constraint on the total ionizing photon production at $z>6$ [135]. More accurate polarization measurements by *Planck* and by continued operation of *WMAP* will further tighten the constraint on the reionization history of the universe, $x_e(z)$ [136]. In the longer term, future radio observations such as the *Square Kilometer Array* will map out the distribution of intergalactic hydrogen in the early universe. The topology of reionization and its evolution will be probed by SKA [44,137].

The first stars in the universe are predicted to be massive, as discussed in this article, and so they are likely progenitors of energetic supernovae and associated GRBs at high redshifts [138]. Infrared color can be utilized to identify supernovae at $z<13$ [132,139]. A realistic 1-year *JWST* survey will discover 1–30 supernovae at $z>5$ [131]. An all-sky near-infrared survey with 26 AB magnitude depth will detect several tens of super-luminous supernovae at $z>10$ [132]. Gamma-ray bursts are the brightest explosions in the universe, and thus are detectable out to redshifts $z>10$ in principle. Recently, the *Swift* satellite has detected a GRB originating at $z>6$ [6,140], thus demonstrating the promise of GRBs as probes of the early universe [141].

Very metal-poor stars—the stellar relics—provide invaluable information on the conditions under which these low-mass stars were formed [142–144]. It is expected that the relics of early-generation stars are orbiting near the centers of galaxies at the present epoch [145]. While, conventionally, halo stars are surveyed to find very metal-poor stars, the APOGEE project is aimed at observing ~ 100 000 stars in the bulge of the Milky Way [146]. The nature of early metal-enrichment must be imprinted in the abundance patterns of the bulge stars.

Altogether, these observations will finally fill the gap in our knowledge of the history of the universe, and thus will end the "Dark Ages".

ACKNOWLEDGEMENTS

The present work is supported in part by Grants-in-Aid from the Ministry of Education, Culture, Sports, Science and Technology of Japan (2168407, 21244021:KO, 20674003:NY). T.H. acknowledges support by Fellowship of the Japan Society for the Promotion of Science for Research Abroad. Portions of this research were conducted at the Jet Propulsion Laboratory, California Institute of Technology, which is supported by the National Aeronautics and Space Administration (NASA).

APPENDIX A: DENSITY FLUCTUATIONS AND MASS VARIANCE

A density perturbation field $\delta(\mathbf{x}) = \rho(\mathbf{x})/\bar{\rho} - 1$ can also be represented by its Fourier transform

$$\delta_{\mathbf{k}} = \frac{1}{V}\int \delta(\mathbf{x}) \exp(-i\mathbf{k}\cdot\mathbf{x}) d^3\mathbf{x}, \qquad \text{(A.1)}$$

where V is the volume of the region under consideration. Note that $\delta_{\mathbf{k}}$ are complex quantities. The second moment, the power spectrum, is often used, and is given by

$$P(k) = V\langle|\delta_{\mathbf{k}}|^2\rangle = V\langle\delta_{\mathbf{k}}\delta_{-\mathbf{k}}\rangle. \qquad \text{(A.2)}$$

The power spectrum gives the probability that the modes $\delta_{\mathbf{k}}$ have amplitudes in the range $|\delta_{\mathbf{k}}|$ and $|\delta_{\mathbf{k}}| + d|\delta_{\mathbf{k}}|$.

The variance of the density field when sampled with randomly placed spheres of radii R is obtained by a weighted integral of the power spectrum as

$$\sigma^2(R) = \frac{1}{2\pi^2}\int P(k)W^2(kR)k^2 dk, \tag{A.3}$$

where the top-hat window function is given by $W(x) = 3(\sin x/x^3 - \cos x/x^2)$. The corresponding mass variance can be obtained by a simple transformation $M = 4\pi/3R^3\bar{\rho}$. For a power law power spectrum with power index n,

$$\sigma^2(R) \propto R^{-(n+3)} \propto M^{-(n+3)/3}. \tag{A.4}$$

Let us define the threshold over-density for gravitational collapse at redshift z as

$$\delta_{\rm crit}(z) = 1.686/D(z), \tag{A.5}$$

where $D(z)$ is the linear growth factor of perturbations to z. Growing perturbations with amplitudes greater than $\delta_{\rm crit}(z)$ at a given epoch z are expected to collapse.

REFERENCES

1. M. Iye et al., Nature 443, 186 (2006).
2. M. D. Lehnert et al., Nature 467, 940 (2010).
3. R. J. Bouwens et al., Nature 469, 504 (2011).
4. X. Fan et al., Astron. J. 125, 1649 (2003).
5. D. J. Mortlock et al., Nature 474, 616 (2011).
6. N. Kawai et al., Nature 440, 184 (2006).
7. R. Salvaterra et al., Nature 461, 1258 (2009).
8. J. Miralda-Escude, Science 300, 1904 (2003).
9. V. Bromm, N. Yoshida, L. Hernquist, and C. F. McKee, Nature 459, 49 (2009).
10. M. Schwarzschild and L. Spitzer, Jr., Observatory 73, 77 (1953).
11. A. A. Penzias and R. W. Wilson, Astrophys. J. 142, 419 (1965).
12. W. C. Saslow and D. Zipoy, Nature 216, 976 (1967).
13. P. J. E. Peebles and R. H. Dicke, Astrophys. J. 154, 891 (1968).
14. T. Hirasawa, K. Aizu, and M. Taketani, Prog. Theor. Phys. 41, 835 (1969).
15. T. Matsuda, H. Sato, and H. Takeda, Prog. Theor. Phys. 42, 219 (1969).
16. T. Yoneyama, Publ. Astron. Soc. Japan 24, 87 (1972).
17. J. B. Hutchins, Astrophys. J. 205, 103 (1976).
18. R. G. Carlberg, Mon. Not. R. Astron. Soc. 197, 1021 (1981).
19. F. Palla, E. E. Salpeter, and S. W. Stahler, Astrophys. J. 271, 632 (1983).
20. Yu. I. Izotov and I. G. Kolesnik, Soviet Astron. 28, 15 (1984).
21. H. Susa, H. Uehara, and R. Nishi, Prog. Theor. Phys. 96, 1073 (1996).
22. D. H. D. Lyth and A. A. Riotto, Phys. Rep. 314, 1 (1999).
23. S. Dodelson, Modern Cosmology (Academic Press, 2003).
24. R. Barkana, Z. Haiman, and J. P. Ostriker, Astrophys. J. 558, 482 (2001).
25. N. Yoshida, A. Sokasian, L. Hernquist, and V. Springel, Astrophys. J. 591, L1 (2003).
26. A. Kosowsky and M. S. Turner, Phys. Rev. D 52, 1739 (1995).
27. R. S. Somerville, J. S. Bullock, and M. Livio, Astrophys. J. 593, 616 (2003).

28. N. Yoshida, A. Sokasian, L. Hernquist, and V. Springel, Astrophys. J. 598, 73 (2003).
29. J. Diemand, B. Moore, and J. Stadel, Nature 433, 389 (2005).
30. J. P. Ostriker and M. J. Rees, Mon. Not. R. Astron. Soc. 179, 541 (1977).
31. T. Matsuda, H. Sato, and H. Takeda, Prog. Theor. Phys. 41, 840 (1969).
32. A. Kashlinsky and M. J. Rees, Mon. Not. R. Astron. Soc. 205, 21 (1983).
33. H. M. P. Couchman and M. J. Rees, Mon. Not. R. Astron. Soc. 221, 53 (1986).
34. L. Gao, N. Yoshida, T. Abel, C. S. Frenk, A. Jenkins, and V. Springel, Mon. Not. R. Astron. Soc. 378, 449 (2007).
35. N. Yoshida, T. Abel, L. Hernquist, and N. Sugiyama, Astrophys. J. 592, 645 (2003).
36. M. Tegmark, J. Silk, M. J. Rees, A. Blanchard, T. Abel, and F. Palla, Astrophys. J. 474, 1 (1997).
37. D. Galli and F. Palla, Astron. Astrophys. 335, 403 (1998).
38. D. Reed, R. Bower, C. S. Frenk, A. Jenkins, T. Theuns, and S. D. M. White, Mon. Not. R. Astron. Soc. 363, 393 (2005).
39. L. Gao and T. Theuns, Science 317, 1527 (2007).
40. G. Jungman, M. Kamionkowski, and K. Griest, Phys. Rep. 267, 195 (1996).
41. D. Spolyar, K. Freese, and P. Gondolo, Phys. Rev. Lett. 100, 1101 (2008).
42. U. Maio, K. Dolag, M. Meneghetti, L. Moscardini, N. Yoshida, C. Baccigalupi, M. Bartelmann, and F. Perrota, Mon. Not. R. Astron. Soc. 373, 869 (2006).
43. M. Zaldarriaga and U. Seljak, Phys. Rev. D 55, 1830 (1997).
44. P. R. Shapiro, K. Ahn, M. A. Alvarez, I. T. Iliev, H. Martel, and D. Ryu, Astrophys. J. 646, 681 (2006).
45. B. W. O'Shea and M. L. Norman, Astrophys. J. 654, 66 (2007).
46. S. Naoz, N. Yoshida, and N. Y. Gnedin, Astrophys. J. 747, 128 (2012).
47. F. Palla, E. E. Salpeter, and S. W. Stahler, Astrophys. J. 52, 155 (1983).
48. K. Omukai and R. Nishi, Astrophys. J. 508, 141 (1998).
49. E. Ripamonti, F. Haardt, A. Ferrara, and M. Colpi, Mon. Not. R. Astron.

Soc. 334, 401 (2002).
50. R. B. Larson, Mon. Not. R. Astron. Soc. 145, 271 (1969).
51. T. Abel, G. L. Bryan, and M. L. Norman, Science 295, 93 (2002).
52. V. Bromm, P. S. Coppi, and R. B. Larson, Astrophys. J. 564, 23 (2002).
53. N. Yoshida, K. Omukai, and L. Hernquist, Science 321, 669 (2008).
54. T. H. Greif, V. Bromm, P. C. Clark, S. C. O. Glover, R. J. Smith, R. S. Klessen, N. Yoshida, and V. Springel Mon. Not. R. Astron. Soc. , (2012).
55. N. Yoshida, K. Omukai, L. Hernquist, and T. Abel, Astrophys. J. 613, 406 (2006).
56. Y. Sabano and Y. Yoshii, Publ. Astron. Soc. Japan 29, 207 (1977).
57. J. Silk, Mon. Not. R. Astron. Soc. 205, 705 (1983).
58. K. Saigo, T. Matsumoto, and M. Umemura Astrophys. J. 615, 65 (2004).
59. M. J. Turk, T. Abel, and B. O'Shea, Science 325, 601 (2009).
60. K. Omukai and Y. Yoshii, Astrophys. J. 599, 746 (2003).
61. S. W. Stahler. F. Palla, and E. E. Salpeter, Astrophys. J. 302, 590 (1986).
62. M. N. Machida, K. Omukai, T. Matsumoto, and S. Inutsuka, Astrophys. J. 677, 813 (2008).
63. A. Stacy, T. H. Greif, and V. Bromm Mon. Not. R. Astron. Soc. 403, 45 (2010).
64. P. C. Clark, S. C. O. Glover, and R. S. Klessen, Astrophys. J. 672, 757 (2008).
65. P. C. Clark et al., Science 331, 1040 (2011).
66. K. Omukai and F. Palla, Astrophys. J. 561, 55 (2001).
67. K. Omukai and F. Palla, Astrophys. J. 589, 677 (2003).
68. C. F. McKee and J. C. Tan, Astrophys. J. 681, 771 (2008).
69. T. Hosokawa and K. Omukai, Astrophys. J. 691, 823 (2009).
70. T. Hosokawa and K. Omukai, Astrophys. J. 703, 1810 (2009).
71. T. Hosokawa, H. W. Yorke, and K. Omukai, Astrophys. J. 721, 478 (2010).
72. T. Hosokawa, K. Omukai, N. Yoshida, and H. W. Yorke, Science 334, 1250 (2011).
73. S. Richling and H. W. Yorke, Astrophys. J. 539, 258 (2000).

74. H. W. Yorke and C. Sonnhalter, Astrophys. J. 569, 846 (2002).
75. T. Nagakura and K. Omukai, Mon. Not. R. Astron. Soc. 364, 1378 (2005).
76. I. D. McGreer and G. Bryan, Astrophys. J. 685, 8 (2008).
77. H. Umeda and K. Nomoto, Astrophys. J. 565, 385 (2002).
78. A. Heger and S. E. Woosley, Astrophys. J. 567, 532 (2002).
79. J. Tumlinson, A. Venkatesan, and J. M. Shull, Astrophys. J. 612, 602 (2004).
80. A. Frebel, J. L. Johnson, and V. Bromm, Mon. Not. R. Astron. Soc. 392, L50 (2009).
81. K. Omukai, Astrophys. J. 534, 809 (2000).
82. K. Omukai, T. Hosokawa, and N. Yoshida, Astrophys. J. 722, 1793 (2010).
83. H. Maki and H. Susa, Publ. Astron. Soc. Japan 59, 787 (2007).
84. D. R. G. Schleicher et al., Astron. Astrophys. 522, A115 (2010).
85. C. Federrath et al., Astrophys. J. 731, 62 (2011).
86. M. J. Turk et al., Astrophys. J. 745, 154 (2012).
87. M. N. Machida, T. Matsumoto, and S. Inutsuka, Astrophys. J. 685, 690 (2008).
88. F. Walter et al., Nature 424, 406 (2003).
89. A. Songaila, Astron. J. 130, 1996 (2005).
90. B. Ciardi, in First Stars III, eds. B. O'Shea et al. (American Institute of Physics, 2008).
91. T. P. Stecher and D. A. Williams, Astrophys. J. 149, 29 (1967).
92. K. Omukai and R. Nishi, Astrophys. J. 518, 64 (1999).
93. Z. Haiman, T. Abel, and M. J. Rees, Astrophys. J. 534, 11 (2000).
94. S.-P. Oh, Astrophys. J. 569, 558 (2001).
95. M. Ricotti and J. P. Ostriker, Mon. Not. R. Astron. Soc. 352, 547 (2004).
96. B. Draine and F. Bertoldi, Astrophys. J. 468, 269 (1996).
97. H. Susa, Astrophys. J. 659, 908 (2007).
98. K. Ahn, Mon. Not. R. Astron. Soc. 375, 881 (2007).

99. B. W. O'Shea and M. L. Norman, Astrophys. J. 673, 14 (2008).
100. M. A. Alvarez, P. R. Shapiro, K. Ahn, and I. T. Iliev, Astrophys. J. 644, 101 (2006).
101. Z. Haiman and G. L. Bryan, Astrophys. J. 650, 7 (2007).
102. V. Bromm and A. Loeb, Astrophys. J. 596, 34 (2003).
103. K. Omukai, R. Schneider, and Z. Haiman, Astrophys. J. 686, 801 (2008).
104. M. E. Machacek, G. L. Bryan, and T. Abel, Mon. Not. R. Astron. Soc. 338, 273 (2003).
105. T. Kitayama, N. Yoshida, H. Susa, and M. Umemura, Astrophys. J. 613, 631 (2004).
106. D. Whalen, T. Abel, and M. L. Norman, Astrophys. J. 610, 14 (2004).
107. H. Susa and M. Umemura, Astrophys. J. 645, L93 (2006).
108. T. Abel, J. H. Wise, and G. L. Bryan, Astrophys. J. 659, 87 (2007).
109. N. Yoshida, S.-P. Oh, T. Kitayama, and L. Hernquist, Astrophys. J. 663, 687 (2007).
110. N. Yoshida, K. Omukai, and L. Hernquist, Astrophys. J. 667, L11 (2007).
111. D. Whalen, B. O'Shea, J. Smidt, and M. L. Norman, Astrophys. J. 679, 925 (2008).
112. V. Bromm, N. Yoshida, and L. Hernquist, Astrophys. J. 596, L135 (2003).
113. K. Wada and A. Venkatesan, Astrophys. J. 591, L38 (2003).
114. T. Kitayama and N. Yoshida, Astrophys. J. 630, 675.
115. T. H. Greif, J. L. Johnson, R. S. Klessen, and V. Bromm, Astrophys. J. 670, 1 (2007).
116. D. Whalen, B. van Veelen, B. W. O'Shea, and M. L. Norman, Astrophys. J. 682, 49 (2008).
117. M. Mori, A. Ferrara, and P. Madau, Astrophys. J. 571, 40 (2002).
118. N. Yoshida, V. Bromm, and L. Hernquist, Astrophys. J. 605, 579 (2004).
119. B. Carr, J. R. Bond, and D. Arnett, Astrophys. J. 277, 445 (1984).
120. J. L Johnson and V. Bromm, Mon. Not. R. Astron. Soc. 374, 1557 (2007).
121. R. Schneider, K. Omukai, A. K. Inoue, and A. Ferrara, Mon. Not. R. Astron. Soc. 369, 1437 (2006).

122. V. Bromm and A. Loeb, Nature 425, 812 (2003).
123. K. Omukai, T. Tsuribe, R. Schneider, and A. Ferrara, Astrophys. J. 667, 117 (2005).
124. J. Wise and T. Abel, Astrophys. J. 665, 899 (2007).
125. T. H. Greif, J. L. Johnson, R. S. Klessen, and V. Bromm, Mon. Not. R. Astron. Soc. 387, 1021 (2008).
126. K. Omukai, Astrophys. J. 546, 635 (2001).
127. C. Shang, G. L. Bryan, and Z. Haiman, Astrophys. J. 402, 1249 (2010).
128. K. Inayoshi and K. Omukai, arXiv:1202.5380.
129. T. Hosokawa, K. Omukai, and H. W. Yorke, arXiv:1203.2613.
130. P. Di Matteo et al., Astrophys. J. 745, L29 (2012).
131. Z. Haiman, in Astrophysics in the Next Decade, eds. H. Thronson et al. (Springer, Dordrecht, 2008).
132. M. Tanaka, T. Moriya, N. Yoshida, and K. Nomoto, Mon. Not. R. Astron. Soc. in press (2012).
133. V. Bromm and N. Yoshida, Ann. Rev. Astron. Astrophys. 49, 373 (2011).
134. E. Komatsu et al., Astrophys. J. 192, 18 (2011).
135. L. Hui and Z. Haiman, Astrophys. J. 596, 9 (2003).
136. G. P. Holder et al., Astrophys. J. 595, 13 (2003).
137. S. Furlanetto, S.-P. Oh, and F. Briggs, Phys. Rep. 433, 181 (2006).
138. S. E. Woosley and J. S. Bloom, Ann. Rev. Astron. Astrophys. 44, 507 (2006).
139. A. Mesinger, B. Johnson, and Z. Haiman, Astrophys. J. 637, 80 (2006).
140. J. Greiner et al., Astrophys. J. 693, 1610 (2009).
141. B. E. Robertson, R. S. Ellis, J. S. Dunlop, R. J. McLure, and D. P. Stark, Nature 468, 49 (2010).
142. N. Christlieb et al., Nature 419, 904 (2002).
143. A. Frebel et al., Nature 434, 871 (2005).
144. E. Caffau et al., Nature 477, 67 (2011).
145. S. D. M. White and V. Springel, in The First Stars, ed. A. Weiss (Springer, Dordrecht, 2000).
146. http://www.sdss3.org

6

Observational Constraints on the Origin of the Elements I. 3D NLTE formation of Mn lines in late-type stars

Maria Bergemann[1], Andrew J. Gallagher[1], Philipp Eitner[1,2], Manuel Bautista[3], Remo Collet[4], Svetlana A. Yakovleva[5], Anja Mayriedl[6], Bertrand Plez[7], Mats Carlsson[8,9], Jorrit Leenaarts[10], Andrey K. Belyaev[5] and Camilla Hansen[1]

[1] Max Planck Institute for Astronomy, 69117 Heidelberg, Germany
[2] Ruprecht-Karls-Universität, Grabengasse 1, 69117 Heidelberg, Germany
[3] Department of Physics, Western Michigan University, Kalamazoo, Michigan 49008, USA
[4] Stellar Astrophysics Centre, Department of Physics and Astronomy, Aarhus University, 8000 Aarhus C, Denmark
[5] Department of Theoretical Physics and Astronomy, Herzen University, St. Petersburg 191186, Russia
[6] Montessori-Schule Dachau, Geschwister-Scholl-Str. 2, 85221 Dachau, Germany
[7] LUPM, UMR 5299, Université de Montpellier, CNRS, 34095 Montpellier, France
[8] Rosseland Centre for Solar Physics, University of Oslo, PO Box 1029 Blindern, 0315 Oslo, Norway
[9] Institute of Theoretical Astrophysics, University of Oslo, PO Box 1029 Blindern, 0315 Oslo, Norway

Citation: Maria Bergemann, Andrew J. Gallagher, Philipp Eitner, Manuel Bautista, Remo Collet, Svetlana A. Yakovleva, Anja Mayriedl, Bertrand Plez, Mats Carlsson, Jorrit Leenaarts, Andrey K. Belyaev and Camilla Hansen "Observational constraints on the origin of the elements - I. 3D NLTE formation of Mn lines in late-type stars" A&A, 631 (2019) A80 DOI: https://doi.org/10.1051/0004-6361/201935811

[10] Institute for Solar Physics, Department of Astronomy, Stockholm University, AlbaNova University Centre, 106 91 Stockholm, Sweden

ABSTRACT

Manganese (Mn) is a key Fe-group element, commonly employed in stellar population and nucleosynthesis studies to explore the role of SN Ia. We have developed a new non-local thermodynamic equilibrium (NLTE) model of Mn, including new photo-ionisation cross-sections and new transition rates caused by collisions with H and H^- atoms. We applied the model in combination with one-dimensional (1D) LTE model atmospheres and 3D hydrodynamical simulations of stellar convection to quantify the impact of NLTE and convection on the line formation. We show that the effects of NLTE are present in Mn I and, to a lesser degree, in Mn II lines, and these increase with metallicity and with the effective temperature of a model. Employing 3D NLTE radiative transfer, we derive a new abundance of Mn in the Sun, A(Mn) = 5.52 ± 0.03 dex, consistent with the element abundance in C I meteorites. We also applied our methods to the analysis of three metal-poor benchmark stars. We find that 3D NLTE abundances are significantly higher than 1D LTE. For dwarfs, the differences between 1D NLTE and 3D NLTE abundances are typically within 0.15 dex, however, the effects are much larger in the atmospheres of giants owing to their more vigorous convection. We show that 3D NLTE successfully solves the ionisation and excitation balance for the RGB star HD 122563 that cannot be achieved by 1D LTE or 1D NLTE modelling. For HD 84937 and HD 140283, the ionisation balance is satisfied, however, the resonance Mn I triplet lines still show somewhat lower abundances compared to the high-excitation lines. Our results for the benchmark stars confirm that 1D LTE modelling leads to significant systematic biases in Mn abundances across the full wavelength range from the blue to the IR. We also produce a list of Mn lines that are not significantly biased by 3D and can be reliably, within the 0.1 dex uncertainty, modelled in 1D NLTE.

Keywords: stars: abundances / Sun: abundances / stars: atmospheres / Sun: atmosphere / line: formation / radiative transfer

INTRODUCTION

Manganese (Mn) is a prominent member of the iron-group family that has interesting connections to several topics in astrophysics. In particular, from the point of view of stellar nucleosynthesis, this element is very sensitive to the physical conditions in supernovae Type Ia (SNIa; Seitenzahl et al. 2013). Hence, the abundances of Mn in metal-poor stars provide powerful constraints on the progenitors and explosion mechanism of this important class of SNe.

Mn displays a large number of Mn I lines spanning a range of excitation potentials in the optical spectra of late-type stars (Bergemann & Gehren 2007). Also a few lines of Mn II can be detected in the blue at ~350 nm, and some strong lines of Mn I are available in the IR at 1.52 μm. Owing to the large number of observable lines, Mn is a useful element to test the excitation and ionisation equilibria in stellar atmospheres. The lines of both ionisation stages are affected by hyperfine splitting (HFS), and some are also very sensitive to stellar activity. For example, the resonance Mn I line at 5394 Å is known to vary across the solar cycle (Vitas et al. 2009; Danilovic et al. 2016).

A large number of studies over the past years have been devoted to the analysis of Mn abundances in the context of stellar population studies and nucleosynthesis. Most of these works have assumed local thermodynamic equilibrium (LTE). There is, however, evidence for the breakdown of the LTE assumption. Johnson (2002) reported a systematic ionisation imbalance of Mn I and Mn II in metal-poor stars. Bonifacio et al. (2009) found a 0.2 dex offset between the abundances of Mn in metal-poor dwarfs and giants. They also observe a significant excitation imbalance, with strong Mn I resonance lines resulting in significantly lower abundances compared to the high-excitation features. Sneden et al. (2016) confirm the excitation imbalance in LTE, but they also find that the ionisation balance is satisfied, if one relies on the high-excitation Mn I lines only. However, that study employed one star only, HD 84937, which can make it difficult to generalise these conclusions to a large sample. Mishenina et al. (2015) also employed LTE models to analyse a large sample of main-sequence stars in the metallicity range from − 1 to + 0.3. Their abundances suggest a modest systematic correlation with T_{eff}, signifying potential departures from LTE and 1D hydrostatic equilibrium.

In earlier studies (Bergemann & Gehren 2007, 2008), we showed that Mn is very sensitive to departures from LTE, also known as non-LTE (NLTE) effects. This is an element of the Fe-group, and is expected to be similar to Fe in terms of line formation properties. However, Mn is prone to stronger NLTE effects than Fe given its lower abundance of two orders of magnitude (in the cosmic abundance scale) compared to Fe, but also significantly higher photo-ionisation cross-sections, and a peculiar atomic structure with a very large number of strong radiative transitions between energy levels with excitation potentials of 2 and 4 eV. In particular, it was shown, on the basis of detailed statistical equilibrium (SE) calculations, that NLTE Mn abundances are significantly higher compared to LTE. This effect increases with decreasing metallicity and log *g* of a star, but also occurs with increasing T_{eff}.

Line formation across the solar granulation has been extensively discussed in the literature, in particular in the series of seminal papers by Dravins et al. (1981); Dravins (1987); Dravins & Nordlund (1990a,b); Nordlund & Dravins (1990), but see also more recent studies of the solar center-to-limb variation (e.g. Lind et al. 2017) and solar abundances (e.g. Caffau et al. 2008, 2009, 2010, 2011; Asplund et al. 2009; Amarsi & Asplund 2017; Amarsi et al. 2018a, 2019). Recently, this work has been extended towards 3D NLTE modelling of spectral line formation in other stars and applied to the lines of H, O, Si, and Al (Amarsi et al. 2016, 2018b, 2019; Nordlander & Lind 2017).

Given the interest in the impact of NLTE and 3D diagnostic on the abundance measurements, we present a re-analysis of Mn abundances in a small sample of well-studied FGK stars using an updated NLTE model atom, and 1D hydrostatic and 3D hydrodynamical model atmospheres. We use new atomic data, including transition probabilities, photo-ionisation cross-sections, and rate coefficients for the transitions caused by the inelastic collisions of Mn I and Mn II ions with H atoms. We compare the results of two 1D statistical equilibrium codes, DETAIL and MULTI2.3 that are both widely used in the community for NLTE analyses of chemical abundances. We also performed full 3D NLTE radiative transfer calculations with the MULTI3D code (Leenaarts & Carlsson 2009) to derive Mn abundance from the high-resolution spectra of the Sun and several metal-poor stars.

The paper is organised as follows: in Sect. 2, we describe the observed spectra. The LTE and NLTE calculations in 1D and 3D, spectrum synthesis, and abundance analysis are documented in Sect. 3. We present a considerable amount of details about the methods of calculations, as this is very important for a judgement of the resulting abundances. The results are presented in Sect. 4. This section includes a discussion of 1D NLTE and 3D NLTE abundance corrections, an analysis of the solar Mn abundance, a comparison between

3D LTE and 3D NLTE line profiles, and the excitation-ionisation balance of Mn I/Mn II in benchmark metal-poor stars. We close with a summary of the results and conclusions in Sect. 5.

OBSERVATIONS

For the Sun, we used the high-resolution flux atlas taken with the KPNO facility (Kurucz et al. 1984). The atlas has a resolving power $R \sim 400\ 000$. Recently, solar spectra taken with the PEPSI instrument at Large Binocular Telescope (LBT; Strassmeier et al. 2018) and with the Fourier transform spectrograph operated by the Institut fuer Astrophysik in Goettingen (Reiners et al. 2016) were released. However, the profiles of Mn lines are very similar in all these atlases. For this reason, we employed the KPNO spectrum in this work.

We also included several metal-poor benchmark stars (HD 84937, HD 140283, and HD 122563) from Bergemann et al. (2012). Their spectra are taken from the UVES-POP database (Bagnulo et al. 2003). These are the *Gaia* benchmark stars with T_{eff} and log g determined using interferometry and astrometry. The estimates of [Fe/H] and micro-turbulence were derived using NLTE radiative transfer for Fe lines (Bergemann et al. 2012). The effective temperatures of two of these stars were recently revised (Karovicova et al. 2018). The new estimates, based on the CHARA angular diameters, are $T_{eff} = 5787 \pm 48$ K for HD 140283 and $T_{eff} = 4636 \pm 37$ K for HD 122563. These estimates are fully consistent with the values we adopted in Bergemann et al. (2012). Creevey et al. (2019) propose a new asteroseismic surface gravity for HD 122563, log$g$ = 1.39 ± 0.01 dex. This is a substantial downward revision of this parameter. However, we tested the effect of log g on the abundance estimates from Mn lines, but found that the abundances change by only 0.05 dex. Hence, we did not recompute the model and used instead the standard models employed in Bergemann et al. (2012).

ANALYSIS

Model Atom and Diagnostic Lines of Mn

The model comprises three ionisation stages and 281 energy levels, with 198 levels of Mn I and 81 levels of Mn II. The model is also closed by the Mn III ground state. The radiative transitions were taken from the Kurucz compilation[1], which includes theoretically predicted and experimental estimates of the oscillator strengths, with the latter given a preference over theoretical estimates. The Mn I part of the model atom is shown in Fig. 1. We do not show the Mn II system in this plot. The ionised species has a very high ionisation potential and the bulk of Mn II lines, which connect the levels do not play any role in the SE of the element (see also the discussion of model atom completeness in Bergemann & Gehren 2008). In contrast to the latter study, we did not include fine structure for most of the Mn II levels, except those which are relevant for the Mn II near-UV lines used in detailed abundance measurements. The full atom is provided in the MULTI2.3 format at the CDS. We include fine structure levels for all energies up to 47 300 cm^{-1}, however, we also test the results using a compact model atom, which is devoid of fine structure for the majority of levels. This is important for our test calculations with full 3D simulations of stellar convection.

The atomic data for the Mn I lines, which we employ in the abundance calculations, are given in Table 1. For all of them, the HFS is included in the model atom directly. That is, the HFS structure of spectral lines was computed for all diagnostic lines of Mn I and Mn II, and included in the radiative transition part of the MULTI model atom. We employed the magnetic dipole constants A and electric quadrupole constants B assembled by Bergemann & Gehren (2008), complementing these with the data from Holt et al. (1999) for the relevant Mn II levels, a ^{5}D and z ^{5}P°. The full HFS patterns for each Mn line is provided in supplementary material. In the SE calculations, we treated the diagnostic Mn I and Mn II lines with Voigt profiles, while all other Mn lines were computed with a Gaussian profile with 13 frequency points. Our tests show that increasing the number of frequency points does not change the statistical equilibrium of the ion, on the other hand with this choice we still have a reasonable frequency quadrature to accurately represent each line profile.

We used the new experimental transition probabilities, where these are available. Most data are from Blackwell-Whitehead & Bergemann (2007) and Den Hartog et al. (2011). For some of the lines, the new log gf values are typically 0.05–0.1 dex lower than the old values, that ultimately leads to slightly higher abundances compared to our earlier work.

The broadening due to elastic collisions with H atoms is adopted from Barklem et al. (2000), where available. These data were derived using the 2nd order Rayleigh–Schroedinger perturbation theory as formulated by Brueckner (1971) and later generalised to transitions between different azimuthal quantum number states by O'Mara (1976), Anstee & O'Mara

(1991, 1995). This theory offers a more accurate representation of the broadening caused by collisions with H atoms than the theory by Unsöld (1927, 1955). The latter theory assumes that only collisions at large separations between atoms can strongly influence the line broadening, thus severely underestimating the line strengths. The main difference in this work with respect to our earlier studies (Bergemann & Gehren 2008) is the implementation of the new photo-ionisation cross-sections for Mn I and the new rates of inelastic collisions due to the interactions of Mn I with H I atoms.

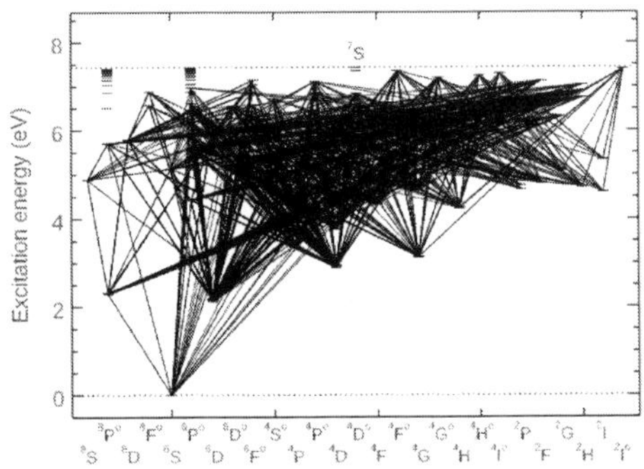

Figure 1. Grotrian diagram of Mn I atomic system. The energy levels of Mn I are shown with black dashes. The levels are connected by radiative transitions (solid black lines).

Table 1. Parameters of Mn I and Mn II lines used for abundance calculations.

λ (Å)	M	N_{HFS}	E_{low} (eV)	Lower level	Upper level	$\log gf$
3488.68	–	3	1.85	$a\,^5D_1$	$z\,^5P^o_1$	−0.937
3496.81	–	2	1.83	$a\,^5D_2$	$z\,^5P^o_3$	−1.779
3497.53	–	3	1.85	$a\,^5D_1$	$z\,^5P^o_2$	−1.418
4018.10	9	5	2.12	$a\,^6D_{9/2}$	$z\,^6D^o_{7/2}$	−0.311
4030.76	4	5	0.00	$a\,^6S_{5/2}$	$z\,^6P^o_{7/2}$	−0.497
4033.07	4	4	0.00	$a\,^6S_{5/2}$	$z\,^6P^o_{5/2}$	−0.647
4034.49	4	3	0.00	$a\,^6S_{5/2}$	$z\,^6P^o_{3/2}$	−0.843
4055.54	9	4	2.14	$a\,^6D_{7/2}$	$z\,^6D^o_{7/2}$	−0.077
4070.28	9	3	2.19	$a\,^6D_{1/2}$	$z\,^6D^o_{1/2}$	−1.039
4451.58	24	3	2.89	$a\,^4D_{7/2}$	$z\,^4D^o_{7/2}$	0.278
4498.90	24	1	2.94	$a\,^4D_{2/2}$	$z\,^4D^o_{5/2}$	0.343
4502.22	24	2	2.92	$a\,^4D_{5/2}$	$z\,^4D^o_{7/2}$	−0.345
4754.03	18	5	2.27	$z\,^8P^o_{5/2}$	$e\,^8S_{7/2}$	−0.086
4761.52	23	4	2.94	$a\,^4D_{1/2}$	$z\,^4F^o_{3/2}$	−0.138
4762.37	23	5	2.88	$a\,^4D_{7/2}$	$z\,^4F^o_{9/2}$	0.425
4765.86	23	3	2.93	$a\,^4D_{3/2}$	$z\,^4F^o_{5/2}$	−0.080
4766.42	23	4	2.91	$a\,^4D_{5/2}$	$z\,^4F^o_{7/2}$	0.100

4783.42	18	5	2.29	$z\,^8P^o_{7/2}$	$e\,^8S_{7/2}$	0.042
4823.51	18	6	2.31	$z\,^8P^o_{9/2}$	$e\,^8S_{7/2}$	0.144
5004.89	22	4	2.91	$a\,^4D_{5/2}$	$z\,^6F^o_{7/2}$	−1.630
5117.93	39	3	3.12	$a\,^4G_{5/2}$	$z\,^4F^o_{3/2}$	−1.140
5255.31	39	6	3.12	$a\,^4G_{11/2}$	$z\,^4F^o_{9/2}$	−0.763
5394.67	2	6	0.00	$a\,^6S_{5/2}$	$z\,^8P^o_{7/2}$	−3.503
5407.42	8	10	2.13	$a\,^6D_{7/2}$	$y\,^6P^o_{7/2}$	−1.743
5420.35	8	9	2.13	$a\,^6D_{7/2}$	$y\,^6P^o_{5/2}$	−1.462
5432.54	2	5	0.00	$a\,^6S_{5/2}$	$z\,^8P^o_{5/2}$	−3.795
5470.63	4	8	2.15	$a\,^6D_{5/2}$	$y\,^6P^o_{5/2}$	−1.702
5516.77	8	8	2.17	$a\,^6D_{3/2}$	$y\,^6P^o_{3/2}$	−1.847
5537.75	4	5	2.18	$a\,^6D_{1/2}$	$y\,^6P^o_{3/2}$	−2.017
6013.49	32	6	3.06	$z\,^6P^o_{3/2}$	$e\,^6S_{5/2}$	−0.251
6016.64	32	6	3.06	$z\,^6P^o_{5/2}$	$e\,^6S_{5/2}$	−0.216
6021.79	32	6	3.06	$z\,^6P^o_{7/2}$	$e\,^6S_{5/2}$	0.034

Photo-Ionisation

We adopted new quantum-mechanical photo-ionisation cross-sections for 84 LS terms of Mn I, which belong to the configurations $3d^6$, $3d^5\,4s$, $3d^5\,4p$, and $3d^4\,4s^2$. The photo-ionisation cross-sections for dipole that allowed transitions in Mn I were computed using the R-matrix method for atomic scattering calculations (Berrington et al. 1987). The solution of the Schrödinger equation for the $N + 1$ electron system is found on the basis of the close-coupling expansion of the wavefunction as

$$\Psi(E; S\,L\pi) = A\sum_i \chi_i\theta_i + \sum_j c_j\phi_j, \quad (1)$$

where A is the anti-symmetrisation operator, χ_i is the target ion wavefunctions in the target state $(S\,L\pi)_i$, θ_j is the wavefunction for the free electron, and ϕ_j are short range correlation functions for the bound (e^-+ion) system.

The calculations were done in LS-coupling and include all states with valence electron excitations up to the principal quantum number $n = 10$. The single-electron orthogonal orbitals that represent the atomic structure of the Mn^+ target, were derived using the AUTOSTRUCTURE code (Badnell 1997). The code employs scaled Thomas–Fermi–Dirac–Amaldi central-field potential. We adopted configuration interaction expansion with spectroscopic orbitals 1s, 2s, 2p, 3s, 3p 3d, 4s, 4p, 4d, 5s, and 5p. The configurations and scaling parameters for the orbitals are presented in Table B.2. LS terms of the target Mn II ion included in the close-coupling expansion are provided in Table B.1.

The cross-sections are sampled at 5000 evenly-spaced energy points between zero and 0.8 Ryd above the first ionisation threshold, followed by 250 points from 0.8 Ryd to 2.0 Ryd. This mesh is also preserved in our NLTE calculations with DETAIL and with MULTI2.3, such that all resonances are fully accounted for in the statistical equilibrium calculations. These resonances are mostly caused by the photo-excitation of the core and dominate the cross-sections of the majority of Mn I states. For the other levels, we employed hydrogenic cross-sections sampled on a regular mesh. The hydrogenic cross-sections were computed using the effective principal quantum number.

Figure 2 shows the total photo-ionisation cross-section for the ground state of Mn I, compared with the central field approximation results of Verner & Yakovlev (1995) and Reilman & Manson (1979). The close coupling expansion accounts forthe photo-ionisation of the outer 4s sub-shell, as well as the open inner 3d sub-shell of the ground state $3d^5\ 4s^2\ {}^6S$ of Mn. The coupling of all relevant photo-ionisation channels results in extensive auto-ionisation structures and ensures that no sharp discontinuity in the cross-section at the 3d inner-shell ionisation edge is present. By contrast, the central field approximation, which misses channel couplings, yields a cross-section that is severely underestimated up to the opening of the 3d sub-shell, where a discontinuity appears. The importance of channel couplings in representing low-energy photo-ionisation cross-sections of iron-peak elements is well known (Bautista & Pradhan 1995). For energies beyond the 3d sub-shell, the central field cross-sections agree very well with our data, which gives additional confidence on the accuracy of our results. Figure A.1 also show the comparison of the cross-sections with the hydrogenic data. Clearly the differences are substantial and amount to several orders of magnitude in the background, but also all resonances that contribute to the over-ionisation at longer wavelengths are missing in the hydrogenic data.

DETAIL does not have a provision for including the partial ionisation channels to specific states of the target ion. Consequently, we adopted the total photo-ionisation cross-sections, computed by adding the partial cross-sections for each Mn I state.

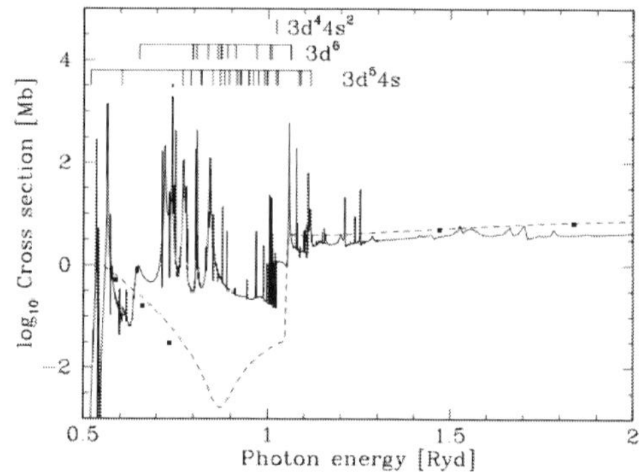

Figure 2. Photo-ionisation cross-section of $3d^5 4s^2$ a ^6S ground term of Mn I. The present R-matrix cross-section is depicted by the solid line. The central-field cross-sections of Verner & Yakovlev (1995) and Reilman & Manson (1979) are indicated by the dashed line and square dots, respectively. Above the figure we show the energies of all $3d^5 4s$, $3d^6$, and $3d^4\ 4s^2$ thresholds.

Inelastic Collisions with H Atoms

The rate coefficients for the bound–bound transitions in Mn I caused by collisions with H atoms, as well as for Mn II collisions with H^-, were taken from Belyaev & Voronov (2017c). We also computed new rates for these processes in this work.

The data from Belyaev & Voronov (2017c) are available for the transitions between 19 levels[2] of Mn I interacting with H and the ground state of Mn II interacting with H^-. They represent collisional excitation, de-excitation, mutual neutralisation, and ion-pair formation[3] due to the transitions between $^7\Sigma^+$ molecular states.

Here we present new calculations of the H I collision rates for 71 additional levels of Mn I interacting with H and the first excited state of Mn II interacting with H^-. The first excited ionic state of the MnH molecule has $^5\Sigma^+$ symmetry and only covalent molecular states of the same symmetry were considered in the non-adiabatic nuclear dynamical calculations. These states are listed in Table B.3.

All calculations were performed within the simplified quantum model (Belyaev & Yakovleva 2017b,a), which allows the identification of a rate coefficient for a particular process using general dependences of the reduced rate coefficients on the electron binding energies. The binding energies are calculated from different ionic limits for the cases of non-adiabatic transitions between $^7\Sigma^+$ molecular states and between $^5\Sigma^+$ states. The rate coefficients for the excitation and de-excitation processes are summed over molecular symmetry, when the initial and the final state of the process have both $^7\Sigma^+$ and $^5\Sigma^+$ symmetries.

Neutralisation rate coefficients for collisions of $Mn^+(3d^54s\ ^5S) + H^-$ are presented in Fig. 3 as a function of the electron binding energy. For the case of the MnH collisional system involving quintet molecular states, the largest rate coefficients at 6000 K correspond to the mutual neutralisation processes at Mn($3d^54s4p\ u^6P^\circ$) + H, Mn ($3d^54s5d\ f^6D$) + H, Mn ($3d^54s4f\ w^6F^\circ$) + H, Mn ($3d^54s6p\ t^6P^\circ$) + H states, having values ~6×10^{-8} cm^3 s^{-1}. The rate coefficients for the (de-)excitation processes are, at least, one order of magnitude lower than the rates for the neutralisation and ion-pair formation processes, as found in previous calculations for other chemical elements (Belyaev et al. 2014, 2017; Yakovleva et al. 2017).

We also derive new rate coefficients for the 42 levels of MnII interacting with H and for the ground state of Mn III interacting with H $^-$. These calculations were performed for the transitions between $^6\Sigma^+$ molecular states as theionic state of MnH^+ has $^6\Sigma^+$ symmetry. The states are presented in Table B.4. Neutralisation rate coefficients for collisions Mn $^{2+}(3d^5\ ^6S) + H^-$ as a function of the electron bound energy are shown in Fig. 4. The largest rate coefficient for the case of MnH^+ collisions at 6000 K with the value of 7.5×10^{-8} cm^3 s^{-1} corresponds to the mutual neutralisation process $Mn^{2+}(3d^5\ ^6S) + H^- \rightarrow Mn^+(3d^55p\ v^5P^\circ) + H$.

Our data apply to J-averaged energy states, but the NLTE model atom includes fine structure. We have tested different recipes that are used in the literature to deal with this case. In particular, Barklem (2007) propose to divide the rate coefficient by the number of the target states. However, we found that the effect of distributing the collision rate coefficients across the target states is virtually null. In particular, for the high-excitation Mn I lines in the model of a metal-poor dwarf, this leads to an error in the line equivalent width of less than 0.1%, which is negligibly small for abundance determinations. We, hence, assigned the same rate coefficient for each fine structure level of a given term[4]. This is analogous to our handling of the photo-ionisation data, which are also provided for a given term. The new rate coefficients are available in the supplementary material.

In the model atom, we tabulated the rates of exothermic processes for the bound-bound reactions, meaning the transitions accompanied by the release of energy ($E_j > E_i$, where the transition occurs from the higher energy level j to the lower energy level i). These rates are almost independent of temperature, which minimises interpolation errors. For the charge transfer, the reverse is true, hence we tabulated the rates of endothermic processes. The reverse rates are computed from the detailed balance internally within the code, see Eq. (5) in Belyaev & Voronov 2017c (note that neither n_H nor n_{H^-} explicitly enter these equations):

$$r_{ij} = r_{ji} \frac{g_j}{g_i} \exp \frac{-\Delta E_{ji}}{k_B T}, \quad (2)$$

where r_{ij} and r_{ji} are rate coefficients for the transitions between i and j; g_i and g_j the statistical weights of these levels; and E_{ji} the energy difference ("energy defect") between the energies of the two states.

Apart from the avoided crossing mechanism, one can estimate additional rate coefficients using the free electron model, which is expected to include other inelastic mechanisms except the long-range ionic covalent mechanism (Barklem 2016; Amarsi et al. 2018c, 2019). We hence also supplemented the model atom with collision rates for all Mn I states computed using the scattering-length approximation[5] according to Eq. (18) of Kaulakys (1991) using the code developed by Barklem (2017). As described by Osorio et al. (2015) and Barklem (2016), the rate coefficients computed in this way need to be redistributed over all possible final spin states. Here, we assume that all transitions have two possible final spin states, and that each final spin state is equally likely, so that the rate coefficients were reduced by a factor of two. The error associated with this assumption is less than a factor of two. The Kaulakys model is developed for application to Rydberg molecular states, hence, our implementation shall be viewed as a limiting case with strong collisional binding, and, hence, might underestimate NLTE effects.

In Fig. 5 we compare the new H collision rates with the data computed using the Steenbock & Holweger (1984) formulation of the classic theory developed by Drawin (1968, 1969). The classic formalism does not cover the mutual neutralisation and ion-pair formation processes. The differences in the H excitation rates for the individual energy levels amount up to 7 orders of magnitude in both directions. The overall distributions as a function of energy difference have similar shapes, with the largest rate coefficients for the transitions between nearby energy levels. The rate coefficients for the charge transfer reactions are also qualitatively similar to the Drawin's bound-free recipe, which describes collisional ionisation, but quantitatively there are major differences of up to 5 orders of magnitude. The quantum-mechanical charge transfer rates are typically lower for the neutral species, whereas for the singly-ionised species these are larger than the Drawin's rates. The Kaulakys rates are typically very low for the low-energy levels, but gradually increase closer to the ionisation threshold and thus somewhat compensate the downturn in the quantum-mechanical data, leading to higher collisional thermalisation. In Sect. 4.1.2 we briefly report on how this impacts the line profiles and the NLTE abundance corrections.

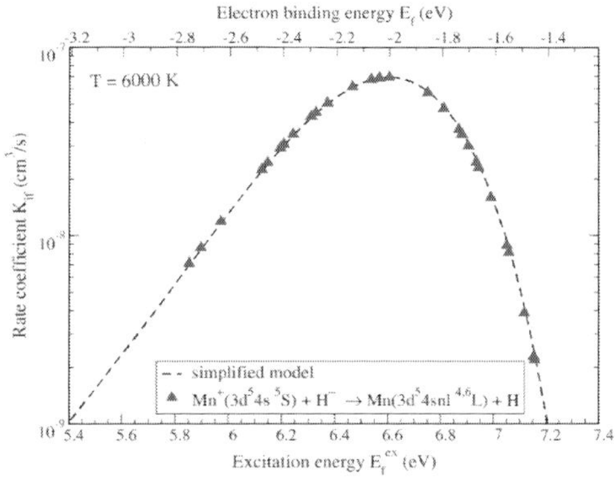

Figure 3. Neutralisation rate coefficients for $Mn^{+}(3d^{5}4s\ ^{5}S) + H^{-}$ collisions asa function of electronic energy in different excited states of Mn I. The dashed line represents the reduced rate coefficient given by the simplified model.

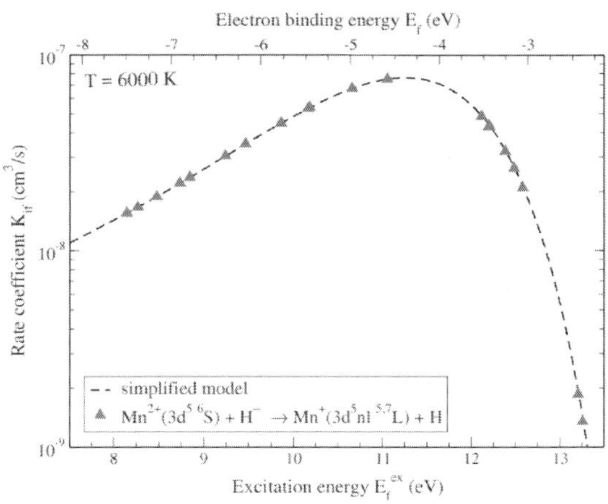

Figure 4. Neutralisation rate coefficients for $Mn^{2+}(3d^{5}\ ^{6}S) + H^{-}$ collisions as a function of electronic energy in different excited states of Mn II. The dashed line represents the reduced rate coefficient given by the simplified model.

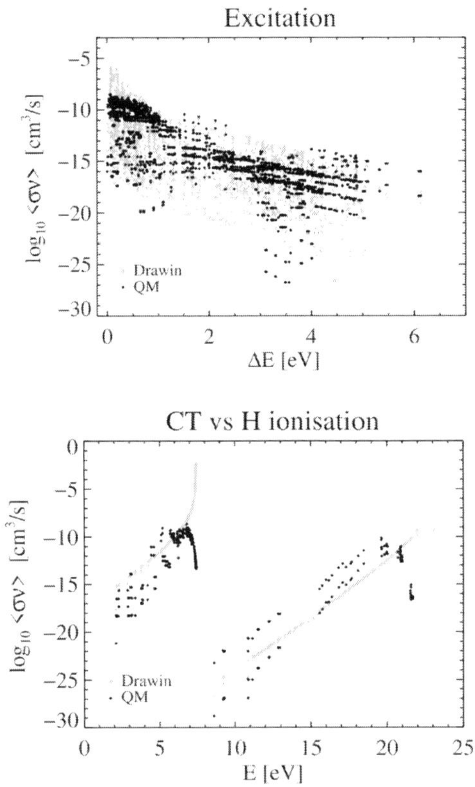

Figure 5. Excitation (*top panel*) and ion-pair formation (*bottom panel*, CT stands for "charge transfer") rate coefficients for Mn I and Mn II as used in this work and compared to Drawin's formulae (Drawin 1968, 1969; Steenbock & Holweger 1984).

Model Atmospheres

As in our previous papers, we used MARCS (Gustafsson et al. 2008) and MAFAGS-OS (Grupp 2004a,b) model atmosphere grids. These are 1D LTE model atmospheres, with certain differences regarding the treatment of convective energy transport (mixing length), opacity, and the solar abundance mixture. The depth discretisation and the vertical extent of the models are also slightly different, as MAFAGS-OS covers the range from − 6 to + 2 in log τ_{5000}, whereas the MARCS models sample the Rosseland optical depths from − 4 to + 2. Nonetheless, the thermodynamic structures of the models for the given input parameters are very similar (Bergemann et al. 2012, 2017).

The 3D model atmospheres are taken from the STAGGER model atmosphere grid (Collet et al. 2011; Magic et al. 2013)and computed with the STAGGER code (Nordlund & Galsgaard 1995). A 3D model consists of a series of computational boxes that represent a time series, which are referred to as snapshots. These snapshots are selected from a larger time series of snapshots that are produced from the STAGGER code and are selected at a time when they have reached dynamical and thermal relaxation. For our purpose and for the sake of time, we have chosen to work with five snapshots. Importantly, and unlike an equivalent 1D model, 3D models provide x, y, and z velocity fields for every voxel. This means that post-processing spectrum synthesis code provides parameter-free description of Doppler broadening, including asymmetric line profiles, which trace these gas flows at each voxel.

Figure 6 depicts the 3D temperature structures (blue 2D histogram) in a representative snapshot for four benchmark stars, along with the 1D MARCS (solid red line) and ⟨3D ⟩ (dashed grey line) profiles. The average temperature of the full 3D model and the 1D hydrostatic model are fairly different in the outermost regions of the atmosphere, as seen by comparing the 1D hydrostatic models with the ⟨3D ⟩ models. In particular, in the outer layers of metal-poor models the 1D hydrostatic models are significantly hotter (up to 500 K) compared to the 3D structures (see also Bergemann et al. 2017). Also the models diverge in deeper regions of the models, where the continuum usually forms. This is mostly due to the treatment of convection between the 1D and 3D model atmospheres.

The average temperature structure of the 1D model of the metal-poor RGB star HD 122563 is not very different from its ⟨3D ⟩ counterpart. Our adopted MARCS models are taken from Bergemann et al. (2012, their Fig. 1). We explore the impact of an adopted 1D model in Sect. 4.6, by performing the abundance analysis with a MAFAGS-OS suit of models. We also note that line formation is not only sensitive to the mean $T(\tau)$ and $P(\tau)$ structures, but also to the horizontal inhomogeneities. The latter play a significant role in the abundance analysis.

The scope of this paper is limited to the analysis of a small sample of 3D models, including that of the Sun, a typical dwarf, and a typical giant (Table 2). We also include tailored 3D models computed for the parameters of the benchmark metal-poor stars HD 122563, HD 140283, and HD 84937 (Sect. 2). To make the NLTE radiative transfer problem computationally tractable, we have to resample the full 3D model cubes onto a less fine, yet equidistant, grid in horizontal coordinates. However, we test the effect of the resolution of the cube for radiative transfer in Sect. 4.2 and find virtually no differences in the resulting atomic number densities and line profiles for horizontal resolutions of $(x,y,z) = 30, 30, 230$ and the original cubes ($(x,y,z) = 240, 240, 230$). Hence, the former is taken to be the default resolution for most of the analysis presented in this work. We solve the 3D NLTE radiative transfer problem for a set of snapshots for each of the 3D model atmospheres listed in Table 2. These snapshots are extracted at regular time intervals from the full simulation that covers roughly two convective turnover timescales (Collet et al. 2011; Magic et al. 2013). The convergence criterion, that is the maximum relative correction in the population numbers, $\max|\delta N/N|$, is set to 10^{-3}. This is fully sufficient according to our experience with 1D NLTE radiative transfer.

Statistical Equilibrium

We used two different codes to compute the SE of Mn. One is MULTI2.3 (Carlsson 1992), the other code is DETAIL (Butler & Giddings 1985). The codes solve the equations of radiative transfer and SE assuming a 1D geometry. The assumption of trace elements is used, that is, the element that is modelled in SE is assumed to have no effect on the model atmosphere. This is a good assumption for Mn as it is not an electron donor, nor does it have a high impact on the overall opacity. Both codes adopt the accelerated lambda iteration (ALI) technique and the operator acting on the source function.

The basic differences between the codes are described in Bergemann et al. (2012). The tests described in the following sections willbe performed imposing the same input conditions (LTE populations) and the same model atmospheres, in order to maximise the consistency. The main difference between the codes are in the handling of thermodynamic parameters and of the background opacities. In particular, DETAIL takes the partial pressures and partition functions from the input model atmosphere, whereas MULTI2.3 includes a package to compute these parameters given the input structures $T(\tau)$ and $P_e(\tau)$ as a function of optical depth or column mass.

In order to maximise the homogeneity of the analysis, we have also computed background opacity tables for MULTI2.3 using the updatedline lists from DETAIL (Bergemann et al. 2015). The MARCS (Gustafsson et al. 2008) and Turbospectrum (Plez 2012) codes were used to generate a table of opacities for a set of temperature and pressure points at more than 10^5 wavelengths, Mn[6] lines being omitted. This table is then interpolated by MULTI2.3 to produce detailed line background opacities.

All calculations with MULTI2.3 are carried out by simultaneously solving the intensity at all angles, using the Feautrier method with all scattering terms included consistently (the ISCAT option set to 1). This is important when scattering in the background opacity is significant, as is the case in the Wien regime. We have tested the line formation disabling this option, but found that this has a very strong effect on the blue and UV lines of Mn I and Mn II, significantly over-estimating the line abundances,because of reduced continuum intensities.

MULTI3D is an MPI-parallelised, domain-decomposed NLTE radiative transfer code that solves the equations of radiative transfer in 3D geometry using the ALI method. The formal solution of radiative transfer is done via the short characteristics method (Kunasz & Auer 1988) that solves the integral form of the radiative transfer equation across one subdomain per time step. Carlson (1963)'s A4 quadrature is employed to compute the angle-averaged radiation field in the SE solution. The approximate operator is constructed using the formulation developed by Rybicki & Hummer (1991, 1992), where only the diagonal of the full Λ operator is used (for discussion of this approximation, see, e.g. Bjørgen & Leenaarts 2017).

MULTI3D will accept three types of 3D model atmosphere formats, including the commonly used Bifrost (Gudiksen et al. 2011) and STAGGER models (Magic et al. 2013). The code will also accept any 3D model providing the temperature, density, electron number density, and x, y, and z velocity fields are supplied on a Cartesian grid that is both horizontally periodic and equidistantly spaced. The code can compute radiative transfer using the 1.5D approximation, which treats each column of grid points in a model as a separate plane-parallel atmosphere, or using full 3D radiative transfer. The rate equations are assumed to be time-independent and the advection term is not included. For more information about the code, we refer to Leenaarts et al. (2012) and Bjørgen & Leenaarts (2017).

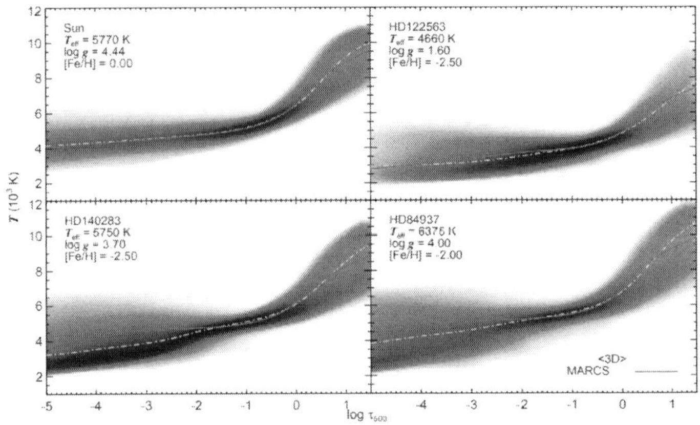

Figure 6. 3D, 1D, and ⟨3D⟩ temperature structures as a function of Rosseland optical depth for benchmark stars. The blue-shaded regions indicate the kinetic temperature distributions in the representative snapshot from the 3D convection simulations. The stellar parameters are given in the inset. We note that ⟨3D ⟩ models are provided only to illustrate the difference between the average structure of the 1D and 3D models, however, the ⟨3D ⟩ are not used in our abundance analysis.

Table 2. Parameters of 3D convective and 1D hydrostatic model atmospheres.

No.	Name	T_{eff} K	log g dex	[Fe/H] dex
1	Sun	5777	4.4	0.0
2	HD 84937	6400	4.0	−2.0
3	HD 140283	5750	3.7	−2.5
4	HD 122563	4600	1.6	−2.5
5	Sub-solar metallicity dwarf	6500	4.0	−1.0
6	Metal-poor dwarf	6500	4.0	−2.0
7	Sub-solar metallicity giant	4500	2.0	−1.0
8	Metal-poor giant	4500	2.0	−2.0

RESULTS

We begin the discussion of results with a brief account of NLTE effects in Mn (Sect. 4.1). The key properties of the statistical equilibrium of Mn are summarised in Sect. 4.1.1. 1D NLTE abundance corrections across a large parameter space are presented in Sect. 4.1.2. Line formation and abundances determined using 3D inhomogeneous atmospheres are the subjects of subsequent sections. Section 4.2.1 deals with the properties of line formation in full 3D solar simulation cubes, that we refer to as photon kinematics. In Sect. 4.2.2, we discuss simplified radiative transfer models and explore how these impact the line profiles compared to the full 3D NLTE solution. The results of the 3D NLTE solar abundance analysis are presented in Sect. 4.3. In Sect. 4.4, we discuss the impact of 3D NLTE on the line profiles, on their equivalent widths (EW), and on the abundance diagnostic for four metal-poor models. The results of LTE and NLTE calculations with 3D models are discussed in Sect. 4.5. Finally, in Sect. 4.6, we use 3D convective models of the benchmark metal-poor stars HD 84937, HD 140383, and HD 122563 to derive 3D NLTE abundances.

1D NLTE

Departures from LTE

This work does not deal extensively with the properties of statistical equilibrium of Mn nor with the details of line formation, as this has been discussed in great detail in our previous work (Bergemann & Gehren 2007). Additionally Bergemann & Gehren (2008) address the details of line transfer in metal-poor stars. It suffices to remind the reader that Mn, similar to other Fe-group elements, is a photo-ionisation dominated ion. Simply stated, the large photo-ioinisation cross-sections of Mn I energy levels imply significant over-ionisation in the atmospheres of cool stars in the more general SE case compared to LTE. Mn I becomes significantly over-ionised (compared to LTE) in metal-poor or in hotter stellar atmospheres due to their strong UV radiation fields. The effect of the radiation field is furthermore amplified in the atmospheres of giants owing to their lower densities and hence, less efficient collisional processes.

The effect of the radiation field is reflected in the behaviour of Mn I level departure coefficients, b_i, which describe the ratio between NLTE and LTE atomic number densities. Departures from LTE take place in the line formation layers. Figure 7 shows that for Mn I, this ratio is typically $\lesssim 1$, as in NLTE the fraction of atoms in a given energy state is less than that predicted by the Saha–Boltzmann formulae. The departure coefficients of the ionic levels, Mn II, are very close to unity for the lower-lying energy levels, but deviate from thermal for the levels of higher excitation energy, $E \gtrsim 1$ eV.

Figure 7 shows that the departures from NLTE in the line-formation region, $-2 \lesssim \log \tau_{5000} \lesssim 0$, are moderate in the solar atmosphere, but only slightly smaller than our previous estimates in Bergemann & Gehren (2007). The differences with respect to the latter study are caused by the use of new quantum-mechanical photo-ionisation rates and H collision rates, as well as the implementation of fine structure for most of the Mn I levels. Collisions with electrons are not important in the physical conditions of the solar atmosphere. On the other hand, inelastic collisions with H atoms have a non-negligible effect on the atomic level populations and significantly contribute to the overall thermalisation of the system. This effect is not linear and may increase or decrease the departures from LTE for individual energy levels, and hence spectral lines, depending on the T_{eff}, log g, and metallicity of a star.

The comparison of the b_i profiles computed using DETAIL and MULTI2.3 (Fig. 7) suggests that the codes are consistent, given the same input conditions, such as the model atom, line opacities, and model atmospheres. MULTI2.3 predicts slightly larger departures from LTE compared to DETAIL. This has also been shown in our earlier paper for Fe (Bergemann et al. 2012), and is likely related to continuum opacities and/or the numerical implementation of the coupled SE and radiative transfer equations.

The behaviour of departure coefficients is very different in the model atmosphere of a metal-poor red giant star (Fig. 8). All Mn I levels show a stronger under-population compared to the solar model, implying larger differences between LTE and NLTE abundances in metal-poor stars. The energy levels of the Mn II lines are also affected by NLTE. In particular, the levels of the upper term $z\,^5P^o$ experience overpopulation caused by the radiative pumping in nine strong near-UV lines of Mn II multiplet Nr. 31 ($a\,^5D$–$z\,^5P^o$) in the deeper layers. However, the levels of $z\,^5P^o$ become under-populated at $\log \tau_{5000} \sim -1.5$, as these lines progressively become optically thin. Consequently, one would expect significant NLTE effects on the formation of Mn II lines, largely driven by the changes in the line source function itself. Comparing the departure coefficients computed using the old atom from Bergemann & Gehren (2008) (Fig. 8, top panel) and the new atom in this work (Fig. 8, middle panel), we find substantial differences. The influence of new quantum-mechanical collisions with hydrogen is admittedly greater in the new atom, despite the larger photo-ionisation cross-sections. On the other hand, contrasting the results obtained using the DETAIL code with MULTI2.3 (Fig. 8, middle and bottom panels) confirms that, similar to the Sun (see Fig. 7), the two codes produce quantitatively similar outputs. In the outer layers, the departures are slightly different, which could be related to the differences in the outer boundary conditions.

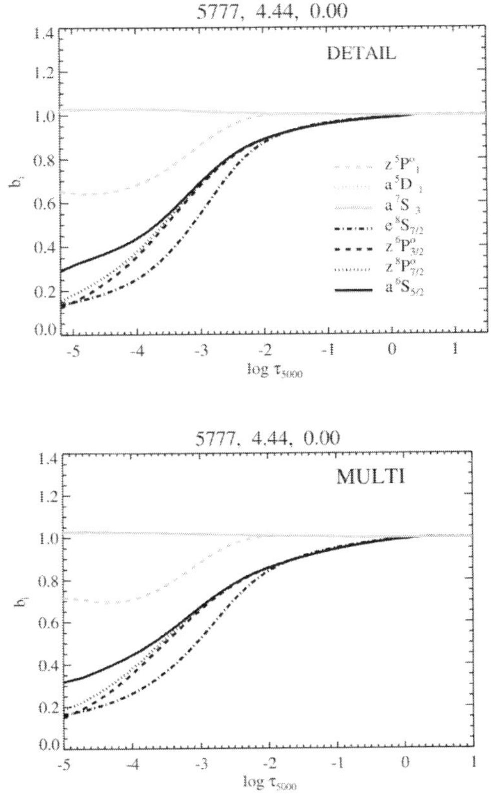

Figure 7. Mn departure coefficients for solar MARCS model atmosphere as a function of optical depth at 5000 Å, computed using DETAIL (*top panel*) and MULTI (*bottom panel*) codes. The surface parameters of the Sun, (T_{eff}, log *g*, and [Fe/H]) are given in the figure details.

NLTE Abundance Corrections

An NLTE abundance correction is the quantity that is commonly used in stellar abundance studies to correct the abundances derived under the assumption of LTE. The abundance correction is defined as $\Delta = A(\text{NLTE}) - A(\text{LTE})$, that is the difference in abundance required to fit a given spectral line assuming 1D LTE or 1D NLTE. We also employ this concept in our 3D NLTE analysis in Sect. 4.2.

Figure 9 illustrates the NLTE abundance correction for several metal-poor models in the metallicity range − 3 to 0. We only plot selected lines to illustrate the key results: the 3488 Å line of Mn II, multiplet 4 (4030, 4033, 4034 Å), multiplet 18 (4783, 4823 Å), and multiplet 32 (6013, 6016, 6021 Å) of Mn I. The behaviour of abundance corrections within a given multiplet is very similar, hence the lines aregrouped by multiplets. The correction is not tailored to any particular star and is computed assuming the reference NLTE [Mn/Fe] of zero. We also explored the abundance corrections computed using LTE [Mn/Fe] abundance of − 0.5 to − 0.8, as it would be typically measured assuming LTE in metal-poor stars and found no significant differences in the corrections. The individual curves represent three possible scenarios that differ in the completeness of the model atom.

In Fig. 9, we illustrate the sensitivity of the corrections to the quantum-mechanical H data. The CH case (note that CH is not an abbreviation) corresponds to the model atom, which is devoid of quantum-mechanical H-collisionalexcitation processes (but includes charge transfer). The CH0 model lacks charge transfer processes, but includes collisional excitation. These two cases are compared to the model, which has both excitation and charge transfer (CH/CH0 included). None of the three cases presented in this figure includes the Kaulakys (1985) collision rates.

The H collisions clearly have a different impact on the line formation in the atmospheres of giants and dwarfs. Based on the NLTE abundance corrections in the figures, it appears that for dwarfs, H collisions serve as a thermalising agent, decreasing the difference between NLTE and LTE. In the atmospheres of red giants, the effect is somewhat counter-intuitive: the lines of multiplet 32 (6013–6021 Å triplet) have smaller NLTE corrections when H collisions are excluded. In fact, this is the indirect effect of over-ionisation, which is more efficiently transferred to higher-excitation states by collisions with H. On the other hand, the lines of multiplet 18 (4783, 4823 Å) behave as expected from the simple considerations of increased rate of collisional thermalisation. Figure 9 also suggests that charge transfer (CT) reactions are more important in the atmospheres of dwarfs. Neglecting CT fully typically leads to abundance corrections being over-estimated by 0.05 dex for the models with [Fe/H] $\gtrsim -3$.

Figure 10 illustrates the influence of collision rates computed using the Kaulakys (1985) recipe. The reference atom includes the Kaulakys (1985) collisions in addition to the data from Belyaev et al. (2017), and we compare this model with the model that is devoid of the Kaulakys data. The differences with our reference model atom are modest, and do not exceed 0.05 dex for the dominant part of the parameter space. The effects that are possibly most significant occur when we neglect Kaulakys (1985) collisions. This leads to slightly over-estimated NLTE abundance corrections for the multiplet 32 Mn I lines (6013 – 6021 Å triplet) in the RGB models. On the other hand, the NLTE abundance corrections for these very high-excitation lines are systematically under-estimated in the model of a dwarf.

The general picture is that the NLTE abundance corrections for the Mn I lines are positive and increase with decreasing metallicity, supporting our earlier study of Mn (Bergemann & Gehren 2007, 2008) and of other similar ions (e.g. Fe, Bergemann et al. 2012; Lind et al. 2012). The corrections are slightly larger for the RGB model, especially at lower [Fe/H]. The higher-excitation lines, such as those of multiplets 18 and 32 are more sensitive to NLTE. Their NLTE abundance corrections typically increase with decreasing metallicity, but this trend slightly flattens below [Fe/H] ~ −2. Another noteworthy feature of these diagrams is the fact that the Mn II lines are also not immune to NLTE. It has been often assumed in the literature that the lines of ionic species do not show NLTE effects. The few strong excited lines of Mn II at 1.85 eV show the classical NLTE effect of photon loss. This effect is small, but shows in the atmospheres of dwarfs and giants (see Fig. 10). It implies that lower abundances would be obtained from Mn II line, especially for the metal-poor stars with [Fe/H] = −2. At lower metallicity, the Mn II lines become sufficiently weak and radiative pumping effects dominate, leading to positive NLTE abundance corrections.

It is interesting, and it possibly presents the main difference with respect to our earlier study, that the strong resonance triplet of Mn I at 4030–4034 Å and the excited lines show very similar NLTE abundance corrections. The NLTE corrections exceed just about 0.4 dex in the atmospheres of RGB stars with [Fe/H] = −3. This is important, as LTE abundances derived from the resonance Mn I lines are known to be significantly lower compared to high-excitation Mn I features (Bonifacio et al. 2009; Sneden et al. 2016). Compared to the high-excitation features, in the 1D NLTE analysis there is no room for differentially larger NLTE corrections for the 4030–4034 Å triplet lines. In our previous work, a higher degree of over-ionisation in Mn I, and, in particular, over-ionisation from the ground state, was achieved by employing a tailored S_H scaling factor to the Drawin collisional (excitation and ionisation) Mn I + H I rates. As a consequence, the NLTE abundance corrections for the resonance triplet lines were significantly higher.

Below we show that 3D NLTE calculations suggest substantial differential effects between Mn I lines of different excitation potentials. These differential effects help to improve the excitation balance (Sect. 4.2), effectively providing the physical basis for the effect, which is mimicked by using inefficient H collisions in 1D models.

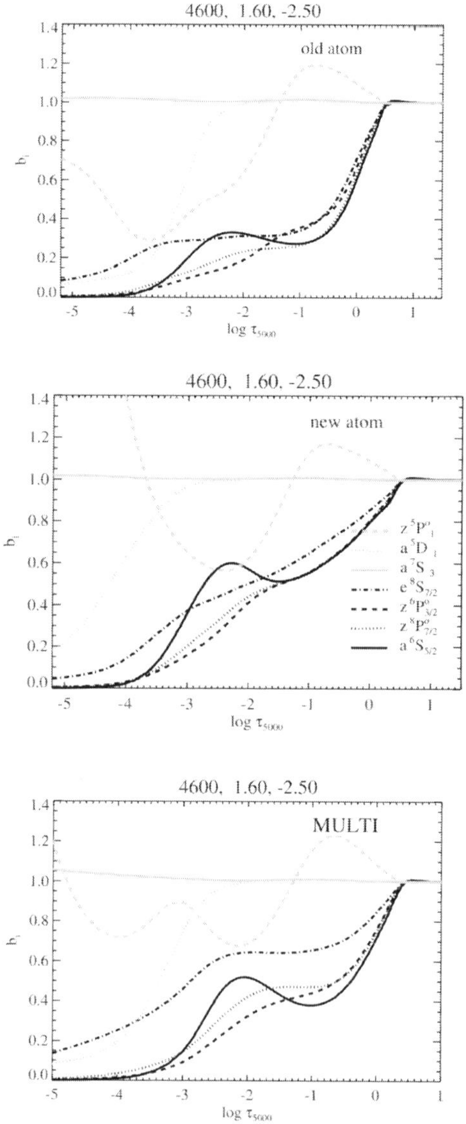

Figure 8. Mn departure coefficients for model atmosphere of a metal-poor red giant. These were computed using the old Mn atom from Bergemann & Gehren

(2007) (*top*) and the new atom from this paper (*middle panel*: DETAIL, *bottom panel*: MULTI). The stellar parameters (T_{eff}, log g, and [Fe/H]) are given in the figure titles.

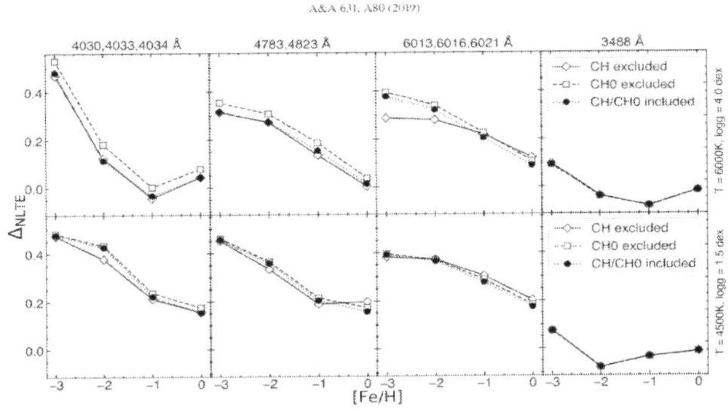

Figure 9. NLTE abundance corrections for Mn I (4030, 4033, 4034, 4783, 4823, 6013, 6016, and 6021 Å) and Mn II (3488 Å) lines. These are computed for a small grid of MARCS model atmospheres representative of dwarfs: T_{eff} = 6000, log g = 4.0 (*top panels*),and red giants: T_{eff} = 4500, log g = 1.5 (*bottom panels*) for a range of metallicities from 0 to − 3 dex. Different curves represent the corrections derived using the model atoms with reduced complexity: (a) CH excluded, ignoring the excitation processes by collisions with H atoms; (b) CH0 excluded, ignoring the charge transfer reactions; and (c) CH/CH0 included with excitation and CT rates from the quantum-mechanical calculations. None of the three models (a–c) include the Kaulakys recipe.

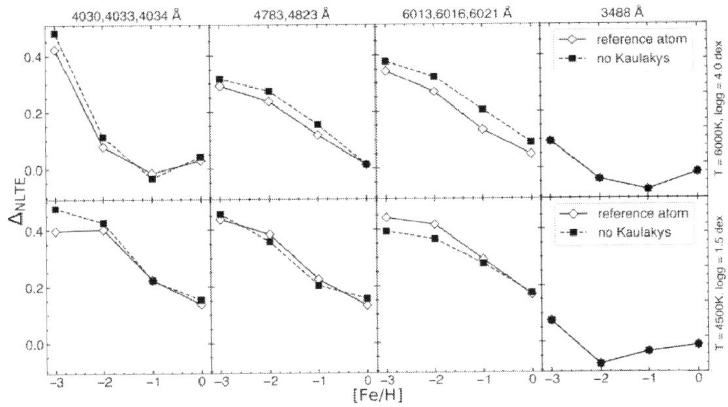

Figure 10. NLTE abundance corrections for Mn I and Mn II lines. These are computed for a small grid of MARCS model atmospheres representative of

dwarfs: T_{eff} = 6000, log g = 4.0 (*top panels*),and red giants: T_{eff} = 4500, log g = 1.5 (*bottom panels*) for a range of metallicities from 0 to − 3 dex. Different curves represent the corrections derived using two model atoms, one with and the other without collisions derived using the Kaulakys recipe.

3D NLTE

Photon Kinematics

Figure 11 illustrates spatially-resolved NLTE intensity profiles of two Mn I lines in the solar model at the disc centre. All profiles are normalised to the average continuum intensity for the corresponding spectral line in the snapshot. The lines were chosen such that the effect of the HFS is minimal, in order to isolate the effect of granular motions on the profiles. The profiles are taken for every fourth point along each horizontal coordinate in the simulation domain (i.e. for 8 × 8 = 64 points out of 900) in order to not overload the figure. The bisectors for each line component are shown in the right-hand side panels. In addition, the solid curves indicate the profiles extracted from the granule and inter-granular lane in Fig. 12.

Overall, the behaviour of the lines is very similar to that described earlier by Dravins & Nordlund (1990a, for example in their Fig. 6 for a Sun-like star α Cen A). The weaker high-excitation Mn I line at 5004 Å shows a strong anti-correlation between the depth of the line core and the line shift (right-hand panel). This is, in fact, the weakest unblended solar Mn I line with the EW of only 13 mÅ. The line profiles with the strongest blue-shift and the highest intensity contrast form above the granules, where the upwards streaming motions of hotter material are characterised by higher velocities, and higher granular temperatures account for the brighter background continuum. The dominant NLTE effect of over-ionisation leads to brightening in the line core. The lack of any pronounced curvature in the bisectors of the blue-shifted components suggests that there is little vertical variation of velocity field in the upflows.

The Mn I line at 4502 Å, which is stronger but has the same lower excitation potential as the 5004 Å line, also shows a very broad distribution of line shifts. This line is close to saturation, as is evidenced by the broad, rectangular inner core in the bluest components. Similar to the 5004 line, the bisectors of the blue-shifted components, which form above granules, are typically *l*-shaped, that is, the line profiles are nearly symmetric. Although the blue-shifted components are very strong, and are characterised by very (vertically) extended line-formation regions, this again suggests that the vertical variation of velocity fields in the granules is small. On the other hand, the bisectors of the profiles that form above the inter-granular lanes tend to approximate a *c*-shape. These line components are highly asymmetric, their cores are very broad, and tend be skewed to the red. Dravins & Nordlund (1990a) found that this is a characteristic feature of the lines that form across inter-granular regions with a larger vertical-velocity gradient with depth.

3D Test Cases

Calculations of 3D NLTE radiative transfer are very computationally expensive. Hence, we explored whether a simplified treatment of NLTE radiative transfer with 3D simulations offers a suitable alternative to full 3D NLTE calculations.

In particular, we illustrate and discuss the results of calculations obtained using more compact model atmosphere cubes and different radiative transfer solvers (1.5D versus full 3D). We also compare the flux profiles obtained using different solar snapshots. In what follows, we limit the discussion to two Mn I lines only. One of them is a strong resonance line at 5394 Å, where the effects of NLTE and 3D convection are most pronounced. The second line is that at 4502 Å, discussed in the previous section. This line is least affected by the HFS, hence its shape is a good test case to explore the effects of NLTE and 3D convective flows.

The top left panel of Fig. 13 shows the line profiles computed in full 3D NLTE versus 1.5D calculations (Sect. 3.3). The difference is negligible and it amounts to less than 0.5% in the flux level, or <0.01 dex in abundance, that is, far less than the uncertainty incurred by the other sources of error. In the case of metal-poor stars, the differences are also modest and do not exceed a few percent in EW. Since the computational time of 3D NLTE scales as $O(n^{4/3})$, where n is the number of grid points (meaning the number of iterations is proportional to the number of grid points in the *z* direction), it is a reasonable approximation to use 1.5D NLTE calculations, at least for the stars with parameters similar to the models tested in this work.

The top right panel of Fig. 13 illustrates the impact of the model atmosphere resolution. We employ three models: one resampled to the resolution of 5 × 5 × 102, the other with the resolution of 30 × 30 × 230 (our default resolution), as well as the maximum resolution of 100 × 100

× 230 that we can afford with the computing time available to us. The comparison of the medium- and maximum- resolution simulation can be found in the appendix. The effects are visible in the line profiles, but they only slightly change the shape of the line and have a very modest effect on the line equivalent width. This effect comes mostly through the impact of velocities on the line opacity, $\kappa(x, \nu \times (1 - v(x)/c))$ as is best seen in lines that are not very sensitive to HFS, such as the 4502 Å line (left panel in Fig. 13). Their shapes are closely approximated by a Gaussian fit, and the effects of velocity fields are easily distinguished. In particular, the red wing of the line appears to be slightly darker in the model with higher resolution. There is also an implicit effect of velocities on the line source function, as it depends on the radiation field at the line frequencies, which is non-local and connected to the velocity distribution at all spatial points along the beam.

We also find (Fig. 13, bottom left panel) that the flux profiles computed using different (regularly-spaced in time) snapshots do not show significant variations. The 5394 Å line changes its depth by ~ 3%. The position of the line centre changes by a few mÅ. The 4502 Å line is almost insensitive to the temporal effects. This suggests that the statistical properties of velocity fields, temperature, and pressure inhomogeneities in each snapshot are representative of some average distribution values for the relaxed simulation given the values of T_{eff}, log g, and [Fe /H]. The temporal variability of spatially resolved profiles has been examined in detail in Dravins (1990); Dravins & Nordlund (1990b, their Sect. 5). They found that the characteristic evolution of line shapes and depths lacks a unique timescale, that was a manifestation of the lack of a unique scale in the spatial power spectra. Hence, the ensemble averages of spatially resolved line profiles are very similar in different snapshots, as we will also demonstrate in Sect. 4.4 for other models.

Finally, the bottom right panel of Fig. 13 illustrates the line profiles computed with 1D LTE and 3D NLTE. This is a major difference that amounts up to 0.2 dex in the abundance (see next section). Similar to the earlier studies for ions with similarly complex electronic structure, and ionisation potential (e.g. Amarsi et al. 2016; Lind et al. 2017, for Fe I), we find that the effects of 3D NLTE are to weaken the line compared to 1D LTE. Mn I is a photo-ionisation-dominated ion, that is the NLTE effects are driven by a strong non-local radiation field across multiple ionisation channels of the ion in the UV and in the optical. Of all Mn I lines, the resonance lines at 5394 and 5432 Å are most sensitive to this effect, as the lines are comparatively weak in the solar spectrum, and their NLTE effect is mainly caused by the change in the line opacity that scales in the first order as the ratio of the NLTE to LTE number densities of the lower energy level of the transition. The higher-excitation lines, such as that at 4502 Å show more modest NLTE effects in the model atmosphere of the Sun, both in 1D and in 3D.

The results of our test calculations indicate that vertical radiative transfer in NLTE with full 3D data cubes at the resolution of $(x,y,z) = 30, 30, 230$ is the most suitable approach and we employed this method in our calculations of Mn abundances for the Sun (Sect. 4.3) and the benchmark stars (Sect. 4.6). The approach allows affordable 3D NLTE calculations within reasonable timescales and provides line profiles consistent with the original resolution of the 3D cubes, as well as, with the full 3D radiative transfer calculations.

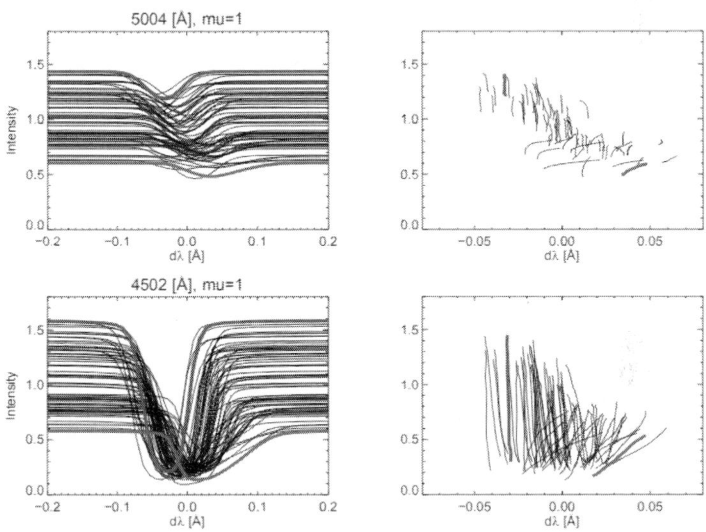

Figure 11. Spatially resolved line profiles (*left panels*) and their bisectors (*right panels*) for solar disc centre in model snapshot 020. The solid blue curve indicates the line profile extracted from the granule, whereas the red curves correspond to the inter-granular lane. The corresponding surface areas are marked with yellow and blue boxes in Fig. 12.

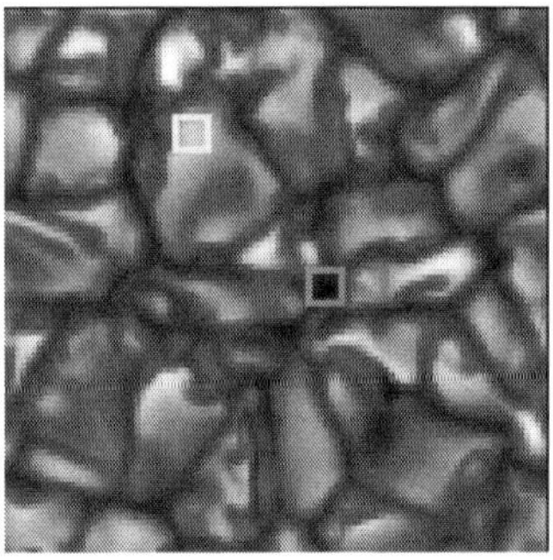

Figure 12. Monochromatic image of solar granulation at wavelength of 5394 Å at solar disc centre.

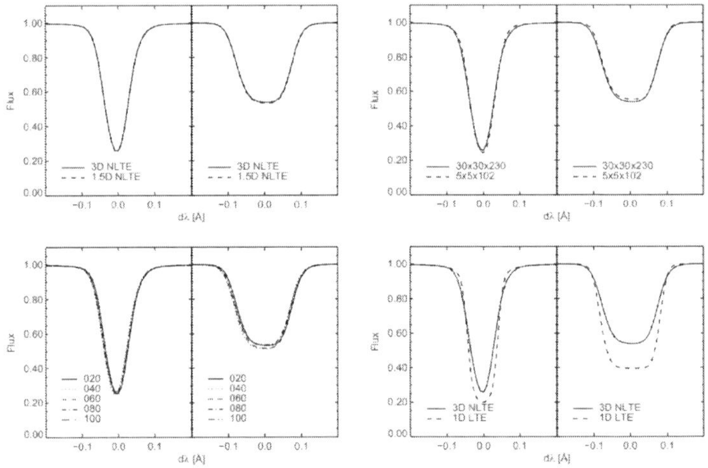

Figure 13. Comparison of theoretical profiles for Mn I lines at 4502 and 5394 Å for solar snapshot 020. *Top left panel*: 1.5D NLTE versus full 3D NLTE. *Top right panel*: line profiles computed with the 3D data cubes scaled down to the resolution of $(x,y,z) = 30, 30, 230$ (solid line) and 5, 5, 102 (dashed line). *Bottom left panel*: full 3D NLTE for the five solar snapshots. *Bottom right panel*: 3D NLTE profiles versus 1D LTE profiles.

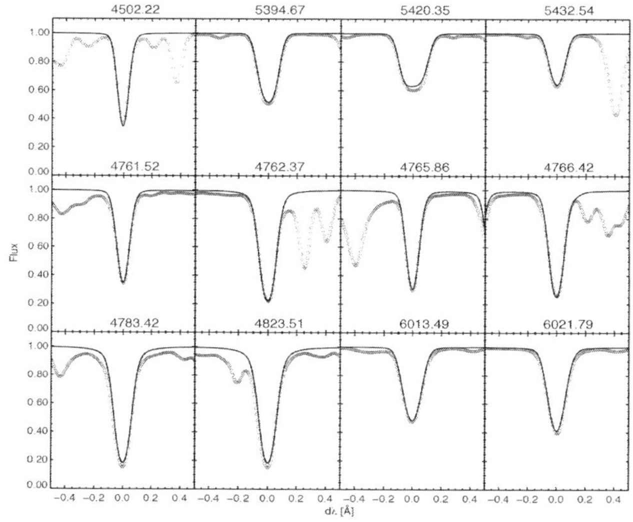

Figure 14. Comparison of 3D NLTE line profiles for selected Mn I lines with observed solar flux spectrum. All model profiles were generated using the same Mn abundance of 5.47 dex.

SOLARMN ABUNDANCE

The Sun is a metal-rich star, and as such, the spectral lines in most cases appear to suffer from blending and/or strong sensitivity to damping or turbulence. As a matter of fact, most Mn I lines are not very useful, being either blended or too strong, with a few exceptions. Nonetheless, we start with the analysis of all Mn I lines that are typically used in the analysis of FGK stars, and later assume a more conservative approach that takes into account only the most reliable features.

The LTE analysis of the solar Mn I lines is carried out using the SIU spectrum synthesis code (Reetz 1999). SIU allows interactive spectral fitting, which is very convenient to test for the presence and effects of blends and asymmetries within a line. This is a very accurate approach, especially when applied to ultra-high-resolution spectra, such as the solar atlases, and has been commonly exploited in the literature.

Figure 14 presents some examples of 3D NLTE line profiles for the solar spectrum. Note that in this plot, we have not adjusted the Mn abundances, but assumed the abundance of 5.47 dex, as recommended by earlier studies of the solar Mn abundance. Most of the Mn I lines are very broad, with the width of 0.2−0.3 Å, which is a consequence of the HFS. It is obvious that the available atomic and HFS data describe the line profiles remarkably well, both in terms of the asymmetries and the multi-component structure. The latter is particularly prominent in the cores of the weak resonance or ground-state lines, such as the 5394 and 5516 Å line. The core of the line at 5420 Å is slightly too weak, compared with the observations. This could be due to its sensitivity to activity (Wise et al. 2018). Note that also the 5394, 5432, and 6013 lines could be affected (Wise et al. 2018), and in particular, the 5394 Å line has a long history of research owing to its variation with the global solar activity cycle (Danilovic & Vince 2005; Livingston et al. 2007). Some studies suggest that the line is not as sensitive to granular motions as other features, because of its wide HFS, and is more sensitive to the intergranular magnetic concentrations, similar to the CH *G*-band (Vitas et al. 2009). According to our 3D NLTE results, however, the 5394 Å line is sensitive to convection and shows a significant difference between modelling with 1D hydrostatic and 3D convective models. Overall, the agreement of the model 3D NLTE line profiles with the observed solar data suggests that the atomic data quality and the physical quality of spectral models are sufficient to satisfactorily describe the properties of Mn lines in the solar spectrum.

The abundances determined from the Mn I lines that we view as reliable photospheric abundance diagnostics are shown in Fig. 15. The major challenge in the analysis of the solar spectrum is the contamination of line profiles by blends. These are present in the wings of the strong Mn I lines, such as the 4761, 4762, 4765, and 4766 Å, but also the lines of multiplet 18 (4783 and 4823 Å). Besides, there are no quantum-mechanical data for elastic collisions with H atoms for the two latter lines (Barklem et al. 2000), hence, we have to resort to the standard Unsöld formalism that carries an additional source of an uncertainty. We disregard such lines when computing the solar abundance of Mn. The 1D NLTE and 3D NLTE Mn abundances are derived by applying the corrections computed using MULTI2.3 and MULTI3D for the individual spectral lines.

In 1D LTE, the low-excitation potential and/or very strong lines give a lower abundance of Mn compared to the high-EP (weaker) lines. The difference is not large, but significant given the very quality of the observed data and well-constrained fundamental parameters of the Sun. The effects on NLTE in the solar atmosphere are moderate and change the average solar abundance by only ~ 0.02–0.07 dex. NLTE alone does not help to improve the solar excitation balance. More important is the combined influence of convection and NLTE. In the case of NLTE calculations with the 3D convective solar model, the effect of over-ionisation is amplified owing to stronger background radiation fields in the granules. The 3D NLTE–1D LTE difference is equivalent to ~ 0.22 dex in abundance for the resonance lines, but 0.15 dex for the higher excitation lines of other multiplets.

Our solar abundances of Mn are 5.34 ± 0.04 dex in LTE, 5.41 ± 0.05 dex in NLTE, and 5.52 ± 0.04 in 3D NLTE. The 1D NLTE and 3D NLTE values are consistent with the meteoritic abundance of Mn 5.50 ± 0.03 dex reported by Lodders (2003). Our 1D LTE result is ~0.1 dex higher compared to the previous estimate reported in Bergemann & Gehren (2007), 5.23 ± 0.1 dex. On the other hand, it matches very well with our LTE estimate obtained using the MAFAGS solar model, the revised transition probabilities, and the SIU spectrum synthesis code in Blackwell-Whitehead & Bergemann (2007), 5.33 ± 0.1 dex. The new 1D NLTE values are also higher than the estimates in Bergemann & Gehren (2007). In both cases, 1D LTE and 1D NLTE, the difference is due to the revision of the transition probabilities and the NLTE model atom.

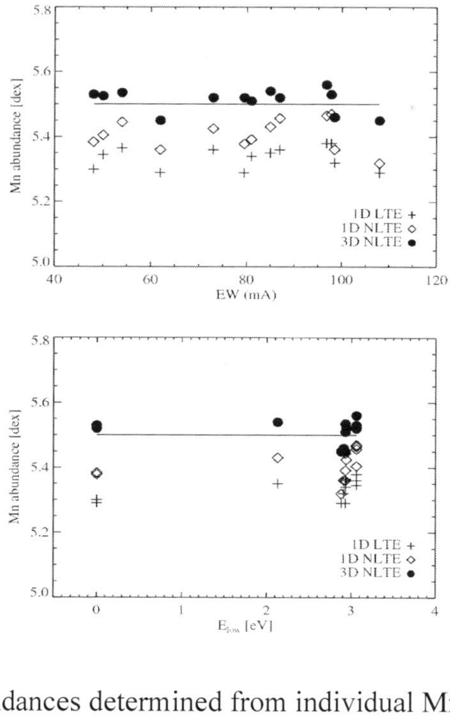

Figure 15. Mn abundances determined from individual Mn I lines in solar flux spectrum using different models: 1D LTE, 1D NLTE, and 3D NLTE. The solid line denotes the Mn abundances measured in CI chondrites, 5.50 ± 0.03 dex (Lodders 2003).

3D NLTE EFFECTS IN METAL-POOR MODEL ATMOSPHERES

Our analysis of line formation with 1D models suggests that the majority of Mn lines are sensitive to NLTE. Also the calculations for the 3D hydrodynamical model of the Sun indicate a strong impact of convection on the line profiles,which does not cancel the NLTE effects, but rather amplifies them for the most spectral features of Mn I.

In this section, we present the analysis of line formation in several model atmospheres of metal-poor stars. We explore the models with parameters with T_{eff} = 6500 K, log g = 4 dex, [Fe/H] = −1 and T_{eff} = 4500 K, log g = 2, [Fe/H] = −1, as well as their metal-poor counterparts with [Fe/H] = −2. For simplicity and didactic clarity, all calculations were performed assuming a scaled solar Mn abundance. Deviations from this assumption may affect the line formation, yet at this stage we are interested in the combined effects of 3D radiative transfer in convective models and NLTE and the impact on Mn I lines with different properties. We explore which lines are least sensitive to these effects, and employ them in the abundance analysis in the follow-up study of a larger stellar sample.

Our results for the four 3D models are illustrated in Fig. 16. We also overplot the predictions of 1D LTE modelling computed using the same Mn model atom and the same Mn abundance. Similar to the solar model, the 3D NLTE line profiles are much weaker than the 1D LTE counterparts. This is the consequence of over-ionisation, which is stronger in metal-poor conditions due to reduced line blanketing and stronger UV radiation fluxes. The topology of convection in FGK stars is set by mass conservation. The hotter rising granules occupy the dominant (~ 2/3) part of the stellar surface, whereas the cooler downdrafts are confined to narrow inter-granular lanes (Stein & Nordlund 1989; Nordlund et al. 2009), although the area subtended by the down-flows increases with depth. Hence, over-ionisation in the granules is the dominant NLTE effect that defines the appearance of the spatially integrated line profiles. In contrast to the Sun, where the effects of convective shifts were marginal, the integrated flux profiles in the metal-poor models show strong asymmetries, with typical blue-shifted cores and skewed wings. This is best seen by comparing the symmetric 1D profiles, convolved with an arbitrary macro-turbulence and the 3D profiles, which do not assume any ad-hoc broadening. Also the line profiles from individual snapshots are remarkably similar to their temporarily averaged counterparts, although a slightly more pronounced temporal variation is seen in the NLTE line profiles computed using the 3D models of dwarfs (as in Fig. 16).

The differences between 1D LTE and 3D NLTE profiles are very large, with only a few notable exceptions. These exceptions are the blue lines of Mn II, such as the 3488 and 3497 Å lines. For dwarfs, these low-excitation Mn II features are nearly invariant to the changes in the model atmosphere structure, and they indeed show very similar line shapes and equivalent widths, regardless of whether one adopts LTE or NLTE radiative transfer. Also a standard choice of the micro-turbulence $\xi_t = 1$ km s^{-1} appears to be satisfactory for a dwarf model, producing a line profile, which closely resembles the full 3D NLTE profile of the Mn II 3488 Å line even in the model with [Fe/H] = −2. For the computational complexity of 3D NLTE modelling, we are unable to explore a larger parameter space of 3D models in this work, but the behaviour of the models with metallicity suggests that this conclusion may hold also for other stars on the main-sequence, as long as [Fe/H] is not in the extremely metal-poor domain.

This fortunate result does not hold for the atmospheres of giants, however, and neither does it generally hold for the Mn I lines. The 3488 Å line of Mn II suffers from non-negligible deviations from NLTE in the RGB models, being significantly stronger in 3D NLTE. On the other hand, the 3D NLTE abundances of all lines of Mn I are significantly higher compared to 1D LTE. This is best illustrated by exploring the differences between 1D LTE and 3D NLTE, and 1D NLTE and 3D NLTE abundances derived from the line of a given equivalent width. This quantity is typically referred to as a 3D NLTE abundance correction (3D NLTE – 1D LTE).

The top panel of Fig. 17 shows that the 3D NLTE abundance corrections amount to 0.25 dex for most Mn I lines in the atmospheres of dwarfs, but in the models of RGB stars the (3D NLTE – 1D LTE) abundance differences tend to be significantly larger. In all cases, 1D analysis tends to substantially underestimate the Mn abundance. In particular, for the lines of multiples 18 and 32 (e.g. 4823 and 6016 Å) we find the abundance corrections of 0.3–0.4 dex in metal-poor models, whereas the blue resonance lines at 4030–4034 Å show even larger 3D NLTE corrections of ~ 0.5 dex at [Fe/H] = −2. This is a very interesting result. On the one hand, our results suggest that the large abundance discrepancies and excitation imbalance seen in the analysis of RGB stars (Bonifacio et al. 2009) may indeed be accounted for by 3D NLTE. On the other hand, we also find that 3D NLTE effects do not significantly impact the excitation balance in the atmospheres of very metal-poor dwarfs, as 3D NLTE corrections are similar for the lines of all Mn I multiplets. This is why this effect may go undetected in 1D LTE studies of stars with similar parameters (e.g. the study of HD 84937 by Sneden et al. 2016, or Mishenina et al. 2015). Also the 3D NLTE corrections for the Mn II lines are very small in the metal-poor atmospheres of dwarfs, and do not exceed − 0.05 dex at [Fe/H] = −2.

Figure 17 also suggests that, whereas 1D LTE always underestimates abundances for the Mn I lines, there are a few Mn I lines that can be reliably modelled with 1D NLTE. These are the lines of multiplets 9 (e.g. 4055, 4070, and 4082 Å), 23 (e.g. 4761, 4762, 4765, and 4766 Å), 24 (e.g. 4436, 4451, 4498, and 4502 Å), and 32 (e.g. 6013, 6016, and 6021 Å). For these features, the differences between 1D NLTE and 3D NLTE are marginal, and do not exceed 0.1 dex in the lowest metallicity ([Fe/H] = − 2) models of both dwarf and RGB stars. The lines of other multiplets show a very significant systematic deviation from 1D NLTE, caused by the impact of convective inhomogeneities. Additionally, for dwarfs, the impact of convection is modest and indeed, most optical Mn I lines, including the resonance Mn I triplet at 403 nm, would be suitable for abundance diagnostic, leading to a modest bias of $\sim^{+0.05}_{-0.15}$ dex. The exact value would also depend on the abundance of Mn itself, and additionally on the assumed value of micro-turbulence in the 1D models. So our estimates of 3D NLTE abundance corrections may not be directly applicable to the 1D LTE measurements in the observed stars. Rather we recommend employing those lines, which according to our tests, show minimal impact of convection and/or NLTE.

The IR lines of Mn I, like those that are used in the H-band abundance analysis (Shetrone et al. 2015), show significant departures from 1D and LTE. However, it is remarkable that all three Mn lines that we tested in this work show very similar 3D NLTE – 1D NLTE differences in dwarf and RGB models. In effect, the abundances for all these models are systematically underestimated by − 0.15 dex with respect to 1D NLTE, and the bias appears to be insensitive to metallicity of the model. It appears reasonable then to apply a positive correction to all H-band Mn 1D NLTE abundance measurements to obtain physically unbiased results.

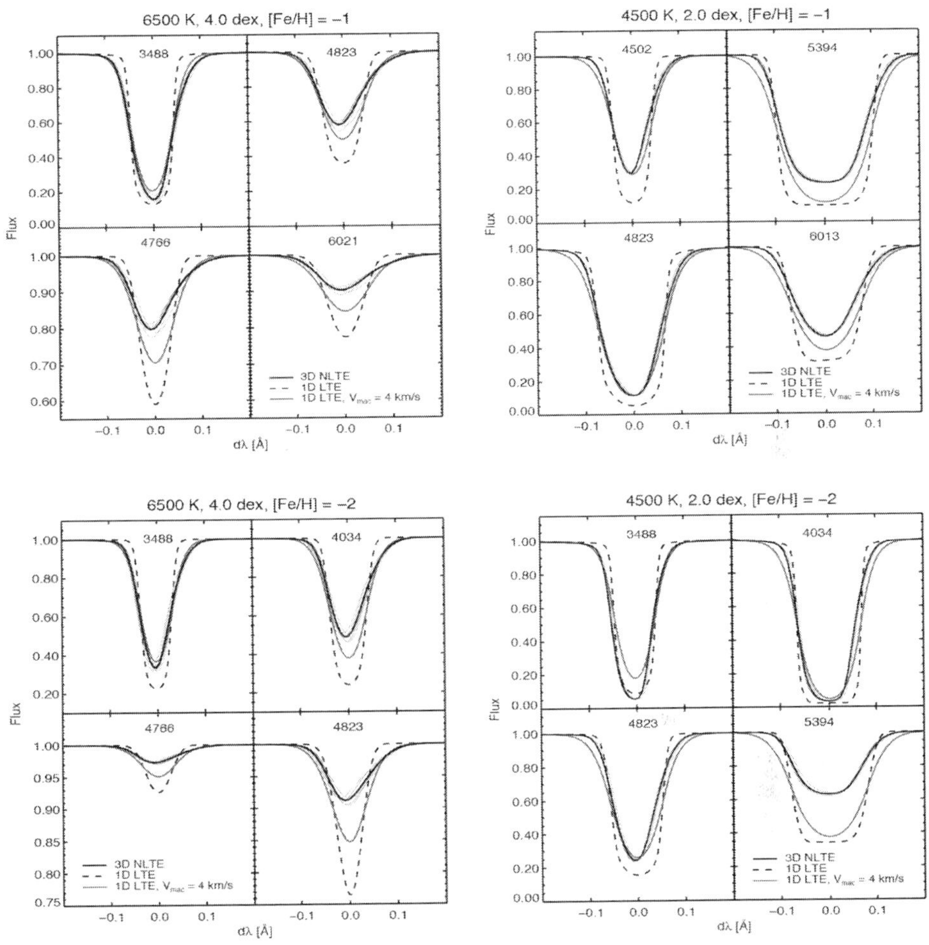

Figure 16. Line profiles of selected Mn I lines in dwarf and RGB models with metallicities [Fe/H] = −1 (*top panels*) and [Fe/H] = − 2 (*bottom panels*). Thick solid curves correspond to the averaged profiles from five snapshots taken at equidistant time steps. Thin lines are the profiles from individual snapshots. 1D LTE profiles are computed assuming ξ_t = 1 km s^{-1}. The 1D LTEprofiles would be stronger, if larger ξ_t values were assumed. Note that different spectral lines of Mn I and Mn II are shown in the panels.

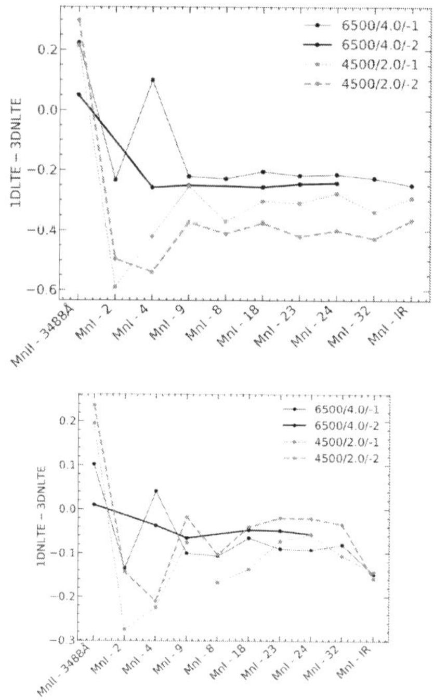

Figure 17. 3D NLTE abundance corrections for selected models.

3D LTE VERSUS 3D NLTE

Figure 18 compares the line profiles computed using 3D LTE, 3D NLTE, and 1D LTE. The latter is convolved with a macro-turbulence of 4 km s^{-1}.

For the [Fe/H] = −1 dwarf model, the 3D LTE line profile of the Mn II line at 3488 Å closely resembles the equivalent 1D LTE profile. The Mn I lines at 4766 and 6021 Å are slightly shallower in 3D LTE compared to 1D LTE, but deeper for the 4823 Å line. On the other hand, all lines of Mn I are much fainter in 3D NLTE compared to 3D LTE, suggesting that a 3D LTE analysis would strongly underestimate the Mn abundances.

This tendency is qualitatively very similar in the dwarf model at [Fe/H] = −2. One notableexception is that the resonance line at 4034 Å is now exceptionally strong in 3D LTE, much stronger than the profiles computed using 1D LTE and 3D NLTE. This behaviour is also known for strong resonance lines of other chemical species. The major implication of this is that 3D LTE modelling is unlikely to solve the excitation imbalance in Mn, which is well-known from 1D LTE studies (e.g. Bonifacio et al. 2009; Sneden et al. 2016). The unfortunate problem is that even in 1D LTE the resonance Mn I lines yield significantly lower abundances, relative to their higher excitation counterparts. The 3D LTE line profiles of the resonance lines are stronger than their 1D LTE counterparts, computed assuming the same Mn abundance, meaning that the 3D LTE abundances of these lines must be reduced to recover the correct equivalent width. For the higher-excitation lines, the differences between 1D and 3D LTE at [Fe/H] = −2 are not large, because the lines are very weak, and form deep in the atmosphere, at $\log \tau_{5000} > -2$, where the temperature inhomogeneities are not pronounced and the average structure of the 1D hydrostatic models closely resembles that of 3D models (see, e.g. Fig. 6).

The behaviour is very similar in the 3D models of giants. 3D LTE line profiles are not too different from 1D LTE at [Fe/H] = −1 despite the modest effect of convection on the line shapes, but they are always stronger compared to 3D NLTE. The differences are exacerbated at low metallicity, [Fe/H] = −2, where 3D LTE calculations greatly overestimate the strength of all Mn I lines compared to 3D NLTE. The UV Mn II line at 3488 Å shows very small NLTE effects in the metal-poor RGB model, in fact, its 3D LTE and 3D NLTE profiles are very similar. On the other hand, the 1D LTE assumption underestimates the strength of the Mn II line, thus overestimating the Mn abundance derived from this feature.

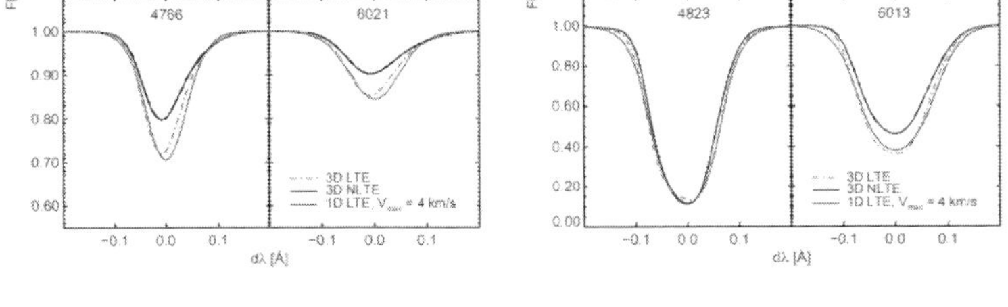

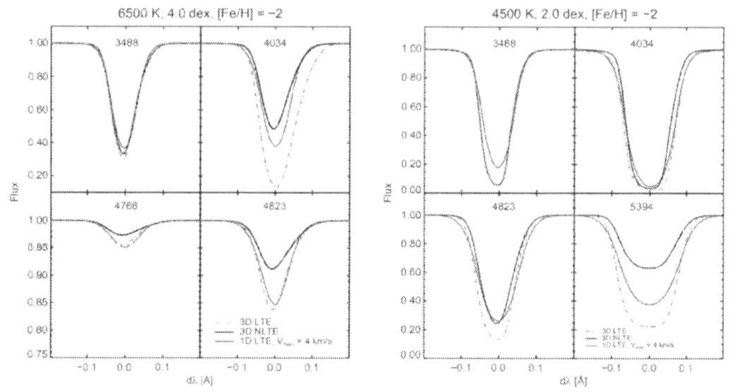

Figure 18. Same as Fig. 16, but thick dotted-dashed brown curves now denote the profiles from 3D LTE calculations for the same five snapshots.

BENCHMARK METAL-POOR STARS

The Mn abundances in the stellar spectra were computed by matching the observed equivalent widths to the grids of 1D LTE and 1D NLTE spectral lines computed using MULTI2.3. The equivalent widths were measured from the UVES-POP spectra of these stars using the SIU code. Unlike in our analysis of the Sun, we have chosen to not employ SIU for abundance measurements in metal-poor stars, as blending is not a problem anymore and a detailed spectrum synthesis is unnecessary. Also MULTI2.3 has an advantage in that it includes background scattering, which is essential for the blue and UV lines. 3D NLTE corrections were computed separately and applied to the 1D LTE abundances derived using the measured EWs.

The results for all three benchmark metal-poor stars are shown in Fig. 19. The error bars correspond to the uncertainties of the EW measurements. We show the abundances derived using the MAFAGS-OS models (left panels) and the MARCS models (right panels) in order to illustrate the impact of the 1D model atmospheres. Overall, the results obtained MAFAGS-OS and MARCS models are in agreement, especially for the model of the metal-poor RGB star HD 122563. The differences can generally be explained by the small differences in the input parameters of the models. In particular, the MAFAGS-OS models assume the following parameters: (a) HD 122563: T_{eff} = 4600 K, log g = 1.60 dex, [Fe/H] = − 2.5 dex; (b) HD 140283: T_{eff} = 5773 K, log g = 3.66 dex, [Fe/H] = − 2.38 dex; and (c) HD 84937: T_{eff} = 6350 K, log g = 4.09 dex, [Fe/H] = −2.15 dex. This is most likely at the origin of the somewhat lower abundances that we get for HD 84937. It is also noteworthy that MARCS models lead to slightly more consistent abundances from Mn I lines for all three reference stars.

Regardless of the choice of 1D models, 1D LTE modelling reveals a significant ionisation imbalance, in that the Mn I lines give systematically lower abundances compared to Mn II lines, confirming the results reported in the literature (Johnson 2002). The difference is most pronounced for the RGB star HD 122563, for which the offset is close to 0.7 dex. We also confirm the Mn I excitation imbalance in all three stars, with strong resonance lines of Mn I producing systematically lower abundances compared to higher excitation lines. For HD 122563, the offset between the lines of multiplet 4 (the 4030–4034 Å triplet) and the lines of other multiplets is ~ 0.2 dex, whereas for the subgiant HD 140283 and turn-off star HD 84937 the offset is slightly larger, of the order ~ 0.3 dex.

The ionisation balance is significantly improved for all three stars in 1D NLTE. In particular, for HD 84937 and HD 140283, the abundances derived from the high-excitation lines are now consistent with the abundances derived from the Mn II lines. For HD 122563, the ionisation balance in 1D NLTE is only partially improved, but there is still a differential effect of − 0.2 dex for the high-excitation lines and − 0.5 dex for the low-excitation lines.

The improvement is most striking in 3D NLTE. The 3D NLTE abundance corrections are very large and positive for both ionisation stages. For HD 122563, this brings Mn I and Mn II lines into agreement. For HD 84937 and HD 140283, the 3D NLTE abundances are higher compared to 1D NLTE, however, the differential effect is not as large. It should be emphasised that stellar parameters of these three stars are well-constrained by independent methods, and the fact that ionisation balance is satisfied in 3D NLTE, without resorting to ad-hoc parameters, like micro-turbulence in 1D, is remarkable. For comparison, to explain the ionisation imbalance in HD 122563 by the error in stellar parameters, its T_{eff} would have to be increased to 4900 K (that would bring the resonance lines up by approximately +0.7 dex to be consistent with Mn II), or alternatively the ξ_t increased to 3.5 km s^{-1} dex that would push the Mn II abundances down by approximately − 0.7 dex. Decreasing the log g might help to increase the abundance from the resonance lines, but a change to log $g < 0.5$ would be necessary. All these changes in stellar parameters can be ruled out, given the robust constraints from asteroseismology and interferometry (Creevey et al. 2012; Karovicova et al. 2018; Creevey et al. 2019). Also micro-turbulence, despite being a free parameter, is relatively well-constrained in the literature (Bergemann et al. 2012; Afşar et al. 2016).

Sneden et al. (2016) recently analysed HD 84937 by employing 1D LTE models. They adopted somewhat different stellar parameters for this metal-poor turn-off star: $T_{eff} = 6300$ K, log $g = 4.0$, [Fe/H] = −2.15, and $\xi_t = 1.5$ km s^{-1}. Their estimates of [Mn/Fe] for this star are approximately − 0.25 for the Mn II lines and approximately −0.3 for the Mn I lines (their Fig. 7). On the other hand, they also tabulate the abundances derived by taking the ratios of the elements derived from the lines in the same ionisation stage (e.g. Mn I to Fe I and Mn II to Fe II). In the latter case, their values are −0.42 for the Mn II lines and −0.46 for the Mn I lines (their Table 5).

Our 1D LTE estimates for HD 84937 are [Mn/Fe] = − 0.24 dex, as derived from the Mn II lines, and [Mn/Fe] = −0.47 dex for the high-excitation Mn I lines. Similar to Sneden et al. (2016), we find that the Mn I triplet lines give the abundances that are ~ 0.2 dex lower compared to the high-excitation lines. Our abundances derived from the Mn I lines may appear to be lower, but this is likely the consequence of the adopted metallicity. Indeed, we adopt [Fe/H] = −2.0 dex in this work. This is consistent with our ⟨3D⟩ NLTE estimate in Bergemann et al. (2012), where we found metallicities in the range from [Fe/H] = −2.04 to −1.95 dex, depending on the choice of the H-scaling factor. Amarsi et al. (2016) find [Fe/H] = −1.96 ± 0.02(stat) ± 0.04(sys) dex from the full 3D NLTE Fe analysis. On the other hand, Sneden et al. (2016) adopt a very low metallicity, [Fe/H] = −2.15 dex, that leads to significantly higher Mn abundances. It is not clear which of the estimates from Sneden et al. (2016) are to be given preference and which of their two methodological approaches is more consistent with our estimate. Contrary to that study, we do not consider lines with EWs $\lesssim$ 3 mÅ in the UVES spectrum of HD 84937 as reliable, such as the lines at 4762, 4765, 4766 Å, and the red triplet at 6013–6021 Å. This, combined with the choice of the atmospheric models, stellar parameters, and methods to determine the abundances, may account for the somewhat different results in our study and in Sneden et al. (2016).

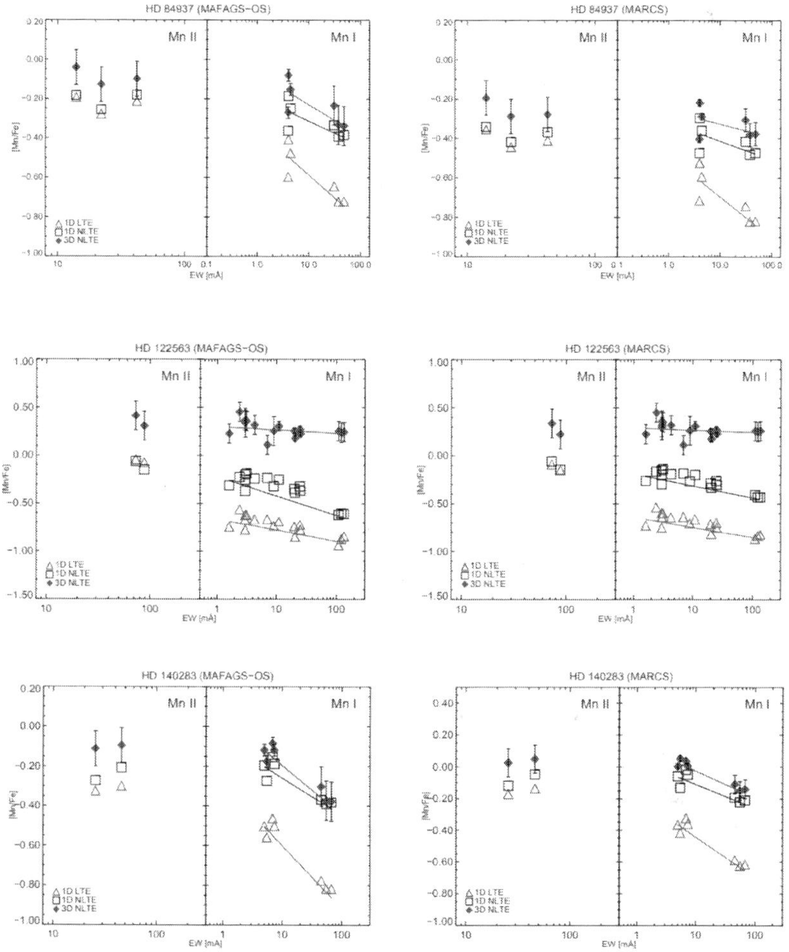

Figure 19. Mn abundances in three metal-poor stars. *Left panels*: the abundances determined using MAFAGS-OS and 3D models. *Right panels*: the abundances determined using the MARCS and 3D models.

CONCLUSIONS

We present the first 3D NLTE analysis of Mn line formation in inhomogeneous model atmospheres. We employed three different statistical equilibrium codes, two of them to compute radiative transfer in 1D geometry (MULTI2.3 and DETAIL), whereas MULTI3D is used for detailed 3D NLTE calculations.

The NLTE model atom was assembled using the new atomic data for different processes. We used the R-matrix method to compute new photo-ionisation cross-sections for 84 terms of Mn I, and employed them in place of hydrogenic cross-sections. We also computed new collision rates for 71 terms of Mn I interacting with H and for the first excited state of Mn II interacting with H. The latter are supplemented with the data for 19 Mn I levels and the ground state of Mn II presented by Belyaev & Voronov (2017c). We also implement the collision rates computed using the scattering-length approximation according to Kaulakys (1985); Barklem (2016). The new photo-ionisation cross-sections and the new rates of inelastic collisions represent the main difference with respect to our earlier results.

We confirm that the NLTE effects in Mn I are driven by over-ionisation. The qualitative behaviour of the departure coefficients and the NLTE abundance corrections is very similar. All Mn I lines in the optical suffer from strong NLTE effects and display positive NLTE abundance corrections, which increase with decreasing metallicity of the model atmosphere. LTE modelling underestimates Mn abundances, by approximately − 0.5 dex in the models of metal-poor red giants, and to a lesser degree in the models of dwarfs. The NLTE abundance corrections are sensitive to the implementation of collision rates that affects the results at the level of ~ 0.1 dex. Departing from Bergemann & Gehren (2008), we find that the new model does not produce large NLTE effects for the resonance lines of Mn I. We attribute this to the use of a tailored, ad-hoc scaling factor on collisions in the previous study that produced qualitatively similar results to our 3D NLTE calculations.

Our 3D modelling of solar Mn lines reveals unique features of line formation in the convective models, known from the previous studies (e.g. Dravins et al. 1981; Dravins 1987; Dravins & Nordlund 1990a,b; Nordlund & Dravins 1990). The spatially-resolved Mn lines show pronounced asymmetries. For the weaker lines, we find a strong anti-correlation between the line core depth and the line shift. The strong lines show a very broad distribution of line shifts, associated with granular motions. The bluer components are rather symmetric, but red components of the spatially-resolved features are very asymmetric and broad, tracing the differences in the line formation in the granules and in the inter-granular lanes.

We perform 1D LTE, 3D LTE, 1D NLTE, and 3D NLTE calculations for a large set of Mn lines, including the Mn II UV lines, but also the commonly-used optical lines, and the IR Mn I lines in the *H*-band. We find large differences between the four scenarios, which can be broadly summarised as follows:

- All lines of Mn I are significantly weaker in 3D NLTE, compared to 1D LTE. As a consequence, 1D LTE underestimates the Mn abundances by approximately −0.6 dex for the blue resonance Mn I lines, but − 0.4 dex for the optical lines in the RGB models with [Fe/H] = − 2. The effect is smaller for the models of dwarfs: the systematic bias incurred by 1D LTE is approximately −0.2 dex and does not change substantially with [Fe/H].
- The commonly used blue lines of Mn II at 3488 and 3497 Å are typically too weak in 1D LTE and 1D NLTE calculations, compared to 3D NLTE modelling. This effect is caused by significant line scattering, and is greatly amplified in the 3D inhomogeneous model atmospheres. Hence, 1D LTE and 1D NLTE typically overestimate Mn abundance derived from the Mn II 3488 and 3497 Å lines. 3D NLTE effects depend strongly on the atmospheric parameters of a star, and, particularly in metal-poor dwarf models with [Fe/H] = −2.0, the 3D NLTE results are close to 1D LTE.
- The impact of convection is modest for the high-excitation Mn I lines with lower excitation potential >2 eV. In particular, the least affected lines are those belonging to multiplets 9, 23, 24, and 32. We recommend using these lines for the abundance analysis in 1D NLTE.
- 3D LTE modelling substantially overestimates the line strength of resonance Mn I lines, compared to 1D LTE and 3D NLTE. As a consequence, the excitation imbalance reported for Mn I lines across a broad metallicity range (Bonifacio et al. 2009) will not be cured, but rather amplified by using 3D LTE.
- All IR *H*-band lines of Mn I suffer from strong systematic bias. 1D LTE underestimates the abundance derived from these lines by 0.2–0.35 dex. On the other hand, the difference between 1D NLTE and 3D NLTE is only − 0.15 dex, and this value does not depend on [Fe/H], T_{eff}, or log g of the model atmosphere. We suggest employing this correction in 1D NLTE studies, in order to account for 3D effects.

We derive a new 3D NLTE solar abundance of Mn, 5.52 ± 0.03 dex, which is consistent with the CI meteoritic abundance, 5.50 ± 0.03 (Lodders

2003). The 1D LTE and 1D NLTE abundances are lower, 5.34 ± 0.04 dex in LTE, and 5.41 ± 0.05 dex in NLTE. Our 3D NLTE value is slightly higher compared to the estimate of 5.42 ± 0.04 dex reported by Scott et al. (2015), however, they did not compute full 3D NLTE, but rather applied the NLTE corrections computed using a 1D model. In particular, in the latter study, we employed an Mn atom with Drawin's collision rates and hydrogenic photo-ionisation, in contrast to the detailed quantum-mechanical data in this work.

Our results for the metal-poor benchmark stars offer a considerably different picture on Mn abundances at low metallicity, contrasting with earlier 1D LTE studies. We find that 3D NLTE abundances in HD 84937, HD 140283, and HD 122563 are significantly higher. For HD 122563, we obtain a perfect ionisation balance in 3D NLTE that cannot be otherwise explained by the atomic data uncertainties and stellar parameters. The 3D NLTE Mn abundances derived from the Mn I lines are ~ 0.5–0.7 (HD 84937, HD 140283)to 1 dex (HD 122563) higher compared to 1D LTE results. Also the 3D NLTE abundance derived from the Mn II lines are ~ 0.15 (HD 84937, HD 140283)to 0.4 dex (HD 122563) larger compared to 1D LTE. Effects of this magnitude are also expected for other stars in this metallicity regime. We strongly recommend applying NLTE, and, if possible, 3D NLTE radiative transfer to the analysis of Mn lines. Alternatively, the Mn I lines of certain multiplets (see above) can be used as a relatively reliable (with a bias of ~ 0.1 dex) diagnostics with hydrostatic models.

In the next paper in the series (Eitner et al., in prep.), we will apply the models developed in this work to a large sample of stars to explore the chemical evolution and nucleosynthesis of Mn in the Galaxy.

ACKNOWLEDGEMENTS

All calculations are run on MPCDF clusters Draco and Cobra. We thank Anish Amarsi for kindly providing us with the H collision rates computed using the Kaulakys recipe. This study is supported by SFB 881 of the DFG (subprojects A05, A10) and by the Research Council of Norway through its Centres of Excellence scheme, project number 262622. S.A.Y. and A.K.B. gratefully acknowledge support from the Ministry for Education and Science (Russian Federation), projects No. 3.5042.2017/6.7, 3.1738.2017/4.6. J.L. has received support through a grant from the Knut och och Alice Wallenberg foundation (2016.0019). Funding for the Stellar Astrophysics Centre is provided by The Danish National Research Foundation (Grant agreement no.: DNRF106). B.P. is partially supported by the CNES, Centre National d'Etudes Spatiales.

APPENDIX A ADDITIONAL FIGURES

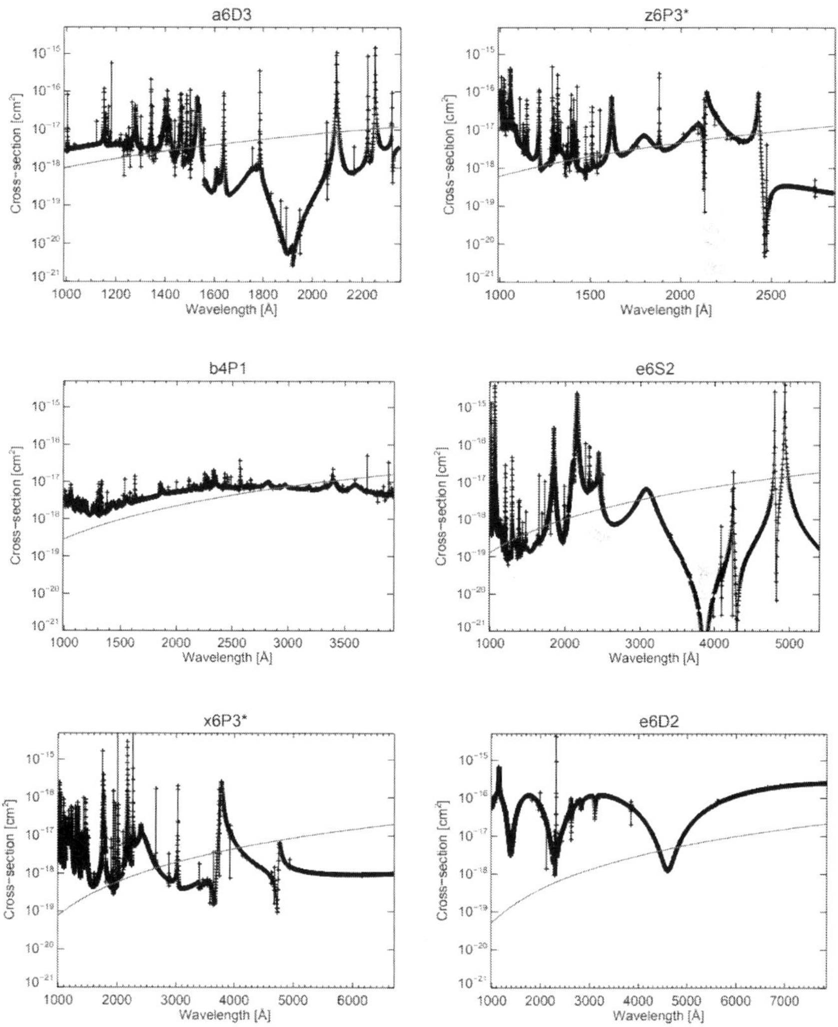

Figure A.1. Photo-ionisation cross-sections for selected Mn I levels with quantum-mechanical data. *Top to bottom*: a6D3, z6P3*, b4P1, e6S2, x6P3*, and e6D2. Red lines illustrate the hydrogenic cross-sections computed using the effective principal quantum number.

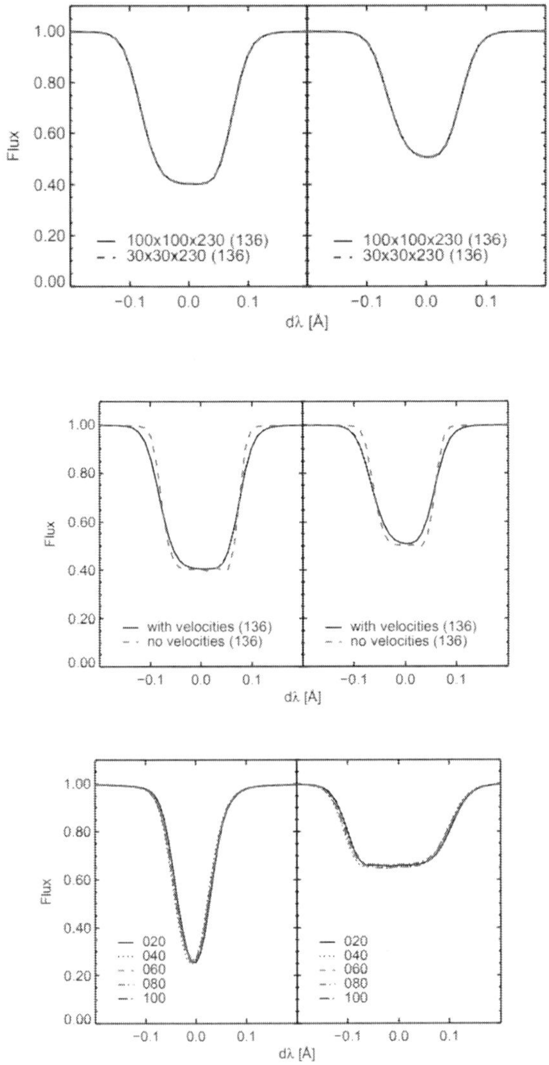

Figure A.2. Comparison of theoretical profiles for Mn I lines at 5394 and 5432 Å in 3D Sun model. Shown are the impact of the resolution of the 3D atmosphere cube (*top panel*) and the radiative transfer calculations with or without velocity field in opacity (*middle panel*). The *bottom panel* **demonstrates the line profiles computed for five different solar snapshots selected at regular time intervals.**

APPENDIX B ADDITIONAL TABLES

Table B.1. LS terms of target Mn II ion included in close-coupling expansion.

Configuration	Term	Energy (Ryd)		Configuration	Term	Energy (Ryd)		Configuration	Term	Energy (Ryd)	
		theory	exp.			theory	exp.			theory	exp.
$3d^5(^6S)4s$	^7S	0.0000	0.0000	$3d^5(^2G)4s$	^3G	0.5430	0.4508	$3d^5(^4D)4p$	$^3D^o$	0.7630	0.6688
$3d^5(^6S)4s$	^5S	0.1048	0.0863	$3d^5(^4F)4s$	^3F	0.5491	0.4544	$3d^5(^4D)4p$	$^3F^o$	0.7678	0.6720
$3d^6$	^5D	0.1560	0.1331	$3d^5(^2F)4s$	^1F	0.5495	0.4290	$3d^5(^2I)4p$	$^3I^o$	0.7847	0.7137
$3d^5(^4G)4s$	^5G	0.2983	0.2513	$3d^5(^2H)4s$	^1H	0.5607	0.4698	$3d^5(^4P)4p$	$^3S^o$	0.7881	0.6735
$3d^6$	^3H	0.3230	0.2793	$3d^5(^2F)4s$	^3F	0.5742	0.4773	$3d^5(^2G)4s$	^1G	0.7973	0.6620
$3d^6$	^3P	0.3358	0.2764	$3d^5(^2G)4s$	^1G	0.5760	0.4798	$3d^5(^4D)4p$	$^3P^o$	0.7976	0.6907
$3d^5(^4P)4s$	^5P	0.3406	0.2726	$3d^6$	^3P	0.5859	0.4853	$3d^5(^2I)4p$	$^1H^o$	0.7978	0.7209
$3d^6$	^3F	0.3436	0.2882	$3d^6$	^3F	0.5929	0.4900	$3d^5(^2I)4p$	$^3H^o$	0.8043	0.7263
$3d^6$	^3G	0.3488	0.3178	$3d^5(^2F)4s$	^1F	0.6204	0.5081	$3d^5(a\ ^2D)4p$	$^1D^o$	0.8247	0.7191
$3d^5(^6S)4p$	$^7P^o$	0.3564	0.3518	$3d^5(^2S)4s$	^3S	0.6371	0.5184	$3d^5(a\ ^2D)4p$	$^3F^o$	0.8259	0.7260
$3d^5(^4D)4s$	^5S	0.3660	0.2992	$3d^5(^4G)4p$	$^5G^o$	0.6402	0.5878	$3d^5(^2I)4p$	$^1I^o$	0.8374	0.7454
$3d^5(^4G)4s$	^3G	0.3828	0.3027	$3d^6$	^1G	0.6486	0.5425	$3d^5(a\ ^2F)4p$	$^1G^o$	0.8449	0.7407
$3d^6$	^3D	0.4022	0.3448	$3d^44s^2$	^5D	0.6552	0.5043	$3d^5(^4F)4p$	$^5G^o$	0.8508	0.7475
$3d^6$	^1I	0.4043	0.3528	$3d^5(^4G)4p$	$^5H^o$	0.6555	0.5986	$3d^5(a\ ^2D)4p$	$^3P^o$	0.8524	0.7409
$3d^5(^4P)4s$	^3P	0.4133	0.3311	$3d^5(^4G)4p$	$^5F^o$	0.6680	0.6077	$3d^5(^2F)4p$	$^3G^o$	0.8528	0.7493
$3d^6$	^1G	0.4133	0.3545	$3d^5(^4P)4p$	$^5D^o$	0.6839	0.6102	$3d^5(^4F)4p$	$^5F^o$	0.8543	0.7473
$3d^5(^6S)4p$	$^5P^o$	0.4331	0.3959	$3d^5(^4P)4p$	$^5S^o$	0.6858	0.6099	3d5.(a ^2D)4p	$^1F^o$	0.8549	0.7401
$3d^5(^2I)4s$	^3I	0.4373	0.3754	$3d^5(^4G)4p$	$^3F^o$	0.6887	0.6181	$3d^5(^2D)4p$	$^3D^o$	0.8591	0.7447
$3d^5(^4D)4s$	^3D	0.4449	0.3628	$3d^5(^4G)4p$	$^3H^o$	0.6925	0.6181	$3d^5(^4F)4p$	$^5D^o$	0.8604	0.7536
$3d^6$	^1S	0.4478	0.3714	$3d^5(^2D)4s$	^3D	0.6977	0.5702	$3d^5(a\ ^2F)4p$	$^3F^o$	0.8650	0.7553
$3d^5(^2I)4s$	^1I	0.4720	0.4038	$3d^5(^4P)4p$	$^5P^o$	0.7002	0.6231	$3d^5(a\ ^2F)4p$	$^3D^o$	0.8673	0.7541
$3d^5(^4F)4s$	^5F	0.4810	0.3972	$3d^5(^4D)4p$	$^5F^o$	0.7185	0.6419	$3d^5(a\ ^2G)4p$	$^3H^o$	0.8674	0.7698
$3d^6$	^1D	0.4812	0.3930	$3d^5(^4P)4p$	$^3P^o$	0.7193	0.6301	$3d^5(^4F)4p$	$^3G^o$	0.8741	0.7646
$3d^5(^2F)4s$	^3F	0.4965	0.4074	$3d^5(^2D)4s$	^1D	0.7243	0.5975	$3d^5(^2H)4p$	$^3I^o$	0.8824	0.7805
$3d^5(^2D)4s$	^3D	0.4990	0.3968	$3d^5(^4G)4p$	$^3G^o$	0.7289	0.6427	$3d^5(a\ ^2D)4p$	$^1P^o$	0.8864	0.7679
$3d^6$	^1F	0.5099	0.4492	$3d^5(^4D)4p$	$^5P^o$	0.7335	0.6501	$3d^5(^2H)4p$	$^3G^o$	0.8890	0.7804
$3d^5(^2D)4s$	^1D	0.5254	0.4274	$3d^5(^4D)4p$	$^5D^o$	0.7412	0.6579	$3d^5(a\ ^2F)4p$	$^1D^o$	0.8973	0.7779
$3d^5(^2H)4s$	^3H	0.5269	0.4408	$3d^5(^4P)4p$	$^3D^o$	0.7431	0.6459	$3d^5(^2I)4p$	$^1H^o$	0.9274	0.7212

Table B.2. AUTOSTRUCTURE configuration expansions for Mn II.

Configuration expansion
$3d^6$, $3d^54s$, $3d^54p$, $3d^54d$, $3d^55s$, $3d^55p$, $3d^44s^2$, $3d^44s4p$, $3d^44s4d$, $3d^44p^2$
$\lambda_{1s} = 1.00000$, $\lambda_{2s} = 1.25438$, $\lambda_{2p} = 1.10633$, $\lambda_{3s} = 1.10274$, $\lambda_{3p} = 1.06738$, $\lambda_{3d} = 1.06340$, $\lambda_{4s} = 1.02844$, $\lambda_{4p} = 1.11707$, $\lambda_{4d} = 1.46349$, $\lambda_{5s} = 1.00000$, $\lambda_{5p} = 1.00000$

Table B.3. Scattering channels correlated to MgH molecular $^5\Sigma^+$ states, asymptotic energies (*J*-average experimental values taken from NIST) with respect to the ground state and electronic bound energies with respect to the ionisation limit $Mn^+(3d^54s\ ^5S) + H$.

j	Scattering channels	Asymptotic energy (eV)	Bound energy (eV)
1	$Mn(3d^54s^2\ a^6S) + H$	0.000000	−8.60854
2	$Mn(3d^64s\ a^6D) + H$	2.145077	−6.46346
3	$Mn(3d^64s\ a^4D) + H$	2.914772	−5.69377
4	$Mn(3d^54s4p\ z^6P^o) + H$	3.073870	−5.53467
5	$Mn(3d^54s^2\ a^4G) + H$	3.134155	−5.47438
6	$Mn(3d^54s^2\ b^4D) + H$	3.768363	−4.84018
7	$Mn(3d^54s4p\ z^4P^o) + H$	3.849324	−4.75921
8	$Mn(3d^54s4p\ y^6P^o) + H$	4.430886	−4.17765
9	$Mn(3d^64s\ b^4G) + H$	4.663140	−3.94540
10	$Mn(3d^54s5s\ e^6S) + H$	5.133436	−3.47510
11	$Mn(3d^54s4d\ e^6D) + H$	5.853606	−2.75493
12	$Mn(3d^54s5p\ w^6P^o) + H$	5.897434	−2.71110
13	$Mn(3d^54s4p\ y^6F^o) + H$	5.972531	−2.63601
14	$Mn(3d^54s5s\ f^6S) + H$	6.126726	−2.48181
15	$Mn(3d^54s5s\ e^4S) + H$	6.148568	−2.45997
16	$Mn(3d^54s4p\ v^6P^o) + H$	6.196287	−2.41225
17	$Mn(3d^54s4p\ z^4H^o) + H$	6.209216	−2.39932
18	$Mn(3d^54s4p\ y^4F^o) + H$	6.243758	−2.36478
19	$Mn(3d^54s4p\ x^6F^o) + H$	6.328611	−2.27993
20	$Mn(3d^54s6s\ g^6S) + H$	6.311380	−2.29716
21	$Mn(3d^54s4p\ x^4P^o) + H$	6.371961	−2.23658
22	$Mn(3d^54s4p\ u^6P^o) + H$	6.466860	−2.14168
23	$Mn(3d^54s5d\ f^6D) + H$	6.537766	−2.07077
24	$Mn(3d^54s4f\ w^6F^o) + H$	6.568422	−2.04012
25	$Mn(3d^54s6p\ t^6P^o) + H$	6.606162	−2.00238
26	$Mn(3d^54s7s\ h^6S) + H$	6.752220	−1.85632
27	$Mn(3d^54s4d\ g^6D) + H$	6.812434	−1.79610
28	$Mn(3d^54s5p\ w^4P^o) + H$	6.868987	−1.73955
29	$Mn(3d^54s5f\ v^6F^o) + H$	6.880172	−1.72837
30	$Mn(3d^54s7p\ r^6P^o) + H$	6.880244	−1.72829
31	$Mn(3d^54s6d\ h^6D) + H$	6.904357	−1.70418
32	$Mn(3d^54s5p\ t^4P^o) + H$	6.934452	−1.67409
33	$Mn(3d^54s5p\ s^6P^o) + H$	6.944067	−1.66447
34	$Mn(3d^65s\ i^6D) + H$	6.989790	−1.61875
35	$Mn(3d^54s6f\ u^6F^o) + H$	7.050626	−1.55791
36	$Mn(3d^54s8p\ q^6P^o) + H$	7.058384	−1.55016
37	$Mn(3d^54s4p\ v^4P^o) + H$	7.116923	−1.49162
38	$Mn(3d^54s7f\ t^6F^o) + H$	7.153545	−1.45499
39	$Mn(3d^54s9p\ p^6P^o) + H$	7.156234	−1.45230
40	$Mn(3d^54s10p\ o^6P^o) + H$	7.220870	−1.38767
41	$Mn(3d^54s11p\ ^6P^o) + H$	7.265446	−1.34309
42	$Mn(3d^54s12p\ ^6P^o) + H$	7.297423	−1.31112
43	$Mn(3d^54s13p\ ^6P^o) + H$	7.321116	−1.28742
44	$Mn(3d^54s14p\ ^6P^o) + H$	7.339133	−1.26941

45	$Mn(3d^5 4s15p\ ^6P^o) + H$	7.351767	−1.25677
46	$Mn(3d^5 4s16p\ ^6P^o) + H$	7.364333	−1.24421
47	$Mn(3d^5 4s17p\ ^6P^o) + H$	7.373379	−1.23516
48	$Mn(3d^5 4s18p\ ^6P^o) + H$	7.380681	−1.22786
49	$Mn(3d^5 4s19p\ ^6P^o) + H$	7.386776	−1.22176
50	$Mn(3d^5 4s20p\ ^6P^o) + H$	7.391888	−1.21665
51	$Mn(3d^5 4s21p\ ^6P^o) + H$	7.396210	−1.21233
52	$Mn(3d^5 4s22p\ ^6P^o) + H$	7.399904	−1.20864
53	$Mn(3d^5 4s23p\ ^6P^o) + H$	7.403078	−1.20546
54	$Mn(3d^5 4s24p\ ^6P^o) + H$	7.405828	−1.20271
55	$Mn(3d^5 4s25p\ ^6P^o) + H$	7.408231	−1.20031
56	$Mn(3d^5 4s26p\ ^6P^o) + H$	7.410336	−1.19820
57	$Mn(3d^5 4s27p\ ^6P^o) + H$	7.412199	−1.19634
58	$Mn(3d^5 4s28p\ ^6P^o) + H$	7.413848	−1.19469
59	$Mn(3d^5 4s29p\ ^6P^o) + H$	7.415318	−1.19322
60	$Mn(3d^5 4s30p\ ^6P^o) + H$	7.416628	−1.19191
61	$Mn(3d^5 4s31p\ ^6P^o) + H$	7.417802	−1.19074
62	$Mn(3d^5 4s32p\ ^6P^o) + H$	7.418876	−1.18966
63	$Mn(3d^5 4s33p\ ^6P^o) + H$	7.419839	−1.18870
64	$Mn(3d^5 4s34p\ ^6P^o) + H$	7.420701	−1.18784
65	$Mn(3d^5 4s35p\ ^6P^o) + H$	7.421504	−1.18704
66	$Mn(3d^5 4s36p\ ^6P^o) + H$	7.422245	−1.18629
67	$Mn(3d^5 4s37p\ ^6P^o) + H$	7.422891	−1.18565
68	$Mn(3d^5 4s38p\ ^6P^o) + H$	7.423498	−1.18504
69	$Mn(3d^5 4s39p\ ^6P^o) + H$	7.424055	−1.18448
70	$Mn(3d^5 4s40p\ ^6P^o) + H$	7.424578	−1.18396
71	$Mn(3d^5 4s41p\ ^6P^o) + H$	7.425049	−1.18349
72	$Mn^+(3d^5 4s\ ^5S) + H^-$	7.854539	−0.754

Table B.4. Scattering channels correlated to MgH^+ molecular $^6\Sigma^+$ states, asymptotic energies (*J*-average experimental values taken from NIST with respect to the ground state and electronic bound energies with respect to the ionisation limit Mn $^{2+}(3d^5\ ^6S)$ + H).

j	Scattering channel	Asymptotic energy (eV)	Bound energy (eV)
1	$Mn^+(3d^5 4s\ a^7S) + H$	0.000000	−15.6400
2	$Mn^+(3d^5 4s\ a^5S) + H$	1.174501	−14.4655
3	$Mn^+(3d^6\ a^5D) + H$	1.808479	−13.8315
4	$Mn^+(3d^5 4s\ a^5G) + H$	3.418429	−12.2216
5	$Mn^+(3d^5 4s\ b^5D) + H$	4.070230	−11.5698
6	$Mn^+(3d^5 4p\ z^7P^o) + H$	4.787521	−10.8525
7	$Mn^+(3d^5 4p\ z^5P^o) + H$	5.386607	−10.2534
8	$Mn^+(3d^5 4p\ z^5H^o) + H$	8.145176	−7.49481
9	$Mn^+(3d^5 4p\ z^5F^o) + H$	8.267108	−7.37288
10	$Mn^+(3d^5 4p\ y^5P^o) + H$	8.476972	−7.16302
11	$Mn^+(3d^5 4p\ y^5F^o) + H$	8.734600	−6.90539
12	$Mn^+(3d^5 4p\ x^5P^o) + H$	8.845407	−6.79458
13	$Mn^+(3d^5 5s\ e^7S) + H$	9.244271	−6.39572
14	$Mn^+(3d^5 5s\ e^5S) + H$	9.469252	−6.17074
15	$Mn^+(3d^5 4d\ e^7D) + H$	9.863794	−5.77620
16	$Mn^+(3d^5 4p\ x^5F^o) + H$	10.16684	−5.47314
17	$Mn^+(3d^5 4d\ e^5D) + H$	10.18466	−5.45532
18	$Mn^+(3d^5 5p\ x^7P^o) + H$	10.66097	−4.97902
19	$Mn^+(3d^5 5p\ v^5P^o) + H$	11.05872	−4.58127
20	$Mn^+(3d^5 6s\ f^7S) + H$	12.11680	−3.52319

21	$Mn^+(3d^56s\ f^5S) + H$	12.20133	−3.43865
22	$Mn^+(3d^54f\ z^7F^o) + H$	12.20296	−3.43702
23	$Mn^+(3d^54f\ e^5F^o) + H$	12.20797	−3.43202
24	$Mn^+(3d^55d\ f^7D) + H$	12.38580	−3.25419
25	$Mn^+(3d^55d\ f^5D) + H$	12.48374	−3.15624
26	$Mn^+(3d^55s\ e^5G) + H$	12.58296	−3.05702
27	$Mn^+(3d^54d\ e^5I) + H$	13.20636	−2.43362
28	$Mn^+(3d^55s\ g^5D) + H$	13.25794	−2.38204
29	$Mn^+(3d^57s\ g^7S) + H$	13.40594	−2.23404
30	$Mn^+(3d^55f\ y^7F^o) + H$	13.44114	−2.19885
31	$Mn^+(3d^55f\ f^5F^o) + H$	13.44490	−2.19509
32	$Mn^+(3d^57s\ g^5S) + H$	13.44576	−2.19423
33	$Mn^+(3d^55g\ e^7G) + H$	13.45920	−2.18079
34	$Mn^+(3d^55g\ f^5G) + H$	13.45925	−2.18074
35	$Mn^+(3d^56d\ g^7D) + H$	13.54475	−2.09524
36	$Mn^+(3d^55p\ s^5F^o) + H$	14.09071	−1.54927
37	$Mn^+(3d^58s\ h^7S) + H$	14.09672	−1.54327
38	$Mn^+(3d^56f\ x^7F^o) + H$	14.11439	−1.52559
39	$Mn^+(3d^56g\ f^7G) + H$	14.12574	−1.51425
40	$Mn^+(3d^56g\ g^5G) + H$	14.12579	−1.51420
41	$Mn^+(3d^56f\ g^5F^o) + H$	14.13750	−1.50249
42	$Mn^+(3d^57d\ h^5D) + H$	14.25117	−1.38882
43	$Mn^{2+}(3d^5\ ^6S) + H^-$	14.88599	−0.754

Table B.5. NLTE abundance corrections for MARCS models with T_{eff} = 4500 K and log g = 1.5 dex.

λ (Å)	[Fe/H]	CH excl.	CH0 excl.	CH/CH0 incl.
4030	0	0.157	0.176	0.153
4033	−1	0.212	0.236	0.222
4034	−2	0.378	0.434	0.426
	−3	0.473	0.483	0.477
	0	0.198	0.173	0.156
4783	−1	0.189	0.214	0.203
4823	−2	0.333	0.366	0.358
	−3	0.452	0.462	0.456
6013	0	0.203	0.184	0.174
6016	−1	0.304	0.288	0.277
6021	−2	0.372	0.370	0.364
	−3	0.382	0.396	0.392
	0	−0.013	−0.014	−0.013
3488	−1	−0.035	−0.035	−0.035
	−2	−0.078	−0.079	−0.079
	−3	0.076	0.074	0.074

Table B.6. NLTE abundance corrections for MARCS models with T_{eff} = 6000 K and log g = 4.0 dex.

λ (Å)	[Fe/H]	CH excl.	CH0 excl.	CH/CH0 incl.
4030	0	0.043	0.078	0.042
4033	−1	−0.043	0.002	−0.033
4034	−2	0.124	0.183	0.113
	−3	0.469	0.529	0.481
	0	0.002	0.038	0.014
4783	−1	0.135	0.186	0.154
4823	−2	0.272	0.308	0.274
	−3	0.315	0.354	0.317
6013	0	0.123	0.109	0.090
6016	−1	0.220	0.226	0.205
6021	−2	0.280	0.341	0.322
	−3	0.288	0.395	0.378
	0	−0.017	−0.017	−0.016
3488	−1	−0.082	−0.080	−0.079
	−2	−0.038	−0.041	−0.040
	−3	0.098	0.093	0.093

Table B.7. NLTE abundance corrections for MARCS models with T_{eff} = 4500 K and log g = 1.5 dex.

λ (Å)	[Fe/H]	Reference atom	No Kaulakys
4030	0	0.138	0.153
4033	−1	0.223	0.222
4034	−2	0.401	0.426
	−3	0.396	0.477
	0	0.132	0.156
4783	−1	0.225	0.203
4823	−2	0.383	0.358
	−3	0.438	0.456
6013	0	0.168	0.174
6016	−1	0.292	0.277
6021	−2	0.416	0.364
	−3	0.442	0.392
	0	−0.012	−0.013
3488	−1	−0.035	−0.035
	−2	−0.078	−0.079
	−3	0.075	0.074

Table B.8. NLTE abundance corrections for MARCS models with T_{eff} = 6000 K and log g = 4.0 dex.

λ (Å)	[Fe/H]	Reference atom	No Kaulakys
4030	0	0.029	0.042
4033	−1	−0.015	−0.033
4034	−2	0.078	0.113
	−3	0.423	0.481
	0	0.013	0.014
4783	−1	0.118	0.154
4823	−2	0.236	0.274
	−3	0.292	0.317
6013	0	0.048	0.090
6016	−1	0.134	0.205
6021	−2	0.269	0.322
	−3	0.344	0.378
	0	−0.016	−0.016
3488	−1	−0.079	−0.079
	−2	−0.040	−0.040
	−3	0.093	0.093

REFERENCES

1. Afşar, M., Sneden, C., Frebel, A., et al. 2016, ApJ, 819, 103
2. Amarsi, A. M., & Asplund, M. 2017, MNRAS, 464, 264
3. Amarsi, A. M., Lind, K., Asplund, M., Barklem, P. S., & Collet, R. 2016, MNRAS, 463, 1518
4. Amarsi, A. M., Barklem, P. S., Asplund, M., Collet, R., & Zatsarinny, O. 2018a, A&A, 616, A89
5. Amarsi, A. M., Nordlander, T., Barklem, P. S., et al. 2018b, A&A, 615, A139
6. Amarsi, A. M., Barklem, P. S., Asplund, M., Collet, R., & Zatsarinny, O. 2018c, A&A, 616, A89
7. Amarsi, A. M., Nissen, P. E., Asplund, M., Lind, K., & Barklem, P. S. 2019, A&A, 622, L4
8. Anstee, S. D., & O'Mara, B. J. 1991, MNRAS, 253, 549
9. Anstee, S. D., & O'Mara, B. J. 1995, MNRAS, 276, 859
10. Asplund, M., Grevesse, N., Sauval, A. J., & Scott, P. 2009, ARA&A, 47, 481
11. Badnell, N. R. 1997, J. Phys. B At. Mol. Phys., 30, 1
12. Bagnulo, S., Jehin, E., Ledoux, C., et al. 2003, The Messenger, 114, 10
13. Barklem, P. S. 2007, A&A, 462, 781
14. Barklem, P. S. 2016, A&ARv, 24, 9
15. Barklem, P. S. 2017, Astrophysics Source Code Library [record ascl:1701.005]
16. Barklem, P. S., Piskunov, N., & O'Mara, B. J. 2000, A&AS, 142, 467
17. Bautista, M. A., & Pradhan, A. K. 1995, J. Phys. B At. Mol. Phys., 28, L173
18. Belyaev, A. K., & Yakovleva, S. A. 2017a, A&A, 608, A33
19. Belyaev, A. K., & Yakovleva, S. A. 2017b, A&A, 606, A147
20. Belyaevc, A. K., & Voronov, Y. V. 2017c, A&A, 606, A106
21. Belyaev, A. K., Yakovleva, S. A., & Barklem, P. S. 2014, A&A, 572, A103
22. Belyaev, A. K., Voronov, Y. V., Yakovleva, S. A., et al. 2017, ApJ, 851, 59
23. Bergemann, M., & Gehren, T. 2007, A&A, 473, 291

24. Bergemann, M., & Gehren, T. 2008, A&A, 492, 823
25. Bergemann, M., Lind, K., Collet, R., Magic, Z., & Asplund, M. 2012, MNRAS, 427, 27
26. Bergemann, M., Kudritzki, R.-P., Gazak, Z., Davies, B., & Plez, B. 2015, ApJ, 804, 113
27. Bergemann, M., Collet, R., Amarsi, A. M., et al. 2017, ApJ, 847, 15
28. Berrington, K. A., Burke, P. G., Butler, K., et al. 1987, J. Phys. B At. Mol. Phys., 20, 6379
29. Bjørgen, J. P., & Leenaarts, J. 2017, A&A, 599, A118
30. Blackwell-Whitehead, R., & Bergemann, M. 2007, A&A, 472, L43
31. Bonifacio, P., Spite, M., Cayrel, R., et al. 2009, A&A, 501, 519
32. Brueckner, K. A. 1971, ApJ, 169, 621
33. Butler, K., & Giddings, J. 1985, Newsletter on Analysis of Astronomical Spectra (London: University of London), 9
34. Caffau, E., Ludwig, H.-G., Steffen, M., et al. 2008, A&A, 488, 1031
35. Caffau, E., Maiorca, E., Bonifacio, P., et al. 2009, A&A, 498, 877
36. Caffau, E., Ludwig, H.-G., Bonifacio, P., et al. 2010, A&A, 514, A92
37. Caffau, E., Ludwig, H.-G., Steffen, M., Freytag, B., & Bonifacio, P. 2011, Sol. Phys., 268, 255
38. Carlson, B. 1963, in Statistical Physics, eds. B. Alder, S. Fernbach, & M. Rotenberg (Berlin: Springer), 1
39. Carlsson, M. 1992, ASP Conf. Ser., 26, 499
40. Collet, R., Magic, Z., & Asplund, M. 2011, J. Phys. Conf. Ser., 328, 012003
41. Creevey, O. L., Thévenin, F., Boyajian, T. S., et al. 2012, A&A, 545, A17
42. Creevey, O., Grundahl, F., Thévenin, F., et al. 2019, A&A, 625, A33
43. Danilovic, S., & Vince, I. 2005, Mem. Soc. Astron. It., 76, 949
44. Danilovic, S., Solanki, S. K., Livingston, W., Krivova, N., & Vince, I. 2016, A&A, 587, A33
45. Den Hartog, E. A., Lawler, J. E., Sobeck, J. S., Sneden, C., & Cowan, J. J. 2011, ApJS, 194, 35
46. Dravins, D. 1987, A&A, 172, 211
47. Dravins, D. 1990, A&A, 228, 218

48. Dravins, D., & Nordlund, A. 1990a, A&A, 228, 184
49. Dravins, D., & Nordlund, A. 1990b, A&A, 228, 203
50. Dravins, D., Lindegren, L., & Nordlund, A. 1981, A&A, 96, 345
51. Drawin, H.-W. 1968, Z. Phys., 211, 404
52. Drawin, H. W. 1969, Z. Phys., 225, 470
53. Grupp, F. 2004a, A&A, 420, 289
54. Grupp, F. 2004b, A&A, 426, 309
55. Gudiksen, B. V., Carlsson, M., Hansteen, V. H., et al. 2011, A&A, 531, A154
56. Gustafsson, B., Edvardsson, B., Eriksson, K., et al. 2008, A&A, 486, 951
57. Holt, R. A., Scholl, T. J., & Rosner, S. D. 1999, MNRAS, 306, 107
58. Johnson, J. A. 2002, ApJS, 139, 219
59. Karovicova, I., White, T. R., Nordlander, T., et al. 2018, MNRAS, 475, L81
60. Kaulakys, B. 1985, J. Phys. B At. Mol. Phys., 18, L167
61. Kaulakys, B. 1991, J. Phys. B At. Mol. Phys., 24, L127
62. Kunasz, P., & Auer, L. H. 1988, J. Quant. Spectr. Rad. Transf., 39, 67
63. Kurucz, R. L., Furenlid, I., Brault, J., & Testerman, L. 1984, Solar Flux Atlas from 296 to 1300 nm (New Mexico: National Solar Observatory)
64. Leenaarts, J.,& Carlsson, M. 2009, in The Second Hinode Science Meeting: Beyond Discovery-Toward Understanding, eds. B. Lites, M. Cheung, T. Magara, J. Mariska, & K. Reeves, ASP Conf. Ser., 415, 87
65. Leenaarts, J., Carlsson, M., & Rouppe van der Voort, L. 2012, ApJ, 749, 136
66. Lind, K., Bergemann, M., & Asplund, M. 2012, MNRAS, 427, 50
67. Lind, K., Amarsi, A. M., Asplund, M., et al. 2017, MNRAS, 468, 4311
68. Livingston, W., Wallace, L., White, O. R., & Giampapa, M. S. 2007, ApJ, 657, 1137
69. Lodders, K. 2003, ApJ, 591, 1220
70. Magic, Z., Collet, R., Asplund, M., et al. 2013, A&A, 557, A26
71. Mishenina, T., Gorbaneva, T., Pignatari, M., Thielemann, F.-K., & Korotin, S. A. 2015, MNRAS, 454, 1585
72. Nordlander, T., & Lind, K. 2017, A&A, 607, A75

73. Nordlund, A., & Dravins, D. 1990, A&A, 228, 155
74. Nordlund, Å., & Galsgaard, K. 1995, A 3D MHD code for parallel computers, Tech. rep., Niels Bohr Institute, University of Copenhagen
75. Nordlund, Å., Stein, R. F., & Asplund, M. 2009, Liv. Rev. Sol. Phys., 6, 2
76. O'Mara, B. J. 1976, MNRAS, 177, 551
77. Osorio, Y., Barklem, P. S., Lind, K., et al. 2015, A&A, 579, A53
78. Plez, B. 2012, Turbospectrum: Code for spectral synthesis, Astrophysics Source Code Library [record ascl:1205.004]
79. Reetz, J. K. 1999, Sauerstoff in kühlen Sternen und die chemische Entwicklung der Galaxis (Munich, Germany: LMU München)
80. Reilman, R. F., & Manson, S. T. 1979, ApJS, 40, 815
81. Reiners, A., Mrotzek, N., Lemke, U., Hinrichs, J., & Reinsch, K. 2016, A&A, 587, A65
82. Rybicki, G. B., & Hummer, D. G. 1991, A&A, 245, 171
83. Rybicki, G. B., & Hummer, D. G. 1992, A&A, 262, 209
84. Scott, P., Asplund, M., Grevesse, N., Bergemann, M., & Sauval, A. J. 2015, A&A, 573, A26
85. Seitenzahl, I. R., Cescutti, G., Röpke, F. K., Ruiter, A. J., & Pakmor, R. 2013, A&A, 559, L5
86. Shetrone, M., Bizyaev, D., Lawler, J. E., et al. 2015, ApJS, 221, 24
87. Sneden, C., Cowan, J. J., Kobayashi, C., et al. 2016, ApJ, 817, 53
88. Steenbock, W., & Holweger, H. 1984, A&A, 130, 319
89. Stein, R. F., & Nordlund, A. 1989, ApJ, 342, L95
90. Strassmeier, K. G., Ilyin, I., & Steffen, M. 2018, A&A, 612, A44
91. Unsöld, A. 1927, Z. Astrophys., 43, 563
92. Unsöld, A. 1955, Physik der Sternatmospharen: MIT Besonderer Berucksichtigung der Sonne (Berlin: Springer)
93. Verner, D. A.,& Yakovlev, D. G. 1995, A&AS, 109, 125
94. Vitas, N., Viticchiè, B., Rutten, R. J., & Vögler, A. 2009, A&A, 499, 301
95. Wise, A. W., Dodson-Robinson, S. E., Bevenour, K., & Provini, A. 2018, AJ, 156, 180
96. Yakovleva, S. A., Barklem, P. S., & Belyaev, A. K. 2017, MNRAS, 473, 3810

SECTION 4:
INTERSTELLAR CHEMISTRY

7

H_2 Formation on Interstellar Dust Grains: The Viewpoints of Theory, Experiments, Models and Observations

Valentine Wakelam[a], Emeric Bron[b,c], Stephanie Cazaux [d,e], Francois Dulieu[f], Cécile Gry[g], Pierre Guillard [h], Emilie Habart[i], Liv Hornekær [j], Sabine Morisset [k], Gunnar Nyman[l], Valerio Pirronello[m], Stephen D. Price[n], Valeska Valdivia[o], Gianfranco Vidali[p], Naoki Watanabe[q]

[a]Laboratoire d'astrophysique de Bordeaux, Univ. Bordeaux, CNRS, B18N, allée Geoffroy Saint-Hilaire, Pessac, 33615, France
[b]Instituto de Ciencias de Materiales de Madrid (CSIC), Madrid, 28049, Spain
[c]LERMA, Obs. de Paris, PSL Research University, CNRS, Sorbonne Universités, UPMC Univ. Paris 06, ENS, F-75005, France
[d]Faculty of Aerospace Engineering, Delft University of Technology, Delft, Netherlands
[e]Leiden Observatory, Leiden University, P.O. Box 9513, Leiden, NL 2300 RA, Netherlands
[f]LERMA, Université de Cergy Pontoise, Sorbonne Universités, UPMC Univ. Paris 6, PSL Research University, Observatoire de Paris, UMR 8112 CNRS, 5 mail Gay Lussac 95000 Cergy Pontoise, France
[g]CNRS, LAM, Laboratoire d'Astrophysique de Marseille, Aix Marseille Univ, Marseille, France
[h]Sorbonne Universités, UPMC Univ. Paris 6 & CNRS, UMR 7095, Institut d'Astrophysique de Paris, 98 bis bd Arago, Paris, 75014, France
[i]Institut d'Astrophysique Spatiale, Univ. Paris-Sud & CNRS, Univ. Paris-Saclay - IAS, bâtiment 121, univ Paris-Sud, Orsay, 91405, France

Citation: V. Wakelam et al. "H2 formation on interstellar dust grains: The viewpoints of theory, experiments, models and observations" Molecular Astrophysics 9 (2017) 1–36 https://doi.org/10.1016/j.molap.2017.11.001

[j]Dept. Physics and Astronomy, Aarhus University, Ny Munkegade 120, Aarhus C, 8000, Denmark
[k]Institut des Sciences Moléculaires d'Orsay, ISMO, CNRS, Université Paris-Sud, Université Paris Saclay, Orsay, F-91405, France
[l]Department of Chemistry and Molecular Biology, University of Gothenburg, Gothenburg, SE 412 96, Sweden
[m]Dipartimento di Fisica e Astronomia, Universitá di Catania, Via S. Sofia 64, Catania, 95123 Sicily, Italy
[n]Chemistry Department, University College London, 20 Gordon Street, London WC1H 0AJ, UK
[o]Laboratoire AIM, Paris-Saclay, CEA/IRFU/DAp - CNRS, Université Paris Diderot, Gif-sur-Yvette Cedex, 91191, France
[p]201 Physics Bldg., Syracuse University, Syracuse, NY, 13244, USA
[q]Institute of Low Temperature Science, Hokkaido University, Sapporo, Hokkaido, 060-0819, Japan

ABSTRACT

Molecular hydrogen is the most abundant molecule in the universe. It is the first one to form and survive photo-dissociation in tenuous environments. Its formation involves catalytic reactions on the surface of interstellar grains. The micro-physics of the formation process has been investigated intensively in the last 20 years, in parallel of new astrophysical observational and modeling progresses. In the perspectives of the probable revolution brought by the future satellite JWST, this article has been written to present what we think we know about the H_2 formation in a variety of interstellar environments.

INTRODUCTION

Molecular hydrogen is, by a few orders of magnitude, the most abundant molecule in the Universe. The first detection of this molecule in the interstellar medium (ISM) was obtained via a rocket flight in 1970 (Carruthers, 1970), three decades after the first interstellar detection of CH, CH+ and CN (see Snow and McCall, 2006, and references therein). Since H_2 is a symmetric and homonuclear diatomic molecule, electric dipole driven ro-vibrational transitions are forbidden and only weak electric-quadrupole transitions are allowed, making its detection extremely difficult in emission[1], unless the emission is from energized environments such as those with, for example, high temperature or high luminosity.

In diffuse molecular clouds, which are regions characterized by molecular fractions $f_{H_2} = 2n_{H_2}/n_H > 0.1$ (n_{H_2} being the number density of H_2 molecules and n_H the total proton number density), the first molecule to form is H_2 (Snow and McCall, 2006). In Photo-Dissociation Regions (PDRs), which are predominantly neutral regions bathed in far ultraviolet light, the emission of H_2 is a tracer of the physical conditions of the cloud e.g. (Hollenbach and Tielens, 1999). In such environments H_2 can be dissociated by ultraviolet radiation, and therefore an efficient route for molecular formation must be present (Jura, 1974, Jura, 1975). Furthermore, molecular hydrogen, either in its neutral or ionized form, controls much of the chemistry in the ISM. In dense clouds where UV penetration is greatly reduced, most of the hydrogen is in molecular form, and most of the Universe's molecular hydrogen resides in these dense clouds.

It has been recognized for a long time that under ISM conditions H_2 cannot be formed efficiently enough in the gas-phase to explain its abundance. Indeed, even in the40′s van de Hulst (1949) had proposed his dirty ice model of dust, where molecules form by combination of atoms on the surface. The link between the presence of H_2 and dust was noted a long time ago (Hollenbach et al., 1971). Indeed, it is now well established that H_2 formation occurs via catalytic reactions on surfaces of interstellar dust grains.

The aim of this paper is to provide the current status of the understanding of the formation of H_2 on interstellar dust grains and identify the important questions that still remain to be answered in this field. This account is motivated by the new observational possibilities that the James Webb Space Telescope (JWST) should provide. In addition, over the last ten years great progress in the modeling of astrophysical media, as well as in the understanding of the associated molecular physics, has been made. Sometimes this progress is directly linked to specific experiments (e.g. Pirronello et al., 1997b; Pirronello, Biham, Liu, Shen, Vidali, 1997, Creighan, Perry, Price, 2006; Watanabe et al., 2010) or calculations and simulations (e.g. Katz, Furman, Biham, Pirronello, Vidali, 1999, Cuppen, Kristensen, Gavardi, 2010, Cazaux, Morisset, Spaans, Allouche, 2011); at other times progress results from an intrinsic change in the treatment of one specific aspect of the formation process, such as stochastic effects (Green et al., 2001; Biham et al., 2001, e.g.). Given the nature of this progress, earlier works in the literature and values used in models can rapidly become outdated, leading to potentially significant differences in the predictions of models if the most up-to-date values are not used. Given this issue, this paper presents, in a unified account, the current viewpoint regarding the formation of molecular hydrogen on interstellar dust grains from the perspective of observers, modelers and chemical physicists. To this end, a group of specialists from these three disciplines gathered for 3 days in Arcachon (France) in June 2016. This paper is the result of this meeting and aims to present the "state of the art" in characterizing and understanding interstellar H_2 formation.

The paper is organized as follows: Section 2 gives an overview of the properties of H_2 and the challenges involved in observing H_2 in space. Section 2 also presents a summary of theoretical and laboratory work aimed at understanding the processes involved in H_2 formation on dust grain analogs (silicates, carbonaceous materials and ices): sticking, diffusion, reaction, desorption and energy the partitioning of the nascent H_2 as it leaves the surface. Several astrophysical models used to study the chemistry of H_2 in various environments are also briefly described in this section. In Section 3, we provide a list of values for the physico-chemical quantities necessary to describe the sticking, diffusion and reactivity of H_2 that can be used in astrochemical models. Section 4 gives an in-depth view of the formation of H_2 in different interstellar environments. A summary and a set of conclusions is then provided at the end of the paper.

STATE OF THE ART

Methods and Tools to Observe H_2 in the Universe

Properties of the H_2 Molecule

Containing two identical hydrogen atoms linked by a covalent bond, the hydrogen molecule is homonuclear and thus highly symmetric. Due to this symmetry, the molecule has no permanent dipole moment and so all the observed ro-vibrational transitions are forbidden electric quadrupole transitions ($\Delta J=\pm 2$) with low values of the spontaneous emission coefficient (A). Since H_2 is the lightest possible molecule it has a low moment of inertia, and hence a large rotational constant (B/k_B=85.25 K), leading to widely spaced energy levels even when the rotational quantum number J is small. In addition, there are no radiative transitions between ortho-H_2 (spins of H nuclei parallel, odd J) and para-H_2 (spins antiparallel, even J), so the ortho and para molecules constitute 2 almost independent states of H_2. The first accessible rotational transition is therefore $J=2\rightarrow 0$, which has an associated energy of $\Delta E/k_B \sim 510$ K. Even so, the lowest excited rotational levels of molecular hydrogen are not easily populated, making H_2 one of the most difficult molecules to detect in space via emission. In absorption, the situation is different since Lyman ($B^1\Sigma_u^1$) and Werner ($C^1\Sigma_u$) electronic bands in the far-UV (from 912 Å to 1155Å) provide a very sensitive tool to detect even very diffuse H_2, down to column densities as low as a few 10^{12} cm−2 – provided a space-born far-UV spectroscopic facility, as well as a UV-bright background source, are available.

Excitation Mechanisms

H_2 may be excited via several mechanisms as described below. The relative population of the H_2 levels depends on the exciting sources and the physical conditions of the gas.

-Inelastic collisions: If the gas density and temperature are high enough, inelastic collisions with H^0, He, H_2 and e− can be the dominant excitation mechanism, at least for the lower rotational energy levels (e.g. Le Bourlot et al., 1999).

-Radiative pumping: In the presence of far-ultraviolet radiation (FUV, $\lambda > 912$ Å), the molecule is radiatively pumped into its electronically excited states. As it decays back into the electronic ground state, it populates the high vibrational levels, and the subsequent cascade to v=0 gives rise to a characteristic distribution of level populations and fluorescent emission in the visible and infrared (IR) regions of the spectrum (e.g. Black, van Dishoeck, 1987, Sternberg, 1989). This excitation mechanism is observed in PDRs where it is the dominant pathway for excitation of ro-vibrational and high rotational levels. UV pumping could also contribute significantly to the excitation of the pure rotational 0-0 S(2)-S(5) lines, since their upper states (v=0, J=4-7) are relatively high in energy and their critical densities are high even at moderate temperatures ($n_{crit} \geq 10^4$ cm−3 for $T \leq 500$ K).

-By formation: The internal energy of the nascent H_2 can also specifically affect the level populations. However, of all the UV photons absorbed by H_2 only 10 to 15% lead to dissociation. Then, for an equilibrium between photo-dissociation and formation, the ratio of the rates of formation pumping and fluorescent pumping of the high-excitation levels in the electronic ground state is $\sim 15/85$. Fluorescent pumping should therefore dominate over formation pumping by a factor five. Thus, unless the level distribution of newly formed H_2 is strongly concentrated toward a small number of high energy levels, the H_2 formation excitation will not specifically affect the H_2 spectrum (see e.g. Black, van Dishoeck, 1987, Le Bourlot, Pineau des

Forets, Roueff, Dalgarno, Gredel, 1995 for models and e.g. Burton et al. 2002 for possible observational signatures). -X-ray photons and cosmic rays: In X-ray emitting environments (such as active galactic nuclei or young stellar objects), X-rays which are capable of penetrating deeply into zones opaque to UV photons, can influence the excitation of H_2 (e.g. Maloney, Hollenbach, Tielens, 1996, Tiné, Lepp, Gredel, Dalgarno, 1997). H_2 excitation may also occur by collisions with secondary electrons generated by cosmic ray ionization.

H_2 Excitation Diagrams: What Information Can We Get?

H_2 excitation diagrams are commonly used to show the population distribution of the molecules across the available levels. Assuming the mid-IR lines are optically thin, the column density of the upper level of each pure rotational transition is measured from the spectral line flux F_v of a given transition according to $N_u = 4\pi F_v/(h\nu A\Omega)$, where h is Planck's constant, ν is the frequency of the transition, A is the Einstein coefficient for the transition, and Ω is the solid angle of the observed region. In Local Thermodynamic Equilibrium (LTE), the upper level column density is related to both the excitation temperature T, and the total column density N_{tot} via, $N_u/g_u = N_{tot}\exp(-E_u/k_B T)/Z(T)$, where E_u is the energy of the upper level of the transition, k_B is the Boltzmann constant and Z(T) is the partition function[2], and $g_u = (2S+1)(2J+1)$ is the degeneracy of the upper level of the transition. In this last expression S is the spin quantum number for a given J transition. The spin value is S=0 for even J (para-H_2), and S=1 for odd J (ortho-H_2). The H_2 excitation diagram is usually presented as a plot of $\log_e(N_u/g_u)$ versus E_u/k (see Fig. 1). For a single excitation temperature the slope of a line fit to these points would be proportional to T^{-1}.

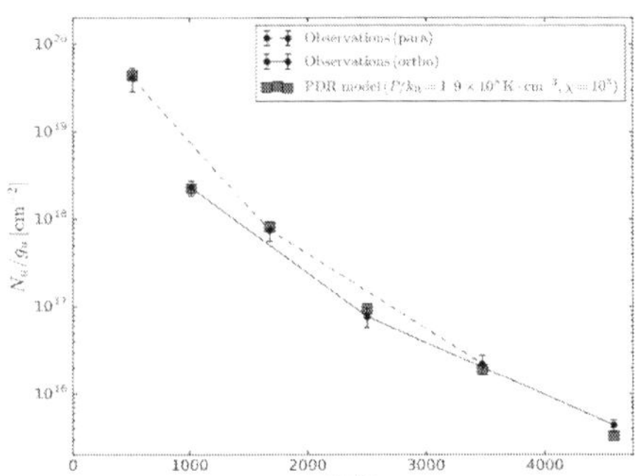

Figure 1. Rotational diagram of H_2 in the NGC7023 NW PDR, comparing the observations (Fuente et al., 1999) with PDR models (with the Meudon PDR

Code, Le Petit et al., 2006). Ortho and para transitions are distinguished to highlight the non-LTE ortho-para ratio.

Two approaches to fit the H_2excitation data referred to above will now be discussed. The first is a traditional method of fitting single or multiple temperature components to the excitation diagrams. This method was first used for the local diffuse ISM detected in absorption in Copernicus spectra of a few bright stars (Spitzer and Cochran, 1973) (see Fig. 2) and has been generalized to many Copernicus (Savage et al., 1977) and FUSE (Rachford, Snow, Destree, Ross, Ferlet, Friedman, Gry, Jenkins, Morton, Savage, Shull, Sonnentrucker, Tumlinson, Vidal-Madjar, Welty, York, 2009, Rachford, Snow, Tumlinson, Shull, Blair, Ferlet, Friedman, Gry, Jenkins, Morton, Savage, Sonnentrucker, Vidal-Madjar, Welty, York, 2002) lines of sight. For translucent lines of sight generally studied in absorption, the excitation diagrams yield mean gas temperatures around 55–80 K from the first excitation levels J=0 to J=2, and excitation temperatures above 180 K from the higher J levels. This method is commonly used to study H_2 studies in other galaxies. It is generally assumed that, for the lower pure rotational H_2 transitions, the ortho and para-H_2 species should be in collisional equilibrium. As shown by Roussel et al. (2007) for H_2 densities $\gtrsim 10^3$ cm^{-3}, most of the lower rotational transitions should be thermalized, and temperatures derived from fits to the ortho- and para-H_2 transitions should yield consistent temperatures. After normalizing by the ortho-para ratio (OPR), significant deviations from LTE would appear as an offset between the odd- and even-J H_2 transitions when plotted on an excitation diagram.

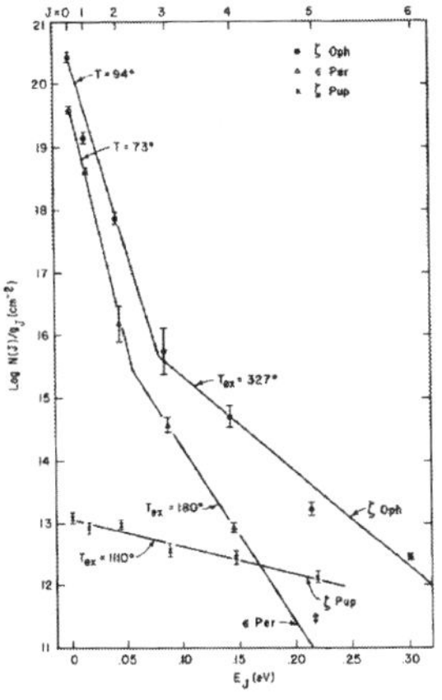

Figure 2. First H_2 excitation diagram published for three stars observed with Copernicus (Spitzer and Cochran, 1973). This diagram illustrates the fact that

two distinct temperatures are needed to fit all J levels, except for low H_2 column densities ($N(H_2) < 10^{15}$ cm^{-2}).

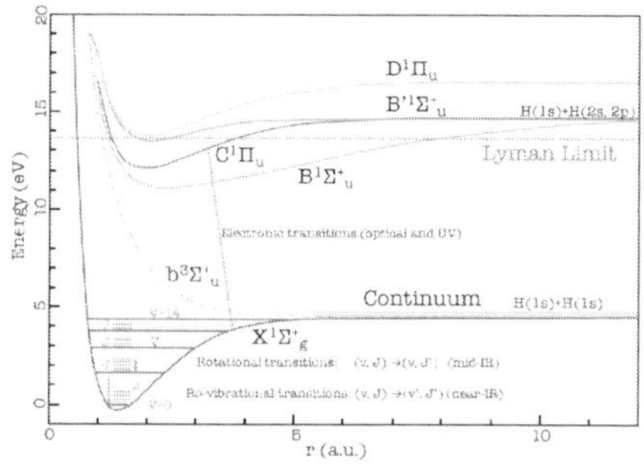

Figure 3. Electronic potentials of H_2 as a function of the separation between both atoms. The subscripts g and u stand for gerade (even) and ungerade (odd) symmetries. Vibrational and rotational levels are indicated schematically for the lowest electronic level. Energy levels are indicated with respect to the ground state (v=0). Levels with vibrational excitation v > 14, in the continuum region, lead to the dissociation of the molecule. Adapted from Le Petit (2002).

A second method of fitting the excitation data is an extension of the first method, by assuming that the molecular gas temperatures can be modeled as a single power-law distribution, again assuming that the gas is in thermal equilibrium (Appleton, Guillard, Togi, Alatalo, Boulanger, Cluver, Pineau des Forêts, Lisenfeld, Ogle, Xu, 2017, Togi, Smith, 2016).

A non-LTE ortho-para ratio appears in excitation diagrams as a systematic offset between the data for ortho and para levels (see Fig. 1). Such non-thermalized OPRs (for the rotational levels) are commonly observed in PDRs (Fleming, France, Lupu, McCandliss, 2010, Fuente, Martín-Pintado, Rodríguez-Fernández, Rodríguez-Franco, de Vicente, Kunze, 1999, Habart, Abergel, Boulanger, Joblin, Verstraete, Compiègne, Pineau Des Forêts, Le Bourlot, 2011, Habart, Boulanger, Verstraete, Pineau des Forêts, Falgarone, Abergel, 2003, Moutou, Verstraete, Sellgren, Leger, 1999), and can either indicate that other conversion mechanisms dominate over reactive collisions (e.g. dust surface conversion, Le Bourlot, 2000, Bron, Le Petit, Le Bourlot, 2016), or that H_2 doesn't have time to thermalize because of fast advection through the dissociation front. Non-LTE OPRs are also commonly seen in the excitation diagrams associated with ro-vibrational transitions, but these ratios are not indicative of the actual OPR of the gas because of preferential pumping effects affecting the populations of the vibrational states (Sternberg and Neufeld, 1999).

H_2 Transitions and Specific Diagnostic Power

The radiative and collision properties of the H_2 molecule make it a diagnostic probe of unique capability in a variety of environments (See Sect. 4 for a discussion of these environments).

-A unique probe of gas excitation: Many competing mechanisms can contribute to the excitation of molecular hydrogen. Since we understand reasonably well the radiative and collisional properties of this molecule we can construct realistic models of the response of H_2 to its surrounding to probe the dominant heating processes taking place in a given environment (e.g., photon heating, shocks, dissipation of turbulence, X-rays).

-A thermometer and mass scale of the warm gas: The lowest rotational transitions of H_2, generally promoted by collisions, provide a wonderful thermometer for the bulk of the gas above $\sim$ 80 K. The rotational excitation of H_2 becomes important only for temperatures $T \gtrsim 80$ K because the J=2 state lies 510 K above the J=0 state (J=3 lies 845 K above J=1). Due to the low A values of the associated optical transitions, any optical depth effects are usually unimportant for these spectroscopic lines. H_2 lines are optically thin up to column densities as high as 10^{23} cm^{-2}. Furthermore, H_2 is the principal constituent of the molecular gas. Thus, these spectral lines provide accurate probes of the mass of the cool/warm ($T \gtrsim 80$ K) gas.

-A unique probe of the warmest photo-dissociation layers subject to photo-evaporation: Self-shielding of H_2 against photo-dissociation is efficient from low H_2 column densities. H_2 can then be present when other molecules, such as CO, would already be photo-dissociated. Thus H_2 can probe, in a unique way, the outer warmest photo-dissociation layers of clouds or proto-planetary disks which are subject to photo-evaporation.

Three types of spectroscopic transitions can be observed for H_2 (shown in Fig. 3, see also Field et al., 1966): the electronic bands in the UV (shown in Fig. 4) , the ro-vibrational transitions in the near-IR (shown inFig. 5), and the pure rotational transitions in the mid-IR (shown in Figs. 5 and 6). Electronic transitions of H_2, in the UV, can be used as probes of two gas regimes: (i) in absorption to probe cold gas ($T \sim$ 50-100 K, such as in the diffuse ISM); (ii) in emission to probe highly excited gas ($T \sim$ few 1000 K such as in outflows or inner disks). UV absorption measurements of vibrationally excited interstellar H_2 can also be used as probes of highly excited gas. H_2 electronic transitions in absorption occur between the ground vibrational level of the ground electronic state ($X^1\Sigma_g^+$) and the vibrational levels of the first ($B^1\Sigma_u^+$) or the second ($C^1\Pi_u$) excited electronic states. In the $X^1\Sigma_g^+$ state the v=1 vibration level is $\approx$ 6000 K above the ground state, so that ro-vibrational excitation (such as that associated with the 2.12 μm line) requires kinetic temperatures T > 1000 K or FUV pumping excitation. The main utility of these near-IR H_2 lines lies in their applicability for probing very small quantities of hot gas. H_2 pure-rotational emission in the mid-IR traces the bulk of the warm gas, generally at temperatures from 100 K up to 1000 K.

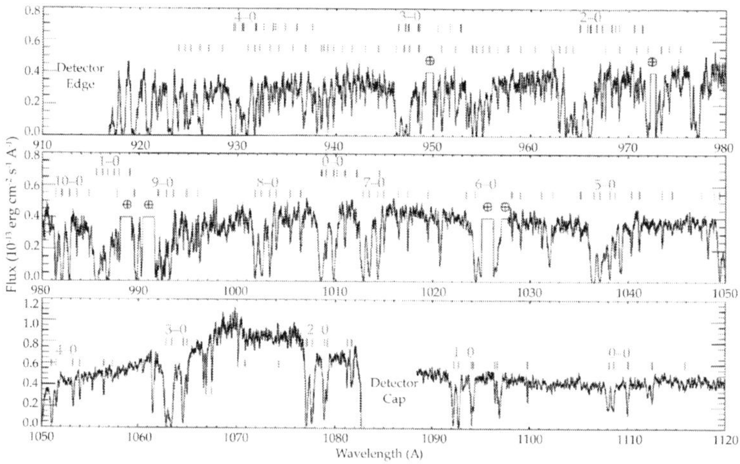

Figure 4. Full FUSE spectrum of ESO 141-G55, which illustrates diffuse Galactic H_2 detected in absorption in the spectrum of a Seyfert galaxy. $N(H_2)=1.910^{19}cm^{-2}$; $N(HI)=3.510^{20}cm^{-2}$. This spectrum has a resolution of R $\approx$ 12,000 and S/N $\approx$ 15 per smoothed (30 km s^{-1}) bin (1040 - 1050 Å) and S/N $\approx$ 25 at 1070 Å. Lower (red) and upper (blue) ticks mark the detected Lyman and Werner lines of H_2, respectively. Bright terrestrial airglow lines superimposed on the interstellar HI lyman absorption lines have been truncated. From Shull et al. (2000). (For interpretation of the references to colour in this figure legend, the reader is referred to the web version of this article.)

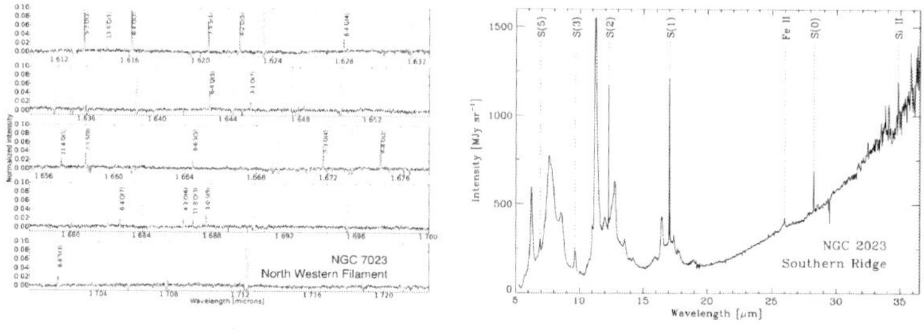

Figure 5. Left pannel: Part of the near-IR spectra from the north western filament

of the reflection nebula NGC 7023, which illustrates H_2 rovibrationnally excited detected in emission in PDRs. This spectrum obtained with the Immersion GRating INfrared Spectrograph (IGRINS), has a resolution of R $\simeq$ 45,000. The spectra show here are into the wavelength ranges 1.610-1.722 μm. The intensity has been normalized by the peak of the 1-0 S(1) line. The dash-red lines display OH airglow emission lines, observed at "off" position120" to the north from the target. Within the1" × 15" slit and the total wavelength coverage 1.45-2.45 μm, 68 H_2 emission lines from rovibrationnally excited H_2 have been detected. From Le et al. (2017). Right pannel: Spitzer mid-IR spectra toward the reflection nebula NGC 2023, which illustrates H_2 rotationally excited detected in emission in PDRs. Full spectral coverage from the four Spitzer/IRS modules (SL2, SL1, SH, and LH with a resolution of R ~ 60-120 and 600), as obtained by averaging 15 pixels that sample the Southern Ridge emission of the nebula. H_2 pure rotational and atomic fine structure emission lines are identified over strong PAH features and dust continuum. From Sheffer et al. (2011). (For interpretation of the references to colour in this figure legend, the reader is referred to the web version of this article.)

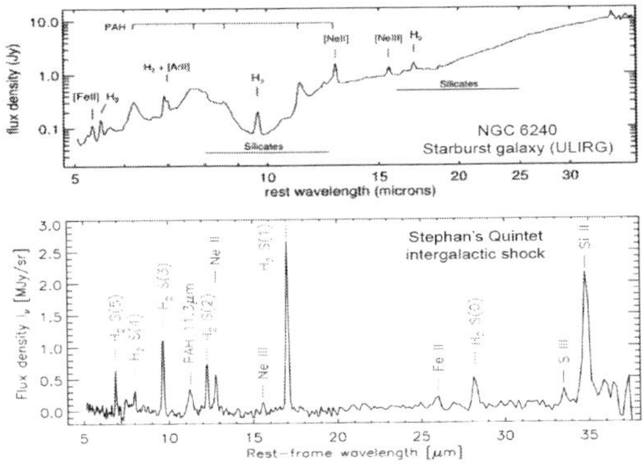

Figure 6. Two examples of extragalactic mid-infrared spectra (taken with the Spitzer IRS) showing prominent rotational lines of H_2. Top panel: spectrum from Armus et al. (2006) of NGC 6240, a nearby (z=0.0245) merging galaxy that has a powerful starburst, a buried (pair of) AGN, and a superwind. Prominent emission lines and absorption bands (horizontal bars) are marked. Bottom: spectrum from Guillard et al. (2010) of the Stephan's Quintet intragroup medium, taken in between two colliding galaxies. The shocked medium is rich in H_2 but with very weak star formation and UV radiation field. Note the strength of the H_2 lines (marked in red) compared to the dust continuum, as opposed to the star-forming galaxy NGC 6240 shown above. (For interpretation of the references to colour in this figure legend, the reader is referred to the web version of this article.)

Observational Challenges: How and Where Can We observe the h_2 Molecule in Space?

As noted above, the electronic transitions of H_2 occur at ultraviolet wavelengths, a region of the spectrum to which the Earth's atmosphere is opaque; hence, observations in this spectral region can only be made from space. The first detection of H_2 beyond the Solar System was made by Carruthers (1970) via UV absorption spectroscopy employing a rocket-borne spectrometer. This discovery was followed by UV observations with the Copernicus space mission that confirmed the presence of the hydrogen molecule in diffuse interstellar clouds (for a first review on this subject see Spitzer and Jenkins, 1975, and references therein). The H_2 absorption lines from the diffuse ISM, i.e. those arising from the low-lying rotational levels of the lowest vibrational level of the ground electronic state (as mentioned in Section 2.1.2), can only be observed in the far UV, below 1130 Å, accessible to Copernicus, ORPHEUS and FUSE (see Fig. 4), as well as HST/COS after 2010 (but only at low resolution with R $\approx$ 2000). Only the excited vibrational levels have lines above 1150 Å, accessible to IUE, and GHRS and STIS on board HST, but they are detected only in a few ISM lines of sight of very high excitation (see Meyer et al., 2001, for an absorption spectrum of vibrationally excited H_2 toward HD 37903, the star responsible for the illumination of NGC 2023). In emission those lines appear only in circumstellar regions like the cited case of the accretion disk observed with HST by France et al. (2010), or in many T Tauri stars observed with IUE, HST or FUSE.

Ro-vibrational and rotational transitions of H_2 are faint because of their quadrupolar origin, as noted above. Moreover, these lines lie, most of the time, on top of a very bright continuum due to the emission of interstellar dust (e.g., see Fig 5); hence, observations at high spectral resolution are needed to disentangle these weak molecular lines. Ground based high-resolution spectrographs (e.g., VLT, Gemini, Subaru) are commonly used to probe the near-IR H_2 ro-vibrational lines. For the case of rotational lines which occur in the mid-IR, the Earth's atmosphere is again, at best, only partially transparent. The mid-IR window with high sky background is a very challenging region of the spectrum in which to perform high sensitivity observations from the ground. Thus, H_2 mid-IR emission studies from the ground (e.g., VISIR, TEXES) are, to date, limited to relatively bright sources (with fluxes typically larger than 1 Jy). Space-based platforms are needed to observe fainter infrared sources in the mid-IR, but here spectral and spatial resolution are limited (e.g. ISO, Spitzer). Finally, most of the interstellar H_2 can lie hidden in cool, shielded regions (e.g. Combes and Pineau des Forêts, 2000) where the molecular excitation could be too low for to H_2 to be seen via emission lines, and the local extinction is too high to allow the detection of lines resulting from UV pumping. In these regions, a way to estimate indirectly the molecular fraction has been proposed by Li and Goldsmith (2003) by measuring the residual atomic hydrogen fraction via HI Narrow Self-Absorption (HINSA) observations.

In the near future, mid-IR instrumentation such as the high-resolution mid-IR spectrograph EXES in the airborne observatory SOFIA, and the mid-IR spectrograph MIRI in the James Webb Space Telescope will greatly increase the critical observational sensitivity, spatial and spectral resolution, and will provide stringent tests of our current understanding of H_2 in space.

In the following section we give a few examples of multi-wavelength observations of H_2 transitions in Galactic and extra-galactic environments.

Galactic Environments

H_2 lines have been detected from Galactic sources as diverse as photo-dissociation regions (PDRs), shocks associated with outflows or supernovae remnants, circumstellar envelopes and proto-planetary disks (PPDs) around young stars, planetary nebulae (PNe), diffuse ISM, and the galactic center. UV absorption lines measured with FUSE, a very sensitive experiment which detected H_2 down to $N(H_2) < 10^{14}$ cm^{-2}, show that H_2 is in fact ubiquitous in our Galaxy (e.g. Shull et al., 2000). UV absorption lines enable us to measure the column densities of H_2 in the rotational J levels of the ground vibrational and electronic states in diffuse and translucent lines of sight, with visual extinctions (A_V) up to about 3 to 5, and to measure the molecular fractions f_{H2}. In the diffuse ISM, with visual extinctions $A_V \leq 1$ mag, the molecular fractions range from 10^{-6} at low HI column density up to $\sim 40\%$ (Gillmon, Shull, Tumlinson, Danforth, 2006, Savage, Bohlin, Drake, Budich, 1977, Spitzer, Jenkins, 1975). In translucent sight lines (which are lines of sight with greater extinction AV=1−5mag), the molecular fraction can be as high as 70% (Rachford, Snow, Destree, Ross, Ferlet, Friedman, Gry, Jenkins, Morton, Savage, Shull, Sonnentrucker, Tumlinson, Vidal-Madjar, Welty, York, 2009, Rachford, Snow, Tumlinson, Shull, Blair, Ferlet, Friedman, Gry, Jenkins, Morton, Savage, Sonnentrucker, Vidal-Madjar, Welty, York, 2002), but is never too close to 1. UV absorption lines also enable us to estimate the H_2 formation rate in the diffuse ISM (e.g. Jura, 1975, Gry,

Boulanger, Nehmé, Pineau des Forêts, Habart, Falgarone, 2002), as well as characterize the gas temperature and excitation (e.g. Spitzer, Cochran, 1973, Gry, Boulanger, Nehmé, Pineau des Forêts, Habart, Falgarone, 2002, Nehmé, Gry, Boulanger, Le Bourlot, Pineau Des Forêts, Falgarone, 2008, Bron, Le Petit, Le Bourlot, 2016). H_2 absorption lines have also been detected towards circumstellar envelopes of young stellar objects, YSOs, (e.g. Martin-Zaidi et al., 2008). The main limitations here are the restricted number of sources against which H_2 can be detected in absorption (which is limited to interstellar gas with $A_V \leq 5$ and intercepting the line of sight toward a bright UV source, thus it prohibits the observation of dense molecular clouds). In practice, UV absorption observations only allow the study of the molecular gas in the Solar Neighborhood or in cirrus and molecular clouds at high Galactic latitude (Gillmon, Shull, 2006, Gillmon, Shull, Tumlinson, Danforth, 2006), in intermediate-velocity clouds in the Galactic halo, in the Magellanic Clouds (Tumlinson et al., 2002), in a few external galaxies, and in objects with a specific geometry. On the other hand, as mentioned above, far-UV H_2 emission lines have unveiled the presence of hot gas in the disks of many T-Tauri stars. In cases where mid-IR CO spectra, or traditional accretion diagnostics, suggest that the inner gas disk has dissipated, far-UV H_2 observations offer unambiguous evidence for the presence of a molecular disk (e.g. France et al., 2012).

Infrared emission of H_2 was first observed via the 2 μm ro-vibrational lines (most notably the 1-0 S(1) line at 2.12 μm) towards Galactic shocks, PDRs, and PNe (e.g., Gatley, Jones, Hyland, Wade, Geballe, Krisciunas, 1986, Pak, Jaffe, Keller, 1996, Burton, Geballe, Brand, Moorhouse, 1990, Lemaire, Field, Maillard, Des Forêts, Falgarone, Pijpers, Gerin, Rostas, 1999, Walmsley, Natta, Oliva, Testi, 2000, Cox, Huggins, Maillard, Habart, Morisset, Bachiller, Forveille, 2002). Luhman et al. (1997) provided the first combined IR/UV picture of an H_2 fluorescence cascade in a single object (a PDR). Deep near-infrared spectra of bright PDRs, taken with high resolution, enable us to detect many emission lines from ro-vibrationally excited molecular hydrogen that arise from transitions out of many upper ro-vibrational levels of the electronic ground state (e.g., Kaplan, Dinerstein, Oh, Mace, Kim, Sokal, Pavel, Lee, Pak, Park, Sok Oh, Jaffe, 2017, Le, Pak, Kaplan, Mace, Lee, Pavel, Jeong, Oh, Lee, Chun, Yuk, Pyo, Hwang, Kim, Park, Sok Oh, Yu, Park, Minh, Jaffe, 2017, see Fig. 5). Since atmospheric transmission in the K band is relatively good, H_2 lines in the near-IR have been searched for in relatively faint objects using large telescopes (such as disks, e.g. Carmona et al., 2011).

ISO and Spitzer provided a fundamental step forward, in that they enabled us to exploit the potential of the H_2 pure rotational lines in the mid-infrared, probing the bulk of the warm gas. This gave access to the H_2 rotational diagram in various sources (PDRs, shocks, YSOs, PNe, SNR, low UV excited clouds, diffuse ISM), as well as its spatial variation in some extended sources (e.g. Neufeld, Melnick, Harwit, 1998, Draine, Bertoldi, 1999, Cesarsky, Cox, Pineau Des Forêts, van Dishoeck, Boulanger, Wright, 1999, Rosenthal, Bertoldi, Drapatz, 2000, van den Ancker, Wesselius, Tielens, 2000, Lefloch, Cernicharo, Cabrit, Noriega-Crespo, Moro-Martin, Cesarsky, 2003, Neufeld, Nisini, Giannini, Melnick, Bergin, Yuan, Maret, Tolls, Güsten, Kaufman, 2009, Maret, Bergin, Neufeld, Green, Watson, Harwit, Kristensen, Melnick, Sonnentrucker, Tolls, Werner, Willacy, Yuan, 2009, Goldsmith, Velusamy, Li, Langer, 2010, Fleming, France, Lupu, McCandliss, 2010, Habart, Walmsley, Verstraete, Cazaux, Maiolino, Cox, Boulanger, Pineau des Forêts, 2005b, Habart, Abergel, Boulanger, Joblin, Verstraete, Compiègne, Pineau Des Forêts, Le Bourlot, 2011, Hewitt, Rho, Andersen, Reach, 2009, Rho, Hewitt, Bieging, Reach, Andersen, Güsten, 2017, Sheffer, Wolfire, Hollenbach, Kaufman, Cordier, 2011, Mata, Ramos-Larios, Guerrero, Nigoche-Netro, Toalá, Fang, Rubio, Kemp, Navarro, Corral, 2016). H_2 data have allowed us to better characterize the shocks (e.g., measure the temperature history, and age) associated with outflows from young stars or supernova remnants and the different possible H_2 excitation mechanisms at different evolutionary stages of young stellar objects and planetary nebulae (e.g., Neufeld, Melnick, Harwit, 1998, Pineau Des Forêts, Flower, Cesarsky, Cox, Pineau Des Forêts, van Dishoeck, Boulanger, Wright, 1999, van den Ancker, Wesselius, Tielens, 2000, Lefloch, Cernicharo, Cabrit, Noriega-Crespo, Moro-Martin, Cesarsky, 2003). This work also showed that H_2 is a major contributor to the cooling of astrophysical media where physical conditions lie in between those of hot molecular gas and cold molecular gas. It has proved possible to estimate the gas temperatures and densities of the transitional layers of the ISM which separate ionized and neutral molecular gas. However, by comparing the observations with the PDR model predictions, the model can account well for the H_2 rotational line intensities and excitation temperature in strongly irradiated PDRs (e.g. Sheffer et al., 2011, see Fig. 5), but underestimates the H_2 excitation temperature and intensity of the excited rotational lines in low/moderate UV irradiated regions (e.g. Goldsmith, Velusamy, Li, Langer, 2010, Habart, Abergel, Boulanger, Joblin, Verstraete, Compiègne, Pineau Des Forêts, Le

Bourlot, 2011). This underlines that our understanding of the warm H_2 gas is incomplete and could suggest additional excitation of H_2 or gas, or out-of-equilibrium processes.

Extra-Galactic Environments

Observations of molecular hydrogen emission from external galaxies started with the detection of near-IR emission coming from hot molecular gas found in photo-dissociation regions or shocks, especially in the central regions hosting AGN or major starbursts (e.g. Wright, Geballe, Graham, 1993, Mouri, 1994, Goldader, Joseph, Doyon, Sanders, 1997). Because they are difficult to observe from the ground, the observations of the pure rotational lines of H_2 from external galaxies started with the Infrared Space Observatory (ISO), and continued with the Spitzer infrared (IR) satellite (e.g. Valentijn, van der Werf, 1999, Lutz, Sturm, Genzel, Spoon, Moorwood, Netzer, Sternberg, 2003, Verma, Charmandaris, Klaas, Lutz, Haas, 2005). In star forming galaxies, rotational H_2 line emission is thought to come from PDRs (Higdon, Armus, Higdon, Soifer, Spoon, 2006, Rigopoulou, Kunze, Lutz, Genzel, Moorwood, 2002, Roussel, Helou, Hollenbach, Draine, Smith, Armus, Schinnerer, Walter, Engelbracht, Thornley, Kennicutt, Calzetti, Dale, Murphy, Bot, 2007). A tight correlation between the H_2 and IR luminosity is inferred for star-forming galaxies and the H_2 to polycyclic aromatic hydrocarbon (PAH) luminosity ratio in this kind of galaxies is within the range of values that are expected from PDR emission, suggesting that UV photons are the main H_2 excitation source. The Active Galactic Nuclei galaxies exhibit a stronger H_2 to PAH ratio than dwarf and star-forming galaxies (Roussel et al., 2007), suggesting that the radiation from the AGN is not sufficient to drive H_2 emission.

More recently Spitzer IRS observations have shown that our census of the warm H_2 gas in galaxies may be severely incomplete, revealing a new class of galaxies (including elliptical galaxies, AGN, galaxy groups, and galaxy clusters) with strongly enhanced H_2 rotational emission lines, while classical star formation indicators (far-infrared continuum emission, ionized gas lines, polycyclic aromatic hydrocarbons, PAHs) are strongly suppressed (Appleton, Xu, Reach, Dopita, Gao, Lu, Popescu, Sulentic, Tuffs, Yun, 2006, Cluver, Appleton, Boulanger, Guillard, Ogle, Duc, Lu, Rasmussen, Reach, Smith, Tuffs, Xu, Yun, 2010, Guillard, Boulanger, Lehnert, Appleton, Pineau des Forêts, 2015a, Guillard, Boulanger, Pineau Des Forêts, Appleton, 2009, Guillard, Ogle, Emonts, Appleton, Morganti,

Tadhunter, Oosterloo, Evans, Evans, 2012b, Ogle, Antonucci, Appleton, Whysong, 2007, Ogle, Boulanger, Guillard, Evans, Antonucci, Appleton, Nesvadba, Leipski, 2010, Ogle, Davies, Appleton, Bertincourt, Seymour, Helou, 2012, Peterson, Appleton, Helou, Guillard, Jarrett, Cluver, Ogle, Struck, Boulanger, 2012). Among the sample of H_2-luminous objects, the Stephan's Quintet is certainly the object where the astrophysical context is clear enough to identify the dominant source of energy that powers the H_2 emission and to associate it with the mechanical energy released in a galactic collision (Appleton, Guillard, Boulanger, Cluver, Ogle, Falgarone, Pineau des Forêts, O'Sullivan, Duc, Gallagher, Gao, Jarrett, Konstantopoulos, Lisenfeld, Lord, Lu, Peterson, Struck, Sturm, Tuffs, Valchanov, van der Werf, Xu, 2013, Appleton, Guillard, Togi, Alatalo, Boulanger, Cluver, Pineau des Forêts, Lisenfeld, Ogle, Xu, 2017, Guillard, Boulanger, Pineau Des Forêts, Appleton, 2009, Guillard, Boulanger, Pineau des Forêts, Falgarone, Gusdorf, Cluver, Appleton, Lisenfeld, Duc, Ogle, Xu, 2012a). In these sources, the luminosity of the H_2 lines cannot be accounted for by UV or X-ray excitation, and their properties suggest that the dissipation of turbulence is the main heating mechanism for the warm H_2 gas. The strong H_2 line emission is a dominant gas cooling channel and traces the turbulent cascade of energy associated with the formation of multiphase gas. The dynamical interaction between gas phases drives a cycle where H_2 gas is formed out of shocked atomic gas (Guillard et al., 2009). In the M82 starburst galaxy, the galactic wind is observed to be loaded with H_2 gas with dust entrained (Beirão et al., 2015). Because the timescale to accelerate molecular material from the galactic disk to tens of kpc in the wind is longer than the dynamical timescale of the outflow, and because the H_2 excitation is consistent with models of slow shocks, it has been argued that the H_2 gas in the outflow is formed by post-shock cooling during the interaction of the wind with the gas in the galactic halo.

Observational Constraints on the H_2 Formation Rate

The determination of the formation rate and abundance of H_2 for a given region of the ISM is crucial, as it controls most of the subsequent development of the chemical complexity, as well as a substantial part of physics of the region, and can allow us to discriminate between the H_2 formation mechanisms that may be operating. Early studies (Gould, Salpeter, 1963, Hollenbach, Salpeter, 1971, Jura, 1975) provided the first estimates of H_2 formation rates in the diffuse ISM, concluding that grain surface chemistry is an unavoidable route for efficient molecular hydrogen formation.

The observationally determined H_2 formation rate coefficient in the diffuse ISM, $R_{H_2} \sim 3 - 4 \times 10^{-17}$ cm^3s^{-1} (Gry et al., 2002), appears to be rather invariant. Nevertheless, Habart et al. (2004) estimate considerably higher H_2 formation rates at high gas temperatures in PDRs. Using as a diagnostic the ratio of the rotational to ro-vibrational lines of H_2, as observed and as predicted by PDR models, they determined H_2 formation rates similar or higher (factor of 5 for moderately excited PDRs) than that measured in diffuse clouds. H_2 appears to form efficiently in PDRs with gas and grains at high temperatures ($T_{gas} \sim 300$ K and $T \sim 30$ K for a grain at thermal equilibrium with the radiation field). However, it must be underlined that these results are based on the assumptions that PDRs are static, in equilibrium, while propagation of the ionization and photo-dissociation fronts will bring fresh H_2 into the zone emitting line radiation. These rate values are thus upper limits. Finally, as we mention later, no observational signatures of the excitation state or ortho-to-para ratio of the newly-formed (nascent) H_2 have been obtained yet. This last point remains observationally challenging.

Experiments and Quantum Calculations

The formation of molecular hydrogen in interstellar space occurs primarily on the surface of dust grains. In diffuse clouds, grains are bare and are usually classified as silicates or carbonaceous materials. In the silicate class we have olivines ($(Mg_x, Fe_{1-x})_2 SiO_4$) and pyroxines ($Mg_x, Fe_{1-x} SiO_3$). Observational evidence shows that these particles are sub-micron in size and are largely in an amorphous form (Draine, 2003, Jones, Fanciullo, Köhler, Verstraete, Guillet, Bocchio, Ysard, 2013). In the carbonaceous class we have sp^3 (nanodiamonds), sp^2 (graphite, PAHs) and mixed sp^3-sp^2 (amorphous carbon) carbon bonded materials. Again, observational evidence shows that these species are nano to sub-micron in size and largely amorphous (Jones et al., 2013). Further information about dust grains and their laboratory analogs can be found in Draine (2011), Henning, 2010a, Henning, 2010b, Jones et al. (2013), Krügel (2007).

Following the first experiments studying H_2 formation on a polycrystalline olivine sample (Pirronello, Biham, Liu, Shen, Vidali, 1997, Pirronello, Liu, Shen, Vidali, 1997), laboratory investigations focused on the various processes involved in H_2 formation on surfaces, in order to find rate limiting process(es) for the specifically chosen conditions (kinetic energy, material, morphology, temperature, etc.). In the next subsection, we list the processes relevant to the formation of molecular hydrogen on dust grains and mention the most common experimental and theoretical techniques that have been used to study such processes. There is a vast literature on these processes, but most of it is for well-characterized systems; that is, for processes occurring on clean and well-characterized surfaces, usually single crystal surfaces (Kolasinski, 2008).

The application of results from the surface science literature to astrophysics environment should be performed very carefully, since the chemical and morphological compositions of actual ISM dust grains are largely unknown, and the processes described below (sticking, binding, diffusion, etc.) depend on many parameters, such as the kinetic energy of the incoming atom and the chemical and morphological composition of the surface. For a recent detailed review of experimental and theoretical work on the formation of H_2 on dust grains, relevant to the ISM, see Vidali (2013). For descriptions of apparatus and measuring methods, see Fraser et al. (2002), Fraser and van Dishoeck (2004), Lemaire et al. (2010), Perry et al. (2002), Vidali et al. (2005), Watanabe and Kouchi (2008) as well as the reviews by Hama and Watanabe (2013), Vidali (2013).

Relevant Processes: Sticking, Binding, Diffusion, Reaction, and Desorption

The formation of molecular hydrogen on a solid surface involves a few key physical processes: trapping, binding, diffusion, reaction and desorption. It helps to make the distinction between weak, long-range interactions with the surface, and strong, localized interactions. In the former, called physisorption, the particle approaching from the gas phase interacts with the surface via long-range van der Waals forces (Bruch et al., 2007a). This binding energy is of the order of tens of meV (Vidali et al., 1991). Experiments studying H atoms interacting with silicate, water ice and graphite surfaces at low kinetic energy and low sample temperatures suggest that physisorption is the class of interaction that is pertinent in these cases, as discussed below.

For the case of a strong interaction between the adsorbate and the surface, chemisorption, a strong bond (~ eV) is formed between the incoming atom and the surface (Kolasinski, 2008). This class of interaction is important, for example, in experiments involving energetic H atoms interacting with the basal plane of graphite (Hornekær, Rauls, Xu, Šljivančanin, Otero, Stensgaard, Lægsgaard, Hammer, Besenbacher, 2006a, Hornekær, Šljivančanin, Xu, Otero, Rauls, Stensgaard, Lægsgaard, Hammer,

Besenbacher, 2006b, Zecho, Güttler, Sha, Lemoine, Jackson, Küppers, 2002b) or of thermal H atoms encountering PAHs (Cazaux, Boschman, Rougeau, Reitsma, Hoekstra, Teillet-Billy, Morisset, Spaans, Schlathölter, 2016, Mennella, Hornekær, Thrower, Accolla, 2012, Rauls, Hornekær, 2008, Snow, Le Page, Keheyan, Bierbaum, 1998, Thrower, Jørgensen, Friis, Baouche, Mennella, Luntz, Andersen, Hammer, Hornekær, 2012).

In the trapping of an atom on a surface, the atom from the gas phase has to lose enough of its kinetic energy to remain confined to the surface. Trapping and sticking are often used interchangeably, but here we will define trapping as the temporary residence of the atom on the surface; that is, the atom is not necessarily fully energetically accommodated. We contrast "trapping" with "sticking" where the particle is fully accommodated (i.e. thermalized) on the surface. The residence time on the surface is then determined by the strength of the bond to the surface and the surface temperature. Sticking has been measured for hydrogen molecules on a variety of surfaces and at different incident kinetic energies (Chaabouni, Bergeron, Baouche, Dulieu, Matar, Congiu, Gavilan, Lemaire, 2012, Matar, Bergeron, Dulieu, Chaabouni, Accolla, Lemaire, 2010). Due to technical difficulties, there are only few experimental measurements of atomic hydrogen sticking on analogs of interstellar dust grains (Pirronello et al., 2000). Computationally, the sticking process has been investigated for H/H_2 on graphite (Cazaux, Morisset, Spaans, Allouche, 2011, Kerwin, Jackson, 2008, Lepetit, Jackson, 2011b, Lepetit, Lemoine, Medina, Jackson, 2011a, Morisset, Ferro, Allouche, 2010, Sha, Jackson, Lemoine, Lepetit, 2005) and on water ice (Al-Halabi, Kleyn, van Dishoeck, Kroes, 2002, Al-Halabi, van Dishoeck, 2007, Buch, 1989, Buch, Czerminski, 1991, Hollenbach, Salpeter, 1970, Masuda, Takahashi, 1997, Masuda, Takahashi, Mukai, 1998, Veeraghattam, Manrodt, Lewis, Stancil, 2014) using methods ranging from molecular dynamics simulations to fully quantum mechanical calculations.

The binding and diffusion of atoms on surfaces are particularly important processes in the formation of molecular hydrogen on dust grains. Binding regulates the time a hydrogen atom or molecule resides on a grain via the relationship

$$\tau = \tau_0 \ \exp \ \frac{E_\mathrm{b}}{k_\mathrm{B} T_\mathrm{dust}} \ ,$$

where τ_0 is related to the fundamental vibrational frequency ν_0 of an atom in a potential energy well describing the motion perpendicular to the surface, E_b

is the binding energy of the atom on the surface, k_B the Boltzmann constant and T_{dust} the temperature of the surface.

For a crystalline surface of a given material, only a handful of binding sites need to be considered. For example, in the case of physisorption of H on the basal plane of graphite, the deepest binding energy site is at the center of the graphitic hexagon (Petucci et al., 2013), while in the case of H chemisorption on graphite, the preferred binding site is on top of a carbon atom (Jeloaica and Sidis, 1999). However, actual dust grains are amorphous, and therefore a range of binding sites, with potentially different binding energies, need to be considered.

For long-distance diffusion, the atoms suffer a larger risk of being trapped at deep potential sites, hence long-distance diffusion tends to be limited by higher activation barriers than short-distance diffusion (see Section 2.2.4). Except for cases where adsorption is activated, the desorption energy of a particle from the surface is the same as the binding energy.

Using the technique of temperature programmed desorption (TPD) in which the temperature of the surface is increased rapidly and the desorbing particles collected, the distribution of desorption energies, and therefore of binding energy sites, has been obtained for many atom/molecule - surface systems (Amiaud, Fillion, Baouche, Dulieu, Momeni, Lemaire, 2006, Amiaud, Fillion, Dulieu, Momeni, Lemaire, 2015, He, Frank, Vidali, 2011). Fig. 8 shows the desorption of D_2 that has been deposited on an amorphous silicate surface. The peaks are rather wide, indicating that D_2 is desorbing from sites with a wide distribution of binding energies. For a comparison with desorption from a single silicate crystal, see (He, Frank, Vidali, 2011, Vidali, Li, 2010). Because of technical limitations in detecting atomic hydrogen, the distribution of the energy sites available on analogs of dust grain surfaces is known primarily for molecular hydrogen and its isotopologues, rather than for atomic hydrogen. In the TPD experiments, information of atomic diffusion is derived from H_2 formation (recombination) rates coupled with H-atom diffusion. Therefore, the obtained activation energy should depend on the initial coverage of H atoms. That is, experiments at high coverage, where only short-distance diffusion is required for recombination, tend to yield lower activation energies for diffusion (see Section 2.2.4).

Given the low surface temperature of dust grains (10-20 K) in some ISM environments where molecules are formed, the motion of an atom that has landed on a grain surface may be restricted. The rate of hydrogen atoms landing on a sub-micron interstellar dust grain is very low. For a sub-micron-sized grain of cross-sectional area $\sigma \sim 10^{-10}$ cm^2, the number of hydrogen atoms landing on the grain per second is given by $\dot{N} = \frac{1}{4} n_H \times v \times \sigma$, where n_H is the H number density in the gas and v its speed. For n=10^4 atoms/cm^3 and v=5×10^4cms^{-1}, $\dot{N} = 10^{-2}$ per second.

For a successful H+H $\rightarrow$ H_2 reaction, either the surface needs to be saturated with H atoms, or an H atom has to sample a large part of the grain before encountering another H atom. Thus, experimental and theoretical works have aimed at characterizing H atom diffusion on morphologically complex surfaces and at finding the conditions required to obtain a high coverage of H atoms on the surface.

For an H atom on a surface, diffusion can proceed via tunneling or thermal hopping. The surface temperature strongly regulates the thermally activated hopping rate, as in the Arrhenius expression $\Gamma = \nu \exp\left[\frac{-E_d}{k_B T_{dust}}\right]$, where E_d is the energy barrier for diffusion and $\nu = \nu_0 \exp\left(\frac{\Delta S}{k_B T}\right)$, with ΔS being the change in entropy between the saddle point and the adsorption site (Tsong, 2005). In practice, the approximation $\nu \sim \nu_0$ (the characteristic vibrational frequency of the particle in the potential well for motion leading to the saddle point) is made.

Tunneling, the rate of which has a very weak dependence on the temperature of the surface, should dominate the diffusion rate at sufficiently low temperatures. However, the tunneling rate depends strongly on the width and height of the energy barrier, as in

$$k_d = \nu_0 \exp \frac{-2a(2m_H E_d)^{\frac{1}{2}}}{\hbar},$$

assuming a rectangular potential, where a and E_d are the width and height of the barrier, and m_H is the mass of the hydrogen atom. In an amorphous solid, there can be a wide distribution of both barrier widths and heights, leading to long residence times of the H atoms in the deep wells of the binding energy landscape. This situation is not unlike the trapping of electrons in amorphous silicon.

The diffusion of a single atom, by random hopping or tunneling, across the surface is called single particle (or tracer) diffusion, as to distinguish it from the concentration driven diffusion (Tsong, 2005) . The direct measurement of tracer diffusion is obtained using visual methods such as field ion microscopy and scanning tunneling microscopy, or with quasi-elastic particle scattering (Miret-Artés and Pollak, 2005). Due to technical restrictions, and because these techniques are applied to conductive surfaces (typically metals and graphite), they find few applications in probing astrochemically relevant surfaces. Indirect methods, such as the ones used in experiments to study HD and H_2 formation (see next Section for details), can provide estimates of the average mobility of H atoms in astrochemically relevant situations.

The activation energy for diffusion is empirically related to the binding energy. From experiments probing atoms weakly adsorbed on well-ordered surfaces, we find $E_d \sim \alpha E_b$ and α=0.3, where E_d and E_b are the energy barrier for thermally activated diffusion and the binding energy, respectively (Bruch et al., 2007b). Analysis of experiments studying H atoms on dust grain analogs give a wider range of values of α, from $\alpha \sim 0.3$ up to $\alpha \sim 0.8$ (Katz, Furman, Biham, Pirronello, Vidali, 1999, Perets, Lederhendler, Biham, Vidali, Li, Swords, Congiu, Roser, Manico, Brucato, Pirronello, 2007, Perets, Biham, Manicó, Pirronello, Roser, Swords, Vidali, 2005), depending largely on the morphology of the surface. For the case of H atoms strongly localized on the surface, as in the case of H chemisorbed on graphite or PAHs, the calculated thermal activation energy to migrate out of the adsorption site can be comparable to the energy for desorption.

Experiments have shown that on the surfaces of a wide variety of solids the formation of H_2 occurs via three main mechanisms: Langmuir-Hinshelwood, Eley-Rideal, and "hot atom" (Kolasinski, 2008). In the Langmuir-Hinshelwood mechanism, atoms from the gas phase first become accommodated on the surface and then, via diffusion, they encounter each other and react. The resulting molecule might or might not leave the surface, depending on how the energy gained in the reaction is partitioned.

In the Eley-Rideal reaction, an incoming atom interacts directly with a partner on the surface; the incident atom is not accommodated on the surface. The resulting molecule is likely to leave the surface retaining much of the energy gained in the reaction. Here, the cross-section for the reaction is of the order of atomic dimensions. The "hot atom" mechanism is similar to the Eley-Rideal one, but here the atom first lands on the surface, without becoming fully accommodated, and proceeds to sample the surface at supra-thermal energy until it finds and reacts with a partner species.

Examples of detections of Eley-Rideal or abstraction reactions relevant to ISM environments are found in works involving experiments of H atoms interacting with PAHs, hydrogenated amorphous carbon and graphite (Cazaux,

Boschman, Rougeau, Reitsma, Hoekstra, Teillet-Billy, Morisset, Spaans, Schlathölter, 2016, Hornekær, Rauls, Xu, Šljivančanin, Otero, Stensgaard, Lægsgaard, Hammer, Besenbacher, 2006a, Hornekær, Šljivančanin, Xu, Otero, Rauls, Stensgaard, Lægsgaard, Hammer, Besenbacher, 2006b, Mennella, 2008, Mennella, Hornekær, Thrower, Accolla, 2012, Snow, Le Page, Keheyan, Bierbaum, 1998, Thrower, Jørgensen, Friis, Baouche, Mennella, Luntz, Andersen, Hammer, Hornekær, 2012, Zecho, Güttler, Sha, Lemoine, Jackson, Küppers, 2002b). The various processes involved in the formation of H_2 are summarized in Fig. 7.

Theoretical calculations, confirmed by experimental results studying H_2 formation on graphite (Islam, Cecchi-Pestellini, Viti, Casu, 2010, Islam, Latimer, Price, 2007, Latimer, Islam, Price, 2008), show that ro-vibrational excitation of H_2 leaving the surface upon formation is concentrated around high values of the vibrational quantum number (3–4) and low values of the rotational quantum number. There have been a few attempts to detect a signature of such H_2 formation, via observations of transitions of H_2 in appropriate regions of the ISM, but with no success to date (Thi, van Dishoeck, Bell, Viti, Black, 2009, Tiné, Williams, Clary, Farebrother, Fisher, Meijer, Rawlings, Davis, 2003).

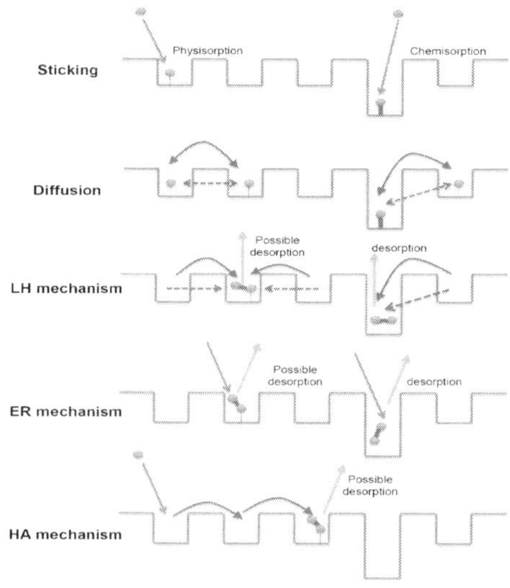

Figure 7. Schematic illustrating key surface processes involved in the formation of H_2.

The ortho (odd rotational quantum number J) to para (even J) ratio of H_2 can yield information concerning the thermal history of the associated cloud, as well as concerning the conditions associated with H_2 formation on the grains of the cloud (Le Bourlot, 2000, Wilgenbus, Cabrit, Pineau des Forêts, Flower, 2000), although presence of other H_2 molecules co-adsorbed on the grain can hinder the release of excited molecules (Congiu et al., 2009).

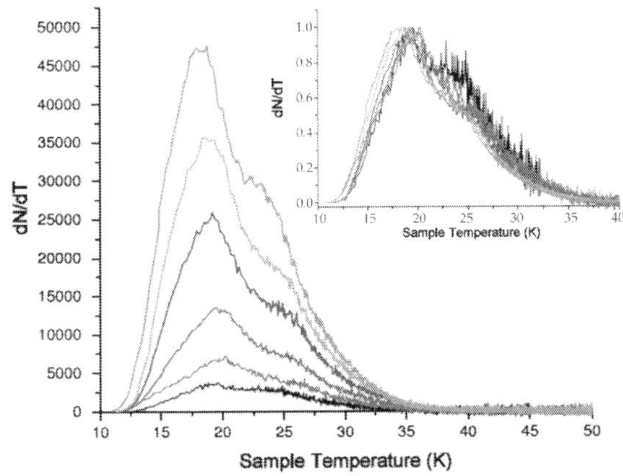

Figure 8. TPD traces of D_2 after irradiation of an amorphous silicate sample at 12 K for different lengths of time (2, 4, 8, 16 and 32 min). Inset: normalized traces; for clarity, the trace of 2 min irradiation (black line) is not shown for T > 32 K. From Vidali and Li (2010) - GV.

In the laboratory, measurements of the ro-vibrational state of nascent (freshly-formed) molecules on surfaces have been used to determine whether the ortho to para ratio of such newly-synthesized molecules would be different from the statistical ratio (Hama and Watanabe, 2013); for temperatures greater that about 200 K, the statistical ratio is 3. Measurements show that for H_2 formed on surfaces of amorphous solid water at low temperature the nascent molecules possess the appropriate statistical value of the OPR (Gavilan, Vidali, Lemaire, Chehrouri, Dulieu, Fillion, Congiu, Chaabouni, 2012, Watanabe, Kimura, Kouchi, Chigai, Hama, Pirronello, 2010).

Experimental and theoretical results of H_2 formation need to be appropriately adapted to the conditions present in the relevant environments of the ISM; this adaption can be performed by using robust experimental data in computer simulations of processes occurring in ISM environments. Specifically, experiments are performed at much higher H atom fluxes than are present in the ISM, and probe chiefly the kinematics of the reactions. Hence, simulations need to translate this laboratory information to reveal its impact under the conditions pertaining in the ISM: such as low fluxes of H atoms impinging on grains and steady-state conditions. For example, Katz et al. (1999) used rate equations to fit experimental data (temperature programmed desorption traces) of H_2 formation on polycrystalline and amorphous silicates and on amorphous carbon. They then used the results to predict the formation of H_2under the conditions of the ISM. Cazaux et al. (2005) considered both physisorption and chemisorption interactions in their rate equations and fitted the same data as in (Katz et al., 1999) but with more parameters. They found that only a physisorption interaction between H atoms and the surface was necessary to explain the data. Cuppen and Herbst (2005) instead used continuous-time, random-walk Monte Carlo code to study the effect of surface roughness on the formation of molecular hydrogen using a model square lattice. These investigators found that roughness increased the grain temperature range over which H_2formation is efficient. Stochastic effects, arising from the fact that the actual size distribution of dust grains in the ISM is skewed to small grains, have also been taken into accounts in models by Biham and Lipshtat (2002), Bron et al. (2014), Cazaux and Spaans (2009), Cuppen et al. (2006), Le Bourlot et al. (2012).

Silicate Surfaces

The ubiquitous observation of molecular hydrogen in widely varying interstellar environments poses significant challenges in explaining its formation. In diffuse clouds, dust grains are largely bare and the formation of H_2 occurs on silicates and amorphous/graphitic carbon (graphite, amorphous carbon, and PAHs). The first experiments studying H_2 formation on dust grain analogs involved a polycrystalline silicate (Pirronello et al., 1997b). In these experiments, the aim was to measure the efficiency of H_2 formation under conditions which simulated the ISM.

Quantifying the formation of H_2 is particularly challenging. For example, in a typical experiment, molecular hydrogen is dissociated and the resulting atoms directed onto a sample surface, see Fig. 9. Although it is possible to dissociate up to nearly 90% of the H_2 molecules in such an H atom source, the remaining un-dissociated species will contaminate the sample, making it impossible to determine if molecules on the surface came from the source or are the product of atomic recombination on the surface. This limitation was lifted in the work of Pirronello et al. (1997b) by using two beamlines directed at the sample, one dosing H atoms and the other dosing for D atoms. In this situation, under the associated experimental conditions, the formation of HD can only occur on the surface of the sample. Another technical limitation of this class of experiments is associated with contamination. Even in a state-of-the-art ultra-high vacuum apparatus (base pressure 10^{-10} torr), the adsorption of background gas (mostly hydrogen) on the surface of the sample limits the sensitivity and duration of experiments studying H_2 formation. Using highly collimated beams, as shown in Fig. 9, allows experimental operating pressures approaching 10^{-10} Torr.

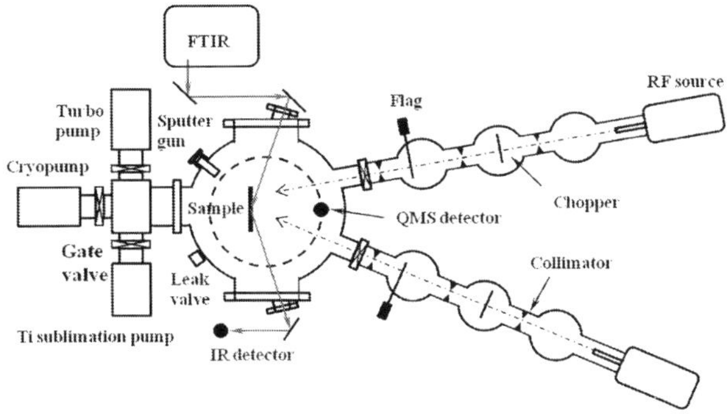

Figure 9. Apparatus at Syracuse University used to study H_2 formation on silicate surfaces. Two independent beam lines converge on a sample mounted on a rotatable flange. A quadrupole mass spectrometer mounted on a rotatable platform can quantify and identify both the products from the surface and the species in the incident beams.

Another technical limitation associated with laboratory experiments is the fact that fluxes of H atoms employed are, for practical reasons, orders of magnitude higher than in the ISM. This mismatch of fluxes cannot be solved directly. However, with careful experimental design and the use of simulations to reproduce the conditions in the ISM, as performed by Katz et al. (1999) and Biham and Lipshtat (2002), the efficiency of H_2 formation in the ISM can be obtained from experimental kinetic data. Further work by the Biham's group studied the effect of particle size (Lipshtat and Biham, 2005) and porosity (Perets and Biham, 2006) on H_2 formation in interstellar environments. Diffusion of H atoms was included in the simulations of the formation kinetics (Katz et al., 1999), revealing that the ratio of the energy barrier for H atom diffusion to the binding energy is considerably higher than the typically assumed value ($\sim$ 0.3) for physisorbed atoms on single crystal surfaces (Bruch et al., 2007b). Because the H atom binding energy could not be well constrained, only an upper limit of 0.7 could be obtained for this ratio in the fitting of Katz et al. (1999). The reason for this unusually high diffusion barrier is likely to be the complex morphology of polycrystalline and amorphous silicates.

Subsequent experiments showed that the efficiency of atomic recombination to form H_2 is dependent on the morphology of the surface, the efficiency being larger on amorphous silicates than on crystalline or polycrystalline silicates. The simulations by Katz et al. (1999) of H_2 formation on polycrystalline silicate and on amorphous carbon, and by Perets et al. (2007) on amorphous silicate, showed that, under the conditions present in diffuse interstellar clouds, the efficiency of atomic recombination to form H_2 is high over only a narrow range of temperatures. In the experiments modeled by Katz et al. (1999) and by Perets et al. (2007), most of the molecules that were formed remained on the surface and only a small minority left the surface following their formation. Both the molecules coming off the surface upon formation, and the ones that remained on it, were detected using a quadrupole mass spectrometer whose sensitivity is inversely proportional to the speed of the particles.

In an experiment studying H_2 formation, in which it was possible to detect molecules leaving the surface in superthermal ro-vibrational states, Lemaire et al. (2010) found that some nascent molecules were formed on, and ejected from, the surface at a temperature as high as 70K. Although the kinetic energy of H_2 was not measured, experimental conditions and analogy with the experiments studying H_2 or HD formation on graphite, by Baouche et al. (2006), Islam et al. (2010), 2007); Latimer et al. (2008), suggest that the kinetic energy was of the order of an eV. Thus, it is possible that earlier experiments underestimated the proportion of nascent molecules immediately leaving the surface following their formation.

The influence of the morphology of the silicate surface on the kinetics of molecular hydrogen formation was studied by He et al. (2011). This work determined the distribution of the binding energy of the species on the surface using TPD. Specifically, the shape of TPD traces, which record the desorption rate (the differential of the desorption yield), as a function of temperature, can be fitted using a distribution of binding energies.

As mentioned before, it is difficult to detect atomic hydrogen in TPD experiments, especially when the surface coverage is less than one layer. However, experiments studying D_2 show a dramatic difference between TPD spectra from a single crystal of forsterite (Mg_2SiO_4) and from an amorphous silicate. The derived binding energy distribution for the amorphous sample is much wider, and centered at a much higher desorption energy, than the one for the single crystal. Experiments studying H+D → HD formation are consistent with the D_2 experiments: they show that HD formation on a silicate crystal occurs at lower temperatures than on amorphous silicate, suggesting that thermally activated diffusion plays an important role in the reaction (He et al., 2011).

Compared with the significant number of theoretical investigations of the interaction of H atoms with carbonaceous surfaces, there are few reports of theoretical investigations of H atoms interacting with silicate surfaces. Such calculations have been performed involving the stable (010) surface of Mg_2SiO_4 as well as the (001) and (110) surfaces which have higher surface energies. Goumans et al. (2009) used an embedded cluster approach where part of the surface was described by DFT and part by analytic potentials, while Garcia-Gil et al. (2013), Navarro-Ruiz et al. (2014), and Navarro-Ruiz et al. (2015) employed Density Functional Theory. H_2 is formed more readily on the (010) surface due to the fact that on the other surfaces H atoms are more strongly adsorbed and the barriers to diffusion are thus higher. Mg atoms are the most favorable sites for physisorption, while chemisorption is on the oxygen site. However, the physisorption energy of H on crystalline silicates, and the energy barriers to diffusion, are calculated to be considerably higher than the values given by experiments on amorphous silicates. Goumans et al. (2009) invoked hydroxilation of the surfaces used in experiments to reconcile the discrepancy between these theoretical and experimental values, while Navarro-Ruiz et al. (2014) pointed out the challenge for computational studies in correctly taking into account the large dispersion energies in weak interactions and open shell systems when using DFT. The calculations show that H_2 formation via the Langmuir-Hinshelwood mechanism is favored on the (010) surface (Navarro-Ruiz, Martínez-González, Sodupe, Ugliengo, Rimola, 2015, Navarro-Ruiz, Sodupe, Ugliengo, Rimola, 2014).

Carbonaceous Surfaces

H_2 Formation on Graphite

The interaction of atomic hydrogen with graphite surfaces, and the pathways to molecular hydrogen formation on these surfaces, have been studied in considerable detail both theoretically and experimentally. Hydrogen atoms can both physisorb and chemisorb on graphite:

Physisorbed H atoms are weakly bound in a shallow potential well with a depth of 43.3 ± 0.5 meV resulting in a ground state binding energy of 31.6 +- 0.2 meV, as determined by scattering experiments (Ghio et al., 1980). The sticking coefficient has been estimated theoretically to be 5–10% for H atoms with translational energies ranging from 0 to 50 meV (Lepetit, Lemoine, Medina, Jackson, 2011a, Medina, Jackson, 2008).

Once in the physisorbed state, the H atom is highly mobile on the surface with a diffusion barrier predicted by theory to be only 4 meV (Bonfanti et al., 2007). This high mobility allows H atoms to scan a large area of the surface and recombine with any other H atoms they encounter; atoms which could be physisorbed, chemisorbed or incident from the gas phase. This reactivity can occur both at low temperatures, where the atom's high surface mobility is assisted by tunneling, and at higher temperatures, where the high thermally induced mobility may allow a significant surface area of a grain to be explored by the atom, even if the atom's lifetime in the physisorbed state is extremely short (Creighan, Perry, Price, 2006, Cuppen, Hornekær, 2008).

Chemisorption of H atoms on the graphite surface is more complex than physisorption. A single H atom can chemisorb above a carbon atom in the graphite surface with a binding energy of 0.7-1.0 eV (Casolo, Løvvik, Martinazzo, Tantardini, 2009b, Hornekær, Šljivančanin, Xu, Otero, Rauls, Stensgaard, Lægsgaard, Hammer, Besenbacher, 2006b, Ivanovskaya, Zobelli, Teillet-Billy, Rougeau, Sidis, Briddon, 2010, Sha, Jackson, Lemoine, 2002). However, this binding requires the associated carbon atom to pucker up, out of the surface, by 0.1 Å. Thus, the chemisorption is associated with a large energy barrier of 0.15-0.2 eV (Hornekær, Šljivančanin, Xu, Otero, Rauls, Stensgaard, Lægsgaard, Hammer, Besenbacher, 2006b, Jeloaica, Sidis, 1999). As a consequence of this barrier, the sticking probability for H atoms into the chemisorbed state is highly energy dependent and has mainly been estimated theoretically (Bonfanti, Jackson, Hughues, Burghardt, Martinazzo, 2015, Karlicky, Lepetit, Lemoine, 2014, Kerwin, Jackson, 2008, Kerwin, Sha, Jackson, 2006, Morisset, Allouche, 2008, Morisset, Ferro, Allouche, 2010, Sha, Jackson, Lemoine, Lepetit, 2005).

The diffusion of isolated chemisorbed H atoms on graphite is highly disfavored with calculated barriers of 0.8-1.1 eV (Ferro, Marinelli, Allouche, 2003, Hornekær, Rauls, Xu, Šljivančanin, Otero, Stensgaard, Lægsgaard, Hammer, Besenbacher, 2006a). The chemisorption of one H atom dramatically changes the reactivity of carbon atoms on specific neighboring graphitic sites, yielding a reduction, or even disappearance, of the barriers for chemisorption of a second incoming H atom nearby (Hornekær, Rauls, Xu, Šljivančanin, Otero, Stensgaard, Lægsgaard, Hammer, Besenbacher, 2006a, Rougeau, Teillet-Billy, Sidis, 2006). As a consequence, sticking probabilities of H atoms on sites in the vicinity of previously chemisorbed H atoms are increased. Thus, hydrogen atoms predominantly adsorb into dimer or cluster structures on the graphite surface (Ferro, Marinelli, Allouche, 2003, Hornekær, Rauls, Xu, Šljivančanin, Otero, Stensgaard, Lægsgaard, Hammer, Besenbacher, 2006a, Hornekær, Šljivančanin, Xu, Otero, Rauls, Stensgaard, Lægsgaard, Hammer, Besenbacher, 2006b). Diffusion barriers of H atoms within such clusters are reduced and values of 0.2-0.4 eV for diffusion into more energetically favorable adsorption structures have been found (Hornekær et al., 2006b).

Molecular hydrogen formation via reactions between chemisorbed H atoms has a high activation barrier of $\sim$ 1.4 eV (Hornekær, Šljivančanin, Xu, Otero, Rauls, Stensgaard, Lægsgaard, Hammer, Besenbacher, 2006b, Zecho, Güttler, Sha, Jackson, Kuppers, 2002a). In contrast, Eley-Rideal reactions between impinging gas-phase H atoms and chemisorbed H atoms have been shown to be barrier-less for the case of abstraction from specific dimer structures (Bachellerie et al., 2007), and, in recent calculations, also for abstraction of hydrogen monomers (Bonfanti et al., 2011). Experimental measurements involving 2000 K H atoms impinging on a hydrogenated graphite surface, held at 300 K, show Eley-Rideal (including hot atom) cross-sections of 17 $Å^2$ at low H coverage decreasing to 4 $Å^2$ at high coverage (Zecho et al., 2002b). However, at low collision energies, theory predicts that quantum reflection effects will limit the Eley-Rideal abstraction cross section (Casolo et al., 2009).

A few measurements regarding the energy partitioning in molecular hydrogen formation on graphite have been reported. The ro-vibrational distribution has been measured for molecular hydrogen formed via reactions involving physisorbed H atoms on a graphite substrate held at 15–50 K. The measurements show high vibrational excitation energies with the vibrational distribution peaking at v=4 (Latimer et al., 2008), while only minimal rotational excitation, corresponding to an excitation temperature of 300 K, was observed (Creighan, Perry, Price, 2006, Latimer, Islam, Price, 2008). Theoretical calculations predict significantly higher vibrational excitation (Morisset et al., 2005).

The kinetic energy for hydrogen molecules formed from reactions between chemisorbed H atoms on graphite has been measured experimentally and yielded an average of 1.4 eV (Baouche et al., 2006). No measurements exist of the energy partitioning for molecular hydrogen formed by Eley-Rideal abstraction reactions with chemisorbed species. However, using relaxed surface, theory predicts that the majority of the released energy goes into vibrational excitation (Bachellerie, Sizun, Aguillon, Teillet-Billy, Rougeau, Sidis, 2009, Martinazzo, Tantardini, 2006, Martinazzo, Tantardini, 2006, Morisset, Aguillon, Sizun, 2003, Morisset, Aguillon, Sizun, Sidis, 2004, Sizun, Bachellerie, Aguillon, Sidis, 2010) . In these calculations geometrical constraints limit information on the partitioning into rotational excitation. However, approximating the surface as rigid, theoretical results (Farebrother, Meijer, Clary, Fisher, 2000, Meijer, Farebrother, Clary, Fisher, 2001) reveal H_2forms with low vibrational excitation and high rotational excitation.

H_2 Formation on Amorphous Carbon Surfaces

Molecular hydrogen formation was studied experimentally on low temperature (5–20 K) compact amorphous carbon surfaces upon irradiation with 200 K H and D atoms. Formation efficiencies above 50% were observed at 5 K, followed by a rapid fall off in efficiency with increasing surface temperature to below 10% at 18 K (Pirronello et al., 1999). This fall off in efficiency was ascribed to the short lifetime of weakly bound physisorbed H/D atoms at increased temperatures, and may also indicate a reduced mobility of the adsorbed atoms, compared with the case of graphite. The height of the diffusion barrier in this situation is expected to be very strongly dependent on the exact nature of the carbonaceous surface.

Molecular hydrogen formation on hydrogenated porous, defective, aliphatic carbon surfaces has also been studied experimentally. H atoms chemisorb strongly to carbon defects with low activation barriers, an experimentally determined activation barrier of 70 K has been reported (Mennella et al., 2006), and typical binding energies in the range of 3–6 eV. Hence, molecular hydrogen formation from two H atoms chemisorbed on such defective carbon surfaces is generally not energetically favorable. However, once the high energy binding sites have been saturated with H atoms, less tightly bound H species may then adsorb and be available for reaction. Furthermore, molecular hydrogen formation involving chemisorbed H atoms on these surfaces can still proceed via Eley-Rideal or hot atom abstraction reactions. The cross-section for such reactions has been determined for a hydrogenated porous, defective, aliphatic carbon surface over a substrate temperature range 13–300 K and for H atom temperatures ranging from 80 to 300 K. The results show that the reaction is barrierless at low surface temperatures, while a small activation barrier of 130 K was found at surface temperatures above ~ 100 K (Mennella, 2008). An abstraction cross-section of 0.03 Å^2, roughly 1/100 of the value found on graphite, was reported for 300 K H atoms impinging on a 300 K sample (Mennella, 2008). The energy partitioning in molecular hydrogen formation on these surfaces is determined by the exoergicity of the reaction, as a C-H bond has to be broken, and by the finding that the majority of the synthesized molecules are retained in the porous structure following their formation. As a result, molecular hydrogen formed on these surfaces is expected to desorb with low ro-vibrational excitation and low kinetic energy, that is with a temperature, effectively, that of the surface.

An alternative molecular hydrogen formation pathway on hydrogenated amorphous carbon is via irradiation with VUV photons. Experimental investigations show that irradiation of hydrogenated amorphous carbon with 6.8–10.5 eV photons leads to breaking of one C-H bond for every 70 incident photons. The majority of the H atoms released by this process (~ 95%) were observed to react to form molecular hydrogen (Alata et al., 2014). The H_2 molecule synthesized in this manner was observed to be retained in the surface structure. These dynamics again indicate that when these H_2 molecules eventually desorb, they will do so with a ro-vibrational distribution and kinetic energies determined by thermal equilibrium with the substrate.

H_2 Formation on PAHs

The ability of PAH molecules to catalyze molecular hydrogen formation has been investigated by several authors (Cazaux, Boschman, Rougeau, Reitsma, Hoekstra, Teillet-Billy, Morisset, Spaans, Schlathölter, 2016, Mennella, Hornekær, Thrower, Accolla, 2012, Rauls, Hornekær, 2008, Snow, Le Page, Keheyan, Bierbaum, 1998, Thrower, Jørgensen, Friis, Baouche, Mennella, Luntz, Andersen, Hammer, Hornekær, 2012). Simple abstraction of H atoms from the PAH, by incident H atoms to generate H_2, is not energetically favorable on un-functionalized PAH molecules, due to the high C-H bond energies of ~ 4.8 eV. However, such reactions can be activated by the initial addition of excess H atoms to the PAH molecule, which then becomes superhydrogenated. Both experimental and theoretical calculations have demonstrated that this superhydrogenation via hydrogen atom addition is possible (Cazaux, Boschman, Rougeau, Reitsma, Hoekstra, Teillet-Billy, Morisset, Spaans, Schlathölter, 2016, Mennella, Hornekær, Thrower, Accolla, 2012, Rauls, Hornekær, 2008, Thrower, Jørgensen, Friis, Baouche, Mennella, Luntz, Andersen, Hammer, Hornekær, 2012). Addition barriers for PAHs vary, depending on charge state and degree of superhydrogenation (Cazaux, Boschman, Rougeau, Reitsma, Hoekstra, Teillet-Billy, Morisset, Spaans, Schlathölter, 2016, Rauls, Hornckær, 2008).

On the cation of coronene (a prototypical PAH), the first H addition to a C atom on the periphery of the molecule, an "edge" carbon, has a very small barrier (10 meV); while a second H atom addition has been calculated to have a barrier of 30 meV (Cazaux et al., 2016). Subsequent H addition reactions have barriers which alternate in magnitude. Some of the barriers are much higher (of the order of 0.1 eV), leading to the predominance of coronene cations with a magic number of H atoms attached (+5, +11 and + 17 extra hydrogens). The barriers for H addition to a coronene cation with an even number of extra H atoms is large, while hydrogenation of a coronene cations with an odd number of extra H atoms is small, and this alternation occurs until full hydrogenation is reached.

On neutral coronene, a barrier of 60 meV for the addition of the first H atom to an edge site has been calculated. Subsequent H addition reactions have lower barriers or are even barrierless (Rauls and Hornekær, 2008). Comparable experiments on coronene show that the first H atom addition, for atoms with energies of 1400–2000 K, has a cross-section of 0.7 ± 0.4 Å^2 (Thrower et al., 2012). Experiments on both cationic and neutral PAHs demonstrate that, for H atom beams with energies ranging from 300 K - 2000 K, high degrees of superhydrogenation (in many cases complete superhydrogenation) with one excess H atom per C atom, are attainable (Skov et al., 2014).

For the neutral coronene molecule, hydrogen addition measurements can be simulated using addition cross-sections ranging from 0.55-2.0 Å^2 (Cazaux, Boschman, Rougeau, Reitsma, Hoekstra, Teillet-Billy, Morisset, Spaans, Schlathölter, 2016, Skov, Thrower, Hornekær, 2014, Thrower, Jørgensen, Friis, Baouche, Mennella, Luntz, Andersen, Hammer, Hornekær, 2012). However, once the molecule is super-hydrogenated, abstraction reactions to form H_2 compete with H atom addition (Rauls and Hornekær, 2008); such abstraction reactions are often barrierless. These abstraction reactions have been directly observed, via H-D exchange, using the coronene molecule (Mennella, Hornekær, Thrower, Accolla, 2012, Thrower,

Jørgensen, Friis, Baouche, Mennella, Luntz, Andersen, Hammer, Hornekær, 2012). For incident H atoms at 300 K, an abstraction cross-section of 0.06 Å^2 per excess H atom was measured (Mennella et al., 2012). For more energetic H atoms (2000 K) a lower cross-section of 0.01 Å^2 per excess H atom was observed to yield better agreement between simulations and measurements (Skov et al., 2014).

Water Ice Surfaces

In cold molecular clouds where the visual extinction is larger than 3, cosmic dust grains are covered by water ice. The abundance of the H_2 molecule in molecular clouds is determined by the balance between destruction in the gas phase and formation of H_2 on the icy dust surfaces. In this section, we briefly overview the experimental work studying the physico-chemical processes involving hydrogen, which lead to H_2 formation on these icy mantles covering the dust grains. As discussed above, H_2 formation generally occurs via a sequence of elementary processes: sticking, surface diffusion, and recombination reactions. Each process has been targeted by various classes of experiments which will be described in the following subsections.

Upon H_2 formation, dissipation of the heat of reaction must occur. Following the recombination event, the excess energy of about 4.5 eV has to be partitioned between the kinetic and internal energy of the nascent H_2 and the grain. The kinetic energy of the molecule is redistributed by collisions and remains within the cloud, while internal excitation followed by deexcitation generates IR photons that may escape from the cloud. Special attention should be paid to the OPR of the nascent H_2 molecules. The energy difference of $\sim$ 14.7 meV (corresponding to 170 K) between the ground states of the two nuclear spin isomers (ortho and para states) is large enough to affect the gas phase chemistry of H_2 in molecular clouds at around 10 K. Furthermore, since the conversion between the nuclear spin states is forbidden in the gas phase, ortho-para interconversion on the ice surface plays a crucial role in determining the OPR of H_2.

Interstellar Ices

Astronomical observations show that the 3 μm-bands seen in the infrared absorption spectra of dust are well explained by the existence of water ice as a mantle in polycrystalline and/or amorphous forms (Smith, Sellgren, Tokunaga, 1988, Whittet, 1993). Amorphous ice (hereafter, Amorphous Solid Water: ASW) dominates at low temperatures over the polycrystalline

(PCI) phase. Therefore, experimental studies began by preparing and characterizing ASW as an analogue of interstellar ice mantles. The quantity of ice observed in such mantles cannot be due to simple freezing out (accretion) of H_2O molecules synthesized in the gas phase; thus, reactions on the grain surface between hydrogen and oxygen are required. Nevertheless, it is generally believed that ASW produced experimentally by the deposition of water vapor onto cold substrates reproduces interstellar ice-mantles fairly well. However, it should be noted that the porosity of ASW strongly depends on the substrate temperature and the deposition method of H_2O vapor (Stevenson et al., 1999). Furthermore, it is known that additional processing due to D-atom exposure on pre-deposited ASW (Accolla et al., 2011) and also synthesis of ASW through D + O_2 reaction on cold surfaces (Oba et al., 2009) result in formation of compact (less-porous) amorphous ice. For the experiments relevant to H_2 formation on icy dust grains, the ASW samples were typically prepared at around 10 K by introducing H_2O vapor into vacuum chambers, sometimes through a capillary plate.

The chemical and physical properties of ASW are an important research target not only for astronomers but also for physicists and chemists (for a review, see Watanabe and Kouchi, 2008). Briefly, the main remarkable features of ASW produced at ~ 10 K are: the porous structure with its large surface area and, because of its irregular morphology, the wide variety of different adsorption sites for hydrogen atoms and molecules. The surface area of ASW was found to be 10 times larger than that of polycrystalline ice when ASW was deposited with the amount of 10^{17} molecules cm^{-2} at 10 K (Hidaka et al., 2008). These features significantly influence the associated physico-chemical processes of hydrogen atoms and molecules, such as adsorption, desorption and diffusion, when they interact with the ice surface.

Sticking and Adsorption Energy

H_2 formation by H-H recombination first requires adsorption of H atoms onto the icy mantle of a dust grain. Therefore, the sticking coefficients and adsorption energies of H atoms on these surfaces are key parameters. Because of significant experimental difficulties in quantifying these processes, these quantities have often been obtained theoretically. In such calculations, both semi-classical and quantum approaches have been employed (Al-Halabi, van Dishoeck, 2007, Buch, Czerminski, 1991, Buch, Zhang, 1991, Hollenbach, Salpeter, 1970, Hollenbach, Salpeter, 1971, Masuda, Takahashi, Mukai, 1998, Takahashi, Masuda, Nagaoka, 1999b). Here adsorption energies of H atoms on crystalline water ice and ASW were determined to be in the range of 300–600 K. Al-Halabi and van Dishoeck (2007) showed that the adsorption energy for H atoms on ASW is higher than that for crystalline ice by approximately 200 K.

Molecular Dynamics calculations (MD) showed that the sticking coefficients of H atoms on ASW are near unity at around 10 K, when the kinetic energy of the impinging atoms is less than about 100 K. In the MD calculations (Al-Halabi, Fraser, Kroes, van Dishoeck, 2004, Al-Halabi, van Dishoeck, 2007, Buch, Zhang, 1991), adsorption is defined as trajectories having stabilized at a total H-atom energy of ~ 100 K for duration of < 10-15 ps, where the total energy consists of the adsorption and kinetic energies of H atom. The sticking coefficient was found to decrease steeply with the increasing incident energy of atoms. However, it should be noted that MD calculations can only follow a very short time period of the H-atom dynamics. Thus, the calculated sticking coefficients may differ from those derived under realistic conditions in experiments and astrophysical environments (Al-Halabi et al., 2004). Buch and Zhang (1991) evaluated the temperature dependence of the sticking coefficient. A very good match between the experimental values of the sticking coefficient for molecular hydrogen and the theoretical value has been found (Matar et al., 2010). The sticking coefficients for H atoms calculated to date are summarized in Veeraghattam et al. (2014). However, direct measurements of the sticking coefficients and adsorption energies of H atoms are not easy to perform and thus have not yet been made.

Diffusion

In molecular interstellar clouds, the accretion of an H atom onto a dust grain occurs about once a day and the coverage of H atoms on the dust grains is considered to be very low. In this situation, the LH mechanism should dominate the recombination process. Since the recombination itself is a radical-radical barrier-less reaction, the surface diffusion of H atoms across the dust surface is the rate-limiting process for H_2 formation.

The diffusion rate, via thermal hopping over the appropriate energy barriers, can be simply derived from the Arrhenius equation. For example, assuming activation energies of 200 and 500 K with a frequency factor of 10^{12} s^{-1}, the hopping rates become 10^3 and 10^{-10} s^{-1}, respectively, at a surface temperature of 10 K. However, since in reality the surfaces of ASW and even those of PCI consist of a variety of adsorption sites accessed by different diffusion barriers, it is not easy to predict the actual diffusion rate.

Furthermore, diffusion by tunneling should also be considered at low surface temperatures.

The tunneling-diffusion rate is much more sensitive to the shape, especially width, of barriers than thermally activated hopping. Although a rectangular potential is often used for estimating a tunneling rate, this is a very simplified model. The tunneling diffusion of H atoms on ice was first studied theoretically by Smoluchowski (1983). He evaluated tunneling diffusion for a non-periodic amorphous structure and indicated that the diffusion on ASW is significantly slower than on PCI, leading to a corresponding recombination rate at least three orders of magnitude lower on ASW.

In recent years, assuming a thermal hopping mechanism, activation energies for H atom surface diffusion have been derived from TPD experiments. Analyzing the TPD spectra obtained experimentally after H or D atoms have been deposited on ASW, Perets et al. (2005) reported the activation energy to be in a range of 41–55 meV. Similarly, applying a simple rate equation model to the results of TPD experiments, Matar et al. (2008) determined the activation energy of D atom diffusion to be about 22 meV. This discrepancy in the results between the two groups could partly originate from different surface coverages of atoms, which require different diffusion lengths to be covered for recombination. Since the energy depths of adsorption sites have a distribution, long-distance diffusion tends to be limited by higher activation barriers than short-distance diffusion. Also, for an irregular surface structure, like that of ASW, TPD experiments, and their analysis by rate equation models with multiple parameters, result in some ambiguity in determining the value of the diffusion barrier.

Recently, using a combination of laser-induced photo-desorption and resonance enhanced multi-photon ionization (REMPI) methods, Watanabe and coworkers clarified how the diffusion mechanisms, either tunneling or thermal hopping, depend on the diffusion distance on ices (Kuwahata et al., 2015), and measured the activation barriers for thermal hopping of H and D atoms on ASW (Hama, Kuwahata, Watanabe, Kouchi, Kimura, Chigai, Pirronello, 2012, Watanabe, Kimura, Kouchi, Chigai, Hama, Pirronello, 2010). Watanabe and coworkers demonstrated that the ASW surface contains various sites that can be categorized into at least three groups: very shallow-, middle-, and deep-potential sites, with associated diffusion activation energies of < 18, 22 (23 meV for D atoms), and > 30 meV, respectively. These values of barriers cover those obtained by TPD experiments, and have been confirmed by recent quantum calculations (Senevirathne et al., 2017).

H_2 Formation on Ices

Formation by Recombination:

H_2 formation by atomic recombination has most commonly been investigated in TPD experiments where the yields of H_2, HD, and D_2 were detected as a result of the recombination of H and/or D atoms which were deposited on ASW (Amiaud, Dulieu, Fillion, Momeni, Lemaire, 2007, Hornekær, Baurichter, Petrunin, Field, Luntz, 2003, Manicò, Ragunì, Pirronello, Roser, Vidali, 2001, Perets, Biham, Manicó, Pirronello, Roser, Swords, Vidali, 2005, Roser, Manicò, Pirronello, Vidali, 2002). In these experiments, recombination efficiencies, as a function of the surface temperature, were determined together with several parameters characterizing elementary processes such as diffusion. These results are well-summarized elsewhere (Vidali, 2013). Another class of experiment was performed by monitoring the yield of H_2 or HD species, formed by recombination, which were then photodesorbed from the ASW surface; no formal heating of the surface was performed (Hornekær, Baurichter, Petrunin, Field, Luntz, 2003, Watanabe, Kimura, Kouchi, Chigai, Hama, Pirronello, 2010).

As noted above, the formation of H_2 by atomic recombination releases an excess energy of about 4.5 eV. The partitioning of this energy resulting from H_2 formation is an important issue in astronomy. This importance arises because ro-vibrational emission observed from H_2 may partly be due to excitation by the energy released by recombination (Takahashi and Uehara, 2001, and references therein). The number of studies probing the energy partitioning upon H atom recombination on ASW is limited. MD calculations showed that H_2 synthesized from H atoms on ASW can be vibrationally excited in v=7-8 (Takahashi et al., 1999a). Roser et al. (2003) demonstrated experimentally that the kinetic energy of the desorbing H_2 molecules, formed by recombination on ASW, is almost equivalent to the surface temperature. Considering the vibrational energy of the nascent H_2, no highly excited species have been detected following atomic recombination on ice (Hama, Kuwahata, Watanabe, Kouchi, Kimura, Chigai, Pirronello, 2012, Hornekær, Baurichter, Petrunin, Field, Luntz, 2003). This lack of vibrational excitation may well be because most of the freshly synthesized H_2 molecules are re-trapped within the ASW layer and thermalized there (Watanabe et al., 2010).

Formation by Energetic Processes:

Energetic agents (ions, electrons and photons) irradiating ices and ice mixtures can dissociate molecules. The fragments of such dissociative processes can recombine to generate new molecules.

- Ion bombardment: Brown et al. (1982) performed the first quantitative measurements of H_2 and O_2 formation following ion irradiation by sending 1.5 MeV He^+ onto layers of ASW a few hundred Angstrom thick, with the surface held at 10 K. The radicals generated, that are mobile in the energized region around the ion track, recombine in a variety of different ways and diffuse out of the ice layer during the irradiation. When investigating effects induced by energetic agents, it is mandatory to use dose rates in the range in which the effects (molecule formation, sputtering and so on) are proportional to fluxes. Only under such conditions single particle effects, not cumulative ones, are measured. Under these conditions experimental results may be applied to interstellar environments, where particle fluxes are orders of magnitude lower than those used in the laboratory. With this "caveat", the yields (i.e. the average number of molecules released by the ice per each impinging ion) were measured as function of the ice temperature (see Fig. 7 of Brown et al., 1982). Values measured in the laboratory, for a narrow range of energies of the impinging ions, may be extrapolated to all other energies encountered in space by taking into account that the formation and release of product species scale with the stopping power dE/dx (the energy lost per unit path length in the solid) of swift ions. The yield then reaches its maximum for kinetic energies of the fast ions around roughly 100 keV per nucleon, in the case of light ions (Brown et al., 1982). Pirronello and Averna (1988) and Averna and Pirronello (1991) used these laboratory results in a Monte Carlo simulation to asses the importance of the production of H_2 by cosmic ray bombardment of icy grain mantles inside, and at the edge of, dense clouds. Their evaluation should be revised due to the fact that the ionization rates involving cosmic rays in dense clouds have been increased by about an order of magnitude since their papers where published (see e.g. Caselli et al., 1998b).
- UV Irradiation: Westley et al. (1995a); 1995b) observed the H_2 yields desorbed from 50 to 100 K ice during irradiation with Ly-α photons. Kouchi and Kuroda (1991) measured the thermal desorption spectra of H_2 molecules from UV-irradiated pure water ice. These were both rather qualitative investigations.

Later, UV irradiation of water ice (Watanabe et al., 2000) showed that the fraction of D_2 molecules formed in D_2O ice, at 12 K, after a dose of 10^{18} cm^{-2} UV photons ($\lambda \sim$ 126, 172 nm) was around 2% of the number of irradiated D_2O molecules. Most of the D_2 molecules produced, in contrast to the case of ion irradiation, remained trapped in the ice layer. The estimated cross section for this molecular synthesis was about 2.4×10^{-18} cm^2, a value that may render such a process non-negligible under the UV conditions pertaining in certain molecular clouds. Yabushita et al. (2008) detected the formation and release of H_2 molecules in highly excited states (v=0-5; J=0-17) following UV (157 nm) irradiation of ASW at 100 K. Using the REMPI technique, they were able to recognize two distinct formation mechanisms of H_2: a hot H atom abstracts an H atom from an H_2O molecule, yielding cold H_2 molecules, and recombination of two H atoms that yields ro-vibrationally hot H_2.

- Electron Irradiation: In H_2 formation induced by electron irradiation of water ice it is important to understand the process of secondary electron generation by ion bombardment of ice. The process of direct injection of free electron from the gas phase is not very relevant to interstellar environments. Kimmel and Orlando (1996) and Kimmel et al. (1994) studied the production of molecular $H(D)_2$ by irradiating ASW with low energy (5–100 eV) electrons. The D_2 yield increased monotonically with the temperature of the substrate in the range 88 - 145 K, doubling at the highest temperatures. The D_2 molecules showed very little translational energy (20 - 50 meV) but were vibrationally (v = 0 - 4) and rotationally (J = 0 - 12) excited. The authors suggested that the dominant mechanisms for production of D_2 at 100 eV incident electron energy are dissociative recombination of holes (D_2O^+ or D_3O^+) with quasi-free or trapped electrons and dissociation of excitons at the vacuum-surface interface.

Ortho to Para Conversion of H_2 on Ice

The ortho-to-para ratio (OPR) of H_2 released into the gas phase affects the chemical evolution of a cloud, including deuterium fractionation, (Bron, Le Petit, Le Bourlot, 2016, Majumdar, Gratier, Ruaud, Wakelam, Vastel, Sipilä, Hersant, Dutrey, Guilloteau, 2017, Pagani, Roueff, Lesaffre, 2011). As noted above this affect of the OPR arises because the energy difference between the ortho and para rotational ground states is ~ 14.7 meV; this energy corresponds to a temperature of 170 K, which is significantly higher than the local temperature of molecular clouds.

The radiative transformation between ortho and para states is forbidden in the gas phase. Therefore, it is highly desirable to understand how the OPR behaves when the molecules are adsorbed on icy dust grains. Watanabe et al. (2010) demonstrated experimentally that the OPR of nascent H_2 formed on ASW, at a surface temperature of around 10 K, is at the statistical value of 3, but diminishes rather quickly when H_2 is adsorbed on ice at 10 K. Similarly, Gavilan et al. (2012) reported that the OPR of D_2 formed on ASW showed a statistical OPR.

In recent years, using the REMPI method, ortho-to-para conversion rates have been measured on ASW. Sugimoto and Fukutani (2011) determined a conversion rate of $\sim 2.7 \times 10^{-3}\ s^{-1}$ for H_2 and found that the conversion rate for D_2 is much slower than that for H_2; a fact later confirmed by another group (Hama et al., 2012). Chehrouri et al. (2011) reported that trace amounts of O_2 on ASW accelerates the nuclear spin conversion for D_2. The conversion mechanism on ASW was first proposed by Sugimoto and Fukutani (2011). However, the role of surface temperature is not considered in their model. Very recently, the conversion rate for H_2 was found to increase steeply, from $\sim 2.4 \times 10^{-4}$ to $\sim 1.5 \times 10^{-3}\ s^{-1}$, over the temperature range of 9.2 - 16 K (Ueta et al., 2016). This temperature dependence of the conversion rate can be explained by an energy dissipation process via phonons.

Astrophysical Models

In the domain of astrophysics, numerical models are of tremendous importance. Astrophysical objects are complex systems coupling a wide range of physical and chemical processes and of spatial and temporal scales where physical conditions (density, UV flux, etc) vary by orders of magnitude. In contrast, laboratory experiments isolate one or a few processes, which they study over a limited range of physical conditions. Numerical models then constitute the indispensable link between our experimental and theoretical knowledge of the elemental processes and astronomical observations, transforming experimental results into observable predictions, and tracing back unexpected observations to specific processes that need to be investigated in the laboratory.

Due to computational limitations, models cannot treat all aspects at the most detailed level. Astrophysical models differ in the way they focus on some parts of the physics and chemistry, which they treat in great detail because they are the most relevant in a given astrophysical environment, while neglecting, or approximating more crudely, other aspects that are less relevant. We first present here the main approaches to modeling the H_2 formation rate itsef, before discussing the broad families of astrophysical models where these approaches are used, defined in term of the areas of physics and chemistry where they place their focus.

Calculating H_2 Formation Efficiency in Models

In order to calculate the H_2 formation rate in models based on laboratory experiments and/or astrophysical observations, different methods (corresponding to different levels of approximation) have been used.

Until relatively recently, most types of models used an approach based entirely on observational measurements of the H_2 formation rate in diffuse molecular clouds (Gry, Boulanger, Nehmé, Pineau des Forêts, Habart, Falgarone, 2002, Jura, 1974). This approach simply parameterizes the formation rate as the product of the atomic hydrogen abundance n(H), the total gas density n_H (as a tracer of the dust abundance), and an efficiency parameter R_f, determined to be roughly $3\times10^{-17}cm^3s^{-1}$. Additional dependencies, like metallicity and gas temperature, were often included, sometimes also taking the experimentally measured sticking efficiency into account.

The simplest approach based only on laboratory data is known as the rate equation approach, treating the surface abundances (here of H and H_2) as continuous variables obeying differential equations (similar to gas phase chemistry), using the experimentally constrained rates for each process (desorption, migration, etc). One such widely used rate equation model of H_2 formation is Cazaux and Tielens (2002), 2004). Despite its advantages of simplicity and low computational cost, several approximations in this approach can be problematic in some regimes so that other numerical techniques have been developed (Herbst and Shematovich, 2003). The first main problem is the fact that surface populations on dust grains can be of the order of unity, which make their discrete nature crucial (e.g. Langmuir-Hinshelwood formation goes from impossible to possible when going from a single adsorbed atom on a grain to two). In addition, this approach misses any spatial effects on the surface (preferential chemisorption on neighboring sites to a chemisorbed atom, different adsorption sites have different binding energies on amorphous surfaces, etc). The methodology also assumes a fixed dust temperature while the temperature of small grains fluctuates due to stochastic heating by UV photons.

The discrete nature of surface population results in random fluctuations (due to, for example, the stochastic nature of atom adsorption and desorption events). This can be accounted for by describing the surface populations. in statistical terms, by a probability distribution function obeying a master equation (again based on the experimentally constrained rate for the elemental processes) and solved to compute the average formation efficiency (Biham, Furman, Pirronello, Vidali, 2001, Green, Toniazzo, Pilling, Ruffle, Bell, Hartquist, 2001). This approach is thus commonly called the master equation approach. Analytical solutions are sometimes possible (Biham and Lipshtat, 2002), otherwise, numerical solution of the master equation incurs a significantly higher numerical cost than rate equation approaches. Approximations allowing faster solution have been proposed, such as the moment equation approach (Le Petit, Barzel, Biham, Roueff, Le Bourlot, 2009, Lipshtat, Biham, 2003). A careful comparison of the master equation methodology to various approximations of this approach, and to the simpler rate equation formalism, was presented in Rae et al. (2003). The master equation approach has also been extended to simultaneously account for dust temperature fluctuations of small grains (Bron et al., 2014).

Finally, Monte Carlo simulations can, in principle, take any level of detail into account, by simulating realizations of the random evolution of the system (again using experimentally constrained rates), over which averages and other statistics can be computed (Charnley, 2001). This approach is, however, very costly in term of computational time, and is thus rarely used inside complete astrophysical models. It has been used to study spatial effects neglected by other approaches, such as variations of binding energies across the surface (Chang, Cuppen, Herbst, 2005, Cuppen, Herbst, 2005), or the effect of dust temperature fluctuations (Cuppen et al., 2006). Vasyunin et al. (2009) and Chang and Herbst (2014) have presented a new generation of Monte Carlo models computing both the gas-phase and surface chemistries for large networks (similar to the ones used by rate equation models). These models are called unified Monte Carlo models. Chang and Herbst (2014) however use a microscopic approach for the surface chemistry and a macroscopic one for the gas whereas Chang and Herbst (2014) use a macroscopic approach for both these environments.

We also note that modifications to the simple rate equation approach have been proposed to approximate effects such as discrete population fluctuations (Cuppen, Garrod, 2011, Garrod, 2008), diffusion on a spherical rather than plane surface (Lohmar, Krug, 2006, Lohmar, Krug, Biham, 2009), or surface inhomogeneity (Chang et al., 2006).

Hydrodynamical and MHD Simulations

Hydrodynamical and Magneto-Hydro-Dynamical(MHD) simulations aiming at reproducing astrophysical systems, such as galaxy formation and star formation within molecular clouds, solve the equations for the hydrodynamics (or the magnetohydrodynamics, when fluids are magnetized) in order to describe the evolution in time of the system. However, the coupling of detailed chemistry to these high-resolution, multi-dimensional time-dependent simulations is still challenging.

While most hydrodynamical and MHD simulations only treat gas dynamics (possibly including thermal balance and gravity), it is only in recent years that computational capabilities have allowed us to include the formation of molecular hydrogen. However, since in astrophysical simulations the scales at which the H_2 molecules form are never resolved, the formation and destruction processes are included as sub-grid models, described by reaction rates derived from observational constraints (Gry, Boulanger, Nehmé, Pineau des Forêts, Habart, Falgarone, 2002, Jura, 1974).

Approximate, less computationally demanding, approaches to compute H_2 abundances are usually stationary (Krumholz, McKee, Tumlinson, 2009, Kuhlen, Krumholz, Madau, Smith, Wise, 2012, Robertson, Kravtsov, 2008). In contrast, a time-dependent and self consistent description of the chemical evolution of the system requires much greater computational resources (Bekki, 2013, Dobbs, Glover, Clark, Klessen, 2008, Glover, Abel, 2008, Glover, Federrath, Mac Low, Klessen, 2010, Glover, Mac Low, 2007a, Glover, Mac Low, 2007b, Gnedin, Tassis, Kravtsov, 2009, Henderson, Bekki, 2016, Hocuk, Cazaux, Spaans, Caselli, 2016, Koyama, Inutsuka, 2002, Mac Low, Glover, 2012, Micic, Glover, Banerjee, Klessen, 2013, Micic, Glover, Federrath, Klessen, 2012, Pelupessy, Papadopoulos, van der Werf, 2006, Smith, Glover, Clark, Klessen, Springel, 2014, Valdivia, Hennebelle, Gérin, Lesaffre, 2016, Walch, Girichidis, Naab, Gatto, Glover, Wünsch, Klessen, Clark, Peters, Derigs, Baczynski, 2015).

The treatments used can involve different degrees of detail in representing the chemical networks. Approaches range from following the evolution of the H_2 abundance only (Valdivia et al., 2016), or that of a few more species in reduced networks (Glover, Federrath, Mac Low, Klessen, 2010, Lim, Rawling, Williams, 2001, Lim, Rawlings, Williams, 1999), to much more complex chemical networks containing hundreds of chemical species and thousands of coupled chemical reactions (Grassi, Bovino, Schleicher, Gianturco, 2013, Ziegler, 2016). The elemental processes of H_2 formation (diffusion, desorption,) are usually not treated in these models, and simplified formulae are used, based observational determinations of H_2 formation efficiency (Gry, Boulanger, Nehmé, Pineau des Forêts, Habart, Falgarone, 2002, Jura, 1974) and scaled by the sticking coefficient of hydrogen atoms and the total dust cross-section per hydrogen nucleon (Bekki, 2013).

The dominant destruction of H_2 by photo-dissociation brings a complication, as it requires some calculation of radiative transfer. Dust extinction and self-shielding by H_2 (in the approximation of Draine and Bertoldi 1996) are usually taken into account, and are crucial in order to obtain the right photo-dissociation rates (Hartwig et al., 2015). Since the computational cost of calculating column densities is high (Clark, Glover, Klessen, Bonnell, 2012, Valdivia, Hennebelle, 2014), simulations use simplified treatments, such as ray-tracing schemes (Inoue and Inutsuka, 2012), logarithmic approximations (Clark, Glover, Klessen, Bonnell, 2012, Valdivia, Hennebelle, 2014), and even constant shielding parameters to avoid modeling radiative transfer in the Lyman and Werner bands.

Hydrodynamical and MHD simulations have shown that the formation of H_2 begins in the initial phases of the molecular cloud formation (Clark, Glover, Klessen, Bonnell, 2012, Glover, Abel, 2008), and that compressive motions within the gas (either driven by the galaxy dynamics or the turbulence) accelerate the formation of molecular hydrogen on galactic scales (Dobbs et al., 2006), as well as on the scale of interstellar clouds (Glover, Mac Low, 2007a, Glover, Mac Low, 2007b, Micic, Glover, Federrath, Klessen, 2012, Valdivia, Hennebelle, Gérin, Lesaffre, 2016). Clark et al. (2012) and Smith et al. (2014) have shown that the formation of CO, which is the main tracer of molecular gas, is significantly delayed with respect to the formation of H_2. Theoretical (Smith, Glover, Clark, Klessen, Springel, 2014, Wolfire, Hollenbach, McKee, 2010) and observational studies (Grenier, Casandjian, Terrier, 2005, Langer, Velusamy, Pineda, Willacy, Goldsmith, 2014, Paradis, Dobashi, Shimoikura, Kawamura, Onishi, Fukui, Bernard, 2012) point out that there is a significant amount of H_2 (between 30 and 70 percent) which is not traced by CO. Simulations have shown that the multiphase nature of molecular clouds (where a cold and dense phase coexists with a diffuse and warm one), and their turbulent motions, have a significant effect on the abundance of H_2. Turbulent motions mix the cold and warm phases, while the multiphase structure allows the survival of molecules in shielded pockets of warm gas (Valdivia et al., 2016). The presence of a hot phase provides a confining pressure that increases the abundance of H_2 in cold gas clumps (Dobbs and Bonnell, 2007). Such a conclusion is consistent with the results of Henderson and Bekki (2016), who find that in disk galaxies a strong ram pressure enhances the formation of H_2.

Since molecular hydrogen also contributes to the cooling of the gas, it is perhaps expected that abundances of H_2 should affect the star formation rate and the spatial distribution of star forming regions (Christensen et al., 2012). However, the recent works of Glover and Clark (2014) and Richings and Schaye (2016) have shown that H_2 has little influence on the star formation process. On the other hand, the abundance of H_2 is important for molecular outflows (Richings and Schaye, 2016), and it plays an important role on the warm chemistry of the ISM (Valdivia et al., 2017a).

Kinetic Models for Complex Chemistry

Astrochemical models aiming at studying the molecular complexity in dense and shielded interstellar regions (cold cores for instance) focus on a detailed description of the complex chemical network at play, and on the time-dependent aspect of this chemistry (as some chemical timescales become comparable to the dynamical lifetime of these clouds). Such models thus solve kinetic differential equations in which each individual chemical and physical process is assigned a rate (Wakelam et al., 2013).

The first kinetic models (Graedel, Langer, Frerking, 1982, Leung, Herbst, Huebner, 1984, Prasad, Huntress, 1980) were based solely on gas-phase chemistry (bimolecular reactions, interactions with UV photons or cosmic-ray particles). In these models, the H_2 formation rate was computed from the collision rate of H atoms with dust grains (Hollenbach, Salpeter, 1971, Hollenbach, Werner, Salpeter, 1971). In the90′s (Hasegawa, Herbst, 1993, Hasegawa, Herbst, 1993, Hasegawa, Herbst, Leung, 1992), models including gas-grain interactions and grain surface chemistry began to be developed. Models of the surface processes were based on the Langmuir-Hinshelwood mechanism. The H accretion rate depends on the grain's geometrical cross section, the concentration of grains, the gas-phase thermal velocity of the H atoms and a sticking probability. The model then computes the time taken for a hydrogen atom to explore the entire grain via diffusion

($t_{diff} = \frac{N_S}{\nu_0 \exp(-E_{diffus}/(k_B T_{dust}))}$) with N_s the total number of surface sites on a grain). The rate of H_2 formation is $2t_{diff,H}^{-1} N_H^2 n_d$ with N_H the number of H atoms on the grains and n_d the number density of grains (see Hasegawa et al., 1992). This approximation assumes that there are at least two hydrogen atoms on the grain at each time and therefore overestimates the H_2 production at the accretion limit (at low density) (Caselli et al., 1998a).

In these models, the precise nature of the grain surface is usually ignored and it is assumed that the grains are covered by water ice; hence, only the H_2 binding energy on H_2O is used. The parameters required are associated with the sticking, diffusion and desorption processes in the physisorption regime. Due to their connection with deuteration, ortho-to-para conversion on grains may also be an important phenomenon to model (Majumdar, Gratier, Ruaud, Wakelam, Vastel, Sipilä, Hersant, Dutrey, Guilloteau, 2017, Pagani, Salez, Wannier, 1992). H_2 formation on these icy surfaces may induce some secondary effects due to the significant amount of energy released in the formation of the H-H bond. Some experimental observations, such as the reduction of the porosity of the ice (Accolla et al., 2011) associated with H_2 formation, are not included in models. Conversely, some proposed phenomena, which have not been observed experimentally (Amiaud, Dulieu, Fillion, Momeni, Lemaire, 2007, Minissale, Dulieu, Cazaux, Hocuk, 2016), such as molecular desorption induced by H_2 formation (Takahashi and Williams, 2000) or the H rejection mechanism, are sometimes included in models (Le Petit et al., 2009).

PDR Models

PDR models focus on the interaction of UV radiation with molecular clouds. They involve simultaneous computation of radiative transfer and the chemical and thermal state of the gas, with a detailed treatment of the micro-physics of both gas and grains. Current major examples include the Meudon PDR code (Le Bourlot, Le Petit, Pinto, Roueff, Roy, 2012, Le Petit, Nehmé, Le Bourlot, Roueff, 2006), the KOSMA-τ code (Röllig, Szczerba, Ossenkopf, Glück, 2013, Störzer, Stutzki, Sternberg, 1996), CLOUDY (Ferland et al., 2013), and the PDR codes of Kaufman et al. (2006), 1999) or of Meijerink and Spaans (2005). Most of these codes are one dimensional and stationary, although some higher dimensional and/or time-dependent codes have been developed recently (e.g., Bisbas, Bell, Viti, Yates, Barlow, 2012, Motoyama, Morata, Shang, Krasnopolsky, Hasegawa, 2015, Hosokawa, Inutsuka, 2006), often at the cost of a simplified micro-physics and chemistry.

In PDR models H_2 formation was initially described empirically by a constant formation efficiency (e.g. using Jura 1974), but some codes have progressively included a more detailed calculation over a full distribution of grain sizes (e.g., Le Bourlot, Le Petit, Pinto, Roueff, Roy, 2012, Röllig, Szczerba, Ossenkopf, Glück, 2013), using a rate equation treatment of the adsorption, migration, reaction and evaporation of physisorbed and chemisorbed hydrogen atoms. Grains are heated by the UV photons and H_2 formation thus occurs on bare silicate and carbonaceous surfaces (in the warm surface layer of the region, where UV photons are not yet absorbed). On these warm bare grains, physisorbed atoms can evaporate quickly and chemisorption can play an important role in allowing an efficient H_2 formation (e.g. Le Bourlot et al. 2012) as observed in PDRs (e.g. Habart et al. 2004).

Very small grains and PAHs represent the majority of the dust surfaces on which H_2 formation can occur. H_2 formation on PAHs could be an important contribution to the total formation rate (Boschman et al., 2015). In addition, small grains undergo large temperature fluctuations on short time scales, as their low heat capacities make their temperatures sensitive to single UV photon absorption events. These fluctuations can significantly increase the efficiency of physisorption-based, Langmuir-Hinshelwood, H_2 formation in the UV-rich environments of PDRs, as was shown by Bron et al. (2014) by coupling a master equation approach to the Meudon PDR Code. In addition, the sticking coefficients at the warm gas temperatures of PDRs (several 100 K) are also important parameters (Cazaux et al., 2011). Finally, the role of surface processes in H_2 ro-vibrational excitation (in the gas) has also been investigated. Ortho-para conversion on the surface was shown to have an important impact on the gas-phase OPR of H_2 in PDRs (Bron, Le Petit, Le Bourlot, 2016, Le Bourlot, 2000). The impact of the internal energy content of the H_2 molecule following its formation has been less explored in models, in part due to the fact that there is little associated observational data to reproduce.

Shock Models

Ubiquitous in the Universe, shocks are produced by violent pressure disturbances such as star-driven (or AGN-driven) jets and winds, supernovae explosions, and collisions between molecular clouds or, at larger scales, collisions between galaxies and between accretion filaments and halos of galaxies.

Due to the coupling between the chemistry and the dynamics in shocked regions, the formation of H_2, and emission from H_2, are among the most difficult astrophysical processes to model (see reviews addressing the physics and chemistry of interstellar shocks by Draine, 1980, McKee, Hollenbach, 1980, Chernoff, 1987, Hollenbach, McKee, 1989, Draine, McKee, 1993, Hartigan, 2003). Since H_2 is one of the main coolants in shocks, its energetics have always been considered in molecular shock models (Flower, Le Bourlot, Pineau des Forêts, Cabrit, 2003, Flower, Pineau des Forêts, 2010, Flower, Pineau des Forets, Hartquist, 1985, Guillard, Boulanger, Pineau Des Forêts, Appleton, 2009). Magneto-hydrodynamic (MHD) shock models have been updated with a benchmark of experimental results associated with H_2 formation on carbonaceous material – see e.g. models by Flower and Pineau des Forêts (2015), Guillet et al. (2009) with data from Cuppen and Hornekær (2008), Cuppen et al. (2010). In this context, the role of chemisorption and the transfer from chemisorbed states to physisorbed states (Cazaux and Tielens, 2004) is a key point, which allows models to bridge the low and the high temperature regimes in these environments. Due to their intrinsic high temperature gradients, reproducing shocked regions without taking into account both the chemisorption and physisorption of hydrogen, and their interplay, seems to be an oversimplification.

In the presence of a magnetic field, ubiquitous in the ISM, the shock wave satisfies not only the equations representing the fluid dynamics, but also Maxwell's equations. The magnetic field (B) does not interact the same way with charged and neutral particles, which can result in a decoupling of the charged (ions and electrons) and neutral fluids. Depending on the intensity of B and the ionization fraction of the gas, we distinguish between two types of MHD shocks, C- and J-types:

***J*-shocks:** if B is weak or absent, or if the ionization fraction of the gas is high, the collisional coupling between charged and neutral particles is strong. These particles behave like a single fluid, coupled to the magnetic field. The properties of these shocks are similar to that of the hydrodynamical shocks. Across the shock front, the variables (pressure, density, velocity, etc.) change abruptly (hence J for Jump. The transition region, which has a scale of the order of the mean free path, is treated as a discontinuity. The pre-shock and post-shock values of the gas density, pressure and temperature are related by the Rankine-Hugoniot jump conditions.

***C*-shocks:** if the magnetic field is present and the ionization fraction low (typically x_e=10^{-7}–10^{-8} in dense n_H=10^3–10^4 cm^{-3} molecular clouds), the neutral and charged fluids are decoupled from each other. These shocks are named multi-fluids or C–shocks (C for Continuous), because in this case the discontinuity is smoothed and the gas parameters vary continuously across the shock front.

In the shock wave, H_2 is collisionally excited (or destroyed) and chemically destroyed, but reforms on grain surfaces in the post-shock gas. In state-of-the art MHD shock codes (e.g. Flower and Pineau des Forêts, 2015), the rate of H_2 formation is simply parametrized as a function of the gas hydrogen density n_H, the grain density n_{gr} and grain size r_{gr}, and a coefficient that incorporates the probability that an H atom sticks to a grain, $1/(1+T_{eff}/T_{cr})^{0.5}$, with T_{eff} being the effective grain temperature (Le Bourlot et al., 2002) and T_{cr}≈30–100K a critical grain temperature below which the sticking probability tends to 1. In the shock codes, the H_2 formation rate [cm^{-3} s^{-1}] is thus often written as $n_H n_{gr} \pi r_{gr}^2 \left(\frac{8\pi T_{eff}}{\pi m_H (1 + T_{eff}/T_{cr})} \right)^{0.5}$.

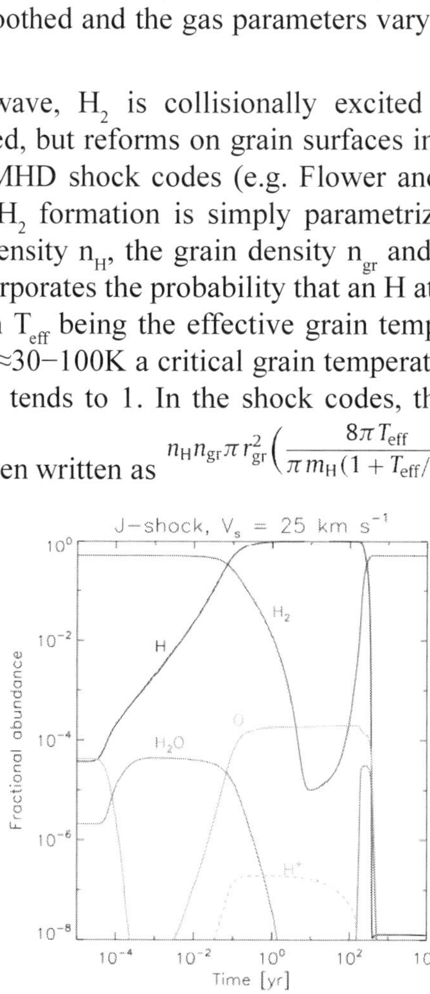

Figure 10. Fractional abundance profiles in a J-shock computed with the (Flower and Pineau des Forêts, 2015) MHD shock code. The shock velocity is 25 km s^{-1} and the pre-shock gas density n_H=10^4 cm^{-3}. The pre-shock magnetic field is B_0=0μG. Abundances are relative to H ($n(X)/n_H$).

Fig. 10 illustrates the impact of a dissociative J-shock wave in a molecular cloud. The passage of a shock wave impacts the physical and chemical quantities upon which the H_2 formation rate depends: the hydrogen density and grain properties (see Section 4.5.3 for a discussion about the effect of shocks on grains). Rapid dissociation occurs, followed by H_2 re-formation in the post-shock gas because a significant number of grains survive the passage of the shock. Fig. 11 illustrates the impact of the nature of the shock (i.e. B field and ionization fraction) on the H_2 excitation. For high energy levels, the diagram is flatter for J-shocks than for C-shocks because the maximum temperature reached in the shock is higher.

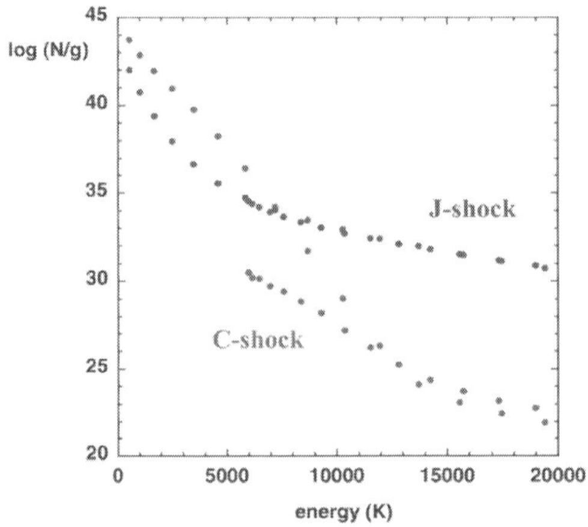

Figure 11. Comparison of the H_2 excitation diagrams of C- and J-type shock waves from the MHD code of Flower and Pineau des Forêts (2010). For both types of models, the shock velocity is 20kms^{-1} and the pre-shock density is n_H=2×10^4 cm^{-3}. For the C- (resp. J) type model, the initial transverse magnetic field intensity is B=140μG (resp. B=14μG).

MHD shock codes have been used to fit the observed H_2 diagrams, which provide clues on the dominant excitation mechanism, and physical conditions (pre-shock densities, shock velocities, magnetic field intensity), both in Galactic (e.g. Gusdorf et al., 2012) and extragalactic environments (Appleton, Guillard, Togi, Alatalo, Boulanger, Cluver, Pineau des Forêts, Lisenfeld, Ogle, Xu, 2017, Guillard, Boulanger, Pineau Des Forêts, Appleton, 2009, Guillard, Ogle, Emonts, Appleton, Morganti, Tadhunter, Oosterloo, Evans, Evans, 2012b, Lesaffre, Pineau des Forêts, Godard, Guillard, Boulanger, Falgarone, 2013).

RECOMMENDED PARAMETERS

The aim of this section is to provide a table of reviewed physical parameters as inputs for models. In the previous parts, we have revealed the heterogeneity of results associated with interstellar H_2 formation and the large variety of systems involved with this process. Even though a physical parameter can be precisely defined and evaluated, it is often not directly usable in models in such a fundamental form. For example, experiments and calculations have shown that for a correct description of H atom diffusion, a single height (and width) of a diffusion barrier is not sufficient; a distribution of barrier heights is a more realistic approach. However, in astrochemical models the use of such distributions is not implemented and therefore one has to choose a reasonable parameter that approximates the real complexity. We are well aware of the imperfections of this approach to parametrization, but at present it is the only approach that is viable. Therefore, each value in the table of physical parameters is the best approximation in the view of the authors. The parameters we present in the table are those commonly used in existing models.

We propose 8 prototypical classes of surface, each class, of course, often representative of a large variety of substrates: (1) amorphous and (2) crystalline silicates, (3) amorphous carbon (including a very large variety of compounds, from HAC to onion rings), (4) graphite, (5) PAHs and (6) PAH cations, and two morphologies of ice, one geometrically flat such as (7) polycrystalline ice, and one with a high degree of molecular disorder, such as (8) porous amorphous ice (ASW) grown at low (< 30 K) temperature.

Sticking Coefficients

For physisorption, the sticking coefficient is predominantly a function of the thermal energy of the incident atoms. Of course, in principle, the surface temperature is also a parameter. However, there is a distinct lack of measurements of sticking as a function of relevant surfacc tcmpcratures, and we have to neglect this parameter. This simplification is supported by the fact that physisorption plays a major role in environments where the surface temperature of grains is low and will not strongly affect sticking. We use the parametrization of Matar et al. (2010), of the sticking probability as a function of T_{gas}=T which depends on 3 parameters (S_0, T_0, β) such that

$$S_{\mathrm{phys}}(T) = S_0 \frac{1 + T/T_0}{(1 + T/T_0)^{\beta}}$$

where β is a geometrical factor which is equal to 2.5 for the case of isotropic deposition, which occurs in the ISM. Thus, this leaves the parameters S_0 and T_0. When no information exists we use unity for S_0 as it is usually close to one. T_0 is the only scaling parameter of the sticking coefficient for physisorption.

For sticking involving chemisorption we use the formalism of Sha et al. (2005) referred to in Cuppen et al. (2010) for the sticking probability. This probability also depends on the surface temperature T_s of the grain;

$$S_{\mathrm{chem}}(T, T_s) = \frac{\exp(-E_{\mathrm{chem}}/T)}{(1 + 5 \times 10^{-2}\sqrt{T + T_s} + 1 \times 10^{-14}T^4}$$

Therefore S_{chem} is only governed by one physical parameter E_{chem} which is the entrance barrier to the chemisorption sites. However, in case of large barriers, this expression cannot be used because H atoms can tunnel through the barrier and transmission coefficients across this barrier should be accounted for, especially at low H atom energies (Cazaux et al., 2011). In this latter study, a mixed classical-quantum dynamics calculation showed that by considering phonons, the resulting sticking probabilities are different than the transmission probabilities of the chemisorption barrier, which are typically used to model sticking probabilities. However, the resulting H_2 formation rate is only slightly influenced by the effect of phonons at low temperatures.

Energy Barriers

In the table of parameters below, we let E_{diffus} correspond to the height of the diffusion barrier between two physisorption sites. A lower value means faster diffusion. $E_{binding}$ is the energy required for an H atom to desorb from a physisorption site. $E_{diffPh\text{-}Chem}$, is the height of the diffusion barrier for going from a physisorption site to a chemisorption site. For the specific case of aromatic carbon rings, the barrier to chemisorption is dependent on whether one or more H atoms are already chemisorbed. In this case we indicate the values for the first and second hydrogen atom adsorption reaction. For graphite and neutral PAHs adsorption events beyond the first H atom often have a significantly lower or even a vanishing barrier.

We have not listed the diffusion barrier between two chemisorption sites because under astrophysical conditions this slow diffusion rarely contributes efficiently to H_2 formation. However calculated values are available for some situations (Ferro et al. (2003); Hornekær et al. (2006b)).

$E_{ReacCh\text{-}Ch}$ is the barrier to the reaction of two adjacent chemisorbed H atoms. $E_{Bind\text{-}Ch}$ is the binding energy of an isolated H atom on the listed surface.

σ_{ER} is the cross-section in Å^2 pr. chemisorbed H atom for H_2 production via Eley-Rideal abstraction reactions with gas phase H atoms

Table of Recommended Values

In Table 1 we typeset values which come directly from a value already described above, or in a paper cited earlier, in regular roman font. Sometimes a range of existing values is indicated. For some other cases, the adopted value is an educated choice and we indicate such values in italic. When several determinations of a given parameter exist, we use bold-face in order to indicate that these values have been confirmed by different methods (experiments or calculations).

Table 1. Selected values characterizing processes of hydrogen interaction with surfaces of dust grain analogs. $S_H(T)$ is the sticking coefficient of H at the temperature T, $E_{diffus.}$ is the activation energy of H diffusion, $E_{binding}$ is the binding energy of H to surfaces; energy values are in meV; 1 meV = 11.6 K/k_B.

PHYSISORPTION								
	1: Amorphous Silicate	2 : Crystalline Silicate	3: Amorphous Carbon	4: Graphite	5: PAH	6: PAH+	7: Crystalline Ice	8: Porous Amorphous Solid Water
A: S_H (T_{gas})	S_0=1.0, T_0=25 K	S_0=1.0, T_0=25 K	S_0=1.0, T_0=125K	S_0=1.0, T_0=24K	S_0=1.0, T_0=24K	S_0=1.0, T_0=24K	S_0=1.0, T_0=52K	S_0=1.0, T_0=52K
B: $E_{diffus.}$	35	8.5	44	4	4	4	< 18;16	**22**
C: $E_{binding}$	44	25	57	40	40	40	30	**50**
CHEMISORPTION								
D: $S_{chem}(T,T_g)$	E_{chem}=300	E_{chem}=260	E_{chem}=6	E_{chem}= 150	E_{chem}=60	E_{chem}=10	n.a	n.a
E: EdifpH−Chem	332	278	6	190;0	100;0	50;70	n.a	n.a
F: ERe-acCh−Ch	48	1510 (74)	> 1400	1400	n.a	n.a	n.a	n.a
G: EBind−Ch	1786	398	> 870;4000	870	1450	2810	n.a	n.a
H: σ_{ER}	n.a	n.a	0.03	4–17	0.06	0.06	n.a	n.a
DYNAMICS AND OTHER PROCESSES								
I: Nascent OPR	-	-	-	-	-	-	-	3 to 1

J: NSC	-	-	-	Yes	-	-	-	Varies with T
K: Int. Energy	medium v, low j	-	No	v 2–5, low j	-	-	No	No
L: Kin. Energy	-	-	No/Low	1.3 eV	-	-	-	No
M: Energetic formation	-	-	Yes	-	-	yes	-	Yes

The lines of the table are labeled by a letter, and rows by a number: number 1 for the amorphous silicate, number 2 for crystalline silicate, number 3 for amorphous carbon, number 4 for graphite, number 5 for PAH, number 6 for PAH cation, number 7 for crystalline ice and number 8 for porous amorphous solid water.

Letter A refers to the sticking to physisorption sites.

A1 : From reference Chaabouni et al. (2012).

A2 : No data, we propose the values of A1.

A3 : No clear data, however due to the possibly high porosity and roughness, and better mass matching of aliphatic groups, the sticking should be higher than on other substrates. The adopted values corresponds to a sticking of 50% at 300 K.

A4 : Best fit from Lepetit and Jackson (2011b). We point out that values for silicates and graphite surfaces seem to be very similar despite the fact that the adopted methods are independent (experiments vs calculations).

A5, A6: No data. We propose the values of A4.

A7 : From reference Matar et al. (2010), although the measurements were performed on compact ASW. This choice was made because compact ASW has the same flat topology as the crystalline ice and differs from the crystalline structure at meso-scale (tens of molecules). Therefore, the sticking properties of this type of ice are closer to crystalline ice than the very disordered amorphous solid water ice.

A8 : We take the values of A7. They are lower limits because it is known that the sticking of light species is enhanced by the porous structures of ice (e.g. Hornekaer et al. (2005)). We do not adopt different and higher values because the degree of porosity varies too much.

Letter B refers to the energy of diffusion $E_{diffus.}$ between two physisorption sites

B1 : From reference Perets et al. (2007)

B2 : From reference He et al. (2011). Note that in the case of polycrystalline silicates there are deeper adsorption sites and that the value of the diffusion barrier has been estimated to be 24 meV (Katz et al., 1999). Furthermore, Navarro-Ruiz et al. (2014) have computed a value of 60 meV using a molecular cluster representative of crystalline forsterite. We keep the low value of barriers from the reference He et al. (2011) because it compares different morphologies of silicates. The set of adopted binding and diffusion barriers may be less than perfect, but it is at least self-consistent.

B3 : From reference Katz et al. (1999). We point out that this value is the highest value for the diffusion barrier of all surfaces. We propose only this one existing value although amorphous carbon can adopt a large variety of morphologies wiht potentially different physical properties.

B4 : From calculations of Bonfanti et al. (2007)

B5, B6 : No data. We propose using the value of B4

B7: Diffusion on crystalline ice is known to be faster than the capabilities of experimental measurements at 8–15 K (Hama et al., 2012) which implies that the diffusion barrier is less than 18 meV. We choose a conservative values of 16 meV, although calculations propose even lower values (Senevirathne et al., 2017).

B8 : There is different experimental and theoretical work relatively convergent to establish a distribution of diffusion barriers (see previous section). The value 22 meV that we finally adopt is an effective value which is a consensus between the authors. However, note the reference (Perets et al., 2005) which gives a higher value (55 meV), the signature of a very deep physisorption site.

Letter C refers to the binding energy $E_{binding}$ of the H atom physisorbed on the surface.

C1 : From reference Perets et al. (2007)

C2: From reference He et al. (2011). See comment in B2

C3: From reference Katz et al. (1999). See comment in B3.

C4: From Ghio et al. (1980) and Bonfanti et al. (2007)

C5,C6: No data. We propose using the value of C4

C7 and C8: Consensus values adopted from the work in four different groups.

Letter D refers to the sticking to chemisorption sites.

D1 : From reference Navarro-Ruiz et al. (2015).

D2 : From reference Navarro-Ruiz et al. (2014). The presence of an adsorbed H on the surface changes the barriers significantly: the reaction between a chemisorbed atom and a physisorbed atom can have a barrier as low as 74 meV.

D3: From reference Mennella et al. (2006).

D4, D5, D6: From references Cazaux et al. (2016); Hornekær et al. (2006b); Rauls and Hornekær (2008).

Letter E refers to diffusion barrier of H atom going from the physisorption site to a chemisorption site $E_{difpH-Chem}$

E1: From reference Navarro-Ruiz et al. (2015).

E2: From reference Navarro-Ruiz et al. (2014). See comment in D2.

E3: Large spread in values expected. We propose using the value from D3.

E4, E5, E6: These values are obtained from the corresponding D values by adding the binding energy of the physisorption state.

Letter F refers to the barrier to the reaction of two adjacent chemisorbed H atoms $E_{ReacCh-Ch}$.

F1: From reference Navarro-Ruiz et al. (2015).

F2: From reference Navarro-Ruiz et al. (2014). See comment in D2.

F3: Large spread in values expected depending on e.g. aliphatic vs. aromatic nature of the amorphous film. We suggest using the value of F4 as a lower limit, since aliphatic structures are expected to offer even higher energy binding sites for H atoms with associated higher recombination barriers. The main reactivity for chemisorbed H atoms is expected to be via Eley-Rideal or hot-atom reactions or via reactions with physisorbed H atoms.

F4: From reference Hornekær et al. (2006b)

Letter G refers to the binding energy of an isolated H atom in chemisorption site $E_{Bind-Ch}$.

G1: From reference Navarro-Ruiz et al. (2015).

G2: From reference Navarro-Ruiz et al. (2014). See comment in D2.

G3: Large spread in values expected depending on e.g. aliphatic vs. aromatic nature of the amorphous film. For predominantly aromatic films we suggest using the value of G4 as a lower limit, while a value of 4 eV is suggested for mainly aliphatic structures following reference Mennella (2008). It is to note that any model using these combined values should also be coherent with the value of the cross section experimentally measured.

G4 : From references Hornekær et al. (2006a); Lehtinen et al. (2004)

G5: Listed value is for the most stable adsorption site of Rauls and Hornekær (2008): the energies of the different adsorption sites are 700, 700, 1450 meV for central, edge and outer-edge sites, respectively.

G6: Listed value if for the most stable absorption site of Cazaux et al. (2011): the energies of the different adsorption sites are 1900, 2140, 2810 meV for central, edge and outer-edge sites, respectively.

Letter H refers to the cross-section σ_{ER} in Å^2 pr. chemisorbed H atom for H_2 production via Eley-Rideal abstraction reactions with gas phase H atoms

H3: From reference Mennella (2008).

H4: From reference Zecho et al. (2002b).

H5: From reference Mennella et al. (2012).

H6: No data. We propose using the value of H5.

The Letters I to M refer to the dynamics and other processes about the H_2 formation on the listed surface. The blank spaces in this part of the table remind us that there are some important issues related to the formation of H_2, like the internal energy of the molecule released, and the capability of the surface to induce nuclear spin conversion, which are still unknown for many of our prototypical surfaces.

Letter I and J refer to the nascent Ortho-Para Ratio (OPR) and to the Nuclear Spin Conversion (NSC), respectively.

I8: from references Gavilan et al. (2012); Watanabe et al. (2010)

J4: from Palmer and Willis (1987)

J8: from Ueta et al. (2016)

Letter K refers to the repartition of the internal energy of the H_2 molecule formed.

K1: from reference Lemaire et al. (2010)

K3: from reference Mennella (2008)

K4: Experiments from Creighan et al. (2006); Islam et al. (2007); Latimer et al. (2008). Many different calculations (e.g. Morisset et al. (2005))

K7: from reference Congiu et al. (2009) when some H_2 is covering the surface, unless probably medium v.

K8: from reference Congiu et al. (2009)

Letter L refers to the informations about kinetic energy of the H_2 molecule formed.

L3:Expected from porous nature of the grains.

L4: from reference Baouche et al. (2006)

L8: Expected from the porous nature of the ice

Letter M refers to the informations about the energetic formation, such as UV, electrons or ions bombardments, that would lead to H_2 formation and release in the gas phase.

M3: from reference Alata et al. (2014)

M6: calculations Allain et al. (1996); Le Page et al. (2001)

M8: Many references with ions, UV and electrons (see text above on water ice section)

H_2 FORMATION IN DIFFERENT ASTROPHYSICAL ENVIRONMENTS

The interstellar medium involves a wide range of physical (temperature and density) and irradiation (UV, X-ray and cosmic-ray particles) conditions. In this section of the paper, we focus on some specific regions of the ISM were H_2 is formed, observed or has a particular role in the chemistry of more complex species. Specifically, we discuss: diffuse clouds, dense photo-dissociation regions (PDRs, for which we distinguish high illumination PDRs, abbreviated HI, and low illumination PDRs, abbreviated LI), cold shielded regions, hot shielded regions and molecular shocks inside the Galaxy and in other galaxies. Table 2 summarizes some key physical parameters of these sources. The first column gives the UV illumination measured by the G_0 parameter, which measures the local energy density in far UV photons (between 912 Å and 2400 Å), normalized to the energy density of the average interstellar UV field u_{Habing}=5.3×10^{-14}erg·cm^{-3} (Habing, 1968):

$$G_0 = \frac{\int_{912A}^{2400A} u_\lambda(\lambda)\, d\lambda}{u_{\text{Habing}}}.$$

The second column gives the proton density n_H of the medium (in cm^{-3}). The third and fourth columns list the gas and dust temperatures. The fifth column gives the fraction of hydrogen in the molecular form f_{H_2} (when this parameter is unity, all the hydrogen is molecular). The sixth column is the observed H_2 ortho-para ratio (OPR). Finally, the last column gives the

H_2 formation efficiency R_{H_2} derived from observations (in cm^3 s^{-1}). In the following, we present these environments and discuss the relevant physical-chemical processes involved in the formation and destruction of H_2 in each of these regions of the ISM.

Table 2. Physical parameters of each type of astrophysical source.

Source	G_0 (Habing unit)	n_H (cm^{-3})	Gas T (K)	Dust T (K)	f_{H_2}	H_2 o/p	R_{H_2} ($\times 10^{-17}$)
Dense PDR (HI)	a few 10^3 - 10^4 (a)	10^4 - 10^6 (a, b, c)	10 - 3000 (a)	25 - 70 (g)	0 - 1	1 - 3 (k, l)	(3 - 15) (r)
Dense PDR (LI)	a few 10 - 100 (a)	10^3 - 10^5 (d, e)	10 - 400 (d)	13 - 30 (g)	0 - 1	1 - 3 (m, l, e)	(10 - 30) (r)
Diffuse clouds	1	1–100	50 - 130	20 (h)	10^{-4}–0.8 (j)	0.4 - 0.9 (n, o)	(3 - 5) (s, n, o)
Cold Cores	1	10^4 - 10^5 (f)	10 (f)	6 - 15 (i)	1	0 - 0.1 (p, q)	?
Hot Cores (f)	1	> 10^6	100 - 300	100 - 300	?	?	?
Shocks (Galactic)	1 - 10(u)	10^3–10^4 (v)	10 - 10^4	?	0.2 - 1	0.5 - 3 (w)	?
Shocks (extra-Galactic)	0 - 100(x)	10 - 10^5	10 - 10^5	?	10^{-5} - 1 (y)	1 - 3 (z)	?

References: (a) Hollenbach and Tielens (1999), (b) Parmar et al. (1991), (c) Köhler et al. (2014a), (d) Habart et al. (2005a), (e) Habart et al. (2011), (f) Bergin and Tafalla (2007), (g) Arab et al. (2012), (h) Planck Collaboration et al. (2014), (i) Hocuk et al. (2016), (j) Snow et al. (2008), (k) Fuente et al. (1999), (l) Fleming et al. (2010), (m) Habart et al. (2003), (n) Gry et al. (2002), (o) Lacour et al. (2005), (p) Troscompt et al. (2009), (q) Pagani et al. (2009), (r) Habart et al. (2004), (s) Jura (1974), (t) Herbst and van Dishoeck (2009), (u) Lesaffre et al. (2013), (v), Gusdorf et al. (2015), (w) Nisini et al. (2010), (x) Guillard et al. (2009); 2012b) and Uzgil et al. (2016), (y) Flower et al. (2003), (z) Harrison et al. (1998).

Diffuse and Translucent Interstellar Medium

Diffuse clouds represent the regime in the ISM where moderate density clouds are exposed to the background UV field of the galaxy (the interstellar radiation field), and consequently where nearly all molecules are rapidly destroyed by photo-dissociation. In diffuse clouds, molecular hydrogen has been observed through its FUV absorption lines (Savage et al. 1977 with Copernicus, Gry, Boulanger, Nehmé, Pineau des Forêts, Habart, Falgarone,

2002, Rachford, Snow, Tumlinson, Shull, Blair, Ferlet, Friedman, Gry, Jenkins, Morton, Savage, Sonnentrucker, Vidal-Madjar, Welty, York, 2002, Rachford, Snow, Tumlinson, Shull, Roueff, Andre, Desert, Ferlet, Vidal-Madjar, York, 2001, Shull, Tumlinson, Jenkins, Moos, Rachford, Savage, Sembach, Snow, Sonneborn, York, Blair, Green, Friedman, Sahnow, 2000, Snow, Rachford, Tumlinson, Shull, Welty, Blair, Ferlet, Friedman, Gry, Jenkins, Lecavelier des Etangs, Lemoine, Morton, Savage, Sembach, Vidal-Madjar, York, Andersson, Feldman, Moos, 2000 with FUSE). In such clouds, from 10^{-6} to about half of the total number of hydrogen nuclei are bound in hydrogen molecules (e.g., Spitzer, Cochran, Hirshfeld, 1974, Shull, Beckwith, 1982). From their H_2 content, these clouds can be categorized as

(1) Diffuse atomic clouds. Here the molecular fraction $f_{H_2} = \frac{2nH_2}{2nH_2+nH} \leq 0.1$. The density ranges from 10 to 100 cm^{-3} and the gas and dust temperatures are around 30–100 K and 15–20 K respectively (Snow and McCall, 2006, for a review).

(2) Diffuse molecular clouds. Here the molecular fraction $f_{H_2} \geq 0.1$, but carbon is still mostly ionized $f_{C^+} = \frac{nC^+}{nC^++nC+nCO} \geq 0.5$. They exist from an extinction of ~ 0.2, and have densities ranging from 100 to 500 cm−3 and gas and dust temperatures around 30–100 K and 15–20 K respectively.

(3) Translucent clouds. With sufficient protection from interstellar radiation (from extinctions of 1–2) carbon begins its transition from an ionized atomic state into a neutral atomic (C) or molecular (CO) form. Clouds in this transition phase have been defined as "translucent" (van Dishoeck and Black, 1989). Here, until Av ~ 2, hydrogen is mostly molecular with densities from 500 to 5000 cm^{-3}. The gas and dust temperatures are about 15–40 K (van Dishoeck and Black, 1989) and 13–19 K (Forbrich, Öberg, Lada, Lombardi, Hacar, Alves, Rathborne, 2014, Launhardt, Stutz, Schmiedeke, Henning, Krause, Balog, Beuther, Birkmann, Hennemann, Kainulainen, Khanzadyan, Linz, Lippok, Nielbock, Pitann, Ragan, Risacher, Schmalzl, Shirley, Stecklum, Steinacker, Tackenberg, 2013, Nielbock, Launhardt, Steinacker, Stutz, Balog, Beuther, Bouwman, Henning, Hily-Blant, Kainulainen, Krause, Linz, Lippok, Ragan, Risacher, Schmiedeke, 2012, Roy, André, Palmeirim, Attard, Könyves, Schneider, Peretto, Men'shchikov, Ward-Thompson, Kirk, Griffin, Marsh, Abergel, Arzoumanian,

Benedettini, Hill, Motte, Nguyen Luong, Pezzuto, Rivera-Ingraham, Roussel, Rygl, Spinoglio, Stamatellos, White, 2014) respectively.

Dust Properties in the Diffuse and Translucent ISM

The dust properties in the diffuse ISM have been the subject of numerous studies; including, for example, the Draine and Li 2007 and Compiègne et al. 2011 models. More recently, theoretical modeling based on laboratory experiments (THEMIS[3], Jones et al. 2013) was applied with great success in the analysis and interpretation of Planck, Herschel and Spitzer observations, even in the most diffuse regions. Ysard et al. (2015) are able, with small variations in the dust properties, to explain most of the variations in the dust emission observed by Planck-HFI in the diffuse ISM.

The dust size distribution in the diffuse ISM, derived by Ysard et al. (2015), can be described as in the following. Small grains (≤ 20 nm) follow a power-law size distribution[4] with α=-5, with a minimum size of a_{min}=40 Å. Large grains follow a log-normal size distribution with a peak at a size of 0.15 μm. Using this dust size distribution and the radiation field and the gas density distribution found in the diffuse ISM, the (mass-weighted) mean temperature of dust grains derived as being $\sim$ 19 K for small and big carbon grains, and $\sim$ 16 K for big silicate grains. The maximum temperature has been derived as being $\sim$ 160 K for small carbon grains, $\sim$ 50 K for big carbon grains, and $\sim$ 16 K for big silicate grains. The temperature of big dust grains has also been estimated observationally from modified blackbody fits as T $\sim$ 20 K (e.g., Planck collaboration XI, 2014), but this value results from a mix of dust at different temperatures along the line of sight.

Main Chemical Processes for the Formation of H_2 Depending On the Environment

Diffuse Clouds with low Radiation Field

In diffuse clouds, H_2 forms with a rate of $1-3\times10^{-17}n_H n(H)cm^{-3}s^{-1}$ where n_H is the total proton density and n(H) is the density of H-atoms (Gry, Boulanger, Nehmé, Pineau des Forêts, Habart, Falgarone, 2002, Hollenbach, Werner, Salpeter, 1971, Jura, 1974). At grain temperatures typical of diffuse cloud conditions (Gry, Boulanger, Nehmé, Pineau des Forêts, Habart, Falgarone, 2002, Hocuk, Cazaux, Spaans, Caselli, 2016, Mathis, Mezger, Panagia, 1983), which is 15–20 K, physisorbed H atoms can still remain attached to dust grains and H_2 can form efficiently through the reaction of 2 physisorbed

H atoms (Langmuir-Hinshelwood mechanism). This formation process is very efficient and depends on the binding energies of the physisorbed H atoms, which have been derived experimentally and are reported in Table 1. Rate equations and Monte Carlo simulations for the formation of H_2 on such surfaces, where both physisorption and chemisorption are considered (Cazaux, Spaans, 2009, Cazaux, Tielens, 2002, Cazaux, Tielens, 2004), or where the roughness of the surface is taken into account (Chang, Cuppen, Herbst, 2006, Cuppen, Herbst, 2005), show that H_2 formation efficiency involving physisorbed H atoms can reach 100% for a surface temperature of 20 K. Therefore, in diffuse environments with low radiation field (i.e. with dust temperatures lower than 20 K), the formation of H_2 is mainly due to the involvement of physisorbed H atoms on dust grains. In Fig. 12, the main mechanisms responsible for the formation of H_2 in diffuse, translucent and molecular clouds are presented. For a diffuse cloud with low G_0, the main process is the encounter of two H physisorbed H atoms: the Langmuir-Hinshelwood mechanism. This mechanism dominates for low dust temperatures ($T_{dust} \leq 20\,K$).

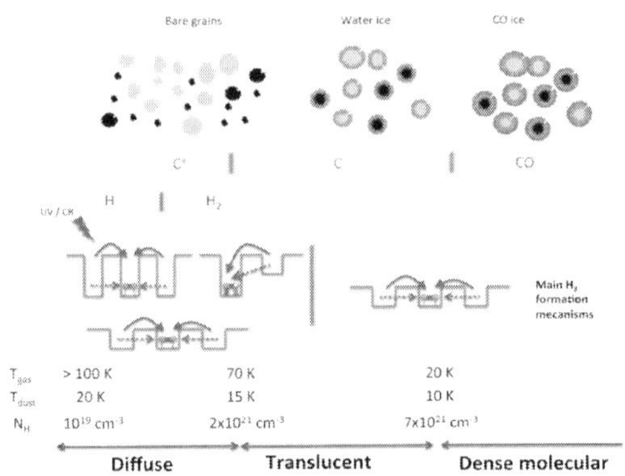

Figure 12. Sketch of the main H_2 formation processes in diffuse to dense interstellar medium.

Diffuse Clouds with High Radiation Field

In environments where the radiation field is significant, but not strong enough to dissociate PAHs, which corresponds to $n_H/G_0 \geq 3\ 10^{-2}$ (Le Page et al., 2009), where G_0 is the radiation field in Draine's units, PAHs could also play a role in the formation of H_2. Under such conditions, PAH cations can contribute to the formation of H_2 with rates comparable with the typical rate in the ISM. However, in regions typical of the diffuse ISM, the radiation field is less important, $n_H/G_0 \sim 1-10^2$, and the formation of H_2 on PAHs is less efficient (Le Page et al., 2009). In these conditions, H_2 forms predominantly on dust grains (Boschman et al., 2015) through the Eley-Rideal mechanism involving chemisorbed H atoms. The possible mechanisms for the formation of H_2 in diffuse clouds with high G_0, shown in Fig. 12, are either through abstraction of H_2 from PAHs, or through Eley Rideal mechanism on dust grains involving chemisorbed H atoms, or by photolysis of hydrogenated amorphous carbons (Alata et al., 2014).

Translucent Gas

In translucent clouds, the temperature of dust grains becomes lower than in diffuse clouds (15-18 K) and, above visual extinctions of around 3, ices start to cover the dust surfaces. The interactions of H atoms with ices are different than those with bare surfaces because (1) H atoms cannot chemisorb on ices and (2) H can only physisorb on icy surfaces with binding energies comparable, or slightly lower, than the binding energies on bare surfaces. These points imply that H_2 formation on icy dust can only involve physisorbed H atoms, and therefore that H_2 in translucent clouds is formed predominantly via the Langmuir-Hinshelwood mechanism. However, the efficiency of the H_2 formation on icy dust strongly depends on the porosity of the ices. If the ices are crystalline, the efficiency of forming H_2 can be 100% up to surface temperatures of $\sim$ 12 K. Above this surface temperature, the efficiency will drop exponentially, which implies that H_2 formation will be very inefficient at higher temperatures. If the ices are porous, the efficiency is 100% until the surface temperature reaches $\sim$ 19K. In this case, the formation of H_2 is efficient for the range of temperatures encountered in translucent clouds. Fig. 12 summarizes the conditions met in translucent clouds as well as the main process for the formation of H_2, which is via the Langmuir-Hinshelwood mechanism involving two physisorbed H atoms.

Dense PDRs

Photo-dissociation regions (PDRs) are regions of neutral interstellar clouds in which the heating and chemistry are dominated by the impact of stellar UV photons (Hollenbach, Tielens, 1997, Hollenbach, Tielens, 1999). This definition encompasses both diffuse neutral clouds (cf. previous subsection) in the ambient galactic UV field, and dense molecular clouds exposed to the radiation of nearby young massive stars. This second kind of environment, commonly called dense PDRs, is the focus of this section.

Physical conditions in PDRs

The physics and chemistry of PDRs is controlled by the local gas density n and the intensity G_0 of the UV radiation field. In dense PDRs, the gas densities can range from $10^3 cm^{-3}$ to $10^7 cm^{-3}$, with UV intensities G_0 from a few units to the order of 10^6 (Hollenbach and Tielens, 1999). The UV field intensities impinging on dense PDRs seem to be correlated to the gas density of the dense structures (Young Owl et al., 2002), and directly proportional to their thermal pressure (Chevance, 2016),Joblin et al., hinting at the role of radiative feedback from the stars in the formation of these dense structures in star forming regions.

The typical physical conditions in dense PDRs are given in Table 2. As the large ranges of UV fields and gas densities result in large variations of the gas conditions and dust properties (which can affect the possible H_2 formation mechanisms at work) we separate low-illumination PDRs ($G_0 < 1000$) from high illumination PDRs ($G_0 > 1000$). Typical examples of low illumination PDRs are the Horsehead (Habart et al., 2005a), and the ρ Oph. (Habart et al., 2003) PDRs. Famous high illumination PDRs include NGC7023 NW (Köhler et al., 2014a), Joblin et al., and the Orion Bar (Parmar et al., 1991,Joblin et al.).

The UV field is progressively extinguished by dust, and by the molecules it dissociates, as it penetrates into the cloud. As a result, the physical conditions vary with the optical depth A_V into the cloud. As H_2 dissociation occurs through line absorption, H_2 self-shielding is important and results in a sharp transition from atomic to molecular hydrogen (see Sternberg et al. 2014 and Bialy and Sternberg 2016 for an analytical theory of the H/H_2 transition), the depth of this transition being highly dependent on the ratio of the gas density to UV intensity (the optical depth of the transition goes from a few 10^{-4} to ~ 1). The resulting PDR structure is layered, with a hot atomic layer before the H/H_2 transition, a warm molecular layer between the H/H_2 transition and the $C^+/C/CO$ transition, and colder molecular gas deeper inside the cloud. This structure is represented in Fig. 13, with a more precise description of the physical conditions in the different layers of the PDR.

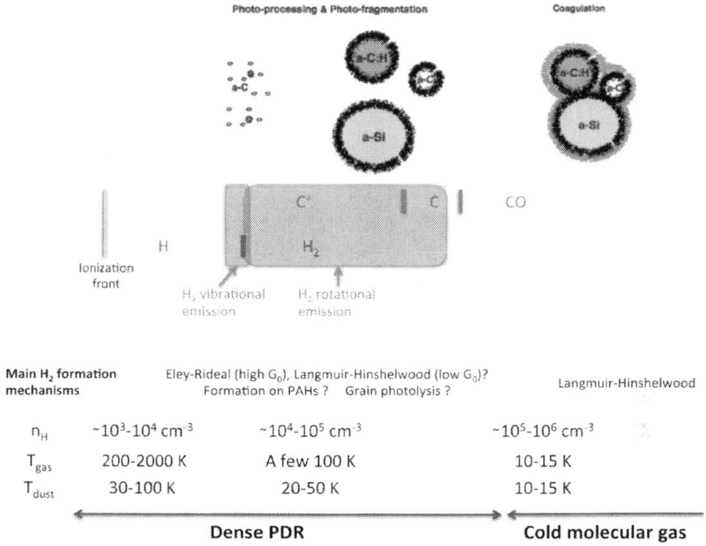

Figure 13. Schematic illustrating the main H_2 formation processes in PDRs.

Dust Evolutionary Processes in PDRs

Due to the strong UV fields and gas density variations associated with PDRs, the nature of the dust in these regions evolves from the edge of the PDR to its center, as well as varying from one PDR to another (e.g., Rapacioli, Joblin, Boissel, 2005, Rapacioli, Calvo, Joblin, Parneix, Toublanc, Spiegelman, 2006, Berné, Joblin, Deville, Smith, Rapacioli, Bernard, Thomas, Reach, Abergel, 2007, Pilleri, Montillaud, Berné, Joblin, 2012, Arab, Abergel, Habart, Bernard-Salas, Ayasso, Dassas, Martin, White, 2012, Arab, 2012, Köhler, Habart, Arab, Bernard-Salas, Ayasso, Abergel, Zavagno, Polehampton, van der Wiel, Naylor, Makiwa, Dassas, Joblin, Pilleri, Berné, Fuente, Gerin, Goicoechea, Teyssier, 2014a, Pilleri, Joblin, Boulanger, Onaka, 2015).

Studies involving the whole dust emission spectrum, from the infrared to the (sub-)mm, indicate significant variations of the dust properties compared with that in the diffuse ISM (e.g., Arab, Abergel, Habart, Bernard-Salas, Ayasso, Dassas, Martin, White, 2012, Arab, 2012). In particular, a decrease in the abundance of small polyaromatic rich carbon grains (denoted as PAHs) by at least a factor of 2 is very likely. This reduction could result from photo-destruction due to the strong UV radiative energy input. Moreover, Pilleri et al. (2012) found that the fraction of carbon locked in very small grains, relative to the total carbon in the IR band carriers (i.e., very small grains and PAHs), decreased with increasing UV radiation field, which was interpreted as evidence for photo-destruction of very small grains. In PDRs, such as NGC 7023, Pilleri et al. (2015) also provide evidence for a change in the aliphatic to aromatic composition, most likely due to the strong UV radiative energy input. This deduction could suggest that photo-processing of very small grains produces PAHs.

Deeper inside the PDRs, with a decreased UV field and an increased gas density, small grains must coagulate onto the surface of big grains. Compared with the diffuse ISM, an increase in the emissivity of the big grains indicates coagulation (e.g., Köhler, Guillet, Jones, 2011, Arab, 2012). Indirect determinations of the dust extinction properties in some PDRs have also been reported (e.g. determination of R_V=5.62 in the NGC7023 NW PDR by Witt et al. 2006), hinting at increased grain sizes compared to the diffuse ISM. However, we must emphasize that dust properties pertinent to the diffuse ISM (such as the Draine and Li 2007 and Compiègne et al. 2011 models) remain used in most PDR models.

Models indicate that the grains remain bare until a visual extinction of $\gtrsim 3$ (Esplugues, Cazaux, Meijerink, Spaans, Caselli, 2016, Hollenbach, Kaufman, Bergin, Melnick, 2009), so that bare surfaces are most relevant for understanding H_2 formation in PDRs.

Dust Temperatures in PDRs and Size Distribution

Dust grains in the PDR are heated by the UV field and their temperature decreases from the edge to the inner part of the cloud. The grain temperature remains significantly lower than the gas temperature in most PDRs. For large grains, temperatures derived from modified blackbody fits and radiative transfer model are given in Table 2.

In high illumination PDRs (such as the Orion Bar, Arab et al., 2012), dust temperatures are found to decrease from 70 K to 35 K, with values of 50−60K in the H/H_2 transition region (where H_2 emission lines peak). Similar gradients but with overall lower temperatures (due to lower UV fields) are found in NGC7023 NW (50−25K with 30 K at the H/H_2 transition, Köhler et al. 2014a), and in the Horsehead PDR (30−13K, 20−30K at the H/H_2 transition).

The dust size distributions in the ISM (including diffuse and dense PDRs) result in most of the available dust surface being associated with small grains. The small grains are thus potentially the most important contributors to H_2

formation, in particular in warm regions. The temperature of small grains has a more complex behavior than that of larger grains. Due to the presence of UV photons, the temperature of small dust grains fluctuates constantly (spikes of a few 100 K for a grain of dimension a few nm absorbing a 912 Å photon). This effect was first investigated in order to understand the IR emission of PAHs and very small grains (Desert, Boulanger, Shore, 1986, Draine, Li, 2001), which is dominated by these high temperature spikes. These temperature fluctuations can also significantly affect the efficiency of the different H_2 formation mechanisms in PDRs, as small grains spend a large fraction of their time at low temperatures (10–20K) between UV photon absorptions, even at the warm edge of PDRs. Here then, surface processes can proceed between temperature spikes, significantly increasing the efficiency of the Langmuir-Hinshelwood mechanism for H_2 formation (Bron, Le Bourlot, Le Petit, 2014, Cuppen, Morata, Herbst, 2006), or of ortho-para conversion of H_2 on grain surfaces (Bron et al., 2016).

Discussion of the Relevant or Possible Formation Mechanisms in PDRs

In PDRs, the Langmuir-Hinshelwood formation mechanism for H_2, where two H atoms physisorbed on a grain surface meet and react, does not appear to be effective on big grains at thermal equilibrium; here, the low physisorption energies lead to very short residence times at the dominant temperatures of the dust in these regions. In order to allow formation of H_2 at higher surface temperatures, mechanisms that involve chemisorbed H atoms have been proposed by several authors (e.g., Cazaux, Tielens, 2004, Le Bourlot, Le Petit, Pinto, Roueff, Roy, 2012). This alternative formation process is usually modeled using the Eley-Rideal (ER) mechanism. Here, H atoms bonded chemically to the surface (chemisorbed), are stationary on the surface until an H atom from the gas-phase interacts with them to form an H_2 molecule.

The ER mechanism, which requires high gas temperatures to allow H atoms to enter the chemisorbed state, is predicted to be efficient only in high illumination PDRs ($G_0 > 10^3$, e.g. Le Bourlot et al., 2012). Boschman et al. (2015) model the influence of this ER route to H_2 formation from PAHs, on total H_2 formation rates in PDR. They find that the photodesorption of H_2 from PAHs can reproduce the high H_2 formation rates derived in moderately/highly excited PDRs (Habart et al., 2004). Nevertheless, the presence of the thermal barrier to chemisorption, observed experimentally (when hydrogenating coronene cations, Boschman et al., 2012), limits this process to high gas temperatures (> 200 K). For PDRs with low/intermediate excitation ($G_0 < 200$ and $n_H > 10^3$ cm^{-3}), Bron et al. (2014) show that the LH mechanism on small grains with fluctuating temperatures could be an efficient route to H_2 formation.

Other processes involving chemisorbed H atoms, such as the photon-processing of grains, are also of great interest in PDRs. As shown experimentally by Alata et al. (2014), UV photon-irradiation of a-C:H leads to very efficient production of H_2 molecules with rates similar to the ones derived in moderately/highly excited PDRs. In line with the interstellar evolution of carbonaceous dust, H_2 formation may occur via such UV photon-driven C-H and C-C bond dissociations in a-C:H (nano-)grains and the associated decomposition of those grains observed in PDRs. This process could also liberate (the precursors to) species such as C_2H, C_3H_2, C_3H^+, C_4H, which have been observed in PDRs (Cuadrado, Goicoechea, Pilleri, Cernicharo, Fuente, Joblin, 2015, Guzmán, Pety, Goicoechea, Gerin, Roueff, Gratier, Öberg, 2015, Pety, Gratier, Guzmán, Roueff, Gerin, Goicoechea, Bardeau, Sievers, Le Petit, Le Bourlot, Belloche, Talbi, 2012, Pety, Teyssier, Fossé, Gerin, Roueff, Abergel, Habart, Cernicharo, 2005),Pilleri et al.. Jones and Habart (2015) theoretically investigate this UV-induced H_2 formation pathway, adopting the dust composition and size distribution from the Jones et al. (2013) dust model, which is specifically tuned to the evolution of the optical and thermal properties of a-C:H grains in the ISM. They conclude that such a process would be sustainable as long as the radiation field is intense enough to photo-dissociate C-H bonds but not intense enough to break the C-C bonds in the aliphatic bridging structure; the latter process would photo-fragment the a-C(:H) grains. Thus, this H_2 formation mechanism will be inefficient deep inside the cloud, because there are too few extreme ultraviolet (EUV) photons left. Neither will this process be important in intense radiation fields, because of the rapid photo-fragmentation of the grains. These ideas appear to be in general agreement with the H_2 observations presented in Habart et al. (2004), who suggest an enhanced H_2 formation rate in moderately-excited PDRs. Because the small grains (a ~ 0.5-5 nm) dominate the dust surface and the C-H bond photo-dissociation efficiency decreases with dust size (other channels such as thermal excitation or fluorescence begin to compete), these grains make the largest contribution to the total H_2 formation rate. However, either fast rehydrogenation of the small grains or advection of unprocessed grains from deeper inside the cloud are necessary to sustain a steady H_2 formation rate.

This aspect of the UV promoted process still has to be investigated in more detail, by going beyond the steady-state approach of the current main PDR models. These possible H_2 formation processes are summarized in Fig. 13.

Impacts of H_2 Formation in PDR Studies

The efficiency of H_2 formation controls the depth within the cloud at which the dominant form of hydrogen changes from atomic to molecular. As the gas temperature decreases, as we go from the PDR edge to the inner part of the cloud, the H_2 formation efficiency thus affects the gas temperature at which molecular gas begins to appear, and hence the amount of warm molecular material in PDRs. The warm molecular layer immediately following the H/H_2 transition is crucial to explain several PDR tracers, such as the rotational emission of H_2 (Habart et al., 2004), high-J CO lines Joblin et al., and abundances of CH^+ (and other species resulting from warm molecular chemistry). These tracers are then used to understand energy transfer (e.g. radiative vs. mechanical) in extragalactic studies (e.g. Kamenetzky, Rangwala, Glenn, Maloney, Conley, 2014, Rosenberg, van der Werf, Aalto, Armus, Charmandaris, Díaz-Santos, Evans, Fischer, Gao, González-Alfonso, Greve, Harris, Henkel, Israel, Isaak, Kramer, Meijerink, Naylor, Sanders, Smith, Spaans, Spinoglio, Stacey, Veenendaal, Veilleux, Walter, Weiß, Wiedner, van der Wiel, Xilouris, 2015). A precise understanding of the H_2 formation mechanisms at play is thus crucial for the interpretation of these tracers.

Cold Shielded ISM

Here we define the cold shielded interstellar medium as any medium characterized by low temperatures (below 15 K), medium species densities (a few 10^4), and high visual extinction (i.e., absence of directly incident UV photons). Such regions, like for example cold dense cores as defined by Bergin and Tafalla (2007), make up one stage of the star formation process. There is no observational evidence of small grains (Rapacioli, Joblin, Boissel, 2005, Tibbs, Paladini, Cleary, Muchovej, Scaife, Stevenson, Laureijs, Ysard, Grainge, Perrott, Rumsey, Villadsen, 2016), while grain growth is suspected to occur early in the process of cold core formation (Pagani, Steinacker, Bacmann, Stutz, Henning, 2010, Steinacker, Andersen, Thi, Paladini, Juvela, Bacmann, Pelkonen, Pagani, Lefèvre, Henning, Noriega-Crespo, 2015, Ysard, Köhler, Jones, Dartois, Godard, Gavilan, 2016).

Observations indicate a log-normal distribution of dust mass, with a peak in the distribution around 0.5-1 μm (Steinacker, Andersen, Thi, Paladini, Juvela, Bacmann, Pelkonen, Pagani, Lefèvre, Henning, Noriega-Crespo, 2015, Ysard, Köhler, Jones, Dartois, Godard, Gavilan, 2016). These grains are covered by mantles of molecules, mostly composed of H_2O (Boogert et al., 2015). Therefore, in these regions, H_2 formation proceeds mostly via physisorption (LH) pathways. Due to low gas temperature, the sticking coefficient is close to unity. The low temperature of the grains prevents H atom desorption, and H atom diffusion allows efficient H_2 recombination. H_2 formed will progressively desorb as its coverage increases and the average binding energy decreases (Amiaud et al., 2006). Due to the interactions with other adsorbed H_2 molecules, the desorbing hydrogen molecules are expected not to be vibrationally excited (Congiu et al., 2009).

H_2 cannot be directly observed in the cold shielded ISM, but it is very likely that most of the hydrogen is in the molecular form. This is because the destruction of H_2 by secondary UV radiation, induced by cosmic rays, is readily compensated for by a high formation efficiency. From the chemical point of view, H_2 is the most abundant molecule in cold cores; however, the influence of molecular hydrogen on the surface chemistry of the cores is limited. Most hydrogenation reactions on the surfaces involving H_2 have significant barriers so that, even though the abundance of H_2 is larger than that of atomic hydrogen, the reactions with H are considered to be much faster. There are a few of these reactions that are, however, exothermic and may proceed via tunneling; for example, H_2+ OH (Meisner, Lamberts, Kästner, Oba, Watanabe, Hama, Kuwahata, Hidaka, Kouchi, 2012).

Recently, astrophysicists have modified gas-grain codes to take into account the competition between reaction and diffusion on the surfaces (Garrod, Pauly, 2011, Ruaud, Wakelam, Hersant, 2016). This diffusion increases significantly the efficiency of reactions with activation barriers. A simple way to understand this effect is to consider that a species (the most mobile one) reaching a site on the surface which is already occupied by another species, will have a certain probability to remain associated with that species for some time, rather than leaving it instantaneously. During this time, the probability for a reaction with barrier to occur increases. With this new generation of models, the abundance of H_2 on the surface is important.

The crucial parameter to determine the surface abundance of H_2 is not its rate of formation (which is very fast in the cold shielded ISM), it is rather its binding energy. The binding energy of H_2 on water ices is such that at high densities (above 10^9 cm^{-3} at 10 K), the molecules begin to get depleted from the gas phase and start to dominate the coverage of the grains (Hincelin, Chang, Herbst, 2015, Wakelam, Ruaud, Hersant, Dutrey, Semenov, Majumdar, Guilloteau, 2016). The sticking of H_2 onto multi-layers, i.e. on to itself, is much less efficient (Cuppen and Herbst, 2007) slowing down the depletion of H_2 from the gas. The details of the sticking of H_2 on the surface has to be carefully considered in gas-grain models because it can affect the general grain chemistry as for instance shown by Wakelam et al. (2016) in protoplanetary disks.

It has been proposed in the literature that the formation of H_2 on dust surfaces would be so exothermic that each reaction forming H_2 would locally heat the surface of the grains, allowing for the evaporation of some light species such as CO (Duley and Williams, 1993). Such non-thermal evaporation processes have been added to several gas-grain models, and desorption by such a route is claimed to be more efficient than desorption by cosmic-ray heating (Duley, Williams, 1993, Roberts, Rawlings, Viti, Williams, 2007, Willacy, Williams, Duley, 1994). However, in contrast, recent experiments have reported a negative effect on desorption rates due to H_2 formation (Minissale et al., 2016). Here the energy release upon H_2 formation changes the morphology of the ice (Accolla et al., 2011). Calculations have shown that H_2 formation can heat nanograins up to 53 K (Navarro-Ruiz et al., 2014). Taking into account that grains in dark clouds are larger and that water ice efficiently dissipate excess energy, it is unlikely that H_2 formation triggers desorption of other species.

Hot Shielded Regions

In the process of forming a star, cold cores evolve towards warm/hot shielded regions (also known as hot cores or hot corinos for massive and low mass protostars respectively) where the dust and gas temperature is a few hundred Kelvin and the density of the gas is above 10^7 cm^{-3}. Gas and dust are inherited from the previous growth phase, so that hydrogen is almost entirely molecular and grains are big. The ices that formed earlier have evaporated, so that the grains are bare. The H_2 molecules cannot be observed in these regions, and their abundance in itself does not have much impact on the chemistry. The H_2 formation rate however determines the abundance of atomic hydrogen, which is involved in many destruction reactions with activation barriers (Harada et al., 2010).

Rate equation simulations including both physisorption and chemisorption, (Cazaux et al. (2005)) have shown that H_2 formation on carbonaceous surfaces could be efficient up to about 1000 K, although this efficiency would decrease with the temperature. Up to 300 K, both Langmuir-Hinshelwood and Eley-Rideal (with chemisorbed H) mechanisms would contribute. At higher temperatures, the chemisorbed H starts to move and the Langmuir-Hinshelwood mechanism dominates.

Iqbal et al. (2012) performed a detailed study of the temperature dependent H_2 formation efficiency, using continuous-time random-walk Monte Carlo simulations on both silicate and carbonaceous surfaces. Their results show that the H_2 formation rate estimated by the rate equation method was overestimated because the mean abundance of H atoms at high temperature can be smaller than 1H per dust grain (see Caselli, Hasegawa, Herbst, 1998a, Cazaux, Caselli, Tielens, LeBourlot, Walmsley, 2005, Cuppen, Karssemeijer, Lamberts, 2013, for more details on this well known problem associated with rate equation models). Assuming deep chemisorption sites, present together with shallower physisorption sites, the H_2 formation remains efficient up to 700 K on both surfaces without the Eley-Rideal mechanism. These models have however only used low H fluxes, simulating ISM regions with densities smaller than 100 cm^{-3}. At higher H fluxes, the efficiency of H_2 formation may remain significant at even higher temperatures.

Shocked Environments

Shocks in the ISM and their Role in h_2 Formation

In the multiphase environment of galaxies, different types of shocks are expected. Fast shocks, ranging over 100–1000 km s^{-1}, heat the gas to high temperatures ($T \approx 10^6 - 10^8$ K), and produce EUV and X-ray photons that photo-ionize the tenuous medium (Allen et al., 2008). These shocks provide one source of energy that produce the Hot Ionized Medium in galaxies, and in the circum-galactic medium when the accreted gas from cosmological filaments is shocked at the halo boundary (Birnboim, Dekel, 2003, Cornuault, Lehnert, Boulanger, Guillard). On the other hand, low-velocity shocks (typically $\lesssim$ 50 km s^{-1}) are known to (i) initiate the formation of molecules in the gas that cools behind the shock (e.g Hollenbach and McKee, 1979), and (ii) be a very efficient process for exciting molecules, especially H_2, via collisions in the dense gas compressed and heated by the shock. Depending on the physical conditions in the shocked and post-shock gas, the different H_2 formation mechanisms discussed above may occur.

Jenkins and Peimbert (1997) describe a steady change in the line profile as J increases from 0 to 5 in high-resolution IMAPS[5] observations of what they interpret as H_2 forming in the post-shock zone of originally atomic gas. They suggest that the H_2 seen in absorption in the higher J levels (which are broader and slightly shifted toward negative velocities relative to the lowest J levels) is initially produced via the formation of a negative hydrogen ion H− in the warm (2000 K < T < 6500 K) and partially ionized medium just behind the shock, through the reactions:

$$H + e^- \longrightarrow H^- + h\nu \text{and} \quad H^- + H \longrightarrow H_2 + e^- \qquad (1)$$

Later, in zones where the post-shock gas has cooled well below 2000 K and is almost fully recombined, the formation of H_2 on grains dominates and is more easily seen in the lowest J levels.

Gas Conditions in Shocks

The typical physical conditions in shocks are given in Table 2. Here we have separated observations of Galactic and extragalactic shocked regions. In Galactic regions, most of the constraints on the physical conditions in shocked molecular regions come from CO, H_2, SiO and H_2O observations in supernovae remnants (Gusdorf, Anderl, Güsten, Stutzki, Hübers, Hartogh, Heyminck, Okada, 2012, Hewitt, Rho, Andersen, Reach, 2009, Nisini, Giannini, Neufeld, Yuan, Antoniucci, Bergin, Melnick, 2010) and outflows from young stellar objects (e.g. Gusdorf, Riquelme, Anderl, Eislöffel, Codella, Gómez-Ruiz, Graf, Kristensen, Leurini, Parise, Requena-Torres, Ricken, Güsten, 2015, Podio, Codella, Gueth, Cabrit, Bachiller, Gusdorf, Lee, Lefloch, Leurini, Nisini, Tafalla, 2015). In extragalactic shocked regions, like AGN- or starburst-driven outflows or galaxy interactions, observations of H_2 line emission show a distribution of temperatures much above the equilibrium temperature set by UV and cosmic ray heating. The same holds, on much smaller scales, in the case of the diffuse ISM in the Solar neighborhood (Gry, Boulanger, Nehmé, Pineau des Forêts, Habart, Falgarone, 2002, Ingalls, Bania, Boulanger, Draine, Falgarone, Hily-Blant, 2011). The range of gas temperatures can only be accounted for if supersonic turbulence dissipates and heats the gas through spatially localized events (e.g. Falgarone, Phillips, 1990, Hily-Blant, Falgarone, 2009, Guillard, Boulanger, Lehnert, Pineau des Forêts, Combes, Falgarone, Bernard-Salas, 2015).

The formation of H_2 in shocked regions greatly affects the gas cooling efficiency (Cuppen et al., 2010). For J-shocks, the cooling is first dominated by H_2 for a very short period of time, at high temperatures where the molecule experiences dissociation. Then, at $\approx$ 8000 K, the [O i]λ=6300Å line takes over. When the gas has cooled down sufficiently, H_2, H_2O, O and C+ dominate the cooling. More precisely, at T $\approx$ 1000 K, the fine-structure line [O i]λ=63.2μm dominates. Other important lines in this temperature range are [C i], [N i]λ=5200Å, [C ii]λ=158μm, [N ii]λ=121.8μm, [S ii]λ=34.82μm and [Fe ii]λ=25.99μm. At lower temperatures, the oxygen is converted into CO, H_2O, and OH, which become the dominant coolants. For C-shocks, H_2 is the main coolant for T > 300 K, which provides excitation of the mid-IR rotational lines. Remarkably, the cooling is dominated by H_2O emission at T < 200 K (Flower and Pineau des Forêts, 2010).

Dust Processing in Shocks and Impact on h_2 Formation

Grain processing in shocks affects the formation of H_2 and arises from interactions between the grain and other "particles" (grain, atom, ion, atomic nucleus, electron, photon). These interactions can be destructive, i.e. leading to a transfer of grain atoms to the gas-phase, and thus to a net mass loss from the grain, or non-destructive where the grain size distribution becomes modified but the total dust mass is conserved. These processes are counterbalanced by grain formation, specifically accretion and nucleation, but also coagulation between grains.

Destructive interactions can directly eject atoms and ions from the grain surface (via sputtering, ion field emission[6] or direct Coulomb explosion via extreme charging effects arising from electron-grain interactions or the photoelectric effect), or vaporize/sublime the grain, in case of $\gtrsim$ 20 km s−1 grain-grain collisions. Vaporisation/sublimation via absorption of energetic photons (UV, γ) or interaction with cosmic rays, may also be an important destruction mechanism for volatile material, such as ice mantles.

Grain-grain collisions at velocities $\gtrsim$1−2 km s^{-1} lead to non-destructive fragmentation (or shattering) of both grains (Tielens et al., 1994). The size distribution of the fragments follows a power-law, suggesting that dust fragmentation in shocks is responsible for the observed power-law size-distribution in the diffuse ISM (Clayton, Wolff, Sofia, Gordon, Misselt, 2003, Jones, Tielens, Hollenbach, 1996, Weingartner, Draine, 2001). By definition, fragmentation is not a destructive process. It re-distributes the dust mass towards smaller grain sizes, increasing the total cross-section of the distribution of grains, and hence their optical and UV extinction.

Dust grains, through gas-grain interactions, have an important impact on the physical and chemical properties of the shock. The grains participate in the formation of molecules and their removal, and the inertia of the grains can modify the propagation of the shock, especially in the case of C-shocks. A detailed review of the processes involving grains in shocks, as well as their impact on the structure of C-shocks, can be found in (Flower and Pineau des Forêts, 2003). For detailed calculations of the evolution of the dust size distribution in MHD shocks, see Guillet et al. (2011).

Impact of H_2 Formation on Shock Dynamics

In shocks, chemistry and hydrodynamics are coupled. Chemical processes, in particular H_2 formation, have a very important impact on the shock structure. H_2 formation plays a role in the thermal balance of the shock, which in turn impacts the number of reactions the shock can initiate and the consequent reaction rates. Without chemistry, the ionization fraction would vary only via the differential compression of the ionized and neutral fluids by the shock wave, and would have the same value in the pre-shock and post-shock regions. Allowing for chemistry, the ionization degree of the post-shock region is one order of magnitude lower. H_2 affects the ionization fraction because it essentially initiates the neutralization of C^+ via the reaction:

$$C^+ + H_2 \longrightarrow CH^+ + H \qquad (2)$$

followed by a chain of reactions, which balance overall to yield

$$CH^+ + e^- \longrightarrow C + H \qquad (3)$$

Consequently, the ion-neutral coupling is weaker, hence broadening the size of the precursor (the width of the shock is ≈ 5 times larger). Since the energy of the shock is dissipated over a larger region, as compared to the hypothetical case without chemistry, the maximum temperature of the shock is lower (Pineau des Forets et al., 1997).

Mixed Regions

Pure shocked regions are difficult to isolate in Galactic and extragalactic sources (e.g. Ingalls, Bania, Boulanger, Draine, Falgarone, Hily-Blant, 2011, Guillard, Boulanger, Pineau Des Forêts, Appleton, 2009). In addition to the shock excitation, a background UV field or cosmic rays are often present, which affect the thermal and chemical characteristics of the gas, in particular

H_2 formation. For instance, protostars generate a strong ultraviolet radiation field that ionizes their surroundings. This field drives powerful shock waves in the neighboring medium in the form of jets and bipolar outflows, whose structure can be partially organized by a local, strong magnetic field. Such an ejection activity locally modifies the interstellar chemistry, contributing to the cycle of matter. Excitation of H_2 gas in galaxies is therefore a mix of processes, and the degeneracy of models makes the identification of the dominant processes a challenge for modern astrophysics (e.g. Appleton et al., 2013).

The treatment of irradiation, which includes UV heating but also its impact on the chemistry through photo-ionization and photo-dissociation, is crucial in shock models to interpret complex regions where a mix of excitation is observed (Lesaffre et al., 2013). Preliminary results comparing steady-state H_2 abundances with and without radiative transfer and shielding effects in high-resolution MHD simulations seem to show very important differences in H_2 abundances (Hartwig, Glover, Klessen, Latif, Volonteri, 2015, Richings, Schaye, Oppenheimer, 2014). When the protecting effect of the shielding is not taken into account, most of the molecular hydrogen is photo-dissociated and only about 3% of the total mass of H_2 (compared to the case that includes shielding effects) survives in the densest regions where the formation is efficient enough to counterbalance the photo-destruction (Valdivia et al., 2017b). More generally, (magneto)-hydrodynamical simulations of the ISM which includes coupling with radiation and radiative transfer effects will be the subject of intense developments in the next years.

CONCLUSIONS AND PERSPECTIVES

In this paper, we present an overview of where we stand on the question: How does H_2 form in the interstellar medium? This report of our understanding of this fundamental scientific problem, of course, is moulded by the opinions and experience of the authors of this paper. These opinions were distilled by gathering together these scientists who are working in different fields of research which impinge on interstellar H_2 formation: experimental and theoretical physico-chemistry, observers and modelers of the interstellar medium.

As far as chemical physics is concerned, the physisorption (LH) mechanism for H_2 formation has been extensively studied, and with some exceptions, conclusions have converged. Amorphous carbonaceous surfaces have been the least studied, and are probably the richest and most complex materials to investigate. When it comes to H_2 formation involving chemisorbed atoms (ER), molecular synthesis by this pathway has been investigated in much less detail ; specifically, no experiments have been performed on silicates although there are calculations that address this process and should be tested with experiments. Formation of molecular hydrogen on graphite surfaces (an idealized representation of an interstellar dust grain) has been well investigated both experimentally and theoretically. This detailed investigation is probably a unique situation, and even then marginal discrepancies can still be found between different studies, especially for the case of computational work. The characterization of the ro-vibrational excitation of nascent H_2 leaving from the surface of the dust grain has been performed for formation on graphite. These studies can guide searches for observational signatures of H_2 formation in space and, hence, should be extended over a wider range of environmental parameters and surface types.

In calculations of the dynamics of the formation process, the sticking coefficient of H atoms on a graphitic surface is difficult to determine. The current calculations have developed models to take into account energy exchange (via phonons) with the surface during the collision, either for the physisorption site or for the chemisorption site. These models are limited in not just the number of degrees of freedom they consider, but also the limited range of geometries investigated and the fact that the surface is modeled as "perfect". The challenge is to include, in the same model, physisorption sites and chemisorption sites when calculating the sticking coefficient of an H atom on a graphitic surface.

More generally, the coupling between physisorption and chemisorption is probably the most challenging to study, and the most pressing problem with which both experimentalists and quantum physicists are battling. It is noteworthy that most of the studies discussed above have focused on only one of these two formation mechanisms. However, the fact that H_2 is formed in a wide variety of astrophysical environments suggests that both these mechanisms operate in parallel. Furthermore, there is a need to transform the results of laboratory studies of surface processes (sticking, diffusion, desorption, energy partition, etc.) to the efficiency of formation of molecular hydrogen in actual ISM environments. This has been done to a certain extent for silicate surfaces, but more work is needed both on silicates and carbonaceous materials.

New routes to H_2 formation, via UV irradiation of hydrogenated carbons, or via other energetic routes, have to be investigated in greater detail. It has been demonstrated that hydrogenated carbons can efficiently produce H_2 following energetic events, but after this carbonaceous grains need to be resupplied with H atoms. This "refilling mechanism" is complex and probably couples diffusion via physisorbed states and storage via chemisorbed states. The existence of these potential H_2 reservoirs underlines the need to studying the links between physisorption and chemisorption on disordered materials. In the specific case of astrophysical shocks, the interaction of high temperature gas with cold surfaces, and the impact of this interaction on H_2 formation, also still needs to be studied.

Finally, on several relevant surfaces many aspects of the H_2 formation process, such as nuclear spin conversion and energy partitioning, remain open questions. The observed trends, revealed mostly by the study of H interaction with graphite, have to be confirmed. In the case of LH processes, the substrate retains a minor part of the energy released by H_2 formation.

For the chemical modeling of cold cores, the amount of H_2 on the surfaces can have a big impact on the abundances of other species. Thus, the study of H_2 formation in the presence of co-adsorbed H_2 (with perhaps also H_2O, CO, CO_2) by experimentalists would be important. Such experiments would perhaps reveal the possibility of trapping nascent H_2 in ices.

In the near future, JWST (launch foreseen in late 2018) will allow medium resolution ($\lambda/\Delta\lambda \approx 3000$) spectroscopy of H_2 lines, with an increase of two orders of magnitude in spatial resolution and sensitivity compared to ISO or Spitzer. This mission will provide a fantastic set of new data which will present stringent tests of our current understanding of H_2 in space. The gain in sensitivity and spatial resolution (0.3" at 10 μm) will generate high quality maps of the H_2 rotational and ro-vibrational transitions. A wealth of moderately/highly excited J levels will be detectable for the first time. Combined with data probing the ro-vibrational levels, these observations will cover a huge range of excitation energies, making them excellent probes of the excitation of H_2 and the associated gas.

In the nearby Galactic sources, the critical H/H_2 transition zones, that could be very sharp because of the H_2 self-shielding and strong gas density gradients, will be resolved via the H_2 ro-vibrational and rotational emission line profiles. This will constrain the possible H_2 formation mechanisms (ER, LH on small grains fluctuating in temperatures or UV processes) and ortho-to-para ratios associated with H_2 formation in a warm gas. The wavelength coverage of JWST will allow us to study in detail the links between H_2 and the properties of the small dust particles. How rapidly the dust properties change at the cloud edge, via photo-processing, and how these changes impact the H_2 formation process, are just two of the key questions that JWST will address.

Being the first instruments able to spatially and spectrally resolve H_2, and forbidden lines of the ionized gas, at rest-frame near-IR and mid-IR wavelengths out to z=1.5−3.5, the JWST spectrometers will also allow the study of H_2 line emission in the distant Universe. These spectrometers will allow us to investigate the physical state, and the kinematics, of the warm (> 50 K) molecular gas. these observations will constrain the impact of galaxy mergers and AGN feedback on the formation and evolution of massive galaxies. In $z \sim 2$ active star-forming galaxies, high gas velocity dispersions (e.g. Lehnert et al., 2009) suggest high turbulence, and we can expect bright H_2 emission (Guillard, Boulanger, Lehnert, Appleton, Pineau des Forêts, 2015a, Guillard, Boulanger, Pineau Des Forêts, Appleton, 2009), as suggested by H_2 detections in $z \approx 2$ infrared-luminous galaxies (Fiolet, Omont, Lagache, Bertincourt, Fadda, Baker, Beelen, Berta, Boulanger, Farrah, Kovács, Lonsdale, Owen, Polletta, Shupe, Yan, 2010, Ogle, Davies, Appleton, Bertincourt, Seymour, Helou, 2012). Identifying the source of turbulence in high-redshift sources is a key step forward for studies of galaxy evolution.

The spectrometer EXES on SOFIA can probe several H_2 rotational lines with diffraction limited resolution of 1–5" and a spectral resolution of $R \sim 10^4 - 10^5$ at 20 μm. This is the only instrument with sufficient spectral resolution and sensitivity to measure a pure rotational H_2 line profile from a relatively faint source. Studies, like that reported by Jenkins and Peimbert (1997), emphasize the interest of observing H_2 at high spectral resolution in order to study the velocity profiles of the different H_2 rotational excitation levels (J) to, for example:

1. Isolate components or cloud regions with different rotational excitation.
2. Determine the role of internal spatial structures in the self-shielding of diffuse H_2 that strongly influences the HI $\rightarrow$ H_2 transition in galaxies.
3. Study the increase in velocity dispersion with J. Specifically, can we confirm the creation of H_2 in warm, partly-ionized post-shock zones proceeds via the formation of H^-?
4. Measure the temperature and turbulence of H_2 in high-J levels. Is

the excess abundance in J > 2 in diffuse clouds due to shocks, or dissipation of turbulence in vortices?

5. Find what fractions of the 4.5 eV of energy released upon H_2 formation are transferred to internal excitation of the molecule, kinetic motion of the molecule, and the thermal bath of the grain.

The possibility of observing the H_2 electronic bands at high spectral resolution in absorption in the UV might be opened (in the distant future) by the LUVOIR project. LUVOIR is one of four Decadal Survey Mission Concept Studies initiated by NASA in 2016, which should include a high-resolution UV spectrometer. This spectrometer would allow to observe the far-UV electronic bands of H_2 at high resolution (R > 120 000) toward many stars in our Galaxy, as well as towards individual stars in the Local Group galaxies. Such a capability would allow us to extend the study of H_2 formation to very different environments.

ACKNOWLEDGEMENTS

All authors thank the EPOC lab, the LAB, and the Institute of Advanced Study of the University of Cergy Pontoise for their support during the meeting that catalyzed the writing of this manuscript. VW's research is funded by an ERC Starting Grant (3DICE, grant agreement 336474). GV acknowledges financial support from the National Science Foundation's Astronomy & Astrophysics Division (Grants No. 1311958 and 1615897). LH acknowledges support from ERC Consolidator Grant GRANN (grant agreement no. 648551). GN acknowledges support from the Swedish Research Council. VW, FD and SM acknowledge the CNRS program "Physique et Chimie du Milieu Interstellaire" (PCMI) co-funded bythe Centre National d'Etudes Spatiales (CNES). SDP acknowledges funding from STFC, UK. FD thanks VW for accepting the difficult task of managing this project and SDP for his efforts in reading and re-writing the text. V.V acknowledges funding from the European Research Council (ERC) under the European Union›s Horizon 2020 research and innovation programme (MagneticYSOS project, grant agreement No 679937).

REFERENCES

1. Accolla, M., Congiu, E., Dulieu, F., Manicò, G., Chaabouni, H., Matar, E., Mokrane, H., Lemaire, J.L., Pirronello, V., 2011. Changes in the morphology of interstellar ice analogues after hydrogen atom exposure. Phys. Chem. Chem. Phys. (Incorporating Faraday Transactions) 13, 8037. doi:10.1039/c0cp01462a, arxiv:1011.5344.
2. Al-Halabi, A., Fraser, H.J., Kroes, G.J., van Dishoeck, E.F., 2004. Adsorption of CO on amorphous water-ice surfaces. aa 422, 777–791. doi:10.1051/0004-6361: 20035939.
3. Al-Halabi, A., Kleyn, A., van Dishoeck, E., Kroes, G., 2002. Sticking of hydrogen atoms to crystalline ice surfaces: dependence on incidence energy and surface temperature. J. Phys. Chem. B 106 (25), 6515.
4. Al-Halabi, A., van Dishoeck, E.F., 2007. Hydrogen adsorption and diffusion on amorphous solid water ice. Mon. Not. R. Astron. Soc. 382, 1648–1656. doi:10.1111/j. 1365-2966.2007.12415.x.
5. Alata, I., Cruz-Diaz, G.A., Muñoz Caro, G.M., Dartois, E., 2014. Vacuum ultraviolet photolysis of hydrogenated amorphous carbons . I. Interstellar H2 and CH4 formation rates. AA 569, A119. doi:10.1051/0004-6361/201323118.
6. Allain, T., Leach, S., Sedlmayr, E., 1996. Photodestruction of PAHs in the interstellar medium. II. influence of the states of ionization and hydrogenation.. AA 305, 616.
7. Allen, M.G., Groves, B.A., Dopita, M.A., Sutherland, R.S., Kewley, L.J., 2008. The MAPPINGS III library of fast radiative shock models. ApJ Supplement 178, 20–55. doi:10.1086/589652.
8. Amiaud, L., Dulieu, F., Fillion, J.-H., Momeni, A., Lemaire, J.L., 2007. Interaction of atomic and molecular deuterium with a nonporous amorphous water ice surface between 8 and 30 K. J. Chem. Phys. 127 (14). doi:10.1063/1.2746323. arxiv: 144709-144709
9. Amiaud, L., Fillion, J.H., Baouche, S., Dulieu, F., Momeni, A., Lemaire, J.L., 2006. Interaction of D2 with H2O amorphous ice studied by temperature-programed desorption experiments. J. Chem. Phys. 124 (9). doi:10.1063/1.2168446. arxiv: 094702-094702
10. Amiaud, L., Fillion, J.-H., Dulieu, F., Momeni, A., Lemaire, J.-L., 2015. Physisorption and desorption of H2, HD and D2 on amorphous solid water ice. Effect on mixing isotopologue on statistical population of adsorption sites. PCCP 17 (44), 30148–30157. doi:10.1039/

c5cp03985a.

11. Appleton, P.N., Guillard, P., Boulanger, F., Cluver, M.E., Ogle, P., Falgarone, E., Pineau des Forêts, G., O'Sullivan, E., Duc, P.-A., Gallagher, S., Gao, Y., Jarrett, T., Konstantopoulos, I., Lisenfeld, U., Lord, S., Lu, N., Peterson, B.W., Struck, C., Sturm, E., Tuffs, R., Valchanov, I., van der Werf, P., Xu, K.C., 2013. Shock-enhanced c+ emission and the detection of h2o from the Stephan's quintet group-wide shock using herschel. Astrophys. J. 777, 66. doi:10.1088/0004-637X/777/1/66, arxiv:1309.1525.
12. Appleton, P.N., Guillard, P., Togi, A., Alatalo, K., Boulanger, F., Cluver, M., Pineau des Forêts, G., Lisenfeld, U., Ogle, P., Xu, C.K., 2017. Powerful H2 line cooling in Stephan's Quintet. II. Group-wide gas and shock modeling of the warm H2 and a comparison with [C II] 157.7 μm emission and kinematics. Astrophys. J. 836, 76. doi:10.3847/1538-4357/836/1/76, arxiv:1701.03226.
13. Appleton, P.N., Xu, K.C., Reach, W., Dopita, M.A., Gao, Y., Lu, N., Popescu, C.C., Sulentic, J.W., Tuffs, R.J., Yun, M.S., 2006. Powerful high-Velocity dispersion molecular hydrogen associated with an intergalactic shock wave in Stephan's quintet. ApJL 639, L51–L54. doi:10.1086/502646.
14. Arab, H., 2012. Ph.d thesis. University of Paris-Sud 11.
15. Arab, H., Abergel, A., Habart, E., Bernard-Salas, J., Ayasso, H., Dassas, K., Martin, P.G., White, G.J., 2012. Evolution of dust in the Orion Bar with Herschel. I. Radiative transfer modelling. A&A 541, A19. doi:10.1051/0004-6361/201118537, arxiv:1202.1624.
16. Armus, L., Bernard-Salas, J., Spoon, H.W.W., Marshall, J.A., Charmandaris, V., Higdon, S.J.U., Desai, V., Hao, L., Teplitz, H.I., Devost, D., Brandl, B.R., Soifer, B.T., Houck, J.R., 2006. Detection of the Buried Active Galactic Nucleus in NGC 6240 with the Infrared Spectrograph on the Spitzer Space Telescope. Astrophys. J. 640, 204–210. doi:10.1086/500040, arxiv:0511381.
17. Averna, D., Pirronello, V., 1991. Synthesis of H2 at the edge of dense cold clouds by cosmic ray bombardment of grain mantles. A&A 245, 239–242.
18. Bachellerie, D., Sizun, M., Aguillon, F., Teillet-Billy, D., Rougeau, N., Sidis, V., 2009. Unrestricted study of the Eley-Rideal formation of H2 on graphene using a new multidimensional graphene-H-H potential: role of the substrate. Phys. Chem. Chem. Phys. (Incorporating Faraday

Transactions) 11, 2715. doi:10.1039/ B818614F.

19. Bachellerie, D., Sizun, M., Teillet-Billy, D., Rougeau, N., Sidis, V., 2007. Eley-Rideal formation of H2 involving one of two para-chemisorbed H atoms on a graphite surface. Chem. Phys. Lett. 448, 223–227. doi:10.1016/j.cplett.2007.09.071.
20. Baouche, S., Gamborg, G., Petrunin, V.V., Luntz, A.C., Baurichter, A., Hornekær, L., 2006. High translational energy release in H2 (D2) associative desorption from H (D) chemisorbed on C(0001). J. Chem. Phys. 125 (8). doi:10.1063/1.2220565. arxiv: 084712-084712
21. Beirão, P., Armus, L., Lehnert, M. D., Guillard, P., Heckman, T., Draine, B., Hollenbach, D., Walter, F., Sheth, K., Smith, J. D., Shopbell, P., Boulanger, F., Surace, J., Hoopes, C., Engelbracht, C., 2015. Spatially Resolved Spitzer-IRS Spectral Maps of the Superwind in M82. arxiv:1505.04069v1
22. Bekki, K., 2013. Coevolution of dust, gas and stars in galaxies - I. Spatial distributions and scaling-relations of dust and molecular hydrogen. Mon. Not. R. Astron. Soc. 432, 2298–2323. doi:10.1093/mnras/stt589, arxiv:1304.1633.
23. Bergin, E.A., Tafalla, M., 2007. Cold Dark Clouds: The Initial Conditions for Star Formation. ARA. A 45, 339–396. doi:10.1146/annurev.astro.45.071206.100404, arxiv:0705.3765.
24. Berné, O., Joblin, C., Deville, Y., Smith, J.D., Rapacioli, M., Bernard, J.P., Thomas, J., Reach, W., Abergel, A., 2007. Analysis of the emission of very small dust particles from spitzer spectro-imagery data using blind signal separation methods. A.A 469, 575–586. doi:10.1051/0004-6361:20066282, arxiv:0703072.
25. Bialy, S., Sternberg, A., 2016. Analytic HI-to-H2 Photodissociation Transition Profiles. Astrophys. J. 822, 83. doi:10.3847/0004-637X/822/2/83, arxiv:1601.02608.
26. Biham, O., Furman, I., Pirronello, V., Vidali, G., 2001. Master Equation for Hydrogen Recombination on Grain Surfaces. Astrophys. J. 553, 595–603. doi:10.1086/320975, arxiv:0012267.
27. Biham, O., Lipshtat, A., 2002. Exact results for hydrogen recombination on dust grain surfaces. Phys. Rev. E 66, 56103.
28. Birnboim, Y., Dekel, A., 2003. Virial shocks in galactic haloes? Mon. Not. R. Astron. Soc. 345, 349–364. doi:10.1046/j.1365-8711.2003.06955.x, arxiv:0302161.

29. Bisbas, T.G., Bell, T.A., Viti, S., Yates, J., Barlow, M.J., 2012. 3D-PDR: a new three-dimensional astrochemistry code for treating photodissociation regions. Mon. Not. R. Astron. Soc. 427, 2100–2118. doi:10.1111/j.1365-2966.2012.22077.x, arxiv:1209.1091.
30. Black, J.H., van Dishoeck, E.F., 1987. Fluorescent excitation of interstellar H2. Astrophys. J. 322, 412–449. doi:10.1086/165740.
31. Bonfanti, M., Casolo, S., Tantardini, G.F., Martinazzo, R., 2011. Surface models and reaction barrier in Eley-Rideal formation of H2 on graphitic surfaces. Phys. Chem. Chem. Phys. (Incorporating Faraday Transactions) 13, 16680. doi:10.1039/ c1cp21900f.
32. Bonfanti, M., Jackson, B., Hughues, K., Burghardt, I., Martinazzo, R., 2015. Quantum dynamics of hydrogen atoms on graphene ii. sticking. J. Chem. Phys. 143, 124704.
33. Bonfanti, M., Martinazzo, R., Tantardini, G.F., Ponti, A., 2007. Physisorption and diffusion of hydrogen atoms on graphite from correlated calculations on the h– coronene model system. J. Phys. Chem C 111 (16), 5825.
34. Boogert, A.C.A., Gerakines, P.A., Whittet, D.C.B., 2015. Observations of the icy universe.. ARA.A 53, 541–581. doi:10.1146/annurev-astro-082214-122348, arxiv:1501.05317.
35. Boschman, L., Cazaux, S., Spaans, M., Hoekstra, R., Schlathölter, T., 2015. H2 formation on PAHs in photodissociation regions: a high-temperature pathway to molecular hydrogen. A.A 579, A72. doi:10.1051/0004-6361/201323165, arxiv:1504.08276.
36. Boschman, L., Reitsma, G., Cazaux, S., Schlathölter, T., Hoekstra, R., Spaans, M., González-Magaña, O., 2012. Hydrogenation of PAH cations: A First step toward H2 formation. ApJL 761, L33. doi:10.1088/2041-8205/761/2/L33, arxiv:1211.4984.
37. Bron, E., Le Bourlot, J., Le Petit, F., 2014. Surface chemistry in the interstellar medium. II. h2 formation on dust with random temperature fluctuations. A.A 569, A100. doi:10.1051/0004-6361/201322101, arxiv:1407.4473.
38. Bron, E., Le Petit, F., Le Bourlot, J., 2016. Efficient ortho-para conversion of H2 on interstellar grain surfaces. A.A 588, A27. doi:10.1051/0004-6361/201527879, arxiv:1601.04356.
39. Brown, W.L., Augustyniak, W.M., Simmons, E., Marcantonio, K.J., Lnazerotti, L.J., Johnson, R.E., Reimann, C.T., Foti, G., Pirronello,

V., 1982. Erosion and molecular formation in condensed gas films by electronic energy loss of fast ions. Nucl. Instrum. Methods in Phys. Res. A 198, 1–8.

40. Bruch, L., Cole, M., Zaremba, E., 2007a. Physical Adsorption: Forces and Phenomena. Dover books on physics. Dover Publications: Mineola, N.Y.
41. Bruch, L.W., Diehl, R.D., Venables, J.A., 2007b. Progress in the measurement and modeling of physisorbed layers. Rev. Mod. Phys. 79, 1381–1454. doi:10.1103/RevModPhys.79.1381.
42. Buch, V., 1989. Sticking probability of light atoms on the graphite (0001) surface. J. Chem. Phys. 91, 4974–4984. doi:10.1063/1.456738.
43. Buch, V., Czerminski, R., 1991. Eigenstates of a quantum-mechanical particle on a topologically disordered surface - H(D) atom physisorbed on an amorphous ice cluster (H2O)115. J. Chem. Phys. 95, 6026–6038. doi:10.1063/1.461571.
44. Buch, V., Zhang, Q., 1991. Sticking probability of h and d atoms on amorphous ice -A computational study. Astrophys. J. 379, 647–652. doi:10.1086/170537.
45. Burton, M.G., Geballe, T.R., Brand, P.W.J.L., Moorhouse, A., 1990. High spectral resolution observations of fluorescent molecular hydrogen in molecular clouds. Astrophys. J. 352, 625.
46. Burton, M.G., Londish, D., Brand, P.W.J.L., 2002. Formation pumping of molecular hydrogen in the messier 17 photodissociation region. Mon. Not. R. Astron. Soc. 333, 721.
47. Carmona, A., van der Plas, G., van den Ancker, M.E., Audard, M., Waters, L.B.F.M., Fedele, D., Acke, B., Pantin, E., 2011. A survey for near-infrared H2 emission in herbig ae/be stars: emission from the outer disks of HD 97048 and HD 100546. A&A 533, A39. doi:10.1051/0004-6361/201116561, arxiv:1107.1994.
48. Carruthers, G.R., 1970. Rocket observation of interstellar molecular Hydrogen. ApJL 161, L81. doi:10.1086/180575.
49. Caselli, P., Hasegawa, T.I., Herbst, E., 1998a. A proposed modification of the rate equations for reactions on grain surfaces. APJ 495, 309–316. doi:10.1086/305253.
50. Caselli, P., Walmsley, C.M., Terzieva, R., Herbst, E., 1998b. The ionization fraction in dense cloud cores. apj 499, 234–249. doi:10.1086/305624.

51. Casolo, S., Løvvik, O., Martinazzo, R., Tantardini, G., 2009b. Understanding adsorption of hydrogen atoms on graphene. J. Chem. Phys. 130, 054704.

52. Casolo, S., Martinazzo, R., Bonfanti, M., Tantardini, G.F., 2009. Quantum dynamics of the eley-Rideal hydrogen formation reaction on graphite at typical interstellar cloud conditions. J. Phys. Chem. A 113, 14545–14553. doi:10.1021/jp9040265.

53. Cazaux, S., Boschman, L., Rougeau, N., Reitsma, G., Hoekstra, R., Teillet-Billy, D., Morisset, S., Spaans, M., Schlathölter, T., 2016. The sequence to hydrogenate coronene cations: A journey guided by magic numbers. Sci. Rep. 6, 19835. doi:10.1038/srep19835, arxiv:1603.02593.

54. Cazaux, S., Caselli, P., Tielens, A.G.G.M., LeBourlot, J., Walmsley, M., 2005. Molecular Hydrogen formation on grain surfaces. J. Phys. Conf. Series 6, 155–160. doi:10. 1088/1742-6596/6/1/016, arxiv:0502291.

55. Cazaux, S., Morisset, S., Spaans, M., Allouche, A., 2011. When sticking influences H2 formation. A.A 535, A27. doi:10.1051/0004-6361/201117220, arxiv:1107.5193.

56. Cazaux, S., Spaans, M., 2009. HD and H2 formation in low-metallicity dusty gas clouds at high redshift. A&A 496, 365–374. doi:10.1051/0004-6361:200811302, arxiv:0901.0514.

57. Cazaux, S., Tielens, A.G.G.M., 2002. Molecular hydrogen formation in the interstellar medium. ApJL 575, L29–L32. doi:10.1086/342607, arxiv:abs/astro-ph/0207035.

58. Cazaux, S., Tielens, A.G.G.M., 2004. H2 formation on grain surfaces. APJ 604, 222–237. doi:10.1086/381775.

59. Cesarsky, D., Cox, P., Pineau Des Forêts, G., van Dishoeck, E.F., Boulanger, F., Wright, C.M., 1999. ISOCAM spectro-imaging of the H2 rotational lines hfill in the supernova remnant IC 443. A.A 348, 945.

60. Chaabouni, H., Bergeron, H., Baouche, S., Dulieu, F., Matar, E., Congiu, E., Gavilan, L., Lemaire, J.L., 2012. Sticking coefficient of hydrogen and deuterium on silicates under interstellar conditions. A.A 538, A128. doi:10.1051/0004-6361/201117409, arxiv:1202.2473.

61. Chang, Q., Cuppen, H.M., Herbst, E., 2005. Continuous-time random-walk simulation of H2 formation on interstellar grains. A.A 434, 599–611. doi:10.1051/ 0004-6361:20041842.

62. Chang, Q., Cuppen, H.M., Herbst, E., 2006. Effective rate coefficients for molecular hydrogen formation in diffuse interstellar clouds. A.A 458, 497–505. doi:10. 1051/0004-6361:20065913.
63. Chang, Q., Herbst, E., 2014. Interstellar simulations using a unified microscopicMacroscopic monte carlo model with a full gas-Grain network including bulk diffusion in ice mantles. Astrophys. J. 787, 135. doi:10.1088/0004-637X/787/2/ 135, arxiv:1405.0840.
64. Charnley, S.B., 2001. Stochastic theory of molecule formation on dust. ApJL 562, L99–L102. doi:10.1086/324753.
65. Chehrouri, M., Fillion, J.-H., Chaabouni, H., Mokrane, H., Congiu, E., Dulieu, F., Matar, E., Michaut, X., Lemaire, J.L., 2011. Nuclear spin conversion of molecular Hydrogen on amorphous solid water in the presence of O2 traces. Phys.
66. Chem. Chem. Phys. (Incorporating Faraday Transactions) 13, 2172. doi:10.1039/ c0cp01322f.
67. Chernoff, D.F., 1987. Magnetohydrodynamic shocks in molecular clouds. Astrophys. J. 312, 143–169. doi:10.1086/164856.
68. Chevance, M., 2016. The physical processes in the interstellar medium of the Magellanic Clouds. Phd thesis. Université Paris 7 Diderot.
69. Christensen, C., Quinn, T., Governato, F., Stilp, A., Shen, S., Wadsley, J., 2012. Implementing molecular hydrogen in hydrodynamic simulations of galaxy formation. Mon. Not. R. Astron. Soc. 425, 3058–3076. doi:10.1111/j.1365-2966.2012.21628.x, arxiv:1205.5567.
70. Clark, P.C., Glover, S.C.O., Klessen, R.S., Bonnell, I.A., 2012. How long does it take to form a molecular cloud? Mon. Not. R. Astron. Soc. 424, 2599–2613. doi:10.1111/ j.1365-2966.2012.21259.x, arxiv:1204.5570.
71. Clayton, G.C., Wolff, M.J., Sofia, U.J., Gordon, K.D., Misselt, K.A., 2003. Dust grain size distributions from MRN to MEM. Astrophys. J. 588 (2), 871–880. doi:10.1086/374316.
72. Cluver, M.E., Appleton, P.N., Boulanger, F., Guillard, P., Ogle, P., Duc, P.-A., Lu, N., Rasmussen, J., Reach, W.T., Smith, J.D., Tuffs, R., Xu, C.K., Yun, M.S., 2010. Powerful H2 line cooling in Stephan's quintet. i. mapping the significant cooling pathways in group-wide shocks. Astrophys. J. 710, 248–264. doi:10.1088/0004-637X/710/ 1/248, arxiv:0912.0282.
73. Combes, F., Pineau des Forêts, G., 2000. Molecular Hydrogen in

Space. Cambridge Contemporary Astrophysics. Cambridge University Press: Cambridge; New York.

74. Compiègne, M., Verstraete, L., Jones, A., Bernard, J.-P., Boulanger, F., Flagey, N., Le Bourlot, J., Paradis, D., Ysard, N., 2011. The global dust SED: tracing the nature and evolution of dust with dustem. A&A 525, A103. doi:10.1051/0004-6361/201015292, arxiv:1010.2769.

75. Congiu, E., Matar, E., Kristensen, L.E., Dulieu, F., Lemaire, J.L., 2009. Laboratory evidence for the non-detection of excited nascent H2 in dark clouds. Mon. Not. R. Astron. Soc. 397, L96–L100. doi:10.1111/j.1745-3933.2009.00692.x, arxiv:0906. 0723.

76. Cornuault, N., Lehnert, M., Boulanger, F., Guillard, P., 2016. Are cosmological gas accretion streams multiphase and turbulent? arXiv.org, arXiv:1609.04405.

77. Cox, P., Huggins, P.J., Maillard, J.-P., Habart, E., Morisset, C., Bachiller, R., Forveille, T., 2002. High resolution near-infrared spectro-imaging of NGC 7027. A.A 384, 603–619. doi:10.1051/0004-6361:20011780.

78. Creighan, S.C., Perry, J.S.A., Price, S.D., 2006. The rovibrational distribution of H2 and HD formed on a graphite surface at 15-50 K. J. Chem. Phys. 124 (11). doi:10. 1063/1.2174878. arxiv: 114701-114701

79. Cuadrado, S., Goicoechea, J.R., Pilleri, P., Cernicharo, J., Fuente, A., Joblin, C., 2015. The chemistry and spatial distribution of small hydrocarbons in UV-irradiated molecular clouds: the orion bar PDR. A.A 575, A82. doi:10.1051/0004-6361/201424568, arxiv:1412.0417.

80. Cuppen, H.M., Garrod, R.T., 2011. Modelling of surface chemistry on an inhomogeneous interstellar grain. A&A 529, A151. doi:10.1051/0004-6361/201016013.

81. Cuppen, H.M., Herbst, E., 2005. Monte carlo simulations of H2 formation on grains of varying surface roughness. Mon. Not. R. Astron. Soc. 361, 565–576. doi:10. 1111/j.1365-2966.2005.09189.x.

82. Cuppen, H.M., Herbst, E., 2007. Simulation of the formation and morphology of ice mantles on interstellar grains. Astrophys. J. 668, 294–309. doi:10.1086/521014, arxiv:0707.2744.

83. Cuppen, H.M., Hornekær, L., 2008. Kinetic monte carlo studies of Hydrogen abstraction from graphite. J. Chem. Phys. 128 (17) 174707-174707 doi:10.1063/1. 2913238. arxiv:0807.0108.

84. Cuppen, H.M., Karssemeijer, L.J., Lamberts, T., 2013. The kinetic monte carlo method as a way to solve the master equation for interstellar grain

chemistry. Chem. Rev. 113 (12), 8840–8871. doi:10.1021/cr400234a, PMID: arxiv:24187949.

85. Cuppen, H.M., Kristensen, L.E., Gavardi, E., 2010. H2 reformation in post-shock regions. Mon. Not. R. Astron. Soc. 406, L11–L15. doi:10.1111/j.1745-3933.2010. 00871.x, arxiv:1005.1847.
86. Cuppen, H.M., Morata, O., Herbst, E., 2006. Monte Carlo simulations of H2 formation on stochastically heated grains. Mon. Not. R. Astron. Soc. 367, 1757–1765. doi:10. 1111/j.1365-2966.2006.10079.x, arxiv:0601554.
87. Desert, F.X., Boulanger, F., Shore, S.N., 1986. Grain temperature fluctuations - A key to infrared spectra. A.A 160, 295–300.
88. Dobbs, C.L., Bonnell, I.A., 2007. Spiral shocks and the formation of molecular clouds in a two-phase medium. Mon. Not. R. Astron. Soc. 376, 1747–1756. doi:10.1111/j.1365-2966.2007.11552.x.
89. Dobbs, C.L., Bonnell, I.A., Pringle, J.E., 2006. The formation of molecular clouds in spiral galaxies. Mon. Not. R. Astron. Soc. 371, 1663–1674. doi:10.1111/j. 1365-2966.2006.10794.x, arxiv:0602103.
90. Dobbs, C.L., Glover, S.C.O., Clark, P.C., Klessen, R.S., 2008. The ISM in spiral galaxies: can cooling in spiral shocks produce molecular clouds? Mon. Not. R. Astron. Soc. 389, 1097–1110. doi:10.1111/ j.1365-2966.2008.13646.x, arxiv:0806.4312.
91. Draine, B., 2011. Physics of the interstellar and intergalactic medium. Princeton Series in Astrophysics. Princeton University Press:, Princeton, N.J..
92. Draine, B.T., 1980. Interstellar shock waves with magnetic precursors. Astrophys. J. 241, 1021–1038. doi:10.1086/158416.
93. Draine, B.T., 2003. Interstellar dust grains. ARA.A 41, 241–289. doi:10.1146/annurev. astro.41.011802.094840, arxiv:0304489.
94. Draine, B.T., Bertoldi, F., 1996. Structure of stationary photodissociation fronts. Astrophys. J. 468, 269. doi:10.1086/177689, arxiv:9603032.
95. Draine, B.T., Bertoldi, F., 1999. Heating the gas in photodissociation regions. In: Cox, P., Kessler, M. (Eds.), The Universe as seen by ISO, p. 553.
96. Draine, B.T., Li, A., 2001. Infrared emission from interstellar dust. i. stochastic heating of small grains. Astrophys. J. 551, 807–824. doi:10.1086/320227, arxiv:0011318.
97. Draine, B.T., Li, A., 2007. Infrared emission from interstellar dust. IV.

the silicateGraphite-PAH model in the post-Spitzer era. Astrophys. J. 657, 810–837. doi:10. 1086/511055, arxiv:0608003.

98. Draine, B.T., McKee, C.F., 1993. Theory of interstellar shocks. ARA.A 31, 373–432. doi:10.1146/annurev.aa.31.090193.002105.
99. Draine, B.T., Salpeter, E.E., 1979. On the physics of dust grains in hot gas. Astrophys. J. 231, 77–94. doi:10.1086/157165.
100. Duley, W.W., Williams, D.A., 1993. The formation of H2 on interstellar dust. Mon. Not. R. Astron. Soc. 260, 37–42. doi:10.1093/mnras/260.1.37.
101. Esplugues, G.B., Cazaux, S., Meijerink, R., Spaans, M., Caselli, P., 2016. Surface chemistry in photodissociation regions. A.A 591, A52. doi:10.1051/0004-6361/201528001, arxiv:1604.06797.
102. Falgarone, E., Phillips, T.G., 1990. A signature of the intermittency of interstellar turbulence - the wings of molecular line profiles. Astrophys. J. 359, 344–354. doi:10.1086/169068.
103. Farebrother, A.J., Meijer, A.J.H.M., Clary, D.C., Fisher, A.J., 2000. Formation of molecular hydrogen on a graphite surface via an eley-rideal mechanism.. Chem. Phys. Lett. 319, 303. doi:10.1016/S0009-2614(00)00128-7.
104. Ferland, G.J., Porter, R.L., van Hoof, P.A.M., Williams, R.J.R., Abel, N.P., Lykins, M.L., Shaw, G., Henney, W.J., Stancil, P.C., 2013. The 2013 release of cloudy. Revista Mexicana de Astronomía y Astrofísica 49, 137–163 arxiv:1302.4485.
105. Ferro, Y., Marinelli, F., Allouche, A., 2003. Density functional theory investigation of the diffusion and recombination of h on a graphite surface. Chem. Phys. Lett. 368, 609–615. doi:10.1016/S0009-2614(02)01908-5.
106. Field, G.B., Somerville, W.B., Dressler, K., 1966. Hydrogen molecules in astronomy. ARA.A 4, 207. doi:10.1146/annurev.aa.04.090166.001231.
107. Fiolet, N., Omont, A., Lagache, G., Bertincourt, B., Fadda, D., Baker, A.J., Beelen, A., Berta, S., Boulanger, F., Farrah, D., Kovács, A., Lonsdale, C., Owen, F., Polletta, M., Shupe, D., Yan, L., 2010. Mid-infrared spectroscopy of spitzer-selected ultraluminous starbursts at z=2. A&A 524, A33. doi:10.1051/0004-6361/201015504.
108. Fleming, B., France, K., Lupu, R.E., McCandliss, S.R., 2010. Spitzer mapping of polycyclic aromatic hydrocarbon and h2 features in photodissociation regions. Astrophys. J. 725, 159–172.

doi:10.1088/0004-637X/725/1/159, arxiv:1010.0007.

109. Flower, D.R., Le Bourlot, J., Pineau des Forêts, G., Cabrit, S., 2003. The contributions of J-type shocks to the h2 emission from molecular outflow sources. Mon. Not. R. Astron. Soc. 341, 70–80. doi:10.1046/j.1365-8711.2003.06463.x.

110. Flower, D.R., Pineau des Forêts, G., 2003. The influence of grains on the propagation and structure of C-type shock waves in interstellar molecular clouds. Mon. Not. R. Astron. Soc. 343, 390–400. doi:10.1046/j.1365-8711.2003.06716.x.

111. Flower, D.R., Pineau des Forêts, G., 2010. Excitation and emission of H2, CO and H2o molecules in interstellar shock waves. Mon. Not. R. Astron. Soc. 406, 1745–1758. doi:10.1111/j.1365-2966.2010.16834.x.

112. Flower, D.R., Pineau des Forêts, G., 2015. Interpreting observations of molecular outflow sources: the MHD shock code mhd_vode. A.A 578, A63. doi:10.1051/ 0004-6361/201525740.

113. Flower, D.R., Pineau des Forets, G., Hartquist, T.W., 1985. Theoretical studies of interstellar molecular shocks. I - General formulation and effects of the ion-molecule chemistry. Mon. Not. R. Astron. Soc. 216, 775–794. doi:10.1093/mnras/216.4.775.

114. Forbrich, J., Öberg, K., Lada, C.J., Lombardi, M., Hacar, A., Alves, J., Rathborne, J.M., 2014. Some like it cold: molecular emission and effective dust temperatures of dense cores in the pipe nebula. A&A 568, A27. doi:10.1051/0004-6361/ 201423913, arxiv:1406.0540.

115. France, K., Linsky, J.L., Brown, A., Froning, C.S., Béland, S., 2010. Metal Depletion and Warm H2 in the Brown Dwarf 2M1207 Accretion Disk. Astrophys. J. 715, 596–605. doi:10.1088/0004-637X/715/1/596, arxiv:1004.1186.

116. France, K., Schindhelm, E., Herczeg, G.J., Brown, A., Abgrall, H., Alexander, R.D., Bergin, E.A., Brown, J.M., Linsky, J.L., Roueff, E., Yang, H., 2012. A Hubble Space telescope survey of H2 emission in the circumstellar environments of young stars. Astrophys. J. 756, 171. doi:10.1088/0004-637X/756/2/171, arxiv:1207.4789.

117. Fraser, H., Collings, M., McCoustra, M., 2002. Laboratory surface astrophysics experiment. Rev. Sci. Instrum. 73, 2161. doi:10.1063/1.1470232.

118. Fraser, H.J., van Dishoeck, E.F., 2004. SURFRESIDE: A novel experiment to study surface chemistry under interstellar and protostellar

conditions. Adv. Space Res. 33, 14–22. doi:10.1016/j.asr.2003.04.003.

119. Fuente, A., Martín-Pintado, J., Rodríguez-Fernández, N.J., Rodríguez-Franco, A., de Vicente, P., Kunze, D., 1999. Infrared space observatory observations toward the reflection nebula NGC 7023: A Nonequilibrium ortho-to-Para-H2 ratio. ApJL 518, L45–L48. doi:10.1086/312063, arxiv:9904136.

120. Garcia-Gil, S., Teillet-Billy, D., Rougeau, N., Sidis, V., 2013. H atom adsorption on a silicate surface: the (010) surface of forsterite. J. Phys. Chem. C 117 (24), 12612. doi:10.1021/jp4025365.

121. Garrod, R.T., 2008. A new modified-rate approach for gas-grain chemical simulations. Astron. Astrophys. 491 (1), 239.

122. Garrod, R.T., Pauly, T., 2011. On the formation of co2 and other interstellar ices. Astrophys. J. 735, 15. doi:10.1088/0004-637X/735/1/15, arxiv:1106.0540.

123. Gatley, I., Jones, T.J., Hyland, A.R., Wade, R., Geballe, T.R., Krisciunas, K., 1986. The spatial distribution and velocity field of the molecular hydrogen line emission from the centre of the galaxy. Mon. Not. R. Astron. Soc. 222, 299–306. doi:10. 1093/mnras/222.2.299.

124. Gavilan, L., Vidali, G., Lemaire, J.L., Chehrouri, M., Dulieu, F., Fillion, J.-H., Congiu, E., Chaabouni, H., 2012. Experimental investigation of the ortho/para ratio of newly formed molecular Hydrogen on amorphous solid water. Astrophys. J. 760, 35. doi:10.1088/0004-637X/760/1/35.

125. Ghio, E., Mattera, L., Salvo, C., Tommasini, F., Valbusa, U., 1980. Vibrational spectrum of h and d on the (0001) graphite surface from scattering experiments. J. Chem. Phys. 73, 556–561. doi:10.1063/1.439855.

126. Gillmon, K., Shull, J.M., 2006. Molecular hydrogen in infrared cirrus. Astrophys. J. 636, 908–915. doi:10.1086/498055, arxiv:0507587.

127. Gillmon, K., Shull, J.M., Tumlinson, J., Danforth, C., 2006. A FUSE survey of interstellar molecular hydrogen toward high-Latitude AGNs. Astrophys. J. 636, 891–907. doi:10.1086/498053, arxiv:0507581.

128. Glover, S.C.O., Abel, T., 2008. Uncertainties in H2 and HD chemistry and cooling and their role in early structure formation. Mon. Not. R. Astron. Soc. 388, 1627–1651. doi:10.1111/j.1365-2966.2008.13224.x, arxiv:0803.1768.

129. Glover, S.C.O., Clark, P.C., 2014. Molecular cooling in the diffuse interstellar medium. Mon. Not. R. Astron. Soc. 437, 9–20. doi:10.1093/

mnras/stt1809, arxiv:1305.7365.

130. Glover, S.C.O., Federrath, C., Mac Low, M.-M., Klessen, R.S., 2010. Modelling CO formation in the turbulent interstellar medium. Mon. Not. R. Astron. Soc. 404, 2–29. doi:10.1111/j.1365-2966.2009.15718.x, arxiv:0907.4081.
131. Glover, S.C.O., Mac Low, M.-M., 2007a. Simulating the formation of molecular clouds. i. slow formation by gravitational collapse from static initial conditions. ApJ Supplement 169, 239–268. doi:10.1086/512238, arxiv:0605120.
132. Glover, S.C.O., Mac Low, M.-M., 2007b. Simulating the formation of molecular clouds. II. rapid formation from turbulent initial conditions. Astrophys. J. 659, 1317–1337. doi:10.1086/512227, arxiv:0605121.
133. Gnedin, N.Y., Tassis, K., Kravtsov, A.V., 2009. Modeling molecular hydrogen and star formation in cosmological simulations. Astrophys. J. 697, 55–67. doi:10.1088/0004-637X/697/1/55, arxiv:0810.4148.
134. Goldader, J.D., Joseph, R.D., Doyon, R., Sanders, D.B., 1997. Spectroscopy of luminous infrared galaxies at 2 microns. Astrophys. J. 474, 104–120. doi:10.1086/303448.
135. Goldsmith, P.F., Velusamy, T., Li, D., Langer, W.D., 2010. Molecular Hydrogen emission from the boundaries of the Taurus molecular cloud. Astrophys. J. 715, 1370–1382. doi:10.1088/0004-637X/715/2/1370.
136. Gould, R.J., Salpeter, E.E., 1963. The interstellar abundance of the Hydrogen molecule. I. basic processes.. Astrophys. J. 138, 393. doi:10.1086/147654.
137. Goumans, T.P.M., Richard, C., Catlow, A., Brown, W.A., 2009. Formation of H2 on an olivine surface: a computational study. Mon. Not. R. Astron. Soc. 393, 1403–1407. doi:10.1111/j.1365-2966.2008.14155.x.
138. Graedel, T.E., Langer, W.D., Frerking, M.A., 1982. The kinetic chemistry of dense interstellar clouds. ApJ Supplement 48, 321–368. doi:10.1086/190780.
139. Grassi, T., Bovino, S., Schleicher, D., Gianturco, F.A., 2013. Chemical complexity in astrophysical simulations: optimization and reduction techniques. Mon. Not. R. Astron. Soc. 431, 1659–1668. doi:10.1093/mnras/stt284, arxiv:1211.5043.
140. Green, N.J.B., Toniazzo, T., Pilling, M.J., Ruffle, D.P., Bell, N., Hartquist, T.W., 2001. A stochastic approach to grain surface chemical

kinetics. A. A 375, 1111–1119. doi:10.1051/0004-6361:20010961.

141. Grenier, I.A., Casandjian, J.-M., Terrier, R., 2005. Unveiling extensive clouds of dark gas in the solar neighborhood. Science 307, 1292–1295. doi:10.1126/science. 1106924.
142. Gry, C., Boulanger, F., Nehmé, C., Pineau des Forêts, G., Habart, E., Falgarone, E., 2002. H2 formation and excitation in the diffuse interstellar medium. A.A 391 (1), 675–680. doi:10.1051/0004-6361:20020793.
143. Guillard, P., Boulanger, F., Cluver, M.E., Appleton, P.N., Pineau Des Forêts, G., Ogle, P., 2010. Observations and modeling of the dust emission from the H2-bright galaxy-wide shock in Stephan's Quintet. AA 518, A59. doi:10.1051/0004-6361/200913430, arxiv:1004.0677.
144. Guillard, P., Boulanger, F., Lehnert, M.D., Appleton, P.N., Pineau des Forêts, G., 2015a. Warm molecular Hydrogen at high redshift with the James Webb Space Telescope. In: Martins, F., Boissier, S., Buat, V., Cambrésy, L., Petit, P. (Eds.), SF2A2015: Proceedings of the Annual meeting of the French Society of Astronomy and Astrophysics, pp. 81–85. arxiv:1510.02107.
145. Guillard, P., Boulanger, F., Lehnert, M.D., Pineau des Forêts, G., Combes, F., Falgarone, E., Bernard-Salas, J., 2015. Exceptional agn-driven turbulence inhibits star formation in the 3c 326n radio galaxy. A.A 574, A32. doi:10.1051/0004-6361/201423612.
146. Guillard, P., Boulanger, F., Pineau Des Forêts, G., Appleton, P.N., 2009. H2 formation and excitation in the Stephan's Quintet galaxy-wide collision. A.A 502, 515–528. doi:10.1051/0004-6361/200811263, arxiv:0904.4239.
147. Guillard, P., Boulanger, F., Pineau des Forêts, G., Falgarone, E., Gusdorf, A., Cluver, M.E., Appleton, P.N., Lisenfeld, U., Duc, P.-A., Ogle, P.M., Xu, C.K., 2012a. Turbulent molecular gas and star formation in the shocked intergalactic medium of Stephan's Quintet. Astrophys. J. 749, 158. doi:10.1088/0004-637X/749/2/158, arxiv:1202.2862.
148. Guillard, P., Ogle, P.M., Emonts, B.H.C., Appleton, P.N., Morganti, R., Tadhunter, C., Oosterloo, T., Evans, D.A., Evans, A.S., 2012b. Strong molecular Hydrogen emission and kinematics of the multiphase gas in radio galaxies with fast jet-driven outflows. Astrophys. J. 747, 95. doi:10.1088/0004-637X/747/2/95, arxiv:1201. 1503.
149. Guillet, V., Jones, A.P., Pineau Des Forêts, G., 2009. Shocks in dense

clouds. II. Dust destruction and SiO formation in J shocks. A.A 497, 145–153. doi:10.1051/0004-6361/200811115.

150. Guillet, V., Pineau Des Forêts, G., Jones, A.P., 2011. Shocks in dense clouds. III. Dust processing and feedback effects in C-type shocks. A.A 527, A123. doi:10.1051/0004-6361/201015973.

151. Gusdorf, A., Anderl, S., Güsten, R., Stutzki, J., Hübers, H.-W., Hartogh, P., Heyminck, S., Okada, Y., 2012. Probing magnetohydrodynamic shocks with high-J CO observations: W28F. A.A 542, L19. doi:10.1051/0004-6361/201218907, arxiv:1203.2519.

152. Gusdorf, A., Riquelme, D., Anderl, S., Eislöffel, J., Codella, C., Gómez-Ruiz, A.I., Graf, U.U., Kristensen, L.E., Leurini, S., Parise, B., Requena-Torres, M.A., Ricken, O., Güsten, R., 2015. Impacts of pure shocks in the BHR71 bipolar outflow. A.A 575, A98. doi:10.1051/0004-6361/201425142, arxiv:1502.00488.

153. Guzmán, V.V., Pety, J., Goicoechea, J.R., Gerin, M., Roueff, E., Gratier, P., Öberg, K.I., 2015. Spatially resolved L-C3H+ emission in the horsehead photodissociation Region: Further Evidence for a Top-Down Hydrocarbon Chemistry. ApJL 800, L33. doi:10.1088/2041-8205/800/2/L33, arxiv:1502.02325.

154. Habart, E., Abergel, A., Boulanger, F., Joblin, C., Verstraete, L., Compiègne, M., Pineau Des Forêts, G., Le Bourlot, J., 2011. Excitation of H2 in photodissociation regions as seen by Spitzer. A.A 527, A122. doi:10.1051/0004-6361/20077327, arxiv:1012. 5324.

155. Habart, E., Abergel, A., Walmsley, C.M., Teyssier, D., Pety, J., 2005a. Density structure of the Horsehead nebula photo-dissociation region. A&A 437, 177–188.

156. Habart, E., Boulanger, F., Verstraete, L., Pineau des Forêts, G., Falgarone, E., Abergel, A., 2003. H2 infrared line emission across the bright side of the rho Ophiuchi main cloud. A&A 397, 623–634.

157. Habart, E., Boulanger, F., Verstraete, L., Walmsley, C.M., Pineau des Forêts, G., 2004. Some empirical estimates of the H2 formation rate in photon-dominated regions. A.A 414, 531–544. doi:10.1051/0004-6361:20031659, arxiv:0311040.

158. Habart, E., Walmsley, M., Verstraete, L., Cazaux, S., Maiolino, R., Cox, P., Boulanger, F., Pineau des Forêts, G., 2005b. Molecular Hydrogen. Space Sci. Rev. 119, 71–91. doi:10.1007/s11214-005-8062-1.

159. Habing, H.J., 1968. The interstellar radiation density between 912 A

and 2400 A. Bull. Astron. Inst. Netherlands 19, 421.

160. Hama, T., Kuwahata, K., Watanabe, N., Kouchi, A., Kimura, Y., Chigai, T., Pirronello, V., 2012. The mechanism of surface diffusion of h and d atoms on amorphous solid water: existence of various potential sites. Astrophys. J. 757, 185. doi:10.1088/0004-637X/757/2/185.
161. Hama, T., Watanabe, N., 2013. Surface processes on interstellar amorphous solid water: adsorption, diffusion, tunneling reactions, and nuclear-Spin conversion. Chem. Rev. 113 (12), 8783–8839. doi:10.1021/cr4000978.
162. Harada, N., Herbst, E., Wakelam, V., 2010. A new network for higher-temperature gas-phase chemistry. I. A preliminary study of accretion disks in active galactic nuclei. Astrophys. J. 721, 1570–1578. doi:10.1088/0004-637X/721/2/1570.
163. Harrison, A., Puxley, P., Russell, A., Brand, P., 1998. The ortho- to para- ratio of H2 in the starburst of NGC253. Mon. Not. R. Astron. Soc. 297, 624–632. doi:10.1046/j.1365-8711.1998.01525.x.
164. Hartigan, P., 2003. Shock Waves in Outflows from Young Stars. Ap&SS 287, 111–122. doi:10.1023/B:ASTR.0000006209.56314.c8.
165. Hartwig, T., Glover, S.C.O., Klessen, R.S., Latif, M.A., Volonteri, M., 2015. How an improved implementation of H2 self-shielding influences the formation of massive stars and black holes. Mon. Not. R. Astron. Soc. 452, 1233–1244. doi:10.1093/mnras/stv1368, arxiv:1505.00263.
166. Hasegawa, T.I., Herbst, E., 1993. New gas-grain chemical models of quiescent dense interstellar clouds - The effects of H2 tunnelling reactions and cosmic ray induced desorption. Mon. Not. R. Astron. Soc. 261, 83–102. doi:10.1093/mnras/261. 1.83.
167. Hasegawa, T.I., Herbst, E., 1993. Three-Phase chemical models of dense interstellar clouds - Gas dust particle mantles and dust particle surfaces. Mon. Not. R. Astron. Soc. 263, 589. doi:10.1093/mnras/263.3.589.
168. Hasegawa, T.I., Herbst, E., Leung, C.M., 1992. Models of gas-grain chemistry in dense interstellar clouds with complex organic molecules. ApJ Suppl. 82, 167–195. doi:10.1086/191713.
169. He, J., Frank, P., Vidali, G., 2011. Interaction of hydrogen with surfaces of silicates: single crystal vs. amorphous. Phys. Chem. Chem. Phys. (Incorporating Faraday Transactions) 13, 15803. doi:10.1039/c1cp21601e.

170. Henderson, B., Bekki, K., 2016. Significant enhancement of H2 formation in disk galaxies under strong ram pressure. ApJL 822, L33. doi:10.3847/2041-8205/822/2/L33, arxiv:1605.03304.

171. Henning, T., 2010a. Cosmic silicates. ARA.A 48, 21–46. doi:10.1146/annurev-astro-081309-130815.

172. Henning, T., 2010b. Laboratory Astrophysics of Cosmic Dust Analogues. In: Lecture Notes in Physics, Berlin Springer Verlag, 815. Springer Berlin Heidelberg, pp. 313–329. doi:10.1007/978-3-642-13259-9_8.

173. Herbst, E., Shematovich, V.I., 2003. New approaches to the modelling of surface chemistry on interstellar grains. APSS 285, 725–735. doi:10.1023/A: 1026173613191.

174. Herbst, E., van Dishoeck, E.F., 2009. Complex Organic Interstellar Molecules. ARA. A 47, 427–480. doi:10.1146/annurev-astro-082708-101654.

175. Herbst, T.M., Beckwith, S.V.W., Glindemann, A., Tacconi-Garman, L.E., Kroker, H., Krabbe, A., 1996. A near-infrared spectral imaging study of T Tau. Astronom. J. 111, 2403. doi:10.1086/117974.

176. Hewitt, J.W., Rho, J., Andersen, M., Reach, W.T., 2009. Spitzer observations of molecular hydrogen in interacting supernova remnants. Astrophys. J. 694, 1266–1280. doi:10.1088/0004-637X/694/2/1266, arxiv:0901.1622.

177. Hidaka, H., Miyauchi, N., Kouchi, A., Watanabe, N., 2008. Structural effects of ice grain surfaces on the hydrogenation of CO at low temperatures. Chem Phys Lett 456, 36–40. doi:10.1016/j.cplett.2008.03.010, arxiv:0805.3383.

178. Higdon, S.J.U., Armus, L., Higdon, J.L., Soifer, B.T., Spoon, H.W.W., 2006. A spritzer space telescope infrared spectrograph survey of warm molecular hydrogen in ultraluminous infrared galaxies. Astrophys. J. 648, 323–339. doi:10.1086/505701.

179. Hily-Blant, P., Falgarone, E., 2009. Intermittency of interstellar turbulence: parsecscale coherent structure of intense, velocity shear. A.A 500 (2), L29–L32. doi:10. 1051/0004-6361/200912296.

180. Hincelin, U., Chang, Q., Herbst, E., 2015. A new and simple approach to determine the abundance of hydrogen molecules on interstellar ice mantles. A.A 574, A24. doi:10.1051/0004-6361/201424807, arxiv:1410.7375.

181. Hocuk, S., Cazaux, S., Spaans, M., Caselli, P., 2016. How chemistry influences cloud structure, star formation, and the IMF. Mon. Not. R. Astron. Soc. 456, 2586–2610. doi:10.1093/mnras/stv2817, arxiv:1511.06696.

182. Hollenbach, D., Kaufman, M.J., Bergin, E.A., Melnick, G.J., 2009. Water, O2, and ice in molecular clouds. Astrophys. J. 690, 1497–1521. doi:10.1088/0004-637X/690/2/1497, arxiv:0809.1642.

183. Hollenbach, D., McKee, C.F., 1979. Molecule formation and infrared emission in fast interstellar shocks. I Physical processes. ApJ Supplement 41, 555–592. doi:10. 1086/190631.

184. Hollenbach, D., McKee, C.F., 1989. Molecule formation and infrared emission in fast interstellar shocks. III - Results for J shocks in molecular clouds. Astrophys. J. 342, 306–336. doi:10.1086/167595.

185. Hollenbach, D., Salpeter, E.E., 1970. Surface adsorption of light gas atoms. J. Chem. Phys. 53, 79–86. doi:10.1063/1.1673836.

186. Hollenbach, D., Salpeter, E.E., 1971. Surface recombination of Hydrogen molecules. Astrophys. J. 163, 155. doi:10.1086/150754.

187. Hollenbach, D.J., Tielens, A.G.G.M., 1997. Dense photodissociation regions (PDRs). ARA. A 35, 179–216. doi:10.1146/annurev. astro.35.1.179.

188. Hollenbach, D.J., Tielens, A.G.G.M., 1999. Photodissociation regions in the interstellar medium of galaxies. Rev. Mod. Phys. 71, 173–230. doi:10.1103/RevModPhys.71. 173.

189. Hollenbach, D.J., Werner, M.W., Salpeter, E.E., 1971. Molecular Hydrogen in H i Regions. Astrophys. J. 163, 165. doi:10.1086/150755.

190. Hornekaer, L., Baurichter, A., Petrunin, V., Luntz, A., Kay, B., Al-Halabi, A., 2005. Influence of surface morphology on d desorption kinetics from amorphous solid water. J. Chem. Phys. 122, 124701.

191. Hornekær, L., Baurichter, A., Petrunin, V.V., Field, D., Luntz, A.C., 2003. Importance of surface morphology in interstellar H2 formation. Science 302, 1943–1946. doi:10.1126/science.1090820.

192. Hornekær, L., Rauls, E., Xu, W., Šljivancanin, ˇ Ž., Otero, R., Stensgaard, I., Lægsgaard, E., Hammer, B., Besenbacher, F., 2006a. Clustering of chemisorbed h(d) atoms on the graphite (0001) surface due to preferential sticking. Phys. Rev. Lett. 97 (18), 186102. doi:10.1103/PhysRevLett.97.186102.

193. Hornekær, L., Šljivancanin, ˇŽ., Xu, W., Otero, R., Rauls, E., Stensgaard,

I., Lægsgaard, E., Hammer, B., Besenbacher, F., 2006b. Metastable structures and recombination pathways for atomic Hydrogen on the graphite (0001) Surface. Phys. Rev. Lett. 96 (15), 156104. doi:10.1103/PhysRevLett.96.156104.

194. Hosokawa, T., Inutsuka, S.-i., 2006. Dynamical expansion of ionization and dissociation front around a massive star. II. On the generality of triggered star formation. Astrophys. J. 646, 240–257. doi:10.1086/504789, arxiv:0511165.

195. Ingalls, J.G., Bania, T.M., Boulanger, F., Draine, B.T., Falgarone, E., Hily-Blant, P., 2011. Spitzer infrared spectrograph detection of molecular Hydrogen rotational emission towards translucent clouds. Astrophys. J. 743, 174. doi:10.1088/0004-637X/743/2/174, arxiv:1109.2544.

196. Inoue, T., Inutsuka, S.-i., 2012. Formation of turbulent and magnetized molecular clouds via accretion Flows of H I clouds. Astrophys. J. 759, 35. doi:10.1088/0004-637X/759/1/35, arxiv:1205.6217.

197. Iqbal, W., Acharyya, K., Herbst, E., 2012. Kinetic Monte Carlo studies of H2 formation on grain surfaces over a wide temperature range. Astrophys. J. 751, 58. doi:10. 1088/0004-637X/751/1/58.

198. Islam, F., Cecchi-Pestellini, C., Viti, S., Casu, S., 2010. Formation pumping of molecular hydrogen in dark clouds. Astrophys. J. 725, 1111.

199. Islam, F., Latimer, E.R., Price, S.D., 2007. The formation of vibrationally excited HD from atomic recombination on cold graphite surfaces. J. Chem. Phys. 127 (6), 064701.

200. Ivanovskaya, V.V., Zobelli, A., Teillet-Billy, D., Rougeau, N., Sidis, V., Briddon, P.R., 2010. Hydrogen adsorption on graphene: a first principles study. Eur. Phys. J. B 76, 481. doi:10.1140/epjb/e2010-00238-7.

201. Jeloaica, L., Sidis, V., 1999. DFT investigation of the adsorption of atomic hydrogen on a cluster-model graphite surface. Chem. Phys. Lett. 300, 157–162. doi:10. 1016/S0009-2614(98)01337-2.

202. Jenkins, E.B., Peimbert, A., 1997. Molecular Hydrogen in the direction of ζ Orionis A. Astrophys. J. 477, 265–280. doi:10.1086/303694, arxiv:9609144.

203. Joblin, C., Bron, E., Pinto, C., Pilleri, P., Gerin, M., Le Bourlot, J., Le Petit, F., Fuente, A., Berné, O., Goicoechea, J., Habart, E., Köhler, M.,

Teyssier, D., Nagy, Z., Montillaud, J., Vastel, C., Cernicharo, J., Röllig, M., Ossenkopf, V., Bergin, E., in prep.Gas morphology and Energetics in bright Photodissociation Regions.

204. Jones,A.P., Fanciullo, L., Köhler, M., Verstraete, L., Guillet, V., Bocchio, M., Ysard, N., 2013. The evolution of amorphous hydrocarbons in the ISM: dust modelling from a new vantage point. A&A 558, A62. doi:10.1051/0004-6361/201321686, arxiv:1411.6293.
205. Jones, A.P., Habart, E., 2015. H2 formation via the UV photo-processing of a-C:h nano-particles. A&A 581, A92. doi:10.1051/0004-6361/201526487, arxiv:1507. 08534.
206. Jones, A.P., Tielens, A.G.G.M., Hollenbach, D.J., 1996. Grain shattering in shocks: The interstellar grain size distribution. Astrophys. J. 469, 740. doi:10.1086/177823.
207. Jura, M., 1974. Formation and destruction rates of interstellar H2.. Astrophys. J. 191, 375–379. doi:10.1086/152975.
208. Jura, M., 1975. Interstellar clouds containing optically thin H2. Astrophys. J. 197, 575–580. doi:10.1086/153545.
209. Kamenetzky, J., Rangwala, N., Glenn, J., Maloney, P.R., Conley, A., 2014. A survey of the molecular ISM properties of nearby galaxies using the herschel FTS. Astrophys. J. 795, 174. doi:10.1088/0004-637X/795/2/174, arxiv:1409.7062.
210. Kaplan, K.F., Dinerstein, H.L., Oh, H., Mace, G.N., Kim, H., Sokal, K.R., Pavel, M.D., Lee, S., Pak, S., Park, C., Sok Oh, J., Jaffe, D.T., 2017. Excitation of molecular Hydrogen in the Orion bar photodissociation region from a deep near-infrared IGRINS spectrum. APJ 838, 152.
211. Karlicky, F., Lepetit, B., Lemoine, D., 2014. Quantum modeling of hydrogen chemisorption on graphene and graphite. J. Chem. Phys. 140, 124702.
212. Katz, N., Furman, I., Biham, O., Pirronello, V., Vidali, G., 1999. Molecular hydrogen formation on astrophysically relevant surfaces. Astrophys. J. 522, 305–312. doi:10.1086/307642, arxiv:9906071.
213. Kaufman, M.J., Wolfire, M.G., Hollenbach, D.J., 2006. [Si II], [Fe II], [C II], And h2 emission from massive star-forming regions. Astrophys. J. 644, 283–299. doi:10. 1086/503596.
214. Kaufman, M.J., Wolfire, M.G., Hollenbach, D.J., Luhman, M.L., 1999. Far-Infrared and submillimeter emission from galactic and extragalactic photodissociation regions. Astrophys. J. 527, 795–813.

doi:10.1086/308102, arxiv:9907255.

215. Kerwin, J., Jackson, B., 2008. The sticking of H and D atoms on a graphite (0001) surface: The effects of coverage and energy dissipation. J. Chem. Phys. 128 (8). 084702–084702. doi: 10.1063/1.2868771

216. Kerwin, J., Sha, X., Jackson, B., 2006. Classical Studies of H Atom Trapping on a Graphite Surface. J. Phys. Chem. B 110 (38), 18811–18817.

217. Kimmel, G.A., Orlando, T.M., 1996. Observation of negative ion resonances in amorphous ice via low-energy (5-40 eV) electron-stimulated production of molecular hydrogen. Phys. Rev. Lett. 77, 3983–3986. doi:10.1103/PhysRevLett.77.3983.

218. Kimmel, G.A., Orlando, T.M., Vézina, C., Sanche, L., 1994. Low-energy electronstimulated production of molecular hydrogen from amorphous water ice. J. Chem. Phys. 101, 3282–3286. doi:10.1063/1.468430.

219. Köhler, M., Guillet, V., Jones, A., 2011. Aggregate dust connections and emissivity enhancements. A.A 528, A96. doi:10.1051/0004-6361/201016379.

220. Köhler, M., Habart, E., Arab, H., Bernard-Salas, J., Ayasso, H., Abergel, A., Zavagno, A., Polehampton, E., van der Wiel, M.H.D., Naylor, D.A., Makiwa, G., Dassas, K., Joblin, C., Pilleri, P., Berné, O., Fuente, A., Gerin, M., Goicoechea, J.R., Teyssier, D., 2014a. Physical structure of the photodissociation regions in NGC 7023. Observations of gas and dust emission with Herschel. A.A 569, A109. doi:10.1051/0004-6361/201322711, arxiv:1410.2081.

221. Köhler, M., Jones, A., Ysard, N., 2014b. A hidden reservoir of Fe/FeS in interstellar silicates? A.A 565, L9. doi:10.1051/0004-6361/201423985, arxiv:1405.4208.

222. Kolasinski, K.W., 2008. Surface Science : Foundations of Catalysis and Nanoscience. Wiley:, Hoboken, N.J..

223. Kouchi, A., Kuroda, T., 1991. Ultraviolet-Induced Amorphization of Cubic Ice and its Implication for the Evolution of Ice Grains. In: Levasseur-Regoud, A.C., Hasegawa, H. (Eds.), IAU Colloq. 126: Origin and Evolution of Interplanetary Dust. In: Astrophysics and Space Science Library, 173, p. 87. doi:10.1007/978-94-011-3640-2_17.

224. Koyama, H., Inutsuka, S.-i., 2002. An Origin of Supersonic Motions in Interstellar Clouds. ApJL 564, L97–L100. doi:10.1086/338978,

arxiv:0112420.

225. Krügel, E., 2007. An Introduction to the Physics of Interstellar dust. Series in astronomy and astrophysics. Taylor & Francis: Oxford and New York.
226. Krumholz, M.R., McKee, C.F., Tumlinson, J., 2009. The star formation law in atomic and molecular Gas. Astrophys. J. 699, 850–856. doi:10.1088/0004-637X/699/1/850, arxiv:0904.0009.
227. Kuhlen, M., Krumholz, M.R., Madau, P., Smith, B.D., Wise, J., 2012. Dwarf galaxy formation with H2-regulated star formation. Astrophys. J. 749, 36. doi:10.1088/0004-637X/749/1/36, arxiv:1105.2376.
228. Kuwahata, K., Hama, T., Kouchi, A., Watanabe, N., 2015. Signatures of quantumTunneling diffusion of hydrogen atoms on water ice at 10 k. Phys. Rev. Lett. 115 (13), 133201. doi:10.1103/PhysRevLett.115.133201.
229. Lacour, S., Ziskin, V., Hébrard, G., Oliveira, C., André, M.K., Ferlet, R., Vidal-Madjar, A., 2005. Velocity dispersion of the high rotational levels of H2. Astrophys. J. 627, 251–262. doi:10.1086/430291, arxiv:0503419.
230. Langer, W.D., Velusamy, T., Pineda, J.L., Willacy, K., Goldsmith, P.F., 2014. A Herschel [C ii] galactic plane survey. II. CO-dark h2 in clouds. A.A 561, A122. doi:10.1051/0004-6361/201322406, arxiv:1312.3320.
231. Latimer, E.R., Islam, F., Price, S.D., 2008. Studies of HD formed in excited vibrational states from atomic recombination on cold graphite surfaces. Chem. Phys. Lett. 455, 174–177. doi:10.1016/j.cplett.2008.02.105.
232. Launhardt, R., Stutz, A.M., Schmiedeke, A., Henning, T., Krause, O., Balog, Z., Beuther, H., Birkmann, S., Hennemann, M., Kainulainen, J., Khanzadyan, T.,
233. Linz, H., Lippok, N., Nielbock, M., Pitann, J., Ragan, S., Risacher, C., Schmalzl, M., Shirley, Y.L., Stecklum, B., Steinacker, J., Tackenberg, J., 2013. The earliest phases of star formation (EPos): a herschel key project. the thermal structure of lowmass molecular cloud cores. A.A 551, A98. doi:10.1051/0004-6361/201220477, arxiv:1301.1498.
234. Le, H.A.N., Pak, S., Kaplan, K., Mace, G., Lee, S., Pavel, M., Jeong, U., Oh, H., Lee, H.I., Chun, M.Y., Yuk, I.S., Pyo, T.S., Hwang, N., Kim, K.M., Park, C., Sok Oh, J., Yu, Y.S., Park, B.G., Minh, Y.C., Jaffe, D.T., 2017. Fluorescent H2 emission lines from the reflection

nebula NGC 7023 observed with Igrins. Astrophys. J. 841, 13.

235. Le Bourlot, J., 2000. Ortho to para conversion of H2 on interstellar grains. A&A 360, 656–662. Le Bourlot, J., Pineau des Forets, G., Roueff, E., Dalgarno, A., Gredel, R., 1995. Infrared diagnostics of the formation of H 2 on interstellar dust. APJ 449, 178L.

236. Le Bourlot, J., Le Petit, F., Pinto, C., Roueff, E., Roy, F., 2012. Surface chemistry in the interstellar medium. I. H2 formation by Langmuir-Hinshelwood and Eley-Rideal mechanisms. A. A 541, A76. doi:10.1051/0004-6361/201118126, arxiv:1202.0374.

237. Le Bourlot, J., Pineau des Forêts, G., Flower, D.R., 1999. The cooling of astrophysical media by H2. Mon. Not. R. Astron. Soc. 305, 802–810. doi:10.1046/j.1365-8711. 1999.02497.x.

238. Le Bourlot, J., Pineau des Forêts, G., Flower, D.R., Cabrit, S., 2002. New determinations of the critical velocities of C-type shock waves in dense molecular clouds: application to the outflow source in Orion. Mon. Not. R. Astron. Soc. 332, 985–993. doi:10.1046/j.1365-8711.2002.05373.x.

239. Le Page, V., Snow, T.P., Bierbaum, V.M., 2001. Hydrogenation and Charge States of PAHS in Diffuse Clouds. I. Development of a Model. ApJ. Suppl. 132, 233–251. doi:10.1086/318952.

240. Le Page, V., Snow, T.P., Bierbaum, V.M., 2009. Molecular hydrogen formation catalyzed by polycyclic aromatic hydrocarbons in the interstellar medium. Astrophys. J. 704, 274–280. doi:10.1088/0004-637X/704/1/274.

241. Le Petit, F., 2002. Etude des propri??t??s de HD dans le Milieu Interstellaire. Observations FUSE et modélisation. Phd thesis. Université Paris 7.

242. Le Petit, F., Barzel, B., Biham, O., Roueff, E., Le Bourlot, J., 2009. Incorporation of stochastic chemistry on dust grains in the meudon pdr code using moment equations. i. application to the formation of h{2} and hd. Astron. Astrophys. 505, 1153. doi:10.1051/0004-6361/200912269.

243. Le Petit, F., Nehmé, C., Le Bourlot, J., Roueff, E., 2006. A model for atomic and molecular interstellar gas: the meudon PDR code. ApJ. Suppl. 164, 506–529. doi:10.1086/503252, arxiv:0602150.

244. Lefloch, B., Cernicharo, J., Cabrit, S., Noriega-Crespo, A., Moro-Martin, A., Cesarsky, D., 2003. Warm molecular hydrogen and ionized

Neon in the HH 2 outflow. APJ 590, L41.

245. Lehnert, M.D., Nesvadba, N.P.H., Tiran, L.L., Matteo, P.D., van Driel, W., Douglas, L.S., Chemin, L., Bournaud, F., 2009. Physical conditions in the interstellar medium of intensely star-forming galaxies at Redshift 2. A.A 699, 1660–1678. doi:10.1088/ 0004-637X/699/2/1660.

246. Lehtinen, P.O., Foster, A.S., Ma, Y., Krasheninnikov, A.V., Nieminen, R.M., 2004. Irradiation-induced magnetism in graphite: A density functional study. Phys. Rev. Lett. 93, 187202. doi:10.1103/ PhysRevLett.93.187202.

247. Lemaire, J.L., Field, D., Maillard, J.P., Des Forêts, G.P., Falgarone, E., Pijpers, F.P., Gerin, M., Rostas, F., 1999. High resolution fourier transform spectroscopy of H2 IR emission in NGC 7023. A.A 349, 253.

248. Lemaire, J.L., Vidali, G., Baouche, S., Chehrouri, M., Chaabouni, H., Mokrane, H., 2010. Competing Mechanisms of Molecular Hydrogen Formation in Conditions Relevant to the Interstellar Medium. ApJL 725, L156–L160. doi:10.1088/2041-8205/725/2/L156.

249. Lepetit, B., Jackson, B., 2011b. Sticking of hydrogen on supported and suspended graphene at low temperature. Phys. Rev. Lett. 107, 236102. doi:10.1103/PhysRevLett.107.236102.

250. Lepetit, B., Lemoine, D., Medina, Z., Jackson, B., 2011a. Sticking and desorption of hydrogen on graphite: a comparative study of different models. J. Chem. Phys. 134 (11) 114705–114705. doi:10.1063/1.3565446.

251. Lesaffre, P., Pineau des Forêts, G., Godard, B., Guillard, P., Boulanger, F., Falgarone, E., 2013. Low-velocity shocks: signatures of turbulent dissipation in diffuse irradiated gas. A.A 550, A106. doi:10.1051/0004-6361/201219928, arxiv:1301.7598.

252. Leung, C.M., Herbst, E., Huebner, W.F., 1984. Synthesis of complex molecules in dense interstellar clouds via gas-phase chemistry - A pseudo time-dependent calculation. ApJ Suppl. 56, 231–256. doi:10.1086/190982.

253. Li, D., Goldsmith, P.F., 2003. H i narrow self-Absorption in dark clouds. Astrophys. J. 585, 823–839. doi:10.1086/346227, arxiv:0206396.

254. Lim, A.J., Rawling, J.M.C., Williams, D.A., 2001. Numerical hydrodynamics with a parallel integrated chemistry. II. the interaction of a stellar jet with a molecular cloud. A. A 376, 336–347.

doi:10.1051/0004-6361:20010925.

255. Lim, A.J., Rawlings, J.M.C., Williams, D.A., 1999. Numerical hydrodynamics with a parallel integrated chemistry - I. A shear boundary layer. Mon. Not. R. Astron. Soc. 308, 1126–1132. doi:10.1046/j.1365-8711.1999.02826.x.
256. Lipshtat, A., Biham, O., 2003. Moment equations for chemical reactions on interstellar dust grains. A. A 400, 585–593. doi:10.1051/0004-6361:20021902, arxiv:0212200.
257. Lipshtat, A., Biham, O., 2005. The effect of grain size distribution on H2 formation rate in the interstellar medium. Mon. Not. R. Astron. Soc. 362, 666–670. doi:10. 1111/j.1365-2966.2005.09356.x, arxiv:0508026.
258. Lohmar, I., Krug, J., 2006. The sweeping rate in diffusion-mediated reactions on dust grain surfaces. Mon. Not. R. Astron. Soc. 370, 1025.
259. Lohmar, I., Krug, J., Biham, O., 2009. Accurate rate coefficients for models of interstellar gas-grain chemistry. Astron. Astrophys. 504 (3), L5.
260. Luhman, M.L., Luhman, K.L., Benedict, T., Jaffe, D.T., Fischer, J., 1997. Near-Infrared Spectra of IC 59/63 and NGC 1535: Comparing infrared and ultraviolet observations of H2. apj 480, L133.
261. Lutz, D., Sturm, E., Genzel, R., Spoon, H.W.W., Moorwood, A.F.M., Netzer, H., Sternberg, A., 2003. ISO spectroscopy of star formation and active nuclei in the luminous infrared galaxy NGC 6240. A&A 409, 867–878. doi:10.1051/0004-6361: 20031165, arxiv:0307552.
262. Mac Low, M.-M., Glover, S.C.O., 2012. The abundance of molecular hydrogen and its correlation with midplane pressure in galaxies: non-equilibrium, turbulent, chemical models. Astrophys. J. 746, 135. doi:10.1088/0004-637X/746/2/135, arxiv:1011.3054.
263. Majumdar, L., Gratier, P., Ruaud, M., Wakelam, V., Vastel, C., Sipilä, O., Hersant, F., Dutrey, A., Guilloteau, S., 2017. Chemistry of TMC-1 with multiply deuterated species and spin chemistry of H2, H2+, H3 + and their isotopologues. Mon. Not. R. Astron. Soc. 466, 4470–4479. doi:10.1093/mnras/stw3360, arxiv:1612.07845.
264. Maloney, P.R., Hollenbach, D.J., Tielens, A.G.G.M., 1996. X-Ray–irradiated Molecular Gas. I. Physical Processes and General Results. Astrophys. J. 466, 561. doi:10. 1086/177532.
265. Manicò, G., Ragunì, G., Pirronello, V., Roser, J.E., Vidali, G., 2001. Laboratory measurements of molecular hydrogen formation on

amorphous water ice. ApJL 548, L253–L256. doi:10.1086/319116.

266. Maret, S., Bergin, E.A., Neufeld, D.A., Green, J.D., Watson, D.M., Harwit, M.O., Kristensen, L.E., Melnick, G.J., Sonnentrucker, P., Tolls, V., Werner, M.W., Willacy, K., Yuan, Y., 2009. Spitzer mapping of molecular Hydrogen pure rotational lines in NGC 1333: A detailed study of feedback in star formation. Astrophys. J. 698, 1244–1260. doi:10.1088/0004-637X/698/2/1244, arxiv:0904.0603.
267. Martin-Zaidi, C., van Dishoeck, E.F., Augereau, J.C., Lagage, P.O., Pantin, E., 2008. Searching for molecular hydrogen mid-infrared emission in the circumstellar environments of herbig be stars. A. A 489, 601.
268. Martinazzo, R., Tantardini, G.F., 2006. Quantum study of eley-rideal reaction and collision induced desorption of hydrogen atoms on a graphite surface. i. h– chemisorbed case. J. Chem. Phys. 124 (12), 124702.
269. Martinazzo, R., Tantardini, G.F., 2006. Quantum study of eley-rideal reaction and collision induced desorption of hydrogen atoms on a graphite surface. ii. h-physisorbed case. J. Chem. Phys. 124 (12), 124703.
270. Masuda, K., Takahashi, J., 1997. The sticking probability of a hydrogen atom on icy mantle. Adv. Space Res. 19, 1019–1022. doi:10.1016/S0273-1177(97)00344-X.
271. Masuda, K., Takahashi, J., Mukai, T., 1998. Sticking probability and mobility of a hydrogen atom on icy mantle of dust grains. A.A 330, 773–781.
272. Mata, H., Ramos-Larios, G., Guerrero, M.A., Nigoche-Netro, A., Toalá, J.A., Fang, X., Rubio, G., Kemp, S.N., Navarro, S.G., Corral, L.J., 2016. Spitzer mid-infrared spectroscopic observations of planetary nebulae. Mon. Not. R. Astron. Soc. 459, 841–853. doi:10.1093/mnras/stw646, arxiv:1603.06667.
273. Matar, E., Bergeron, H., Dulieu, F., Chaabouni, H., Accolla, M., Lemaire, J.L., 2010. Gas temperature dependent sticking of hydrogen on cold amorphous water ice surfaces of interstellar interest. J. Chem. Phys. 133 (10). doi:10.1063/1.3484867, arxiv:1008.4407. 104507–104507.
274. Matar, E., Congiu, E., Dulieu, F., Momeni, A., Lemaire, J.L., 2008. Mobility of D atoms on porous amorphous water ice surfaces under

interstellar conditions. A. A 492, L17–L20. doi:10.1051/0004-6361:200810434, arxiv:0810.2242.

275. Mathis, J.S., Mezger, P.G., Panagia, N., 1983. Interstellar radiation field and dust temperatures in the diffuse interstellar matter and in giant molecular clouds. A. A 128, 212–229.

276. McKee, C.F., Hollenbach, D.J., 1980. Interstellar shock waves. ARA. A 18, 219–262. doi:10.1146/annurev.aa.18.090180.001251.

277. Medina, Z., Jackson, B., 2008. Quantum studies of light particle trapping, sticking, and desorption on metal and graphite surfaces. J. Chem. Phys. 128 (11). doi:10. 1063/1.2890043. 114704–114704.

278. Meijer, A., Farebrother, A., Clary, D., Fisher, A., 2001. Time-dependent quantum mechanical calculations on the formation of molecular hydrogen on a graphite surface via an eley? rideal mechanism. J. Phys. Chem. A 105 (11), 2173.

279. Meijerink, R., Spaans, M., 2005. Diagnostics of irradiated gas in galaxy nuclei. i. a far-ultraviolet and X-ray dominated region code. A.A 436, 397–409. doi:10.1051/ 0004-6361:20042398, arxiv:0502454.

280. Meisner, J., Lamberts, T., Kästner, J., 2017. Atom Tunneling in the Water Formation Reaction H_2 + OH → H_2O + H on an Ice Surface. ArXiv e-prints1708.05559.

281. Mennella, V., 2008. HD formation by abstraction of h/d chemisorbed in carbon grains with d/h atoms under simulated interstellar conditions. APJL 684, L25. doi:10.1086/592001.

282. Mennella, V., Baratta, G.A., Palumbo, M.E., Bergin, E.A., 2006. Synthesis of CO and CO2 molecules by UV irradiation of water ice-covered hydrogenated carbon grains. Astrophys. J. 643, 923–931. doi:10.1086/502965.

283. Mennella, V., Hornekær, L., Thrower, J., Accolla, M., 2012. The catalytic role of coronene for molecular hydrogen formation. ApJL 745, L2. doi:10.1088/2041-8205/745/1/L2.

284. Meyer, D.M., Lauroesch, J.T., Sofia, U.J., Draine, B.T., Bertoldi, F., 2001. The rich ultraviolet spectrum of vibrationally excited interstellar h2 toward HD 37903. ApJL 553, L59–L62. doi:10.1086/320504.

285. Micic, M., Glover, S.C.O., Banerjee, R., Klessen, R.S., 2013. Cloud formation in colliding flows: influence of the choice of cooling function. Mon. Not. R. Astron. Soc. 432, 626–636. doi:10.1093/mnras/stt489, arxiv:1303.4751.

286. Micic, M., Glover, S.C.O., Federrath, C., Klessen, R.S., 2012. Modelling H2 formation in the turbulent interstellar medium: solenoidal versus compressive turbulent forcing. Mon. Not. R. Astron. Soc. 421, 2531–2542. doi:10.1111/j.1365-2966.2012. 20477.x, arxiv:1103.3056.

287. Minissale, M., Dulieu, F., Cazaux, S., Hocuk, S., 2016. Dust as interstellar catalyst: i. Quantifying the chemical desorption process. A&A 585. doi:10.1051/0004-6361/201525981.

288. Miret-Artés, S., Pollak, E., 2005. The dynamics of activated surface diffusion. J. Phys. Condensed Matter 17, S4133–S4150. doi:10.1088/0953-8984/17/49/009.

289. Morisset, S., Aguillon, F., Sizun, M., 2003. The dynamics of H2 formation on a graphite surface at low temperature. Phys. Chem. Chem. Phys. (Incorporating Faraday Transactions) 5, 506–513. doi:10.1039/B208492A.

290. Morisset, S., Aguillon, F., Sizun, M., Sidis, V., 2004. Role of surface relaxation in the Eley-Rideal formation of H2 on a graphite surface. J. Phys. Chem. A 108, 8571–8579. doi:10.1021/jp049969q.

291. Morisset, S., Aguillon, F., Sizun, M., Sidis, V., 2005. Wave-packet study of h[sub 2] formation on a graphite surface through the Langmuir–Hinshelwood mechanism. J. Chem. Phys. 122 (19), 194702.

292. Morisset, S., Allouche, A., 2008. Quantum dynamic of sticking of a H atom on a graphite surface. J Chem Phys 129 (2).

293. Morisset, S., Ferro, Y., Allouche, A., 2010. Study of the sticking of a hydrogen atom on a graphite surface using a mixed classical-quantum dynamics method. J. Chem. Phys. 133 (4).

294. Motoyama, K., Morata, O., Shang, H., Krasnopolsky, R., Hasegawa, T., 2015. A hydrochemical hybrid code for astrophysical problems. i. code verification and benchmarks for a photon-dominated region (PDR). Astrophys. J. 808, 46. doi:10.1088/0004-637X/808/1/46, arxiv:1506.00471.

295. Mouri, H., 1994. Molecular hydrogen emission in galaxies: determination of excitation mechanism. Astrophys. J. 427, 777–783. doi:10.1086/174184.

296. Moutou, C., Verstraete, L., Sellgren, K., Leger, A., 1999. The rich spectroscopy of reflection nebulae. In: Cox, P., Kessler, M. (Eds.), The Universe as Seen by ISO. In: ESA Special Publication, 427, p. 727.

297. Navarro-Ruiz, J., Martínez-González, J.Á., Sodupe, M., Ugliengo,

P., Rimola, A., 2015. Relevance of silicate surface morphology in interstellar H2 formation. Insights from quantum chemical calculations. Mon. Not. R. Astron. Soc. 453, 914–924. doi:10.1093/mnras/stv1628.

298. Navarro-Ruiz, J., Sodupe, M., Ugliengo, P., Rimola, A., 2014. Interstellar H adsorption and H2 formation on the crystalline (010) forsterite surface: a B3LYP-D2* periodic study. Phys. Chem. Chem. Phys. (Incorporating Faraday Transactions) 16, 17447. doi:10.1039/C4CP00819G.
299. Nehmé, C., Gry, C., Boulanger, F., Le Bourlot, J., Pineau Des Forêts, G., Falgarone, E., 2008. Multi-wavelength observations of a nearby multi-phase interstellar cloud. A&A 483, 471–484. doi:10.1051/0004-6361:20078373, arxiv:0802.4059.
300. Neufeld, D.A., Melnick, G.J., Harwit, M., 1998. Infrared space observatory observations of molecular hydrogen in HH 54: Measurement of a nonequilibrium ratio of Ortho- to Para-H2. ApJL 506, L75–L78. doi:10.1086/311636.
301. Neufeld, D.A., Nisini, B., Giannini, T., Melnick, G.J., Bergin, E.A., Yuan, Y., Maret, S., Tolls, V., Güsten, R., Kaufman, M.J., 2009. Spitzer spectral line mapping of protostellar outflows. i. basic data and outflow energetics. Astrophys. J. 706, 170–183. doi:10.1088/0004-637X/706/1/170, arxiv:0910.1107.
302. Nielbock, M., Launhardt, R., Steinacker, J., Stutz, A.M., Balog, Z., Beuther, H., Bouwman, J., Henning, T., Hily-Blant, P., Kainulainen, J., Krause, O., Linz, H., Lippok, N., Ragan, S., Risacher, C., Schmiedeke, A., 2012. The Earliest Phases of Star formation (EPoS) observed with Herschel: the dust temperature and density distributions of B68. A&A 547, A11. doi:10.1051/0004-6361/201219139, arxiv:1208.4512.
303. Nisini, B., Giannini, T., Neufeld, D.A., Yuan, Y., Antoniucci, S., Bergin, E.A., Melnick, G.J., 2010. Spitzer spectral line mapping of protostellar outflows. II. H2 emission in L1157. Astrophys. J. 724, 69–79. doi:10.1088/0004-637X/724/1/69, arxiv:1009.3834.
304. Oba, Y., Miyauchi, N., Hidaka, H., Chigai, T., Watanabe, N., Kouchi, A., 2009. Formation of compact amorphous H2O ice by codeposition of Hydrogen atoms with Oxygen molecules on grain surfaces. Astrophys. J. 701, 464–470. doi:10.1088/0004-637X/701/1/464.
305. Oba, Y., Watanabe, N., Hama, T., Kuwahata, K., Hidaka, H., Kouchi, A., 2012. Water formation through a quantum tunneling surface reaction, oh + h 2, at 10 k. Astrophys. J. 749 (1), 67.

306. Ogle, P., Antonucci, R., Appleton, P.N., Whysong, D., 2007. Shocked molecular hydrogen in the 3C 326 radio galaxy system. Astrophys. J. 668, 699–707. doi:10.1086/521334, arxiv:0707.0896.
307. Ogle, P., Boulanger, F., Guillard, P., Evans, D.A., Antonucci, R., Appleton, P.N., Nesvadba, N., Leipski, C., 2010. Jet-powered molecular hydrogen emission from radio galaxies. Astrophys. J. 724, 1193–1217. doi:10.1088/0004-637X/724/2/1193, arxiv:1009.4533.
308. Ogle, P., Davies, J.E., Appleton, P.N., Bertincourt, B., Seymour, N., Helou, G., 2012. Ultraluminous star-forming galaxies and extremely luminous warm molecular hydrogen emission at z = 2.16 in the PKS 1138-26 radio galaxy protocluster. Astrophys. J. 751, 13. doi:10.1088/0004-637X/751/1/13, arxiv:1203.3019.
309. Pagani, L., Roueff, E., Lesaffre, P., 2011. Ortho-H2 and the age of interstellar dark clouds. ApJL 739, L35. doi:10.1088/2041-8205/739/2/L35, arxiv:1109.6495.
310. Pagani, L., Salez, M., Wannier, P.G., 1992. The chemistry of H2D(+) in cold clouds. Astron. Astrophys. 258, 479–488.
311. Pagani, L., Steinacker, J., Bacmann, A., Stutz, A., Henning, T., 2010. The ubiquity of micrometer-Sized dust grains in the dense interstellar medium. Science 329, 1622. doi:10.1126/science.1193211, arxiv:1110.4180.
312. Pagani, L., Vastel, C., Hugo, E., Kokoouline, V., Greene, C.H., Bacmann, A., Bayet, E., Ceccarelli, C., Peng, R., Schlemmer, S., 2009. Chemical modeling of L183 (L134N): an estimate of the ortho/para H2 ratio. A&A 494, 623–636. doi:10.1051/0004-6361:200810587, arxiv:0810.1861.
313. Pak, S., Jaffe, D.T., Keller, L.D., 1996. H 2 emission from the inner 400 parsecs of the galaxy. ApJL 457, L43. doi:10.1086/309881, arxiv:9604174.
314. Palmer, R., Willis, R., 1987. Rotational states of physisorbed hydrogen on graphite. Surf. Sci. 179 (1), L1–L5. doi:10.1016/0039-6028(87)90105-1.
315. Paradis, D., Dobashi, K., Shimoikura, T., Kawamura, A., Onishi, T., Fukui, Y., Bernard, J.-P., 2012. Dark gas in the solar neighborhood from extinction data. A.A 543, A103. doi:10.1051/0004-6361/201118740, arxiv:1205.3384.
316. Parmar, P.S., Lacy, J.H., Achtermann, J.M., 1991. Detection of low-J

pure-rotational emission from h2 in the orion bar region - Evidence for small-scale clumpiness. ApJL 372, L25–L28. doi:10.1086/186015.

317. Pelupessy, F.I., Papadopoulos, P.P., van der Werf, P., 2006. Incorporating the molecular gas phase in galaxy-sized numerical simulations: first applications in dwarf galaxies. Astrophys. J. 645, 1024–1042. doi:10.1086/504366, arxiv:0603581.
318. Perets, H., Lederhendler, A., Biham, O., Vidali, G., Li, L., Swords, S., Congiu, E., Roser, J., Manico, G., Brucato, J.R., Pirronello, V., 2007. Molecular hydrogen formation on amorphous silicates under interstellar conditions. Astrophys. J. 661, L163. doi:10.1086/518862, arxiv:0703248.
319. Perets, H.B., Biham, O., 2006. Molecular hydrogen formation on porous dust grains. Mon. Not. R. Astron. Soc. 365, 801–806. doi:10.1111/j.1365-2966.2005.09803.x, arxiv:0506492.
320. Perets, H.B., Biham, O., Manicó, G., Pirronello, V., Roser, J., Swords, S., Vidali, G., 2005. Molecular hydrogen formation on ice under interstellar conditions. Astrophys. J. 627, 850–860. doi:10.1086/430435, arxiv:0412202.
321. Perry, J.S.A., Gingell, J.M., Newson, K.A., To, J., Watanabe, N., Price, S.D., 2002. An apparatus to determine the rovibrational distribution of molecular hydrogen formed by the heterogeneous recombination of H atoms on cosmic dust analogues. Meas. Sci. Technol. 13, 1414. doi:10.1088/0957-0233/13/9/306.
322. Peterson, B.W., Appleton, P.N., Helou, G., Guillard, P., Jarrett, T.H., Cluver, M.E., Ogle, P., Struck, C., Boulanger, F., 2012. Detection of powerful mid-IR H2 emission in the bridge between the taffy galaxies. Astrophys. J. 751, 11. doi:10.1088/0004-637X/751/1/11, arxiv:1203.4203.
323. Petucci, J., LeBlond, C., Karimi, M., Vidali, G., 2013. Diffusion, adsorption, and desorption of molecular hydrogen on graphene and in graphite. J. Chem. Phys. 139 (4). 044706–044706. doi: 10.1063/1.4813919
324. Pety, J., Gratier, P., Guzmán, V., Roueff, E., Gerin, M., Goicoechea, J.R., Bardeau, S., Sievers, A., Le Petit, F., Le Bourlot, J., Belloche, A., Talbi, D., 2012. The IRAM-30 m line survey of the horsehead PDR. II. first detection of the l-C3h+ hydrocarbon cation. A.A 548, A68. doi:10.1051/0004-6361/201220062, arxiv:1210.8178.

325. Pety, J., Teyssier, D., Fossé, D., Gerin, M., Roueff, E., Abergel, A., Habart, E., Cernicharo, J., 2005. Are PAHs precursors of small hydrocarbons in photodissociation regions? the horsehead case. A.A 435, 885–899. doi:10.1051/0004-6361:20041170, arxiv:0501339.

326. Pilleri, P., Joblin, C., Boulanger, F., Onaka, T., 2015. Mixed aliphatic and aromatic composition of evaporating very small grains in ngc 7023 revealed by the 3.4/3.3 mum ratio. A.A 577, 16.

327. Pilleri, P., Joblin, C., Fuente, A., Bron, E., Gerin, M., Cernicharo, J., Pety, J., Goicoechea, J. R., Le Petit, F., Treviño-Morales, S. P., González-García, M., Cuadrado, S., Valdivia, V., Evaporation of very small grains and gas-phase hydrocarbon production in the NGC 7023 NW photon-dominated region. To be submitted to A&A.

328. Pilleri, P., Montillaud, J., Berné, O., Joblin, C., 2012. Evaporating very small grains as tracers of the UV radiation field in photo-dissociation regions. A.A 542, A69. doi:10.1051/0004-6361/201015915, arxiv:1204.4669.

329. Pineau Des Forêts, G., Flower, D., 1999. H2 emission from shocks in molecular outflows: the significance of departures from a stationary state.

330. Pineau des Forets, G., Flower, D.R., Chieze, J.-P., 1997. The physical and chemical effects of c-shocks in molecular outflows. In: Reipurth, B., Bertout, C. (Eds.), Herbig-Haro Flows and the Birth of Stars. In: IAU Symposium, 182, pp. 199–212.

331. Pirronello, V., Averna, D., 1988. Production of molecular hydrogen by cosmic rays in interstellar clouds - A relevant process? A.A 196, 201–206.

332. Pirronello, V., Biham, O., Liu, C., Shen, L., Vidali, G., 1997. Efficiency of molecular hydrogen formation on silicates. ApJL 483, L131–L134. doi:10.1086/310746, arxiv:9704236.

333. Pirronello, V., Biham, O., Manicò, G., Roser, J.E., Vidali, G., 2000. Laboratory studies of molecular hydrogen formation on surfaces of astrophysical interest. In: Combes, F., Pineau des Forêts, G. (Eds.), Molecular Hydrogen in Space, p. 71.

334. Pirronello, V., Liu, C., Roser, J.E., Vidali, G., 1999. Measurements of molecular hydrogen formation on carbonaceous grains. A&A 344, 681–686.

335. Pirronello, V., Liu, C., Shen, L., Vidali, G., 1997. Laboratory synthesis

of molecular hydrogen on surfaces of astrophysical interest. ApJL 475, L69–L72. doi:10.1086/310464, arxiv:9611022.

336. Planck Collaboration, Abergel, A., Ade, P.A.R., Aghanim, N., Alves, M.I.R., Aniano, G., Armitage-Caplan, C., Arnaud, M., Ashdown, M., Atrio-Barandela, F., et al., 2014.

337. Planck 2013 results. XI. all-sky model of thermal dust emission. A.A 571, A11. doi:10.1051/0004-6361/201323195, arxiv:1312.1300.

338. Podio, L., Codella, C., Gueth, F., Cabrit, S., Bachiller, R., Gusdorf, A., Lee, C., Lefloch, B., Leurini, S., Nisini, B., Tafalla, M., 2015. The Jet and the Disk of the HH 212 Protostar Imaged by ALMA. In: Iono, D., Tatematsu, K., Wootten, A., Testi, L. (Eds.), Revolution in Astronomy with ALMA: The Third Year. In: Astronomical Society of the Pacific Conference Series, 499, p. 209.

339. Prasad, S.S., Huntress Jr., W.T., 1980. A model for gas phase chemistry in interstellar clouds. I - The basic model, library of chemical reactions, and chemistry among C, N, and O compounds. ApJ Supplement 43, 1–35. doi:10.1086/190665.

340. Rachford, B.L., Snow, T.P., Destree, J.D., Ross, T.L., Ferlet, R., Friedman, S.D., Gry, C., Jenkins, E.B., Morton, D.C., Savage, B.D., Shull, J.M., Sonnentrucker, P., Tumlinson, J., Vidal-Madjar, A., Welty, D.E., York, D.G., 2009. Molecular hydrogen in the far ultraviolet spectroscopic explorer translucent lines of sight: the full sample. ApJ Supplement 180, 125–137. doi:10.1088/0067-0049/180/1/125.

341. Rachford, B.L., Snow, T.P., Tumlinson, J., Shull, J.M., Blair, W.P., Ferlet, R., Friedman, S.D., Gry, C., Jenkins, E.B., Morton, D.C., Savage, B.D., Sonnentrucker, P., Vidal-Madjar, A., Welty, D.E., York, D.G., 2002. A far ultraviolet spectroscopic explorer survey of interstellar molecular hydrogen in translucent clouds. Astrophys. J. 577, 221–244. doi:10.1086/342146, arxiv:0205415.

342. Rachford, B.L., Snow, T.P., Tumlinson, J., Shull, J.M., Roueff, E., Andre, M., Desert, J.M., Ferlet, R., Vidal-Madjar, A., York, D.G., 2001. Far ultraviolet spectroscopic explorer observations of molecular hydrogen in translucent interstellar clouds. II. the line of sight toward HD 110432. Astrophys. J. 555, 839.

343. Rae, J.G.L., Green, N.J.B., Hartquist, T.W., Pilling, M.J., Toniazzo, T., 2003. Diffusive grain-surface chemistry involving the atomsand diatomic molecules of two elements. A&A 405, 387–396. doi:10.1051/0004-6361:20030415, arxiv:0303655.

344. Rapacioli, M., Calvo, F., Joblin, C., Parneix, P., Toublanc, D., Spiegelman, F., 2006. Formation and destruction of polycyclic aromatic hydrocarbon clusters in the interstellar medium. A&A 460, 519–531. doi:10.1051/0004-6361:20065412, arxiv:0609186.

345. Rapacioli, M., Joblin, C., Boissel, P., 2005. Spectroscopy of polycyclic aromatic hydrocarbons and very small grains in photodissociation regions. A&A 429, 193–204. doi:10.1051/0004-6361:20041247.

346. Rauls, E., Hornekær, L., 2008. Catalyzed routes to molecular hydrogen formation and hydrogen addition reactions on neutral polycyclic aromatic hydrocarbons under interstellar conditions. Astrophys. J. 679, 531–536. doi:10.1086/587614.

347. Rho, J., Hewitt, J.W., Bieging, J., Reach, W.T., Andersen, M., Güsten, R., 2017. Discovery of broad molecular lines and of shocked molecular hydrogen from the supernova remnant G357.7+0.3: HHSMT, APEX, spitzer, and SOFIA observations. Astrophys. J. 834, 12. doi:10.3847/1538-4357/834/1/12, arxiv:1610.05459.

348. Richings, A.J., Schaye, J., 2016. The effects of metallicity, UV radiation and nonequilibrium chemistry in high-resolution simulations of galaxies. Mon. Not. R. Astron. Soc. 458, 270–292. doi:10.1093/mnras/stw327, arxiv:1506.08829.

349. Richings, A.J., Schaye, J., Oppenheimer, B.D., 2014. Non-equilibrium chemistry and cooling in the diffuse interstellar medium - II. Shielded gas. Mon. Not. R. Astron. Soc. 442, 2780–2796. doi:10.1093/mnras/stu1046, arxiv:1403.6155.

350. Rigopoulou, D., Kunze, D., Lutz, D., Genzel, R., Moorwood, A.F.M., 2002. An isosws survey of molecular hydrogen in starburst and seyfert galaxies. A&A 389, 374–386. doi:10.1051/0004-6361:20020607, arxiv:0206135.

351. Roberts, J.F., Rawlings, J.M.C., Viti, S., Williams, D.A., 2007. Desorption from interstellar ices. Mon. Not. R. Astron. Soc. 382, 733–742. doi:10.1111/j.1365-2966.2007. 12402.x, arxiv:0708.3374.

352. Robertson, B.E., Kravtsov, A.V., 2008. Molecular hydrogen and global star formation relations in galaxies. Astrophys. J. 680, 1083–1111. doi:10.1086/587796, arxiv:0710.2102.

353. Röllig, M., Szczerba, R., Ossenkopf, V., Glück, C., 2013. Full SED fitting with the KOSMA-τ PDR code. I. Dust modelling. A A. 549, A85. doi:10.1051/0004-6361/201118190, arxiv:1211.3546.

354. Rosenberg, M.J.F., van der Werf, P.P., Aalto, S., Armus, L., Charmandaris, V., DíazSantos, T., Evans, A.S., Fischer, J., Gao, Y., González-Alfonso, E., Greve, T.R., Harris, A.I., Henkel, C., Israel, F.P., Isaak, K.G., Kramer, C., Meijerink, R., Naylor, D.A., Sanders, D.B., Smith, H.A., Spaans, M., Spinoglio, L., Stacey, G.J., Veenendaal, I., Veilleux, S., Walter, F., Weiß, A., Wiedner, M.C., van der Wiel, M.H.D., Xilouris, E.M., 2015. The Herschel Comprehensive (U)LIRG emission survey (HERCULES): CO ladders, fine structure lines, and neutral gas cooling. Astrophys. J. 801, 72. doi:10.1088/0004-637X/801/2/72, arxiv:1501.02985.

355. Rosenthal, D., Bertoldi, F., Drapatz, S., 2000. ISO-SWS observations of OMC-1: H2 and fine structure lines. A.A 356, 705.

356. Roser, J.E., Manicò, G., Pirronello, V., Vidali, G., 2002. Formation of molecular hydrogen on amorphous water ice: influence of morphology and ultraviolet exposure. Astrophys. J. 581, 276–284. doi:10.1086/344200.

357. Roser, J.E., Swords, S., Vidali, G., Manicò, G., Pirronello, V., 2003. Measurement of the kinetic energy of hydrogen molecules desorbing from amorphous water ice. ApJL 596, L55–L58. doi:10.1086/379095.

358. Rougeau, N., Teillet-Billy, D., Sidis, V., 2006. Double h atom adsorption on a cluster model of a graphite surface. Chem. Phys. Lett. 431, 135–138. doi:10.1016/j.cplett. 2006.09.069.

359. Roussel, H., Helou, G., Hollenbach, D.J., Draine, B.T., Smith, J.D., Armus, L., Schinnerer, E., Walter, F., Engelbracht, C.W., Thornley, M.D., Kennicutt, R.C., Calzetti, D., Dale, D.A., Murphy, E.J., Bot, C., 2007. Warm Molecular Hydrogen in the Spitzer SINGS Galaxy Sample. Astrophys. J. 669, 959–981. doi:10.1086/521667, arxiv:0707.0395.

360. Roy, A., André, P., Palmeirim, P., Attard, M., Könyves, V., Schneider, N., Peretto, N., Men'shchikov, A., Ward-Thompson, D., Kirk, J., Griffin, M., Marsh, K., Abergel, A., Arzoumanian, D., Benedettini, M., Hill, T., Motte, F., Nguyen Luong, Q., Pezzuto, S., Rivera-Ingraham, A., Roussel, H., Rygl, K.L.J., Spinoglio, L., Stamatellos, D., White, G., 2014. Reconstructing the density and temperature structure of prestellar cores from Herschel data: A case study for B68 and L1689B. A&A 562, A138. doi:10.1051/0004-6361/201322236, arxiv:1311.5086.

361. Ruaud, M., Wakelam, V., Hersant, F., 2016. Gas and grain chemical composition in cold cores as predicted by the Nautilus three-phase model. Mon. Not. R. Astron. Soc. 459, 3756–3767. doi:10.1093/mnras/

stw887, arxiv:1604.05216.

362. Savage, B.D., Bohlin, R.C., Drake, J.F., Budich, W., 1977. A survey of interstellar molecular hydrogen. i. Astrophys. J. 216, 291–307. doi:10.1086/155471.

363. Senevirathne, B., Andersson, S., Dulieu, F., Nyman, G., 2017. Hydrogen atom mobility, kinetic isotope effects and tunneling on interstellar ices (ih and ASW). Mol. Astrophys. 6, 59–69. doi:10.1016/j.molap.2017.01.005.

364. Sha, X., Jackson, B., Lemoine, D., 2002. Quantum studies of Eley-Rideal reactions between H atoms on a graphite surface. J. Chem. Phys. 116, 7158–7169. doi:10. 1063/1.1463399.

365. Sha, X., Jackson, B., Lemoine, D., Lepetit, B., 2005. Quantum studies of H atom trapping on a graphite surface. J. Chem. Phys. 122 (1). doi:10.1063/1.1827601. 014709–014709

366. Sheffer, Y., Wolfire, M.G., Hollenbach, D.J., Kaufman, M.J., Cordier, M., 2011. PDR Model mapping of physical conditions via spitzer/IRS spectroscopy of H2: theoretical success toward NGC 2023-South. Astrophys. J. 741, 45. doi:10.1088/0004-637X/741/1/45, arxiv:1110.4614.

367. Shull, J.M., Beckwith, S., 1982. Interstellar molecular hydrogen. ARA.A 20, 163–190. doi:10.1146/annurev.aa.20.090182.001115.

368. Shull, J.M., Tumlinson, J., Jenkins, E.B., Moos, H.W., Rachford, B.L., Savage, B.D., Sembach, K.R., Snow, T.P., Sonneborn, G., York, D.G., Blair, W.P., Green, J.C., Friedman, S.D., Sahnow, D.J., 2000. Far Ultraviolet Spectroscopic Explorer Observations of Diffuse Interstellar Molecular Hydrogen. ApJL 538, L73–L76. doi:10. 1086/312782.

369. Sizun, M., Bachellerie, D., Aguillon, F., Sidis, V., 2010. Investigation of ZPE and temperature effects on the eley-Rideal recombination of hydrogen atoms on graphene using a multidimensional graphene-H-H potential. Chem. Phys. Lett. 498, 32–37. doi:10.1016/j.cplett.2010.08.039.

370. Skov, A.L., Thrower, J.D., Hornekær, L., 2014. Polycyclic aromatic hydrocarbons - catalysts for molecular hydrogen formation. Faraday Discuss. 168, 223. doi:10.1039/C3FD00151B.

371. Smith, R.G., Sellgren, K., Tokunaga, A.T., 1988. A study of H2O ice in the 3 micron spectrum of OH 231.8+4.2 (OH 0739-14). Astrophys. J. 334, 209–219. doi:10. 1086/166831.

372. Smith, R.J., Glover, S.C.O., Clark, P.C., Klessen, R.S., Springel, V., 2014. CO-dark gas and molecular filaments in Milky Way-type galaxies. Mon. Not. R. Astron. Soc. 441, 1628–1645. doi:10.1093/mnras/stu616, arxiv:1403.1589.

373. Smoluchowski, R., 1983. Adsorption and mobility on amorphous surfaces. Application to astrophysical problems.. J. Phys. Chem. 87, 4229–4233. doi:10.1021/j100244a050.

374. Snow, T.P., Le Page, V., Keheyan, Y., Bierbaum, V.M., 1998. The interstellar chemistry of PAH cations. Nature 391, 259. doi:10.1038/34602.

375. Snow, T.P., McCall, B.J., 2006. Diffuse Atomic and Molecular Clouds. ARA.A 44, 367– 414. doi:10.1146/annurev.astro.43.072103.150624.

376. Snow, T.P., Rachford, B.L., Tumlinson, J., Shull, J.M., Welty, D.E., Blair, W.P., Ferlet, R., Friedman, S.D., Gry, C., Jenkins, E.B., Lecavelier des Etangs, A., Lemoine, M., Morton, D.C., Savage, B.D., Sembach, K.R., Vidal-Madjar, A., York, D.G., Andersson, B.- G., Feldman, P.D., Moos, H.W., 2000. Far ultraviolet spectroscopic explorer observations of molecular hydrogen in translucent interstellar clouds: the line of sight toward HD 73882. ApJL 538, L65–L68. doi:10.1086/312791, arxiv:0005090.

377. Snow, T.P., Ross, T.L., Destree, J.D., Drosback, M.M., Jensen, A.G., Rachford, B.L., Sonnentrucker, P., Ferlet, R., 2008. A new FUSE survey of interstellar HD. Astrophys. J. 688, 1124–1136. doi:10.1086/592288, arxiv:0808.0926.

378. Spitzer Jr., L., Cochran, W.D., 1973. Rotational excitation of interstellar H_{2}. Astrophys. J. 186, L23. doi:10.1086/181349.

379. Spitzer Jr., L., Cochran, W.D., Hirshfeld, A., 1974. Column densities of interstellar molecular hydrogen. ApJ Suppl. 28, 373–389. doi:10.1086/190323.

380. Spitzer Jr., L., Jenkins, E.B., 1975. Ultraviolet studies of the interstellar gas. ARA.A 13, 133–164. doi:10.1146/annurev.aa.13.090175.001025.

381. Steinacker, J., Andersen, M., Thi, W.-F., Paladini, R., Juvela, M., Bacmann, A., Pelkonen, V.-M., Pagani, L., Lefèvre, C., Henning, T., Noriega-Crespo, A., 2015. Grain size limits derived from 3.6 μm and 4.5 μm coreshine. A.A 582, A70. doi:10. 1051/0004-6361/201425434, arxiv:1508.04691.

382. Sternberg, A., 1989. Ultraviolet fluorescent molecular hydrogen

emission. Astrophys. J. 347, 863–874. doi:10.1086/168177.

383. Sternberg, A., Le Petit, F., Roueff, E., Le Bourlot, J., 2014. H I-to-H2 transitions and H I column densities in galaxy star-forming regions. Astrophys. J. 790, 10. doi:10. 1088/0004-637X/790/1/10, arxiv:1404.5042.

384. Sternberg, A., Neufeld, D.A., 1999. The ratio of ortho- to para-H2 in photodissociation regions. Astrophys. J. 516, 371–380. doi:10.1086/307115, arxiv:9812049.

385. Stevenson, K.P., Kimmel, G.A., Dohnalek, Z., Smith, R.S., Kay, B.D., 1999. Controlling the morphology of amorphous solid water. Science 283, 1505. doi:10.1126/science.283.5407.1505.

386. Störzer, H., Stutzki, J., Sternberg, A., 1996. FUV irradiated molecular clumps: spherical geometry and density gradients. A.A 310, 592–602.

387. Sugimoto, T., Fukutani, K., 2011. Electric-field-induced nuclear-spin flips mediated by enhanced spin-orbit coupling. Nat. Phys. 7, 307–310. doi:10.1038/nphys1883.

388. Takahashi, J., Masuda, K., Nagaoka, M., 1999a. Product energy distribution of molecular Hydrogen formed on icy mantles of interstellar dust. Astrophys. J. 520, 724–731. doi:10.1086/307461.

389. Takahashi, J., Masuda, K., Nagaoka, M., 1999b. The formation mechanism of molecular hydrogen on icy mantles of interstellar dust. Mon. Not. R. Astron. Soc. 306, 22–30. doi:10.1046/j.1365-8711.1999.02480.x.

390. Takahashi, J., Uehara, H., 2001. H2 emission spectra with new formation pumping Models. Astrophys. J. 561, 843–857. doi:10.1086/323364.

391. Takahashi, J., Williams, D.A., 2000. Chemically driven desorption of CO from icy grains in dark clouds. Mon. Not. R. Astron. Soc. 314 (2), 273–278. doi:10.1046/j. 1365-8711.2000.03291.x.

392. Thi, W.-F., van Dishoeck, E.F., Bell, T., Viti, S., Black, J., 2009. ISO observation of molecular hydrogen and fine-structure lines in the photodissociation region IC63. Mon. Not. R. Astron. Soc. 400, 622–628. doi:10.1111/j.1365-2966.2009. 15501.x.

393. Thrower, J.D., Jørgensen, B., Friis, E.E., Baouche, S., Mennella, V., Luntz, A.C., Andersen, M., Hammer, B., Hornekær, L., 2012. Experimental evidence for the formation of highly superhydrogenated polycyclic aromatic hydrocarbons through h atom addition and their catalytic role in H2 formation. Astrophys. J. 752, 3. doi:10.1088/0004-

637X/752/1/3.

394. Tibbs, C.T., Paladini, R., Cleary, K., Muchovej, S.J.C., Scaife, A.M.M., Stevenson, M.A., Laureijs, R.J., Ysard, N., Grainge, K.J.B., Perrott, Y.C., Rumsey, C., Villadsen, J., 2016. Using cm observations to constrain the abundance of very small dust grains in Galactic cold cores. Mon. Not. R. Astron. Soc. 456, 2290–2300. doi:10.1093/mnras/stv2759, arxiv:1511.06626.
395. Tielens, A.G.G.M., McKee, C.F., Seab, C.G., Hollenbach, D.J., 1994. The physics of graingrain collisions and gas-grain sputtering in interstellar shocks. Astrophys. J. 431, 321–340. doi:10.1086/174488.
396. Tiné, S., Lepp, S., Gredel, R., Dalgarno, A., 1997. Infrared response of hH to X-Rays in dense clouds. Astrophys. J. 481, 282–295. doi:10.1086/304048.
397. Tiné, S., Williams, D.A., Clary, D.C., Farebrother, A.J., Fisher, A.J., Meijer, A.J.H.M., Rawlings, J.M.C., Davis, C.J., 2003. Observational Indicators of Formation Excitation of H2. Ap&SS 288, 377–389. doi:10.1023/B:ASTR.0000006062.57984.9c.
398. Togi, A., Smith, J.D.T., 2016. Lighting the dark molecular gas: H2 as a direct tracer. Astrophys. J. 830, 18. doi:10.3847/0004-637X/830/1/18, arxiv:1607.08036.
399. Troscompt, N., Faure, A., Maret, S., Ceccarelli, C., Hily-Blant, P., Wiesenfeld, L., 2009. Constraining the ortho-to-para ratio of H2 with anomalous H2CO absorption. A. A 506, 1243–1247. doi:10.1051/0004-6361/200912770.
400. Tsong, T.T., 2005. Mechanisms and energetic of atomic processes in surface diffusion. Physica A Statistical Mechanics and its Applications 357, 250–281. doi:10. 1016/j.physa.2005.06.029.
401. Tumlinson, J., Shull, J.M., Rachford, B.L., Browning, M.K., Snow, T.P., Fullerton, A.W., Jenkins, E.B., Savage, B.D., Crowther, P.A., Moos, H.W., Sembach, K.R., Sonneborn, G., York, D.G., 2002. A far ultraviolet spectroscopic explorer survey of interstellar molecular Hydrogen in the Small and Large Magellanic Clouds. Astrophys. J. 566, 857–879. doi:10.1086/338112, arxiv:0110262.
402. Ueta, H., Watanabe, N., Hama, T., Kouchi, A., 2016. Surface temperature dependence of Hydrogen ortho-para Conversion on Amorphous Solid Water. Phys. Rev. Lett. 116 (25), 253201. doi:10.1103/PhysRevLett.116.253201.

403. Uzgil, B.D., Bradford, C.M., Hailey-Dunsheath, S., Maloney, P.R., Aguirre, J.E., 2016. Constraining the ISM Properties of the Cloverleaf Quasar Host Galaxy with Herschel SpectroscopyConstraining the ISM properties of the cloverleaf quasar host galaxy with Herschel spectroscopy. Astrophys. J. 832, 209. doi:10.3847/0004-637X/832/2/209, arxiv.1610.00347.

404. Valdivia, V., Godard, B., Hennebelle, P., Gerin, M., Lesaffre, P., Le Bourlot, J., 2017a. Origin of CH+ in diffuse molecular clouds. Warm H2 and ion-neutral drift. A&A 600, A114. doi:10.1051/0004-6361/201629905, arxiv:1612.03030.

405. Valdivia, V., Hennebelle, P., 2014. A fast tree-based method for estimating column densities in adaptive mesh refinement codes. Influence of UV radiation field on the structure of molecular clouds. A&A 571, A46. doi:10.1051/0004-6361/201423720, arxiv:1410.0624.

406. Valdivia, V., Hennebelle, P., Gérin, M., Lesaffre, P., 2016. H2 distribution during the formation of multiphase molecular clouds. A&A 587, A76. doi:10.1051/0004-6361/201527325, arxiv:1512.05523.

407. Valdivia, V., Hennebelle, P., Godard, B., Gerin, M., Lesaffre, P., Le Bourlot, J., 2017b. Chemical Evolution of Turbulent Multiphase Molecular Clouds arxiv:1711.01824.

408. Valentijn, E.A., van der Werf, P.P., 1999. First extragalactic direct detection of largescale molecular hydrogen in the disk of NGC 891. ApJL 522, L29–L33. doi:10. 1086/312208.

409. van de Hulst, H.C., 1949. The solid particles in interstellar space. (Recherches Astronomiques de l'Observatoire d'Utrecht 11 No. 2).

410. van den Ancker, M.E., Wesselius, P.R., Tielens, A.G.G.M., 2000. ISO spectroscopy of young intermediate-mass stars in the BD+40deg4124 group. A&A 355, 194.

411. van Dishoeck, E.F., Black, J.H., 1989. Interstellar C2, CH, and CN in translucent molecular clouds. Astrophys. J. 340, 273–297. doi:10.1086/167391.

412. Vasyunin, A.I., Semenov, D.A., Wiebe, D.S., Henning, T., 2009. A unified Monte Carlo treatment of gas-Grain chemistry for large reaction networks. i. testing validity of rate equations in molecular clouds. Astrophys. J. 691, 1459–1469. doi:10.1088/0004-637X/691/2/1459, arxiv:0810.1591.

413. Veeraghattam, V.K., Manrodt, K., Lewis, S.P., Stancil, P.C., 2014. The

sticking of atomic hydrogen on amorphous water ice. Astrophys. J. 790, 4. doi:10.1088/0004-637X/790/1/4.

414. Verma, A., Charmandaris, V., Klaas, U., Lutz, D., Haas, M., 2005. Obscured Activity: AGN, Quasars, Starbursts and ULIGs Observed by the Infrared Space Observatory. Space Sci. Rev. 119, 355–407. doi:10.1007/s11214-005-8072-z, arxiv: 0507165.
415. Vidali, G., 2013. H2 Formation on Interstellar Grains. Chem. Rev. 113, 8752–8782. doi:10.1021/cr400156b.
416. Vidali, G., Ihm, G., Kim, H.-Y., Cole, M.W., 1991. Potentials of physical adsorption. Surf. Sci. Rep. 12, 135–181. doi:10.1016/0167-5729(91)90012-M.
417. Vidali, G., Li, L., 2010. Molecular hydrogen desorption from amorphous surfaces at low temperature. Journal of Physics Condensed Matter 22, 304012. doi:10.1088/0953-8984/22/30/304012.
418. Vidali, G., Roser, J., Manicó, G., Pirronello, V., Perets, H.B., Biham, O., 2005. Formation of molecular hydrogen on analogues of interstellar dust grains: experiments and modelling. In: Saija, R., Cecchi-Pestellini, C. (Eds.), Journal of Physics Conference Series. In: Journal of Physics Conference Series, 6, pp. 36–58. doi:10.1088/1742-6596/6/1/003, arxiv:0504543.
419. Wakelam, V., Cuppen, H.M., Herbst, E., 2013. Astrochemistry: Synthesis and Modelling. Springer Berlin Heidelberg, Berlin, Heidelberg, pp. 115–143. doi:10.1007/978-3-642-31730-9_4.
420. Wakelam, V., Ruaud, M., Hersant, F., Dutrey, A., Semenov, D., Majumdar, L., Guilloteau, S., 2016. Importance of the H2 abundance in protoplanetary disk icesfor the molecular layer chemical composition. A&A594,A35.doi:10.1051/0004-6361/201628748,arxiv:1609.01471.
421. Walch, S., Girichidis, P., Naab, T., Gatto, A., Glover, S.C.O., Wünsch, R., Klessen, R.S., Clark, P.C., Peters, T., Derigs, D., Baczynski, C., 2015. The SILCC (SImulating the LifeCycle of molecular Clouds) project - I. Chemical evolution of the supernovadriven ISM. Mon. Not. R. Astron. Soc. 454, 238–268. doi:10.1093/mnras/stv1975, arxiv:1412.2749.
422. Walmsley, C.M., Natta, A., Oliva, E., Testi, L., 2000. The structure of the orion bar. Aa 364, 301.
423. Watanabe, N., Horii, T., Kouchi, A., 2000. Measurements of D2 yields from amorphous D2O ice by ultraviolet irradiation at 12 K. Astrophys.

J. 541, 772–778. doi:10.1086/309458.

424. Watanabe, N., Kimura, Y., Kouchi, A., Chigai, T., Hama, T., Pirronello, V., 2010. Direct measurements of hydrogen atom diffusion and the spin temperature of nascent h2 molecule on amorphous solid water. ApJL 714, L233–L237. doi:10. 1088/2041-8205/714/2/L233.
425. Watanabe, N., Kouchi, A., 2008. Ice surface reactions: A key to chemical evolution in space. Prog. Surf. Sci. 83, 439–489. doi:10.1016/j. progsurf.2008.10.001.
426. Weingartner, J.C., Draine, B.T., 2001. Dust grain-size distributions and extinction in the milky way, large magellanic cloud, and small magellanic cloud. Astrophys. J. 548, 296–309. doi:10.1086/318651.
427. Westley, M.S., Baragiola, R.A., Johnson, R.E., Baratta, G.A., 1995a. Photodesorption from low-temperature water ice in interstellar and circumsolar grains. Nature 373, 405–407. doi:10.1038/373405a0.
428. Westley, M.S., Baragiola, R.A., Johnson, R.E., Baratta, G.A., 1995b. Ultraviolet photodesorption from water ice. P&SS 43, 1311–1315. doi:10.1016/0032-0633(95) 00088-M.
429. Whittet, D.C.B., 1993. Observations of Molecular Ices. In: Miller, T.J., Williams, D.A. (Eds.), Dust and Chemistry in Astronomy. The Graduate Series in Astronomy Dust. Philadelphia, PA: Institute of Physics Publishing, p. 9.
430. Wilgenbus, D., Cabrit, S., Pineau des Forêts, G., Flower, D., 2000. The ortho:para-H2 ratio in C- and J-type shocks. Astrophys. J. 356, 1010.
431. Willacy, K., Williams, D.A., Duley, W.W., 1994. Desorption of grain mantles in quiescent dark clouds by H2 formation. Mon. Not. R. Astron. Soc. 267, 949. doi:10.1093/mnras/267.4.949.
432. Witt, A.N., Gordon, K.D., Vijh, U.P., Sell, P.H., Smith, T.L., Xie, R.-H., 2006. The excitation of extended red emission: new constraints on Its carrier from Hubble Space Telescope Observations of NGC 7023. Astrophys. J. 636, 303–315. doi:10.1086/498052, arxiv:0509352.
433. Wolfire, M.G., Hollenbach, D., McKee, C.F., 2010. The dark molecular gas. Astrophys. J. 716, 1191–1207. doi:10.1088/0004-637X/716/2/1191, arxiv:1004.5401.
434. Wright, G.S., Geballe, T.R., Graham, J.R., 1993. Velocity resolved spectroscopy of molecular hydrogen emission in NGC6240. In: Shull, J.M., Thronson, H.A. (Eds.), Evolution of Galaxies and their Environment.

435. Yabushita, A., Hama, T., Iida, D., Kawanaka, N., Kawasaki, M., Watanabe, N., Ashfold, M.N.R., Loock, H.-P., 2008. Measurements of energy partitioning in H2 formation by photolysis of amorphous water ice. ApJL 682, L69. doi:10.1086/590531.

436. Young Owl, R.C., Meixner, M.M., Fong, D., Haas, M.R., Rudolph, A.L., Tielens, A.G.G.M., 2002. Testing Models of Low-Excitation Photodissociation Regions with FarInfrared Observations of Reflection Nebulae. Astrophys. J. 578, 885–896. doi:10. 1086/342647.

437. Ysard, N., Köhler, M., Jones, A., Dartois, E., Godard, M., Gavilan, L., 2016. Mantle formation, coagulation, and the origin of cloud/core shine. II. comparison with observations. A&A 588, A44. doi:10.1051/0004-6361/201527487, arxiv:1602.00541.

438. Ysard, N., Köhler, M., Jones, A., Miville-Deschênes, M.-A., Abergel, A., Fanciullo, L., 2015. Dust variations in the diffuse interstellar medium: constraints on milky way dust from planck-HFI observations. A&A 577, A110. doi:10.1051/0004-6361/201425523, arxiv:1503.07435.

439. Zecho, T., Güttler, A., Sha, X., Jackson, B., Kuppers, J., 2002a. Adsorption of hydrogen and deuterium atoms on the (0001) graphite surface. jcp, keywords = Adsorption, Auger Effect, Desorption, Deuterium, Electron Energy, Electrons, Energy Dissipation, Energy Spectra, Graphite, High Resolution, Hydrogen, Hydrogen Atoms, Kinetics, Potential Energy, Thermal Energy, Adsorption/desorption kinetics, Thermal desorption 117, 8486–8492. doi:10.1063/1.1511729.

440. Zecho, T., Güttler, A., Sha, X., Lemoine, D., Jackson, B., Küppers, J., 2002b. Abstraction of d chemisorbed on graphite (0001) with gaseous h atoms. Chem. Phys. Lett. 366, 188–195. doi:10.1016/S0009-2614(02)01573-7.

441. Ziegler, U., 2016. A chemical reaction network solver for the astrophysics code NIRVANA. A.A 586, A82. doi:10.1051/0004-6361/201527262.

8

The First Steps of Interstellar Phosphorus Chemistry

J. Chantzos[1], V. M. Rivilla[2], A. Vasyunin[3,4], E. Redaelli[1], L. Bizzocchi[1], F. Fontani[2] and P. Caselli[1]

[1] Center for Astrochemical Studies, Max-Planck-Institut für extraterrestrische Physik, Gießenbachstraße 1, 85748 Garching, Germany
[2] INAF – Osservatorio Astrofisico di Arcetri, Largo Enrico Fermi 5, 50125 Firenze, Italy
[3] Ural Federal University, Ekaterinburg, Russia
[4] Visiting Leading Researcher, Engineering Research Institute "Ventspils International Radio Astronomy Centre" of Ventspils University of Applied Sciences, Inženieru 101, Ventspils 3601, Latvia

ABSTRACT

Context. Phosphorus-bearing species are essential to the formation of life on Earth, however they have barely been detected in the interstellar medium. In particular, towards star-forming regions only PN and PO have been identified so far. Since only a small number of detections of P-bearing molecules are available, their chemical formation pathways are not easy to

Citation: J. Chantzos, V. M. Rivilla, A. Vasyunin, E. Redaelli, L. Bizzocchi, F. Fontani and P. Caselli "The first steps of interstellar phosphorus chemistry" A&A, 633 (2020) A54 DOI: https://doi.org/10.1051/0004-6361/201936531

 Open Access funding provided by Max Planck Society.

constrain and are thus highly debatable. An important factor still missing in the chemical models is the initial elemental abundance of phosphorus, that is, the depletion level of P at the start of chemical models of dense clouds.

Aims. In order to overcome this problem, we study P-bearing species in diffuse and translucent clouds. In these objects phosphorus is expected to be mainly in the gas phase and therefore the elemental initial abundance needed in our chemical simulations corresponds to the cosmic one and is well constrained.

Methods. For the study of P-bearing chemistry we used an advanced chemical model. We updated and significantly extended the P-chemistry network based on chemical databases and previous literature. We performed single-pointing observations with the IRAM 30 m telescope in the 3 mm range towards the line of sight to the strong continuum source B0355+508 aiming for the (2–1) transitions of PN, PO, HCP, and CP. This line of sight incorporates five diffuse and/or translucent clouds.

Results. The (2–1) transitions of the PN, PO, HCP, and CP were not detected. We report high signal-to-noise-ratio detections of the (1–0) lines of ^{13}CO, HNC, and CN along with a first detection of $C^{34}S$ towards this line of sight. We have attempted to reproduce the observations of HNC, CN, CS, and CO in every cloud with our model by applying typical physical conditions for diffuse or translucent clouds. We find that towards the densest clouds with $v_{LSR} = -10, -17$ km s^{-1} the best-fit model is given by the parameters (n(H), A_V, T_{gas}) = (300 cm^{-3}, 3 mag, 40 K).

Conclusions. According to our best-fit model, the most abundant P-bearing species are HCP and CP (~10^{-10}). The molecules PN, PO, and PH_3 also show relatively high predicted abundances of ~10^{-11}. We show that the abundances of these species are sensitive to visual extinction, cosmic-ray ionization rate, and the diffusion-to-desorption energy ratio on dust grains. The production of P-bearing species is favored towards translucent rather than diffuse clouds, where the environment provides a stronger shielding from the interstellar radiation. Based on our improved model, we show that the (1–0) transitions of HCP, CP, PN, and PO are expected to be detectable with estimated intensities of up to ~200 mK.

Keywords: astrochemistry / line: identification / molecular processes / ISM: molecules

INTRODUCTION

Phosphorus is essential for biotic chemistry, since it is a fundamental component of many important biological molecules, such as nucleic acids and phospholipids. Phosphorus is therefore vital to life on Earth and can consequently play an important role in exoplanets (Schaefer & Fegley 2011). Despite its importance, the chemistry of P-bearing molecules is in its infancy and remains poorly understood. The aim of this work is to add an important missing piece to the puzzle: unveiling the first steps of P chemistry via observations and chemical simulations of simple P-bearing molecules in diffuse clouds.

The ion P^+ was detected in several diffuse clouds by Jura & York (1978), where an elemental abundance of ~2×10^{-7} with a low P depletion factor of between approximately two and three was derived. However, a more recent study by Lebouteiller et al. (2006) showed that phosphorus remains mostly undepleted towards diffuse clouds. In addition, P has been identified towards dwarf and giant stars (Maas et al. 2017; Caffau et al. 2016), while detections of simple P-bearing molecules (PN, PO, HCP, CP, CCP, NCCP, PH_3) have been done towards the circumstellar material of carbon- and oxygen-rich stars (Agúndez et al. 2007, 2014a,b; Tenenbaum et al. 2007; De Beck et al. 2013; Ziurys et al. 2018). The species PN and only very recently PO are the only P-bearing molecules to have been discovered towards dense star-forming regions (Turner & Bally 1987; Fontani et al. 2016, 2019; Rivilla et al. 2016; Lefloch et al. 2016; Mininni et al. 2018) and molecular clouds in the Galactic Center (Rivilla et al. 2018). The limited number of available observations hinders our understanding of the chemical pathways involved in P chemistry. The main uncertainty in P chemistry is the unknown depletion factor of P in molecular clouds. In general, chemical models of dark clouds start with the so-called "low-metal abundances", where the elemental abundances of heavy elements (such as P, S, Fe, Mg) are reduced by orders ofmagnitude to reproduce molecular observations (e.g., Agúndez & Wakelam 2013), but with poor understanding of the chemical processes at the base of such depletions. In the case of P, the level of depletion is still very uncertain. While Turner et al. (1990) and Wakelam et al. (2015) used high depletion factors of $600–10^4$ with respect to cosmic P abundance, recent works have shown that it could be as low as ~ 100 (Rivilla et al. 2016; Lefloch et al. 2016).

As only a very limited number of P-bearing molecules have been detected in star-forming regions, it is very hard to put constraints on the elemental abundance of P in the gas phase and on the major chemical pathways. In order to elucidate the interstellar P chemistry we focus on diffuse clouds, which represent the first steps of molecular-cloud evolution. Diffuse clouds can provide important constraints on P chemistry, since P in these objects is not strongly affected by depletion, meaning that the initial P abundance that can be used for chemical simulations is well constrained (Lebouteiller et al. 2006). With this approach we are able to remove an important uncertainty in our model and use a reliable starting point for our chemical simulations.

The existing chemical and physical models focus solely on diffuse clouds (e.g., Dalgarno 1988; Le Petit et al. 2004; Cecchi-Pestellini et al. 2012; Godard et al. 2014). For example, in Godard et al. (2014), a model including dissipation of turbulence was applied to reproduce the observed molecular abundances in the diffuse interstellar medium (ISM). The main results showed that chemical complexity is strongly linked to turbulent dissipation, which was able to reproduce the high abundances of CO and other species (such as C^+ and HCO^+) observed towards Galactic diffuse clouds. Le Petit et al. (2004) describe the development of a chemical model of the diffuse cloud towards ζ Persei that was able to reproduce the abundance of H_3^+ and other species, like CN and CO. This was achieved by modeling two phases, namely a small dense phase (~ 100 au) with a density of n(H) = 2 × 10^4 cm^{-3} and a larger diffuse region (4 pc) with n(H) = 100 cm^{-3}. In addition, the reproduction of the CH^+ abundance and that of the rotationally excited H_2 required the inclusion of shocks into the model. Similar results were achieved by Cecchi-Pestellini et al. (2012) when including the injection of hot H_2 into the model.

Previous observations (Corby et al. 2018; Liszt et al. 2018; Thiel et al. 2019) prove the chemical complexity and the wide range of densities, temperatures, and visual extinctions of diffuse and translucent clouds, making them promising targets for observations of P-bearing molecules. Diffuse clouds are characterized by low densities with n(H) = 100–500 cm^{-3} and are therefore more exposed to interstellar radiation, which can destroy molecules. Translucent clouds on the other hand are an intermediate state between diffuse and dense molecular clouds, being more protected from UV radiation (1 mag < A_V < 5 mag). They are denser with typical densities of n(H) = 500–5000 cm^{-3} and are consequently cooler (T_{gas} = 15–50 K), showing higher chemical complexity (Snow & McCall 2006; Thiel et al. 2019). One prominent candidate that has been widely studied in previous works (e.g., Liszt et al. 2018, and references therein) is the gas that lies along the line of sight to the compact extragalactic continuum source

B0355+508. This strong blazar is located at a very low latitude in the outer Galaxy ($b = -1.6037°$), meaning that the way through the Galactic disk is long and therefore gathers a significant amount of distributed Galactic diffuse gas (Pety et al. 2008). Indeed, the line of sight towards B0355+508 shows a complex kinematic structure which incorporates several diffuse and translucent clouds. The detections of numerous molecules like S- and CN-bearing species as well as small hydrocarbons towards B0355+508 also indicate the rich chemistry present in this diffuse and translucent gas (e.g., Liszt et al. 2018, and references therein). The substantial velocity structure coupled with a high chemical complexity of this line of sight enables us to adjust our chemical and physical model to every cloud component and find which physical conditions most favor the abundances of P-bearing molecules. Other background sources that have previously been studied are either lacking the chemical (like B0224+671) or the velocity (such as B0415+479) features which are essential for the present work.

In this paper, we present single-pointing observations of the (2–1) transitions of HCP, CP, PN, and PO and chemical simulations of their molecular abundances towards the line of sight to B0355+508 in order to investigate P-bearing chemistry within diffuse and translucent clouds, the precursors of molecular clouds. In Sect. 2 we describe the observational details. Section 3 summarizes the results of the observations. In Sect. 4 we describe our updated phosphorus chemical network as well as the grid of models that we apply in order to reproduce the observations of HNC, CN, CS, and CO towards every cloud component along the line of sight. Furthermore, in Sect. 5 we focus on the P-bearing chemistry based on our best-fit model (which was determined in Sect. 4). In particular, we report the predicted molecular abundances of HCP, CP, PN, PO, and PH_3 and we study their dependence on visual extinction, cosmic-ray ionization rate, and diffusion-to-desorption energy ratio on dust grains. A future outlook and conclusions are summarized in Sects. 6 and 7.

OBSERVATIONS

The observations of the HCP (2–1), CP (2–1), PN (2–1), and PO (2–1) transitions in the 3 mm range were carried out at the IRAM 30 m telescope located at Pico Veleta (Spain) towards the line of sight to the compact extragalactic quasar B0355+508. Table 1 lists the observed transitions, the spectroscopic constants, and the telescope settings at the targeted frequencies: the upper state energy is described by E_{up}, the upper state degeneracy is given by g_u, while A_{ul} stands for the Einstein coefficient of the transition $u \rightarrow l$. The main beam efficiency and the main beam size of the telescope at a given frequency are denoted by the parameters B_{eff} and θ_{MB}, respectively.For our observations we used the EMIR receiver with the E090 configuration (3 mm atmospheric window). We applied three observational setups, in which every setup covered a total spectral coverage of 7.2 GHz (each sub-band covered 1.8 GHz). As a backend we used the Fast Fourier Transform Spectrometer with a frequency resolution of 50 kHz (0.15 km s^{-1} at 100 GHz). In addition, we applied the wobbler switching mode with an amplitude offset of ± 90″. Pointing and focus of the telescope was performed every 2 h on the background source B0355+508 itself and was found to be accurate to within 2″.

The intensity of the obtained spectra was converted from antenna (T^*_A) to main beam temperature (T_{mb}) units, using the following relation: $T_{mb} = \frac{F_{eff}}{B_{eff}} \times T^*_A$, where F_{eff} is the forward efficiency. F_{eff} is equal to 95% in the targeted frequency range.

RESULTS

The compact extragalactic source B0355+508 is located at α = 3h 59m 29.73s, δ = 50°57′50.2″ with a low galactic latitude of $b = -1.6037°$, incorporating a large amount of Galactic gas along the line of sight that harbors up to five diffuse and/or translucent clouds at velocities of −4, − 8, − 10, − 14 and −17 km s^{-1} (e.g., Liszt et al. 2018, and references therein). The flux of the blazar B0355+508 is variable over time and has been measured at ~ 3 mm to be on average equal to (4.62 ± 1.02) Jy, after averaging the flux of 76 different observations (Agudo 2017). This corresponds to a temperature T_c of (0.96 ± 0.21) K at a beam size of 27″ by taking into account the Rayleigh-Jeans-Approximation. The obtained spectra were reduced and analyzed by the GILDAS software (Pety 2005). Every detected line was fitted via the standard CLASS Gaussian fitting method. For the derivation of the peak opacity we use the radiative transfer equation

$$T_{mb} = (J_\nu(T_{ex}) - J_\nu(T_{bg}) - J_\nu(T_c)) \times (1 - \exp(-\tau)) \Rightarrow$$

$$\tau = -\ln\left(1 - \frac{T_{mb}}{J_\nu(T_{ex}) - J_\nu(T_{bg}) - J_\nu(T_c)}\right), \quad (1)$$

where T_{ex} is the excitation temperature, T_{bg} is the cosmic background temperature, and $J(T) = (\frac{h\nu}{k_B})(e^{\frac{h\nu}{k_B T}} - 1)^{-1}$ describes the Rayleigh-Jeans temperature in Kelvin[1]. After obtaining the peak opacity τ, the column density N is then estimated by following the relation:

$$N = \tau \sqrt{\frac{16\pi^3}{\ln 2} \frac{\nu^3 Q_{\rm rot}(T_{\rm ex}) \Delta v \, e^{E_u/k_B T_{\rm ex}}}{c^3 A_{ul} \, g_u \, (e^{h\nu/k_B T_{\rm ex}} - 1)}}, \quad (2)$$

with k_B being the Boltzmann constant, Δv the line width (FWHM), ν the transition frequency, c the speed of light, and h the Planck constant. $Q_{rot}(T_{ex})$ gives the partition function of a molecule at a given excitation temperature T_{ex}.

The (2–1) transitions of HCP, CP, PN, and PO were not detected within our observations (see Fig. 1). We derive 3σ upper limits for the opacities and column densities of the P-bearing species using Eqs. (1) and (2). Due to the low densities in diffuse clouds, molecules are expected to show no collisional excitation. The column densities were calculated assuming $T_{ex} = T_{bg} = 2.7$ K, which simplifies Eq. (1) to:

$$\tau = -\ln\left(1 + \frac{T_{\rm mb}}{J_\nu(T_{\rm c})}\right). \quad (3)$$

The results are summarized in Table 2.

We detected the HNC (1–0), CN (1–0), and $C^{34}S$ (2–1) transitions in absorption as well as the ^{13}CO (1–0) in emission at the 3 mm range with a high signal-to-noise ratio (S/N), ranging from 6 to 80[2]. Figure 2 shows all the detected spectra towards the line of sight to B0355+508. In the case of CN we were able to detect and resolve four hyperfine components from 113.123 to 113.191 GHz (see Fig. 3). Every hyperfine component was detected in the three velocity components at − 8, − 10, and − 17 km s^{-1} except for the one weak transition at 113.123 GHz, which was identified only in two clouds (at − 10, − 17 km s^{-1}). The molecule HNC was identified in all five cloud components, while C^{34} S (in absorption) and ^{13} CO (in emission) were detected solely towards the densest features, at − 10 and − 17 km s^{-1}. Table 3 lists the identified species and the corresponding spectroscopic parameters.

For estimating the CN column density we use the hyperfine component at 113.170 GHz. Our derived opacities and column densities of CN agree within a factor of two to three with previous results (Liszt & Lucas 2001), while the HNC results are well reproduced within a factor of 1.5. Table 4 summarizes the derived opacities and column densities of the detected species, as well as the obtained line intensities and rms levels.

Lucas & Liszt (2002) reported the detection of the main isotopolog $C^{32}S$ with the IRAM Plateau de Bure interferometer (PdBI) and estimated a column density of $(4.27 \pm 0.16) \times 10^{12}$ cm^{-2} at the − 10 km s^{-1} component and $(3.06 \pm 0.32) \times 10^{12}$ cm^{-2} at − 17 km s^{-1}. With the above values and the column densities of $C^{34}S$ calculated in this work, we derive a sulfur isotopic ratio $^{32}S/^{34}S$ of 12.8 ± 4.8 and 18.7 ± 9.5 for the components at − 10 and − 17 km s^{-1}, respectively. The latter value is in good agreement with the $^{32}S/^{34}S$ ratio for the local ISM of 24 ± 5 (Chin et al. 1996). However, the isotopic ratio determined for $v = -10$ km s^{-1} is significantly lower than the local interstellar value, which could be the result of opacity effects of the $C^{32}S$ line. In addition, the determination of the sulphur isotopic ratio was based on just one spectral line of the main species $C^{32}S$ and its isotopolog, which also yields a high uncertainty. To our knowledge this is the first detection of $C^{34}S$ towards this line of sight, owing to the high spectral resolution of ~50 kHz and high sensitivity (rms of ~4 mK) achieved with our observations.

In Liszt & Lucas (1998), detections of the main species ^{12}CO and its isotopolog ^{13}CO are reported, which were obtained with the PdBI as well as the NRAO 12 m telescope. The single-dish observations covered a large beam of 60″, thus seeing CO and its isotopolog in emission, while the interferometric observations were sensitive only to the very narrow column of gas towards the strong background blazar, giving rise to absorption lines. For deriving the excitation temperatures and the column densities, both emission (single-dish data) and absorption lines (interferometric data) were considered.

In Fig. 2 it is clearly visible that the strong ^{13} CO emission line at − 10 km s^{-1} overlaps with an absorption feature at around − 8 km s^{-1}. This is probably due to the fact that absorption is present close to the background source, meaning that emission and absorption lines are merged together in our observations with the IRAM 30 m telescope. This contamination effect is influencing the line profile at − 10 km s^{-1} which subsequently results in an unreliable fit. This could possibly explain why the ^{13} CO column density derived at − 10 km s^{-1} deviates by a factor of about four from previous results (Liszt & Lucas 1998), while towards − 17 km s^{-1}, $N(^{13}CO)$ is well reproduced within 10% (see Table 4)[3]. Our derived isotopic ratio $^{12}CO/^{13}CO$ at − 17 km s^{-1} is equal to 16.7 ± 1.4. For this calculation, we used the column density of ^{12}CO derived in Liszt & Lucas (1998) with $N(^{12}CO) = (6.64 \pm 0.47) \times 10^{15}$ cm^{-2}. The resulting CO isotopic ratio is almost a factor of four lower than the local interstellar ratio $^{12}C/^{13}C = 60$ (Lucas & Liszt 1998). This was already confirmed by previous studies (Liszt 2007, 2017) that show an increased insertion of ^{13} C into CO towards clouds in the translucent regime with elevated densities and/or smaller radiation fields, which lead to an enhanced abundance of ^{13}CO by a factor of two to four. Under these conditions isotope exchange fractionation ($^{13}C^{+} + {}^{12}CO \rightarrow {}^{12}C^{+} + {}^{13}CO$ + 35 K) is more dominant than selective photodissociation.

Table 1. Spectroscopic parameters of the observed species and telescope settings.

Species	Transitions	E_{up} (K)	Frequency (GHz)	A_{ul} (10^{-5} s^{-1})	g_u	B_{eff} (%)	θ_{MB} ('')	References
HCP	$J = 2-1$	5.8	79.90329	0.04	5	83	31	1
PN	$J = 2-1$	6.8	93.97977	2.92	5	80	26	2
CP	$N = 2-1, J = 3/2-1/2, F = 2-1$	6.8	95.16416	0.33	5	80	26	3
PO	$J = 5/2-3/2, \Omega = 1/2, F = 3-2$, e	8.4	108.99845	2.13	7	78	23	4
PO	$J = 5/2-3/2, \Omega = 1/2, F = 2-1$, e	8.4	109.04540	1.92	5	78	23	4
PO	$J = 5/2-3/2, \Omega = 1/2, F = 3-2$, f	8.4	109.20620	2.14	7	78	23	4
PO	$J = 5/2-3/2, \Omega = 1/2, F = 2-1$, f	8.4	109.28119	1.93	5	78	23	4

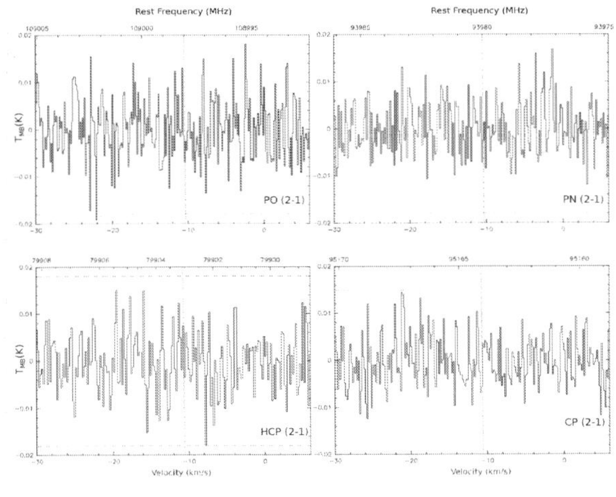

Figure 1. Spectra of the nondetected (2–1) transitions of PO, PN, HCP, and CP. The upper x-axis shows the rest frequency (in MHz) and the lower one is a velocity axis (in km s^{-1}). The red dashed line indicates the 3 level and the blue dashed line shows the transition frequency of the corresponding molecule. In the case of PO, we show as an example one of the observed transitions at 108.998 GHz.

Table 2. Derived upper limits for the opacity and the column density of HCP, CP, PN, and PO.

Species	Frequency (GHz)	τ	N (10^{11} cm^{-2})	T_{MB} (K)	rms (mK)
HCP	79.90329	<0.02	<22.7	<0.02	6
CP	95.16416	<0.02	<12.6	<0.02	5
PN	93.97977	<0.02	<0.42	<0.02	6
PO	108.99845	<0.02	<4.29	<0.02	6
	109.04540	<0.02	<6.70	<0.02	6
	109.20620	<0.02	<4.34	<0.02	6
	109.28119	<0.02	<6.69	<0.02	6

Notes. The upper limits are 3σ.

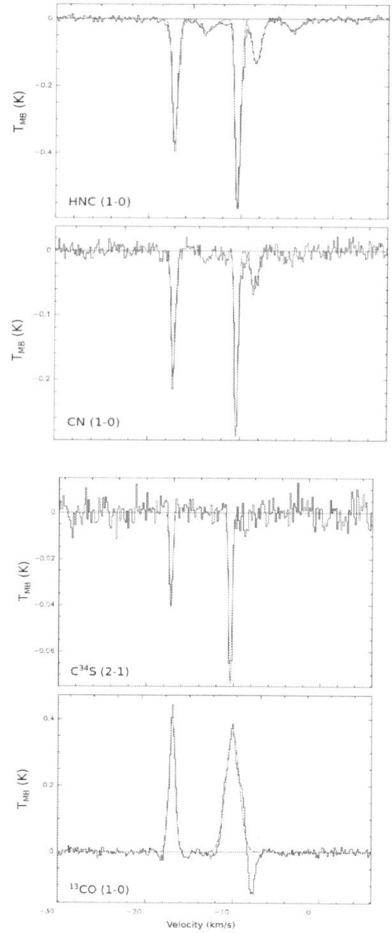

Figure 2. Spectra of the detected species HNC, CN, $C^{34}S$, and ^{13} CO in the 3 mmrange towards the line of sight to the extragalactitc source B0355+508. The red line represents the CLASS Gaussian fit.

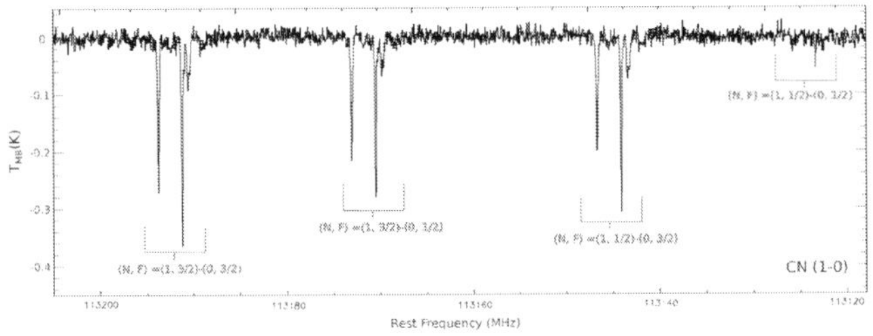

Figure 3. Detected hyperfine components of the CN (1–0) transition between 113.12 and 113.20 GHz. The three strongest hyperfine components were detected in the three clouds with v_{LSR} =–8, – 10, – 17 km s^{-1} except for the one weak transition (N, F) = (1, 1⁄2) - (0, 1⁄2), which was identified only in the two densest clouds (at – 10, – 17 km s^{-1}).

Table 3. Spectroscopic parameters of the detected species and telescope settings.

Species	Transitions	E_{up} (K)	Frequency (GHz)	A_{ul} 10^{-5} s^{-1}	g_u	B_{eff} (%)	θ_{MB} (″)	Reference
HNC	J = 1–0	4.4	90.66357	2.69	3	81	27	1
CN	N = 1–0, F = 3/2–1/2	5.5	113.17049	0.51	4	78	22	2
$C^{34}S$	J = 2–1	6.9	96.41295	1.60	5	80	26	3
^{13}CO	J = 1–0	5.3	110.20135	0.006	6	78	22	4

Table 4. Gaussian fitting results of CN, HNC, $C^{34}S$, ^{13} CO.

Species	Velocity (km s^{-1})	Δv (km s^{-1})	τ	T_{MB} (K)	rms (mK)	Spectral resolution (km s^{-1})	N (cm^{-2}) (this work)	N (cm^{-2}) (previous work) [(1)]
CN								
	−17.23 ± 0.01	0.54 ± 0.02	0.23 ± 0.07	0.20 ± 0.02	9	0.13	(0.87 ± 0.28) × 10^{13}	(2.14 ± 0.24) × 10^{13}
	−10.41 ± 0.01	0.49 ± 0.01	0.36 ± 0.09	0.29 ± 0.02	9	0.13	(1.23 ± 0.34) × 10^{13}	(3.34 ± 0.37) × 10^{13}
	−8.47 ± 0.04	0.97 ± 0.09	0.06 ± 0.03	0.06 ± 0.01	9	0.13	(0.43 ± 0.20) × 10^{13}	(0.76 ± 0.21) × 10^{13}
HNC								
	−17.197 ± 0.004	0.73 ± 0.01	0.50 ± 0.10	0.37 ± 0.01	6	0.17	(0.69 ± 0.16) × 10^{12}	(0.74 ± 0.02) × 10^{12}
	−10.392 ± 0.003	0.73 ± 0.01	0.87 ± 0.16	0.56 ± 0.01	7	0.17	(1.20 ± 0.23) × 10^{12}	(1.14 ± 0.04) × 10^{12}
	−8.37 ± 0.01	1.14 ± 0.04	0.15 ± 0.04	0.13 ± 0.01	6	0.17	(0.32 ± 0.15) × 10^{12}	(0.39 ± 0.02) × 10^{12}
	−13.68 ± 0.07	2.30 ± 0.23	0.04 ± 0.01	0.04 ± 0.01	6	0.17	(0.16 ± 0.08) × 10^{12}	(0.10 ± 0.02) × 10^{12}
	−4.32 ± 0.06	1.71 ± 0.15	0.04 ± 0.02	0.04 ± 0.01	6	0.17	(0.12 ± 0.06) × 10^{12}	(0.14 ± 0.01) × 10^{12}
$C^{34}S$								
	−17.20 ± 0.02	0.43 ± 0.04	0.04 ± 0.02	0.04 ± 0.01	4	0.16	(1.64 ± 0.82) × 10^{11}	<3.2 × 10^{11} [(2)]
	−10.39 ± 0.01	0.48 ± 0.02	0.08 ± 0.03	0.07 ± 0.01	4	0.16	(3.33 ± 1.23) × 10^{11}	<4.4 × 10^{11} [(2)]
^{13}CO								
	−16.935 ± 0.004	0.82 ± 0.01	0.154 ± 0.004	0.41 ± 0.01	8	0.14	(3.98 ± 0.16) × 10^{14}	(4.34 ± 0.51) × 10^{14} [(3)]
	−10.03 ± 0.01	1.82 ± 0.03	0.135 ± 0.004	0.36 ± 0.01	13	0.14	(7.73 ± 0.38) × 10^{14}	(1.79 ± 0.26) × 10^{14} [(3)]

CHEMICAL MODELING

The goal of the present study is to constrain and improve our model of diffuse and translucent clouds to make reliable predictions regarding the abundances of P-bearing species (and also others). For this reason, we used the observations of HNC, CN, CS, and CO in order to constrain the physical parameters in our model. The chemical code that we applied was developed by Vasyunin & Herbst (2013) with an updated grain-surface chemistry (Vasyunin et al., in prep.). The model includes a gas-grain chemical network with 6000 gas-phase reactions, 200 surface reactions, and 660 species. Accretion and desorption processes regulate and connect the gas-phase and grain surface chemistry. The code numerically solves coupled differential equations (chemical rate equations) and computes a set of time-dependent molecular abundances. Since the observations were carried out towards diffuse and translucent clouds, we considered as initial elemental abundances the standard Solar elemental composition (see Table 5). We note that our initial elemental abundances are significantly different compared to the low metal abundances used in Wakelam et al. (2015) for dark clouds (200 times more abundant S and up to 10^4 more abundant Fe, Cl, P, and F). In particular, the initial abundance of P is 2.6×10^{-7} and is therefore well constrained unlike in dense molecular clouds. This approach will help us better elucidate the chemistry ofP since a key parameter for the chemical model is well determined. In addition, we begin our chemical simulations with the entirety of hydrogen being in its atomic form in order to start with pure atomic diffuse cloud conditions.

The Chemical Network of Phosphorus

The phosphorus chemical network that has been used in previous studies (Fontani et al. 2016; Rivilla et al. 2016) has been extended with new available information in the literature (new reactions, updated reaction rates, desorption energies, etc.). In particular, chemical reactions of several P-bearing species, such as PN, PO, HCP, CP, and PH_3, were included and/or updated in our chemical network. The reaction rates were taken from the online chemical databases KInetic Database for Astrochemistry (Wakelam et al. 2015, KIDA)[4] and the UMIST Database for Astrochemistry (McElroy et al. 2013, UDfA)[5], as well as from numerous previous papers (Thorne et al. 1984; Adams et al. 1990; Millar 1991; Anicich 1993; Charnley & Millar 1994; Jiménez-Serra et al. 2018). In particular we included several reactions involving the formation and destruction of PH_n ($n = 1, 2, 3$) and their cationic species from Charnley & Millar (1994) and Anicich (1993), along with the chemical network proposed by Thorne et al. (1984) that contains production and loss routes for P, PO, P^+, PO^+, PH^+, HPO^+ and H_2 PO^+. In addition, weextended the PN chemical network based on the work by Millar et al. (1987), and we took into account the gas-phase reaction P + OH → PO + H proposed byJiménez-Serra et al. (2018), as well as two formation routes of PN in the gas phase, N +CP → PN + C and P + CN → PN + C, by Agúndez et al. (2007). Finally, we included the photodissociation reactions of PN, PO, HCP, and PH_n (n = 1, 2, 3) based on the reaction rates given in KIDA and UDfA. The reaction rates of the photodissociation of PH_n were assumed to be equal to the analogous reactions for NH_n.

Concerning the chemistry taking place on grain surfaces, we took into account the hydrogenation reactions of P-bearing species (where the letter "g" denotes a grain surface species) as well as their corresponding desorption reactions:

- gH + gP → gPH,
- gH + gPH → gPH_2,
- gH + gPH_2 → gPH_3.

The desorption energy of PH_3 was calculated based on that of NH_3 and amounts to ~ 5800 K. This corresponds to an evaporation temperature of ~100 K, which is in good agreement with the value of ~90 K proposed by Turner et al. (1990)[6]. The reactive desorption efficiency in our chemical model is set equal to 1%. An increased reactive desorption of 10% changes the predicted abundances of the aforementioned P-bearing molecules by less than a factor of two. Another nonthermal desorption mechanism included in our model is the cosmic-ray desorption, which is fully described in Hasegawa & Herbst (1993). Based on this study, dust grains are heated upon impact with cosmic rays reaching a peak temperature T_{dust} of 70 K, which subsequently leads to preferential desorption of molecules from grain surfaces. This type of desorption is however negligible in diffuse clouds, where photodesorption dominates. In our model we adopt for all species a photodesorption rate of 3×10^{-3} molecules per incident UV photon, as was determined in Öberg et al. (2007) based on laboratory measurements of pure CO ice. This stands in good agreement with the photodesorption yield of ~ 10^{-3} molecules per UV photon found for other species, such as H_2O, O_2 and CH_4 (Öberg et al. 2009; Fayolle et al. 2013; Dupuy et al. 2017).

Table 5. Assumed solar initial elemental abundances (Asplund et al. 2006).

Species	Abundances
H	1.0
He	8.5×10^{-2}
N	6.8×10^{-5}
O	4.9×10^{-4}
C^+	2.7×10^{-4}
S^+	1.3×10^{-5}
Si^+	3.2×10^{-5}
Fe^+	3.2×10^{-5}
Na^+	1.7×10^{-6}
Mg^+	3.9×10^{-5}
Cl^+	3.2×10^{-7}
P^+	2.6×10^{-7}
F^+	3.6×10^{-8}

Comparison to Observations

In order to reproduce the observed abundances of HNC, CN, CS, and CO in every cloud towards the line of sight to B0355+508 we produce a grid of models applying typical physical conditions for diffuse or translucent gas (Snow & McCall 2006; Thiel et al. 2019). We note here that, since our chemical model is not treating isotopic species, we are using as a reference for our comparison the main species ^{12}CO and C^{32} S instead of ^{13}CO and $C^{34}S$. For the fractional abundances of ^{12}CO and $C^{32}S$, we are adopting the column densities determined in Liszt & Lucas (1998) and Lucas & Liszt (2002). In addition, for the clouds at − 14 and − 4 km s^{-1} we use for CN and CS the upper limits derived in this work (N(CN) < 10^{12} cm^{-2}) and in Lucas & Liszt (2002). The parameter space that we investigate is listed below:

- n(H) = 100–1000 cm^{-3}, spacing of 100 cm^{-3},
- A_V = 1–5 mag, spacing of 1 mag,
- T_{gas} = 20–100 K, spacing of 10 K.

The chemical evolution in each model is simulated over 10^7 yr (100 time steps) assuming static physical conditions. For the cosmic-ray ionization rate ζ(CR) we use a value of 1.7×10^{-16} s^{-1} (Indriolo & McCall 2012, see Sect. 5 for further explanation). This also corresponds to the values applied in Godard et al. (2014) and Le Petit et al. (2004), where the best-fit models provided a ζ(CR) of 10^{-16} s^{-1} and 2.5×10^{-16} s^{-1}, respectively.

Given the above parameter space, we calculate the level of disagreement $D(t, r)$ between modeled and observed abundances (for the species HNC, CN, CS and CO), which, following Wakelam et al. (2010) and Vasyunin et al. (2017), we define as

$$D(t, r) = \sum_{j=1}^{N_{species}} |\log(x^j_{mod}(t, r)) - \log(x^j_{obs})|, \quad (4)$$

with $r = (n(\mathrm{H}), A_V, T_{gas})$ and $x^j_{obs,mod}$ is the observed or modeled abundance of species j, respectively. We then determine the minimal value of $D(t, r)$, (noted as $D_{min}(t, r)$), that corresponds to the best-fit model; the parameters (t, r) provided by the best-fit model are the ones giving the smallest deviation between observations and predictions. The smaller the $D_{min}(t, r)$, the better the agreement.

According to Pety et al. (2008) and references therein, the clouds with $v_{LSR} = -10$ and -17 km s^{-1} towards B0355+508 show strong ^{12}CO emission lines that originate from dense regions (with $n(\mathrm{H}) = 300–500$ cm^{-3} and $N(H_2) > 10^{21}$ cm^{-2}) which are just outside the synthesized beam (combination of the 30 m and PdBI telescopes), but still within the IRAM 30 m beam (Pety et al. 2008). Based on the CO (2–1) maps shown in Pety et al. (2008) with 22″ and 5.8″ resolutions, respectively, the cloud at $v_{LSR} = -10$ km s^{-1} shows the most pronounced, dense substructure. This is confirmed by the fact that this particular cloud component produces the most detectable amounts of observed species in the data presented in this work and previous works by Liszt & Lucas (2001) and Lucas & Liszt (2000), and is therefore chemically the most complex one. The component at $v_{LSR} = -8$ km s^{-1} shows a similar structure to the one at $v_{LSR} = -10$ km s^{-1}, incorporating a dense region as well. Pety et al. (2008) suggest that the two components are part of the same cloud, even though they are distinguishable in absorption and show different levels of chemical complexity. With a higher spatial resolution of 5.8″, the CO emission at $v_{LSR} = -8$ km s^{-1} is separated from the one at $v_{LSR} = -10$ km s^{-1} and is clearly visible. The diffuse gas seen at velocities -14 and -4 km s^{-1} shows barely any ^{13}CO (Liszt & Lucas 1998, and this work) or ^{12}CO (Pety et al. 2008; Liszt & Lucas 1998) in emission, which suggests that the density of these clouds is too low to sufficiently excite CO. Pety et al. (2008) estimated a low to moderate density of these clouds to be ~ 64–256 cm^{-3} with $A_V < 2$ mag.

Since we are performing single-dish observations with a beam of ~ 22″, we also cover the high-density regions that produce significant ^{12}CO emission (Pety et al. 2008). For this reason we constrain our grid of models to high densities of ≥ 300 cm^{-3} for the denser clouds (-8, -10, -17 km s^{-1}), while for the low-density objects at -14 km s^{-1} and -4 km s^{-1} we restrict our input parameters to $n(\mathrm{H}) \leq 200$ cm^{-3} and $A_V < 2$ mag. The calculation of the molecular abundances was done with respect to the H_2 column densities ($N(H_2) \sim 4-5 \times 10^{20}$ cm^{-2}) that were derived towards every cloud by Liszt et al. (2018). However, our model provides the fractional abundance of a species X with respect to the total number of hydrogen nuclei, as in $n(X)/n(H)$, with the total volume density of hydrogen defined as $n(H) = n(H\,I + 2 \times H_2)$. The surface mobility parameters that we set as default values in our model (see Sect. 5.3) enable fast and effective formation of H_2 on the surfaces of grains. At the end of our simulations (at $t = 10^7$ yr), the H_2 abundance reaches a value of 40–50% (depending on the set of parameters). This means that almost the entire hydrogen is predicted to be in its molecular form at the late phases of the chemical evolution. Following this consideration, we divide all the observed abundances and their upper limits mentioned in this paper by a factor of two, since $n(X)/n(H\,I + 2 \times H_2) \simeq n(X)/2 \times n(H_2) = 0.5 \times n(X)/n(H_2)$. We note that this expression applies to the high-density parts of the clouds and does not account for the low-density (and H I rich) gas along the line of sight.

Figure 4 shows the results of the grid of models that was applied to reproduce the observations towards $v_{LSR} = -17$ km s^{-1}. In particular, we plot $D(t_{best}, r)$, which describes the deviation between observed and modeled abundances at the time of best agreement t_{best}, versus the density, temperature, and visual extinction. Between an A_V of 1 and 3 mag the smallest level of disagreement $D(t_{best}, r)$ reduces by 13%. The main discrepancy between observed and modeled abundances at low A_V comes from the fact that a high visual extinction results in higher molecular abundances and is therefore able to reproduce the chemical complexity seen towards the translucent clouds. For models with $A_V > 3$ mag the minimal $D(t_{best}, r)$ barely changes (less than 1% of increase). The smallest $D(t_{best}, r)$ increases with respect to the density and temperature up to 3 and 2%, respectively. This is a clear indication that the most influential physical parameter in our analysis is the visual extinction.

For the cloud component at $v_{LSR} = -17$ km s^{-1} the best-fit model with $D_{min}(t, r)$ is reached ata time $t_{best} = 6.2 \times 10^6$ yr and has the parameters: $r_{best} = (n(H), A_V, T_{gas}) = (300$ cm^{-3}, 3 mag, 40 K$)$. At $t_{best} = 6.2 \times 10^6$ yr we also fulfill the assumption of having most of the hydrogen in molecular form, as the H_2 abundance reaches a value of 0.45. Based on this model, we show in Fig. 5 the time dependent abundances of CO, CN, CS and HNC over 10^7 yr as well as the corresponding observed abundances towards $v_{LSR} = -17$ km s^{-1}. Our chemical model reproduces the observed species CO, CN, and CS very well within a factor of ~ 1−1.4 at t_{best} = 6.2 × 10^6 yr. As can be seen in Fig. 5, the predicted abundances follow the order of the observed quantities. The most significant discrepancy is found in case of HNC, where the chemical model underestimates the observed abundance by a factor of four at the time of best agreement (see Table 6). According to the model, one of the main destruction mechanisms of HNC is: C^+ + HNC → C_2N^+ + H; based on the online chemical databases, its reaction rate remains uncertain. In UDfA the given reaction rate was determined theoretically (Leung et al. 1984) and therefore entails a high uncertainty. Experimental studies of the above chemical route are still needed to make reliable predictions of the HNC abundance.

The smallest deviation with the observations towards the cloud with v_{LSR} = −10 km s^{-1} was produced by the same set of parameters: r_{best} = (n(H), A_V, T_{gas}) = (300 cm^{-3}, 3 mag, 40 K). However, in this case the best-fit model gives a $D_{min}(t, r)$ that is slightly larger by a factor of approximately 1.4. The molecular abundances observed towards v_{LSR} = −8 km s^{-1} are best reproduced with an A_V of 5 mag, a density of 400 cm^{-3} and a gas temperature of 40 K. The smallest level of disagreement between an A_V of 3 and 5 mag differs by less than 1%.

For the two remaining clouds with v_{LSR} = −4 km s^{-1} and v_{LSR} = −14 km s^{-1} the best-fit model in both cases is given by the parameters: (n(H), A_V, T_{gas}) = (200 cm^{-3}, 1 mag, 30 K) at t_{best} = 10^7 yr. Here, the discrepancy in both clouds arises mostly from the fact that the model underestimates the CS abundance by a factor of between approximately six and nine. In our model, CS is being effectively destroyed via photodissociation due to the low A_V. Table 7 lists the best-fit parameters that were determined towards every cloud component.

We note that towards the same line of sight there have been detections of several other molecules, as reported in Liszt et al. (2008). The best-fit model determined towards v_{LSR} = −17 km s^{-1} and v_{LSR} = −10 km s^{-1} is able to reproduce within one order of magnitude the species OH, C_2H, H_2 CO, NH_3 and CH, while other species such as HCN, SO, H_2S and C_3H_2 are strongly underestimated by up to two orders of magnitude. This is a clear indication that the chemical network of certain molecules (other than P-bearing ones) still needs to be extended and updated. This however will be addressed in future work, as the present paper focuses mainly on P chemistry.

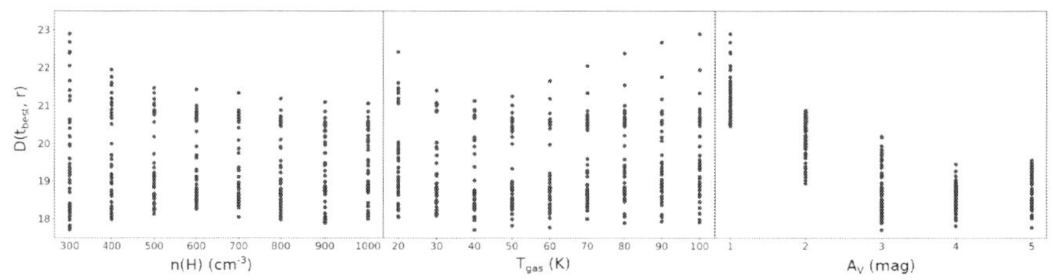

Figure 4. Results of the grid of models applying typical conditions for diffuse or translucent clouds in order to reproduce the observations towards the cloud at v_{LSR} =−17 km s^{-1}. The deviation between observations and model at the time of best agreement t_{best} is given by $D(t_{best}, r)$, which is plotted versus density, temperature, and visual extinction. The best-fit model is given at a time t_{best} = 6.2 × 10^6 yr and has the following parameters: (n(H), A_V, T_{gas}) = (300 cm^{-3}, 3 mag, 40 K).

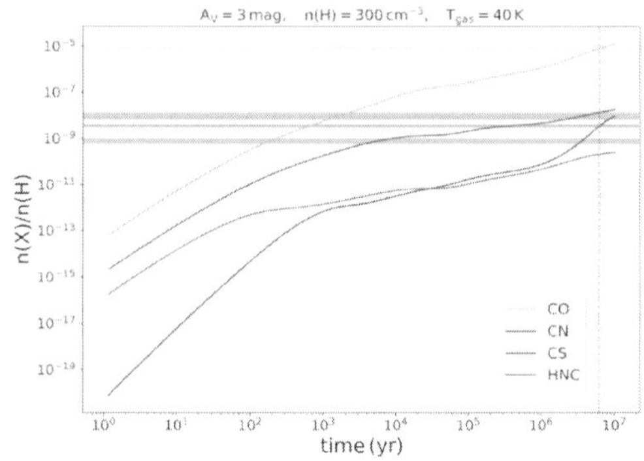

Figure 5. Chemical evolution of the abundances of CO, CN, CS and HNC over 10^7 yr predicted by our best-fit model with the parameters (n(H), A_V, T_{gas}) = (300 cm^{-3}, 3 mag, 40 K). The colored horizontal bands correspond to the observed abundances towards the cloud with v_{LSR} =−17 km s^{-1}, including the inferred uncertainties. The vertical dashed line indicates the time of best agreement (t = 6.2 × 10^6 yr) between observations and model results.

Table 6. Observed abundances for the cloud component with $v_{LSR} = -17$ km s^{-1} and predictions of the species HNC, CO, CS, and CN based on our best-fit model at the time of best agreement $t = 6.2 \times 10^6$ yr.

Species	Observed abundance	Predicted abundance	Ratio (observed/predicted)
HNC	$8.0(1.9) \times 10^{-10}$	2.1×10^{-10}	3.8
CN	$1.0(0.3) \times 10^{-8}$	1.4×10^{-8}	0.7
CS	$3.6(0.4) \times 10^{-9}$	3.4×10^{-9}	1.1
CO	$7.7(0.6) \times 10^{-6}$	8.0×10^{-6}	1.0

Table 7. Set of physical parameters that give the best agreement between model results and observations towards every cloud component.

Velocity (km s^{-1})	n(H) (cm^{-3})	A_V (mag)	T_{gas} (K)	t_{best} (10^6 yr)
−17	300	3	40	6.2
−14	200	1	30	10
−10	300	3	40	6.2
−8	400	5	40	2.3
−4	200	1	30	10

DISCUSSION: THE CHEMISTRY OF PHOSPHORUS

Based on the above results we can conclude that the molecular abundances observed at $v_{LSR} = -8, -10, -17$ km s^{-1} can be best reproduced by a more "shielded" ($A_V > 1$ mag) interstellar medium that allows the build-up of molecules to occur more efficiently. The resulting visual extinction A_V of 3 mag should be viewed as an average value over the region covered by our beam (~ 22″). Within this region the denser clumps are most likely translucent in nature. Hence, the observed cloud components are probably heterogeneous clouds, incorporating diffuse and translucent material, filled with relatively abundant molecules. This result stands in good agreement with a study by Liszt (2017), which involved modeling the CO formation and fractionation towards diffuse clouds. One of the main results of this latter study was that strong ^{13}CO absorption lines observed in the millimeter- and UV-range can be explained by higher densities (≥ 256 cm^{-3}) and weaker radiation (and thus higher visual extinction), as already mentioned in Sect. 3. Our conclusions also agree well with the work done by Thiel et al. (2019) in which the physical and chemical structure of the gas along the line of sight to SgrB2(N) was studied; here, complex organic molecules such as

NH_2CHO and CH_3CHO were detected in the majority of the clouds, which at the same time proved to have relatively high visual extinctions ($A_V = 2.5–5$ mag with $N(H_2) > 10^{21}$ cm^{-2}), thus consisting mainly of translucent gas. According to Thiel et al. (2019) the column density of H_2 that corresponds to an A_V of 3 mag is ~3×10^{21} cm^{-2}. This is also consistent with the study by Pety et al. (2008), which states that the bright ^{12}CO emission originates from dense regions with $N(H_2) > 10^{21}$ cm^{-2}. The gas observed at velocities of − 14 and − 4 km s^{-1} on the other hand, corresponds mainly to a "classical" diffuse cloud with a visual extinction of ~ 1 mag according to the above analysis. These clouds also yielded the smallest amounts of the detected molecular abundances. Since chemical complexity seems to be favored towards translucent rather than diffuse gas, for the following discussion we use the model that provided the best fit towards the dense clouds with $v_{LSR} = -17$ km s^{-1} and $v_{LSR} = -10$ km s^{-1} as a reference.

According to our best-fit model, P^+ has a gas-phase abundance of 1.8×10^{-7} at the end of our simulations, being a factor of approximately 1.4 lower than its cosmic value, which indicates that little depletion takes place. The main reservoir of phosphorus other than P^+ is atomic P, having an abundance of 7.4×10^{-8} at 10^7 yr. Atomic P is formed mainly through the electronic recombination of P^+. For our models with elevated densities (10^3 cm^{-3}) we reach high elemental depletion (such as for C^+, S^+ and P^+) after running the code for 10^7 yr. This is consistent with the results presented by Fuente et al. (2019), which show significant depletion of C, O, and S happening already towards translucent material at the edge of molecular clouds (3–10 mag) with $1 - 5 \times 10^3$ cm^{-3}. In Appendix A we investigate further the expected depletion of phosphorus when transitioning from diffuse- to dense-cloud conditions. We find that there is a significant depletion of atomic P on dust grains after the final density of 10^5 cm^{-3} is reached. This in turn leads to a strong increase of gPH_3, that becomes the main carrier of phosphorus in the dense phase. We also find a considerable decrease of the molecules HCP, CP, PN, PO and PH_3 due to freeze-out on grains and their destruction route with H_3^+ after the final density is attained at $t \sim 10^6–10^7$ yr.

The most abundant P-bearing molecules in the gas phase are HCP and CP, with maximal abundances of 3.4×10^{-10} and 2.1×10^{-10}, respectively[7]. The formation and destruction pathways of both HCP and CP are strongly related to the electron fraction, as they are mainly produced (throughout the entire chemical evolution) by dissociative recombination of the protonated species PCH_2^+ and destroyed by reacting with C^+, the main carrier of positive charge in diffuse clouds. Two additional P-bearing species that are predicted by our model to have "observable" abundances in the gas phase are PN and PO, with respective maximal abundances of 4.8×10^{-11} and 1.4×10^{-11}. The most productive formation pathways for PN start with P + CN → PN + C, N + PH → PN + H and end with N + CP → PN + C. In the late stage of evolution (~ $0.5 \times 10^6 - 10^7$ yr) PN is primarily being destroyed by He^+ + PN → P^+ + N + He. The species PO is mainly produced over the entire chemical evolution of 10^7 yr by the dissociative recombination of HPO^+: $HPO^+ + e^- \rightarrow$ PO + H, and is mostly destroyed by reactions with C^+ and H^+. On the other hand, HPO^+ is efficiently formed via $P^+ + H_2O \rightarrow HPO^+ + H$[8]. An additional reaction that becomes relevant at progressive times (~ 10^6–10^7 yr) is O + PH → PO + H with a ~ 10% reaction significance.

Another relatively abundant P-bearing species in the gas-phase based on our best-fit model is phosphine, PH_3, with a maximal abundance of ~1.6×10^{-11} at a late time of 10^7 yr. We note here that the species PH is also predicted to be detectable with a maximal abundance of ~ 3.6×10^{-11}. Unlike PH however, PH_3 has already been detected in circumstellar envelopes of evolved stars (Agúndez et al. 2014a), indicating that it could be an important P-bearing species in interstellar environments such as diffuse and translucent clouds. We therefore focus in the following sections on the PH_3 rather than the PH chemistry. Based on our chemical model, PH_3 is formed most efficiently on dust grains in the early phase, being released to the gas-phase via reactive desorption: gH + $gPH_2 \rightarrow PH_3$. Its formation proceeds after 1.4×10^3 yr with the photodesorption process $gPH_3 \rightarrow PH_3$ being the most effective reaction. Since the evaporation temperature of PH_3 lies at ~ 100 K, the main mechanism driving the desorption of PH_3 at low temperatures is photodesorption (instead of thermal desorption). Switching off the photodesorption in our model leads to a decrease of the PH_3 gas-phase abundance of two orders of magnitude. Once in the gas phase, PH_3 is mostly destroyed by reactions with C^+ and H^+ as well as through the photodissociation reaction: $PH_3 + h\nu \rightarrow PH_2 + H$. The most abundant species on grains is gPH_3, with a maximal abundance of 7.2×10^{-10}. Almost all the atomic P that depletes onto the dust grains reacts with gH and forms gPH (gP + gH → gPH), which subsequently forms gPH_3 through further hydrogenation. Table 8 summarizes all the main formation and destruction pathways for the molecules PN, PO, HCP, CP and PH_3 at three different times ($t = 10^3, 10^5, 10^7$ yr). The last column shows the significance of the given reaction in the total formation or destruction rate of the species of interest.

Figure 6 depicts the time-dependent abundances of PN, PO, HCP, CP and PH_3 over 10^7 yr predicted by the best-fit model along with the computed 3σ upper limits. The predicted abundance for PO lies a factor of about 40 below the observational upper limit at $t = 10^7$ yr, while the current upper limits of HCP and CP are about one order of magnitude higher than the model predictions. Finally, for PN the modeled abundance almost reaches the observed value at the end of our simulations. This means that in all cases the predicted abundances of P-bearing species are lower than the derived upper limits. Future observations of the ground-energy transitions (1–0) will help us to constrain these upper limits even more (see Sect. 6 for further justification). Table 9 lists the predicted abundances of the above species given by our chemical model at $t = 10^7$ yr along with the corresponding upper limits. In the case of PO we show only the lowest value of the four upper limits that were derived for each transition.

In the following discussion we focus on how deviations from our best-fit model can affect the chemistry of P-bearing species. In particular, we examine the dependence of the abundances of HCP, CP, PN, PO and PH_3 on increasing visual extinction A_V, increasing cosmic-ray ionization rate ζ(CR), and alternating the surface mobility constants (diffusion-to-desorption ratio E_b/E_D and possibility of quantum tunneling for light species).

Effects of Visual Extinction on the P-Bearing Chemistry

In this section, we analyze how an increase of A_V is affecting the predicted abundances of P-bearing species. For this purpose we consider the parameters of the best-fit model with $n(H) = 300$ cm^{-3} and $T_{gas} = 40$ K, while varying the A_V from 1 to 10 mag. By keeping the density constant, we avoid high levels of elemental depletion. The increase in visual extinction can then be explained by a figurative increase of the size of the source. Figure 7 (left panel) shows the predicted abundances of P-bearing species at the end of our simulations ($t = 10^7$ yr) under the effect of varying the visual extinction. All species reach a maximal abundance at an of 4 mag. The abundances of HCP, CP, and PN barely change for $A_V > 4$ mag, while for the rest of the molecules the abundances drop; especially in the case of PH_3 where we see a substantial decrease of almost two orders of magnitude. As already mentioned in Sect. 5, the most effective formation process of PH_3 is the photodesorption $gPH_3 \rightarrow PH_3$. Thus, a high visual extinction attenuates the incoming UV-field and therefore the desorption of gPH_3.

In order to better understand the A_V dependence of the remaining molecular abundances, we plotted in Fig. 7 the predicted abundances of the species that mainly form and destroy HCP, CP, PN, and PO (see Table 8) as a function of the visual extinction. In particular, we simulated the abundances of PCH_2^+, C^+, P^+, He^+, and H^+ as well as HPO^+. In the case of HCP (and also CP) its abundance increases up to an A_V of 4 mag and then remains constant above that value. This behavior is correlated with the increase of the PCH_2^+ abundance up to an A_V of 3 mag as well as the decreaseof C^+ up to a visual extinction of 4 mag. The species PO seems to be more strongly affected by the increasing A_V. Its abundance will also increase for $A_V \leq 4$ mag which again stands in correlation with the decrease of the C^+ abundance (the main "destroyer" of PO), followed by a drop in abundance up to 7 mag. This on the other hand results from the decrease in HPO^+ abundance (the main precursor of PO) in the same A_V range. An increase in A_V will decrease the P^+ abundance (due to the decrease of the total ionization rate), as can be seen in Fig. 7. In addition, an enhanced A_V slightly decreases the H_2O abundance (by a factor of two), since the most effective formation for H_2O at late times is the photodesorption $gH_2O \rightarrow H_2O$ (see footnote 7). Therefore, for higher A_V, both P^+ and H_2O decrease, meaning that HPO^+ and subsequently PO reduce in abundance as well.

Effects of the Cosmic-Ray Ionization Rate on the Chemistry of P-Bearing Species

As already mentioned in Sect. 4.2, for all the applied models we use a value of 1.7×10^{-16} s^{-1} for the cosmic-ray ionization rate ζ(CR), as was derived by Indriolo & McCall (2012). This is also consistent with previous work in which diffuse and translucent clouds were studied as well (Fuente et al. 2019; Godard et al. 2014; Le Petit et al. 2004). However, we should note here that in Indriolo & McCall (2012) several cosmic-ray ionization rates were derived towards 50 diffuse lines of sight, ranging from 1.7×10^{-16} to 10.6×10^{-16} s^{-1} with a mean value of 3.5×10^{-16} s^{-1}. Due to the complex and not yet fully understood nature of our observed clouds, we test our chemical model by also applying the elevated values of ζ(CR) = 3.5×10^{-16} s^{-1} and 10.6×10^{-16} s^{-1} in order to examine the influence of the cosmic-ray ionization rate on P-bearing chemistry. As for the remaining parameters of the code (such as A_V and T_{gas}) we use the values given by our best-fit model (see Sect. 4.2). Figure 8 shows the chemical evolution of the species PN, PO, HCP, CP and PH_3 over 10^7 yr for ζ(CR) = 1.7×10^{-16} s^{-1} and 10.6×10^{-16} s^{-1} in the left and right panels, respectively.

Table 10 summarizes the predicted abundances of P-bearing species for the three different cosmic-ray ionization rates given in Indriolo & McCall (2012). As one can see that PN shows the most substantial decrease in abundance with increasing ζ(CR). From the lowest (ζ(CR) = 1.7×10^{-16} s^{-1}) to the highest (ζ(CR) = 10.6×10^{-16} s^{-1}) cosmic-ray ionization rate, the PN abundance decreases by a factor of approximately 730, while for HCP, CP, and PH_3 we see a drop by a factor of about 85 and 50, respectively. As already mentioned, PN is heavily destroyed by He^+ with a ~40% reaction significance. An increase of ζ(CR) up to a value of 10.6×10^{-16} s^{-1} significantly enhances the ionization of He and H by a factor of about 20 and 30, respectively, via cosmic-ray-induced secondary UV photons: He + CRP → He^+ + e^- and H + CRP → H^+ + e^- (the abundance of C^+ increases by a factor of ~ 6). Therefore, the destruction path with H^+ becomes relevant for all P-bearing species showing a 10–40% loss efficiency. The effect is the strongest in the case of PN, because PN is mainly formed through CP which drastically decreases and is also efficiently destroyed by both He^+ and H^+. The PO abundance is only reduced by a factor of about 20 after increasing ζ(CR) up to 10.6×10^{-16} s^{-1}, despite being heavily destroyed by H^+. On the other hand, the significance of the dissociative recombination of HPO^+ increases up to 50% which in turn counterbalances the loss through H^+. An increased ζ(CR) of 10.6×10^{-16} s^{-1} enhances the abundance of P^+ up to ~2.5 × 10^{-7}, nearly reaching its cosmic value of ~2.6×10^{-7} (Asplund et al. 2006), while the abundance of atomic P decreases down to ~ 9.5×10^{-9} via the enhanced reaction with C^+ and H^+.

Table 8. Main formation and destruction mechanisms for the species PN, PO, HCP, CP, and PH_3 based on the best-fit chemical model at times: $t = 10^3, 10^5, 10^7$ yr.

Species	Time (yr)	Reaction type	Reaction	Reaction importance (%)
PN	10^3	Formation	$P + CN \rightarrow PN + C$	47
		Formation	$N + PH \rightarrow PN + H$	25
		Formation	$N + PO \rightarrow PN + O$	17
	10^5	Formation	$N + CP \rightarrow PN + C$	27
		Destruction	$H^+ + PN \rightarrow PN^+ + H$	−27
		Destruction	$He^+ + PN \rightarrow P^+ + N + He$	−23
		Formation	$N + PH \rightarrow PN + H$	10
	10^7	Destruction	$He^+ + PN \rightarrow P^+ + N + He$	−41
		Formation	$N + CP \rightarrow PN + C$	38
		Destruction	$H^+ + PN \rightarrow PN^+ + H$	−8
PO	10^3	Formation	$HPO^+ + e^- \rightarrow PO + H$	48
		Destruction	$C^+ + PO \rightarrow PO^+ + C$	−44
	10^5	Formation	$HPO^+ + e^- \rightarrow PO + H$	47
		Destruction	$H^+ + PO \rightarrow PO^+ + H$	−37
		Destruction	$C^+ + PO \rightarrow PO^+ + C$	−9
	10^7	Formation	$HPO^+ + e^- \rightarrow PO + H$	36
		Destruction	$C^+ + PO \rightarrow PO^+ + C$	−26
		Formation	$O + PH \rightarrow PO + H$	13

HCP	10^3	Formation	$PCH_2^+ + e^- \rightarrow HCP + H$	55
		Destruction	$C^+ + HCP \rightarrow CCP^+ + H$	−22
		Destruction	$C^+ + HCP \rightarrow HCP^+ + C$	−22
	10^5	Formation	$PCH_2^+ + e^- \rightarrow HCP + H$	50
		Destruction	$H^+ + HCP \rightarrow HCP^+ + H$	−39
	10^7	Formation	$PCH_2^+ + e^- \rightarrow HCP + H$	50
		Destruction	$C^+ + HCP \rightarrow CCP^+ + H$	−17
		Destruction	$C^+ + HCP \rightarrow HCP^+ + C$	−17
CP	10^3	Destruction	$C^+ + CP \rightarrow CP^+ + C$	−45
		Formation	$PCH_2^+ + e^- \rightarrow CP + H_2$	34
	10^5	Destruction	$H^+ + CP \rightarrow CP^+ + H$	−38
		Formation	$PCH_2^+ + e^- \rightarrow CP + H_2$	29
		Formation	$HCP^+ + e^- \rightarrow CP + H$	18
	10^7	Formation	$PCH_2^+ + e^- \rightarrow CP + H_2$	35
		Destruction	$C^+ + CP \rightarrow CP^+ + C$	−32
		Destruction	$H^+ + CP \rightarrow CP^+ + H$	−9
PH_3	10^3	Destruction	$C^+ + PH_3 \rightarrow PH_3^+ + C$	−45
		Formation	$gH + gPH_2 \rightarrow PH_3$	31
		Formation	$gPH_3 \rightarrow PH_3$	23
	10^5	Formation	$gPH_3 \rightarrow PH_3$	49
		Destruction	$H^+ + PH_3 \rightarrow PH_3^+ + H$	−26
		Destruction	$C^+ + PH_3 \rightarrow PH_3^+ + C$	−20
	10^7	Formation	$gPH_3 \rightarrow PH_3$	49
		Destruction	$C^+ + PH_3 \rightarrow PH_3^+ + C$	−36
		Destruction	$PH_3 + h\nu \rightarrow PH_2 + H$	−6

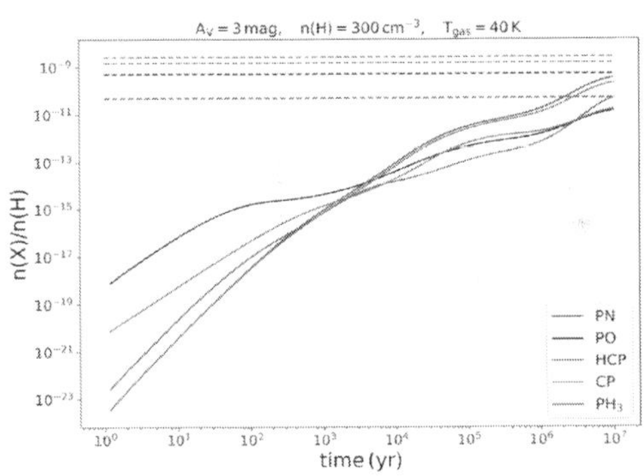

Figure 6. Variation of the predicted abundances of PN, PO, HCP, CP and PH_3 over 10^7 yr in our best-fit model. The dashed lines represent the 3 upper limits derived from the observations at $v_{LSR} = -17$ km s^{-1}. In the case of PO we use 5 × 10^{-10} as an upper limit (see Table 9 and text for explanation).

Table 9. Observed and predicted abundances at time $t = 10^7$ yr for the species PO, PN, HCP, CP, and PH_3 given by our best-fit model.

Species	Observed abundance	Predicted abundance
PN	$<4.9 \times 10^{-11}$	4.8×10^{-11}
PO	$<5.0 \times 10^{-10}$	1.4×10^{-11}
HCP	$<2.6 \times 10^{-9}$	3.4×10^{-10}
CP	$<1.5 \times 10^{-9}$	2.1×10^{-10}
PH_3	–	1.6×10^{-11}

Effects of the Diffusion-to-Desorption Ratio on the Chemistry of P-Bearing Species

The chemistry in the ISM is heavily influenced by the presence of dust grains (Caselli & Ceccarelli 2012). The mobility of the depleted species on the surface of dust grains depends on two mechanisms: thermal hopping and quantum tunneling for the lightest species H and H_2 through potential barriers between surface sites (Hasegawa et al. 1992). Without the possibility of tunneling, the species are not able to scan the grain surface quickly at low temperatures and the total mobility decreases. The parameters that strongly determine the surface chemistry are the diffusion-to-desorption energy ratio E_b/E_D as well as the thickness of the potential barrier between adjacent sites. Based on previous studies (Hasegawa et al. 1992; Ruffle & Herbst 2000; Garrod & Herbst 2006), Vasyunin & Herbst (2013) proposed three different values for the E_b/E_D ratio: 0.3, 0.5, and 0.77. In the case of low ratios ($E_b/E_D = 0.3$) we activate in our model the possibility of quantum tunneling for light species, while for the other two cases, surface mobility is only controlled by thermal hopping (and quantum tunneling is deactivated). The potential barriers are assumed to have rectangular shape and a thickness of 1 Å (Vasyunin & Herbst 2013). In our model we utilize the first set of parameters ($E_b/E_D = 0.3$, with tunneling), nevertheless, since the chemistry of P-bearing species is still highly uncertain, we examine how the remaining two sets of parameters ($E_b/E_D = 0.5$, 0.77, no tunneling) influence the predicted abundances. Table 11 lists the predictions for PN, PO, HCP, CP, and PH_3 as well as H_2 at $t = 10^7$ yr for the three different sets of surface mobility parameters proposed in Vasyunin & Herbst (2013). As Table 11 shows, the H_2 abundance decreases by a factor of four by switching from setup 1 (E_b/E_D = 0.3 with tunneling) to setup 2 (E_b/E_D = 0.5 no tunneling), and finally experiences a dramatic drop of a factor 50 when increasing the E_b/E_D up to 0.77 (overall change of a factor 200 between setups 1 and 3).

The reduction of the H_2 abundance has a significant impact on the formation of PCH_2^+ and PH, which affects the PN, PO, HCP and CP abundances through the following reactions:

PN

- $P^+ + H_2 \rightarrow PH_2^+$
- $PH_2^+ + e^- \rightarrow PH + H$
- $N + PH \rightarrow PN + H$

PO

- $P^+ + H_2 \rightarrow PH_2^+$
- $PH_2^+ + e^- \rightarrow PH + H$
- $O + PH \rightarrow PO + H$

HCP

- $HCP^+ + H_2 \rightarrow PCH_2^+$
- $PCH_2^+ + e^- \rightarrow HCP + H$

CP

- $HCP^+ + H_2 \rightarrow PCH_2^+$
- $PCH_2^+ + e^- \rightarrow CP + H_2$

Both PH and PCH_2^+ decrease by a factor of approximately 20 when increasing the E_b/E_D up to 0.77. In addition, the abundance of H^+ is increased by a factor of about 25, since the reduction of H_2 formation leads to more atomic hydrogen and subsequently more H^+. The enhanced H^+ abundance results in a stronger destruction of all P-bearing species through their reaction with H^+. The species HCP and CP are also strongly affected by changing the surface mobility parameters, with an overall decrease by a factor of about 70 and 65 in abundance, respectively. In both cases the dissociative recombination of PCH_2^+ is essential during the whole chemical evolution for the formation of HCP and CP showing a reaction significance of 30–99%. A decrease of PCH_2^+ due to lower H_2 abundance therefore results in reduced HCP and CP formation. The largest effect is seen for PN, where a diffusion-to-desorption ratio of 0.77and no quantum tunneling of light species reduces the PN abundance by a factor of 300 (see Fig. 9).

Besides the effective loss through H^+, the substantial decrease in PN is also related to the reduction of CP, which is the main precursor of PN at late times. In addition, the reaction N + PH → PN + H is important for PN formation over the entire chemical evolution of 10^7 yr with a 10–50% formation efficiency (for E_b/E_D = 0.77 and no tunneling). This means that the reduction of the H_2 abundance decreases PH, which in turn produces less PN. In the case of PO however, the change in abundance between the two extreme cases is only a factor of about 20. Here, the route O +PH → PO + H increases in significance only up to 3% at late times (4 × 10^6–10^7 yr), indicating that the decrease of PH will not considerably affect PO production. Furthermore, the reduction of PO due to H^+ is compensated through its effective formation via the dissociative recombination of HPO^+. Finally, the abundance of PH_3 decreases only by a factor of 13 in total when changing the surface chemistry constants. Despite being heavily destroyed by H^+, PH_3 is still sufficiently formed through the photodesorption of gPH_3.

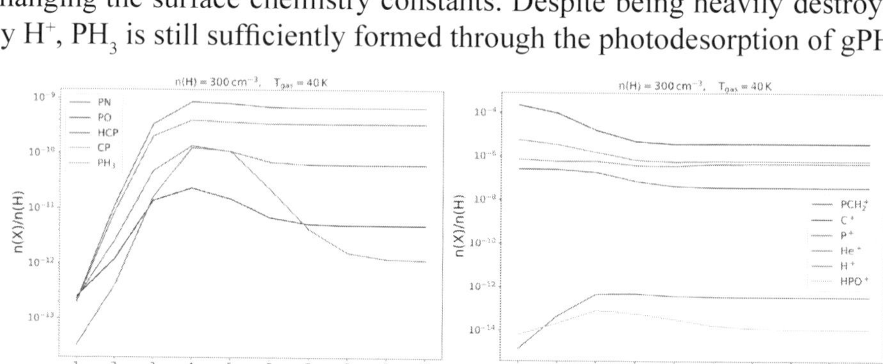

Figure 7. Predicted abundances of P-bearing molecules as a function of visual extinction A_V. The molecular abundances shown here are computed at $t = 10^7$ yr. *Right panel*: predicted abundances of PCH_2^+, C^+, P^+, He^+, H^+, and HPO^+ as they are contributing the most to the formation and destruction of HCP, CP, PN, PO, and PH_3 (*left panel*).

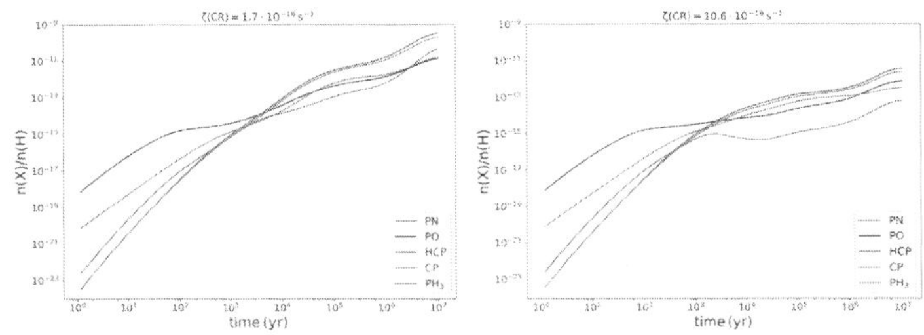

Figure 8. Chemical evolution of P-bearing molecules as a function of time under the effects of cosmic-ray ionization rates of $\zeta(CR) = 1.7 \times 10^{-16}$ s^{-1} (*left panel*) and 10.6×10^{-16} s^{-1} (*right panel*).

Table 10. Predicted abundances of the species PN, PO, HCP, CP, and PH_3 at $t = 10^7$ yr for three different cosmic-ray ionization rates (see text for explanation).

Species	Predicted abundances ($\zeta(CR) = 1.7 \times 10^{-16}$ s^{-1})	Predicted abundances ($\zeta(CR) = 3.5 \times 10^{-16}$ s^{-1})	Predicted abundances ($\zeta(CR) = 10.6 \times 10^{-16}$ s^{-1})
PN	4.8×10^{-11}	2.9×10^{-12}	6.6×10^{-14}
PO	1.4×10^{-11}	4.4×10^{-12}	7.5×10^{-13}
HCP	3.4×10^{-10}	6.7×10^{-11}	3.8×10^{-12}
CP	2.1×10^{-10}	4.2×10^{-11}	2.5×10^{-12}
PH_3	1.6×10^{-11}	2.7×10^{-12}	3.2×10^{-13}

Table 11. Predicted abundances of the species PN, PO, HCP, CP, and PH_3 as well as H_2 at $t = 10^7$ yr for three different sets of surface mobility parameters (see text for explanation).

Species	Predicted abundances ($E_b/E_D = 0.3$ with tunneling)	Predicted abundances ($E_b/E_D = 0.5$ no tunneling)	Predicted abundances ($E_b/E_D = 0.77$ no tunneling)
PN	4.8×10^{-11}	5.0×10^{-13}	1.6×10^{-13}
PO	1.4×10^{-11}	1.2×10^{-12}	6.7×10^{-13}
HCP	3.4×10^{-10}	1.3×10^{-11}	4.6×10^{-12}
CP	2.1×10^{-10}	9.5×10^{-12}	3.2×10^{-12}
PH_3	1.6×10^{-11}	1.8×10^{-12}	1.2×10^{-12}
H_2	4.8×10^{-1}	1.3×10^{-1}	2.4×10^{-3}

FUTURE OBSERVATIONS

Thanks to the sensitive observations (rms of ~6 mK) of the (2–1) transitions of HCP, CP, PN and PO we were able to obtain good upper limits for the column densities and abundances of the above species (see Tables 2 and 9) and thus constrain P chemistry. The observations of HNC, CN, CS, and CO helped us to put important constraints on the main physical parameters of the targeted diffuse and translucent clouds, that is, the visual extinction, the density, and the gas temperature. For the prospect of future observations we want to estimate the expected line intensities of the (1–0) transitions of HCP, CP, PN and PO (at ~ 40−65 GHz) based on our new and improved diffuse-cloud model. Since the densities present in diffuse and translucent clouds are too low to show any collisional excitation ($T_{ex} = T_{bg} = 2.7$ K), the (1–0) transitions are expected to be more strongly populated than the higher-energy transition levels. For these calculations, we take into account the nonthermal nature of the blazar emission, meaning that the flux increases with decreasing frequency. In particular, we apply a power law to the emission of the blazar with $\frac{F}{F_0} = (\frac{\nu}{\nu_0})^{-\alpha}$, where F is the flux, ν is the corresponding frequency, and α is the spectral index. By using the fluxes determined in Agudo (2017) at 3 and 1.3 mm we infer a spectral index of $\alpha \sim 1.06$. Following this, we determine the flux at 7 mm to be ~11 Jy, which in turn corresponds to a temperature of ~26 K with a beam size of 17″ (at 7 mm with the Green Bank Telescope). As Table 12 shows, the derived peak intensities of the species PN, PO, HCP and CP vary from 10 to 200 mK, making these lines "detectable" with radio telescopes, such as the Green Bank Telescope (GBT) and the Effelsberg Telescope. The capabilities of these instruments will allow us to reach rms levels down to 4 mK and enable possible detections up to the 50σ level. The only exception is PH_3 with a (1–0) transition at 266.944 GHz. The flux of the background source at that frequency based on the above power law is equal to 1.91 Jy. This corresponds to a background temperature T_c of 0.4 K with a beam size of9″ (with the IRAM telescope), which in the end results in a very weak, nondetectable absorption line.

CONCLUSIONS

The aim of this work is to understand through observations and chemical simulations which physical conditions favor the production of P-bearing molecules in the diffuse ISM and to what degree. Observing diffuse clouds offers us the opportunity to constrain animportant parameter in our chemical simulations, namely the depletion level of phosphorus (and in general the initial elemental abundances).

We performed single-pointing observations (IRAM 30 m telescope) of the (2–1) transitions of the species PN, PO, HCP and CP at 3 mm towards the line of sight to the bright continuum source B0355+508. None of the above transitions were detected. Nevertheless, the sensitive observations yielding an rms level of ~ 6 mK allowed us to obtain reliable upper limits (see Tables 2 and 9).

We have obtained high S/N detections of the (1–0) lines of HNC, CN, and ^{13}CO between 80 and 110 GHz. We also show a first detection of $C^{34}S$ (2–1) at 96 GHz towards the two densest cloud components at − 10 and − 17 km s^{-1}. Following this, we were able to derive a sulfur isotopic ratio $^{32}S/^{34}S$ of 12.8 ± 4.8 and 18.7 ± 9.5 towards the − 10 and − 17 km s^{-1} features, with the latter being close to the local interstellar value of 24 ± 5 (Chin et al. 1996). The detected molecular species show the highest abundances towards the two components at − 10 and − 17 km s^{-1}, as already shown in previous studies (e.g., Liszt et al. 2018, and references therein).

Based on the detected molecular abundances, we updated our chemical model in order to provide reliable predictions of abundances and line intensities of P-containing molecules that will serve as a guide for future observations. For this purpose we ran a grid of chemical models, with typical physical conditions of diffuse or translucent clouds, trying to reproduce the observed abundances and upper limits of HNC, CN, CO, and CS in every cloud component along the line of sight (at − 4, −8, −10, −14 and −17 km s^{-1}). For the clouds with $v_{LSR} = -10$ km s^{-1} and −17 km s^{-1}, the best agreement between observed and modeled abundances is reached at a time $t_{best} = 6.2 \times 10^6$ yr and at $r_{best} = (n(\mathrm{H}), A_V, T_{gas}) = (300\ \mathrm{cm}^{-3}$, 3 mag, 40 K). We chose this set of parameters as a reference for modeling the phosphorus chemistry.

According to our best-fit model mentioned above, the most abundant P-bearing species are HCP and CP (~ 10^{-10}) at a time of $t = 10^7$ yr. The species PN, PO, and PH_3 also show relatively high predicted abundances of 4.8 × 10^{-11} to 1.4 × 10^{-11} at the end of our simulations. All species are effectively destroyed through reactions with C^+, H^+, and He^+. The molecules HCP, CP, and PO are efficiently formed throughout the entire chemical evolution via the dissociative electron recombination of the protonated species PCH_2^+ and HPO^+, respectively. In addition, the species PH_3 is mainly formed on dust grains through subsequent hydrogenation reactions of P, PH, and PH_2 and then released to the gas-phase via photodesorption. Finally, PN is formed at late times (10^5 –10^7 yr) mainly through the reaction N + CP → PN + C.

We also examined how the visual extinction A_V, the cosmic-ray ionization rate ζ(CR), and the surface mobility on dust grains affect the chemistry of P-bearing species. We found that all P-bearing species are strongly sensitive to the visual extinction: low A_V values of 1 and 2 mag lead to very low P-bearing molecular abundances of ~ 10^{-14}–10^{-12}, indicating that a translucent region rather than a diffuse one is needed to produce observable amounts of P-containing species. All examined species in our study are influenced by the cosmic-ray ionization rate as well. An increasing ζ(CR) enhances the abundance of He^+, H^+ and C^+, which in turn are effectively destroying all P-bearing species. A similar conclusion was found when changing the diffusion-to-desorption ratio to $E_b/E_D = 0.77$ and deactivating the possibility of quantum tunneling of light species on grain surfaces. This setup increases the H^+ abundance, which in turn efficiently reacts with and destroys PN, PO, HCP, CP, and PH_3. Finally, we performed a study of the P-depletion level by tracing the phosphorus chemistry from a diffuse to a dense cloud with the application of a dynamical model that varies the density, the gas and dust temperature, the cosmic-ray ionization rate, and the visual extinction with time (see Appendix A). We came to the main conclusion that at high densities of ~ 10^5 cm^{-3}, atomic P is strongly depleted through freeze-out on dust grains, resulting in a significant increase of the gPH_3 abundance. The molecules PN, PO, HCP, CP, and PH_3 are also affected by freeze-out on grains and are destroyed by their reaction with H_3^+ when reaching the dense phase at timescales of ~10^6–10^7 yr.

Based on the predictions of our improved diffuse-cloud model, the (1–0) transitions of HCP, CP, PN, and PO are expected to be detectable with estimated intensities ranging from 10 to 200 mK. A possible detection of the above species will help us to further constrain the physical and chemical properties of our model and help us to better understand interstellar phosphorus chemistry.

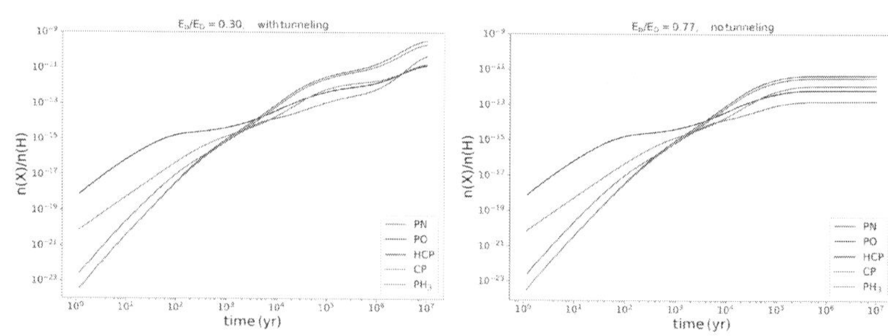

Figure 9. Chemical evolution of P-bearing molecules as a function of time for a diffusion-to-desorption ratio E_b/E_D of 0.3 (with quantum tunneling) shown in the *left panel* and for a E_b/E_D of 0.77 (without quantum tunneling) in the *right panel*.

Table 12. Estimated absorption line intensities for the (1–0) transitions of HCP, CP, PN, and PO towards B0355+508 for T_{ex} = 2.73 K, a FWHM linewidth of Δv = 0.5 km s^{-1} and based on the predicted abundances given by our best-fit model at $t = 10^7$ yr.

Species	Transitions	E_{up} (K)	Frequency (GHz)	A_{ul} (10^{-6} s^{-1})	g_u	Estimated intensities (mK)	Reference
HCP	$J = 1-0$	1.9	39.95190	0.04	3	23	1
PN	$J = 1-0$	2.3	46.99028	3.04	3	214	2
CP	$N = 1-0, J = 3/2-1/2, F = 2-1$	2.3	47.98288	0.43	5	51	3
PO	$J = 3/2-1/2, \Omega = 1/2, F = 2-1$, e	3.2	65.31224	3.83	5	12	4

ACKNOWLEDGEMENTS

We thank the anonymous referee for his/her comments that significantly improved the present manuscript. The authors also wish to thank the IRAM Granada staff for their help during the observations. V.M.R. has received funding from the European Union's Horizon 2020 research and innovation programme under the Marie Skłodowska-Curie grant agreement No 66493. Work by A.V. is supported by the Latvian Science Council via the project lzp-2018/1-0170. J.C. acknowledges Dr. J. C. Laas for his support with the Python programming.

APPENDIX A THE DEPLETION OF PHOSPHORUS

The main advantage of studying the early phases of star formation is to avoid high levels of elemental depletion and thus to constrain the initial abundances used in our model to their cosmic values. This is crucial especially for phosphorus, as the small number of detections of P-bearing species in the ISM makes the determination of the P-depletion level quite difficult. In order to obtain an approximate estimation of the expected depletion level, we apply a dynamical model with time-dependent physical conditions that allows us to follow the chemical evolution of P-bearing species from a diffuse to a dense cloud. In particular we simulate a "cold" stage in which a free-fall collapse takes place within 10^6 yr (Vasyunin & Herbst 2013; Garrod & Herbst 2006). During that time the density increases from n(H) = 300 to 10^5 cm^{-3} and the visual extinction rises from 1 to 40 mag. The gas temperature decreases from 40 to 10 K, while the dust temperature drops slightly, from 20 to 10 K. Finally, the cosmic-ray ionization rate also changes from 1.7×10^{-16} s^{-1} to 1.3×10^{-17} s^{-1}. We note here that the changes in the above-mentioned physical constants happen within 10^6 yr, while the total chemical evolution is over 10^7 yr. This means that between 10^6 and 10^7 yr the model becomes static with the above parameters retaining the values they reached at 10^6 yr. That way, we simulate a long-lived collapse that provides enough time for chemical processes such as depletion to evolve.

As a first step, we plot the chemical evolution of the sum of abundances of gas-phase and solid-phase P-bearing species separately (see lower left panel of Fig. A.1). It is clearly visible how at late times, the gas-phase species decrease, and in return the grain species increase in abundance due to depletion. In particular, the sum of the gas-phase abundances of P-bearing species reduces by a factor of ~ 3000 at $t = 10^7$ yr. This does not correspond to the elemental depletion, but it indicates the redistribution of phosphorus between the gas phase and the dust grains. The right-hand panel of Fig. A.1 shows the time-dependent abundances of the main carriers of phosphorus in the gas phase and on grains. The species that experience the largest change during the transition from diffuse to dense cloud are P^+ and gPH_3. The P^+ abundance strongly decreases down to $\sim 10^{-16}$, mainly through its destruction reactions with OH, CH_4, S, and H_2. Atomic P decreases significantly because of freeze-out on dust grains, which is also evident through the increase in gP. According to the model almost all P that freezes out, quickly reacts with hydrogen on grains, and finally forms gPH_3 (after successive hydrogenation), which reaches a high abundance of $\sim 2.5 \times 10^{-7}$ at the end of our simulations.

Finally, Fig. A.2 shows the time-dependent abundances of PN, PO, HCP, CP, PH_3 in the gas phase (left panel) and the corresponding grain species (right panel). All species reach their peak abundances at around 10^6 yr, followed by a strong decrease due to freeze-out on dust grains as well as through their reaction with H_3^+ (at $t = 10^6$–10^7 yr). The species PN, PH_3, and HCP show a more significant freeze-out than CP and PO, as they are the most abundant molecules in the gas-phase at $t = 10^6$ yr. The freeze-out process is also clearly evident from the substantial increase of the corresponding grain species once high densities of ~ 10^4–10^5 cm^{-3} are reached (see right panel of Fig. A.2.)

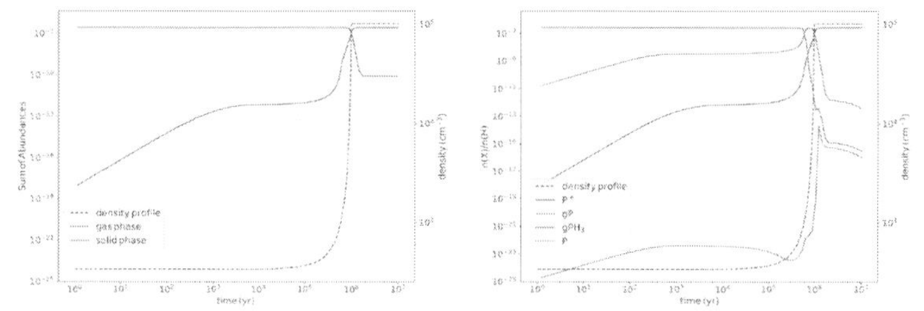

Figure A.1. Results of our dynamical model that simulates the transition from a diffuse to a dense cloud. *Left panel*: sum of abundances of all P-bearing species in the gas phase (red line) and the solid phase (blue line) as a function of time. *Right panel*: chemical evolution of the main carriers of phosphorus in the gas and solid phase: P^{+}, P, gP and gPH_3. In both figures the density profile of the free-fall collapse is depicted as a black dashed line.

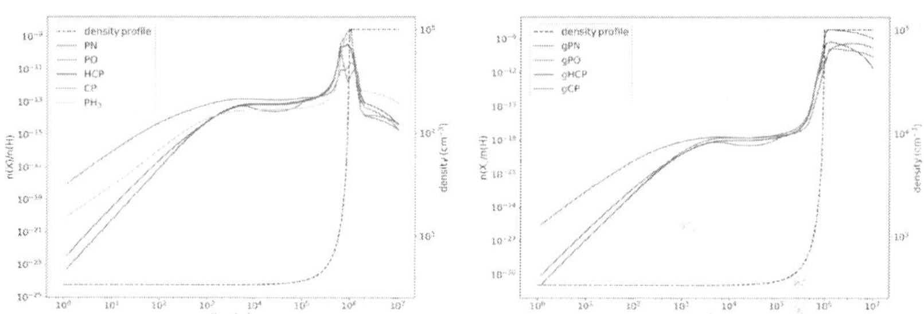

Figure A.2. Chemical evolution of PN, PO, HCP, CP and PH_3 (*left panel*) and the corresponding grain species (*right panel*) as a function of time based on our dynamical model (diffuse to dense cloud). The black dashed line illustrates the density profile of the free-fall collapse. The gPH_3 abundance is shown in Fig. A.1.

REFERENCES

1. Adams, N. G., McIntosh, B. J., & Smith, D. 1990, A&A, 232, 443
2. Agudo, I. 2017, in Submm/mm/cm QUESO Workshop 2017 (QUESO2017), 1
3. Agúndez, M., & Wakelam, V. 2013, Chem. Rev., 113, 8710
4. Agúndez, M., Cernicharo, J., & Guélin, M. 2007, ApJ, 662, L91
5. Agúndez, M., Cernicharo, J., Decin, L., Encrenaz, P., & Teyssier, D. 2014a, ApJ, 790, L27
6. Agúndez, M., Cernicharo, J., & Guélin, M. 2014b, A&A, 570, A45
7. Anicich, V. G. 1993, ApJS, 84, 215
8. Asplund, M., Grevesse, N., & Jacques Sauval, A. 2006, Nucl. Phys. A, 777, 1
9. Bailleux, S., Bogey, M., Demuynck, C., Liu, Y., & Walters, A. 2002, J. Mol. Spectr., 216, 465
10. Bizzocchi, L., Esposti, C. D., Dore, L., & Puzzarini, C. 2005, Chem. Phys. Lett., 408, 13
11. Caffau, E., Andrievsky, S., Korotin, S., et al. 2016, A&A, 585, A16
12. Caselli, P., & Ceccarelli, C. 2012, A&ARv, 20, 56
13. Cazzoli, G., Puzzarini, C., & Lapinov, A. V. 2004, ApJ, 611, 615
14. Cazzoli, G., Cludi, L., & Puzzarini, C. 2006, J. Mol. Struct., 780, 260
15. Cecchi-Pestellini, C., Duley, W. W., & Williams, D. A. 2012, ApJ, 755, 119
16. Charnley, S. B., & Millar, T. J. 1994, MNRAS, 270, 570
17. Chin, Y. N., Henkel, C., Whiteoak, J. B., Langer, N., & Churchwell, E. B. 1996, A&A, 305, 960
18. Corby, J. F., McGuire, B. A., Herbst, E., & Remijan, A. J. 2018, A&A, 610, A10
19. Dalgarno, A. 1988, Astrophys. Lett. Commun., 26, 153
20. De Beck, E., Kamiński, T., Patel, N. A., et al. 2013, A&A, 558, A132
21. Dixon, T. A., & Woods, R. C. 1977, J. Chem. Phys., 67, 3956
22. Dupuy, R., Bertin, M., Féraud, G., et al. 2017, A&A, 603, A61
23. Fayolle, E. C., Bertin, M., Romanzin, C., et al. 2013, A&A, 556, A122
24. Fontani, F., Rivilla, V. M., Caselli, P., Vasyunin, A., & Palau, A. 2016, ApJ, 822, L30

25. Fontani, F., Rivilla, V. M., van der Tak, F. F. S., et al. 2019, MNRAS, 489, 4530
26. Fuente, A., Navarro, D. G., Caselli, P., et al. 2019, A&A, 624, A105
27. Garrod, R. T., & Herbst, E. 2006, A&A, 457, 927
28. Godard, B., Falgarone, E., & Pineau des Forêts, G. 2014, A&A, 570, A27
29. Gottlieb, C. A., Myers, P. C., & Thaddeus, P. 2003, ApJ, 588, 655
30. Hasegawa, T. I., & Herbst, E. 1993, MNRAS, 261, 83
31. Hasegawa, T. I., Herbst, E., & Leung, C. M. 1992, ApJS, 82, 167
32. Indriolo, N., & McCall, B. J. 2012, ApJ, 745, 91
33. Jiménez-Serra, I., Viti, S., Quénard, D., & Holdship, J. 2018, ApJ, 862, 128
34. Jura, M., & York, D. G. 1978, ApJ, 219, 861
35. Klapper, G., Lewen, F., Gendriesch, R., Belov, S. P., & Winnewisser, G. 2000, J. Mol. Spectr., 201, 124
36. Le Petit, F., Roueff, E., & Herbst, E. 2004, A&A, 417, 993
37. Lebouteiller, V., Kuassivi, V., & Ferlet, R. 2006, ASP Conf. Ser., 348, 480
38. Lefloch, B., Vastel, C., Viti, S., et al. 2016, MNRAS, 462, 3937
39. Leung, C. M., Herbst, E., & Huebner, W. F. 1984, ApJS, 56, 231
40. Liszt, H. S. 2007, A&A, 476, 291
41. Liszt, H. S. 2017, ApJ, 835, 138
42. Liszt, H. S., & Lucas, R. 1998, A&A, 339, 561
43. Liszt, H., & Lucas, R. 2001, A&A, 370, 576
44. Liszt, H. S., Pety, J., & Lucas, R. 2008, A&A, 486, 493
45. Liszt, H., Gerin, M., Beasley, A., & Pety, J. 2018, ApJ, 856, 151
46. Lucas, R., & Liszt, H. 1998, A&A, 337, 246
47. Lucas, R., & Liszt, H. S. 2000, A&A, 358, 1069
48. Lucas, R., & Liszt, H. S. 2002, A&A, 384, 1054
49. Maas, Z. G., Pilachowski, C. A., & Cescutti, G. 2017, ApJ, 841, 108
50. McElroy, D., Walsh, C., Markwick, A. J., et al. 2013, A&A, 550, A36
51. Millar, T. J. 1991, A&A, 242, 241 [NASA ADS] [Google Scholar]
52. Millar, T. J., Bennett, A., & Herbst, E. 1987, MNRAS, 229, 41P

53. Mininni, C., Fontani, F., Rivilla, V. M., et al. 2018, MNRAS, 476, L39
54. Öberg, K. I., Fuchs, G. W., Awad, Z., et al. 2007, ApJ, 662, L23
55. Öberg, K. I., Linnartz, H., Visser, R., & van Dishoeck, E. F. 2009, ApJ, 693, 1209
56. Pety, J. 2005, in SF2A-2005: Semaine de l' Astrophysique Francaise, eds. F. Casoli, T. Contini, J. M. Hameury, & L. Pagani (Les Ulis, France: EDP Sciences), 721
57. Pety, J., Lucas, R., & Liszt, H. S. 2008, A&A, 489, 217
58. Rivilla, V. M., Fontani, F., Beltrán, M. T., et al. 2016, ApJ, 826, 161
59. Rivilla, V. M., Jiménez-Serra, I., Zeng, S., et al. 2018, MNRAS, 475, L30
60. Ruffle, D. P., & Herbst, E. 2000, MNRAS, 319, 837
61. Saito, S., Yamamoto, S., Kawaguchi, K., et al. 1989, ApJ, 341, 1114
62. Saykally, R. J., Szanto, P. G., Anderson, T. G., & Woods, R. C. 1976, ApJ, 204, L143
63. Schaefer, L., & Fegley, Jr. B. 2011, ApJ, 729, 6
64. Snow, T. P., & McCall, B. J. 2006, ARA&A, 44, 367
65. Tenenbaum, E. D., Woolf, N. J., & Ziurys, L. M. 2007, ApJ, 666, L29
66. Thiel, V., Belloche, A., Menten, K. M., et al. 2019, A&A, 623, A68
67. Thorne, L. R., Anicich, V. G., Prasad, S. S., & Huntress, W. T. J. 1984, ApJ, 280, 139
68. Turner, B. E., & Bally, J. 1987, ApJ, 321, L75
69. Turner, B. E., Tsuji, T., Bally, J., Guelin, M., & Cernicharo, J. 1990, ApJ, 365, 569
70. Vasyunin, A. I., & Herbst, E. 2013, ApJ, 762, 86
71. Vasyunin, A. I., Caselli, P., Dulieu, F., & Jiménez-Serra, I. 2017, ApJ, 842, 33
72. Wakelam, V., Herbst, E., Le Bourlot, J., et al. 2010, A&A, 517, A21
73. Wakelam, V., Loison, J.-C., Herbst, E., et al. 2015, ApJS, 217, 20
74. Ziurys, L. M., Schmidt, D. R., & Bernal, J. J. 2018, ApJ, 856, 169

SECTION-5:
LABORATORY-BASED ASTROCHEMISTRY

9

Grain Surface Models and Data for Astrochemistry

H.M. Cuppen[1] · C. Walsh[2,3] · T. Lamberts[1,4] · D. Semenov[5] · R.T. Garrod[6] · E.M. Penteado[1] · S. Ioppolo[7]

[1] Institute for Molecules and Materials, Radboud University Nijmegen, Heyendaalseweg 135, 6525 AJ Nijmegen, The Netherlands
[2] Leiden Observatory, Leiden University, PO Box 9513, 2300 RA Leiden, The Netherlands
[3] School of Physics and Astronomy, University of Leeds, Leeds LS2 9JT, UK
[4] Computational Chemistry Group, Institute for Theoretical Chemistry, University of Stuttgart, Pfaffenwaldring 55, 70569 Stuttgart, Germany
[5] Max Planck Institute for Astronomy, Königstuhl 17, 69117 Heidelberg, Germany
[6] Depts. of Astronomy & Chemistry, University of Virginia, McCormick Road, PO Box 400319, Charlottesville, VA 22904, USA
[7] Department of Physical Sciences, The Open University, Walton Hall, Milton Keynes MK7 6AA, UK

ABSTRACT

The cross-disciplinary field of astrochemistry exists to understand the formation, destruction, and survival of molecules in astrophysical

Citation: H.M., Walsh, C., Lamberts, T. et al. "Grain Surface Models and Data for Astrochemistry". Space Sci Rev 212, 1–58 (2017). https://doi.org/10.1007/s11214-016-0319-3

environments. Molecules in space are synthesized via a large variety of gas-phase reactions, and reactions on dust-grain surfaces, where the surface acts as a catalyst. A broad consensus has been reached in the astrochemistry community on how to suitably treat gas-phase processes in models, and also on how to present the necessary reaction data in databases; however, no such consensus has yet been reached for grain-surface processes. A team of ~25 experts covering observational, laboratory and theoretical (astro) chemistry met in summer of 2014 at the Lorentz Center in Leiden with the aim to provide solutions for this problem and to review the current state-of-the-art of grain surface models, both in terms of technical implementation into models as well as the most up-to-date information available from experiments and chemical computations. This review builds on the results of this workshop and gives an outlook for future directions.

INTRODUCTION

The very presence of anything but atoms and obscuring minuscule dust grains in the interstellar medium (ISM) was inconceivable by astronomers merely a hundred years ago. Even the brightest minds of the time, such as Sir Arthur Eddington, were doubtful about the existence of molecules in the vast interstellar void. In his Bakerian lecture he pointed out that "…it is difficult to admit the existence of molecules in interstellar space because when once a molecule becomes dissociated there seems no chance of the atoms joining up again" (Eddington 1926).

However, around one decade later, absorption electronic transitions of the first interstellar molecular species, CN, CH, and CH^+, were identified (Swings and Rosenfeld 1937; McKellar 1940; Douglas and Herzberg 1941). The rapid development of radio and infrared detectors following World War II has since allowed the discovery of ~190 molecules in the ISM, as of March 2016 (see http://www.astro.uni-koeln.de/cdms/molecules). These interstellar species have a multitude of orbital electronic configurations and include stable molecules, radicals, open-shell molecules, cations, and anions.

Many interstellar molecules are recognizable from terrestrial and atmospheric chemistry. Among those are relatively stable species, e.g., water (H_2O), molecular hydrogen and nitrogen (H_2 and N_2), and carbon dioxide (CO_2), all of which consist of just a few atoms. More complex, hydrogen-rich saturated organic molecules are also present in space, e.g., formaldehyde (H_2CO), glycolaldehyde ($HCOCH_2OH$), methanol (CH_3OH), formic acid (HCOOH), and dimethyl ether (CH_3OCH_3) (Ehrenfreund and Charnley 2000; Herbst and van Dishoeck 2009). These species "bridge the gap" between the simple species listed previously and those considered of prebiotic and biological importance, e.g., amino acids. Other interstellar molecules are more exotic and unique to space. These include highly-unsaturated carbon chains and cages, e.g., $HC_{11}N$ (Bell et al. 1997), and the fullerenes, C_{60}, $C_{60}+$, and C_{70} (Cami et al. 2010; Berné et al. 2013; Campbell et al. 2015), the latter of which are also the largest molecular species discovered to date in the ISM. Even larger macromolecules, polyaromatic hydrocarbons (PAHs), consisting of between tens and hundreds of carbon atoms, are identifiable in space as a distinct class of species through their characteristic infrared bands (see the review by Tielens 2008). In summary, it is now known that the interstellar matter out of which stars and planets form has a substantial molecular component, which plays a pivotal role in the thermal balance of the ISM and its evolution (Tielens 2010).

The first theoretical models that successfully explained the presence and abundances of early observed molecular species were developed by Bates and Spitzer (1951), Watson and Salpeter (1972b), Herbst and Klemperer (1973), and Watson (1974), and thereafter significantly extended. The common perception in modern astrophysics is that many interstellar molecules, including complex unsaturated molecules, can be readily formed through purely gas-phase kinetics. Ion-molecule reactions and dissociative recombination reactions are of particular importance. Such processes typically do not have activation barriers and thus the rate coefficients are large, and possibly even enhanced, at low temperatures (~10–20 K, Adams et al. 1985; Herbst and Leung 1986). However, gas-phase chemistry alone cannot efficiently synthesize saturated organic species. The available reaction pathways typically require high temperatures and/or three-body reactions—conditions that are not usually met in the ISM.

Another efficient route towards increasing molecular complexity in the ISM is the chemical kinetics that occurs on dust-grain surfaces. Intriguingly, the most abundant molecule in space, molecular hydrogen, is formed almost exclusively via surface chemistry in the local universe (Gould and Salpeter 1963; Hollenbach and Salpeter 1971; Watson and Salpeter 1972a). The dust-grain surface has several roles. Firstly, the surface serves as a local "meeting point" for molecules or atoms that become bound to the dust grain via electrostatic or van der Waals forces, so-called physisorption, or by forming chemical bonds with its surface, so-called chemisorption. Secondly, the dust grain lattice can accommodate a portion of the excess energy usually generated during surface-mediated association reactions, stabilizing the product, and thus allowing large polyatomic species to be efficiently synthesized. Thirdly, in dense ($\gtrsim 10^4$ cm^{-3}) and cold ($\ll 100$ K) interstellar environments, thick ice mantles can grow on dust grains (~100~100 monolayers). Molecular species trapped in the ice mantle under these conditions are protected from further processing by the FUV interstellar radiation field (ISRF), although they are exposed to the significantly lower strength ambient radiation field generated internally by the interaction of cosmic rays with molecular hydrogen. Eventually, over long timescales ($\gtrsim 10^4$ yr) these pristine ices can be processed by the heating and enhanced irradiation associated with the star-formation process, potentially forming even more complex volatile and refractory organic compounds, including amino acids (Kvenvolden et al. 1970; Ehrenfreund and Charnley 2000; Elsila et al. 2009). These molecules may then be delivered to young protoplanets and planets via accretion early in the evolution of the planetary system, or at a later time via bombardment by pristine icy bodies (Anders 1989; Chyba et al. 1990; Cooper et al. 2001).

Most modern astrochemical models of ISM chemistry simulate dust-grain surface chemical kinetics processes with various degrees of complexity (Tielens and Hagen 1982; Hasegawa and Herbst 1993a; Garrod 2013). Models using large reaction networks typically adopt the rate-equation approach as is done for the gas-phase chemistry where the time evolution of surface species' abundances is described by a set of coupled ordinary differential equations, and the abundances considered "averaged" over the entire dust grain population, i.e., the mean-field approximation. One of the major challenges in these models, is the accurate treatment of the stochasticity of diffusive surface processes. This becomes critical when abundances of reactants on the dust-grain surface becomes very low, i.e., $\ll 1 \ll 1$ reactant per dust grain, and fluctuates with time, thus rendering the rate-equation approach unfeasible (Gillespie 1976; Green et al. 2001; Charnley 2001). This is the so-called "accretion-limited" case. A number of approximate or precise micro- and macroscopic Monte Carlo techniques have been proposed to overcome this issue (e.g., Biham et al. 2001; Charnley 2001; Lipshtat and Biham 2004; Stantcheva and Herbst 2004; Chang et al. 2005; Garrod 2008; Vasyunin and Herbst 2013). Another challenge is to account for the multilayered nature of dust-grain ice mantles, and to take all relevant processes into account in the modeling, e.g., inter-lattice diffusion, mobility/immobility of reactants, desorption, porosity trapping (see Cuppen and Herbst 2007; Chang et al. 2007; Kalvāns and Shmeld 2010; Wolff et al. 2011; Taquet et al. 2012; Vasyunin and Herbst 2013). A further obstacle in both approaches (rate equation and stochastic) is the lack of appropriate laboratory data on binding energies and desorption efficiencies of molecular ices of astrophysical interest, as well as the energy barriers and branching ratios for surface reactions. Large reaction networks can treat up to a few hundred different surface species; however, only a handful of reaction systems and molecules have been theoretically or experimentally studied. Moreover, the underlying molecular mechanism is not always fully understood, which makes it hard to scale up to astrophysically relevant timescales.

A wealth of evidence suggests that dust-grain surface processes are important over a wide range of interstellar conditions and star-formation environments, while models and observations are rapidly advancing to trace this chemical evolution through to at least the protoplanetary disk phase (Henning and Semenov 2013; Dutrey et al. 2014; Walsh et al. 2014). The new Atacama Large Millimeter Array (ALMA), with its orders-of-magnitude increase in sensitivity and resolving power, is expected to give us an unprecedented view of potentially pre-biotic and biologically-relevant molecules in various astrophysical environments over the coming years. The analysis of these new data will require much more elaborate, and more diverse, gas-grain astrochemical models than have been developed so far. Unfortunately, a major stumbling-block in our understanding of pre-biotic chemistry in the ISM is the lack of a standardized and comprehensive approach to simulate grain-surface chemistry. In the case of gas-phase chemistry, several publicly available databases with reactions and the corresponding rate data exist, of which the UMIST Database for Astrochemistry (UDfA) and the KInetic Database for Astrochemistry (KIDA) are the most widely used (McElroy et al. 2013; Wakelam et al. 2012). Consensus on how these data should be used, including how the rate coefficient is calculated and over which temperature ranges it is viable, has been reached, and the quality of the data in these databases is regularly reviewed. However, for grain-surface chemistry, this is not yet the case. Modelers often compile their own grain-surface reaction networks, and most are not publicly available, primarily due to the lack of an agreed and standardized approach.

Fortunately, many of the assumptions within the models can now be tested using surface science techniques with interstellar ice analogs. Over the past few years, substantial progress has been made on the understanding of various grain-surface reaction systems, including which processes are dominant and under what conditions, as well as the underlying mechanisms. In the summer of 2014, astronomers, experimentalists and theoretical chemists came together during a Lorentz Center Workshop ("Grain-Surface

Networks and Data for Astrochemistry") to identify the needs of modelers for their models, the appropriate formalisms to use, and to identify how recent experimental techniques and results can help to test and improve the models. In this paper we summarize the key findings of this workshop and relay our recommendations for the treatment of grain-surface chemistry to the astrochemical community. First we describe the outline of a typical gas-grain model (Sect. 2) and in Sects. 3 to 8 we discuss, in turn, the various processes that need to be considered in astrochemical models of surface chemistry: accretion, desorption, surface reactions, diffusion (thermal diffusion in the surface and bulk and quantum tunneling), and photoprocesses. We also address the more technical aspects of writing and executing code such as numerical precision in chemical models (Sect. 9) and finally end with a discussion on a test case of CO hydrogenation to form complex molecules (Sect. 10) and a future outlook (Sect. 11).

OUTLINE OF A GENERIC GAS-GRAIN CODE

Gas-grain astrochemical models typically use the rate-equation approach to describe both the gas-phase and grain-surface chemistry using chemical kinetics. This generates a set of stiff ordinary differential equations that can be numerically solved using a multi-step integrator, e.g., via Runge-Kutta or Adams algorithms. In chemistry, rate equations are often applied to describe macroscopic experimental effects and account for many-body effects with a mean-field approach. As we mentioned above, this may not be the case on a dust-grain surface under particular conditions; hence, using the rate-equation approach to describe interstellar surface chemistry can lead to large errors when compared with results using more realistic stochastic techniques. The main reason why modelers persist with such a method is the convenience, stability, and the rather fast numerical performance of the pure chemical kinetics codes, even for reaction networks which consist of thousands of reactions involving hundreds of molecules (e.g., Dalgarno and Black 1976; Leung et al. 1984; D'Hendecourt et al. 1985; Brown and Charnley 1990; Hasegawa et al. 1992; Bergin et al. 1995; Millar et al. 1997; Aikawa et al. 1996; Willacy et al. 1998; Semenov et al. 2010; Wakelam et al. 2010; Agúndez and Wakelam 2013; Albertsson et al. 2013; McElroy et al. 2013; Grassi et al. 2014). As an indication, rate equations require CPU time of ~1--60~1--60 seconds. Typically, the addition of the modified rate approach to the rate equation model, which use the same numerical scheme, slows it by a factor of several due to the performance penalty for accounting for probabilities of reactants to be on the grain surface (Garrod et al. 2009). A multiphase model (Furuya et al. 2016) without bulk ice chemistry, but with swapping only, takes typically ~30–60 minutes per trajectory, using a full gas-ice network with deuterium chemistry. It is hence computationally feasible to use such a model to simulate a collapsing core model; tracing 35000 parcels from the prestellar core phase to the circumstellar disk phase results in ~35000 CPU hours or ~2 weeks on a ~100-core machine. When bulk ice chemistry is included, the CPU time increases by a factor of 10–100. Adding in bulk chemistry increases this to months and hence a multi-phase model with bulk ice chemistry coupled with 2-D/3-D physical models remains a computationally challenging problem. On the other hand, macroscopic Monte Carlo models like presented in Vasyunin et al. (2009) require much more CPU time, from hours to days, for a simulation of a TMC-1 type cloud. What is more important, Monte Carlo models usually have a rather limited range of physical conditions that can be considered due to their slow performance. Microscopic Monte Carlo models (Lamberts et al. 2014b; Cuppen et al. 2009; Chang and Herbst 2014, 2016) are restricted to an even smaller chemical network and require days to weeks. The method of choice is hence highly dependent on the available computer power, the problem that one would like to address which dictates the level of detail in the grain description required, and the heterogeneity and complexity of the astrophysical object that one is interested in. In recent times, efforts have been made to simulate both laboratory and astrophysical conditions with the same model, thus using the laboratory simulation as a benchmark (e.g., Lamberts et al. 2013). Especially for these cases, microscopic Monte Carlo methods are worth the extra computational effort since they allow to include more surface complexity that might be crucial to gain insight in the physical and chemical processes occurring in the experiments. At the same time, laboratory environments typically deal with well-constrained physical conditions and a limited chemical network.

Here we present a recipe for the construction of a chemical kinetics model based on the rate-equation approach (see also Semenov et al. 2010).

The chemical system consists of two major phases: the gas-phase and the dust-grain surface ice mantle. If all the necessary kinetic data are provided (e.g., from a database) and the initial abundances are assumed (e.g., from observations) or generated (e.g., using a similar model), a chemical kinetics code numerically solves the equations of first- and second-order kinetics and returns time-dependent molecular concentrations. Under typical ISM conditions, i.e., low densities, three-body reactive collisions are usually irrelevant and hence ignored. Here, we focus solely on the grain-surface chemistry aspect of the code. The treatment of gas-phase chemistry has been described in a number of papers, including McElroy et al. (2013) and Wakelam et al. (2010).

As schematically shown in Fig. 1, species on a grain surface generally experience four types of processes: (i) accretion (or adsorption) onto the surface, (ii) desorption from the surface, (iii) diffusion across the surface or on/within the ice mantle, and (iv) reaction. When grain-surface ice mantles are still exposed to far-UV radiation, species contained within can also be photodissociated. This leads to the following expression for the change in surface abundance:

$$\begin{aligned}\frac{d}{dt}n_s(A,t) &= \sum_{i,j} f_{react}(i+j \longrightarrow A) - \sum_{i} f_{react}(i+A \longrightarrow) \\ &\quad + \sum_{i} f_{UV\ diss,i} - f_{UV\ diss,A} \\ &\quad + f_{acc,A} - [f_{evap,A} + f_{nonth,A}].\end{aligned} \tag{1}$$

The first four terms in this expression account for the gain and loss of species A due to grain-surface reactions or photodissociation reactions, respectively. The fifth term expresses the accretion of species A from the gas phase onto the grain, and the final term denotes the desorption of species A from the grain back into the gas phase. This latter process can occur via thermal desorption or by non-thermal processes, whereby desorption is trigged by the input of external energy in the form of far-UV or X-ray photons or high energy particles or by energy released during in-situ exothermic reactions. In the subsequent Sections we will discuss in detail the functional forms usually adopted for each of these chemical processes.

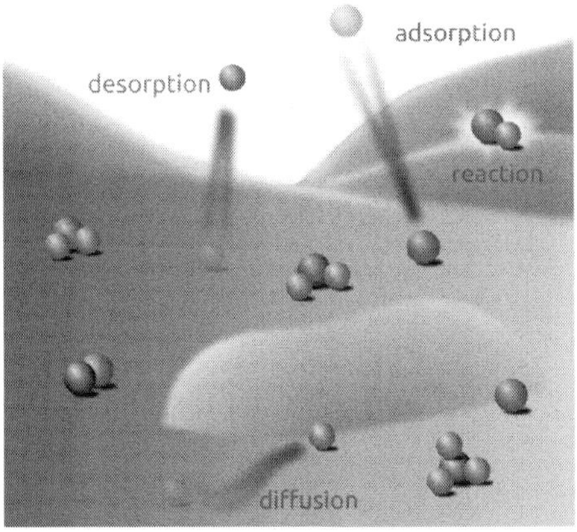

Figure. 1: Overview of the most important grain surface processes that will be covered in this review

As previously mentioned, for low surface abundances, the mean-field assumption inherent in the rate-equation approach breaks down, and several stochastic methods have been developed to overcome this issue. Although the description of the chemistry is intrinsically more accurate, stochastic models are computationally much more demanding than rate equations, and for the purpose of this review we will limit ourselves to rate-equation models. Modifications to the rate-equation approach can be made to better treat the surface chemistry in the accretion-limited case. Caselli et al. (1998) were the first to propose such an adjustment. They applied a semi-empirical approach to scale down the reaction rates for those cases where the surface migration of atomic hydrogen is significantly faster than its accretion rate onto grains (Caselli et al. 1998). This method gave good agreement with stochastic methods for a number of cases; however, it was not clear how applicable the method was outside of the tested regime. More recently, a new modified-rate approach was suggested by Garrod (2008) which improves upon the original.

ACCRETION

The accretion term $f_{acc,A}$ in Eq. (1) accounts for the adsorption of gas-phase species onto the dust grains. It is determined by the collisional frequency of a gas-phase species with a grain, times a sticking efficiency, S_A:

$$f_{\mathrm{acc,A}} = S_{\mathrm{A}} v_{\mathrm{A}} n_{\mathrm{grain}} \pi r_{\mathrm{grain}}^2 n_{\mathrm{g}}(\mathrm{A}), \tag{2}$$

where ng(A) is the number density of species A in the gas phase, r_{grain} is the average radius of a dust grain (~0.1 μm for ISM-like grains) with number density, ngrainngrain, and v_A is the average gas-phase thermal velocity,

$$v_{\mathrm{A}} = \sqrt{\frac{8kT_{\mathrm{gas}}}{\pi m_{\mathrm{A}}}}. \tag{3}$$

This in turn depends on the gas temperature, T_{gas}, the mass of the species, m_A, and Boltzmann's constant k. The sticking efficiency or probability, S_A, of species A to the surface is determined by how well it can dissipate its kinetic energy. This depends on the dust-grain and gas temperature, on the relative masses of the substrate molecules and the incoming species, and on the presence of a barrier for sticking, typically restricted to chemisorption. For most species at low gas and grain temperatures, this results in a sticking fraction near unity, with the exception of hydrogen. Computationally, sticking fractions have been determined by Molecular Dynamics (Buch and Czerminski 1991; Al-Halabi et al. 2002, 2003, 2004; Batista et al. 2005; Veeraghattam et al. 2014), perturbation and effective Hamiltonian theories, close coupling wavepacket, and reduced density matrix approaches (Lepetit et al. 2011), or by the much-more-approximate soft-cube method (Logan and Keck 1968; Burke and Hollenbach 1983). These studies typically focus on the accretion of a single atom or molecule on an otherwise bare surface, whereas the sticking coefficient could be coverage dependent, especially for chemisorption where for high coverage there are simply fewer sites available for sticking. Experimentally it was found that the sticking coefficient of physisorbed H_2 increases linearly with the number of deuterium molecules already adsorbed on the surface (Amiaud et al. 2007; Chaabouni et al. 2012).

DESORPTION

The desorption term in Eq. (1) represents the desorption of the species from the grain surface back into the gas phase. Various desorption processes are possible and usually a particular distinction is made between thermal desorption and non-thermal desorption. For the latter process, a multitude of different mechanisms is possible, such as photodesorption (Westley et al. 1995b; Öberg et al. 2007, 2009a,b), sputtering by cosmic rays or grain heating by cosmic rays (Hasegawa and Herbst 1993a; Herbst and Cuppen 2006), and reactive or chemical desorption, where the excess heat generated upon reaction allows desorption of the products (Garrod et al. 2006, 2007; Dulieu et al. 2013). Here, thermal and reactive desorption are briefly discussed. Photodesorption is discussed in a separate section on photoprocesses since photodesorption and photodissociation are parallel processes which require a different treatment.

Thermal Desorption

The residence time of a species on the dust-grain surface is predominantly determined by its desorption rate. The thermal desorption rate, in turn, depends on the binding energy of the species to the surface, $E_{\mathrm{bind,A}}$,

$$k_{\mathrm{evap,A}} = \nu \exp\left(-\frac{E_{\mathrm{bind,A}}}{kT}\right), \tag{4}$$

where ν is a characteristic attempt frequency. Here ν and $E_{\mathrm{bind,A}}$ are important input parameters for astrochemical models. Usually the following equation for the characteristic frequency (Tielens and Allamandola 1987) is assumed,

$$\nu = \sqrt{\frac{2N_s E_{\mathrm{bind,A}}}{\pi^2 m_A}}, \tag{5}$$

where N_s is the surface density of binding sites and mAmA is the mass of species A. Tielens and Allamandola (1987) derived this expression assuming that the vibrational frequency perpendicular to the surface equals the vibrational frequency parallel to the surface and that the binding can be described by a harmonic potential, which might not be an accurate assumption for a physisorbed species. They also derived an expression including rotational degrees of freedom and one for the frequency of a free particle $\nu = \frac{kT}{h}$. Together this leads to an estimation of $\nu = 10^{12}\ \mathrm{s}^{-1}$ 1, in accordance with all three approaches.

The binding energy, $E_{\mathrm{bind,A}}$, and the thermal desorption rate, $k_{\mathrm{evap,A}}$, can be experimentally obtained using Temperature Programmed Desorption (TPD). These experiments are usually performed under ultra-high-vacuum conditions (base pressure better than ~10^{-9}) coupled with a quadrupole mass spectrometer. The temperature of the substrate can be carefully controlled using a cryostat. A TPD experiment consists of two phases: (i) the substrate is brought to a constant low temperature and a known quantity of one or more species is deposited, and (ii) the temperature is linearly increased and the desorption monitored using the mass spectrometer.

Different types of analysis methods can be applied to obtain kinetic parameters, such as desorption energy, desorption order, and the so-called "prefactor", which is analogous to (although not entirely equivalent to) the characteristic frequency, ν. Sometimes the latter two parameters are assumed and only the first is obtained from the analysis, other groups use for instance "leading edge fitting" to obtain all three simultaneously. Whichever method is applied, the three parameters are not completely independent and therefore a desorption energy derived from experiment should be used in combination with its corresponding prefactor. In most gas-grain codes, the computationally convenient description of the prefactor by Eq. (5) is adopted, although sometimes not physically accurate. It is recommended that experimentalists always quote the desorption energy derived in combination with the prefactor, and a fixed integer value for the desorption order, so that the binding energies are used in an appropriate manner in astrochemical models.

The desorption order is an important consideration worthy of further discussion. Zeroth-order desorption, i.e., a constant desorption rate, generally occurs when multiple layers of the same species are deposited. The number of surface species available for desorption (limited to the top few monolayers) remains the same; hence, the desorption rate is independent of the number of total species on the surface. In the sub-monolayer regime, first-order desorption is observed. Second-order desorption, i.e., a quadratic dependence of the desorption rate on the number of surface species, is also seen. This can occur in two cases: (i) when the surface exhibits a distribution of binding sites, and (ii) through chemical desorption of species that are formed via a second-order surface reaction.

In many astrochemical models, the first-order thermal desorption rate is assumed,

$$f_{\mathrm{thermal,A}} = k_{\mathrm{evap,A}} n_{\mathrm{s}}(\mathrm{A}) \tag{6}$$

where $n_s(A)$ is the number density of species A adsorbed on the grain surface. As mentioned above, this only strictly occurs in the sub-monolayer regime. In two-phase gas-grain astrochemical models there is no positional information on the various species, so it is not known which species occupy the top layers of the ice mantle. However, it is possible to apply a fix to the thermal desorption rate to account for the fractional composition of the ice mantle, as well as treating thermal desorption as a zeroth-order process in the multilayer regime. This involves counting the number of monolayers present within the ice mantle,

$$N_{\text{mono}} = \frac{\sum_i n_{s,i}}{N_s \sigma_g n_{\text{grain}}}, \quad (7)$$

where the numerator is the total number density of surface species per unit volume, and the denominator is the total number of surface sites available per unit volume. For $N_{mono} > N_{act,}$ where, N_{act} is the assumed number of "active" monolayers (typically ~2--4), the thermal desorption rate is given by,

$$f_{\text{thermal,A}} = k_{\text{evap,A}} N_{\text{act}} \chi_A N_s \sigma_g n_{\text{grain}}, \quad (8)$$

Where $\chi_A = n_s(A)/\sum_i n_{s,i}$ is the fractional abundance of species A in the ice mantle. For $N_{\text{mono}} \leq N_{\text{act}}$, the rate switches to the first-order desorption rate.

Table 1 lists the binding energies of a wide collection of stable species that have been determined using the TPD technique and are therefore relatively well constrained. The binding energies have been mostly determined for the desorption of pure ices from different substrates. The differences between the different substrates are rather small and become negligible in the multilayer regime (Green et al. 2009). The uncertainties on the binding energies quoted in Table 1 can have different origins: experimental errors, errors in the fit, or—especially for the amorphous silicate surfaces—they can represent a range of binding energies which is an intrinsic property of the substrate.

Table 1 List of experimentally determined binding energies

Species	EbindEbind **(K)**	**Prefactor ($cm^{-2} s^{-1}$)**	**Comment**	**Ref.**
H_2	440		amorphous water ice	A
	480±10		silicate	B
	555±35		mixed $H_2O:CH_3OH:H_2$	C

H_2O	4800±96	$10^{23\pm1}$	graphite	D
	4800		silicate	E
	4825±5		unannealed water ice	F
	5027±87		annealed water ice	F
	5600	10^{30}	amorphous water ice	G
	5770±60	10^{30}	crystalline water ice	G
	5930±240	10^{28}	amorphous silica	H
H_2CO	3260±60	10^{28}	mixed H_2CO:H_2O = 1:14	I
	3730±110	10^{28}	silicate	I
CH_3OH	4235±15		pure	J
	4930±98	$6\times10^{21\pm3}$	graphite/multilayer	D
	5770±95	$9\times10^{9\pm3}$ [a]	graphite/monolayer	D
CO	826±24	$(7\pm2)\times10^{26}$	pure	K
	828±28	7.1×10^{26}	non-porous water ice	L
	830±40	7.1×10^{26}	silicate	L
	850±55	7.1×10^{26}	crystalline water ice	L
	858±15	7.2×10^{26}	pure	M
	856±15	7.2×10^{26}	layered CO–O_2	M
	865±18	7.2×10^{26}	mixed CO:O_2 = **1:1/multilayer**	M
	955±18	7.6×10^{11}	mixed CO:O_2 = **1:1/monolayer**	M
	855±25		pure	N
	855±25	$7\times10^{26\pm1}$	pure	O
	880±36		layered	H
	955±4		pure	F
	1180±24	$(5\pm1)\times10^{14}$ [a]	amorphous water ice	K
CO_2	2235±50		mixed SO_2:CO_2 = **20:1**	J
	2270±71	9.3×10^{26}	non-porous water ice	L
	2270±80	9.3×10^{26}	silicate	L
	2360±83	9.3×10^{26}	crystalline water ice	L
	2690±150		pure	P
	2860±150		mixed H_2O:CO_2 = **20:1**	P
O	1100		amorphous silicate	E
	1660±60		porous amorphous water ice	Q
	1765±232		amorphous silicate/monolayer	R
	1850±90		bare amorphous silicate	Q

O_2	912±15	6.9×10^{26}	pure	M
	904±15	6.9×10^{26}	layered CO–O_2	M
	896±18	6.8×10^{26}	mixed O_2:CO = 1:1	M
	895±36	6.9×10^{26}	silicate	L
	898±30	6.9×10^{26}	non-porous water ice	L
	902±24		layered	H
	904		amorphous silicate/monolayer	R
	925±25		pure	O
	936±40	7×10^{26}	water/crystalline	L
	1250		amorphous silicate	E
N_2	790±25		pure	O
	830±36		pure	H
NH_3	2790±144	$(8\pm3)\times10^{21}$	graphite	D
	3075±25		pure	J

Desorption rates depend exponentially on binding energies and uncertainties in these binding energies can have a large effect, even at dark cloud conditions where the temperature is well below the desorption temperature of the vast majority of the surface species (Penteado et al. 2016). Since in most systems the diffusion barrier is calculated by taking a fixed ratio with the binding energy, changing binding energies not only affects the temperature at which species desorb, i.e., the temperature at which species cannot participate in the grain surface chemistry, but also the onset temperature at which species start to diffuse. A sensitivity analysis of grain surface chemistry under dark cloud conditions to binding energies of ice species showed that the model results appear particularly sensitive to the binding energy of H_2 (Penteado et al. 2016). The dust temperatures in molecular cloud cores are relatively well constrained to precision of ~1 K by a number of *Herschel* studies (e.g., Stutz et al. 2010; Launhardt et al. 2013; Lippok et al. 2016).

The experiments show that the molecules indeed desorb with a (close to) zeroth-order rate in the multilayer regime whereas they desorb with a (close to) first-order rate in the monolayer regime, as explained above (Fraser et al. 2001). It is difficult to experimentally obtain similar results for radical species due to their high reactivity (and correspondingly short lifetime). Binding energies for radicals can only be obtained in an indirect manner, usually involving the simulation of experimental data, and an exploration of the possible parameter space, using stochastic chemical models. However, there are recent experimental results reporting the experimental determination of the binding energy of atomic oxygen on a range of surfaces (Dulieu et al. 2013; He et al. 2014, 2015), showing that for some species, direct measurements are possible.

Most TPD experiments are performed using pure ices to allow an unambiguous interpretation of the results and to minimize the chance of contamination. Some studies on mixed and layered ice have been done to better mimic the composition of interstellar ice mantles. The introduction of more species in the ice immediately increases the complexity of the desorption process. The binding energy of individual species will vary depending on its surrounding material, and the dominant ice-mantle species can prevent other species from desorbing. Collings et al. (2004) showed, for instance, that a fraction of molecules like CO and CO_2 can become trapped in an ice mantle which consists predominantly of water ice. The trapped CO and CO_2 are then released at the higher temperatures expected for water desorption. However, laboratory timescales are significantly shorter than those in the ISM; hence, trapped species may have sufficient time to escape the ice mantle since they will also have had sufficient time to segregate (Öberg et al. 2009d). This process depends on a large number of parameters including surface temperature, ice composition and mixing ratio. Two-phase astrochemical models can include the effects of trapping in a somewhat empirical manner by allowing a fraction of volatile species such as CO to have a binding energy similar to the species within which they are trapped (e.g., CO_2 or H_2O, see, e.g., Viti et al. 2004). Three-phase and multilayer models can simulate the effects of trapping by allowing diffusion of surface species into the bulk ice mantle (and vice versa). We discuss the treatment of bulk diffusion in Sect. 7.

REACTIVE/CHEMICAL DESORPTION

Chemical desorption is desorption of reaction products from the grain surface by excess reaction energy. This type of desorption is also referred to as reactive desorption. Garrod et al. (2006) were the first to suggest this mechanism to explain, e.g., the gas-phase detection of methanol in cold dark clouds. They based their initial model on the Rice-Ramsperger-Kessel (RRK) theory, which relates the excess energy and the binding energy of species to a desorption probability. They modified this theory by adding an unconstrained aa parameter which they chose to be 0.1. In a follow-up study, Garrod et al. (2007) showed that chemical desorption may play an important role in explaining the observed abundances of different gas-phase chemical species, particularly in dark molecular clouds. Later Cazaux et al. (2010) came to similar conclusions when they included this mechanism in their model for water formation on grains.

The first constraints on the probability of this mechanism were obtained using Molecular Dynamics simulations (Andersson et al. 2006; Arasa et al. 2010, 2011). In these studies, the fate of photoproducts of water ice photodissociation—OH and H—were monitored in time. In some cases, the photoproducts were found to recombine to form water which subsequently escaped from the ice mantle: this can loosely be described as reactive desorption driven by photodissociation. However, as will be discussed later, this can also be thought of as "photodesorption" (see Sect. 8). In these simulations, the desorption probability was highly dependent on the location of the dissociated molecule in the ice mantle. Recombination events in the fourth layer of the ice or further below almost exclusively resulted in trapping of the reformed water molecule. These results are limited to a water-rich environment and they may not be applicable to the formation of the first monolayers of the water ice mantle.

What remains to be quantified is the efficiency of reactive desorption which is not driven by photoprocessing. The first experimental study by Dulieu et al. (2013) measured the chemical desorption of reaction products through sequential O_2 hydrogenation experiments on an amorphous silicate or a graphite surface, where the amount of deposited O_2 remained in the (sub)monolayer regime. They find substantial desorption of the formed H_2O molecules, which is caused, at least in part, by the lack of binding opportunities with surrounding molecules. Moving to the multilayer regime, they find that the desorption probability for the O+OO+O recombination reaction drops to negligible values (Minissale and Dulieu 2014). Expanding their studies to other reaction systems (e.g., CO+HCO+H, H2CO+HH2CO+H), they determined relatively low reactive desorption rates in the (close-to) sub-monolayer regime ($\lesssim$10 %, Minissale et al. 2016).

Despite the lack of conclusive experimental evidence for chemical desorption driven purely by exothermicity of reactions (and not by photoprocessing), especially in the multi-layer regime, astrochemical models typically still account for such a process by implementing the Garrod et al. (2007) prescription

$$P = \frac{a\Xi}{1 + a\Xi} \tag{9}$$

with

$$\Xi = \left[1 - \frac{E_{\text{bind}}}{E_{\text{exo}}}\right]^{s-1} \tag{10}$$

where E_{exo} is the exothermicity of the reaction, and ss is the number of vibrational modes in the molecule/surface-bond system. This number is s=2 for diatomic species; for all others, s=3N−5, where NN is the number of atoms in the molecule, which is assumed to be non-linear and forms an extra "bond" to the surface. The efficiency parameter aa is not well constrained and is generally used as universal input parameter with a value between 0.01 and 0.1. Figure 2 shows the sensitivity of gas phase and ice abundances to this parameter in a laminar solar nebula model by changing aa from 0.05 to 0.01. The figure shows that changes can locally be several orders of magnitude, but integrated over the height of the disk the changes are relatively small for several species. Ices in disks concentrate around the midplane and are nearly absent in upper layers due to thermal evaporation and photodesorption. The changes in column densities are hence mainly determined by the changes in abundance around the midplane. Interestingly, the abundances of CH_3OH and CH_4 increase both in the gas phase and in the ice by lowering the reactive desorption. This is presumably since these species are formed in several steps and a lower reactive desorption efficiency keeps the intermediate species on the grain, enabling the full reaction route to proceed.

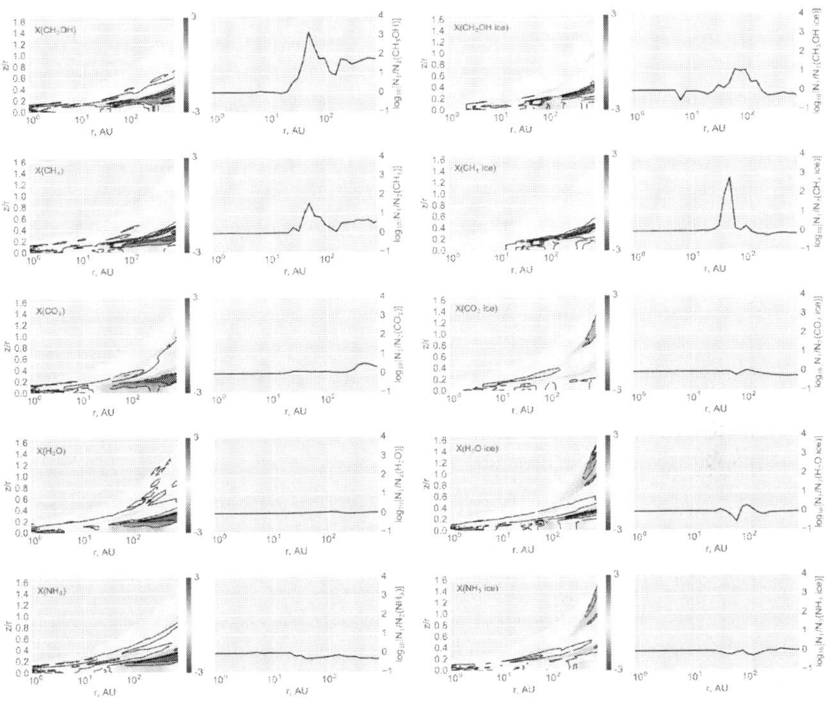

Figure. 2: The change in molecular abundance of a selection of species in a laminar solar nebula at 1 Myr, when using a chemical desorption efficiency of 0.01 (N_1) instead of 0.05 (N_2). The log of relative ratios are given both as function of location (height z and radius r) and integrated over z as a function of r. Relative gas phase abundances are in the *left panels*, the corresponding ice abundances in the panels on the *right-hand side*

REACTIONS

Generally, surface reactions are thought to occur via one of three mechanisms: the diffusive *Langmuir-Hinshelwood* mechanism, where both species move over the surface and react upon meeting, the stick-and-hit *Eley-Rideal* mechanism where one (stationary) reactant is hit by another species from the gas phase, and the *hot-atom* mechanism (which is a combination of both) where non-thermalized species travel some distance over the surface before finding a fellow reactant. Photodissociation is usually treated separately and will be discussed in Sect. 8. Under astrophysical conditions where ices are abundant, the gas and dust grains typically have similar temperatures, and the chemical timescales tend to be significantly longer than the thermalization timescale; hence, the hot-atom mechanism is often considered not important. The analytical expression to describe the Langmuir-Hinshelwood mechanism on a surface is

$$f_{\text{react,LH}}(i + j \longrightarrow \text{A}) = P_{\text{react,LH},i,j}(k_{\text{scan},i} + k_{\text{scan},j})n_s(i,t)n_s(j,t) \tag{11}$$

where ns(i,t) is the number of species ii present on the surface at time t and k_{scan},ikscan,i the rate by which species i scans the grain surface. The scanning rate is given by

$$k_{\text{scan}} = \frac{k_{\text{hop}}}{N_{\text{sites}}} \tag{12}$$

where N_{sites} is the number of binding sites per grain. For amorphous solid water with a density of 0.94 gcm−30.94 gcm^{-3}, the site density is 1×10^{15} molecules cm^{-2}. Simulations of CO_2, which is a bulkier molecule, on top of water ice showed a site density of 6×10^{14} molecules cm^{-2} (Karssemeijer et al. 2014a). Both lead to approximately $\approx10^6$ binding sites per grain for a standard grain of 0.1 μm.

The scanning rate determines the meeting frequency of the two particles i and j due to the mobility of one, or both, reactant(s). The Langmuir-Hinshelwood mechanism is dependent upon the abundances of both reactant and hence is a second-order process. The hopping rate, khopkhop, will be discuss in Sect. 6.

The rate coefficient, Preact,$_{LH,i,j}$, accounts for the probability that a possible reaction barrier will be crossed during the encounter. This probability is assumed to be 1 for a reaction with zero activation energy, and 0.5 if the two reactants are the same species. For reactions with an activation barrier, EaEa, that occur on a dust-grain surface with a temperature, TT, this probability is

$$P_{\text{react,LH},i,j} = \exp\left(-\frac{E_a}{kT}\right) \tag{13}$$

when the barrier is crossed through thermal activation as schematically depicted in Fig. 3 for the reaction H + H_2CO. In this case there is a clear transition state that determines the rate limiting energy barrier. Tunneling through the reaction barrier is also possible, greatly increasing the probability of reaction (see, e.g., Hasegawa et al. 1992). This occurs through delocalization of the transition state. As can be seen for the H--H_2CO complex, light species are much more delocalized and quantum-mechanical tunneling is hence of main importance for reactions where light species, e.g., H, D, are involved in the bond breaking or forming. Tunneling is discussed in more detail in Sect. 5.2.

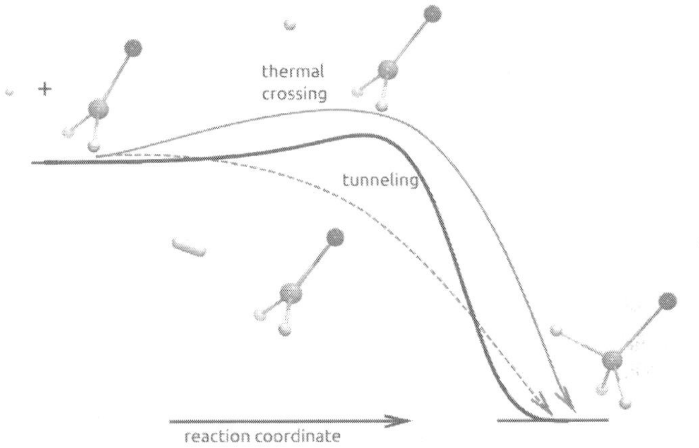

Figure. 3: Schematic representation of crossing a reactive barrier either through tunneling or through thermal activation. The $H+H_2CO \longrightarrow H_3CO$ reaction is used as an example.

Although conceptually simple, in reality the situation is more complex. First, a surface reaction may have several exit channels leading to a number of various products, similar to reactions in the gas phase. For most examples each of these channels will have its own transition state and corresponding (temperature-dependent) rate. The reaction constant does not need to include a special scaling to account for this effect, but the branching ratios αα are a natural outcome of the model

$$\alpha_l = \frac{k_l}{\sum_{j=1,m} k_j} \tag{14}$$

for the all possible mm reaction channels. In constructing a reaction network, one should be very careful when adding new product channels especially when the reaction rates come from very different sources (surface vs. gas phase experiments, computations) since some product channels might be heavily suppressed. For some reactions, only a destruction rate is known and branching ratios are determined separately. Individual product rate should in this case be adjusted accordingly.

Second, for a diffusive surface reaction to happen, the two molecules must remain adsorbed in close vicinity until they react, otherwise they can migrate away from each other or even desorb as schematically depicted in Fig. 4. Therefore, a surface reaction process is in competition with diffusion and desorption (Herbst and Millar 2008). Consequently, the reaction constant for product channel kk takes the following expression (see Equation 6 in Garrod and Pauly 2011),

$$P^{\mathrm{full},k}_{\mathrm{react,LH},i,j} = P^{k}_{\mathrm{react,LH},i,j} \frac{\nu \sum_{l=1,m} P^{l}_{\mathrm{react,LH},i,j}}{\nu \sum_{l=1,m} P^{l}_{\mathrm{react,LH},i,j} + k_{\mathrm{hop},i} + k_{\mathrm{hop},j} + k_{\mathrm{evap},i} + k_{\mathrm{evap},j}}, \quad (15)$$

where the k_{hop} and k_{evap} are the thermal hopping (scanning) and evaporation rates for the reactants i and j, consequently. In the majority of astrophysically relevant situations the evaporation terms are small in magnitude compared with the hopping terms and can be safely neglected.

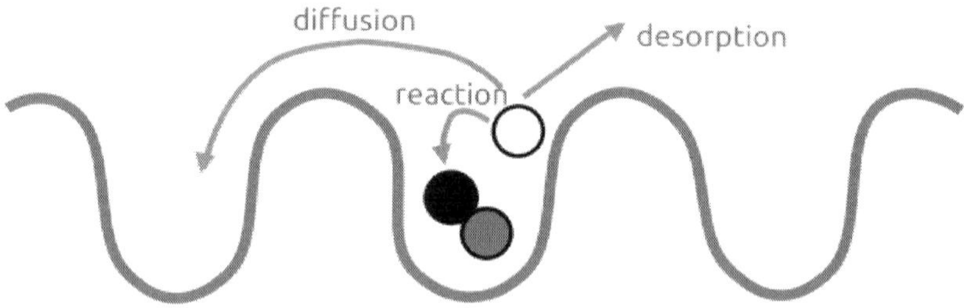

Figure. 4: Surface reactions can be attempted as long as the reactions are each others vicinity. Hence, reaction competes with diffusion and desorption. Faster diffusion will lead to a higher meeting rate but also in a shorter reaction time

The Eley-Rideal mechanism is considered to be important only for high surface coverages or low surface mobility (Ruffle and Herbst 2001). An example where this can become important is during catastrophic freeze-out of CO in prestellar cores. During this phase, a layer of reactive CO ice forms on the grains (Pontoppidan 2006). Under these circumstances, Eley-Rideal could be an important mechanism in the formation of methanol. It can be included in models by using the following expressions

$$f_{\mathrm{react,ER}}(i + j \longrightarrow \mathrm{A}) = P_{\mathrm{react,ER},i,j} f_{\mathrm{acc},i} n_{\mathrm{s}}(j,t) + P_{\mathrm{react,ER},i,j} f_{\mathrm{acc},j} n_{\mathrm{s}}(i,t). \quad (16)$$

The reaction constant is different for the Eley-Rideal mechanism. Here the two reactants have only one attempt to cross the reaction barrier E_a, so diffusion and desorption competition is of no importance. The corresponding rate coefficient is much simpler than in the LH-case (Eq. (15)):

$$P_{\mathrm{react,ER},i,j} = \alpha_l \exp\left(-\frac{E_a}{kT}\right). \tag{17}$$

Surface Experiments

Surface reactions can be monitored in the laboratory using an ultra-high vacuum setup (better than $\sim 10^{-9}$ mbar and H_2 as main gas residue in the chamber) and experiments are generally performed in two ways. Either the reactants are deposited in sequence, referred to as *pre-deposition* experiments, or in tandem in a so-called *co-deposition* experiment. The first gives a better control over the total dose and the predeposited amount can either be in the monolayer regime on top of an astrophysically relevant surface (e.g., a silicate or carbonaceous substrate), or a thicker ice if the reactant is a stable species, e.g., CO. For pre-deposition ice experiments, the final yield of the newly formed species for a selected radical fluence and ice temperature is largely limited by the penetration depth of the reactants in the ice. For the case of hydrogenation of CO ice, the maximal penetration depth of H atoms is four monolayers; therefore, only the upper layers of the ice are hydrogenated. Co-deposition experiments generally give a higher signal as they do not suffer from such penetration effects. Moreover, they are particularly useful in experiments involving radical species other than H atoms. Radicals are generally formed in a microwave discharge or RF (radio frequency) source in which the stable precursor is injected and subsequently dissociated. The dissociation products are then piped to the substrate, usually through a cooling pipe that thermalizes or cools the species to room temperature or even below. Generally, the dissociation is not 100 % efficient and recombinations of the radical species can occur in the cooling pipe. Depending on the initial stable precursor gas that is used, the desired radical might not be the only radical formed during the discharge which can lead to a beam of mixed composition. For hydrogen gas, this is H and H_2 where the latter sticks to the surface only in the monolayer regime. For oxygen gas, this is O and O_2 and in a predeposition experiment where first an ice is deposited which is then exposed to O atoms, O_2 has been seen to form a thick layer on top of the predeposited species depending on the surface temperature, which can inhibit any reaction.

Surface reactions are generally monitored by means of Fourier transform infrared spectroscopy and gas-phase molecules, generated by performing a post-experiment TPD, by mass spectrometry using a quadrupole mass spectrometer. With an *in-situ* technique like infrared spectroscopy, the amount of species formed can be obtained as a function of time if the band strength of the species is known. If the dose is also known, a formation rate in the experiment can then be determined. However, this depends on a combination of processes and their competition, i.e., diffusion, desorption, and reaction. Moreover, often several reactive species are present and a multitude of different reactions are possible at the same time. Thus, the reaction pathways and associated branching ratios can become nontrivial to disentangle as the systems studied increase in complexity. In addition, the characteristics of the substrate—its composition and surface structure—can affect reaction pathways and rates in the sub-monolayer regime. However, these data (rates, pathways, and branching ratios) are critical for advancing surface-chemical networks for use in astrochemistry. Simulating laboratory conditions can aid in extracting or constraining reaction data.

Although quantitative information on the reaction barrier or rate is not easily obtainable from surface experiments, experiments are extremely useful in detecting whether specific reactions can proceed under circumstances close to those present in the ISM and this is the aim of most experimental studies. It is also not trivial to *calculate* rates of surface reactions by quantum chemical methods. Surface reactions require many atoms in the calculation, which unequivocally increases the computational time. One of the approximations that can be made is the use of a model surface, e.g., a coronene molecule to model a carbonaceous grain surface (Adriaens et al. 2010), or a water cluster to mimic a thick ice layer (Rimola et al. 2014). Another promising method that can be used to include the effect of the surface on reaction rates is quantum mechanics/molecular mechanics (QM/MM, Goumans et al. 2009). Although such methods are robust for studying the effect of the surface on the reaction barrier height and shape, they cannot account for the adsorption of energy through phonon excitation. Therefore, they can result in different dynamical behavior of the products and thus, different rates.

Much more data is available for gas-phase reactions and a reasonable assumption for the construction of a surface reaction network is to lend from gas-phase data. As far as we are aware, there have been no reports of reactions that are efficient on the surface and not in the gas phase (or vice versa), except for association reactions of the form, A+B⟶AB. This type of reaction is very inefficient in the gas phase without the presence of a third body, since it has to proceed through radiative association. For surface reactions, the grain acts as a third body. For reactions on clusters, the efficiency of association reactions likely lies somewhere in between. Examples of reactions that have a high barrier in the gas phase but nevertheless proceed in the solid phase at low temperatures are $CO + H$, $H_2CO + H$, $H_2 + OH$ and $H + H_2O_2$ (see Tables 2 and 3). This is thanks to the possibility of many crossing attempts, energy dissipation for reactions of the type A+B⟶AB, and tunneling; although the latter is relevant in the gas phase as well.

Table 2 Astrochemically relevant surface reactions for which tunneling has been confirmed experimentally

	Reaction	Comment		Ref.
1	$H+CO \longrightarrow HCO$	H-addition		A
2	$H+CH_2O \longrightarrow CH_3O/CH_2OH$	H-addition		B
3	H+CkHl⟶CkHl+1H+CkHl⟶CkHl+1	H-addition	C_2H_2, C_2H_4, C_2H_6	C
		H-addition	C_6H_6	D[a]
4	H+CkHlOmNn⟶H2+CkHl−1OmNnH +CkHlOmNn⟶H2+CkHl−1OmNn	H-abstraction by H	CH_3OH[b]	E
			CH_3NH_2 [c]	F[a]
			Glycine[b]	G
5	$H_2+OH \longrightarrow H+H_2O$	H-abstraction by OH		H
6	$H+H_2O_2 \longrightarrow H_3O_2 \longrightarrow H_2O+OH$	H-addition		I[d]
7	$H+SiH_4 \longrightarrow SiH_3+H_2$ $SiH_3 \longrightarrow$ Si–Si⟶a–SiH[e]	H-abstraction		J[a]

[a]Based on high efficiency at low-T for high barrier process; no T-dependence or KIE studies

1. [b]Only on C-H sites
2. [c]On both C-H and N-H sites, but likely via a different mechanism
3. [d]See also additional comments in the Faraday Discussions
4. [e]Proposed mechanism
5. A—Hiraoka et al. (2002), Watanabe and Kouchi (2002), Hidaka et al. (2007); B—Watanabe and Kouchi (2002), Fuchs et al. (2009), Hidaka et al. (2009); C—Hiraoka et al. (1999, 2000); D—Hama et al. (2014); E—Nagaoka et al. (2007), Hidaka et al. (2009); F—Oba et al. (2014a); G—Oba et al. (2015); H—Oba et al. (2012); I—Miy-

auchi et al. (2008), Oba et al. (2014b); J—Hiraoka et al. (2001)

Table 3 Astrochemically relevant surface reactions for which tunneling has been studied theoretically

	Reaction	Comment	Method	TminTmin	Ref.
1	$H+CO \rightarrow HCO$	H-addition	HQTST	5 K	A
2	$CO+O \rightarrow CO_2$	O-addition	HQTST	10 K	B
3	$CO+OH \rightarrow CO_2+H$		SCTST	200 K	C
4	$H+CH_3OH \rightarrow CH_2OH+H_2$	H-abstraction by H	HQTST	30 K	D
5	$H+CH_4 \rightarrow CH_3+H_2$	H-abstraction by H	RPMD	200 K	E
6	$H+C_6H_6 \rightarrow C_6H_7$	H-addition	HQTST	40 K	F
7	$O+CH_4 \rightarrow OH+CH_3$	H-abstraction by O	RPMD, VTST/MT	200 K	G
8	$OH+H_2O \rightarrow H_2O+OH$	H-abstraction by OH	Q-RPH	200 K	H[a]
9	$OH+CH_4 \rightarrow H_2O+CH_3$	H-abstraction by OH	RPMD, VTST/MT	200 K	I
10	$OH+NH_3 \rightarrow H_2O+NH_2$	H-abstraction by OH	CVT/μOMT	200 K	J
11	$OH+H_2 \rightarrow H_2O+H$	H-abstraction by OH	SCTST	200 K	K
12	$OH+OH \rightarrow H_2O+O$		SCTST	200 K	L

[a]Small water clusters, (H2O)n(H2O)n, $n=1–3$

1. A—Andersson et al. (2011); B—Goumans and Andersson (2010); C—Nguyen et al. (2012), Weston et al. (2013); D—Goumans and Kästner (2011); E—Li et al. (2013); F—Goumans and Kästner (2010); G—Gonzalez-Lavado et al. (2014); H—Gonzalez et al. (2011); I—Suleimanov and Espinosa-Garcia (2015); J—Monge-Palacios et al. (2013a,b); K—Nguyen et al. (2011); L—Nguyen and Stanton (2013)

TUNNELING

Experiments have shown that a number of surface reactions proceed through quantum-mechanical tunneling, a small summary of which is shown in Table 2. An extensive explanation can be found in Hama and Watanabe (2013). Briefly, both hydrogen-addition and hydrogen-abstraction reactions are important processes that can occur through tunneling.

When tunneling is involved in "crossing" the reaction barrier, the description of the rate coefficient changes. To accurately describe tunneling, a full quantum-mechanical calculation is preferred and there are numerous methods available to calculate tunneling rates (summarized recently by Kästner 2014). However, such methods typically lie beyond the scope of most astrochemical models. Instead, the most common way to account for tunneling in models is by using the Wentzel-Kramers-Brillouin approximation and the (crude) assumption of a rectangular barrier, which leads to a rate constant,

$$k_{\mathrm{react},i,j} = \nu \exp\left(-\frac{2a}{h}\sqrt{2\mu E_{\mathrm{a}}}\right), \tag{18}$$

where ν is an attempt frequency (as described previously), aa is the width of the barrier, h is Planck's constant, μ is the effective mass of the system and E_a is the reaction barrier. Temperature no longer plays a role, in contrast with the rate coefficient for thermally-activated reactions, and the reaction probability increases with decreasing reduced mass and barrier width.

For a particular reaction system, below the so-called cross-over temperature, the contribution of quantum-mechanical tunneling to the reaction rate dominates over the thermal contribution. Tunneling can only occur through the exothermic part of the barrier. For an endothermic reaction, the rate coefficient described by Eq. (18) should be scaled by a Boltzmann factor accounting for the difference in energy between the reactant and product states, following arguments of detailed balance

$$k_{\mathrm{react},j,i} = \nu \exp\left(-\frac{2a}{h}\sqrt{2\mu E_{\mathrm{a}}}\right) \exp\left(-\frac{\Delta E}{kT}\right) \tag{19}$$

where E_a and ΔE are defined in Fig. 5. This decreases the reaction rate considerably, and can potentially significantly alter the outcome of large astrochemical models (e.g., Lamberts et al. 2014a).

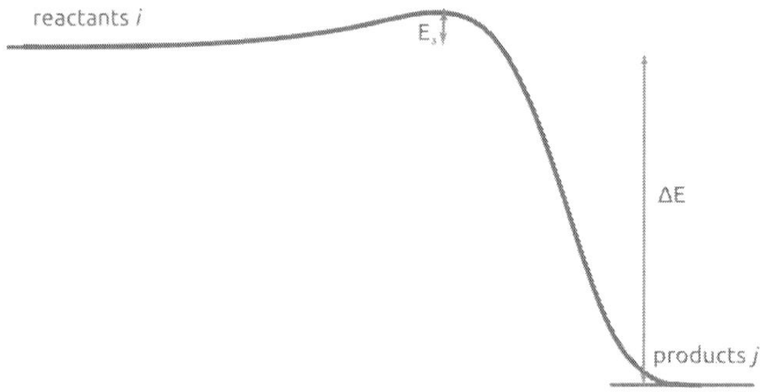

Figure. 5: Energy profile of a reaction the forward reaction is exothermic; the backward reaction is endothermic and its rate should be described by Eq. (19)

First of all, the width of the barrier, aa, in the simple approximation mentioned above, is often not known and therefore commonly approximated as 1–2 Å (Garrod and Herbst 2006; Furuya et al. 2013; Walsh et al. 2014), initiated by older literature such as Tielens and Hagen (1982) and Hasegawa et al. (1992). However, the shape of the barrier can also have a significant effect on the tunneling transmission coefficient. Eq. (18) assumes a rectangular barrier, i.e., with equal initial and final energies, which is not the case for most chemical reaction systems. A computationally cheap method to improve on this rectangular shape is using the Eckart model for the shape of the potential (Eckart 1930). However, this requires knowledge of the forward and backward reaction barriers, as well as the imaginary frequency of the transition state. Therefore, (quantum) calculations need to be available, which is typically the case only for the gas phase. Recently, Taquet et al. (2013) demonstrated the significance of using the Eckart model versus a rectangular barrier on the calculation of the rates for a set of surface reactions, using gas-phase theoretical data as input.

Secondly, let us consider the mass dependence, which can be intuitively understood by considering the uncertainty principle $\Delta x \Delta p > \hbar/2$.. A large mass will result in a large Δp, causing the accuracy of the position, ΔxΔx, to be better defined. Hence, a particle is more localized and is less likely to "leak" some of its probability density into or through a barrier. Eq. (18) requires the effective mass, μμ, along the reaction coordinate, which for a simple bond forming reaction can be approximated by,

$$\mu = \left(\frac{1}{m_1} + \frac{1}{m_2}\right)^{-1} = \frac{m_1 m_2}{m_1 + m_2},$$
(20)

where m_1 and m_2 are the masses of the *individual atoms* forming the bond. However, in astrochemical models, generally the reduced mass of the *system* is used,

$$\mu = \left(\frac{1}{m_A} + \frac{1}{m_B}\right)^{-1} = \frac{m_A m_B}{m_B + m_A},$$
(21)

where are A and B are the reacting species, regardless of the reaction mechanism. For abstraction reactions A+XB⟶AX+BA+XB⟶AX+B, the situation becomes more complex and depends on the incoming angle. The expression that can be applied to calculate the reduced mass for linear systems A–X–B is

$$\mu = \frac{m_A m_X (1 + c)^2 + m_X (m_A + c^2 m_B)}{(m_A + m_X + m_B)(1 + c^2)}.$$
(22)

Here, cc is the ratio between the infinitesimal change in bond length of R_{AX} and R_{XB}. For reactions where species A and B have equal mass, or both have masses much larger than that of X, μ is between $\frac{1}{2}m_X$ and m_X. In general, however, this depends on the forces acting on the system. More extensive theoretical literature can be found in Bell (1980).

As an example, the reduced mass for abstraction reactions by OH, e.g.,

$$A\text{–}H + OH \longrightarrow A \cdot + H_2O,$$
(23)

should reflect that the tunneling species is the hydrogen atom and the effective mass should be μ≈mH. However, using Eq. (21) will yield a much higher mass and this has a large influence on the reaction rate. Consider, for example, a reaction with a barrier E_a of 2000 K and a width of 1 Å. Using Eq. (18) with two different values for μ ($\mu_1 = 10$ and $\mu_2 = 1$ a.u.) yields $k_1/k_2 = 2 \times 10^{-3}$.

Typically, gas grain codes use information on barrier height and the masses of the products to calculate the rates. We recommend to include the tunneling reaction rate as obtained from high-level gas-phase calculations as an additional input parameter in an astrochemical model. When such rates are unavailable, the original approximate Eq. (18) can be reinstalled.

The importance of tunneling can be experimentally confirmed in two ways: either by studying the temperature dependence—to determine whether this confirms the predicted behavior of either Eq. (15), (17) or (18)—or by making use of the mass dependence of tunneling. In this latter case, an experiment is performed (at least) twice, using different isotopes of the same species between the two (or more) runs. Since isotopes are chemically equivalent, but have a different mass, this is a convenient way to detect the difference in tunneling efficiency. If the reaction products scale with the effective mass of the reaction, this is usually interpreted as a proof of tunneling and is called the Kinetic Isotope Effect (KIE). For instance, a reaction with a hydrogen atom, H+X⟶HX, can be compared to the same reaction with a deuterium atom, D+X⟶DX. In both cases, the experiments are performed at temperatures as low as possible, which is typically setup-dependent.

In order to study the influence of tunneling on a reaction with a relatively high precision from a computational perspective, it is necessary to use methods that encompass more than only an E_{ckart} calculation of the barrier, or a Wigner correction (Wigner 1938) to the rate. For instance, a variety of Transition State Theories (TST) have been applied, such as Variational TST (or Canonical Variational Theory, CVT) combined with a Multidimensional Tunneling correction (VTST/MT), Semi-Classical TST (SCTST), and Harmonic Quantum TST (HQTST). Other methods, such as Quantum-Reaction Path Hamiltonian method (Q-RPH), Ring Polymer Molecular Dynamics (RPMD), and Free Energy Instanton Theory (FEIT) have also been employed. Several recent papers comment on the differences and accuracy of these various methods with respect to each other (Nyman 2014; Zhang et al. 2014; Hele et al. 2015). In Table 3, a number of gas-phase studies are listed that focus on reactions that may be important in surface astrochemistry. However, this is by no means an exhaustive list and focuses only on recent papers. Note that we have excluded the reaction H_2+O⟶OH from this table, based on its endothermicity (Lamberts et al. 2014a) and on the high barrier and, consequently, low reaction rate (Nguyen and Stanton 2014).

Temperature Dependence of Surface Reactions

Although many reaction rates are temperature independent, because they are either barrierless or occur through quantum tunneling, reactions still have a temperature window within which they are most effective. This is a consequence of the temperature dependence of the diffusion and desorption of reactants. Diffusion is required for reactants to meet in a Langmuir-Hinshelwood reaction and determines the lower bound of the temperature window; desorption, on the other hand, sets the upper bound. As a result, at low temperatures, surface chemistry will mostly be dominated by hydrogenation reactions,

$$X\cdot + H \longrightarrow X\text{--}H \qquad (24)$$

effectively converting all radical species to more stable species such as H_2O, CH_4, NH_3, and CH_3OH. In parallel, hydrogen abstraction can recreate surface radicals,

$$X\text{--}H + Y\cdot \longrightarrow X\cdot + Y\text{--}H, \qquad (25)$$

but most likely with a much lower efficiency. Whereas hydrogenation reactions have been extensively studied, for hydrogen abstraction reactions, the reaction rates and the importance of tunneling remain unknown.

At higher temperatures, the residence time of hydrogen atoms on the surface becomes too short for hydrogenation reactions to dominate, and other surface reaction types increase in importance. For example, high-mass star-forming cores, that are in the so-called "hot core" phase, are found to be rich in gas-phase complex organic molecules, including alcohols, aldehydes, carboxylic acids, esters, and ethers (Bisschop et al. 2007b). While several of these classes of molecules may have viable gas-phase formation mechanisms, the larger species are thought to be formed primarily (or exclusively) on the surfaces of dust grains, or within/upon dust-grain ice mantles (Garrod and Herbst 2006; Garrod et al. 2008). The main reaction mechanism for the formation of complex organics on the grains is the creation of functional-group radicals on/within the ice mantle, such as CH_3 and CH_3O, both of which may be formed via the photodissociation of methanol (CH_3OH). As the temperature in the core increases to above 20 K20 K or so, these radicals become mobile, diffuse across or through the ice mantle, thereby allowing radical-radical association reactions to become competitive with hydrogenation of radicals by abundant H atoms. As temperatures rise yet further ($\gtrsim$100 K), the grain-surface-formed molecules desorb into the gas phase, where they are detected typically via (sub)mm rotational spectroscopy.

Ubiquitous molecules, including methyl formate ($HCOOCH_3$) and dimethyl ether (CH_3OCH_3), appear to have efficient dust-grain surface formation routes. Astronomical observations and astrochemical modeling (e.g., Belloche et al. 2014) indicate that surface/ice-mantle formation mechanisms may also be efficient for very large molecules like propyl cyanide (C_3H_7CN); however, in such cases, the earlier formation of smaller homologues (e.g. ethyl cyanide, C_2H_5CN) is usually required. The removal of a hydrogen atom from these molecules produces the necessary large radicals, to which other functional groups may be added, further increasing chemical complexity.

The most abundant molecule in the ices is water itself, which may be photodissociated to form the highly reactive OH radical. At relatively low temperatures, most OH produced recombines with H, or with mobile, abundant radicals like CH_3 and HCO. However, models suggest that above around 60 K (Garrod 2013), OH becomes sufficiently mobile to find and react with large stable molecules, to abstract an H atom and produce a large radical. At this stage, H-abstraction by OH (along with NH_2, which is also formed by H-abstraction from ammonia) becomes the dominant formation mechanism for the molecular radicals that are the precursors to even larger species on the grains.

While relatively few of the H-abstraction reactions of OH invoked in this dust-grain chemistry have been directly measured, even in the gas phase, those for which rates have been determined display very small activation energy barriers to H abstraction by OH, typically less than 1000 K. Crucially, barriers of this size are comparable to the expected diffusion barrier for surface OH, meaning that for these reactions, the surface diffusion of OH is the rate-limiting step. Thus, as soon as temperatures are sufficiently high for OH diffusion, OH becomes the key instigator of radical production.

Unfortunately, few of these barriers to H-abstraction by OH from large, saturated molecules have been determined, and the calculation of rates in models is further complicated by the fact that the abstracted hydrogen atom may be able to tunnel through the barrier, meaning that information about the barrier shape is required. Alternatively, if tunneling is efficient, then in many cases the key quantity needed in astrochemical gas-grain calculations will be the diffusion barrier for OH. However, an accurate determination of each one of these quantities will be necessary for a full understanding of how complex organic molecules form in different temperature regimes in interstellar regions.

The rather elegant "warm-up" scenario for the origin of so-called "hot core" molecules has been muddied by recent observational and laboratory results. High sensitivity observations have shown that complex organic molecules are also present in the gas phase in cold environments (~ 10 K), albeit at a low level with respect to molecular hydrogen (Öberg et al. 2010, 2011; Cernicharo et al. 2012; Bacmann et al. 2012; Vastel et al. 2014). Bisschop et al. (2007a) investigated the laboratory hydrogenation of solid acetaldehyde, CH_3CHO, under interstellar relevant conditions, using both RAIRS and TPD to analyze the results. The experiments showed that the hydrogenation of CH_3CHO leads to the formation of 20 % of ethanol, C_2H_5OH, showing for the first time that surface hydrogenation of unsaturated complex species can be responsible for the abundances of more complex saturated species detected in dense interstellar clouds. In addition, it has now been demonstrated experimentally that molecules such as glycolaldehyde ($HOCH_2CHO$) and ethylene glycol ($HOCH_2CH_2OH$) can form at 10 K *without* the need for irradiation (Fedoseev et al. 2015). Here, the reaction scheme also relies mainly on hydrogenation reactions which are known to be efficient at low temperatures; however, the scheme also involves dimerization of the HCO radical which allow the formation of the C–C backbone (see also Woods et al. 2013). These results suggest that complex molecules are already present in the ice mantles prior to warm up in the environs of young stars.

A few alternatives to the UV photo-induced surface chemistry hypotheses involve gas-phase reactions (Rawlings et al. 2013; Balucani et al. 2015). The so called "Rapid-Radical Association" mechanism proposes that COMs are formed by three-body gas-phase reactions between radicals in warm high density gas (Rawlings et al. 2013). This environment exists, for a very short period of time, following the sudden and total sublimation of dust-grain ice mantles driven by the catastrophic recombination of trapped hydrogen atoms, and other radicals, in the ice. More recently, Balucani et al. (2015) proposed a new gas-phase model to form some of the COMs, such as dimethyl ether and methyl formate, starting from methanol in the gas phase. In the proposed scheme, dimethyl ether is the precursor of methyl formate via an efficient reaction overlooked by previous models. Very recently, electronic structure and kinetic calculations showed that the gas-phase reaction NH_2+H_2CO leading to formamide is barrierless. Hence, for some species, there is no need to invoke grain-surface chemistry provided that the necessary precursors are available in the gas phase (Barone et al. 2015). The main complicating factor in deciphering the chemistry of COMs is that their abundances and abundance ratios usually vary significantly from source to source. Moreover, not all the detected COMs and their abundances can be explained by gas-phase reactions alone. Therefore, it is likely that surface reactions on ice grains still play an important role in the dense regions of the ISM.

Figure 6 taken from Linnartz et al. (2015) summarizes the extensive laboratory work that investigated the surface formation of several simple and more complex species through atom-addition and radical-radical recombination reactions under dense cloud conditions. The figure shows how the surface reaction routes to H_2O, CO_2, HCOOH, CH_3OH, $(CH_2OH)_2$, HNCO, NH_3, and NH_2OH, are interlinked and are initiated by the H/O/N-atom addition to simpler precursor species, like CO molecules. The efficiency of these reaction routes, and their contribution to the molecular abundances observed in space, depends on a number of physico-chemical parameters. For instance, in atom addition/abstraction reactions, the height of an activation barrier determines whether or not a reaction can proceed at 10 K. Radical-radical recombination reactions are barrierless and therefore their inefficient thermal diffusion at 10 K is the limiting factor. Another important parameter is the molecular environment. In the case of exothermic reactions, polar (water-rich) ices can promote surface chemistry through the dissipation of extra energy in their H-bond network.

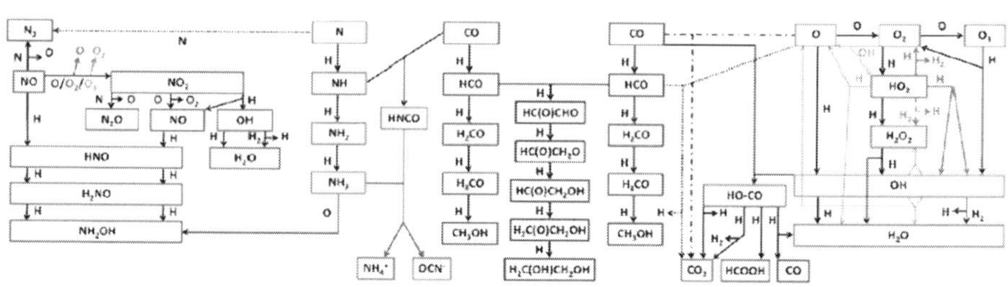

Figure. 6: A summary of non-energetic surface chemistry through atom-addition and radical-radical recombination reactions based on experimental evidence. The *arrows* indicate possible pathways, but other (energetic) processes are at play as well. The figure clearly shows the complexity of non-energetic ice chemistry and the possibility for this type of chemistry to create complex molecules without additional energy input. Figure reproduced from Linnartz et al. (2015)

DIFFUSION

As discussed in the previous section, diffusion rates are key to determining the rate for the Langmuir-Hinshelwood reaction mechanism, because they regulate the meeting frequency between reactants. Models require as input diffusion barriers and binding energies for each of the surface species included in the reaction network to determine the hopping rate

$$k_{\mathrm{hop,th},i} = \nu \exp\left(-\frac{E_{\mathrm{diff}}}{kT}\right). \tag{26}$$

The scanning rate kscankscan is determined by either thermal hopping or quantum mechanical tunneling, depending on the mass of the diffusing species.

In early models, H atoms were assumed to diffuse via quantum tunneling. After experimental studies showed that the diffusion of H atoms is rather slow, thermal hopping became favored in models (Pirronello et al. 1997b,a, 1999; Katz et al. 1999). More recently, the discussion on the nature of the diffusion mechanism for atoms was reopened, with experiments suggesting H atoms can quantum tunnel under particular conditions (Watanabe et al. 2010; Kuwahata et al. 2015). It has also been postulated that atoms as heavy as oxygen can diffuse via quantum tunneling (Congiu et al. 2014). Heavier species are most likely to diffuse through thermal hopping.

Obtaining diffusion barriers experimentally remains challenging. Diffusion rates cannot be measured directly and have to be inferred from experiments using a model. Often diffusion barriers are measured through reaction, where a known dose of both mobile reactants and relatively immobile (stationary) reactants are deposited. If the reaction between species is diffusion limited, i.e., there exists no reaction barrier and in the regime of low surface coverage, the diffusion rate can be inferred from the disappearance of the reactants as a function of temperature, dose, and/or time. This method is limited to reactive species and, because one works within the submonolayer regime, sensitivity of detection is an issue. Usually products are measured using mass spectrometry during TPD. Another method is to deposit a layer of ASW (amorphous solid water) on top of an ice consisting of the species of interest. If the temperature is raised above the desorption temperature of this species, the rate limiting step for desorption is the diffusion of the species through the ASW layer. In the case of porous ASW, pore-wall diffusion is probed which mimics surface diffusion. For compact ASW, bulk diffusion is more likely studied. Here, the disappearance of the diffusing species is monitored, usually by IR spectroscopic techniques. This method is limited to stable species with a desorption temperature below that of ASW. Surface diffusion rates for CO molecules have been studied this way (Mispelaer et al. 2013; Karssemeijer et al. 2014b; Lauck et al. 2015).

For both methods of study, the model that is used to extract the diffusion rates is crucial. In the case of reaction, one has to ensure that no other reactions play a role. Radical beams are often not 100 % pure and other species will also be present on the surface. Moreover, species might not be instantaneously thermalized when deposited on the substrate, so that they are able to move some distance superthermally. One has to be sure that the second reactant is indeed stationary. Furthermore, the effect of the warm-up phase during TPD needs to be considered which can thermally enhance diffusion (and subsequent reaction). In the case of the two-layer experiments, care should be taken when considering to which type of diffusion the system is limited: bulk diffusion, pore wall diffusion, or surface diffusion upon a layer of the same species adsorbed on the ASW (see Fig. 7).

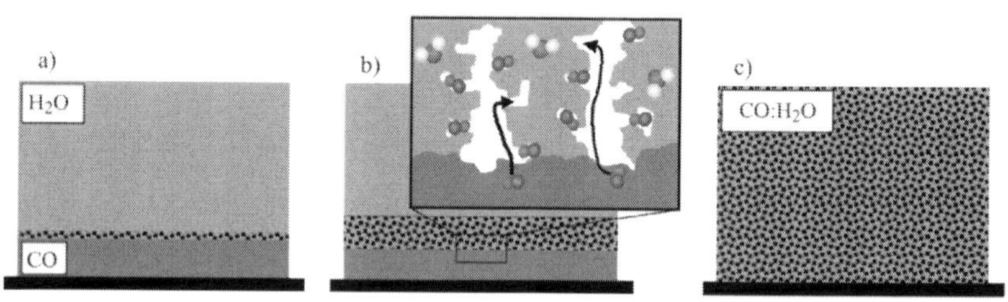

Figure. 7: Schematic of the mixing process. (**a**) Layered system at t=0t=0. (**b**) Occurrence of mixing, with the inset showing the CO molecules diffusing along the micropore surfaces into the strong-binding nanopore sites. (**c**) After some period of time, the layer becomes fully mixed. Figure taken from Lauck et al. (2015)

The species that has been most studied to determine its surface diffusion is atomic hydrogen. Because it is a radical, it is studied through reaction, either with itself to form H_2, with D to form HD, or with O_2 to from H_2O_2, and ultimately H_2O. However, the results between different experimental groups are not in agreement (see Hama and Watanabe 2013, for a review). The latest results on an ASW surface show that very shallow-potential binding sites ($(E_{diff} \leq 18 \pm 2$ meV)) are dominant, while there are also middle- ($E_{diff} = 22 \pm 1$ meV) and deep- ($E_{diff} = 30$ meV) potential binding sites (Watanabe et al. 2010; Hama et al. 2012). These particular results were obtained with a new technique for the detection of product species. The remaining species are photodesorbed by a laser as a function of delay time between deposition and laser desorption and the detection occurs through REMPI (resonance-enhanced multi-photon ionization). This eliminates the effect of warm-up which can be an issue for TPD experiments. These experiments support the following picture (Watanabe et al. 2010; Hama et al. 2012; Kuwahata et al. 2015): H atoms land on the ASW surface, diffuse rapidly using the abundant shallow and middle sites, and are finally trapped in the deepest sites. Once a significant number of H atoms become trapped, the subsequent H atoms recombine with the trapped H atoms. Therefore, the effective diffusion rate becomes dependent upon coverage and is also affected by the number of deep binding sites present on the ice surface.

This picture is also observed at an atomistic level through adaptive kinetic Monte Carlo simulations of CO diffusion on an ASW surface (Karssemeijer et al. 2014b). Here, for a single CO molecule on an ASW surface, the diffusion is limited by diffusion out of the deep binding sites (ECOdiff=84--114 meVEdiffCO=84--114 meV), whereas when the deep sites are filled with CO molecules, the diffusion rate is increased (ECOdiff=48--79 meVEdiffCO=48--79 meV). In the simulations, a large spread in diffusion barriers between different surfaces is found, in addition to the large spread observed for each specific surface. However, given a specific surface and surface coverage, the temperature-dependent diffusion constant follows an Arrhenius-like behavior, i.e., it can be described by a single diffusion barrier, which is consistent the diffusion barrier for moving out of the deepest available binding site. This is good news for astrochemical gas-grain models because, for simplicity, they typically do not account for local chemical or structural differences on the grain. Instead, these models aim to provide a macroscopic view. For this purpose, a single diffusion barrier per species likely suffices. A collection of different barriers and binding energies obtained in this way is presented in Table 4.

Table 4 Desorption and diffusion energy barriers for CO and CO_2 **on the proton-disordered and the ordered Fletcher phase of ice Ih**

Substrate	**Adsorbate**	EbindEbind [a] **(meV)**	EdiffEdiff **(meV)**	*f*	**Approach**	**Method**	**Ref.**
Disordered	CO	143	49	0.34	comp.	AKMC	A
(frozen)	CO_2	299	127	0.42	comp.	AKMC	B
	H_2O	670	220	0.33	comp.	AKMC	C
Fletcher	CO	133	39	0.29	comp.	AKMC	B

(frozen)	CO_2	291	110	0.38	comp.	AKMC	B
Fletcher	CO	137	42	0.31	comp.	AKMC	B
(free)	CO_2	323	122	0.38	comp.	AKMC	B
ASW (frozen)[b]	CO (1+3)[c]	96	147	0.42	comp.	AKMC	D
ASW (frozen)[b]	CO (1+6)[c]	86	134	0.36	comp.	AKMC	D

[a]Binding energies are time-averaged over the kinetic Monte Carlo runs at temperatures of 25 and 70 K for CO and CO_2

[b]Amorphous solid water (ASW) values are calculated from simulations published in Karssemeijer et al. (2014b) on substrate Sc2S2c. Values are also at T=25 KT=25 K

[c]There is only one mobile CO molecule on the substrate. The remaining 3 or 6 CO molecules are immobilized in strong binding pore sites on the substrate

A—Karssemeijer et al. (2012); B—Karssemeijer and Cuppen (2014); C—Pedersen et al. (2015); D—Karssemeijer et al. (2014b)

Table 4 is far from complete, and to obtain the remaining diffusion barriers required for the astrochemical models, the diffusion energy of a species is assumed to be a universal, fixed fraction, f, of the desorption energy: $E_{\text{diff},i} = f E_{\text{bind},i}$. There is no fundamental physical argument for such a universal ratio to exist and it is used solely due to the lack of data. This ratio is most likely dependent upon the species, the substrate material and structure, and on the surface coverage (which determines the likely binding "partner"). There already exist problems with the concept of using a single diffusion and a single desorption barrier even for a single species because they both vary strongly from site to site, especially for amorphous ices. As shown by Vasyunin and Herbst (2013) the value of f influences the outcome of the models. Due to the lack of diffusion information, the existence and possible value of fraction f is very poorly constrained and values between 0.3 and 0.8 are used by the modeling community and is often treated as a free parameter within these typical bounds (Hasegawa et al. 1992; Ruffle and Herbst 2000; Cuppen et al. 2009). An accurate value of f is also important because it affects the conditions under which the accretion limit is reached and thereby whether or not modelers should turn to stochastic models. The data in Table 4 suggests that, at least for stable species, there is a more or less constant ratio and that this ratio, f, is more likely to lie around 0.3--0.40.3--0.4. It is not possible, currently, to give a definitive recommended value. We encourage the experimental and theoretical community to continue to work towards filling this large gap in the necessary input data for astrochemical models, especially for radical species, which, to date, have remained largely unstudied.

BULK PROCESSES

Although the chemistry on interstellar grains starts out with processes on bare carbonaceous or silicate grains, when the ice thickness increases to more than a few monolayers, it becomes important to differentiate between surface and bulk processes. The earliest models of grain-surface chemistry primarily considered the grain surface as a substrate for the formation of molecular hydrogen or other simple species, which could then rapidly desorb back into the gas phase upon recombination (e.g., Watson and Salpeter 1972a; Allen and Robinson 1977; Tielens and Hagen 1982). Hence, originally, ice chemistry was modeled using rate equations that consider only the averaged abundance of a species throughout the entire mantle (Hasegawa et al. 1992). This is a rather crude approximation because bulk species cannot diffuse as easily as those on the surface. "Bulk ice" implies that the species involved in "bulk" processes are fully surrounded by neighboring molecules and are therefore rather tightly bound, leading to the assumption that diffusion within the bulk ice at low temperature is inefficient and therefore, chemistry is also inhibited. The ice mantle can thus be conceptually divided into an ice surface and bulk ice. To this end, three-phase models have been introduced, where the distinction is made between gas-phase species, reactive surface species within the top (few) monolayer(s), and fully inert bulk species in the core of the ice mantle, with terms that allow surface material to be incorporated into the bulk (or vice versa), as the ice thickness increases or decreases (Hasegawa and Herbst 1993b; Garrod and Pauly 2011). Each phase is treated mathematically as an independent entity. More recent models, informed by the discovery of substantial ice mantles on interstellar dust grains via infrared absorption observations, have involved a more concerted effort to treat the build-up of surface ices (e.g., Hasegawa et al. 1992). It is in the extension of these models to the formation of multiple layers of ice that a significant conceptual error should be identified. The absolute reaction rates for surface processes are commonly formulated thus:

$$f_{\text{react}}(A+B) = n_s(A)n_s(B)\left[k_{\text{hop}}(A)+k_{\text{hop}}(B)\right]/N_{\text{sites}} = n_s(A)n_s(B)k_{\text{scan}}(AB) \tag{27}$$

where k_{hop} is the site-to-site hopping rate for either species A or B, and N_{sites} is the number of surface sites present on the grain surface. $n_s(A)$ and $n_s(B)$ are used here as a shorthand to represent the mean populations of each species, $\langle n_s(A)\rangle$ and $\langle n_s(B)\rangle$ This formulation is often further simplified, absorbing all coefficients into a single rate, k_{scan}(AB). The efficiency factor used for reactions involving activation energy barriers is omitted here for simplicity.

Equation (27) is valid in the case where the sum of all surface reactants, which we label nall, is smaller than N_{sites}, i.e. that there is less than one layer of particles on the grain surface. However, if $n_{\text{all}} > N_{\text{sites}}$, the equation is necessarily invalid. This may be seen more clearly if Eq. (27) is re-arranged:

$$f_{\text{react}}(A+B) = n_s(A)k_{\text{hop}}(A)n_s(B)/N_{\text{sites}} + n_s(B)k_{\text{hop}}(B)n_s(A)/N_{\text{sites}} \tag{28}$$

It may be seen that the reaction rate is composed of two analogous parts. The first represents the rate at which all surface species of type A may hop into an adjacent site, multiplied by the probability that the neighboring site is occupied by a species of type B. The second term may be described similarly. This latter probability is the simple ratio $n_s(B)/N_{\text{sites}}$.

Clearly, if $n_s(B)$ exceeds N_{sites}, such a probability is meaningless. More rigorously, this is also true where $n_{\text{all}} > N_{\text{sites}}$.

A majority of two-phase models (i.e. gas-phase and grain surface) retain the usage of NsitesNsites in the reaction rates. In the case of an ice mantle with a thickness of, say, 100 monolayers, this could lead to a rate that is inaccurate by two to four orders of magnitude.

The simplest fix for this problem is to exchange N_{sites} for n_{all}, which may be calculated easily within the model code, for cases where $n_{\text{all}} > N_{\text{sites}}$. This approach removes the immediate inconsistency in the probabilities; however, it retains another assumption implicit in the equations above, namely that not only are the species in all layers of the ice chemically active, but that all such reactants may diffuse within the bulk ice at the surface diffusion rate. Neither of these assumptions should necessarily be used, as explained elsewhere in this paper.

To remove the latter problem while retaining the structure of a two-phase model, one may adjust Eq. (28) further, by retaining NsitesNsites, but substituting ns(A)ns(A) for nmax(A)nmax(A), where:

$$n_{\max}(A) = n_s(A); \quad \text{for } n_{\text{all}} \leq N_{\text{sites}} \tag{29}$$

$$n_{\max}(A) = n_s(A) N_{\text{sites}} / n_{\text{all}}; \quad \text{for } n_{\text{all}} > N_{\text{sites}} \tag{30}$$

And similarly for $n_{\max}(B)$ giving

$$f_{\text{react}}(A + B) = n_{\max}(A) n_{\max}(B) \left[k_{\text{hop}}(A) + k_{\text{hop}}(B) \right] / N_{\text{sites}} \tag{31}$$

This solution removes the error in the probabilities, while limiting the reactive portion of the ice mantle to a single layer, on the assumption that a layer is composed of a total of NsitesNsites particles. Such an approach may be considered a quasi-three-phase model, although the chemically-active portion of the mantle, which is assumed to be the surface layer itself, nevertheless represents the composition of the entire mantle. Only a fully realized three-phase model, consisting of entirely separate phases for the gas, ice mantle, and ice-surface layer can resolve this point. However, the added complexity of the physical treatment, combined with technical challenges (see below), makes three-phase modeling of astrophysically complex objects such as protoplanetary disks including bulk chemistry a more distant prospect, as outlined in Sect. 2.

Another approximation that can be made is the multilayer approach by Taquet et al. (2012, 2013), where the abundances of each single layer are stored in the simulation and these compositions are used to determine the binding energy of the species in the overlying layer. Again only the top monolayer is considered chemically active. Furthermore, recently another multilayer model (using a Monte Carlo approach) has been reported where the chemically active surface is extended to the uppermost four monolayers (Vasyunin and Herbst 2013). Both models account for a correct description of thermal desorption, but still leaves the bulk reactively inert. In microscopic Kinetic Monte Carlo routines, on the other hand, bulk reactivity can be included, for instance through allowing two neighboring molecules to swap position (Öberg et al. 2009d) or incorporating interstitial positions between the molecules in the lattice (Lamberts et al. 2013; Chang and Herbst 2014).

A number of experimental and theoretical studies indicate that bulk processes can be important, both at low and high ice temperatures. Here we make the distinction between bulk diffusion and pore-wall diffusion, since the latter is essentially surface diffusion. Experimentally it is not always straightforward to distinguish the two effects. Bulk diffusion appears to play a role in the segregation of mixed H_2O:CO_2 and H_2O:CO ices which leads to the occurrence of "islands" of both mixture components at temperatures of 30 K for CO and 60 K for CO_2 (Öberg et al. 2009d). Exposure of an O_2 ice to hydrogen atoms yields large quantities of H_2O_2 and H_2O, corresponding to tens of monolayers depending on the temperature (Ioppolo et al. 2010). For hydrogenation of CO, a maximum exposure of four monolayers was found experimentally (Fuchs et al. 2009). Furthermore, Molecular Dynamics simulations performed by Andersson et al. (2006) and Arasa et al. (2010) of photodissociation and photodesorption of water ice at low temperature (10–90 K) have shown that the desorption of species (H, OH, H_2O) depends strongly on the distance of the excited molecule to the top monolayer of the ice. The first 4–5 monolayers are, however, taking part in the desorption process, which indicates again that the assumption of a single reactive top monolayer cannot be valid. The desorption depth dependence has been observed for a range of excitation energies, ice temperatures and isotope compositions (Andersson et al. 2006; Andersson and van Dishoeck 2008; Arasa et al. 2010; Koning et al. 2013; Arasa et al. 2015). The studies further show that radical species which are created through photodissociation remain in the bulk of the ice, can use their excitation energy to move some short distance before they thermalize. This will most likely generally hold for radicals formed through an energetic process like photodissociation, exothermic reactions and through cosmic rays.

An analog for hydrogen diffusion at low temperature can also be brought about by reactions that have a net neutral effect, of the type $\mathrm{H{-}A + A \longrightarrow A + H{-}A}$ as mentioned during the Lorentz Center Workshop by Fedoseev. This could be of interest, for instance, in the case of water ices where photodissociation can create OH radicals that may react with neighboring H_2O molecules, $H_2O + OH \longrightarrow OH + H_2O$ (Nguyen and Stanton 2013; Lamberts et al. 2016).

Recently, the discovery of a new class of thin films, *spontelectrics*, may also have interesting implications for diffusion and ordering in ISM ices (Field et al. 2013). Upon deposition onto a substrate at low temperatures, dipolar molecules were seen to spontaneously align, creating electric fields >108 eVm−2>108 eVm^{-2}. The impact of such spontaneous ordering within ISM-like ices is yet to be quantified. An alternative mechanism that could work at high temperatures, is to allow the reactants to reach each other before the ice is close to evaporating, leaving the mantle molecules more freedom to diffuse.

Diffusion through the walls of macropores is less relevant for interstellar ices with respect to laboratory analogs. Interstellar water ice is mostly formed through reactions instead of deposition resulting in compact ice (Oba et al. 2009) by the exothermicity of surface reactions (Accolla et al. 2011). Moreover, laboratory experiments of deposited ices indicate that the macropores present in these ices collapse at least to a certain extent (Bossa et al. 2012; Mispelaer et al. 2013; Isokoski et al. 2014), making pores less relevant. Moreover, there is a class of reactions that are not (merely) diffusion limited, since one of the reactants is one of the main mantle species, but (mainly) thermally activated at higher temperatures. This is possible in particular for reactions between two neighboring species, for instance, in a water-rich environment. Examples of such reactions are $H_2O + D_2O \longrightarrow 2HDO$ (Lamberts et al. 2015), and $H_2O + HNCO \longrightarrow H_3O^+ + OCN^- \longrightarrow HOCN + H_2O$ (Theule et al. 2011). Other thermally activated reactions in an ice are possible leading to the formation of salts (e.g., Bossa et al. 2009; Noble et al. 2014; Bergner et al. 2016). An example is the following network of equilibrium reactions.

$$NH_3 + CO_2 \rightleftharpoons NH_3 + COO^- \tag{32}$$

$$NH_3 + NH_3 + COO^- \rightleftharpoons NH_4 + NH_2COO^- \tag{33}$$

$$NH_4 + NH_2COO^- \rightleftharpoons NH_3 + NH_2COOH \tag{34}$$

leading to the formation of carbamic acid in a protic environment, i.e. a NH_3– or H_2O-dominated ice, where the *reaction* barrier is sufficiently low for proton transfer to occur thermally without any external source of processing. However, experimentally only mixtures with a NH_3:H_2O ratio greater than 1 were considered. In a more dilute environment, which is more representative for interstellar ices these processes become bulk diffusion limited and they require a suitable bulk diffusion mechanism in astrochemical gas-grain models.

Observations of interstellar ices show absorption bands attributed to various cations and anions (OCN^-, $HCOO^-$, NH_4^+, Boogert et al. 2015), and hence these processes should occur. Moreover, it is known that interstellar (and circumstellar) grains are likely charged (e.g., Draine 2003). Grains can become the main carrier of negative charge, for instance in dark and dense midplanes, where the gas is weakly ionized. As grain charge affects the chemistry, it is important to consider. Ionic chemistry in most grain surface models is currently limited to cation-grain recombination, where an ion recombines with a free electron and dissociates rather than sticking to the grain, applying gas-phase products and branching ratios analogous to those from dissociative recombination

$$XY^{+} + GRAIN^{-} \longrightarrow X + Y + GRAIN. \quad (35)$$

The rate of this process is determined by the cation-grain collision rate and is analogous to the accretion rate (Eq. (2)) but can include an enhancement (e.g., Draine and Sutin 1987).

Photoprocesses

It has long been known experimentally that FUV irradiation of interstellar ice analogues induces complex chemistry (e.g., Hagen et al. 1979; D'Hendecourt et al. 1986; Allamandola et al. 1988; Grim et al. 1989; Gerakines et al. 1996; Öberg et al. 2009c). The interaction of FUV photons with ices has both primary and secondary chemical effects which influence both the ice and gas composition. Which of these effects dominates depends upon the FUV flux and shape of the FUV spectrum, the photoexcitation/photodissociation cross sections of the molecular components which make up the ice mixture, and the temperature and structure of the ice. One of these effects is termed "photodesorption", whereby absorption of FUV (or X-ray) photons by an ice leads to desorption (or sublimation) of ice species into the gas phase. Photodesorption was originally thought to occur as a single-step process: the FUV photon excites the molecule into a state which is antibonding (or repulsive) with the surface and the species is ejected. As will be discussed later, it is now known that it can also occur as a multi-step process where the excited species decays back to the ground state and imparts kinetic energy to a surrounding molecule which is then "kicked out". Another effect is photodissociation, whereby the absorption of a FUV photon by a molecule induces dissociation, e.g., H2O+hν⟶OH+HH2O+hν⟶OH+H. It is this latter process which induces chemical changes in FUV-irradiated ices. As might be expected, photodesorption and photodissociation can occur in parallel. Photodissociation has primarily been induced to explain the origin of complex molecules in FUV-irradiated ices with an initial interstellar-like composition (e.g., CO, CO_2, H_2O, CH_4, N_2, NH_3, and CH_3OH).

Photodesorption is now considered an important process for releasing molecules into the gas phase in cold ($\lesssim$ 100 K) astrophysical regions where they would otherwise be frozen out onto icy grain mantles (and thus wholly depleted from the gas, e.g., Hartquist and Williams 1990). The origin of the FUV radiation depends upon the nature of the region. In cold and dark molecular clouds, FUV radiation is generated via the excitation of H_2 molecules by cosmic rays (Prasad and Tarafdar 1983), whereas in weak photodissociation regions (PDRs), the origin is the external, often enhanced, interstellar radiation field (ISRF, Hollenbach and Tielens 1997). In the envelopes surrounding forming stars, the innermost regions are sufficiently hot that thermal desorption is thought to dominate; however, molecules can be released back into the gas in the extremities of the envelope by photodesorption by the ISRF, similar to PDRs (e.g., van Dishoeck and Blake 1998). Recent high-sensitivity observations have suggested that the origin of gas-phase molecules at offset positions from the location of the central forming star may be due to photodesorption by stellar FUV photons along the outflow cavity walls (e.g., Öberg et al. 2010). In protoplanetary disks around young stars, there are multiple origins, including FUV photons from the central star and the external interstellar radiation field, and those generated internally via the excitation of H_2 by cosmic rays or X-rays (e.g., Willacy and Langer 2000). Which source dominates the FUV radiation field depends on the radial distance away from the central star and the vertical distance away from the shielded disk midplane.

Given the potential importance of photodesorption when interpreting observations, this process has routinely been included in astrochemical models for some decades. However, it is only relatively recently that experiments have successfully quantified this process, and also that the underlying physical mechanisms at work have begun to be understood. Photoprocesses have only been included in models relatively recently once it was demonstrated that molecules, such as methanol (CH_3OH), were unable to form efficiently via gas-phase chemistry alone.

In this Section, we give a brief overview of the experimental studies specific to ice photodesorption and photodissociation (Sect. 8.1), and those theoretical calculations which have offered insight into the mechanisms at play (Sect. 8.2). We also provide a recommendation for the implementation of experimental and theoretical data into the computation of photodesorption and photodissociation rates in astrochemical models (Sect. 8.3).

Experimental Studies

The measured photodesorption yields for pure ices are listed in Table 5. Early experiments on the photodesorption of molecules of astrophysical interest resulted in yields spanning from $\sim 10^{-6}$ photon^{-1} (Greenberg 1973), to $\sim 10^{-1}$ photon^{-1} (Nishi et al. 1984; Hartquist and Williams 1990). The large range of values is likely due to the wide array of setups adopted and the challenges of direct detection of the photodesorbed molecules. The former experiments were conducted in the 100–275 nm range at 77 K using a Hg-Xe high pressure arc lamp, whereas the latter were conducted using an ArF 193 nm laser at relatively high fluence between 90 and 135 K. It was not until the experiments by Westley et al. (1995a) and Westley et al. (1995b) that a robust quantification of water ice photodesorption at low temperatures (35--100 K35--100 K) was reported. They determined a yield of (3--8×10−3) photon−1(3--8×10−3) photon−1 at Lyman-αα (121.6 nm). Photodesorption was revisited much later by (Öberg et al. 2007, 2009a,b) who investigated the photodesorption yields of pure CO, N_2, CO_2, and H_2O ices under ultra-high vacuum conditions (base pressure better than 10−9 mbar10−9 mbar) at low temperatures using a broadband hydrogen microwave discharge lamp (120--170 nm120--170 nm). CO, CO_2, and H_2O were found to desorb from thick ices (≫10≫10 monolayers) at low temperatures (<20 K<20 K) with similar yields: (2.7±1.7×10−32.7±1.7×10−3), (2.3±1.4×10−32.3±1.4×10−3), and (1.3±0.8×10−3) photon−1(1.3±0.8×10−3) photon−1. Pure N_2 ice, on the other hand, had a significantly lower yield (<2×10−4) photon−1(<2×10−4) photon−1. Öberg et al. (2009a) also provided evidence that photodesorption occurs primarily from the top 1–2 monolayers of the ice mantle. For a CO ice covered in a thin layer of N_2, the signal for CO photodesorption is decreased whilst that for N_2 is enhanced, and above that expected for pure N_2 ice. This was the first evidence of *co-desorption* which we now term "kick out". Öberg et al. (2009a,b) were also able to determine that photodesorption is a parallel process to photodissociation. For the pure H_2O ice experiments, they observe a product yield for H_2O:OH of 0.7:1.0 at 20 K to 1.7:1.0 at 100 K. The temperature dependence arises from the increased mobility of the photodissociation products at a higher temperature allowing more efficient reformation of the water ice. For pure CO_2 ice, they find CO_2 desorbs as CO with a yield between 20–50 % that of CO_2. Again, at higher temperatures, the ratio of CO_2:CO increases due to increased mobility of the fragments. Hence, for both H_2O and CO_2, the primary process is photodissociation, followed by (i) desorption of the energetic fragments (OH or CO), (ii) recombination of the fragments which have excess energy to desorb, or (iii) "kick out"' of a neighboring molecule by an energetic fragment. Pure O_2 ice was also recently studied by Zhen and Linnartz (2014) under similar experimental conditions, and was found to have a desorption yield (6.0±2.0×10−4) photon−1(6.0±2.0×10−4) photon−1. Ozone (O_3) was also detected (arising from photodissociation of O_2 and reaction between O_2 and O) and similar to CO_2, the yield had a weak temperature dependence where the yield increased at higher temperatures.

Table 5 Measured photodesorption yields for pure ice species and products (where determined)

Ice species	Temperature (K)	Irradiation source	Products	Yield (molecules photon^{-1})	Ref.
H_2O	35–100	H_2 **lamp**	–	$3–8 \times 10^{-3}$	A
	18–100	H_2 **lamp**	OH/H_2O	1.3 ± 0.4–$4.5 \pm 1.2 \times 10^{-3}$	B[a]
CO	15–27	H_2 **lamp**	–	$2.7 \pm 1.3 \times 10^{-3}$	C, D
	7–15	H_2 **lamp**	–	3.5 ± 0.5–$6.4 \pm 0.5 \times 10^{-2}$	E[b]
	14	H_2 **lamp**	–	6.4 ± 0.2–$21.2 \pm 0.3 \times 10^{-2}$	F[c]
	18	Synchrotron	–	$4.1–16 \times 10^{-3}$	G[d], H
N_2	16	H_2 **lamp**	–	$\approx 2 \times 10^{-4}$	C,D
	14	Synchrotron	–	$1.5–5.3 \times 10^{-3}$	I[d], J
CO_2	16–60	H_2 **lamp**	CO/CO_2	$2.3 \pm 1.4 \times 10^{-3}$	C[e]
	10–40	Synchrotron	CO/CO_2	$1–3 \times 10^{-3}$	K
O_2	14–21	H_2 **lamp**	O_2/O_3	$6 \pm 2 \times 10^{-4}$	L
	14	Synchrotron	–	$2.1–3.3 \times 10^{-3}$	I[d]
CH_3OH	20–70	H_2 **lamp**	$C_2H_6/CH_3CHO/C_2H_5OH/CH_3OCH_3$	$2.1 \pm 1.0 \times 10^{-3}$	M[f]
			$HCOOCH_3/HCOCH_2OH/(CH_2OH)_2$		
	8–130	H_2 **lamp**	$CH_4/CO/HCOOH/CH_3CHO/X–CHO$	$<3 \times 10^{-5}$	N[g]
	10	Synchrotron	$CO/CH_3/OH/H_2CO/CH_3O$	$<10^{-5}$	O[h]

[a]The stated range reflects the temperature range, with higher yields at higher temperatures (see Öberg et al. 2009b, for more details)

[b]The stated range reflects the temperature range, with a slightly lower yield at the higher temperature (see Muñoz Caro et al. 2010, for more details)

[c]The stated range reflects the range of relative contributions from Lyman-*α* **and H_2 molecular bands (see Chen et al.** 2014, for more details)

[d]The stated range corresponds to the wavelength-dependent yield integrated over various radiation fields (see Fayolle et al. 2012, 2013, for more details)

[e]Includes the contribution from both CO and CO_2 **photodesorption**

[f]Products detected in situ via RAIRS and following TPD, and photodesorption yield determined indirectly by following loss of CH_3OH ice upon irradiation (see Öberg et al. 2009c, for details)

[g]CO and CH_4 **detected in the gas-phase following irradiation with yields** $\sim 2–60 \times 10^{-3}$ **molecules photon^{-1} (see Cruz-Diaz et al.** 2016, for details)

[h]Photofragments only detected in the gas-phase following irradiation with yields $\lesssim 10^{-4}$ **molecules photon^{-1} (see Bertin et al.** 2016, for details)

A—Westley et al. (1995b); B—Öberg et al. (2009b); C—Öberg et al. (2007) D—Öberg et al. (2009a); E—Muñoz Caro et al. (2010); F—Chen et al. (2014); G—Fayolle et al. (2011); H—Bertin et al. (2012); I—Fayolle et al. (2013); J—Bertin et al. (2013); K—Fillion et al. (2014); L—Zhen and Linnartz (2014); M—Öberg et al. (2009c); N—Cruz-Diaz et al. (2016); O—Bertin et al. (2016)

Later experimental efforts moved towards wavelength-dependent studies. These measurements are more challenging and are conducted at synchrotron facilities where the FUV flux is lower, requiring longer irradiation times to match the fluence typical of interstellar conditions. On the other hand, microwave discharge lamps have a significantly higher flux; however, independent measurements of CO photodesorption have shown that the derived yield is dependent on having a well-calibrated lamp spectrum. Muñoz Caro et al. (2010) determined a desorption yield for pure CO ice of $(3.5\pm0.5\times10^{-2})$ photon^{-1}$(3.5\pm0.5\times10^{-2})$ photon−1 which is one order of magnitude higher than that from Öberg et al. (2009a). Recent investigations showed that the FUV spectrum, in particular the ratio of Lyman-αα (121.6 nm) to H_2 molecular emission bands (110–180 nm), is sensitive to the gas pressure, the type of lamp, the gas composition (pure H_2 versus a mixture of H_2 and He) and the window properties (Chen et al. 2014; Ligterink et al. 2015). Chen et al. (2014) determined CO desorption yields as high as $(2.12^{\pm}0.03\times10^{-1})$ photon−1$(2.12\pm0.03\times10^{-1})$ photon−1 for a lamp setup with pure H_2 and a window giving a ratio of 20 % flux of Lyman-αα relative to the molecular emission bands. This showed that desorption of CO at Lyman-αα is less efficient than at wavelengths covering the H_2 emission bands and turned out to be the first step in quantifying the wavelength dependence of photodesorption. The same group have recently reported VUV (120–160 nm) absorption cross-sections for a plethora pure ices (Cruz-Diaz et al. 2014a,b). All data are publicly available for download.[Footnote1]

Fayolle et al. (2011) and Bertin et al. (2012) presented the first wavelength-dependent measurement of CO photodesorption (between 90 and 160 nm) conducted at the SOLEIL[Footnote2] synchrotron facility. The results showed a strong wavelength dependence corresponding to the vibronic transition spectrum of CO ice. CO was thus confirmed to desorb via the DIET mechanism: Desorption Induced by Electronic Transitions. The results also confirmed the low desorption yield at Lyman-αα wavelengths. Peak yields approaching ~5×10^{-2} photon−1~5×10^{-2} photon−1 were found to correspond with the $A^1\Pi - X^1\Sigma^+$ electronic transitions of CO (~130--160 nm~130--160 nm). Followup studies on pure N_2, O_2, and CO_2 ices showed interesting results (Fayolle et al. 2013; Fillion et al. 2014). The pure N_2 photodesorption yield was found to peak at significantly shorter wavelengths (≲100 nm≲100 nm) with a wavelength dependence corresponding to the $A^1\Pi - X^1\Sigma^+$ electronic transition. A peak yield of 3.77±1.13×10^{-2} photon−13.77±1.13×10^{-2} photon^{-1} was found at 100.8 nm. Photodesorption yields at longer wavelengths were consistent with previous measurements using broad-band hydrogen lamps (Öberg et al. 2009a). Pure O_2 ice photodesorption was also studied and, in contrast to CO and N_2, was found to have a non-negligible contribution due to photodissociation (Fayolle et al. 2013), as also seen later in the broad-band photodesorption experiments (Zhen and Linnartz 2014). The peak photodesorption rates (~0.5×10−2 photon−1~0.5×10−2 photon−1) overlapped with the Schumann-Runge band for O_2 $(\tilde{X}^3\Sigma_g^- - \tilde{B}\Sigma_u^-)$ which leads to dissociation into O(^{3}P) and O(^{1}D) below ~176 nm~176 nm (Parker 2000; Mason et al. 2006). Fillion et al. (2014) reported results from pure CO_2 ice. CO_2 showed a more complicated behavior than the more volatile species discussed above. Both CO and CO_2 molecules were found to photodesorb at wavelengths below 118 nm only, with yields ranging from ~1--3×10^{-3} photon−1~1--3×10^{-3} photon−1. The steep increase in photodesorption signal at 118 nm is associated with the first dipole-allowed transition $((3p\sigma_u)^1\Sigma_u^+ -^1\Sigma_g^+)$ and thereafter follows the photoabsorption continuum. That both CO and CO_2 were seen with similar photodesorption yields suggests that the mechanism likely has multiple paths: (i) photoexcitation of CO_2 which "kick out" surface CO_2 (or CO) molecules (DIET mechanism), (ii) photodissociation of CO_2 generating energetic fragments (CO∗+OCO∗+O) which either "kick out" surface CO_2 (or CO) molecules (DIET mechanism), or desorb directly, or recombine to reform CO_2 which then desorbs. In all cases, the photodesorption signature of CO_2 and CO were similar and as described above. However, for pre-irradiated CO_2 ice, the photodesorption signatures of CO_2 and CO were found to match that of pure CO ice (Fillion et al. 2014). Here, the DIET mechanism is driven by photoexcitation of CO (as opposed to photoexcitation or photodissociation of CO_2).

Wavelength-dependent results from binary N_2:CO ice mixtures confirmed the desorption of surface species via sub-surface excitation (Bertin et al. 2013). The resulting photodesorption signature of surface molecules was found to correspond with the dominant component of the bulk ice composition, e.g., a thin surface layer of N_2 on top of CO ice photodesorbs with a signature of CO photodesorption whereas for a 1:1 mixture, both molecules photodesorb with the signature of both species. Also, Bertin et al. (2013) found that a thin layer (~3 monolayers) of N_2/CO on top of CO/N_2 bulk ice was sufficient to completely quench the CO/N_2 photodesorption signal.

The quantification and understanding of the process of photodesorption has made leaps and bounds in the past few years and for species other than CO and N_2, it has been shown that irradiation also induces photodissociation, that is, photodesorption and photodissociation are parallel processes. For pure (and relatively) simple ices, the resultant products and reaction pathways are straightforward to ascertain, as described above. Many experiments have shown that FUV irradiation of mixed ices induce chemistry; however, these experiments often couple irradiation with thermal processing and it can be difficult to disentangle reaction pathways specific to either FUV or heat processing (e.g., Hagen et al. 1979; D'Hendecourt et al. 1986; Allamandola et al. 1988).

Several experiments have conducted in-situ monitoring of ice composition via infrared spectroscopy during irradiation at low temperatures. Gerakines et al. (1996) irradiated numerous pure ices composed of known interstellar molecules with a broad-band hydrogen lamp and monitored changes in the ice composition with a Transmission Fourier-transform infrared spectrometer. Gerakines et al. (1996) reported a multitude of identified products, even for relatively simple ices (e.g., H_2O, NH_3, and CH_4). They also studied more complex ices: formaldehyde (H_2CO) and methanol (CH_3OH). In the former experiment, products such as polyoxymethylene (POM, (--H2CO--)n(--H2CO--)n), CO, HCO, CO_2, CH_4, CH_3OH, and methyl formate ($HCOOCH_3$), were identified after 1 hour of irradiation. For pure CH_3OH irradiation, similar products were identified (excluding POM). Gerakines et al. (1996) were also the first to extract quantitative information from the experiments and provided photodissociation cross sections (averaged over the lamp spectrum) for those species for which photodissociation was deemed efficient, and proposed potential reaction pathways.

The irradiation of pure and mixed methanol ices was revisited by Öberg et al. (2009c). They also studied the influence of temperature and FUV fluence and monitored the composition of the ice in-situ via RAIRS (Reflection-Absorption Infrared Spectroscopy). In all experiments conducted, complex molecules were observed to form and included CH_3CHO (acetaldehyde), CH_3CH_2OH (ethanol), CH_3OCH_3 (dimethyl ether), $HCOOCH_3$ (methyl formate), $HOCH_2CHO$ (glycolaldehyde), and $(CH_2OH)_2$ (ethylene glycol). The exact products and relative ratios were found to have a weak dependence on temperature during irradiation and a stronger dependence on ice composition (pure CH_3OH versus CH_3OH mixed with CO or CH_4). For example, CO:CH_3OH ices were more enriched with --HCO--HCO molecules while CH_4:CH_3OH ice were more enriched in --CH3--CH3 molecules. Öberg et al. (2009c) also observed a large increase in the abundances of several complex molecules (around an order of magnitude) upon warming the irradiated sample from 20 to 70 K, confirming that radical mobility is temperature dependent and is important for building additional complexity. An analysis of the products allowed them to identify potential reaction pathways and formation cross sections. They also estimated the branching ratios for CH_3OH ice photodissociation at 20 K as $5:1:<1$ for $CH_2OH + H$, $CH_3O + H$, and $CH_3 + OH$. The authors plan an extraction of fundamental data from the experimental results, for example, photodissociation rates, products, and branching ratios, diffusion rates, and binding energies of radical species. At the time of writing (March 2016), this challenging analysis was still ongoing.

As a final note in this Section, Öberg et al. (2009c) also report a photodesorptionyieldfor$CH_3OH$of$(2.1\pm1.0\times10^{-3})$photon−1(2.1±1.0×10−3) photon−1 which was determined by modelling the loss of methanol ice during irradiation as a combination of photodesorption (a zeroth-order process) and photodissociation (a first-order process). Very recently, several groups have revisited this system in light of recent revelations regarding the importance of robust lamp characterization for accurate determinations of photodesorption yields (Chen et al. 2014). The general consensus from several independent groups, is that pure methanol ice does not photodesorb intact. Photofragments only were observed, with a deduced upper limit to intact methanol photodesorption $\sim10^{-6}$–$10^{-5}$$\sim10^{-6}$–$10^{-5}$ molecules per incident photon, i.e., at least two orders of magnitude less than that found for water ice (Cruz-Diaz et al. 2016; Bertin et al. 2016). This low photodesorption yield was found for both pure and mixed (with CO) methanol ice, showing also that the DIET mechanism appears not to work for large molecules such as methanol. Note that this low photodesorption yield for methanol now calls into question the origin of gas-phase methanol in both cold and irradiated environments (e.g., dark clouds, PDRs, and protoplanetary disks), providing an intriguing challenge for molecular astrophysicists and astrochemical modelers.

Theoretical Calculations

Photodesorption has been studied theoretically via classical molecular dynamics (MD) simulations. Most work to date has been conducted on water ice (Andersson et al. 2005, 2006; Andersson and van Dishoeck 2008; Arasa et al. 2010), including its deuterated analogues, HDO and D_2O (Koning et al. 2013). Andersson et al. (2005, 2006) studied the photodesorption mechanisms of H_2O in both crystalline and amorphous water ice at 10 K following photoexcitation of H_2O into the first excited state with photons with wavelengths between 130--150 nm130--150 nm. The MD calculations predicted a H_2O photodesorption efficiency between 0.5–1 % per absorbed photon in top few monolayers of an amorphous water ice surface (Andersson et al. 2005, 2006), which was in reasonable agreement with experimental data available at that time (e.g., Westley et al. 1995a). Andersson and van Dishoeck (2008) went a step further than earlier work and determined the relative contributions of all possible outcomes as function of depth into the ice. Upon photodissociation into H+OHH+OH, the following outcomes are possible: (i) H desorbs and OH is trapped, (ii) H is trapped and OH desorbs, (iii) H and OH are trapped, (iv) H and OH recombine and desorb as H_2O, and (v) H and OH recombine and are trapped as H_2O. There is also a parallel process whereby the excited H product had sufficient energy to "kick out" a neighboring water molecule. Outcomes (i)–(iv) and the "kick out" processes all have the potential to affect the chemical balance of both the ice and gas. Andersson and van Dishoeck (2008) determined that outcome (i) has the highest probability in the surface monolayer (~90 %~90 %), and decreases in efficiency rapidly with depth into the ice reaching ~10 %~10 % efficiency by the sixth monolayer. On the other hand, trapping of either the fragments or the reformed molecule increased in probability with depth, becoming the most probable outcome by the forth monolayer. The MD calculations also provided insight into the behavior of the photodissociation fragments post dissociation. For outcomes in which no desorption occurred, H and OH had sufficient excess energy to travel up to 70 and 60 Å, respectively, on or within the uppermost monolayers of an amorphous ice surface (Andersson and van Dishoeck 2008). Thus photodissociation of ices also creates mobile radicals at 10 K which can take part in further surface reactions.

Arasa et al. (2010) extended the calculations to higher temperatures (up to 90 K). Similar results were found: photodesorption has the highest efficiency in the top few monolayers, and trapping increases in probability with depth into the ice. The efficiency of OH and H_2O photodesorption was found to have weak dependence on temperature with an increase in probability of around 30 %30 %. The probability of the parallel "kick out" mechanism was found to have a more complex behavior with temperature which the authors attribute to the amorphous nature of the ice. Koning et al. (2013) studied isotopic effects, and looked at the astrophysically relevant cases of HDO and D_2O photodissociation in an amorphous H_2O water ice. D atoms were found to photodesorb with a slightly lower efficiency than H atoms. On the other hand, the "kick out" probability of D atoms was found to be higher than that for H atoms due to more efficient energy transfer by the higher mass atom to the surrounding water molecules. Arasa et al. (2015) revisited these results to determine whether photodissociation of HDO (into either OH+D or OD+H) could affect the deuterium fractionation of water in the gas phase. The conclusion was that the effects are likely minimal; however, the lower probability for D atom photodesorption may lead to fractionation effects within the ice. This latter prediction remains to be tested in astrochemical models. Probabilities for each outcome as a function of monolayer and temperature have now been made available for use in astrochemical models (Arasa et al. 2015). Also available are values averaged over the top few monolayers for use in those models which are unable to treat the ice chemistry in a stochastic manner.

Very recent MD calculations have also attempted to understand the underlying mechanism for CO photodesorption via the DIET process (van Hemert et al. 2015). van Hemert et al. (2015) excited selected CO molecules with 8.7--9.5 eV8.7--9.5 eV photons contained within both crystalline and amorphous CO clusters (containing up to 1200 CO molecules). They determined two potential mechanisms. In the first, the excited CO molecule decays to the ground state in an unfavorable configuration and subsequently ejected, and the second corresponds to the aforementioned "kick out" mechanism. Photodesorption yields of (4.0×10^{-3}) photon^{-1} agreed well with the lower end of the range of values currently constrained by experiment.

Implementation in Astrochemical Models

To date, the implementation of experimentally-determined photodesorption yields (Y_X, typically expressed in units of molecules photon^{-1}) in astrochemical models has been fairly pragmatic. The photodesorption rate of a molecule is dependent upon the flux of FUV photons striking the grain surface, FUV photonscm^{-2}s^{-}the cross-sectional area of grain surface available per unit volume, n_g cm2cm−3σgng cm2cm−3, and the surface coverage fraction of the specific molecule in the top 1–2 monolayers of the ice mantle, $\theta_X = n_X/n_S$. In the above expressions, σ_g is the geometrical cross section of a grain surface, n^g is the number density of grains, n_X is the number density of molecules of species X in the surface layers of the ice mantle, and n_S is the total number of species in the surface layers of the ice mantle. In two-phase models, there is no distinction between the surface coverage factor and the bulk coverage factor. However, these numbers will likely differ in three-phase and multilayer models.

The photodesorption rate is thus given by,

$$k_X^{pd} = \theta_X Y_X F_{UV} \sigma_g n_g \text{ cm}^{-3}\text{ s}^{-1}. \tag{36}$$

Derivation of the rate beginning instead with the cross-sectional area of an individual grain surface site leads to an identical expression due to the relation between number density of surface sites, grain surface area, and total number of species per monolayer. Note that the experimentally determined yield implicitly includes the effects of ice opacity to FUV radiation (since the yield is given per incident photon) and the fact that photodesorption only occurs from the top few monolayers of the ice. This results in a zeroth-order rate coefficient: the photodesorption rate is independent of ice thickness (or number density of ice molecules) for thick ices. Within this framework, it is also possible to include branching ratios of the expected products, for those species for which photodissociation is the primary process (e.g., H_2O and CO_2). Several of the experimental groups also provide analytical fits to the photodesorption yields as a function of ice thickness (in monolayers) and temperature (Öberg et al. 2009a,c). This allows the inclusion of a more realistic treatment of the photodesorption rate at low surface coverage (important in models which consider evolution from diffuse to dense conditions).

To show the effect of photodesorption, Fig. 8 again shows a simulation of a laminar solar nebula where we have lowered the photodesorption yield from $10^{-2}$to 10^{-5}. Locally the abundance can change by several orders of magnitudes, but integrated over zz the changes are small, except for gas phase CO_2 which is strongly enhanced in the outer regions.

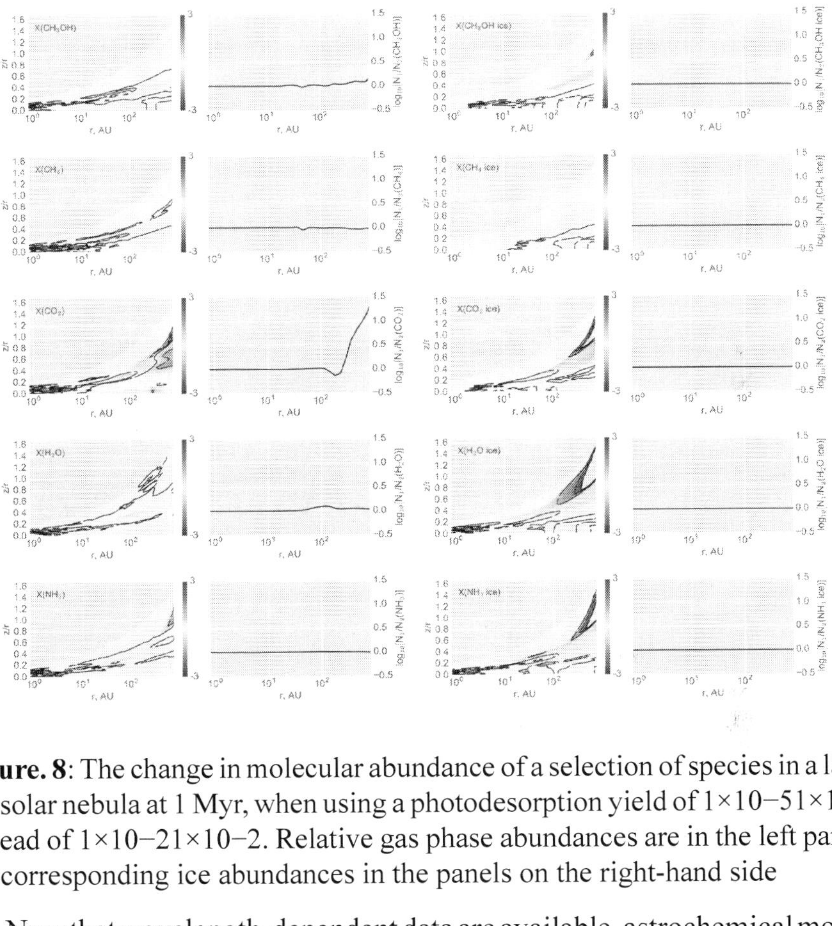

Figure. 8: The change in molecular abundance of a selection of species in a laminar solar nebula at 1 Myr, when using a photodesorption yield of 1×10−51×10−5 instead of 1×10−21×10−2. Relative gas phase abundances are in the left panels, the corresponding ice abundances in the panels on the right-hand side

Now that wavelength-dependent data are available, astrochemical models for which the FUV spectrum is included explicitly (e.g., in protoplanetary disks or PDRs), the total photodesorption yield per species can be found by integrating over the experimentally determined yield and the FUV spectrum,

$$Y_X = \int_\lambda Y_X(\lambda) F_{UV}(\lambda) d\lambda. \qquad (37)$$

These data are now available for pure CO, N_2, O_2, and CO_2 ices (Fayolle et al. 2011, 2013; Fillion et al. 2014). One caveat of this method is that one needs to take into account that the wavelength dependence of the desorption yield will depend upon the dominant molecular component of the ice mantle in the *subsurface* layers of the ice (Bertin et al. 2013). This is possible to track in three-phase and multilayer models, but non trivial in two-phase models. This is because additional assumptions are required, for instance, whether the ice is sequestered into layers and if so, with which layers (H_2O-dominated or CO-dominated) more minor species are most likely to be associated. In the case of H_2O-dominated ices, wavelength dependence has not been determined explicitly but is likely to follow the photodissociation (or absorption) cross section of amorphous solid water (see, e.g., Cruz-Diaz et al. 2014a). It is also worth bearing in mind that, in the regime where water ice is the dominant ice component, the ice temperature could be such that thermal desorption will dominate, except for those molecules which are comparable in volatility with water (e.g., CH_3OH).

In current gas-grain chemical models, photodissociation is treated as a parallel process to photodesorption. Because interstellar ices are optically thin to FUV radiation, photodissociation can occur throughout the ice mantle (in contrast to photodesorption which occurs only from the top few monolayers). Photodissociation rates for specific species are computed by integrating over the photodissociation cross section multiplied by the FUV spectrum, exactly as is done in the gas phase (e.g., van Dishoeck et al. 2006),

$$k_X^{ph} = \int_\lambda \sigma_X(\lambda) F_{UV}(\lambda) d\lambda, \tag{38}$$

with products and associated branching ratios usually adopted from the equivalent gas-phase measurements or estimates. Photoabsorption cross sections in the VUV range are now available for several molecules in the solid phase and are usually a good proxy for computing the photodissociation rates (e.g., Cruz-Diaz et al. 2014a,b). For other species, the gas-phase data is typically used and should be accurate to within a factor of a few. The derivation of photodissociation cross sections and branching ratios in the solid phase remains a complex process requiring careful modelling of ice experiments which explore the full range of multi-dimensional parameter space (Öberg et al. 2009c).

By treating photodissociation as a parallel process to photodesorption in full gas-grain models, models can implicitly account for the production of reactive radicals within and on the ice mantle. The models are also able to treat the fate of such radicals including loss to the gas phase via desorption (whether photo driven or thermally driven), or reaction with other radicals and atoms present in the ice, as discussed in the introduction to this Section. However, in models in which chemical desorption is also included, one runs into the issue in which photodesorption is accounted for twice: one explicitly as outlined above, and one implicitly, where radicals formed via photodissociation recombine to reform the molecule which then has an assumed efficiency for chemical (or reactive) desorption. This latter process is essentially direct photodesorption. Given the inherent uncertainties which exist in large astrochemical models due mainly to the lack of available experimental data, the effect of this "double counting" is likely minor. A potential solution to the problem is to treat photodissociation products as a separate (and likely) excited species in the ice mantle which have a timescale for thermalization guided by the MD calculations.

PRECISION CONSIDERATIONS IN RATE-BASED MODELING

A basic, two-phase model of interstellar gas and grain surface chemistry may be solved easily using the standard rate-equation approach that is commonly used for pure gas-phase chemistry. However, as explained earlier, this method can sometimes lead to unphysically high grain surface production rates, under conditions where two reactants on the surface have mean populations per dust grain that are less than unity. In a mathematically exact treatment, if the reaction between two such species is sufficiently rapid once both particles are present on the grain surface, then the surface populations of those species should become anti-correlated. The rate-equation method cannot reproduce such behavior, as it calculates only the mean, individual populations $\langle n_s(A) \rangle$ and $\langle n_s(B) \rangle$, rather than the mean of the product, $\langle n_s(A) n_s(B) \rangle$.

Garrod (2008) proposed a modification scheme (i.e. a so-called "modified rate" approach) to take account of the inaccuracies of the rate-equation method, which Garrod et al. (2009) showed to be effective over a wide range of physical conditions. However, the full implementation of this method involves the calculation of quantities such as $(1-\langle n_s(A) \rangle)$ In a numerical simulation, the value of this differential becomes highly imprecise as $\langle n_s(A) \rangle$ approaches 1, due to the limited fractional precision available in the calculations (for double precision calculations, the fractional precision is on the order of 10^{-16}). This imprecision can induce the solver to use very small timesteps, significantly slowing the calculations. For this reason, the maximum available precision should be used for modified-rate calculations wherever possible. Several Fortran compilers now include quadruple precision as a standard feature. Converting a chemical code and differential equation solver to quadruple precision is a straight-forward task.

The use of a three-phase gas-grain model can also introduce precision problems. As explained by Hasegawa and Herbst (1993b), the three-phase treatment calculates a net deposition rate onto the dust grains that represents the difference between the total accretion rate of all gas-phase species onto the grains, and the total desorption rate of all surface species back into the gas (as well as any processes such as reaction that may also decrease the total number of surface particles). This net rate is used to determine the rate at which material in the surface layer is pushed down into the bulk ice (in the case of net deposition) or brought up from the bulk into the surface layer (net desorption), in order to maintain the same total population of particles in the surface layer. The net deposition rate therefore strongly couples the calculated abundances of all surface and mantle species with the abundances of whichever gas-phase or surface species have the greatest influence over that net deposition rate. Improving the overall precision of the calculations (i.e. moving to quadruple precision) allows the net rate to be more precisely determined, improving model run-times.

Unfortunately, in the case of three-phase models, simply increasing the overall precision of the calculations is often not enough to avoid a failure of the solver routine. A more careful consideration of the error associated with individual abundances is also required. We may illustrate this point with an example in which the net deposition rate onto dust grains is solely determined by the difference between (i) the rate of deposition of gas phase H_2 and (ii) the rate of desorption of grain surface H_2. If one rate strongly dominates over the other, then the *fractional error* on the net deposition rate is mainly dependent upon the solver-enforced fractional error (i.e. "EPS", error per step) on either the gas-phase or surface H_2 abundance. However, if the two rates are nearly equal, then the error on the net rate will depend upon the error on both abundances. Crucially, in this latter case, if the net rate is very much smaller than either the deposition or desorption rates, then the absolute error on the net rate may become much larger even than the rate itself. Because the calculated gas-phase and grain surface H_2 abundances may fluctuate anywhere within their allowed error limits, this situation can produce wild fluctuations in the net deposition rate. This prompts a shift by the solver routine to smaller and smaller timesteps and ultimately a failure to integrate over manageable timescales.

To avoid this problem, we must enforce an error value that is more stringent than EPS for certain species in the model that affect the net deposition rate. The general EPS value used in a model may be, say, 10−410−4. The solver compares this EPS with the fractional error on each species resulting from the time-integration of the differential equations. If the EPS exceeds the computed error value for all chemical species, then convergence is assumed, and the timestep is complete. In order to impose stronger error handling on the net deposition rate, we may bias this comparison by some factor, according to the sensitivity of the net deposition rate to changes in the abundance of a particular species. If a small, fractional change in a chemical abundance induces a *larger* fractional change in the net deposition rate, then the ratio of these values is used to bias the error comparison to enforce smaller errors. The sensitivity of the net deposition rate to the abundance of each species can be determined at each Jacobian call by the solver. Typically only a few chemical abundances are important to the error dependence of the net deposition rate, such as gas-phase and grain surface H, H_2 and CO.

This more stringent error-handling approach is necessary for the use of three-phase models in all cases where the net deposition rate changes sign during the course of the integration, because such conditions induce extreme fractional errors as the net rate approaches zero.

Test Case: Hydrogenation of CO Ice Leading to Complex Molecules

Catastrophic freeze-out of CO molecules from the gas phase leads to the formation of an apolar (CO-rich) ice. The surface hydrogenation of CO in inter- and circumstellar ices drives the formation of more complex species. Here we will use the hydrogenation of CO as an example of how a combined experimental and modeling approach can result in an increased understanding of complex molecule formation. In 1992, Charnley et al. proposed that formaldehyde (H_2CO) and methanol (CH_3OH) form in the solid phase by the hydrogenation of CO ice. Formaldehyde is the first organic polyatomic molecule detected in the interstellar medium (ISM) (Snyder et al. 1969). Methanol has been observed abundantly in the ISM (Belloche et al. 2013) and is considered to be a resource for the formation of more complex molecules through surface reactions and after its sublimation in the gas phase (Balucani et al. 2015). Originally, gas-phase complex molecules were postulated to arise from dissociative recombination of protonated methanol, which itself was formed via radiative association of smaller ions and molecules (e.g., $CH_3^+ + H_2O \longrightarrow CH_3OH_2^+ + h\nu$ followed by

$CH_3OH_2^+ + e^- \longrightarrow CH_3OH + H$) Pioneering measurements with CRYRING, a low-temperature ion-storage ring, showed that dissociative recombination led predominantly to fragmentation of the protonated species, for instance, for the example listed above, only 3 % of dissociative recombinations where found to lead to methanol (Geppert et al. 2006). On the other hand, the surface formation of methanol has been long under debate, as the first sets of laboratory data gave conflicting results. In their study, Hiraoka et al. (2002) reported only the formation of formaldehyde, whereas Watanabe and Kouchi (2002) confirmed an efficient formation of formaldehyde and methanol ice. This controversy was then solved in 2009 by an independent investigation of the same reaction scheme performed for a wider range of H-atom fluxes and ice temperatures (Fuchs et al. 2009), which showed the different results from the two groups were due to the use of different H-atom fluxes and fluences.

The first (CO+HCO+H) and third (H2CO+HH2CO+H) reaction steps in this H-atom addition scheme presents barriers that have to be overcome, see also Fig. 3. The efficiency of the process therefore depends on the barrier height and how much excess energy is available to cross over. Figure 9 shows the time evolution of the surface abundance (in molecules cm−2cm−2) of CO, H_2CO and CH_3OH during H-atom bombardment of CO ice with an H-atom flux of $5 \times 10^{13}\ cm^{-2}\ s^{-1}$ at different temperatures. This figure is taken from Fuchs et al. (2009). At all the temperatures, the amount of CO decreases as the abundance of H_2CO increases. After an H-atom fluence of 1×10^{17} H atoms cm−2, the formation of methanol starts to increase at the expense of the H_2CO abundance. Similar trends are also reported in Watanabe et al. (2006). For temperatures higher than 15 K, a clear drop in the production rate of methanol is observed by Fuchs et al. (2009). This is mainly due to a lower lifetime of H atoms on the surface of the ice that causes the H-surface abundance to drop substantially and therefore reduce the probability of hydrogenation reactions occurring in the ice at temperatures higher than 15 K.

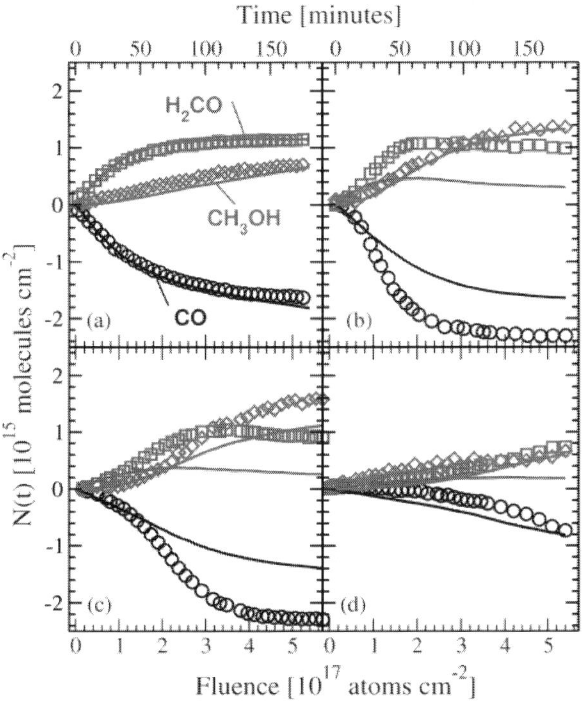

Figure. 9: Time evolution of the surface abundance (in molecules cm^{-2}) of CO, H_2CO and CH_3OH during H-atom bombardment of CO ice with a H-atom flux of 5×10^{13} $cm^{-2}s^{-1}$ at surface temperatures of 12.0 K (**a**), 13.5 K (**b**), 15.0 K (**c**), and 16.5 K (**d**). Experimental data (*symbols*) and Monte Carlo simulation results (*solid lines*) are shown as well. Figure reproduced from Fuchs et al. (2009). Credit: G.W. Fuchs, H.M. Cuppen, S. Ioppolo, C. Romanzin, S.E. Bisschop, S. Andersson, E.F. van Dishoeck, and H. Linnartz: Astron. Astrophys. **505**, 629 (2009), reproduced with permission, copyright ESO

A detailed interpretation of laboratory measurements in terms of understanding processes like deposition of H atoms, diffusion, reaction and desorption is only possible if laboratory work is supported by simulations. Fuchs et al. (2009) used a microscopic kinetic Monte Carlo algorithm to fit effective reaction barriers for $CO + H$ and $H_2CO + H$ to a large set of H_2CO and CH_3OH production and CO loss data as a function of H-atom fluence and temperature indicated by the solid lines in Fig. 9. The outcome of the model yielded typical values in the range of 400 to 500 K. In general, simulations and experimental data are in very good agreement at 12 K. The agreement for 13.5, 15, and 16.5 K is far less good, probably because of missing mechanisms that promote the penetration into the ice. The reaction rates for $H + CO$ and $H + H_2CO$ have only a minor temperature dependence, which suggests that there is a clear tunneling component for the reaction at low temperature. Values from Fuchs et al. (2009) are in good absolute agreement with the barriers found by Awad et al. (2005), who also found a similar temperature behavior. Their values were obtained using a rate equation analysis for T = 10, 15, and 20 K using the data from Watanabe et al. (2006).

Temperature-dependent values are important - within their experimental accuracies - for inclusion in astrochemical models that predict astronomical column densities. The Fuchs et al. Monte Carlo model is further extended to interstellar conditions to compare with observational H_2CO/CH_3OH data. The H_2CO and CH_3OH interstellar abundances obtained in the simulations do not scale directly with H-atom fluence because of the different relative importance of H_2 production and CO hydrogenation in space compared with the laboratory. However, laboratory experiments are required to derive the necessary rates that serve as input to the Monte Carlo program. The obtained H_2CO/CH_3OH ratios for the interstellar simulations are in closer agreement with observational limits than a direct translation of the experimental results. Cuppen et al. (2009) extended the Fuchs et al. (2009) model to simulate in detail the formation of both H_2CO and CH_3OH under interstellar conditions. Unlike the master equation (Stantcheva et al. 2002), rate equation (Ruffle and Herbst 2000), and macroscopic Monte Carlo (Ruffle and Herbst 2000) studies of methanol formation, the continuous-time, random-walk Monte Carlo technique accounts for the layering of the CO. This layering is crucial, because the constant addition of fresh CO tends to protect the underlying layers from hydrogenation and limits the time available for reaction.

The link between solid CO and CH_3OH was also recently confirmed observationally. Using VLT data on the CO band of a large collection of YSOs (Pontoppidan et al. 2003), peak width, peak position and upper limits on the 2152 cm−12152 cm−1 band were extracted and compared to laboratory spectra. This proved that the astronomical spectra are fully consistent with CO ice intimately mixed with CH_3OH, in agreement with a common chemical history as described above (Fraser et al. 2004; Cuppen et al. 2011). A more recent study expands this proposition by comparing literature laboratory spectra of different ice mixtures containing CO to high-resolution Keck/NIRSPEC spectra, of the massive YSO AFGL 7009S and of the low mass YSO L 1489 IRS (Penteado et al. 2015). Accurate band profile comparisons of both the ^{12}CO and ^{13}CO ice bands were carried out.

Moreover, Penteado et al. gave evidence that a gradient of different ratios of CO: CH_3OH mixtures might exist in the grain mantles and that the spectrum of high-mass YSOs requires a higher content of methanol than the low-mass YSOs to reproduce the observed CO band profile.

In the past decade, to fully characterize the methanol formation reaction network, deuteration experiments were also performed on CO ices (Hidaka et al. 2007), which resulted in both deuterated formaldehyde and methanol. In this study, it was found that deuteration rate of CO is 12.5 times lower than the rate of CO hydrogenation under the same experimental conditions. A similar CO deuteration rate was also recently confirmed by Fedoseev et al. (2015). In addition to CO+DCO+D experiments, deuteration studies of H_2CO and CH_3OH ices have also been performed (Nagaoka et al. 2005; Hidaka et al. 2009; Watanabe and Kouchi 2008). Figure 8 of Hidaka et al. (2009) presents the complete surface reaction network as obtained when CO is exposed to H and D atoms based on several studies (Watanabe et al. 2006; Watanabe and Kouchi 2008; Hidaka et al. 2009). Reaction rate constants are shown relative to those of CO+HCO+H at 15 K (Hidaka et al. 2004). The general conclusion here is that although D-atom additions to CO/H_2CO proceed slower than H-atom additions to CO/D_2CO, D-atom induced substitution reactions in H_2CO and CH_2OH take place faster than the corresponding H-atom induced substitutions in D_2CO and CD_2OD.

Recently, Fedoseev et al. (2015) showed the first laboratory evidence that surface hydrogenation of CO ice leads to the formation of molecules with more than one carbon atom along with the formation of H_2CO and CH_3OH. In their experiments, Fedoseev et al. confirmed the surface formation of two larger molecules of astrobiological importance, i.e., glycolaldehyde (HC(O)CH_2OH) and ethylene glycol (H_2C(OH)CH_2OH). The key step in this process is believed to be the recombination of two HCO radicals followed by the formation of a C-C bond as also indicated by modelling work (Woods et al. 2013). The dimerization of HCO is considered by Woods et al. (2013) as one of the formation routes of glycolaldehyde. Their astrophysical model matches the observed estimates in the hot molecular core G31.41+0.31 and low-mass binary protostar IRAS 1629-32422. Fedoseev et al. (2015) experimentally determined that the recombination reaction between two HCO radicals yields glyoxal (HC(O)CHO) at 13 K; sequential hydrogenation by 2 or 4 H-atoms turns HC(O)CHO into glycoladehyde and ethylene glycol, respectively. This is an important experimental finding, as so far surface hydrogenation reactions were mainly shown to be effective in the formation of relatively simple interstellar relevant species (e.g., H_2O,

NH_3, and CH_3OH), while radical-radical recombination reactions were considered efficient in the ISM only at temperatures higher than 20–30 K. The latter assumption comes from the fact that radicals are commonly assumed to become mobile around 30 K (e.g., Garrod 2013; Butscher et al. 2015). Therefore, diffusion of radicals is assumed to be the limiting factor to the formation of COMs through radical-radical recombination reactions. Fedoseev et al. (2015) gave the first laboratory evidence that diffusion is not necessarily needed for radicals to find each other. In CO-rich ices, HCO radicals can be formed next to one another through the hydrogenation of CO molecules and react without diffusing in the ice.

To further strengthen the conclusions of Fedoseev et al. (2015), experimental results are implemented into the model used by Cuppen et al. (2009) based on the continuous-time, random walk Monte-Carlo method described before. This model allows for the simulation of microscopic grain-surface chemistry for the long timescales typical in interstellar space, including the layering of ice during the CO freeze out. The choice of microscopic simulations is because HCO radicals have to stay in close vicinity to each other in order for a reaction to occur. Therefore, a method that accounts for the position of the species in the ice lattice is required for the best representation of the system. In the simulations reported by Fedoseev et al. (2015), five new species, including glycolaldehyde and ethylene glycol, are introduced in the simulations $HC(O)CHO$, $HC(O)CH_2O/HC(O)CHOH$, $HC(O)CH_2OH$, $H_2C(O)CH_2OH/HC(OH)CH_2OH$, and $H_2C(OH)CH_2OH$) The glycolaldehyde abundance was found to correlate with the formaldehyde abundance, while the abundance of ethylene glycol correlates with the abundance of methanol. This is not surprising since both CH_3OH and $H_2C(OH)CH_2OH$ are hydrogen saturated species while both H_2CO and $HC(O)CH_2OH$ are not. Finally, the correlation found between the abundances of $HC(O)CH_2OH$ and $H_2C(OH)CH_2O$ with those of formaldehyde and methanol is as expected typically in ratios of the order of a few percent and in good agreement with observations (Jørgensen et al. 2012; Schöier et al. 2002).

More recently, Chuang et al. (2016) presented new laboratory experiments on the low-temperature solid-state formation of three complex molecules, i.e., methyl formate ($HC(O)OCH_3$), glycolaldehyde and ethylene glycol, through recombination of free radicals formed via H-atom addition and abstraction reactions that occur during the hydrogenation of CO ice at 15

K. The experiments include the co-deposition of H atoms and pure H_2CO, H atoms and pure CH_3OH as well as H atoms and mixtures of H_2CO with CO, H_2CO with CH_3OH, and CH_3OH with CO, in fact extending previous CO hydrogenation studies (e.g., Fuchs et al. 2009; Fedoseev et al. 2015). The aim of these experiments is to resemble the physical-chemical conditions typical of the CO freeze-out stage in dark molecular clouds, when H_2CO and CH_3OH form by the recombination of accreting CO molecules and H atoms on ice grains. The occurrence of H-atom abstraction reactions in ice mantles leads to the formation of a higher number of intermediates (HCO, CH_3O, and CH_2OH) than previously thought, when assuming sequential H-atom addition reactions only. This enhances the probability to form COMs through radical-radical recombination without the need of UV photolysis or cosmic rays as external triggers.

Figure 10 shows the resulting COMs formation network based on the investigated reaction routes. From top to bottom, a sequence of H-atom addition and abstraction tunneling reactions is shown to lead to the formation of CH_3OH. The figure shows that H_2CO can undergo an abstraction reaction induced by H atoms to form HCO radicals, which can be further dehydrogenated to form CO. This pathway increases the total number of HCO formation events and its lifetime in the ice mantle. H_2CO also participates in addition reactions with H-atoms. Formation of CH_3O radicals is confirmed by the detection of solid methyl formate, a species formed most likely under the detection limit in Fedoseev et al. (2015). However, formation of CH_2OH radicals is not fully excluded by Chuang et al. (2016). In contrast, H-atom induced abstraction reaction involving CH_3OH likely yield CH_2OH while no proof for CH_3O formation is found by the authors. The barrier-less recombination of HCO intermediates yielding glyoxal followed by consequent hydrogenation (i.e., the mechanism investigated in Fedoseev et al. 2015) yields $HC(O)CH_2OH$ and $H_2C(OH)CH_2OH$ and is presented in the right panel of the reaction scheme in Fig. 10. Alternatively, the intermediate HCO radicals can directly recombine with CH_3O or CH_2OH to form $HC(O)OCH_3$ and $HC(O)CH_2OH$, respectively. According to the authors, the recombination of two CH_2OH radicals (dash arrow) seems to contribute less to the formation of $H_2C(OH)CH_2OH$.

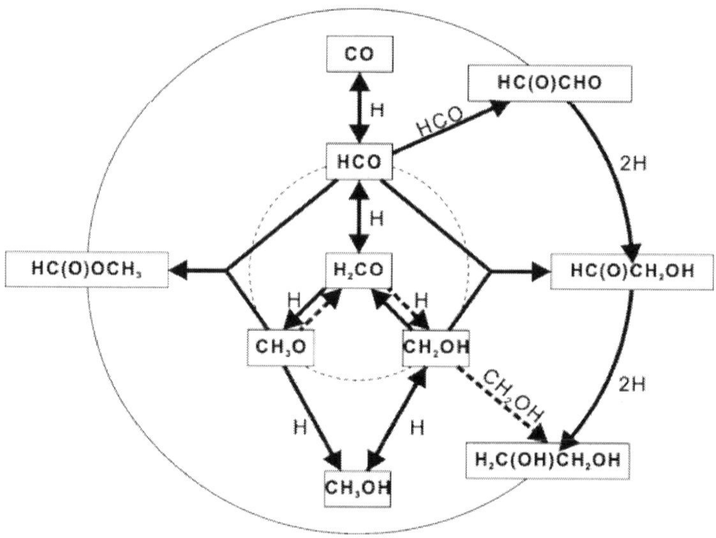

Figure. 10: Extended COM formation network as obtained from the CO, H_2CO, and CH_3OH hydrogenation experiments by Fedoseev et al. (2015) and Chuang et al. (2016). *Solid arrows* indicate the reaction less efficient pathways. Figure reproduced from Chuang et al. (2016)

Chuang et al. (2016) showed that both H-atom addition and abstraction reactions play an important role in effectively increasing the number of interaction events between reactants as well as the lifetime of radicals in the ice. This way, more COMs are formed through recombination of reactive intermediates than previously assumed at 10 K under interstellar conditions. It is important to note that Fuchs et al. (2009) did not include hydrogen abstraction reactions when calculating reaction rates. Clearly, the final products they observed in their experiments were the result of an equilibrium between addition and abstraction reactions. Therefore, the rates inferred by Fuchs et al. (2009) should be regarded as effective rates. The recent laboratory and modeling advances in our understanding of interstellar surface reactions discussed in this section give a strong indication that the COMs observed in cold dense molecular clouds have a solid phase non-energetic origin and are most likely linked to a combination of hydrogenation addition/abstraction reactions and radical-radical recombination reactions at 10 K. CO-rich ices formed upon the catastrophic freeze-out of CO molecules onto interstellar grains seem to be the right environment for these surface reaction to occur efficiently in the ISM.

OUTLOOK

In this review, we have summarized the "state-of-the-art" in terms of the methods used and the data available for treating grain-surface chemistry under astrophysical conditions. Since the first conceptual models of interstellar grain-surface chemistry, now more than 40 years old (e.g., Watson and Salpeter 1972a; Allen and Robinson 1977; Tielens and Hagen 1982), the general structure of a gas-grain code has not significantly changed. What has come on leaps and bounds is the advances in experimental techniques that have enabled the first quantification of many of the parameters required for models including binding energies, reaction barriers, diffusion rates, and desorption yields. Such experiments remain challenging to perform, and as reaction systems become more complex, the various potential chemical pathways and branching ratios remain difficult to disentangle. As discussed here, using sophisticated stochastic astrochemical models to also simulate laboratory data can help extract this information from the experiments which is then input into the larger rate-equation astrochemical models for simulating astrophysical environments.

Despite vast advances in the quantification of interstellar grain-surface chemistry in the past decades, there remains much more to learn. Typical grain-surface networks consist of sometimes hundreds of grain-surface species, the vast majority of which are reactive radicals. Such species are transient by nature and remain difficult to isolate and thus study in the laboratory. We urge the experimental community to continue to push the boundaries of what is possible in the laboratory and develop innovative methods to quantify the necessary data for radicals. In astrochemical models used to simulate interstellar and circumstellar environments, the data adopted for radicals remains the most uncertain.

Thanks to the effort over the past decades, we have a reasonably clear picture of the different processes important for ice chemistry, summarized in each of the Sections in this review. However, one chemical process we have only touched upon is ion-molecule chemistry in ice mantles (e.g., Woon 2011), the gas-phase equivalent of which is arguably the most important for building molecular complexity in the gas phase in space.

A second process we have not covered in detail here, but debate remains on its relative importance, is the direct impact of cosmic-ray particles on ice-covered dust grains. Due the high energies attained by cosmic-ray particles (~ Mev–GeV), such a process is thought to have catastrophic effect on the composition of the ice due to the creation of an ionized particle track through the grain which can also destroy, ionize, and sublime molecules in an explosive manner from the ice mantle (e.g., D'Hendecourt et al. 1982; Leger et al. 1985). Such a process is also stochastic by nature, and remains challenging to account for in a "mean-field" approach, since each impact can have a different track on each individual grain, and not every dust grain will be impacted by a cosmic-ray particle. This process had fallen out of favor as it became increasingly evident that photodesorption was an efficient (and well understood) mechanism for releasing molecules from the ice mantle at low temperatures. Recent revisitations of this process suggest that the overall effect is of the same order of magnitude as other processes considered to influence the gas-grain balance in dark clouds such as photodesorption (e.g., Ivlev et al. 2015).

It is clear that much work remains in linking the observed gas-phase composition in dark clouds with the known ice composition, and astrochemical models remain critical for constraining the potential processes for determining the gas-grain balance. The outcome of this Lorentz Centre Workshop shows what can be achieved when scientists from various fields and disciplines (modelers, experimentalists, and observers) meet to intensively discuss a naturally interdisciplinary field, such as astrochemistry. We have presented the first detailed review of the data and methods used for interstellar/circumstellar grain-surface chemistry and made recommendations, based on a legacy of past and present experience and research, for the treatment of the reactions rates. We have also compiled in a systematic manner the ever-increasing pool of experimental and theoretical data currently available in the literature. As we move into a new era of observational astronomy, with ALMA and JWST both expected to have a transformational impact on the field of astrochemistry, we anticipate regular future meetings on this topic to review, compile, and update the current "state-of-the-art" for the treatment of grain-surface chemistry in astrochemical models.

REFERENCES

1. M. Accolla, E. Congiu, F. Dulieu, G. Manicò, H. Chaabouni, E. Matar, H. Mokrane, J.L. Lemaire, V. Pirronello, Changes in the morphology of interstellar ice analogues after hydrogen atom exposure. Phys. Chem. Chem. Phys. **13**, 8037 (2011). doi:10.1039/C0CP01462A
2. K. Acharyya, Laboratory study of sticking and desorption of H2H2 and its significance in the chemical evolution of dense interstellar medium. Mon. Not. R. Astron. Soc. **443**, 1301–1309 (2014). doi:10.1093/mnras/stu1219
3. K. Acharyya, G.W. Fuchs, H.J. Fraser, E.F. van Dishoeck, H. Linnartz, Desorption of CO and O2O2 interstellar ice analogs. Astron. Astrophys. **466**, 1005–1012 (2007). doi:10.1051/0004-6361:20066272
4. N.G. Adams, D. Smith, D.C. Clary, Rate coefficients of the reactions of ions with polar molecules at interstellar temperatures. Astrophys. J. **296**, 31 (1985). doi:10.1086/184543
5. D.A. Adriaens, T.P.M. Goumans, C.R.A. Catlow, W.A. Brown, Computational study of carbonyl sulphide formation on model interstellar dust grains. J. Phys. Chem. C **114**(4), 1892–1900 (2010). doi:10.1021/jp9083212
6. M. Agúndez, V. Wakelam, Chemistry of dark clouds: databases, networks, and models. Chem. Rev. **113**, 8710 (2013). doi:10.1021/cr4001176
7. Y. Aikawa, S.M. Miyama, T. Nakano, T. Umebayashi, Evolution of molecular abundance in gaseous disks around young stars: depletion of CO molecules. Astrophys. J. **467**, 684 (1996). doi:10.1086/177644
8. A. Al-Halabi, A.W. Kleyn, E.F. van Dishoeck, G.J. Kroes, Sticking of hydrogen atoms to crystalline ice surfaces: dependence on incidence energy and surface temperature. J. Phys. Chem. B **106**, 6515–6522 (2002)
9. A. Al-Halabi, A.W. Kleyn, E.F. van Dishoeck, M.C. van Hemert, G.J. Kroes, Sticking of hyperthermal CO to the (0001) face of crystalline ice. J. Phys. Chem. A **107**, 10615–10624 (2003)
10. A. Al-Halabi, H.J. Fraser, G.J. Kroes, E.F. van Dishoeck, Adsorption of CO on amorphous water-ice surfaces. Astron. Astrophys. **422**, 777–791 (2004). doi:10.1051/0004-6361:20035939
11. T. Albertsson, D.A. Semenov, A.I. Vasyunin, T. Henning, E. Herbst, New extended deuterium fractionation model: assessment at dense

ISM conditions and sensitivity analysis. Astrophys. J. **207**, 27 (2013). doi:10.1088/0067-0049/207/2/27

12. L. Allamandola, S. Sandford, G. Valero, Photochemical and thermal evolution of interstellar/precometary ice analogs. Icarus **76**, 225 (1988). doi:10.1016/0019-1035(88)90070-X
13. M. Allen, G.W. Robinson, The molecular composition of dense interstellar clouds. Astrophys. J. **212**, 396 (1977). doi:10.1086/155059
14. L. Amiaud, F. Dulieu, J.-H. Fillion, A. Momeni, J.L. Lemaire, Interaction of atomic and molecular deuterium with a nonporous amorphous water ice surface between 8 and 30 K. J. Chem. Phys. **127**(14), 144709 (2007). doi:10.1063/1.2746323
15. E. Anders, Pre-biotic organic matter from comets and asteroids. Nature **342**, 255 (1989). doi:10.1038/342255a0
16. S. Andersson, E.F. van Dishoeck, Photodesorption of water ice. A molecular dynamics study. Astron. Astrophys. **491**, 907–916 (2008). doi:10.1051/0004 6361:200810374
17. S. Andersson, G. Kroes, E.F. van Dishoeck, Photodissociation of water in crystalline ice: a molecular dynamics study. Chem. Phys. Lett. **408**, 415 (2005). doi:10.1016/j.cplett.2005.04.073
18. S. Andersson, A. Al-Halabi, G.-J. Kroes, E.F. van Dishoeck, Molecular-dynamics study of photodissociation of water in crystalline and amorphous ices. J. Chem. Phys. **124**(6), 064715 (2006). doi:10.1063/1.2162901
19. S. Andersson, T.P.M. Goumans, A. Arnaldsson, Tunneling in hydrogen and deuterium atom addition to CO at low temperatures. Chem. Phys. Lett. **513**, 31–36 (2011). doi:10.1016/j.cplett.2011.07.073
20. C. Arasa, S. Andersson, H.M. Cuppen, E.F. van Dishoeck, G. Kroes, Molecular dynamics simulations of the ice temperature dependence of water ice photodesorption. J. Chem. Phys. **132**(18), 184510 (2010). doi:10.1063/1.3422213
21. C. Arasa, S. Andersson, H.M. Cuppen, E.F. van Dishoeck, G.J. Kroes, Molecular dynamics simulations of D2OD2O ice photodesorption. J. Chem. Phys. **134**(16), 164503 (2011). doi:10.1063/1.3582910
22. C. Arasa, J. Koning, G. Kroes, C. Walsh, E.F. van Dishoeck, Photodesorption of H2OH2O, HDO, and D2OD2O ice and its impact on fractionation. Astron. Astrophys. **575**, 121 (2015). doi:10.1051/0004-6361/201322695

23. Z. Awad, T. Chigai, Y. Kimura, O.M. Shalabiea, T. Yamamoto, New rate constants of hydrogenation of CO on H2OCOH2OCO ice surfaces. Astrophys. J. **626**, 262 (2005). doi:10.1086/429856
24. A. Bacmann, V. Taquet, A. Faure, C. Kahane, C. Ceccarelli, Detection of complex organic molecules in a prestellar core: a new challenge for astrochemical models. Astron. Astrophys. **541**, 12 (2012). doi:10.1051/0004-6361/201219207
25. N. Balucani, C. Ceccarelli, V. Taquet, Formation of complex organic molecules in cold objects: the role of gas-phase reactions. Mon. Not. R. Astron. Soc. **449**, 16 (2015). doi:10.1093/mnrasl/slv009
26. V. Barone, C. Latouche, D. Skouteris, F. Vazart, N. Balucani, C. Ceccarelli, B. Lefloch, Gas-phase formation of the prebiotic molecule formamide: insights from new quantum computations. Mon. Not. R. Astron. Soc. **453**, 31–35 (2015). doi:10.1093/mnrasl/slv094
27. D.R. Bates, L.J. Spitzer, The density of molecules in interstellar space. Astrophys. J. **113**, 441 (1951). doi:10.1086/145415
28. E.R. Batista, P. Ayotte, A. Bilić, B.D. Kay, H. Jónsson, What determines the sticking probability of water molecules on ice? Phys. Rev. Lett. **95**, 223201 (2005). doi:10.1103/PhysRevLett.95.223201
29. R.P. Bell, *The Tunnel Effect in Chemistry* (Chapman and Hall, London, 1980)
30. M.B. Bell, P.A. Feldman, M.J. Travers, M.C. McCarthy, C.A. Gottlieb, P. Thaddeus, Detection of HC[TINF]11[/TINF]N in the Cold Dust Cloud TMC-1. Astrophys. J. **483**, 61 (1997). doi:10.1086/310732
31. A. Belloche, H.S.P. Müller, K.M. Menten, P. Schilke, C. Comito, Complex organic molecules in the interstellar medium: IRAM 30 m line survey of Sagittarius B2(N) and (M). Astron. Astrophys. **559**, 47 (2013). doi:10.1051/0004-6361/201321096
32. A. Belloche, R.T. Garrod, H.S.P. Muller, K.M. Menten, Detection of a branched alkyl molecule in the interstellar medium: iso-propyl cyanide. Science **345**, 1584 (2014). doi:10.1126/science.1256678
33. E.A. Bergin, W.D. Langer, P.F. Goldsmith, Gas-phase chemistry in dense interstellar clouds including grain surface molecular depletion and desorption. Astrophys. J. **441**, 222 (1995). doi:10.1086/175351
34. J.B. Bergner, K.I. Öberg, M. Rajappan, E.C. Fayolle, Kinetics and mechanisms of the acid-base reaction between NH_3 and HCOOH in interstellar ice analogs. Astrophys. J. **829**, 85 (2016). doi:10.3847/0004-

637X/829/2/85

35. O. Berné, G. Mulas, C. Joblin, Interstellar C60+C60+. Astron. Astrophys. **550**, 4 (2013). doi:10.1051/0004-6361/201220730
36. M. Bertin, E.C. Fayolle, C. Romanzin, K.I. Öberg, X. Michaut, A. Moudens, L. Philippe, P. Jeseck, H. Linnartz, J. Fillion, UV photodesorption of interstellar CO ice analogues: from subsurface excitation to surface desorption. Phys. Chem. Chem. Phys. **14**, 9929 (2012). doi:10.1039/c2cp41177f
37. M. Bertin, E.C. Fayolle, C. Romanzin, H.A.M. Poderoso, X. Michaut, L. Philippe, P. Jeseck, K.I. Öberg, H. Linnartz, J. Fillion, Indirect ultraviolet photodesorption from CO:N_2 binary ices an efficient grain-gas process. Astrophys. J. **779**, 120 (2013). doi:10.1088/0004-637X/779/2/120
38. M. Bertin, C. Romanzin, M. Doronin, L. Philippe, P. Jeseck, N. Ligterink, H. Linnartz, X. Michaut, J. Fillion, UV photodesorption of methanol in pure and co-rich ices: desorption rates of the intact molecule and of the photofragments. Astrophys. J. **817**, 12 (2016). doi:10.3847/2041-8205/817/2/L12
39. O. Biham, I. Furman, V. Pirronello, G. Vidali, Master equation for hydrogen recombination on grain surfaces. Astrophys. J. **553**, 595–603 (2001). doi:10.1086/320975
40. S.E. Bisschop, G.W. Fuchs, E.F. van Dishoeck, H. Linnartz, H-atom bombardment of CO_2, HCOOH, and CH_3CHO containing ices. Astron. Astrophys. **474**, 1061 (2007a). doi:10.1051/0004-6361:20078210
41. S.E. Bisschop, J.K. Jørgensen, E.F. van Dishoeck, E.B.M. de Wachter, Testing grain-surface chemistry in massive hot-core regions. Astron. Astrophys. **465**, 913–929 (2007b). doi:10.1051/0004-6361:20065963
42. A.S. Bolina, W.A. Brown, Studies of physisorbed ammonia overlayers adsorbed on graphite. Surf. Sci. **598**, 45–56 (2005). doi:10.1016/j.susc.2005.08.025
43. A.S. Bolina, A.J. Wolff, W.A. Brown, Reflection absorption infrared spectroscopy and temperature programmed desorption investigations of the interaction of methanol with a graphite surface. J. Chem. Phys. **122**(4), 044713 (2005a). doi:10.1063/1.1839554
44. A.S. Bolina, A.J. Wolff, W.A. Brown, Reflection absorption infrared spectroscopy and temperature-programmed desorption studies of the adsorption and desorption of amorphous and crystalline water

on a graphite surface. J. Phys. Chem. B **109**, 16836–16845 (2005b). doi:10.1021/jp0528111

45. A.A. Boogert, P.A. Gerakines, D.C. Whittet, Observations of the Icy Universe. Annu. Rev. Astron. Astrophys. **53**, 541 (2015). doi:10.1146/annurev-astro-082214-122348
46. J.B. Bossa, P. Theule, F. Duvernay, T. Chiavassa, NH_2CH_2OH thermal formation in interstellar ices contribution to the 5–8 μm region toward embedded protostars. Astrophys. J. **707**, 1524–1532 (2009). doi:10.1088/0004-637X/707/2/1524
47. J. Bossa, K. Isokoski, M.S. de Valois, H. Linnartz, Thermal collapse of porous interstellar ice. Astron. Astrophys. **545**, 82 (2012). doi:10.1051/0004-6361/201219340
48. W.A. Brown, A.S. Bolina, Fundamental data on the desorption of pure interstellar ices. Mon. Not. R. Astron. Soc. **374**, 1006–1014 (2007). doi:10.1111/j.1365-2966.2006.11216.x
49. P.D. Brown, S.B. Charnley, Chemical models of interstellar gas-grain processes. I. Modelling and the effect of accretion on gas abundances and mantle composition in dense clouds. Mon. Not. R. Astron. Soc. **244**, 432–443 (1990)
50. V. Buch, R. Czerminski, Eigenstates of a quantum-mechanical particle on a topologically disordered surface—H(D) atom physisorbed on an amorphous ice cluster $(H_2O)115$. J. Chem. Phys. **95**, 6026–6038 (1991). doi:10.1063/1.461571
51. J.R. Burke, D.J. Hollenbach, The gas-grain interaction in the interstellar medium—thermal accommodation and trapping. Astrophys. J. **265**, 223–234 (1983). doi:10.1086/160667
52. T. Butscher, F. Duvernay, P. Theule, G. Danger, Y. Carissan, D. Hagebaum-Reignier, T. Chiavassa, Formation mechanism of glycolaldehyde and ethylene glycol in astrophysical ices from HCO and CH_2OH recombination: an experimental study. Mon. Not. R. Astron. Soc. **453**, 1587 (2015). doi:10.1093/mnras/stv1706
53. J. Cami, J. Bernard-Salas, E. Peeters, S.E. Malek, Detection of C60C60 and C70C70 in a Young Planetary Nebula. Science **329**, 1180 (2010). doi:10.1126/science.1192035
54. E.K. Campbell, M. Holz, D. Gerlich, J.P. Maier, Laboratory confirmation of C60+C60+ as the carrier of two diffuse interstellar bands. Nature **523**, 322 (2015). doi:10.1038/nature14566

55. P. Caselli, T.I. Hasegawa, E. Herbst, A proposed modification of the rate equations for reactions on grain surfaces. Astrophys. J. **495**, 309 (1998). doi:10.1086/305253
56. S. Cazaux, V. Cobut, M. Marseille, M. Spaans, P. Caselli, Water formation on bare grains: when the chemistry on dust impacts interstellar gas. Astron. Astrophys. **522**, 74 (2010). doi:10.1051/0004-6361/201014026
57. J. Cernicharo, N. Marcelino, E. Roueff, M. Gerin, A. Jiménez-Escobar, G.M. Muñoz Caro, Discovery of the methoxy radical, CH_3O, toward B1: dust grain and gas-phase chemistry in cold dark clouds. Astrophys. J. **759**, 43 (2012). doi:10.1088/2041-8205/759/2/L43
58. H. Chaabouni, H. Bergeron, S. Baouche, F. Dulieu, E. Matar, E. Congiu, L. Gavilan, J.L. Lemaire, Sticking coefficient of hydrogen and deuterium on silicates under interstellar conditions. Astron. Astrophys. **538**, 128 (2012). doi:10.1051/0004-6361/201117409
59. Q. Chang, E. Herbst, Interstellar simulations using a unified microscopic-macroscopic Monte Carlo model with a full gas-grain network including bulk diffusion in ice mantles. Astrophys. J. **787**, 135 (2014). doi:10.1088/0004-637X/787/2/135
60. Q. Chang, E. Herbst, Unified microscopic–macroscopic Monte Carlo simulations of complex organic molecule chemistry in cold cores. Astrophys. J. **819**, 145 (2016). doi:10.3847/0004-637X/819/2/145
61. Q. Chang, H.M. Cuppen, E. Herbst, Continuous-time random-walk simulation of H2H2 formation on interstellar grains. Astron. Astrophys. **434**, 599–611 (2005). doi:10.1051/0004-6361:20041842
62. Q. Chang, H.M. Cuppen, E. Herbst, Gas-grain chemistry in cold interstellar cloud cores with a microscopic Monte Carlo approach to surface chemistry. Astron. Astrophys. **469**, 973–983 (2007). doi:10.1051/0004-6361:20077423
63. S.B. Charnley, Stochastic theory of molecule formation on dust. Astrophys. J. Lett. **562**, 99–102 (2001). doi:10.1086/324753
64. S.B. Charnley, A.G.G.M. Tielens, T.J. Millar, On the molecular complexity of the hot cores in Orion A—grain surface chemistry as 'The last refuge of the scoundrel'. Astrophys. J. **399**, 71 (1992). doi:10.1086/186609
65. Y. Chen, K. Chuang, G.M. Muñoz Caro, M. Nuevo, C. Chu, T. Yih, W. Ip, C.R. Wu, Vacuum ultraviolet emission spectrum measurement of a

microwave-discharge hydrogen-flow lamp in several configurations: application to photodesorption of CO ice. Astrophys. J. **781**, 15 (2014). doi:10.1088/0004-637X/781/1/15

66.

67. K.J. Chuang, G. Fedoseev, S. Ioppolo, E.F. van Dishoeck, H. Linnartz, H-atom addition and abstraction reactions in mixed CO, H_2CO and CH3OH ices—an extended view on complex organic molecule formation. Mon. Not. R. Astron. Soc. **455**, 1702 (2016). doi:10.1093/mnras/stv2288

68. C. Chyba, P. Thomas, L. Brookshaw, C. Sagan, Cometary delivery of organic molecules to the early Earth. Science **249**, 366 (1990). doi:10.1126/science.11538074

69. M.P. Collings, J.W. Dever, H.J. Fraser, M.R.S. McCoustra, Laboratory studies of the interaction of carbon monoxide with water ice. Astrophys. Space Sci. **285**, 633–659 (2003). doi:10.1023/A:1026144806831

70. M.P. Collings, M.A. Anderson, R. Chen, J.W. Dever, S. Viti, D.A. Williams, M.R.S. McCoustra, A laboratory survey of the thermal desorption of astrophysically relevant molecules. Mon. Not. R. Astron. Soc. **354**, 1133–1140 (2004). doi:10.1111/j.1365-2966.2004.08272.x

71. M.P. Collings, V.L. Frankland, J. Lasne, D. Marchione, A. Rosu-Finsen, M.R.S. McCoustra, Probing model interstellar grain surfaces with small molecules. Mon. Not. R. Astron. Soc. **449**, 1826–1833 (2015). doi:10.1093/mnras/stv425

72. E. Congiu, M. Minissale, S. Baouche, H. Chaabouni, A. Moudens, S. Cazaux, G. Manicò, V. Pirronello, F. Dulieu, Efficient diffusive mechanisms of O atoms at very low temperatures on surfaces of astrophysical interest. Faraday Discuss. **168**, 151 (2014). doi:10.1039/C4FD00002A

73. G. Cooper, N. Kimmich, W. Belisle, J. Sarinana, K. Brabham, L. Garrel, Carbonaceous meteorites as a source of sugar-related organic compounds for the early Earth. Nature **414**, 879 (2001). doi:10.1038/414879a

74. G.A. Cruz-Diaz, G.M. Muñoz Caro, Y. Chen, T. Yih, Vacuum-UV spectroscopy of interstellar ice analogs. Astron. Astrophys. **562**, 119 (2014a). doi:10.1051/0004-6361/201322140

75. G.A. Cruz-Diaz, G.M. Muñoz Caro, Y. Chen, T. Yih, Vacuum-UV spectroscopy of interstellar ice analogs. Astron. Astrophys. **562**, 120

(2014b). doi:10.1051/0004-6361/201322621

76. G.A. Cruz-Diaz, R. Martín-Doménech, G.M. Muñoz Caro, Y. Chen, Negligible photodesorption of methanol ice and active photon-induced desorption of its irradiation products. Astron. Astrophys. **592**, 68 (2016). doi:10.1051/0004-6361/201526761
77. H.M. Cuppen, E. Herbst, Simulation of the formation and morphology of ice mantles on interstellar grains. Astrophys. J. **668**, 294–309 (2007). doi:10.1086/521014
78. H.M. Cuppen, E.F. van Dishoeck, E. Herbst, A.G.G.M. Tielens, Microscopic simulation of methanol and formaldehyde ice formation in cold dense cores. Astron. Astrophys. **508**, 275–287 (2009). doi:10.1051/0004-6361/200913119
79. H.M. Cuppen, E.M. Penteado, K. Isokoski, N. van der Marel, H. Linnartz, CO ice mixed with CH_3OH: the answer to the non-detection of the 2152 cm−12152 cm−1 band? Mon. Not. R. Astron. Soc. **417**, 2809 (2011). doi:10.1111/j.1365-2966.2011.19443.x
80. A. Dalgarno, J.H. Black, Molecule formation in the interstellar gas. Rep. Prog. Phys. **39**, 573 (1976). doi:10.1088/0034-4885/39/6/002
81. L.B. D'Hendecourt, L.J. Allamandola, F. Baas, J.M. Greenberg, Interstellar grain explosions—molecule cycling between gas and dust. Astron. Astrophys. **109**, 12–14 (1982)
82. L.B. D'Hendecourt, L.J. Allamandola, J.M. Greenberg, Time dependent chemistry in dense molecular clouds. I. Grain surface reactions, gas/grain interactions and infrared spectroscopy. Astron. Astrophys. **152**, 130–150 (1985)
83. L.B. D'Hendecourt, L.J. Allamandola, R.J.A. Grim, J.M. Greenberg, Time-dependent chemistry in dense molecular clouds. II. Ultraviolet photoprocessing and infrared spectroscopy of grain mantles. Astron. Astrophys. **158**, 119–134 (1986)
84. A.E. Douglas, G. Herzberg, Note on CH+CH+ in Interstellar Space and in the Laboratory. Astrophys. J. **94**, 381 (1941). doi:10.1086/144342
85. B.T. Draine, Interstellar dust grains. Annu. Rev. Astron. Astrophys. **41**, 241–289 (2003). doi:10.1146/annurev.astro.41.011802.094840
86. B.T. Draine, B. Sutin, Collisional charging of interstellar grains. Astrophys. J. **320**, 803 (1987). doi:10.1086/165596
87. F. Dulieu, E. Congiu, J. Noble, S. Baouche, H. Chaabouni, A. Moudens, M. Minissale, S. Cazaux, How micron-sized dust particles determine

the chemistry of our Universe. Sci. Rep. **3**, 1338 (2013). doi:10.1038/srep01338

88. A. Dutrey, D. Semenov, E. Chapillon, U. Gorti, S. Guilloteau, F. Hersant, M. Hogerheijde, M. Hughes, G. Meeus, H. Nomura, V. Piétu, C. Qi, V. Wakelam, Physical and chemical structure of planet-forming disks probed by millimeter observations and modeling, in *Protostars and Planets VI* (2014), pp. 317–338. doi:10.2458/azu_uapress_9780816531240-ch014
89. C. Eckart, The penetration of a potential barrier by electrons. Phys. Rev. **35**, 1303–1309 (1930). doi:10.1103/PhysRev.35.1303
90. A.S. Eddington, Bakerian lecture. Diffuse matter in interstellar space. Proc. R. Soc. A, Math. Phys. Eng. Sci. **111**, 424 (1926). doi:10.1098/rspa.1926.0076
91. P. Ehrenfreund, S.B. Charnley, Organic molecules in the interstellar medium, comets, and meteorites: a voyage from dark clouds to the early Earth. Annu. Rev. Astron. Astrophys. **38**, 427 (2000). doi:10.1146/annurev.astro.38.1.427
92. J.E. Elsila, D.P. Glavin, J.P. Dworkin, Cometary glycine detected in samples returned by Stardust. Meteorit. Planet. Sci. **44**, 1323 (2009). doi:10.1111/j.1945-5100.2009.tb01224.x
93. E.C. Fayolle, K.I. Öberg, H.M. Cuppen, R. Visser, H. Linnartz, Laboratory H_2O:CO_2 ice desorption data: entrapment dependencies and its parameterization with an extended three-phase model. Astron. Astrophys. **529**, 74 (2011). doi:10.1051/0004-6361/201016121
94. E.C. Fayolle, K.I. Öberg, H.M. Cuppen, R. Visser, H. Linnartz, Laboratory H_2O:CO_2 ice desorption: entrapment and its parameterization with an extended three-phase model, in *European Conference on Laboratory Astrophysics*, ECLA. EAS Publications Series, vol. 58 (2012), pp. 327–331. doi:10.1051/eas/1258053
95. E.C. Fayolle, M. Bertin, C. Romanzin, H.A.M. Poderoso, L. Philippe, X. Michaut, P. Jeseck, H. Linnartz, K.I. Öberg, J. Fillion, Wavelength-dependent UV photodesorption of pure N_2 and O_2 ices. Astron. Astrophys. **556**, 122 (2013). doi:10.1051/0004-6361/201321533
96. G. Fedoseev, H.M. Cuppen, S. Ioppolo, T. Lamberts, H. Linnartz, Experimental evidence for glycolaldehyde and ethylene glycol formation by surface hydrogenation of CO molecules under dense molecular cloud conditions. Mon. Not. R. Astron. Soc. **448**, 1288–

1297 (2015). doi:10.1093/mnras/stu2603

97. D. Field, O. Plekan, A. Cassidy, R. Balog, N. Jones, J. Dunger, Spontaneous electric fields in solid films: spontelectrics. Int. Rev. Phys. Chem. **32**, 345 (2013). doi:10.1080/0144235X.2013.767109
98. J. Fillion, E.C. Fayolle, X. Michaut, M. Doronin, L. Philippe, J. Rakovsky, C. Romanzin, N. Champion, K.I. Öberg, H. Linnartz, M. Bertin, Wavelength resolved UV photodesorption and photochemistry of CO_2 ice. Faraday Discuss. **168**, 533 (2014). doi:10.1039/C3FD00129F
99. H.J. Fraser, M.P. Collings, M.R.S. McCoustra, D.A. Williams, Thermal desorption of water ice in the interstellar medium. Mon. Not. R. Astron. Soc. **327**, 1165–1172 (2001)
100. H.J. Fraser, M.P. Collings, J.W. Dever, M.R.S. McCoustra, Using laboratory studies of CO-H_2O ices to understand the non-detection of a 2152 cm−12152 cm−1 (4.647 μm) band in the spectra of interstellar ices. Mon. Not. R. Astron. Soc. **353**, 59–68 (2004). doi:10.1111/j.1365-2966.2004.08038.x
101. G.W. Fuchs, K. Acharyya, S.E. Bisschop, K.I. Öberg, F.A. van Broekhuizen, H.J. Fraser, S. Schlemmer, E.F. van Dishoeck, H. Linnartz, Comparative studies of O_2 and N_2 in pure, mixed and layered CO ices. Faraday Discuss. **133**, 331 (2006). doi:10.1039/b517262b
102. G.W. Fuchs, H.M. Cuppen, S. Ioppolo, C. Romanzin, S.E. Bisschop, S. Andersson, E.F. van Dishoeck, H. Linnartz, Hydrogenation reactions in interstellar CO ice analogues. A combined experimental/theoretical approach. Astron. Astrophys. **505**, 629–639 (2009). doi:10.1051/0004-6361/200810784
103. K. Furuya, Y. Aikawa, H. Nomura, F. Hersant, V. Wakelam, Water in protoplanetary disks: deuteration and turbulent mixing. Astrophys. J. **779**, 11 (2013). doi:10.1088/0004-637X/779/1/11
104. K. Furuya, M.N. Drozdovskaya, R. Visser, E.F. van Dishoeck, C. Walsh, D. Harsono, U. Hincelin, V. Taquet, Water delivery from cores to disks: deuteration as a probe of the prestellar inheritance of H2OH2O. Astron. Astrophys. (2016, in press). doi:10.1051/0004-6361/201629269
105. R.T. Garrod, A new modified-rate approach for gas-grain chemical simulations. Astron. Astrophys. **491**, 239–251 (2008). doi:10.1051/0004-6361:200810518
106. R.T. Garrod, A three-phase chemical model of hot cores: the formation of glycine. Astrophys. J. **765**, 60 (2013). doi:10.1088/0004-

637X/765/1/60

107. R.T. Garrod, E. Herbst, Formation of methyl formate and other organic species in the warm-up phase of hot molecular cores. Astron. Astrophys. **457**, 927–936 (2006). doi:10.1051/0004-6361:20065560
108. R.T. Garrod, T. Pauly, On the formation of CO_2 and other interstellar ices. Astrophys. J. **735**, 15 (2011). doi:10.1088/0004-637X/735/1/15
109. R. Garrod, I.H. Park, P. Caselli, E. Herbst, Are gas-phase models of interstellar chemistry tenable? The case of methanol. Faraday Discuss. **133**, 51 (2006). doi:10.1039/b516202e
110. R.T. Garrod, V. Wakelam, E. Herbst, Non-thermal desorption from interstellar dust grains via exothermic surface reactions. Astron. Astrophys. **467**, 1103–1115 (2007). doi:10.1051/0004-6361:20066704
111. R.T. Garrod, S.L.W. Weaver, E. Herbst, Complex chemistry in star-forming regions: an expanded gas-grain warm-up chemical model. Astrophys. J. **682**, 283–302 (2008). doi:10.1086/588035
112. R.T. Garrod, A.I. Vasyunin, D.A. Semenov, D.S. Wiebe, T. Henning, A new modified-rate approach for gas-grain chemistry: comparison with a unified large-scale Monte Carlo simulation. Astrophys. J. Lett. **700**, 43–46 (2009). doi:10.1088/0004-637X/700/1/L43
113. W.D. Geppert, M. Hamberg, R.D. Thomas, F. Österdahl, F. Hellberg, V. Zhaunerchyk, A. Ehlerding, T.J. Millar, H. Roberts, J. Semaniak, M. af Ugglas, A. Källberg, A. Simonsson, M. Kaminska, M. Larsson, Dissociative recombination of protonated methanol. Faraday Discuss. **133**, 177 (2006). doi:10.1039/b516010c
114. P.A. Gerakines, W.A. Schutte, P. Ehrenfreund, Ultraviolet processing of interstellar ice analogs. I. Pure ices. Astron. Astrophys. **312**, 289–305 (1996)
115. D.T. Gillespie, A general method for numerically simulating the stochastic time evolution of coupled chemical reactions. J. Comp. Physiol. **22**, 403 (1976). doi:10.1016/0021-9991(76)90041-3
116. J. Gonzalez, M. Caballero, A. Aguilar-Mogas, M. Torrent-Sucarrat, R. Crehuet, A. Solé, X. Giménez, S. Olivella, J.M. Bofill, J.M. Anglada, The reaction between HO and (H2O)n(H2O)n (n=1,3n=1,3) clusters: reaction mechanisms and tunneling effects. Theor. Chem. Acc. **128**(4–6), 579–592 (2011). doi:10.1007/s00214-010-0824-5
117. E. Gonzalez-Lavado, J.C. Corchado, Y.V. Suleimanov, W.H. Green, J. Espinosa-Garcia, Theoretical kinetics study of the o(3p)+ch4/

cd4o(3p)+ch4/cd4 hydrogen abstraction reaction: the role of anharmonicity, recrossing effects, and quantum mechanical tunneling. J. Phys. Chem. A **118**(18), 3243–3252 (2014). doi:10.1021/jp5028965

118. R.J. Gould, E.E. Salpeter, The interstellar abundance of the hydrogen molecule. I. Basic processes. Astrophys. J. **138**, 393 (1963). doi:10.1086/147654
119. T.P.M. Goumans, S. Andersson, Tunnelling in the O+COO+CO reaction. Mon. Not. R. Astron. Soc. **406**, 2213–2217 (2010). doi:10.1111/j.1365-2966.2010.16836.x
120. T.P.M. Goumans, J. Kästner, Hydrogen-atom tunneling could contribute to h_2 formation in space. Angew. Chem. **49**(40), 7350–7352 (2010). doi:10.1002/anie.201001311
121. T.P.M. Goumans, J. Kästner, Deuterium enrichment of interstellar methanol explained by atom tunneling. J. Phys. Chem. A **115**(39), 10767–10774 (2011). doi:10.1021/jp206048f
122. T.P.M. Goumans, C.R.A. Catlow, W.A. Brown, J. Kästner, P. Sherwood, An embedded cluster study of the formation of water on interstellar dust grains. Phys. Chem. Chem. Phys. **11**, 5431 (2009). doi:10.1039/b816905e
123. T. Grassi, S. Bovino, D.R.G. Schleicher, J. Prieto, D. Seifried, E. Simoncini, F.A. Gianturco, KROME—a package to embed chemistry in astrophysical simulations. Mon. Not. R. Astron. Soc. **439**, 2386 (2014). doi:10.1093/mnras/stu114
124. N.J.B. Green, T. Toniazzo, M.J. Pilling, D.P. Ruffle, N. Bell, T.W. Hartquist, A stochastic approach to grain surface chemical kinetics. Astron. Astrophys. **375**, 1111–1119 (2001). doi:10.1051/0004-6361:20010961
125. S.D. Green, A.S. Bolina, R. Chen, M.P. Collings, W.A. Brown, M.R.S. McCoustra, Applying laboratory thermal desorption data in an interstellar context: sublimation of methanol thin films. Mon. Not. R. Astron. Soc. **398**, 357–367 (2009). doi:10.1111/j.1365-2966.2009.15144.x
126. L.T. Greenberg, Photodesorption from interstellar dust grains, in *Interstellar Dust and Related Topics*, ed. by J.M. Greenberg, H.C. van de Hulst. IAU Symposium, vol. 52 (1973), p. 413
127. R.J.A. Grim, J.M. Greenberg, M.S. de Groot, F. Baas, W.A. Schutte, B. Schmitt, Infrared spectroscopy of astrophysical ices—new insights in

the photochemistry. Astron. Astrophys. Suppl. Ser. **78**, 161–186 (1989)

128. W. Hagen, L.J. Allamandola, J.M. Greenberg, Interstellar molecule formation in grain mantles—the laboratory analog experiments, results and implications. Astrophys. Space Sci. **65**, 215–240 (1979). doi:10.1007/BF00643502
129. T. Hama, N. Watanabe, Surface processes on interstellar amorphous solid water: adsorption, diffusion, tunneling reactions, and nuclear-spin conversion. Chem. Rev. **113**(12), 8783–8839 (2013). doi:10.1021/cr4000978
130. T. Hama, K. Kuwahata, N. Watanabe, A. Kouchi, Y. Kimura, T. Chigai, V. Pirronello, The mechanism of surface diffusion of H and D atoms on amorphous solid water: existence of various potential sites. Astrophys. J. **757**, 185 (2012). doi:10.1088/0004-637X/757/2/185
131. T. Hama, H. Ueta, A. Kouchi, N. Watanabe, H. Tachikawa, Quantum tunneling hydrogenation of solid benzene and its control via surface structure. J. Phys. Chem. Lett. **5**(21), 3843–3848 (2014). doi:10.1021/jz5019948
132. T.W. Hartquist, D.A. Williams, Cosmic-ray induced desorption and high depletions in dense cores. Mon. Not. R. Astron. Soc. **247**, 343 (1990)
133. T.I. Hasegawa, E. Herbst, New gas-grain chemical models of quiescent dense interstellar clouds—the effects of H_2 tunneling reactions and cosmic ray induced desorption. Mon. Not. R. Astron. Soc. **261**, 83–102 (1993a)
134. T.I. Hasegawa, E. Herbst, Three-phase chemical models of dense interstellar clouds—gas dust particle mantles and dust particle surfaces. Mon. Not. R. Astron. Soc. **263**, 589 (1993b)
135. T.I. Hasegawa, E. Herbst, C.M. Leung, Models of gas-grain chemistry in dense interstellar clouds with complex organic molecules. Astrophys. J. Suppl. Ser. **82**, 167–195 (1992). doi:10.1086/191713
136. J. He, D. Jing, G. Vidali, Atomic oxygen diffusion on and desorption from amorphous silicate surfaces. Phys. Chem. Chem. Phys. **16**, 3493 (2014). doi:10.1039/c3cp54328e
137. J. He, J. Shi, T. Hopkins, G. Vidali, M.J. Kaufman, A new determination of the binding energy of atomic oxygen on dust grain surfaces: experimental results and simulations. Astrophys. J. **801**, 120 (2015). doi:10.1088/0004-637X/801/2/120

138. T.J.H. Hele, M.J. Willatt, A. Muolo, S.C. Althorpe, Communication: relation of centroid molecular dynamics and ring-polymer molecular dynamics to exact quantum dynamics. J. Chem. Phys. **142**(19), 191101 (2015). doi:10.1063/1.4921234

139. T. Henning, D. Semenov, Chemistry in protoplanetary disks. Chem. Rev. **113**, 9016 (2013). doi:10.1021/cr400128p

140. E. Herbst, H.M. Cuppen, Interstellar chemistry special feature: Monte Carlo studies of surface chemistry and nonthermal desorption involving interstellar grains. Proc. Natl. Acad. Sci. USA **103**, 12257–12262 (2006). doi:10.1073/pnas.0601556103

141. E. Herbst, W. Klemperer, The formation and depletion of molecules in dense interstellar clouds. Astrophys. J. **185**, 505–534 (1973). doi:10.1086/152436

142. E. Herbst, C.M. Leung, Effects of large rate coefficients for ion-polar neutral reactions on chemical models of dense interstellar clouds. Astrophys. J. **310**, 378 (1986). doi:10.1086/164691

143. E. Herbst, T.J. Millar, *Low Temperatures and Cold Molecules* (Imperial College Press, London, 2008)

144. E. Herbst, E.F. van Dishoeck, Complex organic interstellar molecules. Annu. Rev. Astron. Astrophys. **47**, 427–480 (2009). doi:10.1146/annurev-astro-082708-101654

145. H. Hidaka, N. Watanabe, T. Shiraki, A. Nagaoka, A. Kouchi, Conversion of H_2CO to CH_3OH by reactions of cold atomic hydrogen on ice surfaces below 20 K. Astrophys. J. **614**, 1124 (2004). doi:10.1086/423889

146. H. Hidaka, A. Kouchi, N. Watanabe, Temperature, composition, and hydrogen isotope effect in the hydrogenation of CO on amorphous ice surface at 10–20 K. J. Chem. Phys. **126**(20), 204707 (2007). doi:10.1063/1.2735573

147. H. Hidaka, M. Watanabe, A. Kouchi, N. Watanabe, Reaction routes in the CO-H2CO-dn-CH3OH-dmCO-H2CO-dn-CH3OH-dm system clarified from H(D) exposure of solid formaldehyde at low temperatures. Astrophys. J. **702**, 291–300 (2009). doi:10.1088/0004-637X/702/1/291

148. K. Hiraoka, K. Yamamoto, Y. Kihara, T. Takayama, T. Sato, Reaction of H atoms with solid C_2H_4 and C_2H_6 at 13 K. Astrophys. J. **514**, 524–528 (1999). doi:10.1086/306942

149. K. Hiraoka, T. Sato, T. Takayama, in *Astrochemistry: From Molecular*

Clouds to Planetary Systems, ed. by Y.C. Minh, E.F. van Dishoeck (Astron. Soc. Pacific, Chelse, 2000), p. 283

150. K. Hiraoka, K. Sato, S. Sato, S. Hishiki, K. Suzuki, Y. Takahashi, T. Yokoyama, S. Kitagawa, Formation of amorphous silicon by the low-temperature tunneling reaction of H atoms with solid thin film of SiH_4 at 10 K. J. Phys. Chem. B **105**, 6950–6955 (2001). doi:10.1021/jp010553b
151. K. Hiraoka, T. Sato, S. Sato, N. Sogoshi, T. Yokoyama, H. Takashima, S. Kitagawa, Astrophys. J. **577**, 265 (2002)
152. D. Hollenbach, E.E. Salpeter, Surface recombination of hydrogen molecules. Astrophys. J. **163**, 155–164 (1971)
153. D.J. Hollenbach, A.G.G.M. Tielens, Dense Photodissociation Regions (PDRs) 1. Annu. Rev. Astron. Astrophys. **35**, 179 (1997). doi:10.1146/annurev.astro.35.1.179
154. S. Ioppolo, H.M. Cuppen, C. Romanzin, E.F. van Dishoeck, H. Linnartz, Water formation at low temperatures by surface O_2 hydrogenation I: characterization of ice penetration. Phys. Chem. Chem. Phys. **12**, 12065 (2010). doi:10.1039/c0cp00250j
155. K. Isokoski, J. Bossa, T. Triemstra, H. Linnartz, Porosity and thermal collapse measurements of H_2O, CH_3OH, CO_2, and H_2O:CO_2 ices. Phys. Chem. Chem. Phys. **16**, 3456 (2014). doi:10.1039/c3cp54481h
156. A.V. Ivlev, T.B. Röcker, A. Vasyunin, P. Caselli, Impulsive spot heating and thermal explosion of interstellar grains revisited. Astrophys. J. **805**, 59 (2015). doi:10.1088/0004-637X/805/1/59
157. J.K. Jørgensen, C. Favre, S.E. Bisschop, T.L. Bourke, E.F. van Dishoeck, M. Schmalzl, Detection of the simplest sugar, glycolaldehyde, in a solar-type protostar with ALMA. Astrophys. J. **757**, 4 (2012). doi:10.1088/2041-8205/757/1/L4
158. J. Kalvāns, I. Shmeld, Subsurface chemistry of mantles of interstellar dust grains in dark molecular cores. Astron. Astrophys. **521**, 37 (2010). doi:10.1051/0004-6361/201014190
159. L.J. Karssemeijer, H.M. Cuppen, Diffusion-desorption ratio of adsorbed CO and CO_2 on water ice. Astron. Astrophys. **569**, 107 (2014). doi:10.1051/0004-6361/201424792
160. L.J. Karssemeijer, A. Pedersen, H. Jónsson, H.M. Cuppen, Long-timescale simulations of diffusion in molecular solids. Phys. Chem. Chem. Phys. **14**, 10844–10852 (2012). doi:10.1039/C2CP41634D

161. L.J. Karssemeijer, G.A. de Wijs, H.M. Cuppen, Interactions of adsorbed CO_2 on water ice at low temperatures. Phys. Chem. Chem. Phys. **16**, 15630–15639 (2014a). doi:10.1039/C4CP01622J
162. L.J. Karssemeijer, S. Ioppolo, M.C. van Hemert, A. van der Avoird, M.A. Allodi, G.A. Blake, H.M. Cuppen, Dynamics of CO in amorphous water-ice environments. Astrophys. J. **781**, 16 (2014b). doi:10.1088/0004-637X/781/1/16
163. J. Kästner, Theory and simulation of atom tunneling in chemical reactions. WIREs Comput. Mol. Sci. **4**(2), 158–168 (2014). doi:10.1002/wcms.1165
164. N. Katz, I. Furman, O. Biham, V. Pirronello, G. Vidali, Molecular hydrogen formation on astrophysically relevant surfaces. Astrophys. J. **522**, 305–312 (1999). doi:10.1086/307642
165. J. Koning, G.J. Kroes, C. Arasa, Isotope effects on the photodesorption processes of X_2O (X=H,DX=H,D) and HOD ice. J. Chem. Phys. **138**, 104701 (2013). doi:10.1063/1.4793733
166. K. Kuwahata, T. Hama, A. Kouchi, N. Watanabe, Signatures of quantum-tunneling diffusion of hydrogen atoms on water ice at 10K. Phys. Rev. Lett. **115**, 133201 (2015). doi:10.1103/PhysRevLett.115.133201
167. K. Kvenvolden, J. Lawless, K. Pering, E. Peterson, J. Flores, C. Ponnamperuma, I.R. Kaplan, C. Moore, Evidence for extraterrestrial amino-acids and hydrocarbons in the murchison meteorite. Nature **228**, 923 (1970). doi:10.1038/228923a0
168. T. Lamberts, H.M. Cuppen, S. Ioppolo, H. Linnartz, Water formation at low temperatures by surface O_2 hydrogenation III: Monte Carlo simulation. Phys. Chem. Chem. Phys. **15**, 8287 (2013). doi:10.1039/c3cp00106g
169. T. Lamberts, H.M. Cuppen, G. Fedoseev, S. Ioppolo, K.-J. Chuang, H. Linnartz, Relevance of the H2+OH2+O reaction pathway for the surface formation of interstellar water. Combined experimental and modeling study. Astron. Astrophys. **570**, 57 (2014a). doi:10.1051/0004-6361/201424252
170. T. Lamberts, X. de Vries, H.M. Cuppen, The formation of ice mantles on interstellar grains revisited "the effect of exothermicity". Faraday Discuss. **168**, 327 (2014b). doi:10.1039/C3FD00136A
171. T. Lamberts, S. Ioppolo, H.M. Cuppen, G. Fedoseev, H. Linnartz, Thermal H/D exchange in polar ice—deuteron scrambling in space.

Mon. Not. R. Astron. Soc. **448**, 3820–3828 (2015). doi:10.1093/mnras/stv278

172. T. Lamberts, G. Fedoseev, F. Puletti, S. Ioppolo, H.M. Cuppen, H. Linnartz, Low-temperature chemistry between water and hydroxyl radicals: H/D isotopic effects. Mon. Not. R. Astron. Soc. **455**, 634 (2016). doi:10.1093/mnras/stv2305
173. T. Lauck, L. Karssemeijer, K. Shulenberger, M. Rajappan, K.I. Öberg, H.M. Cuppen, CO diffusion into amorphous H_2O ices. Astrophys. J. **801**, 118 (2015). doi:10.1088/0004-637X/801/2/118
174. R. Launhardt, A.M. Stutz, A. Schmiedeke, T. Henning, O. Krause, Z. Balog, H. Beuther, S. Birkmann, M. Hennemann, J. Kainulainen, T. Khanzadyan, H. Linz, N. Lippok, M. Nielbock, J. Pitann, S. Ragan, C. Risacher, M. Schmalzl, Y.L. Shirley, B. Stecklum, J. Steinacker, J. Tackenberg, The Earliest Phases of star Formation (EPoS): a Herschel key project. Astron. Astrophys. **551**, 98 (2013). doi:10.1051/0004-6361/201220477
175. A. Leger, M. Jura, A. Omont, Desorption from interstellar grains. Astron. Astrophys. **144**, 147–160 (1985)
176. B. Lepetit, D. Lemoine, Z. Medina, B. Jackson, Sticking and desorption of hydrogen on graphite: a comparative study of different models. J. Chem. Phys. **134**(11), 114705 (2011). doi:10.1063/1.3565446
177. C.M. Leung, E. Herbst, W.F. Huebner, Synthesis of complex molecules in dense interstellar clouds via gas-phase chemistry—a pseudo time-dependent calculation. Astrophys. J. **56**, 231 (1984). doi:10.1086/190982
178. Y. Li, Y.V. Suleimanov, J. Li, W.H. Green, H. Guo, Rate coefficients and kinetic isotope effects of the X+CH4→CH3+HX (X=H,D,Mu) X+CH4→CH3+HX (X=H,D,Mu) reactions from ring polymer molecular dynamics. J. Chem. Phys. **138**(9), 094307 (2013). doi:10.1063/1.4793394
179. N.F.W. Ligterink, D.M. Paardekooper, K. Chuang, M.L. Both, G.A. Cruz-Diaz, J.H. van Helden, H. Linnartz, Controlling the emission profile of an H_2 discharge lamp to simulate interstellar radiation fields. Astron. Astrophys. **584**, 56 (2015). doi:10.1051/0004-6361/201526930
180. H. Linnartz, S. Ioppolo, G. Fedoseev, Atom addition reactions in interstellar ice analogues. Int. Rev. Phys. Chem. **34**, 205 (2015). doi:10.1080/0144235X.2015.1046679

181. N. Lippok, R. Launhardt, T. Henning, Z. Balog, H. Beuther, J. Kainulainen, O. Krause, H. Linz, M. Nielbock, S.E. Ragan, T.P. Robitaille, S.I. Sadavoy, A. Schmiedeke, Earliest phases of star formation (EPoS). Astron. Astrophys. **592**, 61 (2016). doi:10.1051/0004-6361/201525792
182. A. Lipshtat, O. Biham, Efficient simulations of gas-grain chemistry in interstellar clouds. Phys. Rev. Lett. **93**(17), 170601 (2004). doi:10.1103/PhysRevLett.93.170601
183. R.M. Logan, J.C. Keck, Classical theory for the interaction of gas atoms with solid surfaces. J. Chem. Phys. **49**, 860–876 (1968). doi:10.1063/1.1670153
184. N.J. Mason, A. Dawes, P.D. Holtom, R.J. Mukerji, M.P. Davis, B. Sivaraman, R.I. Kaiser, S.V. Hoffmann, D.A. Shaw, VUV spectroscopy and photo-processing of astrochemical ices: an experimental study. Faraday Discuss. **133**, 311 (2006). doi:10.1039/b518088k
185. D. McElroy, C. Walsh, A.J. Markwick, M.A. Cordiner, K. Smith, T.J. Millar, The UMIST database for astrochemistry 2012. Astron. Astrophys. **550**, 36 (2013). doi:10.1051/0004-6361/201220465
186. A. McKellar, Evidence for the molecular origin of some Hitherto unidentified interstellar lines. Publ. Astron. Soc. Pac. **52**, 187 (1940). doi:10.1086/125159
187. T.J. Millar, P.R.A. Farquhar, K. Willacy, The UMIST database for astrochemistry 1995. Astron. Astrophys. **121**, 139 (1997). doi:10.1051/aas:1997118
188. M. Minissale, F. Dulieu, Influence of surface coverage on the chemical desorption process. J. Chem. Phys. **141**, 014304 (2014). doi:10.1063/1.4885847
189. M. Minissale, F. Dulieu, S. Cazaux, S. Hocuk, Dust as interstellar catalyst I. Quantifying the chemical desorption process. Astron. Astrophys. **585**, 24 (2016). doi:10.1051/0004-6361/201525981
190. F. Mispelaer, P. Theulé, H. Aouididi, J. Noble, F. Duvernay, G. Danger, P. Roubin, O. Morata, T. Hasegawa, T. Chiavassa, Diffusion measurements of CO, HNCO, H_2CO, and NH_3 in amorphous water ice. Astron. Astrophys. **555**, 13 (2013). doi:10.1051/0004-6361/201220691
191. N. Miyauchi, H. Hidaka, T. Chigai, A. Nagaoka, N. Watanabe, A. Kouchi, Formation of hydrogen peroxide and water from the reaction of cold hydrogen atoms with solid oxygen at 10 K. Chem. Phys. Lett.

456, 27–30 (2008). doi:10.1016/j.cplett.2008.02.095

192. M. Monge-Palacios, J.C. Corchado, J. Espinosa-Garcia, Dynamics study of the OH+NH3OH+NH3 hydrogen abstraction reaction using QCT calculations based on an analytical potential energy surface. J. Chem. Phys. **138**(21), 214306 (2013a). doi:10.1063/1.4808109
193. M. Monge-Palacios, C. Rangel, J. Espinosa-Garcia, Ab initio based potential energy surface and kinetics study of the OH+NH3OH+NH3 hydrogen abstraction reaction. J. Chem. Phys. **138**(8), 084305 (2013b). doi:10.1063/1.4792719
194. G.M. Muñoz Caro, A. Jiménez-Escobar, J.Á. Martín-Gago, C. Rogero, C. Atienza, S. Puertas, J.M. Sobrado, J. Torres-Redondo, New results on thermal and photodesorption of CO ice using the novel InterStellar Astrochemistry Chamber (ISAC). Astron. Astrophys. **522**, 108 (2010). doi:10.1051/0004-6361/200912462
195. A. Nagaoka, N. Watanabe, A. Kouchi, H-D substitution in interstellar solid methanol: a key route for D enrichment. Astrophys. J. Lett. **624**, 29–32 (2005). doi:10.1086/430304
196. A. Nagaoka, N. Watanabe, A. Kouchi, Effective rate constants for the surface reaction between solid methanol and deuterium atoms at 10 K. J. Phys. Chem. A **111**, 3016–3028 (2007). doi:10.1021/jp068978r
197. T.L. Nguyen, J.F. Stanton, Ab initio thermal rate calculations of HO+HO=O(3P)+H2OHO+HO=O(3P)+H2O reaction and isotopologues. J. Phys. Chem. A **117**(13), 2678–2686 (2013). doi:10.1021/jp312246q
198. T.L. Nguyen, J.F. Stanton, Accurate ab initio thermal rate constants for reaction of o(3p) with h2 and isotopic analogues. J. Phys. Chem. A **118**(27), 4918–4928 (2014). doi:10.1021/jp5037124
199. T.L. Nguyen, J.F. Stanton, J.R. Barker, Ab initio reaction rate constants computed using semiclassical transition-state theory: HO+H2→H2O+HHO+H2→H2O+H and isotopologues. J. Phys. Chem. A **115**(20), 5118–5126 (2011). doi:10.1021/jp2022743
200. T.L. Nguyen, B.C. Xue, R.E. Weston, J.R. Barker, J.F. Stanton, Reaction of HO with CO: tunneling is indeed important. J. Phys. Chem. Lett. **3**(11), 1549–1553 (2012). doi:10.1021/jz300443a
201. N. Nishi, H. Shinohara, T. Okuyama, Photodetachment, photodissociation, and photochemistry of surface molecules of icy solids containing NH3 and pure H_2O ices. J. Chem. Phys. **80**, 3898

(1984). doi:10.1063/1.447172

202. J.A. Noble, P. Theulé, F. Mispelaer, F. Duvernay, G. Danger, E. Congiu, F. Dulieu, T. Chiavassa, The desorption of H_2CO from interstellar grains analogues. Astron. Astrophys. **543**, 5 (2012a). doi:10.1051/0004-6361/201219437
203. J.A. Noble, E. Congiu, F. Dulieu, H.J. Fraser, Thermal desorption characteristics of CO, O2O2 and CO_2 on non-porous water, crystalline water and silicate surfaces at submonolayer and multilayer coverages. Mon. Not. R. Astron. Soc. **421**, 768–779 (2012b). doi:10.1111/j.1365-2966.2011.20351.x
204. J.A. Noble, P. Theule, F. Duvernay, G. Danger, T. Chiavassa, P. Ghesquiere, T. Mineva, D. Talbi, Kinetics of the NH_3 and CO_2 solid-state reaction at low temperature. Phys. Chem. Chem. Phys. **16**, 23604–23615 (2014). doi:10.1039/C4CP02414A
205. G. Nyman, Computational methods of quantum reaction dynamics. Int. J. Quant. Chem. **114**(18), 1183–1198 (2014). doi:10.1002/qua.24661
206. Y. Oba, N. Miyauchi, H. Hidaka, T. Chigai, N. Watanabe, A. Kouchi, Formation of compact amorphous H_2O ice by codeposition of hydrogen atoms with oxygen molecules on grain surfaces. Astrophys. J. **701**, 464–470 (2009). doi:10.1088/0004-637X/701/1/464
207. Y. Oba, N. Watanabe, T. Hama, K. Kuwahata, H. Hidaka, A. Kouchi, Water formation through a quantum tunneling surface reaction, OH+H2OH+H2, at 10 K. Astrophys. J. **749**, 67 (2012). doi:10.1088/0004-637X/749/1/67
208. Y. Oba, T. Chigai, Y. Osamura, N. Watanabe, A. Kouchi, Hydrogen isotopic substitution of solid methylamine through atomic surface reactions at low temperatures: a potential contribution to the D/H ratio of methylamine in molecular clouds. Meteorit. Planet. Sci. **49**, 117–132 (2014a). doi:10.1111/maps.12096
209. Y. Oba, K. Osaka, N. Watanabe, T. Chigai, A. Kouchi, Reaction kinetics and isotope effect of water formation by the surface reaction of solid H2O2H2O2 with H atoms at low temperatures. Faraday Discuss. **168**, 185–204 (2014b). doi:10.1039/C3FD00112A
210. Y. Oba, N. Watanabe, Y. Osamura, A. Kouchi, Chiral glycine formation on cold interstellar grains by quantum tunneling hydrogen–deuterium substitution reactions. Chem. Phys. Lett. **634**, 53–59 (2015). doi:10.1016/j.cplett.2015.05.070

211. K.I. Öberg, F. van Broekhuizen, H.J. Fraser, S.E. Bisschop, E.F. van Dishoeck, S. Schlemmer, Competition between CO and N_2 desorption from interstellar ices. Astrophys. J. Lett. **621**, 33–36 (2005). doi:10.1086/428901

212. K.I. Öberg, G.W. Fuchs, Z. Awad, H.J. Fraser, S. Schlemmer, E.F. van Dishoeck, H. Linnartz, Photodesorption of CO ice. Astrophys. J. Lett. **662**, 23–26 (2007). doi:10.1086/519281

213. K.I. Öberg, E.F. van Dishoeck, H. Linnartz, Photodesorption of ices I: CO, N_2, and CO_2. Astron. Astrophys. **496**, 281–293 (2009a). doi:10.1051/0004-6361/200810207

214. K.I. Öberg, H. Linnartz, R. Visser, E.F. van Dishoeck, Photodesorption of ices. II. H_2O and D_2O. Astrophys. J. **693**, 1209–1218 (2009b). doi:10.1088/0004-637X/693/2/1209

215. K.I. Öberg, R.T. Garrod, E.F. van Dishoeck, H. Linnartz, Formation rates of complex organics in UV irradiated CH_3OH-rich ices. I. Experiments. Astron. Astrophys. **504**, 891–913 (2009c). doi:10.1051/0004-6361/200912559

216. K.I. Öberg, E.C. Fayolle, H.M. Cuppen, E.F. van Dishoeck, H. Linnartz, Quantification of segregation dynamics in ice mixtures. Astron. Astrophys. **505**, 183–194 (2009d). doi:10.1051/0004-6361/200912464

217. K.I. Öberg, S. Bottinelli, J.K. Jørgensen, E.F. van Dishoeck, A cold complex chemistry toward the low-mass protostar B1-b: evidence for complex molecule production in ices. Astrophys. J. **716**, 825–834 (2010). doi:10.1088/0004-637X/716/1/825

218. K.I. Öberg, A.C.A. Boogert, K.M. Pontoppidan, S. van den Broek, E.F. van Dishoeck, S. Bottinelli, G.A. Blake, N.J. Evans II., The Spitzer ice legacy: ice evolution from cores to protostars. Astrophys. J. **740**, 109 (2011). doi:10.1088/0004-637X/740/2/109

219. D.H. Parker, Laser photochemistry of molecular oxygen. Acc. Chem. Res. **33**, 563 (2000). doi:10.1021/ar980012d

220. A. Pedersen, L. Karssemeijer, H.M. Cuppen, H. Jónsson, Long-timescale simulations of H2OH2O admolecule diffusion on Ice Ih(0001) surfaces. J. Phys. Chem. C **119**(29), 16528–16536 (2015). doi:10.1021/acs.jpcc.5b01629

221. E.M. Penteado, A.C.A. Boogert, K.M. Pontoppidan, S. Ioppolo, G.A. Blake, H.M. Cuppen, Spectroscopic constraints on CH_3OH formation: CO mixed with CH_3OH ices towards young stellar objects. Mon. Not.

R. Astron. Soc. **454**, 531–540 (2015). doi:10.1093/mnras/stv1987

222. E.M. Penteado, C.W. Walsh, H.M. Cuppen, Sensitivity analysis of grain surface chemistry to binding energies of ice species (2016, submitted)
223. V. Pirronello, O. Biham, C. Liu, L. Shen, G. Vidali, Efficiency of molecular hydrogen formation on silicates. Astrophys. J. Lett. **483**, 131 (1997a). doi:10.1086/310746
224. V. Pirronello, C. Liu, L. Shen, G. Vidali, Laboratory synthesis of molecular hydrogen on surfaces of astrophysical interest. Astrophys. J. Lett. **475**, 69 (1997b). doi:10.1086/310464
225. V. Pirronello, C. Liu, J.E. Roser, G. Vidali, Measurements of molecular hydrogen formation on carbonaceous grains. Astron. Astrophys. **344**, 681–686 (1999)
226. K.M. Pontoppidan, Spatial mapping of ices in the Ophiuchus-F core. A direct measurement of CO depletion and the formation of CO_2. Astron. Astrophys. **453**, 47–50 (2006). doi:10.1051/0004-6361:20065569
227. K.M. Pontoppidan, H.J. Fraser, E. Dartois, W. Thi, E.F. van Dishoeck, A.C.A. Boogert, L. d'Hendecourt, A.G.G.M. Tielens, S.E. Bisschop, A 3–5 μm VLT spectroscopic survey of embedded young low mass stars I. Structure of the CO ice. Astron. Astrophys. **408**, 981–1007 (2003). doi:10.1051/0004-6361:20031030
228. S.S. Prasad, S.P. Tarafdar, UV radiation field inside dense clouds—its possible existence and chemical implications. Astrophys. J. **267**, 603 (1983). doi:10.1086/160896
229. J.M.C. Rawlings, D.A. Williams, S. Viti, C. Cecchi-Pestellini, W.W. Duley, Episodic explosions in interstellar ices. Mon. Not. R. Astron. Soc. **430**, 264 (2013). doi:10.1093/mnras/sts601
230. A. Rimola, V. Taquet, P. Ugliengo, N. Balucani, C. Ceccarelli, Combined quantum chemical and modeling study of CO hydrogenation on water ice. Astron. Astrophys. **572**, 70 (2014). doi:10.1051/0004-6361/201424046
231. D.P. Ruffle, E. Herbst, New models of interstellar gas-grain chemistry. I. Surface diffusion rates. Mon. Not. R. Astron. Soc. **319**, 837–850 (2000). doi:10.1046/j.1365-8711.2000.03911.x
232. D.P. Ruffle, E. Herbst, New models of interstellar gas-grain chemistry. III. Solid CO_2. Mon. Not. R. Astron. Soc. **324**, 1054–1062 (2001). doi:10.1046/j.1365-8711.2001.04394.x
233. S.A. Sandford, L.J. Allamandola, The condensation and vaporization

behavior of H_2O:CO ices and implications for interstellar grains and cometary activity. Icarus **76**, 201–224 (1988). doi:10.1016/0019-1035(88)90069-3

234. S.A. Sandford, L.J. Allamandola, The volume- and surface-binding energies of ice systems containing CO, CO2, and H_2O. Icarus **87**, 188–192 (1990). doi:10.1016/0019-1035(90)90028-8
235. S.A. Sandford, L.J. Allamandola, Condensation and vaporization studies of CH3OH and NH3 ices: major implications for astrochemistry. Astrophys. J. **417**, 815–825 (1993a). doi:10.1086/173362
236. S.A. Sandford, L.J. Allamandola, H_2 in interstellar and extragalactic ices—infrared characteristics, ultraviolet production, and implications. Astrophys. J. Lett. **409**, 65–68 (1993b). doi:10.1086/186861
237. F.L. Schöier, J.K. Jrgensen, E.F. van Dishoeck, G.A. Blake, Does IRAS 16293-2422 have a hot core? Chemical inventory and abundance changes in its protostellar environment. Astron. Astrophys. **390**, 1001 (2002). doi:10.1051/0004-6361:20020756
238. D. Semenov, F. Hersant, V. Wakelam, A. Dutrey, E. Chapillon, S. Guilloteau, T. Henning, R. Launhardt, V. Piétu, K. Schreyer, Chemistry in disks. Astron. Astrophys. **522**, 42 (2010). doi:10.1051/0004-6361/201015149
239. L.E. Snyder, D. Buhl, B. Zuckerman, P. Palmer, Microwave detection of interstellar formaldehyde. Phys. Rev. Lett. **22**, 679 (1969). doi:10.1103/PhysRevLett.22.679
240. T. Stantcheva, E. Herbst, Models of gas-grain chemistry in interstellar cloud cores with a stochastic approach to surface chemistry. Astron. Astrophys. **423**, 241–251 (2004). doi:10.1051/0004-6361:20040433
241. T. Stantcheva, V.I. Shematovich, E. Herbst, On the master equation approach to diffusive grain-surface chemistry: The H, O, CO system. Astron. Astrophys. **391**, 1069–1080 (2002). doi:10.1051/0004-6361:20020838
242. A. Stutz, R. Launhardt, H. Linz, O. Krause, T. Henning, J. Kainulainen, M. Nielbock, J. Steinacker, P. André, Dust-temperature of an isolated star-forming cloud: Herschel observations of the Bok globule CB244. Astron. Astrophys. **518**, 87 (2010). doi:10.1051/0004-6361/201014537
243. Y.V. Suleimanov, J. Espinosa-Garcia, Recrossing and tunneling in the kinetics study of the OH+CH4→H2O+CH3OH+CH4→H2O+CH3 reaction. J. Phys. Chem. B (2015). doi:10.1021/acs.jpcb.5b02103

244. P. Swings, L. Rosenfeld, Considerations regarding interstellar molecules. Astrophys. J. **86**, 483 (1937). doi:10.1086/143880
245. V. Taquet, C. Ceccarelli, C. Kahane, Multilayer modeling of porous grain surface chemistry. I. The GRAINOBLE model. Astron. Astrophys. **538**, 42 (2012). doi:10.1051/0004-6361/201117802
246. V. Taquet, P.S. Peters, C. Kahane, C. Ceccarelli, A. López-Sepulcre, C. Toubin, D. Duflot, L. Wiesenfeld, Water ice deuteration: a tracer of the chemical history of protostars. Astron. Astrophys. **550**, 127 (2013). doi:10.1051/0004-6361/201220084
247. P. Theule, F. Duvernay, A. Ilmane, T. Hasegawa, O. Morata, S. Coussan, G. Danger, T. Chiavassa, Kinetics of the OCN−OCN− and HOCN formation from the HNCO+H2OHNCO+H2O thermal reaction in interstellar ice analogs. Astron. Astrophys. **530**, 96 (2011). doi:10.1051/0004-6361/201016051
248. A. Tielens, Interstellar polycyclic aromatic hydrocarbon molecules. Annu. Rev. Astron. Astrophys. **46**, 289 (2008). doi:10.1146/annurev.astro.46.060407.145211
249. A.G.G.M. Tielens, *The Physics and Chemistry of the Interstellar Medium* (2010)
250. A.G.G.M. Tielens, L.J. Allamandola, Composition, structure, and chemistry of interstellar dust, in *Interstellar Processes*, ed. by D.J. Hollenbach, H.A. Thronson Jr. Astrophysics and Space Science Library, vol. 134 (1987), pp. 397–469
251. A.G.G.M. Tielens, W. Hagen, Model calculations of the molecular composition of interstellar grain mantles. Astron. Astrophys. **114**, 245–260 (1982)
252. E.F. van Dishoeck, G.A. Blake, Chemical evolution of star-forming regions. Annu. Rev. Astron. Astrophys. **36**, 317 (1998). doi:10.1146/annurev.astro.36.1.317
253. E.F. van Dishoeck, B. Jonkheid, M.C. van Hemert, Photoprocesses in protoplanetary disks. Faraday Discuss. **133**, 231 (2006). doi:10.1039/b517564j
254. M.C. van Hemert, J. Takahashi, E.F. van Dishoeck, Molecular dynamics study of the photodesorption of CO ice. J. Phys. Chem. A **119**, 6354 (2015). doi:10.1021/acs.jpca.5b02611
255. C. Vastel, C. Ceccarelli, B. Lefloch, R. Bachiller, The origin of complex organic molecules in prestellar cores. Astrophys. J. **795**, 2 (2014).

doi:10.1088/2041-8205/795/1/L2

256. A.I. Vasyunin, E. Herbst, A unified Monte Carlo treatment of gas-grain chemistry for large reaction networks. II. A multiphase gas-surface-layered bulk model. Astrophys. J. **762**, 86 (2013). doi:10.1088/0004-637X/762/2/86

257. A.I. Vasyunin, D.A. Semenov, D.S. Wiebe, T. Henning, A unified Monte Carlo treatment of gas-grain chemistry for large reaction networks. I. Testing validity of rate equations in molecular clouds. Astrophys. J. **691**, 1459–1469 (2009). doi:10.1088/0004-637X/691/2/1459

258. V.K. Veeraghattam, K. Manrodt, S.P. Lewis, P.C. Stancil, The sticking of atomic hydrogen on amorphous water ice. Astrophys. J. **790**, 4 (2014). doi:10.1088/0004-637X/790/1/4

259. S. Viti, M.P. Collings, J.W. Dever, M.R.S. McCoustra, D.A. Williams, Evaporation of ices near massive stars: models based on laboratory temperature programmed desorption data. Mon. Not. R. Astron. Soc. **354**, 1141–1145 (2004). doi:10.1111/j.1365-2966.2004.08273.x

260. V. Wakelam, I.W.M. Smith, E. Herbst, J. Troe, W. Geppert, H. Linnartz, K. Öberg, E. Roueff, M. Agúndez, P. Pernot, H.M. Cuppen, J.C. Loison, D. Talbi, Reaction networks for interstellar chemical modelling: improvements and challenges. Space Sci. Rev. **156**, 13–72 (2010). doi:10.1007/s11214-010-9712-5

261. V. Wakelam, E. Herbst, J.-C. Loison, I.W.M. Smith, V. Chandrasekaran, B. Pavone, N.G. Adams, M.-C. Bacchus-Montabonel, A. Bergeat, K. Béroff, V.M. Bierbaum, M. Chabot, A. Dalgarno, E.F. van Dishoeck, A. Faure, W.D. Geppert, D. Gerlich, D. Galli, E. Hébrard, F. Hersant, K.M. Hickson, P. Honvault, S.J. Klippenstein, S. Le Picard, G. Nyman, P. Pernot, S. Schlemmer, F. Selsis, I.R. Sims, D. Talbi, J. Tennyson, J. Troe, R. Wester, L. Wiesenfeld, A KInetic Database for Astrochemistry (KIDA). Astrophys. J. Suppl. Ser. **199**, 21 (2012). doi:10.1088/0067-0049/199/1/21

262. C. Walsh, T.J. Millar, H. Nomura, E. Herbst, S. Widicus Weaver, Y. Aikawa, J.C. Laas, A.I. Vasyunin, Complex organic molecules in protoplanetary disks. Astron. Astrophys. **563**, 33 (2014). doi:10.1051/0004-6361/201322446

263. N. Watanabe, A. Kouchi, Efficient formation of formaldehyde and methanol by the addition of hydrogen atoms to CO in H_2O-CO ice at 10 K. Astrophys. J. Lett. **571**, 173–176 (2002). doi:10.1086/341412

264. N. Watanabe, A. Kouchi, Ice surface reactions: a key to chemical evolution in space. Prog. Surf. Sci. **83**, 439 (2008). doi:10.1016/j.progsurf.2008.10.001
265. N. Watanabe, A. Nagaoka, H. Hidaka, T. Shiraki, T. Chigai, A. Kouchi, Dependence of the effective rate constants for the hydrogenation of CO on the temperature and composition of the surface. Planet. Space Sci. **54**, 1107–1114 (2006). doi:10.1016/j.pss.2006.05.019
266. N. Watanabe, Y. Kimura, A. Kouchi, T. Chigai, T. Hama, V. Pirronello, Direct measurements of hydrogen atom diffusion and the spin temperature of nascent H2H2 molecule on amorphous solid water. Astrophys. J. Lett. **714**, 233–237 (2010). doi:10.1088/2041-8205/714/2/L233
267. W.D. Watson, Ion-molecule reactions, molecule formation, and hydrogen-isotope exchange in dense interstellar clouds. Astrophys. J. **188**, 35 (1974). doi:10.1086/152681
268. W.D. Watson, E.E. Salpeter, Molecule formation on interstellar grains. Astrophys. J. **174**, 321 (1972a). doi:10.1086/151492
269. W.D. Watson, E.E. Salpeter, On the abundances of interstellar molecules. Astrophys. J. **175**, 659 (1972b). doi:10.1086/151587
270. M.S. Westley, R.A. Baragiola, R.E. Johnson, G.A. Baratta, Photodesorption from low-temperature water ice in interstellar and circumsolar grains. Nature **373**, 405 (1995a). doi:10.1038/373405a0
271. M.S. Westley, R.A. Baragiola, R.E. Johnson, G.A. Baratta, Ultraviolet photodesorption from water ice. Planet. Space Sci. **43**, 1311–1315 (1995b). doi:10.1016/0032-0633(95)00088-M
272. R.E. Weston, T.L. Nguyen, J.F. Stanton, J.R. Barker, HO+COHO+CO reaction rates and H/D kinetic isotope effects: master equation models with ab initio SCTST rate constants. J. Phys. Chem. A **117**, 821–835 (2013). doi:10.1021/jp311928w
273. E. Wigner, The transition state method. Trans. Faraday Soc. **34**, 29 (1938). doi:10.1039/TF9383400029
274. K. Willacy, W.D. Langer, The importance of photoprocessing in protoplanetary disks. Astrophys. J. **544**, 903 (2000). doi:10.1086/317236
275. K. Willacy, H.H. Klahr, T.J. Millar, T. Henning, Gas and grain chemistry in a protoplanetary disk. Astron. Astrophys. **338**, 995–1005 (1998)
276. A. Wolff, I. Lohmar, J. Krug, O. Biham, Diffusion-limited reactions on disordered surfaces with continuous distributions of binding

energies. J. Stat. Mech. Theory Exp. **10**, 29 (2011). doi:10.1088/1742-5468/2011/10/P10029

277. P.M. Woods, B. Slater, Z. Raza, S. Viti, W.A. Brown, D.J. Burke, Glycolaldehyde formation via the dimerization of the formyl radical. Astrophys. J. **777**, 90 (2013). doi:10.1088/0004-637X/777/2/90
278. D.E. Woon, Ion-ice astrochemistry: barrierless low-energy deposition pathways to HCOOH, CH_3OH, and CO_2 on icy grain mantles from precursor cations. Astrophys. J. **728**, 44 (2011). doi:10.1088/0004-637X/728/1/44
279. Y. Zhang, T. Stecher, M.T. Cvitas, S.C. Althorpe, Which is better at predicting quantum-tunneling rates: quantum transition-state theory or free-energy instanton theory? J. Phys. Chem. Lett. **5**, 3976–3980 (2014). doi:10.1021/jz501889v
280. J. Zhen, H. Linnartz, UV-induced photodesorption and photochemistry of O_2 ice. Mon. Not. R. Astron. Soc. **437**, 3190 (2014). doi:10.1093/mnras/stt2106

10

Experimental Study of the Penetration of Oxygen and Deuterium Atoms into Porous Water Ice

M. Minissale[1], T. Nguyen[2] and F. Dulieu[2]

[1] Aix-Marseille Université, CNRS, PIIM, Marseille, France
[2] LERMA, Université de Cergy Pontoise, Sorbonne Université, PSL Research University, Observatoire de Paris, UMR 8112 CNRS, 5 mail Gay Lussac, 95000 Cergy Pontoise, France

ABSTRACT

Context. Many interstellar molecules are thought to form on dust grains. In particular, hydrogenation is one of the major mechanisms of the formation of mantle ice. To date it is not clear if H atoms can penetrate the bulk of the ice mantle or if it only has chemical activity on the accessible surface of grains.

Aims. We wish to study the efficiency of atoms deposited on the outer surface of the amorphous solid water to penetrate into the ice bulk.

Citation: M. Minissale, T. Nguyen and F. Dulieu "Experimental study of the penetration of oxygen and deuterium atoms into porous water ice" A&A, 622 (2019) A148 DOI: https://doi.org/10.1051/0004-6361/201833830

Methods. NO molecules react with O and H atoms. They are easily detected by infrared (IR) spectroscopy. These two properties make this molecule an ideal chemical tracer for the penetration of O and H atoms through water ice. In our experiments we first deposited a NO undercoat and covered this layer (at 40 K) with a variable amount of water ice. Then, we exposed this undercoat to D (10 K) or O (40 K) atoms, and we followed the NO consumption and the products that appeared via IR signatures, and we finally analyzed the desorption of all species through a temperature-programmed desorption technique. We experimentally characterize the accessible surface of the ice and provide a model to interpret quantitatively our measurements.

Results. Water ice limits the destruction of tracer NO molecules. The thicker the ice, the more NO remains unreacted. H and O atoms lead to the same amount of NO consumption, pointing out that access to reactants for these two different atoms is identical. We discuss different possible scenarios of NO localization (in and/or on the ice) and determine how this affects our observables (IR data and desorption profiles).

Conclusions. In our experimental conditions, it is not possible to measure any atom penetration through the bulk of the ice. The surface diffusion followed by reaction with NO or by self-reaction (i.e., $H + H \rightarrow H_2$) is faster than bulk diffusion. We propose lower limit values for penetration barriers. Therefore the building of astrophysical ice mantles should be mostly driven by surface reactivity.

Key words: ISM: molecules / molecular processes / astrochemistry / diffusion / atomic processes / solid state: volatile

INTRODUCTION

The formation of certain molecules (such as H_2, H_2O or CO_2) in different astrophysical environments is possible thanks to the catalytic effect of dust grains (e.g., Anders et al. 1974; Pirronello et al. 1999; Cazaux et al. 2010, and references therein). In dense clouds, where the UV photons flux is rapidly attenuated, atoms and molecules accumulate on dust grains and react to form increasingly complex molecules (Caselli & Ceccarelli 2012). The diffusion of the lightest species, in particular hydrogen and oxygen atoms, is often considered as the trigger of chemical reactions that are at the base of the growth of these molecular mantles and differ in composition from the gaseous phase. It was shown very early (Tielens & Hagen 1982) that grain chemistry was mainly limited by diffusion (Cuppen et al. 2017). Experimentally it has been demonstrated that H, O, and N atoms can effectively diffuse on the surface of iced grains at low temperature (~10 K; Matar et al. 2008; Watanabe et al. 2010; Hama et al. 2012; Minissale et al. 2013a, 2016; Congiu et al. 2014; Wakelam et al. 2017).

This type of catalysis dominated by surface diffusion is very effective. The synthesis of methanol from the CO-hydrogenation is an emblematic example, since this molecule is abundantly produced and detected (e.g., Kristensen et al. 2010; Ceccarelli et al. 2017) despite the reaction barriers (Hiraoka et al. 2002; Hidaka et al. 2007; Rimola et al. 2014). It is not yet well understood if once the molecular mantle is built as a consequence of surface reactions, this mantle can evolve because of reactions in its interior. It is known for example that UV penetrates many molecular layers (e.g., 1 μm at 163 nm, Orzechowska et al. 2007) creating radicals in the inner layers of the mantle leading to chemical recomposition. It is therefore necessary to understand what is the diffusion in the bulk. Several studies have focused on the experimental determination of bulk diffusion rates of molecules such as NH_3, CO, and CO_2 (Mispelaer et al. 2013; Lauck et al. 2015; He et al. 2017; Cooke et al. 2018; Ghesquière et al. 2018) on relevant astrophysical ice, but little is known about the bulk diffusion of atoms. However, it is very important to determine if and how the diffusion of atoms takes place in ice, since some increase of the mobility may be expected for atoms, especially H. An efficient diffusion in bulk can indeed allow a rich chemistry in the ice of dense clouds thus favoring the dynamic and temporal evolution of the ice in its whole thickness. A lack of bulk mobility, in the absence of subsequent processing, causes the ice composition to be governed by the surface diffusion and the composition of the accreted gas until the ice is heated or processed.

Water ice is the main compound of astrophysical ices (Boogert et al. 2015), but it is a multiform matrix. In astrophysical environments, water ice can be crystalline or amorphous, referred to as amorphous solid water (ASW), but in dense clouds water iced is thought to be amorphous because some energy processes must take place before it crystallizes (Papoular 2005; Palumbo 2006). Furthermore amorphous ice can be porous or compact: if it is deposited at low temperature (<70 K), it is porous (Stevenson et al. 1999), but it can be recompacted by ionic or UV irradiation (Palumbo et al. 2010) or by simple chemical activity such as hydrogen recombination (Accolla et al. 2011). Finally, if the ice forms directly on the dust grains, it is amorphous and compact (Oba et al. 2009; Accolla et al. 2013). Of course, the diffusion rates of species in the bulk of ice depend on the ice morphology (amorphous or crystalline) and its degree of porosity.

The reactivity of atoms with another compound inserted in the ice (used as a chemical tracer) can be used to experimentally measure atomic diffusion on/in the water ice. If the reaction is barrier-free, the disappearance of the reactive species allows us to infer the access of the atoms to the reactants, and hence their mobility. This is the method we used in Matar et al. (2008) using O_2 consumption as a marker for surface diffusion of D. If we now deposit water ice on top of a reactive partner, by examining the consumption and formation of species according to the thickness of the water ice deposited on top of it, we must be able to deduce the penetration of atoms into the ice. In the present case we use NO, which has no reaction barrier with H (Congiu et al. 2012) and O (Minissale et al. 2014b) and is detectable in the IR, contrary to O_2. However this approach presents certain constraints from an experimental point of view. Compact water ice (amorphous or crystalline) can only grow at sample temperatures above 100 K. Such temperatures are sufficiently high to induce immediate desorption of the "marker", the reactive compound NO. We are therefore limited by the desorption temperature of the chemical marker (i.e., ~ 40 K for NO). This temperature does not allow for the formation of compact ice. The amorphous and porous water ice is what can be studied by this method.

There is a general consensus that water ice in dark clouds is mainly amorphous, but the nature of its porosity degree still remains poorly known (Öberg et al. 2009, 2011; Accolla et al. 2013). For this reason we decided to perform a first experimental study of volumetric diffusion rates of H and O atoms on porous ASW.

Strictly speaking, diffusion on porous ice is both a surface diffusion and a bulk diffusion. Figure 1 depicts this situation. There is surface diffusion on the accessible surface of the pores of the ice, which can be located "inside the ice" and therefore this surface diffusion allows the ice to be swept within its volume by the diffusive atoms. This scenario is sometimes called percolation. Bulk diffusion may occur, but refers to the crossing of a compact/closed ice layer. In particular, if closed pores are present, the atoms must penetrate through the wall of the pores before possibly continuing the surface diffusion on the inner surface of the closed pore.

For the rest of the article we call "surface diffusion" the diffusion on both inner and outer surfaces. Thus, the atoms can penetrate inthe depth of the ice through surface diffusion, which may appear misleading. This is because of the complex structure of the ice that we used. The bulk diffusion corresponds to the diffusion that requires crossing a water layer.

Our article is organized as follows: the following section describes the experimental setup and ice characterization. In Sect. 3 we presents the experimental results. In Sect. 4, we propose a model of rate equations that simulates our experiments. In the last section, we discuss the main conclusions and astrophysical implications of this study.

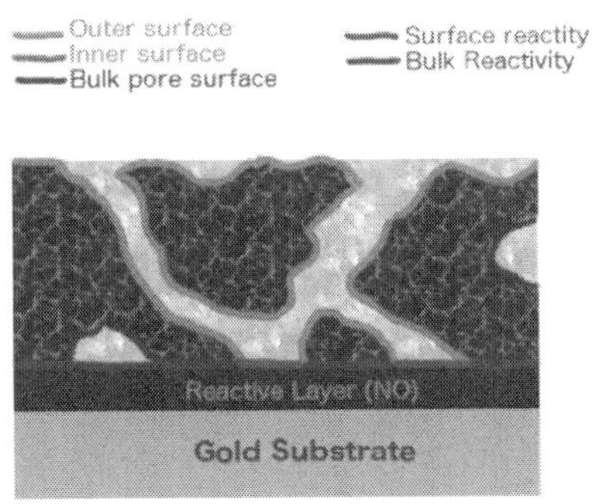

Figure 1. Schematic drawing of the porous amorphous ice grown on top of a NO layer deposited on a gold surface. The outer surface (in green) is the surface directly accessible from the gas phase (with a normal incidence). The inner surface (in blue) is the surface accessible via surface diffusion. Bulk diffusion is required to access closed pores (dark blue). The reactive layer can be accessed via surface diffusion (both inner and outer surfaces) along the cracks or pores. This case is represented in light red. Part of the reactive layer that required bulk diffusion is drawn in dark red.

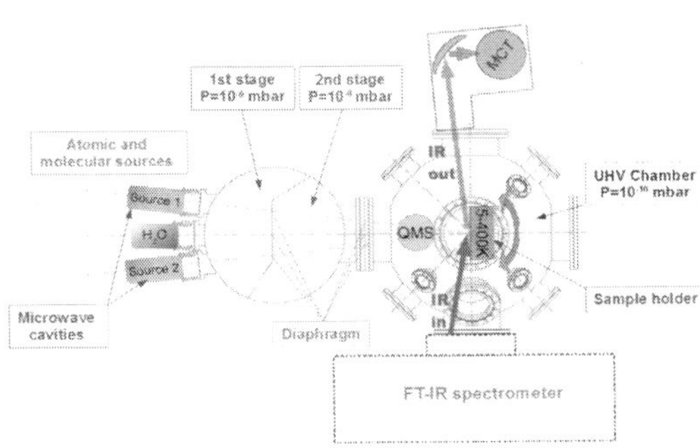

Figure 2. Schematic top view of the Venus setup and the FT-RAIRS facility.

EXPERIMENTAL METHODS

Experimental Setup

The experiments were conducted with the VENUS setup located at LERMA-Cergy lab in Cergy-Pontoise, France. It consists of an ultra high vacuum (UHV) main chamber with a base pressure of 10^{-10} mbar. A golden mirror, placed at the center of UHV chamber, is held at temperatures between 8 and 400 K. The temperature is controlled by a calibrated silicon-diode sensor clamped on the sample holder. Atomic species are directed toward the sample through three diaphragms located a double vacuum chamber to ensure a good differential pumping and keep the pressure in the main chamber very low (less than 5×10^{-10} mbar, during multibeams exposure). The scheme of the setup is shown in Fig. 2.

Adsorbates and products were probed in situ through a Fourier Transform Reflection Absorption Infrared Spectrometer (FT-RAIRS), and a quadrupole mass spectrometer (QMS) used for measuring the beams fluxes and beams compositions and for performing the temperature-programmed desorption (TPD) experiments.

All the experiments were performed following these steps:

(a) *NO deposition*. One monolayer (mL = 10^{15} molecules cm^{-2}) of NO on gold sample held at 40 K.

(b) *H_2O deposition*. *X* mL of H_2O on top of NO ice at 40 K ($X = 0, 2, 3, 6$, and 10 mL).

(c) *Sample temperature*. Cool down to 10 K in the case of D atoms study. The surface is kept at 40 K for O study.

(d) *Beam exposure*. NO-H_2O ice are exposed to a D- or O-atom beam. RAIRS spectra recorded during exposure.

(e) TPD performed after D or O exposure.

This protocol isschematized in Fig. 3.

Deuterium is used in place of hydrogen to increase the signal-to-noise ratio (S/N). In the case of D atoms exposure, we held at 10 K after the same procedure of ice formation at 40 K. The ice is not supposed tochange upon cooling, but can change the morphology if heated. During the O atoms exposure, NO-H_2O ice is held at 40 K to enhance O_2 desorption and prevent an efficient formation of O_3 (Minissale et al. 2014a). These temperatures are chosen so that the diffusion of atoms is fast enough and their desorption is negligible. Both D and O should be in the thermal diffusion regimes in our experimental conditions (Hama et al. 2012; Senevirathne et al. 2017; Congiu et al. 2014).

Atomicspecies, D and O atoms, were generated by dissociating D_2 and O_2 molecules in a quartz tube placed within a Surfatron cavity, which can deliver a maximum microwave power of 200 W at 2.45 GHz. Atoms and undissociated molecules were cooled and instantaneously thermalized upon surface impact with the walls of the quartz tube. Deuterium and oxygen beams were free of electronic excited atoms (Congiu et al. 2009; Minissale et al. 2015). The dissociation efficiency of D_2 and O_2 was $\tau = 50 \pm 5\%$, where τ represents the percentage of dissociated molecules. The value τ defines the atom/molecule ratio in the beam. For example if τ is 0.5, every ten molecules we have ten atoms and five undissociated molecules.

We calibrated the molecular beam as described in Amiaud et al. (2007) and Noble et al. (2012). The first monolayer of NO was reached after an exposure time of about 15 min, which give a flux of $\phi_{NO} = (1.0 \pm 0.3) \times 10^{12}$ molecules cm^{-2} s^{-1}. The D_2 and O_2 molecular fluxes were $\phi_{D_2,O_2} = (1.05 \pm 0.3) \times 10^{12}$ molecules cm^{-2} s^{-1} while D and O atoms fluxes were $\phi_{D,O} = (2.1 \pm 0.3) \times 10^{12}$ atoms cm^{-2} s^{-1}.

Adsorbates and products were probed continuously through FT-RAIRS during atoms exposure. TPD was performed after D or O atoms exposure and the sample was heated up to 250 or 190 K, respectively. Those temperatures allow the desorption of NO-H_2O ice and of newly formed molecules (ND_2OD and N_2O in the case of D, and NO_2 in the case of O).

We use the mL unit, which corresponds to 10^{15} molecules cm^{-2} exposed to the surface. For a typical solid substrate, like crystalline water ice, it also corresponds to the number of adsorption sites (NAS). But in the case of a reactive species, like O atoms, exposing 1 mL does not mean that one layer of O atoms is deposited, since O atoms diffuse and react quickly. If only O_2 is formed (like in the case of exposure on a nonreactive surface held at 30 K), and if the desorption (thermally or chemically induced) is neglected, it would lead to 50% of occupancy of the surface binding sites by O_2 molecules, because two O atoms are recombining into one O_2 molecule. Moreover, in the case of light species, such as H, H_2, D, the sticking coefficient is far to be unity (Chaabouni et al. 2012), so the exposed dose does not correspond to the number of molecules adsorbed on the surface. Finally, when the surface is porous or in presence of clusters, the NAS can greatly vary, and in general, even if the sticking coefficient is unity, 1mL of deposition is not sufficient to cover the full accessible surface, and therefore the saturation of the surface can occur after few mL of exposure.

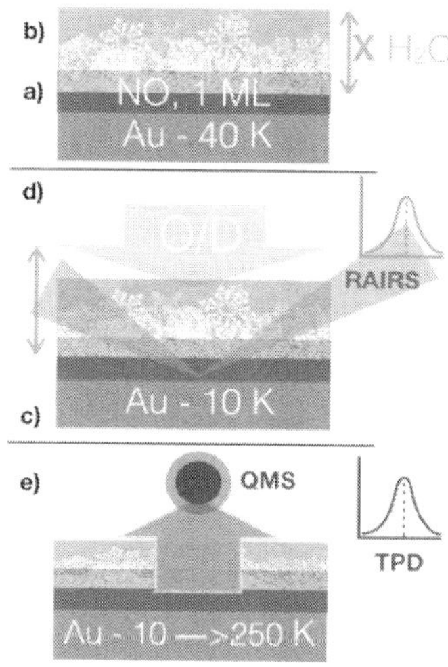

Figure 3. Scheme of the experimental protocol. *Panel a*: deposition of 1 mL of NO at 40 K. *Panel b*: deposition of X mL of H_2O at 40 K. *Panel c*: cool down at 10 K in the case of D-exposure experiments. *Panel d*: deposition of D atoms at 10 K or of O atoms at 40 K, recording continuously IR spectra. *Panel e*: TPD profiles of the species of interest.

Evaluation of the Number of Adsorption Sites with the Water Ice Thickness

Water ice deposited at 40 K is porous (Stevenson et al. 1999). The presence of cavity and pores on the surface increases the NAS and thus the effective surface. We studied how the effective surface seen by the atoms evolves as a function of water ice layer. We used a standard King and Wells method (King & Wells 1972) by depositing D_2 as a function of water ice thickness. We monitor the partial pressure of D_2 in the chamber while a D_2 beam is sent to the surface held at 10 K. Typical measured curves can be seen in Fig. 3 of Amiaud et al. (2007). For each ice substrate there is a specific saturation time that scales to the NAS. The method does not measure directly the NAS since the D_2 saturation of compact ASW occurs at 0.4 mL (of exposure) at 10 K, which corresponds to an occupancy of ~15% of the surface sites taking into account the sticking coefficient (Amiaud et al. 2007). However, the assumption that the saturation time of D_2 scales to the NAS is quantitatively verified with one experimental direct measurement at 5 mL of overlayers of H_2O, using N_2 and CO as direct probes (Nguyen et al. 2018).

The results are shown in Fig. 4. Without addition of amorphous porous water, the surface is considered as planar. The initial surface is a compact ASW of 15 mL thickness grown at 110 K. It is by definition equal to 1 mL. With the addition of extra layers of porous water, the saturation time increases, which corresponds to an increase of the NAS. The effective surface reaches 1.6 for 11 mL of extra water ice deposited at 40 K. In order to take into account the evolution of NAS in the model presented in Sect. 4, we fitted the experimental results with the following exponential law

$$NAS = 1.67 - 0.67 \cdot e^{-\frac{H_2O}{4.43}}. \qquad (1)$$

We stress that this is an empirical law and it is far from being a general law to describe the evolution of NAS as a function of water ice thickness. It is only valid in our specific experimental conditions.

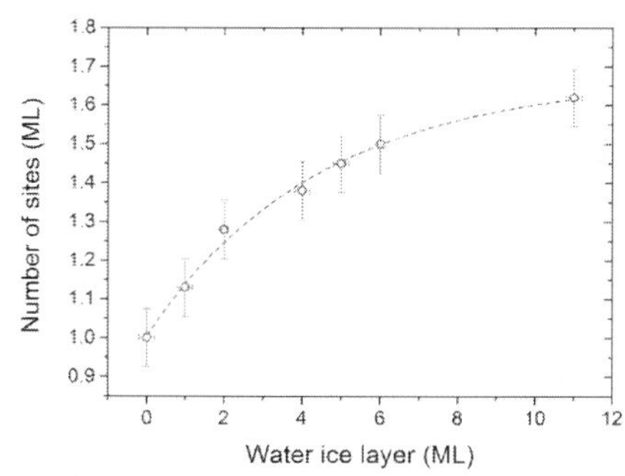

Figure 4. Number of adsorption sites expressed in mL as a function of water ice thickness, derived from a King and Wells method. The initial surface is a compact ASW of 15 mL thickness grown at 110 K covered with the addition of extra layers of porous water.

EXPERIMENTAL RESULTS

Oxygenation of NO Ices

The diffusion of oxygen in ASW is studied using nitric oxide as a reactive marker. Oxygenation of NO can occur through NO reactions with the three oxygen allotropes O_x (O, O_2, and O_3). Solid state reactions generally have two limiting factors, the reagent diffusion barrier and the reaction barrier. As the NO + O reaction is barrier-free, it represents a system suitable for the study of O diffusion. We note that the presence of atomic oxygen on a surface easily leads to the formation of diatomic oxygen and ozone therefore, reactions NO + O_2 and NO + O_3 could occur in solid phase and would make the interpretation of experimental results more difficult. To avoid this problem, the presence of diatomic oxygen and ozone can be limited by performing experiments at a sample temperature of 40 K. At this temperature, O_2 is efficaciously released in the gas phase and, therefore, O_3 cannot be effectively formed in the solid phase (Minissale et al. 2014a).

Figure 5 shows typical IR spectra obtained by exposing 1 mL of NO held at 40 K to O atoms.The red spectrum is taken before O-atom exposure. This curve does not present any IR signature since it has been taken as reference spectrum. The other spectra are taken after 0.2, 0.4, 0.8, and 2.0 mL of O-atom exposure, respectively. We can assign three main signatures: the NO dimer at 1770 cm^{-1} and the NO_2 at 1605 and 1315 cm^{-1}. From the decreasing of NO peak as a function of O-atom exposure, we evaluate the kinetics of NO consumption. Similarly, by integrating the increasing of NO_2 signatures, we evaluate the formation rate of NO_2.

The kinetics of NO consumption and NO_2 formation are shown in Fig. 6 as a function of O-atom exposure for different thickness of H_2O ice (0, 2, 3, 6, and 10 mL). Each curve shows a decrease of the reagent (NO, top panel) correlated with an increase of one of the product (NO_2, bottom panel). As expected for a barrierless reaction the kinetics reach a steady-state plateau after an exposure about equal to one mL, which corresponds to a perfect stochastic reaction.

As evident from experimental data in Fig. 6, the level of the steady state of NO consumption (or NO_2 formation) depends on the thickness of water ice. When no water ice is deposited on top of NO, 1 mL of oxygen atoms is able to react with all the NO previously grown (~1 mL). Two mL of porous water ice are able to prevent a complete consumption of NO ice since less than 0.7 mL are consumed. For 10 mL of water ice, only 0.1 mL of NO reacts with O atoms. We note that once the steady-state plateau is reached, there is no measurable slower increase within our experimental uncertainties

As explained in the experimental procedure, the NO ice deposit is followed by a H_2O deposit on the NO ice. However, because of the porous structure of the water ice and diffusion, NO molecules may be present in an "uncovered" adsorption site after water ice deposition.

The following two populations of NO can therefore be considered:

- covered NO, accessible only by pure bulk diffusion of O atoms (hereinafter NO_b)
- uncovered NO, present on the surface or in the pores of water and accessible by surface diffusion of O atoms (hereinafter NO_s).

In the NO_s population, there are two different subcategories, which could be called "outer surface", and "inner accessible surface". They are depicted in green and blue, respectively, in Fig. 1. The first category corresponds to NO located on the outer surface, and thus can interact directly with the gas. Even though water ice is deposited on top of the NO layer, we cannot neglect the possibility for NO to diffuse during the ice building at 40 K. The second category of NO_s corresponds to NO located in a pore, and it requires surface diffusion of O (or D) to react. However NO molecules organize themselves as dimers (Minissale et al. 2013b, 2014b; Ioppolo et al. 2014), so their mobility should be strongly reduced compared to any other molecules of similar range of volatility (such as CO). Figure 1, for the sake of simplicity, shows only the case in which NO dimers are unable to diffuse, which is an oversimplification. However, the important point here is the following: do O or D have to cross a water layer to react with NO (i.e., need bulk diffusion)?

The details of TPD profiles can help to discuss these possible locations: outer surface, inner surface, or bulk. In Fig. 7 panel a presents the TPD traces of mass 30 (in red) and mass 46 (in green) after the oxidation of NO without a water overlayer. As we have shown in Minissale et al. (2013c), TPD desorption curves of NO_2 ices are characterized by two peaks at masses 30 and 46 (with a constant ratio m46/m30 ~ 13%) at around 125 K. We stress that only the presence of both mass 30 and 46 represents the desorption of NO_2. In panels b and c we can see three peaks at mass 30, named A, B, and C for the sake of clarity. The peak A at 125 K is related to the desorption of surface NO_2 since both masses 30 and 46 are present. Peak B at 148 K has no mass 46 components and it is therefore due to the desorption of NO. It is indeed the NO unreacted that is released during the rearrangement of the ASW, prior to the crystallization. This peak corresponds to NO_b in water bulk (or inaccessible pores). Peak C shows masses 30 and 46 and it corresponds to the release of NO_2 trapped in the pores of the water. The origin of this peak can be either NO_b or NO_s trapped during the reconstruction of the ice during the TPD. On panel d we plot the shape of the water desorption. It shows a two-peaks- desorption. The first peak corresponds to desorption of ASW.

The ASW desorption is reduced when the crystallization occurs. The second peak corresponds to desorption of crystalline water (Smith et al. 1997). We can see that the NO desorption (peak B) occurs during the ASW desorption, whereas the NO_2 desorption exactly matches with the water crystallization. This is known as the so-called volcano effect (May et al. 2013); when the water crystal is expanding, it expels all the impurities, which are NO_2 in our present case. From panels a to d in Fig. 7, we note a decrease of the peak at A; it is no longer visible in panel d, meaning that all NO_2 is formed from NO trapped inside the ice.

We now return to the different scenarios of NO reactivity depending on their initial location. NO_2 desorbing from peak A comes from the surface or from an open pore. The C peak indicates an origin within the volume, in bulk, or in open pores. We can clearly see the change in the balance from peak A to C, corresponding to more and more NO that reacted in the pores or in the bulk. The increase of the B peak corresponds to the unreacted NO, which perfectly matches the plateau values observed in the IR spectra. The TPD traces show that the thicker the ice, the higher the proportion of reaction inside the pores. This was not certain before the study. It is possible to have imagined that the NO film would remain at the surface because it is completely repelled by water growth (water molecules impinging with the kinetic energy of room temperature) or segregated forming islands. Moreover, it seems that NO diffusion during the growth of water ice over layer is not efficient. This could be owing to the ability of NO to form dimers (Minissale et al. 2014b) that may have a low diffusion rate.

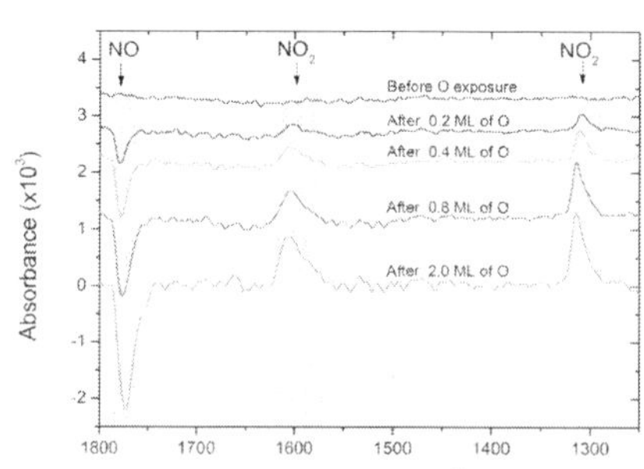

Figure 5. Infrared spectra obtained by exposing to 1 mL of NO held at 40 K to different amount of O atoms: 0, 0.2, 0.4, 0.8, and 2.0 mL. The red curve does not present any IR signature since it has been considered as the reference spectrum. Curves are offset for clarity.

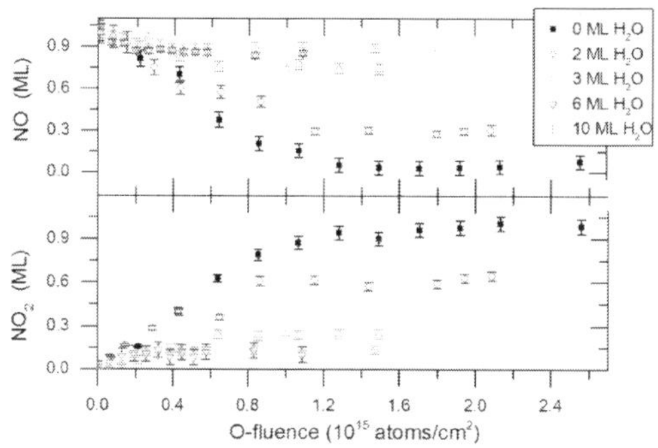

Figure 6. Kinetics of NO consumption and NO_2 formation obtained by integrating IR peaks of NO and NO_2 as a function of O-atom fluence for five different thickness of water ices (0, 2, 3, 6, and 10 mL) held at 40 K.

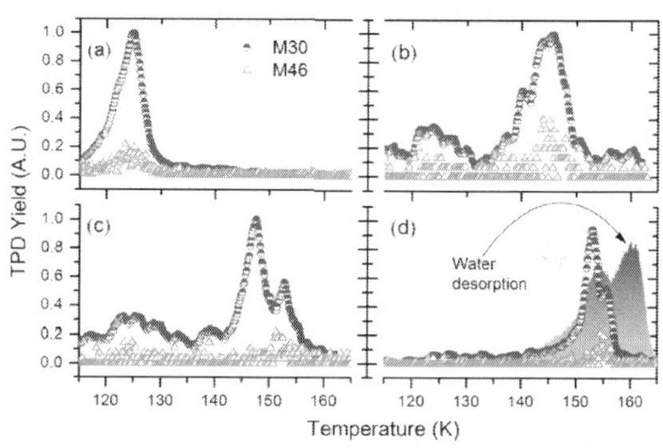

Figure 7. Curves of TPD at mass 30 and 46 after O-exposure of NO/H_2O ice for different thickness of water ices held at 40 K. *Panel a*: 0 mL. *Panel b*: 3 mL. *Panel c*: 6 mL. *Panel d*: 10 mL.

Deuteration of NO Ices

In this section we describe experimental results of deuteration of NO. As studied by Congiu et al. (2012), the deuteration of NO brings to the formation of deuterated hydroxilamine (ND_2OD) via the three subsequent D-atom additions to nitrogen monoxide. Side reactions may lead to some N_2O formation. The first step (NO consumption) is barrierless. In this case, we exposed NO ices held at 10 K to different amount of D atoms (up to 2.1 mL). We studied three different chemical kinetics by varying the amount of water ice deposited on top of NO, i.e., 0, 3, and 6 mL of porous water ice. As in the case of NO + O ice, we can easily follow the kinetics of NO consumption by looking to the decrease of IR band at 1770 cm^{-1}. Unlike the NO + O case, we cannot follow the kinetics of reaction products (ND_2OD) since their IR signatures are hidden by the noise. Nevertheless by performing TPD experiments, we checked the presence of deuterated hydroxilamine after each experiment. NO kinetics is shown in Fig. 8; NO consumption reaches a plateau after a D-fluence of ~ 1 mL in the case of bare NO. This could be explained by a cover effect of NO by the newly formed hydroxilamine: at very beginning of experiments, NO can react efficiently with D through Eley-Rideal mechanism; once hydroxilamine starts to form on the surface, it can partially cover NO, and D atoms have to diffuse to react with it. This means that surface density of D atoms increases and facilitates D_2 formation through D + D reaction. Somehow hydroxilamine could play the same role of water ice preventing a full NO consumption. The role of water ice is clearly visible for the experimental data plotted in Fig. 8. When 3 mL of H_2O are deposited, 0.3 mL of NO are consumed, while only 0.1 mL of NO reacts when 6 mL of H_2O are deposited on top of it.

If we now compare the amount of NO unreacted (in mL) for the O and D experiments, we find that it is 0.67 ± 0.09 mL (for D) and 0.74 ± 0.04 mL (for O) for the case of 3 mL of water overlayer and 0.88 ± 0.07 and 0.89 ± 0.02 mL for 6 mL of water. This amount is equal within our experimental error bars. We can conclude at this stage that the accessibility to NO for Oatoms and D atoms seems to be identical.

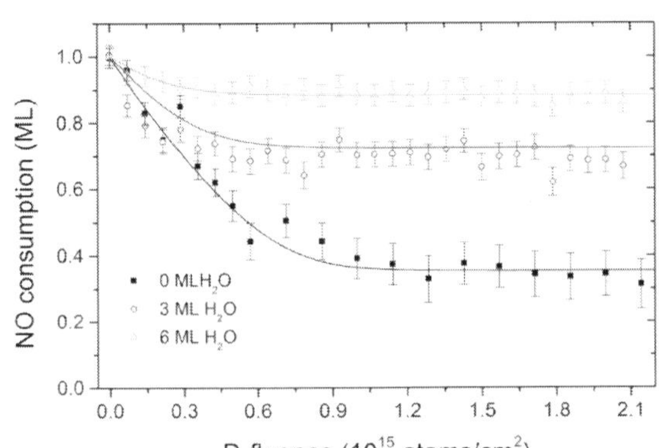

Figure 8. Comparison between model (solid lines) and experimental data (dots). Experimental data are obtained by integrating IR peak of NO following to different fluences of D atoms for three different thickness of water ices (0, 3, and 6 mL) deposited on top of NO held at 40 K. Simulated curves are obtained by using the α_0 law for a deuterium bulk diffusion of 280 K.

MODEL AND DISCUSSION

The model used to fit our experimental data is very similar to that described in Minissale et al. (2013c, 2015). In the case of NO oxygenation we follow the density of six chemical species: O-surface atoms (O_s), coming exclusively from the beam and diffusing only on the surface; O-bulk atoms (O_b), reacting and diffusing only in the ice bulk; NO-surface molecules (NO_s), deposited on the surface and reacting only with O_s; NO-bulk molecules (NO_b), deposited on the surface and reacting only with O_b; NO_2 formed via $NO_s + O_s$ or $NO_b + O_b$; and finally O_2, coming both from the beam and formed by solid-state reactions. Each differential equation is composed of different terms:

- positive terms contribute to the increase of species amount, i.e., a molecule (or atom) arrives from the gas phase or it is formed on the surface;
- negative terms indicate a decrease of species amount, i.e., a molecule (or atom) is consumed on the surface or it desorbs and returns to the gas phase.

The terms involving the Eley-Rideal (ER; gas-surface reactions) and Langmuir-Hinshelwood (LH) mechanisms (reaction induced by surface diffusion) are independent. For this reason we can determine the amount of a species formed (or consumed) via the ER or the LH mechanism. Below we show the set of differential equations built to simulate NO consumption and NO_2 formation:

$$O_s'(mL) = \phi_O(1 - 2O_s - O_2) - \phi_{O_2} O_s - 4k_{O_s,Dif} O_s O_s - k_{O_s,Dif} O_s O_2 - \phi_O NO_s - k_{O_s,Dif} O_s NO_s + k_{ObDif} (O_b - O_s) - O_s Des_{O_s}, \quad (2)$$

$$O_b'(mL) = -k_{O_b,Dif} (O_b - O_s) - k_{O_b,Dif} O_b NO_b, \quad (3)$$

$$O_2'(mL) = \phi_{O_2} (1 - O_s) - \phi_O O_2 + \phi_O O_s + 2k_{O_s,Dif} O_s O_s - k_{O_s,Dif} O_s O_2 - O_2 Des_{O_2}, \quad (4)$$

$$NO_s'(mL) = -\phi_O NO_s - k_{O_s,Dif} O_s NO_s - NO_s Des_{NO_s}, \quad (5)$$

$$NO_b'(mL) = -k_{O_b,Dif} O_b NO_b, \quad (6)$$

$$NO_2'(mL) = \phi_O NO_s + k_{O_s,Dif} O_s NO_s + k_{O_b,Dif} O_b NO_b. \quad (7)$$

O_s, O_b, O_2, NO_s, NO_b, and NO_2 are the densities (expressed in fractions of mL) of the different species, ϕ_O is the flux of O atoms, and ϕ_{O_2} is the nondissociated fraction of O_2 defined in Sect. 2. We define

$$\mathrm{Des}_{O_s} = \nu e^{\frac{-E_{O_s,des}}{T}}, \text{ with } E_{O_s,des} = 1400 \text{ K} \quad (8)$$

$$\mathrm{Des}_{O_2} = \nu e^{\frac{-E_{O_2,des}}{T}}, \text{ with } E_{O_2,des} = 1100 \text{ K} \quad (9)$$

$$\mathrm{Des}_{NO} = \nu e^{\frac{-E_{NO,des}}{T}}, \text{ with } E_{NO,des} = 1300 \text{ K} \quad (10)$$

$$k_{O_s,Dif} = \nu e^{\frac{-E_{O_s,dif}}{T}}, \text{ with } E_{O_s,dif} = 750 \text{ K} \quad (11)$$

$$k_{O_b,Dif} = \nu e^{\frac{-E_{O_b,dif}}{T}}, \text{ with } E_{O_b,dif} \text{ a free parameter,} \quad (12)$$

where T is the sample temperature and $\nu = 10^{12}s^{-1}$ is the trial frequency for attempting a new event; $\mathrm{Des}_{O_s,O_2,NO}$ are the desorption probabilities of O_s, O_2, and NO, respectively; $k_{O\_s,dif}$ and $k_{O\_b,dif}$ are the thermal diffusion probability of O atoms on surface (s) and in the bulk (b). We point out that for the diffusion probability we only consider a pure thermal diffusion barrier since quantum effects are negligible at high temperatures (>30 K). We stress that the only free parameters are $E_{O\_b,dif}$ and the initial amounts of NO_s and NO_b.

As described before water ice is deposited on top of NO molecules. Depending on the amount of water ice deposited, NO molecules can be accessible by bulk (NO_b) or surface (NO_s) paths. We use a single parameter, called α, to take into account the initial amount of NO_b (α) and NO_s (1-α) molecules.

The main difficulty of the analysis lies is the fact that we do not know the amount of NO initially present in the bulk (NO_b). The consumption of NO can occur via surface reactions of NO_s or via diffusion through the ice. We have a unique observable NO, but two competing and potentially compensating pathways. Actually we can obtain the same level of consumption if we increase both α and bulk diffusion. In other words, more NO can be initially covered if the bulk diffusion is more efficient. Therefore, we need to explore the whole domain of bulk diffusion and initial α coefficient to simulate the full kinetic of NO consumption.

Let us first focus on the possible values of α (NO_b). The amounts of NO_b and NO_s are related by the obvious closure relation $NO = NO_b + NO_s$. Because NO= 1 (1 mL of NO is initially deposited) in all our experiments, $NO_s = (1\text{-}\alpha)$. At the present stage, we ignore the dependency of α with the thickness of ASW. By definition α = 0 when there is no water deposition. We can suppose that α is monotonically increasing with the water ice layers (WIL) and that α cannot be larger than the WIL. Here we suppose that at least one water molecule is required to occupy one adsorption site. The value α tends toward 1 when the ASW thickness tends to infinity.

Moreover we know that the NO + O reaction is a barrierless reaction and so all the NO accessible via surface paths should react. This means that the amount of NO_s (i.e., 1-α) cannot be greater than the amount of NO reacted. This hypothesis is reinforced by the fact that in all our experiments, we reach a plateau, indicating that all the possible reactants have been consumed. In Fig. 9 we plot in blue dots the amount of NO unreacted as a function of WILs deposited on top of the NO layer. The possible values of α are represented by the colored zone. The values of α, whatever is the WIL are comprised of between 0 for WIL = 0 and 1 for WIL = ∞. Moreover, taking into account the previous remark that NO_s (i.e., 1-α) cannot be greater than the amount of NO reacted, therefore α is necessarily greater than the unreacted NO. In other words, the contribution of bulk diffusion cannot be negative. So the α domain is restricted to the portion of the Fig. 9 above the experimental values (α_0). To materialize the inferior limit of the possible α, we took an exponential saturation law $\alpha(\mathrm{WIL}) = 1 - e^{-\frac{\mathrm{WIL}}{P_1}}$ where P_1 is the fitting parameters. The curve α_1(WIL) is represented in blue in Fig. 9 and it corresponds to the situation of a low number of NO initially present in the bulk, α_2 (red dotted curve) to an intermediatenumber of initial NO_b, and α_3 (green dashed curve) represents the case where NO molecules are perfectly covered with water ice, and therefore the reactivity can only proceed through bulk diffusion. The three laws are given by

$$\alpha_1(\mathrm{WIL}) = 1 - e^{-\frac{\mathrm{WIL}}{3.447}}, \qquad (13)$$

$$\alpha_2(\mathrm{WIL}) = 1 - e^{-\frac{\mathrm{WIL}}{2.247}}, \qquad (14)$$

$$\alpha_3(\mathrm{WIL}) = 1 - e^{-\frac{\mathrm{WIL}}{0.515}}. \qquad (15)$$

We stress that these are empiric laws that have been defined to explore the space parameter of the possible α values as we can see in Fig. 9.

Figure 10 presents typical comparison of our model and the experiments. We use the kinetic of the NO consumption (and NO_2 formation) for different O fluences. We can very satisfactorily reproduce our experimental data using the α_0 values, which are indeed the values of α equal to theunreacted NO obtained from the plateaus observed in IR spectra. In this case, the best fit is obtained for any value ofbulk diffusion greater than 950 K as can be seen on the right panel of the Fig. 10. This value is to be compared with the value of surface diffusion taken as 750 K. It corresponds to a ratio of ~150 of the two diffusion rates.

Other scenarios of initial covering of NO are tested using different α_x laws. We can see on the χ^2 graph that there exists a minimum for which it is possible to compensate an initial coverage with a lower diffusion barrier. These minima have higher absolute values with the increase of the initial NO coverage (from 1 to 3), and logically show values with lower and lower diffusion barriers, in line with what is expected. The values of χ^2 are higher but, taking into account thatthe α_0 has the best possible absolute values of χ^2 because it ismade from the right plateau values, it is not a strong reason to reject these possibilities of initial coverage. We can for example argue that our α_x laws are not exactly representative of the initial coverages and that another repartition would probably get a lower, and therefore better, minimal value of χ^2. However, on the right side of Fig. 10 we can see that the best fits correspond to unrealistic situations. In particular, the values predicted at high fluences are always decreasing and therefore there is no plateau, as observed in experimental data. Therefore, the best solution is the case of α_0, which corresponds the scenario of the minimum of NO initially covered with the water ice, and thus to a slow bulk diffusion, which needs to be at least two order of magnitude slower than the surface diffusion.

We alsobuilt a similar model to simulate the NO + D kinetics. We considered eight species NO_s, NO_b, D_s, D_b, D_2, NOD, NDOD, and N_2OD. The desorption energy of D and D_2 was fixed to 600 K and the diffusion energy for D_s to 210 K (Wakelam et al. 2017). Even in this case we considered three laws for α and different values of bulk diffusion. We obtained results similar to the case of NO oxygenation. The best fit, shown in Fig. 8, has been found for the α_0 law, ($NO_s = NO_{reacted}$) and the minimum value of bulk diffusion is $E_{Dbdif} = 280$ K. In this case, the ratio of surface diffusion rates to bulk diffusion rates is larger than 1000.

In both cases, NO oxygenation and deuteration, the chemistry seems to be driven by surface reactions (including the pores). Lowering the value of E_{bdif} to values closer to E_{sdif} induces a faster kinetics, that would slowly consumed the NO_b, which is not observed experimentally. A good agreement between experimental and simulation results can be found using $\frac{E_{bulk-diffusion}}{E_{surface-diffusion}} \geq 1.3$.

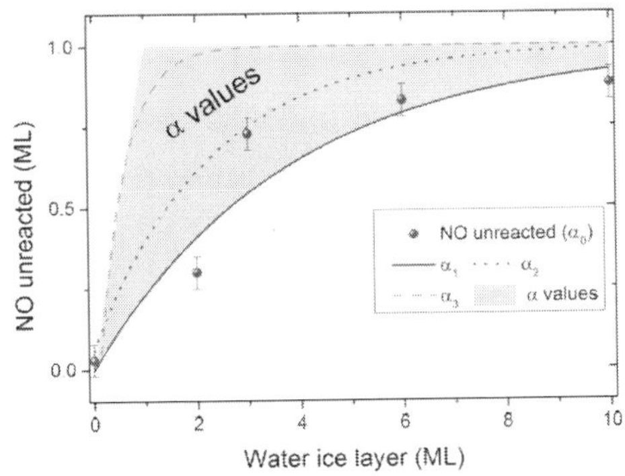

Figure 9. Colored zone: possible values of NO located inside the bulk of ice ($NO_b = \alpha$) as a function of the WIL thickness. Blue points: measured amounts of unreacted NO. Blue, red, and green lines represent three possible parametrized repartitions of initially covered NO molecules

ASTROPHYSICAL IMPLICATIONS AND CONCLUSIONS

In this work, we present an experimental comparison of the bulk diffusion (i.e., the crossing of a compact/closed ice layer) and surface diffusion (including in the pores) of oxygen and deuterium that can occur in conditions mimicking the interstellar ices. In particular we evaluated the ratio of surface to bulk diffusion, where the surface diffusion is the diffusion occurring on the top surface and on the surface of the (open) pores. Potentially surface diffusion can take place in the closed pores too, but it can be neglected as a first approximation since it has to be triggered by bulk diffusion.

We use NO as a chemical tracer. Under O or D exposure, NO is rapidly consumed, down to a plateau indicating that NO can no longer bereached by atoms in our experimental conditions. To mimic the initial fast consumption and the following plateau, only a bulk diffusion (through water ice) two orders of magnitude slower than surface diffusion for O and three orders of magnitude for D can fit our data. We found a lower limit for bulk diffusion of 950 and 280 K for O and D, respectively, and a $E_{diff\,bulk}/E_{diff\,surf} \geq 1.3$ in both cases. We stress that we did not perform experiments using H atom. Since the diffusion of H could be different from D, the conclusion regarding D atoms could not be totally valid for H atoms. In the interstellar medium, the existence of ice mantle is ubiquitous as soon as visual extinction reaches a value of 3. This corresponds to regions where density raises to ~10^{3-4} mol

cm^{-3}. In these conditions, one layer of the mantle takes about 1000 yr to build up for a total of 10^5 yr to reach a 100-layer thick mantle. After this point, most of the gas species should already have accreted on grains, with few notable exceptions such as H, He, H_2, and some of close related species such as H_3^+ and N_2H^+. This is why interstellar ices enter a second stage of their chemical evolution, whose duration depends on the mass of the to-be-born star, before ongoing processes linked to the more energetic next phases of star formation. During the growth of the mantle, one H atom lands on grains every $\sim 2 \times 10^4$ s (see Fig. 8 of Congiu et al. 2014). It can completely scan the surface of a grain (radius $\simeq 0.1$ μm) in about 10^3 s. In the case of D atoms, the lower limit of the bulk diffusion (280 K) corresponds to a life time on the surface of ~ 1 s at 10 K $(\tau = k^{-1}_{D_b,Dif})$. It means that H atoms can scan the surface before penetrating the bulk of the ice, but it could penetrate the ice before the arrival of another H atom. This implies that the H diffusion through the bulk of the ice cannot be fully ruled out with our lower limit of H bulk diffusion. Only a factor of more than 10^6 between the surface and bulk diffusion rates would prevent bulk penetration, the latter being too slow to compete with the H accretion of the gas phase, and its subsequent transformation into H_2. However, since the surface diffusion rate is still at least 1000 times faster than the bulk diffusion, any surface reactant with a concentration greater than a thousandth would statistically react with the H atoms. In other words, H could penetrate deeper in the ice mantle only if very few reactants are present on the outer surface of grains. Thus, we conclude that the surface hydrogenation, in our experiments as well as in the interstellar medium, is dominant over bulk diffusion, which cannot be totally excluded only on the basis of a direct extrapolation of our experiments. In such considerations, we neglect isotopic effects that are probably important but hard to take into account since both surface and bulk diffusions should be affected. For O atoms, crossing the first water bulk layer should take $\sim 10^{29}$ s, which means that it is not possible (within the frame of the classical formalism we use).

However, we point out our values are lower limits for the bulk diffusion barriers as shown by the flat shape of the ^{2} curve of Fig. 10. The present experiments do not show directly whether H diffusion through the ice bulk can reduce the lifetime of radicals formed inside the volume of the ice mantle during UV irradiation or during other energetic processes. The presence of radicals during the initial building of the ice mantle is probably excluded thanks to the high mobility of H atoms at any temperature (Senevirathne et al. 2017). Our experiments show that the only very first layers of the mantle are reacting and models considering that surface reactions rule the ice building (e.g., Vasyunin et al. 2017)seem thus coherent with our laboratory experiments. Hydrogenation or oxidation experiments (Hiraoka et al. 2002; Watanabe et al. 2004; Fuchs et al. 2009; Jing et al. 2011; Ward & Price 2011) have used similar or greater atomic fluxes. This means in a first approximation that the surface to bulk diffusion efficiency is the same or greater. Actually, the higher the flux, the higher the self reactivity, the lower the possibility to penetrate inside the molecular substrate. With the present study, we can understand why, by construction, laboratory experiments favor surface instead of bulk reactivity. However, we provide a lower limit confirming that surface diffusion of H is an order of magnitude faster than bulk diffusion through the ice, even though it certainly changes with the type of molecular layer used.

Finally some works suggest the existence of an efficient ice segregation mechanism of both interstellar (Öberg et al. 2011; Allodi et al. 2014) and cometary ices (Dello Russo et al. 2011; McKay et al. 2018). Nevertheless ice segregation is still an openquestion in astrochemistry (Boogert et al. 2015). The possible H atoms migration inside the bulk can strongly change the chemical composition in the whole ice thickness, but simple thermal differentiation could also be the reason for the observed segregation. This aspect could now be tested thanks to Monte Carlo simulations (Cazaux et al. 2010; Cuppen et al. 2017)using our boundary limits of bulk diffusion barriers.

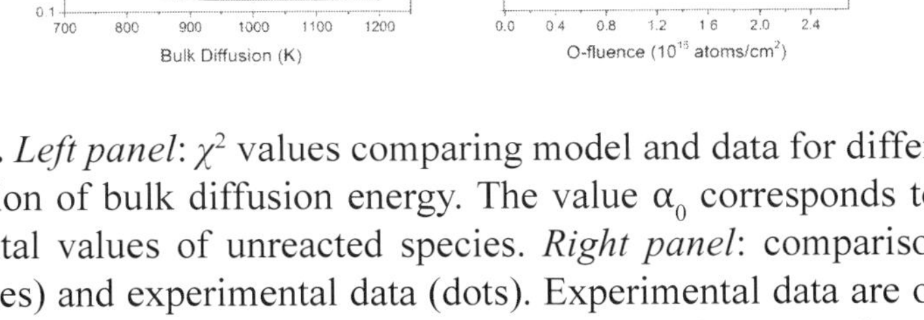

Figure 10. *Left panel*: χ^2 values comparing model and data for different α laws as a function of bulk diffusion energy. The value α_0 corresponds to the exact experimental values of unreacted species. *Right panel*: comparison between model (lines) and experimental data (dots). Experimental data are obtained by integrating IR peaks of NO and NO_2 as a function of O-atom fluence for five

different thickness of water ices (0, 2, 3, 6, and 10 mL) held at 40 K. Simulated curves are obtained by using three laws for the α parameter (α_1, α_2, and α_3) and measured experimental values (α_0) for the best case corresponding to the minimum χ^2 of left panel. For α_0 any choice of diffusion parameter greater than 950 K gives similar results.

ACKNOWLEDGEMENTS

T.N. is a Ph.D. student granted by the Labex MICHEM program. This work was supported by the Program DIM-ACAV, and now DIM-ACAV+, of the Region Ile de France, the CNRS – INSU program Physique et Chimie du Milieu Interstellaire (PCMI), supported by the CNES.

REFERENCES

1. Accolla, M., Congiu, E., Dulieu, F., et al. 2011, Phys. Chem. Chem. Phys., 13, 8037
2. Accolla, M., Congiu, E., Manicò, G., et al. 2013, MNRAS, 429, 3200
3. Allodi, M. A., Ioppolo, S., Kelley, M. J., McGuire, B. A., & Blake, G. A. 2014, Phys. Chem. Chem. Phys., 16, 3442
4. Amiaud, L., Dulieu, F., Fillion, J.-H., Momeni, A., & Lemaire, J. L. 2007, J. Chem. Phys., 127, 144709
5. Anders, E., Hayatsu, R., & Studier, M. H. 1974, ApJ, 192, L101
6. Boogert, A. A., Gerakines, P. A., & Whittet, D. C. 2015, ARA&A, 53, 541
7. Caselli, P., & Ceccarelli, C. 2012, A&ARv, 20, 56
8. Cazaux, S., Cobut, V., Marseille, M., Spaans, M., & Caselli, P. 2010, A&A, 522, A74
9. Ceccarelli, C., Caselli, P., Fontani, F., et al. 2017, ApJ, 850
10. Chaabouni, H., Bergeron, H., Baouche, S., et al. 2012, A&A, 538, A128
11. Congiu, E., Matar, E., Kristensen, L. E., Dulieu, F., & Lemaire, J. L. 2009, MNRAS, 397, L96
12. Congiu, E., Chaabouni, H., Laffon, C., et al. 2012, J. Chem. Phys., 137, 054713
13. Congiu, E., Minissale, M., Baouche, S., et al. 2014, Faraday Discuss., 168, 151
14. Cooke, I. R., Öberg, K. I., Fayolle, E. C., Peeler, Z., & Bergner, J. B. 2018, ApJ, 852, 75
15. Cuppen, H. M., Walsh, C., Lamberts, T., et al. 2017, Space Sci. Rev., 212, 1
16. Dello Russo, N., Vervack, Jr. R. J., Lisse, C. M., et al. 2011, ApJ, 734, L8
17. Fuchs, G. W., Cuppen, H. M., Ioppolo, S., et al. 2009, A&A, 505, 629
18. Ghesquière, P., Ivlev, A., Noble, J. A., & Theulé, P. 2018, A&A, 614, A107
19. Hama, T., Kuwahata, K., Watanabe, N., et al. 2012, ApJ, 757, 185
20. He, J., Emtiaz, S. M., & Vidali, G. 2017, ApJ, 837, 65

21. Hidaka, H., Kouchi, A., & Watanabe, N. 2007, J. Chem. Phys., 126, 204707
22. Hiraoka, K., Sato, T., Sato, S., et al. 2002, ApJ, 577, 265
23. Ioppolo, S., Fedoseev, G., Minissale, M., et al. 2014, Phys. Chem. Chem. Phys., 16, 8270
24. Jing, D., He, J., Brucato, J., et al. 2011, ApJ, 741, L9
25. King, D. A., & Wells, M. G. 1972, Surf. Sci, 29, 454
26. Kristensen, L. E., van Dishoeck, E. F., van Kempen, T. A., et al. 2010, A&A, 516, A57
27. Lauck, T., Karssemeijer, L., Shulenberger, K., et al. 2015, ApJ, 801, 118
28. Matar, E., Congiu, E., Dulieu, F., Momeni, A., & Lemaire, J. L. 2008, A&A, 492, L17
29. May, R. A., Smith, R. S., & Kay, B. D. 2013, J. Chem. Phys., 138, 104501
30. McKay, A. J., Cochran, A. L., DiSanti, M. A., et al. 2018, Icarus, 309, 1
31. Minissale, M., Congiu, E., Baouche, S., et al. 2013a, Phys. Rev. Lett., 111, 053201
32. Minissale, M., Congiu, E., Baouche, S., et al. 2013b, Chem. Phys. Lett, 565, 52
33. Minissale, M., Congiu, E., Manicò, G., Pirronello, V., & Dulieu, F. 2013c, A&A, 559, A49
34. Minissale, M., Congiu, E., & Dulieu, F. 2014a, J. Chem. Phys., 140, 074705
35. Minissale, M., Fedoseev, G., Congiu, E., et al. 2014b, Phys. Chem. Chem. Phys., 16, 8257
36. Minissale, M., Loison, J.-C., Baouche, S., et al. 2015, A&A, 577, A2
37. Minissale, M., Congiu, E., & Dulieu, F. 2016, A&A, 585, A146
38. Mispelaer, F., Theulé, P., Aouididi, H., et al. 2013, A&A, 555, A13
39. Nguyen, T., Baouche, S., Congiu, E., et al. 2018, A&A, 619, A111
40. Noble, J. A., Congiu, E., Dulieu, F., & Fraser, H. J. 2012, MNRAS, 421, 768
41. Oba, Y., Miyauchi, N., Hidaka, H., et al. 2009, ApJ, 701, 464
42. Öberg, K. I., Fayolle, E. C., Cuppen, H. M., van Dishoeck, E. F., & Linnartz, H. 2009, A&A, 505, 183

43. Öberg, K. I., Boogert, A. C. A., Pontoppidan, K. M., et al. 2011, ApJ, 740, 109
44. Orzechowska, G. E., Goguen, J. D., Johnson, P. V., Tsapin, A., & Kanik, I. 2007, Icarus, 187, 584
45. Palumbo, M. E. 2006, A&A, 453, 903
46. Palumbo, M., Baratta, G., Leto, G., & Strazzulla, G. 2010, J. Mol. Struct., 972, 64
47. Papoular, R. 2005, MNRAS, 626, 489
48. Pirronello, V., Liu, C., Roser, J. E., & Vidali, G. 1999, A&A, 344, 681
49. Rimola, A., Taquet, V., Ugliengo, P., Balucani, N., & Ceccarelli, C. 2014, A&A, 572, A70
50. Senevirathne, B., Andersson, S., Dulieu, F., & Nyman, G. 2017, Mol. Astrophys., 6, 59
51. Smith, R. S., Huang, C., Wong, E. K. L., & Kay, B. D. 1997, Phys. Rev. Lett., 79, 909
52. Stevenson, K. P., Kimmel, G. A., Dohnálek, Z., Smith, R. S., & Kay, B. D. 1999, Science, New Ser., 283, 1505
53. Tielens, A. G. G. M., & Hagen, W. 1982, A&A, 114, 245
54. Vasyunin, A., Caselli, P., Dulieu, F., & Jiménez-Serra, I. 2017, ApJ, 842
55. Wakelam, V., Bron, E., Cazaux, S., et al. 2017, Mol. Astrophys., 9, 1
56. Ward, M. D., & Price, S. D. 2011, ApJ, 741, 121
57. Watanabe, N., Nagaoka, A., Shiraki, T., & Kouchi, A. 2004, ApJ, 616, 638
58. Watanabe, N., Kimura, Y., Kouchi, A., et al. 2010, ApJ, 714, L233

SECTION 6:
CHEMISTRY OF THE SOLAR SYSTEM

11

Constraining the Evolutionary History of the Moon and the Inner Solar System: A Case for New Returned Lunar Samples

Romain Tartèse[1], Mahesh Anand[2,3], Jérôme Gattacceca[4], Katherine H. Joy[1], James I. Mortimer[2], John F. Pernet-Fisher[1], Sara Russell[3], Joshua F. Snape[5], Benjamin P. Weiss[6]

[1] Department of Earth and Environmental Sciences, The University of Manchester, Manchester, M13 9PL, UK
[2] Planetary and Space Sciences, School of Physical Sciences, The Open University, Walton Hall, Milton Keynes, MK7 6AA, UK
[3] Department of Earth Sciences, The Natural History Museum, Cromwell Road, London, SW7 5BD, UK
[4] CNRS, Aix-Marseille Univ, IRD, Coll France, INRA, CEREGE, Aix-en-Provence, France
[5] Faculty of Sciences, Department of Earth Sciences, Vrije Universiteit, Amsterdam, The Netherlands
[6] Department of Earth, Atmospheric and Planetary Sciences, Massachusetts Institute of Technology, Cambridge, MA, USA

Citation: Tartèse, R., Anand, M., Gattacceca, J. et al. "Constraining the Evolutionary History of the Moon and the Inner Solar System: A Case for New Returned Lunar Samples". Space Sci Rev 215, 54 (2019). https://doi.org/10.1007/s11214-019-0622-x

ABSTRACT

The Moon is the only planetary body other than the Earth for which samples have been collected *in situ* by humans and robotic missions and returned to Earth. Scientific investigations of the first lunar samples returned by the Apollo 11 astronauts 50 years ago transformed the way we think most planetary bodies form and evolve. Identification of anorthositic clasts in Apollo 11 samples led to the formulation of the magma ocean concept, and by extension the idea that the Moon experienced large-scale melting and differentiation. This concept of magma oceans would soon be applied to other terrestrial planets and large asteroidal bodies. Dating of basaltic fragments returned from the Moon also showed that a relatively small planetary body could sustain volcanic activity for more than a billion years after its formation. Finally, studies of the lunar regolith showed that in addition to containing a treasure trove of the Moon's history, it also provided us with a rich archive of the past 4.5 billion years of evolution of the inner Solar System. Further investigations of samples returned from the Moon over the past five decades led to many additional discoveries, but also raised new and fundamental questions that are difficult to address with currently available samples, such as those related to the age of the Moon, duration of lunar volcanism, the lunar paleomagnetic field and its intensity, and the record on the Moon of the bombardment history during the first billion years of evolution of the Solar System. In this contribution, we review the information we currently have on some of the key science questions related to the Moon and discuss how future sample-return missions could help address important knowledge gaps.

OVERVIEW OF OUR LUNAR SAMPLE COLLECTION

A total of ~382 kg of rock and soil samples were brought back to the Earth by the Apollo and Luna programmes between 1969 and 1976. This sample suite has advanced our scientific understanding of the Moon in a profound manner. Over the last five decades, there are many fundamental questions that have only been addressed through our existing lunar sample collection. These are questions that not only have implications for our understanding of the Earth-Moon system, but also for our understanding of the wider Solar System, such as how planets are built and the nature of the impact flux in the inner Solar System over geological time (e.g., National Research Council 2007). In many cases, this has been possible as a result of the high precision and accuracy of geochemical measurements that can be made in laboratories on Earth relative to geochemical analyses that can be achieved by rovers on the Moon's surface or by remote sensing instruments from lunar orbit.

Despite the many advances the lunar sample collection has enabled, the major limitation of this collection stems from the limited areal extent of the Moon represented by the Apollo and Luna landing sites. Remote sensing orbital spacecraft over the last two decades have highlighted that our existing sample collections come from a compositionally anomalous area of the Moon (Fig. 1). This was first hinted by orbital gamma-ray measurements collected by the Apollo 15 and 16 service modules (Metzger et al. 1977, see also Plate 10.1 of the Lunar Sourcebook—Heiken et al. 1991), and observed in more detail by the 1998 Lunar Prospector and 1994 Clementine missions (Jolliff et al. 2000). This area, referred to as the Procellarum-KREEP terrane (PKT), is enriched in incompatible elements such as potassium (K), rare earth elements (REE), phosphorus (P), uranium (U), and thorium (Th). It also coincides with the extent of the Imbrium basin continuous ejecta blanket (Fig. 1). Consequently, without future samples from other areas such as the feldspathic highlands terrane, or the farside South Pole-Aitken (SPA) basin, there are a number of key lunar science questions that cannot be fully addressed (Table 1).

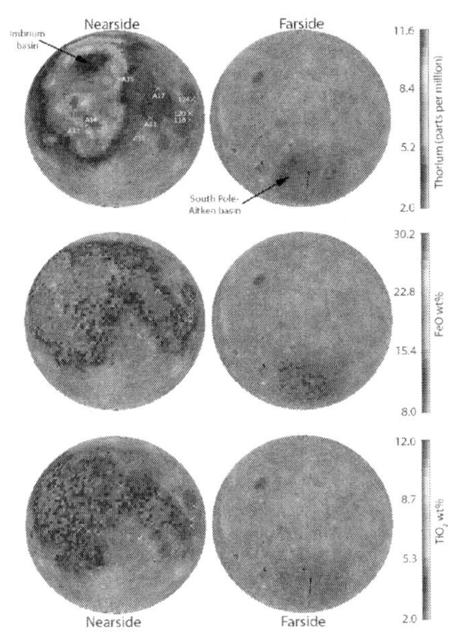

Figure 1. Nearside and farside of the Moon showing distribution of Th, FeO and TiO_2 abundances obtained by the Lunar Prospector Gamma Ray Spectrometer

(2° per pixel data from Prettyman et al. 2006). Data for these elements have been selected to μg g gw abundances of Th>2μgg^{-1}, FeO > 8 wt.% and TiO_2>2 wt.%. These data are overlain on a Clementine albedo map of the Moon in an orthographic projection. Locations of the Apollo (A) and Luna (L) landing sites are from Wagner et al. (2017). The locations of the Imbrium basin and South Pole-Aitken basin are also indicated.

Table 1. Major outstanding science questions that lunar sample-return missions could address. Adapted from Pernet-Fisher et al. (2019b), based on National Research Council (2007) and Crawford and Joy (2014)

Science questions	Opportunities offered by sample-return	Sampling location
Age of the Moon	Multi-system high precision geochronology; Detailed major and trace element geochemistry; Noble gas geochemistry	Primary crust exposures (ideally pure anorthosite—PAN), non PKT
Primordial crust of the Moon: homogeneous vs. heterogeneous ferroan anorthosite (FAN); Relationship between anorthosites and ancient magmatic (i.e., Mg-suite) rocks	Examine petrogenesis of lunar crustal rocks; Multi-system geochronology; Detailed major and trace element geochemistry	Magnesian anorthosites (MAN) outcrops; Farside PAN
Bombardment history of the inner Solar System	Multi-system high precision geochronology; Detailed major and trace element geochemistry; Search for impactor fragments and/or chemical signatures	Impact-melt material from major basin-forming impacts (SPA, Nectaris, Crisium, etc.); Key stratigraphic marker craters (Copernicus, Tycho); Young lava surfaces
Lunar volcanism: timing, nature, duration, volume, emplacement mechanism	Examine source regions; Multi-system geochronology; High precision stable isotope measurements (e.g., O); Detailed major and trace element geochemistry	Young lava flows; Cryptomare lava flows (access ancient volcanism); Mare basalt bedrock (especially within stacked lava flows that would be accessible in lava pit walls); KREEP-rich basalt flows; Pyroclastic glass bead sites

Structure and composition of the lunar interior	Nearside vs. farside crustal thickness dichotomy; Examine mare basalt source regions; Petrology; Multi-system geochronology; Detailed major and trace element geochemistry; Volatile inventory of the lunar mantle	Crustal rocks from the farside; Deep crustal rocks and possible mantle xenoliths from deep basins that breached the crust
Lunar polar volatiles: composition, physical state, form, distribution, context	Volatile release studies; Stable isotope measurements (H, C, N, O, etc.); Detailed major and trace element geochemistry	Volatile-bearing materials in pristine condition from multiple polar locations
Flux and composition of ancient solar wind composition recorded in lunar regolith	Volatile release studies; Stable isotope measurements (H, C, N, O, etc.); Noble gas geochemistry	Ancient paleoregolith sites; Present day regolith; Regolith samples from swirl regions; Core samples
Galactic record archived in lunar surface samples	Analysis of cosmogenic nuclides (^{3}He, ^{21}Ne, ^{38}Ar, ^{83}Kr, ^{126}Xe, ^{10}Be, ^{36}Cl); analysis of supernova-related isotopes (e.g., ^{60}Fe)	Ancient paleoregolith sites; Present day regolith at various sites; Core samples
Lunar magnetic field and anomalies	High-precision magnetic measurements; Multi-system geochronology	Ancient magmatic bedrock (samples collected with known orientations and from multiple stratigraphic layers); Samples from swirl regions

Attempts have been made to answer many of these questions by lunar orbiter remote sensing missions, and lander and rovers missions. For instance, remote sensing instruments (such as the Moon Mineralogy Mapper (Pieters et al. 2009a; Green et al. 2011)—M^3—on Chandrayaan-1 and Diviner (Paige et al. 2010a, 2010b) on the Lunar Reconnaissance Orbiter) have geologically mapped the Moon globally in high-spatial and spectral detail. Spectroscopic imaging has identified lithologies that are rare or absent in the Apollo or Luna collections, such as dunites and pink spinel troctolites (e.g., Corley et al. 2018; Pieters et al. 2011). However, M^3 has a 70 m/pixel maximum resolution (Green et al. 2011) and Diviner infrared spectrometer has an average resolution of 250 m/pixel (Paige et al. 2010a). Therefore, each reported lithology and associated chemical features represent an average over the entire pixel, only painting a broad-scale view of these complex geological features.

Complementary to remote-sensing datasets, rovers on the lunar surface have the potential to make detailed, localised, *in situ* geological investigations. For instance, China's rover Yutu-2 (part of the Chang'E-4 mission) recently reported orthopyroxene- and olivine-bearing lithologies within the Von Kármán crater (situated within the SPA basin; Li et al. 2019). Such a mineral assemblage, together with the geological context of this crater (i.e. material on the Von Kármán crater floor is thought to have been excavated from 72 km below the SPA floor; Huang et al. 2018), led Li et al. (2019) to a suggestion that these lithologies could represent excavated mantle material. This is significant because no mantle rocks have conclusively been identified in the Apollo or Luna collections. However, without laboratory-based geochemical measurements of returned samples, the true provenance of these lithologies and what they tell us about the Moon (for instance, testing predicted lunar magma ocean crystallisation sequences) cannot be unambiguously established.

Undoubtedly, remote sensing and rover missions have made invaluable contributions to our understanding of the Moon by conducting a wide range of science investigations such as identifying new lithologies, globally mapping the chemical heterogeneity of the lunar crust, estimating gravity anomalies, and measuring surface temperature. However, there are some important questions that can only be addressed by detailed geochemical and physical experiments on rock samples in labs on Earth. In this regard, our current best tools for understanding the geology of the Moon outside of the nearside PKT comes from lunar meteorites. Even though lunar meteorites are limited by the fact that they lack exact geographical provenance, precise geological context, and have unknown original orientation, they have uniquely advanced our understanding of the Moon (e.g., Korotev 2005; Joy and Arai 2013). For instance, the lunar meteorite collection contains fragments of the youngest and oldest mare basalts reported to date (e.g., Borg et al. 2009; Snape et al. 2018a; Curran et al. 2019). However, for this example, without known geographic and geological context of where these samples originated on the Moon, it is hard to fully interpret how these mare basalts fit into the wider picture of lunar volcanism.

In this chapter, we detail some of the major outstanding lunar science questions that can only be resolved by the collection of new lunar samples from a geographically diverse set of locations for investigation in labs on Earth (see summary in Table 1).

WHEN DID THE MOON FORM?

Overview

The formation of the Moon is commonly linked with a catastrophic impact between the proto-Earth and a large (potentially Mars-sized or larger) impactor (e.g., Hartmann and Davis 1975). Given this unique relationship between the Earth and the Moon, the timing of this event represents not just a fundamental starting point for lunar evolution, but a key stage in the early evolution of the Earth. Since the return of the first Apollo samples, there have been numerous efforts to determine the age of the Moon (e.g., Tatsumoto 1970; Tera and Wasserburg 1974; Tilton and Chen 1979; Carlson and Lugmair 1988), and significant differences exist among recently published age estimates (e.g., Borg et al. 2011; Gaffney and Borg 2014; McLeod et al. 2014; Avice and Marty 2014; Connelly and Bizzarro 2016; Barboni et al. 2017; Kruijer and Kleine 2017; Thiemens et al. 2019). These age estimates will often be divided into those proposing either an "old Moon" forming at around 4.5 Ga or earlier, and those advocating a "young Moon" age of <4.5 Ga.

Estimates for the Time of Lunar Formation

Because the Moon is not a primitive planetary body, and has had a complex geological history of differentiation, impact modification and extensive volcanic activity, there is no single rock whose age can be measured in order to obtain a direct determination for the time of lunar formation. Nonetheless, perhaps the most straightforward approach employed for constraining the time of the Moon's formation has been to determine the ages of the oldest available lunar samples. For example, Borg et al. (2011) determined an age of 4.360±0.003 Ga4.360±0.003 Ga for the FAN sample 60025, using three independent chronometers (Pb-Pb, ^{147}Sm-^{143}Nd and ^{146}Sm-^{142}Nd). The FAN rocks are believed to represent the remnants of the ancient lunar crust, formed as flotation cumulates in the lunar magma ocean (LMO). As such, the age reported by Borg et al. (2011) can be interpreted as the time at which the LMO crystallised, and by inference suggests a relatively young age for the Moon.

An alternative approach has been to calculate model ages for the mantle sources of a variety of lunar rock types, using different long-lived isotope systems (e.g. U-Pb, Rb-Sr, Sm-Nd and Lu-Hf). While the rocks used in such studies (e.g., basalts formed between ~3.1–4.0 Ga) will often have relatively young crystallisation ages compared to the FAN sample studied by Borg et al. (2011), they will retain an initial isotopic composition from the time at which they crystallised. If these isotopic fingerprints can be determined, it is then possible to calculate the amount of time necessary for radiogenic ingrowth in the mantle sources to generate the initial isotopic composition. Notably, many of these model ages for different lunar lithologies appear to cluster around an age range similar to that of the 60025 FAN sample, as well as that of a number of other ancient lunar crustal rocks (~4350–4400 Ga; e.g., Lugmair and Carlson 1978; Nyquist et al. 1995; Rankenburg et al. 2006; Brandon et al. 2009; Gaffney and Borg 2014; McLeod et al. 2014; Carlson et al. 2014; Snape et al. 2016a). Once again, these model ages have been interpreted as marking the final stages of LMO crystallisation, and potentially suggestive of a relatively young Moon. By contrast, evidence for an older Moon (~4.5 Ga) has been provided by a date of 4.417±0.006 Ga4.417±0.006 Ga obtained on a zircon grain in Apollo 17 breccia 72215 (Nemchin et al. 2009) and by model ages calculated for zircon grains in lunar breccias based on their initial Hf isotopic compositions and $^{207}Pb/^{206}Pb$ crystallisation ages (Taylor et al. 2009; Barboni et al. 2017).

Frustratingly, even studies looking at the same isotope system have produced different estimates for the age of the Moon. A good example, is the ^{182}Hf-^{182}W system, which can potentially provide unique insights due to the short half-life of ^{182}Hf (*ca.* 9 Ma) and the fact that Hf (lithophile) and W (siderophile) are fractioned during planetary core formation. Therefore, if the formation of a planet's core has occurred during the lifetime of ^{182}Hf, the remaining silicate fraction of the planet will develop an excess of ^{182}W. Analyses of lunar samples have identified a relative excess of ^{182}W (Touboul et al. 2015; Kruijer et al. 2015) and higher Hf/W ratios (Thiemens et al. 2019) compared to terrestrial values. These observations can be explained by radiogenic ingrowth of ^{182}W in the bulk silicate Moon (BSM), provided lunar differentiation occurred no later than ~4.51 Ga (Lee et al. 1997; Halliday and Lee 1999; Righter and Shearer 2003; Kleine et al. 2005; Thiemens et al. 2019). Conversely, it has also been suggested that the Moon could have formed later than 4.5 Ga, with the ^{182}W excess in lunar samples being explained via a disproportionate accretion of meteoritic material to the Earth-Moon system during the so-called "late veneer" (Touboul et al. 2015; Kruijer et al. 2015; Kruijer and Kleine 2017).

A number of studies looking at terrestrial rocks have also provided constraints for the timing of the Moon's formation. For example, Avice and Marty (2014) used I-Pu-Xe isotope systematics to determine an age for the Earth's atmosphere of 40+20−1040−10+20 Ma after Solar System formation, which they suggest might provide a lower limit for the timing of the last giant impact event that could have formed the Moon (i.e. the giant impact occurred prior to *ca*. 40 Ma after Solar System formation). Meanwhile, a much younger estimate for the Moon's age was derived by Connelly and Bizzarro (2016). Based on the assumption that the bulk silicate Earth (BSE) originally had a much lower $^{238}U/^{204}Pb$ ratio (μμ-value) than it does today and that the subsequent increase was the result of Pb loss during the giant impact, the authors calculated that this event must have occurred between 4.426–4.417 Ga, in order to explain modern day BSE Pb isotopic compositions.

How Can New Lunar Samples Help Address This Issue?

The lack of agreement with regard to the time of lunar formation is primarily due to different assumptions that have been made in calculating various ages (e.g., the duration of different stages of planetary differentiation, the occurrence and effect of a late veneer), and how these ages are interpreted (e.g., are they representative of global processes and can they be tied to the LMO model?). It is quite possible that new samples will not completely resolve these uncertainties. Furthermore, they will almost certainly raise new questions. However, a wider range of lunar samples has the potential to address a number of key questions that have arisen from previous studies.

Determining ages for the oldest fragments of the lunar crust is particularly challenging, partly due to the age of these samples and the fact that they have been subjected to more than 4 billion years of impact modification. Furthermore, the Apollo samples were all collected from locations on the nearside of the Moon and near to, and in some cases within, the geochemically anomalous PKT region (Fig. 1). Even the Apollo 16 landing site, which provided many of the known FAN samples (such as 60025), is likely not representative of "typical" lunar highlands as it has a fairly elevated KREEP component (e.g., Korotev et al. 2003). Therefore, returning samples from more representative regions of the lunar highlands, particularly on the lunar farside well away from the PKT, will provide a better chance to determine ages for the most ancient crustal lithologies on the Moon.

One of the most striking features about the model ages derived from the Apollo samples is the preponderance of ~4.35–4.4 Ga ages. If similar model ages are derived from samples collected from a wider range of locations (particularly on the lunar farside), it would provide much stronger support for the age representing final crystallisation stages of a global LMO. This would suggest either that the Moon is significantly younger than 4.5 Ga, or that the LMO cooled more slowly than most models predict (Elkins-Tanton et al. 2011). If, on the other hand, samples returned from the lunar farside indicate that the ~4.35–4.4 Ga model ages are an artefact of the lunar nearside or the PKT, it would support the suggestion that previous model ages represent a later large-scale (but not global) resetting of isotope systems. A potential cause of this event could be a particularly large impact event, such as the one previously suggested to have generated the hypothetical "Procellarum basin" (Cadogan 1974). Remote sensing data have failed to detect the typical gravity anomaly associated with impact basins around the Oceanus Procellarum (Andrews-Hanna et al. 2014). Nevertheless, such a large impact event would not only provide a way to explain observed asymmetries between the lunar nearside and farside (Zhu et al. 2019), but could also provide explanation for the ~4.35–4.4 Ga model ages.

A key goal for future sample-return missions is to investigate predicted exposures of material derived from the lunar mantle, particularly in the SPA basin (Yamamoto et al. 2010; Melosh et al. 2017). Our current understanding of the lunar mantle and its composition is based on inferences from samples such as the lunar mare basalts. While it is generally agreed that these rocks have their origins in partial melts of mantle cumulates, they are not a direct representation of the lunar mantle. Having more direct determinations of lunar mantle compositions, particularly regarding the isotopic systematics of mantle material, could address a number of uncertainties associated with models discussed above, which rely on assumptions regarding the nature of the lunar mantle and the BSM.

In case of the Sm-Nd and Lu-Hf model ages described above, an important constraint is whether or not the mantle sources of the Apollo samples evolved from a BSM with a chondritic composition. The ^{182}Hf-^{182}W studies used to constrain the Moon's age rely on estimates for the ratio of Hf/W in the early BSM (Kruijer and Kleine 2017; Thiemens et al. 2019). Similarly, although recent studies of lunar basalts have made it possible to construct a multistage model of Pb isotopic evolution in the Moon (Snape et al. 2016a), using this model to determine the age of the Moon would require a better understanding of the BSM ^{238}U/^{204}Pb ratio (known as the μμ-value) during the Moon's early differentiation. If lunar mantle material indicated that the BSM had very high μμ-values (>1000), it would argue strongly in favour of a young Moon (<4.5 Ga), while evidence for lower BSM μμ-values more typical of those determined for many mare basalt sources (<500), would imply that the Moon is old (~4.5 Ga) (Snape et al. 2016a).

PETROLOGY AND FORMATION HISTORY OF THE LUNAR ANORTHOSITIC CRUST

As a geologically ancient surface, the Moon can inform us about both its own early history and more generally about the expected evolution of the terrestrial planets. The first reports of anorthite-rich lithologies in samples returned from the Apollo missions led to a very simple view of the evolution of the Moon crystallising from a LMO (Wood et al. 1970). In this model, the first minerals to solidify, olivine followed by pyroxene, sank downwards to make a mafic Mg-rich mantle. After solidification of ~75–80% of the LMO, plagioclase began to crystallise (e.g., Snyder et al. 1992). Being less dense than the melt, it rose to form a crust on the lunar surface, bringing with it co-crystallising Fe-bearing mafic minerals. The relative homogeneity of the anorthite composition (An_{94-98}) in Apollo FAN samples supports the LMO model (see discussion in Gross and Joy 2016, and references therein).

Evidence for formation of the different silicate layers in planetary bodies from a magma ocean is most obvious for the Moon, but a fully molten state may be a ubiquitous stage in the evolution of planetary sized bodies and learning about the LMO has implications for planetary evolution more generally. The howardite, eucrite and diogenite (HED) meteorites, thought to originate from asteroid 4-Vesta (e.g., McCord et al. 1970; McSween et al. 2013), have genetic similarities to each other that may be explained by formation from a magma ocean (Righter and Drake 1997). This is likely to be replicated in many large rocky asteroids (McCoy et al. 2006). Mars may also have had a magma ocean, although the wide variety of mineral compositions observed in martian meteorites requires a post-magma ocean period of considerable overturn, mixing and reprocessing (Elkins-Tanton et al. 2003).

The Earth's structure and composition is also consistent with it having a magma ocean stage (Righter et al. 1997). A recent model suggests that the Earth may have been partially molten when it was impacted by a large impactor to produce the Moon. Its molten state may have enabled most of the lunar material to have originated from the Earth rather than from the impactor (Hosono et al. 2019). Thus, magma oceans can have critical implications for how bodies evolve and indeed, how the Moon itself formed.

The study of lunar meteorites, which originate from random locations on the lunar surface, and the comparison of these meteorites to returned Apollo samples, have increased our understanding of crustal heterogeneities and, therefore, of the evolution of the LMO. Most lunar anorthositic meteorites have compositions clearly different from those of the majority of returned Apollo samples in both major and trace element compositions, leading to recent questions about the lunar magma ocean evolution. In terms of major element composition, around 80% of lunar meteorites are significantly more magnesian (and named MAN for magnesian anorthosites) than Apollo FAN samples (Gross et al. 2014). Gross et al. (2014) interpreted this as an indication that much of the lunar crust cannot be the product of a primary magma ocean, in which more ferroan compositions are expected, but likely formed from complex multiple melting events. Trace element analyses of lunar meteorite minerals show a similar picture of complexity. Anorthite in lunar meteorites differs from that in Apollo samples in REE abundances that cannot be explained by crystallisation from a single magma composition (Russell et al. 2014). More recently, FAN clasts in Apollo 15 and 16 samples have been demonstrated to exhibit wide variations in REE abundances that overlap those observed in lunar meteorites, and which cannot be explained by fractional crystallisation from a single source (Pernet-Fisher et al. 2019a) (Fig. 2). The Nd isotope compositions of lunar meteorites and FAN samples likewise suggest that their parent magmas vary in REE abundances, sometimes being enriched in HREE and sometimes in LREE, whereas according to the LMO model the residual melt should always trend towards LREE enrichment, as observed in KREEP samples (Nyquist et al. 2010).

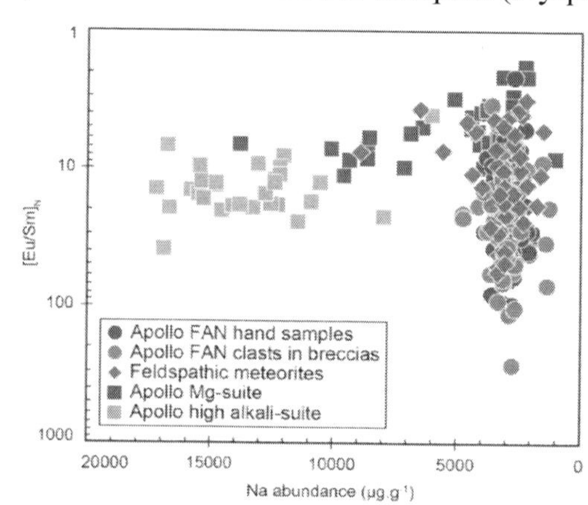

Figure 2. CI chondrite-normalised (CI chondrite REE abundances from Barrat et al. 2012) Eu/Sm vs. Na abundance in lunar plagioclase from Apollo 15

and 16 hand samples and clasts in breccias (after Joy 2013; Pernet-Fisher et al. 2019a), feldspathic lunar meteorites (from Cahill et al. 2004; Treiman et al. 2010; Joy et al. 2011a; Russell et al. 2014), and Apollo Mg-suite and high alkali-suite samples (from Papike et al. 1996; Shervais and McGee 1998, 1999).

All these recent results acquired on lunar meteorites and clasts within Apollo suite samples require a more complex geological history than that provided by simple LMO solidification models. While the Moon is indeed likely to have experienced a magma ocean stage, it seems unlikely that the majority of the lunar surface is comprised mostly of unmodified primordial crust formed about 4.5 billion years ago. Sample-return missions at selected locations could help answer some outstanding questions about the degree of remelting of the lunar crust post-LMO. In order to understand this very early history of the Moon, we require more samples from a variety of lunar locations. The far side of the Moon may be more magnesium-rich than the nearside FAN lithologies (Arai et al. 2008; Ohtake et al. 2012) and, therefore, some anorthosite outcrops from these terrains (Ohtake et al. 2009; Donaldson Hanna et al. 2014; Yamamoto et al. 2015; Lemelin et al. 2015) may be a good focus for a sample-return landing site. Returning samples from anorthosite areas from regions remote from Apollo and Luna landing sites will enable investigating these outstanding issues related to the LMO formation, crystallisation, and the subsequent lunar igneous evolution.

THE BOMBARDMENT HISTORY OF THE INNER SOLAR SYSTEM

Impact bombardment is one of the main geological processes affecting all bodies throughout the Solar System. Because it is not affected by geological reprocessing and hydrospheric/atmospheric weathering, the Moon has preserved a near complete record of the inner Solar System impact history since its formation (e.g., Joy et al. 2016 and references therein) around 4.5 billion years ago (see Sect. 2: When did the Moon form?). The traces left on the Moon by impact bombardment range in size from several thousands of kilometres across (e.g., ~2000×2400 km for the SPA basin; Garrick-Bethell and Zuber 2009) down to micron-sized crater pits observed on pyroclastic glass beads (e.g., McKay et al. 1991). Deciphering the lunar bombardment record can provide crucial information on (i) the secular evolution of bombardment in the inner Solar System, and (ii) the source of material involved and possible changes of impactor types through time.

Timing of Lunar Bombardment

Constraining the temporal evolution of impactor delivery to the Moon is one of the top priorities of research in Solar System science (e.g., National Research Council 2007). This is because the lunar cratering chronology anchors the cratering chronology of all other inner Solar System bodies, and is used to derive crater count model ages for geological units at their surface. Surface observations and gravity data enable deciphering the relative ages of the different lunar basins and craters compared to one another (e.g., Shoemaker and Hackman 1962; Hartmann and Wood 1971; Wilhelms 1987; Stöffler et al. 2006; Fassett et al. 2012; Neumann et al. 2015). However, only samples can provide absolute age constraints derived from precise analysis of parent-daughter radiometric chronometers (e.g., Turner 1970a, 1970b; Tera et al. 1974; Cohen et al. 2000; Norman et al. 2006; Fernandes et al. 2013; Snape et al. 2016a).

The temporal evolution of the flux of impactors to the Moon for its first billion years of evolution is still highly controversial (see Bottke and Norman 2017; Morbidelli et al. 2018 for recent reviews). It is generally believed that immediately after the formation of the Moon the impact flux was high and dominated by leftovers from the planetary accretion epoch, a time period commonly referred to as 'late accretion' during which impactors may have imprinted their chemical signatures into the crystallising LMO (e.g., Walker 2009; Day and Walker 2015; Barnes et al. 2016). Once the crust was formed and the LMO fully solidified, it is unlikely that any extraneous sources could have been added and efficiently mixed into the lunar mantle. The Moon then suffered an intense period of bombardment, as recorded by *ca.* 30–40 impact basins larger than 300 km in diameter scattered at its surface (Fig. 3 and Table 2) (e.g., Fassett et al. 2012; Neumann et al. 2015). However, the formation ages of most of these basins are not precisely known. The SPA basin is probably the oldest—crater counting studies and $^{40}Ar/^{39}Ar$ data obtained on samples that are postulated to have originated from SPA, based on their formation depth for example, have tentatively dated formation of SPA at *ca.* 4.25–4.3 Ga (e.g., Hiesinger et al. 2012; Fernandes et al. 2013; Garrick-Bethell and Miljković 2018). Orientale is probably the youngest basin but its age is indirectly constrained at *ca.* 3.9–3.7 Ga (e.g., Stöffler et al. 2006; Whitten et al. 2011). The Imbrium basin is one of the youngest basins, and its age has recently been proposed to be 3.926±0.002 Ga3.926±0.002 Ga based on U-Pb dating of zircon in an Apollo 12 impact melt breccia and in lunar meteorite Sayh al Uhaymir 169, and of phosphates in four Apollo 14 impact melt breccias (e.g., Snape et al. 2016b). About 30 impact basins larger than 300 km in diameter are proposed to have formed over a time window of ~350–400 million years between the formation of SPA and Imbrium (see Table 2).

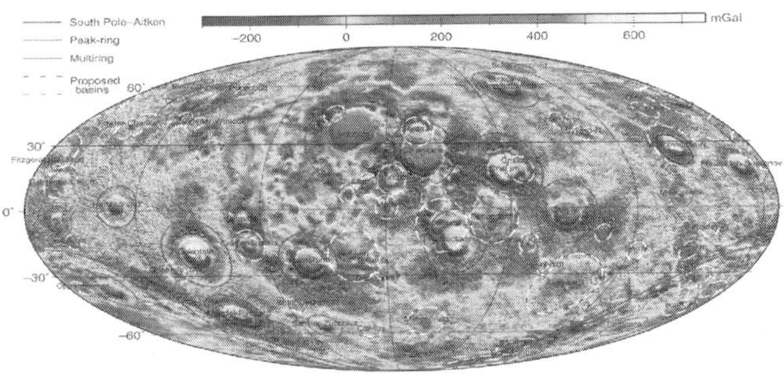

Figure 3. Bouguer anomaly map of the Moon showing the location of large mass concentrations under impact basins (from Neumann et al. 2015). Bouguer-corrected GRAIL gravity anomaly in Mollweide equal-area projection centred on the nearside at 7°E longitude.

Table 2. Sequence of lunar basins with diameters greater than 300 km (after Fassett et al. 2012). Basins in brackets are those whose diameters and/or stratigraphic relationships are uncertain because of observational challenges. Main ring and Bouguer anomaly diameters are from Neumann et al. (2015); ring diameters italicised are probable/possible

Name	Period	Main ring diameter (km)	Bouguer anomaly diameter (km)	Superposition relationships
South Pole-Aitken	Pre-Nectarian	2400		no basin older
Coulomb-Sarton	Pre-Nectarian	*672*	330±18	> Birkhoff
Dirichlet-Jackson	Pre-Nectarian	*452*	220	> Korolev
Cruger-Sirsalis	Pre-Nectarian	*430*	268	
Smythii	Pre-Nectarian	878	438±62	> Crisium
[Schiller-Zucchius]	Pre-Nectarian	361	210	
[Amundsen-Ganswindt]	Pre-Nectarian	378	170	> Schrödinger
Nubium	Pre-Nectarian	*690*	416	> Humorum
[Poincaré]	Pre-Nectarian	312	188	

Lorentz	Pre-Nectarian	351	156	
[Fitzgerald-Jackson]	Pre-Nectarian	*600*	334	> Freundlich-Sharonov
[Birkhoff]	Pre-Nectarian	334	130	
[Ingenii]	Pre-Nectarian	342	154	
[Serenitatis]	Pre-Nectarian?	*923*	556±64	> Nectaris (?)
Apollo	Pre-Nectarian/ Nectarian	492	264	> Korolev, Hertzsprung
Freundlich-Sharonov	Pre-Nectarian/ Nectarian	582	318	> Moscovi-ense
Nectaris	Start of Nectar-ian	885	440±61	
Korolev	Nectarian/Pre-Nectarian	417	202	> Hertz-sprung (?)
[Mendeleev]	Nectarian/Pre-Nectarian	331	156	
Hertzsprung	Nectarian/Pre-Nectarian	571	254±38	
[Grimaldi]	Nectarian/Pre-Nectarian	460	220	> Mendel-Rydberg
Mendel-Rydberg	Nectarian/Pre-Nectarian	650	328±26	
[Planck]	Nectarian/Pre-Nectarian	321	128	> Schröding-er
Moscoviense	Nectarian	421		
Crisium	Nectarian	1076	498±31	> Humbold-tianum
Humorum	Nectarian	816	360±21	
Humboldtianum	Nectarian	618	312±27	
Imbrium	Start of Imbrian	1321	684±45	
Schrödinger	Imbrian	326	154	
Orientale (no buff.)	Imbrian	937	436±20	

Whether impact events during this basin-forming period largely occurred as a cataclysmic spike at *ca.* 3.9 Ga ago (e.g., Tera et al. 1974; Ryder 1990; Cohen et al. 2000; Marchi et al. 2012) or decayed monotonically after the end of planetary accretion from ~4.3–4.2 Ga until ~3.9–3.8 Ga ago (e.g., Hartmann 1970; Neukum et al. 2001; Morbidelli et al. 2012; Werner 2014; Morbidelli et al. 2018), or even after (e.g., Fernandes et al. 2013; Zellner 2017), remains highly debated. One of the explanations for the apparent concentration of chronometric dates at *ca.* 3.9 Ga links it to a bias in largely sampling Imbrium ejecta in the Apollo sample collection (Fig. 1) (e.g.,

Haskin 1998; Spudis et al. 2011; Fernandes et al. 2013; Zellner 2017).

Sources of Impactors

The lunar record contains direct and indirect evidence that projectiles have hit the Moon throughout its history (see Joy et al. 2016 for a recent review). During the 'late accretion' window, a time period of up to around 200 million years between the segregation of the lunar core and the end of the LMO crystallisation (e.g., Elkins-Tanton et al. 2011), impactors may have been efficiently mixed into the lunar mantle and modified its chemical composition. We only have indirect, chemical and isotopic, evidence to estimate the flux and sources of impactors during this early period. The highly siderophile element (HSE) abundances and Os isotope signatures of mare basalts suggest that *ca.* 0.02 wt.% of the mass of the Moon has been added by chondritic impactors during late accretion (e.g., Day et al. 2007; Day and Walker 2015). These chondritic projectiles may have also delivered the bulk of the indigenous lunar H and N budgets (Barnes et al. 2016).

Projectiles continued hitting the Moon after crystallisation of the LMO and formation of a thick anorthosite crust. Volatile elements (H, C, N, S) and HSE abundances and isotopic compositions of lunar regolith and impact-melt samples provide indirect evidence for the types of impactors involved. Analysis of Apollo 14, 15, 16 and 17 samples show a broadly linear correlation between HSE/Ir and $^{187}Os/^{188}Os$ signatures (Puchtel et al. 2008; Fischer-Gödde and Becker 2012; Sharp et al. 2014; Liu et al. 2015; Gleißner and Becker 2017), reflecting the addition of chondritic and non-chondritic components to the lunar crust, the non-chondritic component HSE composition resembling that of iron meteorites, and contributing up to ~30 wt.% HSE in the most fractionated Apollo 16 impact-melt samples (e.g., Gleißner and Becker 2017). Siderophile element abundances in mature regolith samples that have been exposed at the lunar surface for hundreds of millions of years suggest the addition of *ca.* 1–3 wt.% of a CI chondrite-like micrometeorite component (e.g., Keays et al. 1970; Morgan et al. 1972; Korotev et al. 2003; Wolf et al. 2009) often present in the form of micrometre-sized metallic particles, with, on average, about 70% of the total metallic iron content of lunar soils being derived from micrometeorite impactors (Morris 1980).

The lunar regolith also contains relict fragments of various types of projectiles. These are dominated by asteroidal materials, ranging from enstatite chondrite-like to carbonaceous chondrites and iron meteorites (see Joy et al. 2016 for a full list of recovered meteoritic debris). To date no cometary fragments have been positively identified in lunar regolith samples, although there are suggestions that some lunar regolith samples have volatile element signatures consistent with an added cometary component (Gibson and Moore 1973). Finally, a recent investigation of a *ca*. 4.0 billion year old clast from Apollo 14 breccia 14321 suggests that it formed at oxygen fugacity and pressure conditions incompatible with a lunar crustal environment, potentially representing the first positive identification of a terrestrial meteorite in the lunar regolith (Bellucci et al. 2019).

Outstanding Questions and the Potential of Future Sample-Return

Determining the Magnitude and Duration of the Basin-Forming Period

To test the magnitude and duration of the postulated late heavy bombardment, we need samples to be collected and returned to Earth for analysis in our laboratories collected from one, or ideally several, of the older, pre-Nectarian lunar basins. Accessing samples will require careful assessment of site geology and sampling of specific targets to provide well-understood geological context (Norman 2009; Pernet-Fisher et al. 2019b). Site selection ideally needs to consider (i) accessing basins of different ages to test different key stratigraphic markers throughout the early bombardment interval, and (ii) where possible collecting material directly from impact melt deposits, or reworked impact melt deposits to be able to access material formed within the basin event rather than re-processed by later superimposed impact events. Key periods, with targets providing opportunities for sampling multi-ring basin targets, include:

- *Pre-Nectarian age:* The SPA basin, on the lunar farside, which is the largest (>2000 km diameter) and oldest preserved basin on the Moon (Wilhelms 1987). The formation age of SPA is unknown from direct geological sampling. However, it is inferred from crater size-frequency distribution (CSFD) age models to be >4 Ga, with estimates ranging from ~4.25 Ga to ≤4.33 Ga (Hiesinger et al. 2012; Morbidelli et al. 2012; Fernandes et al. 2013; Marchi et al. 2013; Garrick-Bethell and Miljković 2018). Anchoring the age of SPA will constrain the onset of bombardment (Jolliff et al. 2017) after the lunar crust formed and had the material strength to be able to retain a large basin structure (Conrad et al. 2018).

- *Pre-Nectarian or Nectarian age:* The Serenitatis basin (~900 km diameter), for which there is significant debate whether it is an older Pre-Nectarian basin, or a Nectarian basin younger than Nectaris and only just older than Imbrium at *ca.* 3.93 Ga (Spudis et al. 2011; Fassett et al. 2012). This debate is important as the lunar sample community is currently challenged about how to interpret the ages of samples collected from the North and South Massif at the Apollo 17 landing site, and notably if these samples represent Imbrium, Nectaris or Serenitatis ejecta (e.g., Hurwitz and Kring 2016; Thiessen et al. 2017).
- *Nectarian age:* The Nectaris basin (~900 km diameter) is a key stratigraphic marker that gives its name to the Nectarian period (Stuart-Alexander and Wilhelms 1975). It has been postulated that Nectarian ejecta was sampled at the Apollo 16 landing site at North Ray Crater, with basin age estimates ranging from 3.9 Ga to 4.2 Ga (e.g., Fischer-Gödde and Becker 2011, 2012). Collecting various impact melt samples from this nearside highlands site (e.g., Norman 2009; Cohen et al. 2016) is of key interest for potentially determining the timing of formation of about ten basins formed between Nectaris and Imbrium (Table 2). This would provide crucial constraints on whether the putative late heavy bombardment was a true cataclysm (Ryder 1990), followed a longer duration, less intense, 'sawtooth' model (e.g., Turner 1979; Morbidelli et al. 2012), or consisted in a monotonic decline over time (e.g., Hartmann 1970; Neukum et al. 2001; Morbidelli et al. 2018).

Anchoring the Lunar Cratering Chronology

Samples collected from known locations on the Moon, and in particular the mare basalts, have made it possible with CSFDs to quantify the relationship between the age of a geologic unit on a planetary surface and the density of impact craters on that surface (Fig. 4) (Neukum et al. 2001; Stöffler and Ryder 2001; Robbins 2014; Fassett 2016). Returning and dating any new samples that can be confidently linked with particular locations and geologic units on the Moon would provide valuable new data-points for future cratering chronology studies. The importance of this development extends beyond just understanding the impact history of the Moon, since the lunar cratering chronology model is commonly extrapolated to studies of other terrestrial planets (Neukum et al. 2001; Marchi et al. 2009).

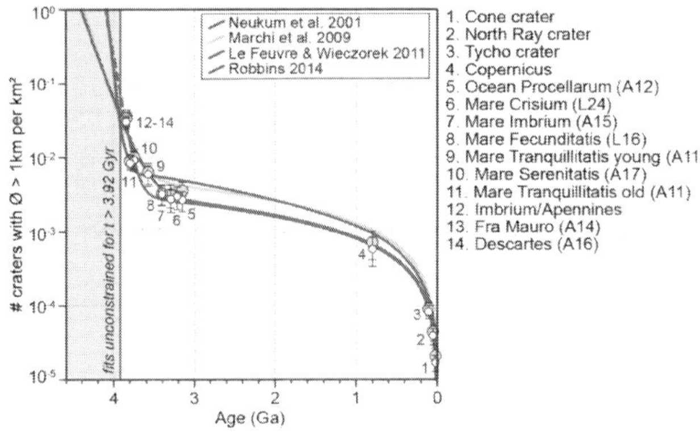

Figure 4. Examples of different lunar crater chronology calibration curves and the Apollo sample age calibration points used (redrawn after Le Feuvre and Wieczorek 2011). Note that (4) Copernicus and (3) Tycho are from indirect sampling at the Apollo 12 and Apollo 17 landing sites, respectively.

The current plans for the China National Space Administration (CNSA) Chang'E-5 mission, planned for late 2019—early 2020, involve returning samples from the Mons Rümker region in northern Oceanus Procellarum (Qian et al. 2018; Zhao et al. 2017), with potential target landing sites close to the young (~1.2–1.5 Ga) basalt flows identified in previous remote sensing studies (Hiesinger et al. 2003; Stadermann et al. 2018). Likewise there are NASA mission proposals to sample *ca.* 1 Ga-old lava flows adjacent to the Aristarchus region of the nearside of the Moon (Draper et al. 2019). Given that current lunar crater chronology studies still rely predominantly on samples with ages between 3.0–4.0 Ga (Fig. 4), samples from these young basalt flows would be immensely valuable, helping to calibrate parts of the lunar cratering curve more relevant to Mars and Venus surface geological processes. Similarly, as meteorite bombardment is thought to have been so much more intense during the first billion years of solar system history, returning samples from any exposed basaltic flows older than 4.0 Ga would place better constraints at the other end of the timescale.

Secular Evolution of Types of Projectiles Hitting the Moon

A thorough investigation of potential changes in the origin(s) of projectiles hitting the Moon requires precise age constraints on (i) impact-melt lithologies in which chemical and isotopic signatures of impactors are measured, and of (ii) regolith and soil samples from which projectile fragments are recovered. Projectiles in regolith samples >3.5 Ga-old could, for example, contain ancient terrestrial and martian crustal fragments that would be key for tackling questions linked with habitability of terrestrial planets and the origin of life on Earth (Armstrong et al. 2002; Crawford et al. 2008; Armstrong 2010; Bellucci et al. 2019).

There are different proxies for determining the duration of exposure of regolith samples to the space environment (see Joy et al. 2016), including (i) the ferromagnetic resonance maturity index (I_s/FeO—this corresponds to the intensity of the ferromagnetic resonance caused by non-interacting fine-grained metal particles ratioed to the FeO abundance, since the amount of fine-grained metal is proportional to both the duration of surface exposure and the amount of FeO available for reduction; Morris 1976, 1978) used to classify surface exposure of regolith samples into immature (I_s/FeO $\leq$ 29), submature (I_s/FeO = 30–59) and mature (I_s/FeO $\geq$ 60) (Morris 1976, 1978), (ii) the amount of 'trapped' solar wind-derived ^{36}Ar (e.g., Wieler and Heber 2003), or (iii) the agglutinate content (McKay et al. 1972). Quantitative exposure ages can be derived from measurement of the abundance of cosmogenic noble gases (^{3}He, ^{21}Ne, ^{38}Ar, ^{81}Kr, ^{126}Xe) produced during regolith interaction with cosmic rays in the top few meters of regolith (e.g., Eugster et al. 2001). However, none of these proxies allow estimating when the samples were exposed at the lunar surface, and exposure ages could even be influence by several discrete episodes of exposure due to the constant gardening of the lunar regolith. A semi-quantitative method of estimating regolith closure ages can be found using the $^{40}Ar/^{36}Ar$ antiquity indicator, which relates the amount of solar wind-derived ^{36}Ar to the amount of 'parentless' ^{40}Ar derived from the lunar exosphere (e.g., McKay et al. 1986; Eugster et al. 2001; Joy et al. 2011b; Fagan et al. 2014; Wieler 2016). The model closure ages for 74 Apollo, Luna and lunar meteorite regolith breccias range from *ca*. 0.1 Ga for some Apollo 15 samples up to *ca*. 3.8–3.9 Ga for some Apollo 14 samples, displaying a rough bimodal distribution with broad peaks between ~3.4–3.8 Ga and 1–2 Ga (Fig. 5), indicating that most regolith breccia samples in our collections formed around 1–2 Ga ago and 3.5–3.8 Ga ago, and that none is older than *ca*. 4 Ga.

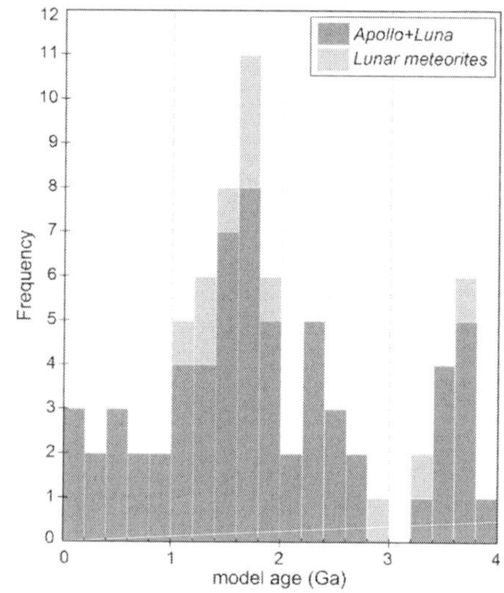

Figure 5. Histogram of regolith breccias model closure ages (data and sources can be found in Fagan et al. 2014).

Sample-return missions targeting regolith breccias with lithification ages older than *ca.* 3.8–3.9 Ga would greatly complement the record present in our collections, potentially providing crucial information on the types of projectiles hitting the Earth-Moon system during the basin-forming epoch before 3.9 Ga ago. The most scientifically important sample sites would be 'palaeoregolith' horizons, which are ancient regolith layers or lenses trapped between datable geological units, such as lava flows or impact melt flows (Crawford et al. 2007, 2010; Fagents et al. 2010; Crawford and Joy 2014). Whilst deep drilling might be necessary to access deeply seated palaeoregolith horizons, some natural access might be provided through layers exposed in the walls of lava tube skylights or in layered boulders in which palaeoregolith horizons may be preserved between volcanic layers (Fig. 6). For instance, Lunar Reconnaissance Orbiter Camera (LROC) Narrow Angle Camera (NAC) images of a Mare Tranquillitatis pit show that several tens of meters of layered units are present below the surface (Robinson et al. 2012a). The basaltic units at the surface where the pit is located are ~3.6 Ga old (Hiesinger et al. 2000), so the deepest palaeoregolith layers in the Mare Tranquillitatis pit may well be over 4 Ga old, considering typical regolith formation rates on the order of 1–2 mm/Ma (Hörz et al. 1991).

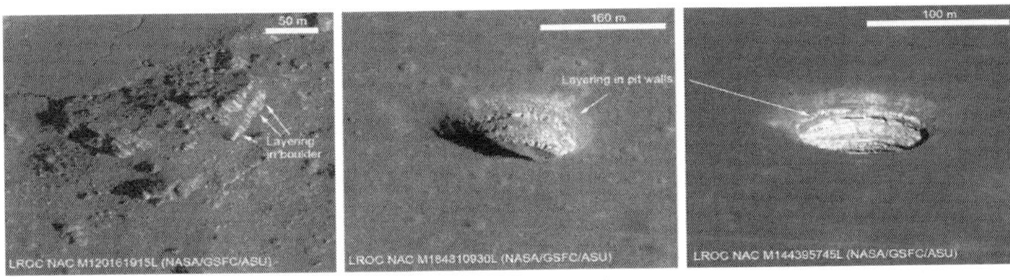

Figure 6. Oblique LROC NAC view of layered boulders found on the lunar surface within Aristarchus crater (left), and of layered pits in Mare Ingenii (centre) and Mare Tranquillitatis (right). Scale bars in all cases have been estimated from the pixel resolution of NAC images (images: NASA/GSFC/Arizona State University, modified by K.H. Joy).

NATURE, TIMING, AND DURATION OF LUNAR MARE VOLCANISM

Overview

The most obvious expressions of volcanic activity on the Moon are the mare basalt infillings of impact basins, particularly on the lunar nearside. The majority of these maria appear to have been emplaced between approximately 3.8–3.1 Ga (Nyquist and Shih 1992; Hiesinger et al. 2003, 2010). Older rocks identified in both the Apollo collection and in lunar meteorites, indicate that some form of basaltic volcanism was occurring on the Moon as early as approximately 4.35 Ga (Terada et al. 2007a; Snape et al. 2018a; Curran et al. 2019). It is not clear to what extent this early volcanism resembled the large scale eruptions of the younger mare basalts, or if these samples resulted from more localised volcanic activity. Meanwhile, the presence of less heavily cratered mare basalt flows indicates that mare volcanism continued until approximately 1.0 Ga in restricted geographical locations (Hiesinger et al. 2003, 2011; Stadermann et al. 2018), but examples of such young lunar basalts have yet to be identified in the sample collection. In this section, we review the current understanding of the Moon's volcanic history based on the Apollo, Luna and meteoritic samples, and highlight outstanding issues that could be addressed by future sample-return missions.

The Nature and Origin of Lunar Volcanism

Discussions of volcanism on the Moon have typically divided samples into those formed as a result of mare and non-mare (or pre-mare) volcanism (Nyquist and Shih 1992). The focus of this paper is primarily on the geochemical nature of samples generated by volcanic processes. The readers interested in recent studies on magma ascent from the lunar mantle and on the possible eruption mechanisms that produced these samples and associated volcanic structures are referred to Head and Wilson (2017) and Wilson and Head (2017, 2018). In addition to basaltic rocks, the mare volcanic samples also include glass beads, produced during pyroclastic "fire fountain" eruptions (Rutherford et al. 2017). Moving beyond these top-level classifications, petrologic and geochemical analyses of samples returned by the Apollo and Luna missions have revealed a compositional diversity in the Moon's basaltic rocks that necessitates more specific classification schemes (Papike et al. 1976; Neal and Taylor 1992). Using the classification proposed by (Neal and Taylor 1992), the lunar basalts are defined first by their bulk TiO_2 content (where: >6 wt% = high-Ti; 1–6 wt% = low-Ti; <1 wt% = very low-Ti [VLT]), then by Al_2O_3 content (>11 wt% = high-Al; <11 wt% = low-Al) and finally by K content (>2000 ppm = high-K; <2000 ppm = low-K). At a more detailed level, even within the mare basalt samples collected from individual landing sites, it is possible to define multiple different basaltic suites based on petrologic texture, relative abundance of minerals, chemical composition and isotope systematics (Rhodes et al. 1976, 1977; Beaty and Albee 1978; Warner et al. 1979; Paces et al. 1991; Jerde et al. 1994; Neal et al. 1994; Snyder et al. 1994). One notable group of samples that are not described by the above classification scheme are the KREEP-rich basalts, which are best represented by samples collected during the Apollo 15 mission (Nyquist et al. 1975; Carlson and Lugmair 1979; Nyquist and Shih 1992), and are characterised by elevated trace element compositions similar to the KREEP signature identified in other Apollo samples (Warren and Wasson 1979).

All of these volcanic samples are generally accepted to have been derived from 5–10% partial melting in the Moon's interior (e.g. Shearer et al. 2006 and references therein), with a common starting point for most petrogenetic models being mafic cumulates formed during the LMO crystallisation. Simply varying the proportions of these different cumulates cannot adequately explain the variability observed in lunar basalts, and so a variety of mechanisms have been proposed to address this issue. This includes the presence of more evolved residual LMO melt enriched in incompatible trace elements (ITE) in the lunar mantle (Snyder et al. 1997; Borg et al. 2004; Hallis et al. 2014), the introduction of ilmenite-bearing material during density-driven overturn of the lunar mantle (Snyder et al. 1997), and the assimilation of KREEP and more evolved plutonic rocks (e.g. granite) by the magmas (Neal and Kramer 2006). Additionally, it is not clear what the heat source for the melting of this material would have been. Trapped ITE-rich material in the lunar mantle provides one potential heat source, but external heat sources, such as meteoritic bombardment and tidal heating, have also been proposed (Borg et al. 2004; Elkins-Tanton et al. 2004; Garrick-Bethell et al. 2006; Shearer et al. 2006).

Timing of Lunar Volcanism

Currently, evidence for the oldest basaltic volcanism on the Moon comes from two lunar meteorites: the basaltic breccia Kalahari (Kal) 009 (4.369±0.007 Ga4.369±0.007 Ga; Terada et al. 2007a; Sokol et al. 2008; Snape et al. 2018a) and the basalt-bearing regolith breccia, Miller Range (MIL) 13317 (4.332±0.002 Ga4.332±0.002 Ga; Snape et al. 2018a) (Fig. 7). Prior to the discovery of these meteorites, the oldest samples of lunar basalt were the high-Al basalts collected during the Apollo 14 mission. Early studies of the volcanic rocks returned from the Apollo 14 landing site yielded ages of approximately 3.9–4.1 Ga, based on internal Rb-Sr isochrons (Compston et al. 1971; Papanastassiou and Wasserburg 1971a; Compston et al. 1972), but further investigations of basaltic clasts in several Apollo 14 breccias determined ages of up to 4.3 Ga (Taylor et al. 1983; Dasch et al. 1987) (Fig. 7). The KREEP basalts collected during the Apollo 15 and 17 missions appear to have formed soon after (and potentially overlapping with) the eruption of the last high-Al basalts, with the largest of these samples (15382 and 15386) having crystallised at *ca.* 3.9 Ga (Stettler et al. 1973; Turner et al. 1973; Nyquist et al. 1975; Carlson and Lugmair 1979; Snape et al. 2016a) (Fig. 7).

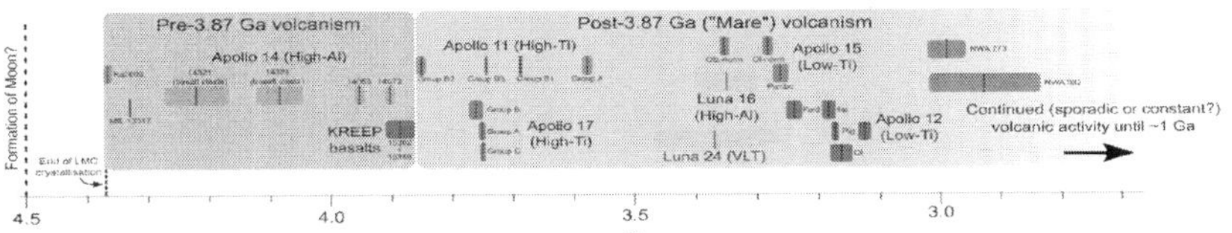

Figure 7. Compilation of crystallisation ages determined for lunar basaltic samples. The boxes are coloured to correspond to different groups of basaltic

samples. The widths of the boxes represent 2 standard deviation uncertainties of either individual sample ages (in the case of the four lunar meteorites; Borg et al. 2004, 2009; Terada et al. 2007a; Sokol et al. 2008; Snape et al. 2018a), or weighted averages of multiple age determinations (including data from: Turner 1970a, 1970b; Papanastassiou and Wasserburg 1971a, 1971b; Compston et al. 1972; Stettler et al. 1973; Turner et al. 1973; Nyquist et al. 1975, 1976; Murthy and Coscio 1976; Guggisberg et al. 1979; Carlson and Lugmair 1979; Paces et al. 1991; Snyder et al. 1994, 1997; Cohen et al. 2001; Fernandes and Burgess 2005; Snape et al. 2016a, 2018b, 2019).

The formation of the high-Al and KREEP basalts was followed by the emplacement of the mare basalts. Based on the samples currently available, this period of lunar volcanism appears to have begun with the eruption of high-Ti basalts. Samples collected at the Apollo 11 landing site have ages varying from ~3.6–3.85 Ga (Turner 1970a, 1970b; Papanastassiou and Wasserburg 1971b; Guggisberg et al. 1979; Snyder et al. 1994; Snape et al. 2016a, 2019), while those sampled by the Apollo 17 mission have a more narrow range of ages, between ~3.75–3.8 Ga (Murthy and Coscio 1976; Nyquist et al. 1976; Paces et al. 1991; Snape et al. 2019) (Fig. 7). The oldest low-Ti mare basalts (~3.28–3.35 Ga; Snape et al. 2019) collected during the Apollo missions were returned from the Hadley-Appennine province by Apollo 15 (Papanastassiou and Wasserburg 1973; Husain 1974; Snape et al. 2019) (Fig. 7). By comparison, the low-Ti basalts collected during the Apollo 12 mission have ages between ~3.24–3.13 Ga (Snyder et al. 1997; Snape et al. 2018b), making them the youngest lunar basalts to have been returned by human sampling (Fig. 7).

Sample-based evidence for more recent magmatic activity on the Moon has been identified in the meteorites Northwest Africa (NWA) 773 (Borg et al. 2004) and NWA 032 (Borg et al. 2009). Both of these samples have ages of ~2.9 Ga (determined by internal Sm-Nd and Rb-Sr isochrons) (Fig. 7). In the case of NWA 773, it is notable that the sample appears to be strongly enriched in ITEs, and exhibits a KREEP-rich geochemical signature, despite it being significantly younger than any of the KREEP-rich samples returned by Apollo, while NWA 032, by contrast, has relatively low ITE concentrations. While these are currently the youngest identified volcanic samples from the Moon, crater size-frequency distribution studies have identified basaltic flows that appear to have been emplaced at ~1.0 Ga (Hiesinger et al. 2003; Stadermann et al. 2018), and potentially some even younger "irregular mare patches" for which ages of <0.1 Ga have been proposed (Braden et al. 2014).

Outstanding Questions and the Potential of Future Sample-Return

Duration and Flux of Lunar Volcanism

Identifying and sampling the youngest lunar basalts would provide new insight into how long small terrestrial planetary bodies can sustain volcanic activity. This issue is connected to a more general issue of the changing flux of lunar volcanism with time (e.g., Hiesinger et al. 2011; Robinson et al. 2012b). Based on the ages of the Apollo, Luna and meteorite basaltic samples, together with crater size-density frequency studies of exposed mare basalt units, the majority of the lunar maria appear to have been emplaced between 3.8–3.1 Ga (Nyquist and Shih 1992; Hiesinger et al. 2003, 2010). Upcoming missions such as the CNSA Chang'E-5 or the Inner SOlar System CHRONology (ISOCHRON) missions could constitute a first step in this direction, since it may involve returning samples from very young basalt flows (~1.2–1.5 Ga) identified in previous remote sensing studies (Hiesinger et al. 2003; Qian et al. 2018; Stadermann et al. 2018; Draper et al. 2019).

Assessing the amount of recent volcanic activity should be a relatively simple issue of identifying and mapping exposed volcanic units. By contrast, determining how common ancient (>4.0 Ga) lunar volcanism was, is likely to be more challenging due to the potential for older units to have been obscured by subsequent eruptions and impact processes. In addition to the ancient volcanic lunar samples discussed previously, remote sensing studies have identified evidence for ancient (>4.0 Ga) mare volcanism in the form of "dark-haloed" craters. These were interpreted as instances where basaltic flows had been buried by the ejecta deposits from large impact craters and basins, and then subsequently re-exposed by smaller impacts (Schultz and Spudis 1979, 1983). Buried basaltic flows of this kind have since been designated by the name "cryptomare" (Head and Wilson 1992; Whitten and Head 2015a, 2015b). Terada et al. (2007a) proposed that the Kal 009 meteorite represented a fragment from such a deposit, and it is true that the bulk-rock composition and mineral chemistry data from the sample are consistent with mare basalts collected during the Apollo and Luna missions (Sokol et al. 2008). Nonetheless, it remains unclear just how much basaltic material could have been buried by impact ejecta.

Sample-return missions to identified dark-haloed craters would provide a valuable first step in testing the cryptomare theory. By performing detailed petrologic, geochemical and isotopic studies of potential cryptomare basalts it would be possible to determine how similar they are to previously studied lunar basalts, as well as when they crystallised. However, individual sample-return missions will almost certainly struggle to provide a definitive answer to the larger question of how common such deposits are. This will likely require a more prolonged human presence on the Moon, enabling protracted and focused geologic sorties with the potential to obtain core samples and map the extent of cryptomare deposits.

Petrogenesis of Lunar Basalts

Returning samples of young and old lunar basalts, together with an increased knowledge of the locations and extent of cryptomare deposits, will be vital to gaining a more complete understanding of mantle sources for lunar volcanism, and addressing questions related to the petrogenesis of lunar basalts. One of the most striking aspects about the lunar mare basalts is their assymetric distribution between the lunar nearside and farside, as well as the spatial correlation between the exposed mare units, the nearside lunar basins and the PKT (Jolliff et al. 2000) (see Fig. 1). Determining the significance of these correlations appears to be fundamental to understanding the sources of lunar basalts and the mechanism(s) that led to partial melting in the lunar mantle.

One suggested explanation for the asymmetric distribution of mare basalts is a propencity for more extrusive volcanism on the thinner nearside crust compared with the thicker crust of the lunar farside (Head and Wilson 2017). Given the location of so many mare basalt flows within large impact basins, there is also a temptation to try and link the period of apparently more intense volcanic activity on the Moon (from ~3.8–3.1 Ga) with the end of the basin-forming period of lunar history and the proposed cataclysm (at ~3.9 Ga). One suggestion is that the formation of the large impact basins on the nearside of the Moon could have led to adiabatic melting in the lunar mantle (e.g., Solomon and Head 1980; Elkins-Tanton et al. 2004). For example, in the model proposed by Elkins-Tanton et al. (2004), partial melting occurs in two stages: (i) *in situ* melting first occurs in response to the decompression associated with basin excavation; (ii) a second stage of adiabatic melting then occurs in response to convection at the lithosphere-asthenosphere boundary, triggered by the isostatic rebound of the lithosphere after basin formation. In the case of the Orientale basin, a significant (60–

100 Ma) time gap has been estimated between the formation of the basin (~3.68 Ga) and the subsequent eruption of the Mare Orientale units (~3.58 Ga), which has been cited as evidence against such impact-induced melting (Whitten et al. 2011). However, melts produced during the second stage of adiabatic melting of the Elkins-Tanton et al. (2004) model could form as late as *ca.* 350 million years after the impact event (Elkins-Tanton et al. 2004). Identifying the distribution of ancient (>4.0 Ga) volcanic units, and determining the times at which they were emplaced by dating returned samples, would be a significant step towards testing the link between lunar volcanism and impact events.

The connection between impact basins and basalts does not, in itself, explain the correlation between the mare basalts and the PKT or the extent to which lunar volcanism is connected with KREEP. Returning basaltic samples further away from the PKT than the Apollo basalts, would help address this issue, and a hint of what might be expected from such samples may be seen in several basaltic lunar meteorites (Yamato-793169, Asuka-881757, Meteorite Hills 01210 and MIL 05035). These samples have lower concentrations of ITEs than most Apollo mare basalts and appear to have been derived from mantle sources with low ratios of U/Pb and Rb/Sr, and high Sm/Nd ratios (Misawa et al. 1993; Terada et al. 2007b; Joy et al. 2008; Liu et al. 2009; Arai et al. 2010). These features have been interpreted as evidence that these rocks may be derived from partial melts of more primitive lunar mantle cumulates than those from which the majority of Apollo basalts originated (Joy et al. 2008; Liu et al. 2009; Arai et al. 2010). The low-ITE concentrations and lack of a KREEP signature in these samples also indicate that their parental magmas did not assimilate significant amounts of KREEP-rich material. Furthermore, while the presence of ITE-rich material in the lunar mantle may have triggered the partial melting that eventually produced many of the Apollo basalts, these samples indicate that the same mechanism may not be sufficient to explain the petrogenesis of all lunar basalts. If these characteristics were found to be common in basalt samples collected away from the PKT, it could provide important insights into the petrogenesis of lunar basalts, the distribution of the proposed urKREEP layer, and the early magmatic evolution and differentiation of the Moon.

STRUCTURE AND COMPOSITION OF THE LUNAR INTERIOR

Geophysical Insights

The Apollo Lunar Surface Experiments Packages (ALSEP) deployed across the Moon's surface by astronauts on the Apollo 12, 14, 15, 16 and 17 missions are one of the key legacies of the Apollo programme. These included active and passive seismic experiments, gravimeters, magnetometers, and heat flow probes at the Apollo 15, 16 and 17 landing sites, for example (e.g., Eichelman and Lauderdale 1974; Wieczorek et al. 2006; Jaumann et al. 2012; Weber 2014). This lunar network of seismometers recorded *ca.* 13000 catalogued events between 1969 and 1977 (e.g., Nakamura et al. 2008), including non-natural events related to impacts of booster rockets and lunar modules from Apollo spacecraft. Natural seismic sources on the Moon have been divided into three categories: meteoroid impacts, shallow moonquakes and deep moonquakes. Deep moonquakes are more frequent than shallow moonquakes. About 7,000 deep moonquakes have been recorded, originating from distinct source regions *ca.* 700–1200 km deep (e.g., Jaumann et al. 2012; Weber 2014). The precise location and periodicity of deep moonquakes are not well understood, but it seems that most deep moonquakes occur at monthly intervals, suggesting that they could be triggered with tidal stress (e.g., Latham et al. 1971). Shallow moonquakes are less frequent than deep moonquakes—28 events have been recorded between 1969 and 1977 (Nakamura 1980)—and more powerful, with magnitudes of *ca.* 1.5 to 5 on a Richter-equivalent scale (e.g., Lammlein 1977; Nakamura 1980). Estimates for the depth of their epicentres range from ~10 km down to ~100 km deep (e.g., Gillet et al. 2017), the shallowest ones being potentially linked with recent thrusting activity on fault scarps suggesting that the Moon is still tectonically active (Watters et al. 2019). Seismic data also point to possible compositional stratification within the lunar mantle—modelling results suggest that there may be a seismic discontinuity *ca.* 500–600 km deep, below which P- and S-wave velocity seems to increase with depth (Nakamura et al. 1982; Khan et al. 2000). The presence of a zone of mineralogical phase transition and change in mantle composition toward more magnesian and aluminous compositions below ~500 km may be interpreted as indicating (i) the base of the LMO, if the whole Moon was not fully melted, (ii) the limit between olivine-dominated early cumulates and orthopyroxene-rich later cumulates in a whole Moon melted scenario, or (iii) the maximum depth of partial melting of the basaltic melt source regions (see Wieczorek et al. 2006 for further discussion).

One of the key findings that arose from interpretation of lunar seismic data is that the Moon is a differentiated body, consisting of a crust, a mantle and a *ca.* 400–500 km radius metallic core comprising a solid inner core and a fluid outer core (e.g., Weber et al. 2011). Gravity data also provide valuable constraints on the Moon's internal structure. The gravity field of the Moon has been mapped with increasing levels of details by successive orbiting spacecraft (NASA Lunar Prospector in the 1990s, JAXA SELENE in the 2000s and NASA GRAIL in the 2010s), the latest maps produced by the GRAIL mission being actually the highest resolution gravity maps for any Solar System body including the Earth (e.g., Zuber et al. 2013). Combined with constraints from Apollo seismic experiments, GRAIL data suggest that the lunar crust has an average thickness of 34–43 km, with the farside crust being *ca.* 10–15 km thicker compared to that of the nearside (Wieczorek et al. 2013). The other main feature revealed by gravity data is the presence of anomalous mass concentrations, known as 'mascons' (Fig. 3). These are associated with large impact basins and thought to originate from the excavation and collapse of impact basins, followed by isostatic adjustment and cooling and contraction of voluminous melt pools (e.g., Melosh et al. 2013).

Sample-Based Insights

Samples of the lunar mantle have yet to be found in our Apollo, Luna and lunar meteorite collections. Our knowledge of the composition of the lunar mantle thus mostly relies on petrological and geochemical studies of mare basalts and pyroclastic glasses produced via mantle partial melting. Importantly, their compositions only reflect that of their mantle source regions at the time of their extraction from the mantle *ca.* 4.3 to 3 billion years ago (see Sect. 5: Nature, timing, and duration of lunar mare volcanism). Numerous primitive lunar basaltic melts, sampled by pyroclastic glasses, are rich in FeO and TiO_2 (e.g., Delano 1986). Experimental work suggests that most of these primitive melts formed at pressures of *ca.* 1.5 to 3 GPa, roughly corresponding to depths of ~300–600 km (e.g., Delano and Livi 1981; Longhi 1992; Grove and Krawczynski 2009; Brown and Grove 2015), while crystalline mare basalts seem to originate from partial melting at depth shallower than ~350 km (e.g., Wieczorek et al. 2006). This thus provides evidence for the presence of Fe- and Ti-rich sources deep in the lunar mantle, which is not predicted by LMO crystallisation models (e.g.,

Snyder et al. 1992; Charlier et al. 2018; Rapp and Draper 2018). To account for this observation, it has been proposed that Rayleigh-Taylor instability causes dense ilmenite-rich cumulates and underlying Fe-rich upper mantle layers formed at the end of the LMO crystallisation to sink and mix in with earlier-formed Mg-rich lower mantle (e.g., Hess and Parmentier 1995). This mantle overturn mechanism also provides a way to bring radioactive incompatible elements such as K, U and Th, enriched in late stage LMO liquids, deep into the mantle. However, it remains unclear whether heating via radioactive decay of these ITEs is a viable mechanism for sustaining mantle partial melting during the main period of mare basalt volcanism ~3.8 to 3.0 Ga ago or if the influx of hotter deep mantle material was required to trigger partial melting in cold mare basalt and pyroclastic glasses source regions (e.g., Spohn et al. 2001; Grove and Krawczynski 2009).

Compositionally, the mare basalt source regions are characterised by (i) low alkali and siderophile element abundances, (ii) Sm-Nd and Lu-Hf isotope compositions fractionated compared to chondritic compositions, and (iii) very low oxygen fugacities, as indicated by the presence of Fe metal and reduced valence state of Fe, Ti and Cr in mare basalts (see Wieczorek et al. 2006 et references therein). For decades after the return of Apollo samples, it has been assumed that the lunar mantle was extremely dry, with estimates of lunar mantle water content lower than 1 part per billion (Taylor et al. 2006). However, with technological advances in lab instrumentation detection limits for species such as water, carbon or halogens such as fluorine and chlorine greatly improved, which permitted the successful detection in 2008 of indigenous water in lunar pyroclastic glasses (Saal et al. 2008). This was followed by a wealth of studies reporting the abundance of water and other volatile species (e.g., C, N, F, S, and Cl) in trapped melt inclusions and mineral phases such as apatite in basaltic lava products (see recent reviews by Anand et al. 2014; McCubbin et al. 2015; Hauri et al. 2017). Reconstructing mantle volatile abundances from those measured in volcanic products is not trivial, notably because volatile elements can be modified by numerous processes such as magmatic degassing, assimilation of solar wind-enriched soils, spallogenic reactions at the surface, or contamination during sample preparation on Earth. Therefore, current estimates for the water abundance in the mantle source regions of mare basalts and pyroclastic glasses range between *ca.* 1 and 100 parts per million (e.g., McCubbin et al. 2015; Hauri et al. 2017). It remains unclear whether this fairly large range of estimates reflects uncertainties in back-calculating water contents for mare basalt source regions, heterogeneous distribution of water in the lunar mantle, or a combination of both.

Outstanding Questions and the Potential of Future Sample-Return

The location of Apollo landing sites within or just around the anomalous PKT area (Fig. 1) suggests that Apollo samples provide us with a biased view of the Moon (see Sect. 5: Nature, timing, and duration of lunar mare volcanism). This is also where ALSEP packages were deployed by Apollo astronauts, suggesting that inferences about the stratification of the lunar interior based on seismic data might also be biased. A global geophysical network (e.g., Neal 2011; Smith et al. 2012; Neal et al. 2019) with stations distributed all over the lunar surface would provide invaluable information for better constraining the structure of the Moon's interior, including the possible discontinuity at ~500 km depth, the presence of partially molten layers deep in the lower mantle, and the size and structure of the core. Global geophysical data such as the gravity data obtained by GRAIL support the crustal thickness dichotomy, the farside crust being *ca.* 10–15 km thicker compared to the nearside crust (Wieczorek et al. 2013). Several scenarios have been proposed to explain such dichotomy, including (i) asymmetric convection during LMO crystallisation and crustal growth (Loper and Werner 2002; Ohtake et al. 2012), (ii) asymmetric impact cratering (Wood 1973), (iii) ejecta deposition from SPA (Zuber et al. 1994), (iv) inhomogeneous early tidal heating (Garrick-Bethell et al. 2010), (v) accretion of a companion moon (Jutzi and Asphaug 2011), or (vi) giant impact of a ~500–800 km diameter impactor onto the lunar nearside soon after the Moon's formation (e.g., Zhu et al. 2019). Returning samples from the farside crust would certainly help in disentangling between these different scenarios, in addition to providing opportunities to evaluate the possible chemical differences between farside magnesian anorthosites and nearside ferroan anorthosites (see Sect. 3: Petrology and formation history of the lunar anorthitic crust).

Sampling pieces of the lunar mantle would prove crucial to our understanding of its mineralogical and chemical nature. In particular, this would allow direct measurement of its volatile and water inventory, which has key implications for investigations of how the Moon formed (e.g., Hauri et al. 2017), and of LMO crystallisation and thermal evolution of the Moon's interior (e.g., Evans et al. 2014; Khan et al. 2014). Impact cratering processes may have breached the crust periodically and could have brought up the pieces of lunar mantle at the surface; SELENE/Kaguya spectral data indeed suggest that numerous olivine-rich (mantle) exposures are found around large impact basins characterised by thin crustal thickness, including Mare Moscoviense, Crisium, Imbrium, Humorum, Nectaris, Serenitatis and basins within SPA such as Schrödinger (Yamamoto et al. 2010; see location of basins on Fig. 3). The Yutu-2 rover, which landed successfully in Von Kármán crater within the SPA basin on the lunar farside (bottom right corner on Fig. 3) on-board the CNSA Chang'e 4 lander, may indeed have identified, remotely through near infrared spectroscopy, samples excavated from the lunar mantle (Li et al. 2019). Such impact basins with very thin crust thus represent key targets for sample-return missions of pieces of the Moon's mantle.

ICY DEPOSITS AT THE LUNAR POLES

Evidence for the Presence of Volatiles at the Moon's Poles

The occurrence of water ice at the lunar poles has been postulated for decades (e.g., Watson et al. 1961; Arnold 1979) because of the presence of deep topographic lows at the poles and the low obliquity of the Moon. In the 1990s, possible detection of water ice in polar regions of Mercury using Earth-based radar observations (Harmon and Slade 1992) refuelled the speculation that the lunar poles might also act as a trap for water ice and other volatiles. At the same time, data returned by instruments on board the two NASA lunar orbiters Clementine and Lunar Prospector supported the presence of water ice in permanently shadowed regions (PSR) at the lunar poles (Nozette et al. 1996, 1997; Feldman et al. 1998, 2000), although subsequent Earth-based radar investigations challenged the existence of lunar polar water ice, which became a topic of strong scientific controversy (see Campbell et al. 2006, and references therein). Further remote observations carried out by a broad range of instruments on-board India's Chandrayaan-1 and NASA's Deep Impact, Cassini, and Lunar Reconnaissance Orbiter (LRO) missions, coupled with the direct detection of *ca.* 5 wt.% water and other volatile species in the plume formed by the LCROSS impact-experiment in the Cabeus crater, strongly suggested that there is indeed OH/H_2O molecules and water ice at the surface or sub-surface of lunar polar regions (Clark 2009; Pieters et al. 2009b; Sunshine et al. 2009; Colaprete et al. 2010; Gladstone et al. 2010; Hayne et al. 2010, 2015; Mitrofanov et al. 2010; Paige et al. 2010b; Schultz et al. 2010; Spudis et al. 2013; Fisher et al. 2017; Li and Milliken 2017; Li et al. 2018) (Fig. 8). However, we have little insights on how ice occurs (i.e. solid lumps of ice vs. ice mixed in with lunar soils), on its vertical and lateral distributions, and on its origin(s). For example, the production of OH/H_2O species at the surface could result from interaction of solar-wind protons with the regolith (e.g., Farrell et al. 2015). Cometary and volatile-rich asteroidal objects that have impacted the Moon for billions of years are obvious sources for accumulation of ice at the lunar poles. The migration of water molecules from lower latitudes to polar cold traps in permanently shadowed craters is also possible (e.g., Schorghofer and Taylor 2007), implying that magmatic water degassed onto the lunar surface during volcanic eruptions could have made its way to polar traps.

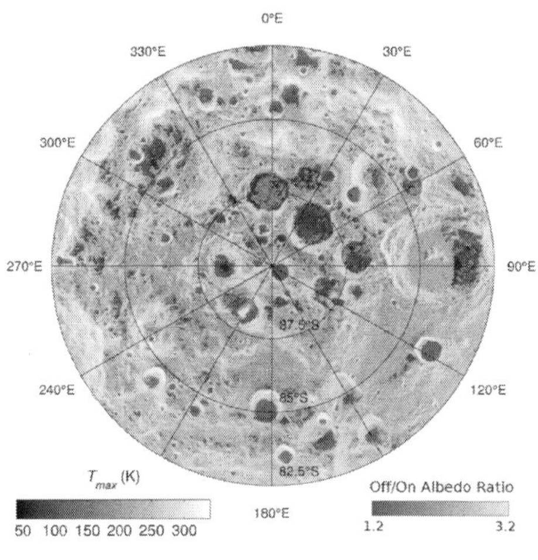

Figure 8. Locations of anomalous UV albedo consistent with water ice (figure from Hayne et al. 2015). Colours indicate points with off/on-band albedo ratio values >1.2 and Lyman-αα albedo <0.03, increasing from deep blue (1.2) to white (>3.2). The average Moon outside of cold traps has a ratio of 0.9. Ratio values in the range 1.2–4.0 are consistent with water ice concentrations of 0.1–2.0 wt.%. If patchy exposures of pure water ice are mixed with dry regolith, the abundance could be up to 10 wt.%.

Prospecting Missions to the Poles

At this stage only future prospecting missions to the lunar surface can provide firmer constraints on the abundance, form, distribution, and origin(s) of water ice and other lunar polar volatiles. Landers with drilling capabilities should be able to provide crucial information on the composition, abundance and vertical distribution of polar volatiles at a given location.

In addition to providing important scientific constraints on the source and processing of these polar volatiles, each single point measurement will represent ground truth for orbital measurements. PROSPECT (Package for Resource Observation and in-Situ Prospecting for Exploration Commercial Exploitation and Transportation) is a payload currently developed by the European Space Agency (ESA) to carry out these types of activities (Carpenter et al. 2016). Broadly, PROSPECT can be deployed on a mobile platform or a fixed lander, and its technical capabilities can be tailored to different objectives. Currently, the first deployment of PROSPECT on the lunar surface is planned on the Russian Luna-27 mission that should land in a lunar south polar region. PROSPECT is comprised of a drill (ProSEED) that will extract regolith samples from up to *ca.* 1.2 m depth. The samples will be delivered to a miniature chemical laboratory (ProSPA) where volatile species (e.g., H_2, H_2O, CO, CO_2, N_2, noble gases) will be degassed in order to measure their abundances and stable isotope compositions (Barber et al. 2017). Thanks to their mobility, rovers could provide complementary information on the composition, abundance and lateral distribution of subsurface polar volatiles across traverses up to a few kilometres long, which would also reveal key scientific information and ground truth for orbital measurements. One such example is the NASA Volatiles Investigating Polar Exploration Rover (VIPER) mission concept (Colaprete et al. 2019). To characterise the budget of polar volatiles, VIPER would rove across the lunar surface and locate areas of high hydrogen concentration using its neutron spectrometer. Once the presence of subsurface volatiles is confirmed, regolith samples would be drilled and delivered to a miniature chemical laboratory in which volatile species (e.g., H_2, H_2O, CO, CO_2, N_2, noble gases) would be degassed and their abundances quantified (Colaprete et al. 2019).

If the presence of polar volatiles at the lunar subsurface is confirmed by prospecting missions, the next logical step would then be to collect and return samples to the Earth for further characterisation of polar volatiles. Only sample-return missions allow fully comprehensive scientific investigations using present day state-of-the-art analytical equipment. In the future, technological improvements should allow addressing new questions provided the returned samples are curated properly. The paradigm shift seen over the last decade on the issue of indigenous lunar water, through re-investigation of Apollo samples using 21st century lab equipment, perfectly illustrates the scientific importance of collection and curation of samples (e.g., Anand et al. 2014; McCubbin et al. 2015). In addition to providing ground truth for orbital measurements, a sample-return mission from a polar site where cold trapped volatiles occur would establish a comprehensive inventory of what species are present. Analysis of the isotopic composition of species such as H, C, N, O, S, and noble gases would provide key constraints on the source(s) of these volatiles and the processes that may have helped to concentrate them. To this end great care would be needed during sample collection and handling on the lunar surface, in the returned capsule to Earth, and on Earth. Temperature increase during sample collection through drilling for example could lead to ice and gas loss through sublimation. Therefore, the thermal and mechanical history of the samples collected will need to be precisely recorded.

THE LUNAR REGOLITH AS A RECORDER OF THE SUN'S COMPOSITION, ACTIVITY AND GALACTIC ENVIRONMENT

For half a century, the lunar regolith record has been used to decipher the elemental and isotopic composition of the solar wind (SW). In fact the first experiment deployed on the Moon by Apollo astronauts was designed to directly collect SW (Fig. 9). A 30 cm × 140 cm, 15-micron thick, aluminium foil sheet was exposed to solar rays at the lunar surface for 77 minutes in order to determine the SW He, Ne and Ar abundances and isotopic compositions (Geiss et al. 1969).

Figure 9. Astronaut Buzz Aldrin deploying the Solar-Wind Experiment about 5 meters north-northwest of the Apollo 11 lunar module. Credit NASA photograph AS11-40-5872.

The SW is dominated by H (~95%) and He (~4%), with heavier elements such as C, N, O, Ne, Mg, Si and Fe comprising <1% (e.g., Anders and Grevesse 1989; Ogilvie and Coplan 1995). If the outermost layers of recently irradiated (i.e. in the last 100 Ma or so) lunar soil grains preserve an accurate record of the present-day composition of the SW, it has been proposed that systematic variations of isotopic ratios of some elements such as He, Ne, Ar, N or C, in soils irradiated billions of years ago could reflect changes of the SW composition through time (e.g., Becker and Clayton 1975; Kerridge 1975; Becker 1980; Clayton and Thiemens 1980; Becker and Pepin 1989; Geiss and Bochsler 1982, 1991; Kerridge et al. 1991; Benkert et al. 1993). Several hypotheses have been proposed to account for such changes, including (i) variations in acceleration conditions in the solar corona for light species such as He, (ii) contamination of the composition of the solar convective zone by interstellar material with exotic compositions, or (iii) change of spallation reaction rates and/or depth of occurrence in the photosphere with time (e.g., Kerridge 1980). However, coordinated analyses of multiple species in ilmenite from Apollo 12, 14 and 17 soil and regolith breccia samples via progressive etching (e.g., Heber et al. 2003), combined with high precision measurement of the noble gas and nitrogen isotope compositions of the SW directly collected in ultrapure targets at the L1 Lagrangian point for more than 2 years by the NASA Genesis spacecraft, seem to indicate that no such secular variations of the SW isotopic composition are recorded in lunar regolith samples (see Wieler 2016 for an extensive discussion on this topic).

Solar Wind Isotope Compositions Estimated from Lunar Sample Analysis

Estimates of the solar wind isotope compositions for the noble gases He, Ne and Ar, and for H, C, N, and O, obtained through analysis of lunar samples are summarised in Table 3.

Table 3. Summary of some lunar sample-based estimates for solar wind isotope compositions and compositions measured on Genesis samples or estimated from absorption lines in the solar photosphere

Ratio	Value	±	Analysed target	References
$^{3}He/^{4}He$ ($\times 10^{4}$)	4.64	0.18	Genesis DoS	1
	4.26	0.22	Apollo SWE	2
	4.57	0.16	Ilmenite in "young" soil 71501	3
	4.47	0.26	Average of ilmenite from soils and breccias	4
$^{20}Ne/^{22}Ne$	13.78	0.06	Genesis DoS	1
	13.97	0.05	Genesis AloS	5
	14.00	0.04	Genesis Au foil, AuoS and Al collectors	6
	13.70	0.30	Apollo SWE	2
	13.80	0.20	Ilmenite in "young" soil 71501	3
$^{36}Ar/^{38}Ar$	5.47	0.02	Genesis DoS	1
	5.50	0.02	Genesis Si collectors	7
	5.50	0.01	Genesis AloS	5
	5.50	0.01	Genesis Au foil, AuoS and Al collectors	6
	5.40	0.30	Apollo SWE	2
	5.48	0.10	Ilmenite in "young" soil 71501	3
	5.53	0.06	From surface-correlated Ar from Apollo 15 soil feldspar	8
δD_{VSMOW} (‰)	<−999		Genesis Si collector	9
	<−970		Bulk Apollo 11 breccia samples	10
	<−950		Depth profile in soil grains in sample 79035	11
	<−840		Lowest δD in Apollo 11 and 17 agglutinates	12
$\delta^{13}C_{VPDB}$ (‰)	?		Genesis collectors	–
	−105	40	Depth profile in soil grains in sample 79035	13
	−85	25	SW estimates from solar photosphere δ13Cδ13C value obtained by analysing CO absorption lines	14

$\delta^{15}N_{AIR}$ (‰)	−390	180	Genesis Au cross arms	15
	−407	7	Genesis concentrator	16
	−417	100	Genesis DoS	17
	<−240		Depth profile in soil grains in sample 79035	11
$\delta^{18}O_{VSMOW}$ (‰)	<−40	15	Depth profile in metal grains in sample 79035	18, 19
	−102.3	6.6	Genesis SiC collector	20

Abbreviations: DoS—diamond-like carbon on silicate substrate; AloS—aluminium deposited on sapphire; AuoS—gold deposited on sapphire. References: 1—Heber et al. (2009); 2—Geiss et al. (2004); 3—Benkert et al. (1993); 4—Heber et al. (2003); 5—Meshik et al. (2007); 6—Pepin et al. (2012); 7—Vogel et al. (2011); 8—Frick et al. (1975); 9—Huss et al. (2012a); 10—Friedman et al. (1970); 11—Hashizume et al. (2000); 12—Liu et al. (2012); 13—Hashizume et al. (2004); 14—Lyons et al. (2018); 15—Marty et al. (2010); 16—Marty et al. (2011); 17—Huss et al. (2012b); 18—Hashizume and Chaussidon (2005); 19—Hashizume and Chaussidon (2009); 20—McKeegan et al. (2011)

Hydrogen, Carbon, Nitrogen and Oxygen

Both bulk and secondary ion mass spectrometry (SIMS) analysis of H isotope ratios in lunar soils, regolith breccia samples, soil minerals, and glassy agglutinates show that SW is an important source of H in the lunar regolith, as indicated by very low δD_{VSMOW} signatures[Footnote1] down to *ca.* −850‰ to −970‰ (e.g., Epstein and Taylor 1970; Friedman et al. 1970; Hashizume et al. 2000; Liu et al. 2012). The lowest δD_{VSMOW} values measured in lunar samples are slightly higher than the SW δD_{VSMOW} value of *ca.* −999‰ estimated for a Genesis Si collector (Huss et al. 2012a), indicating that lunar surface soils may contain some non-solar deuterium that may have formed via spallation reactions for example. The SW D/H ratio does not seem to have evolved much, if any, over time since sample 79035, which has a $^{40}Ar/^{36}Ar$-derived antiquity age of around 2 Ga (e.g., Wieler et al. 1996; Heber et al. 2003), yielded $(\delta D_{VSMOW})_{SW}$ <−950‰ (Hashizume et al. 2000).

Depth profile analyses in grains from soil 79035, which contains surface H dominated by SW sources (Hashizume et al. 2000), have yielded very low $\delta^{13}C_{VPDB}$ values[Footnote2] of *ca.* −105±40\textperthousand−105±40\textperthousand at a depth of 15–30 nm (Hashizume et al. 2004). Such values cannot be compared to Genesis data since the $(^{13}C/^{12}C)_{SW}$ has not yet been determined. However, it matches fairly well with recent estimates for $(\delta^{13}C_{VPDB})_{SW}$ of *ca.* −75±15 to −90±15\textperthousand−90±15\ textperthousand derived from a new estimate for the $\delta^{13}C_{VPDB}$ of the Sun's photosphere of −48±15\textperthousand−48±15\textperthousand and modelling of C isotope fractionation during transport from the photosphere to the corona via Coulomb drag (Lyons et al. 2018). Overall this suggests that the SW is depleted in ^{13}C compared to most planetary and asteroidal materials that are characterised by $\delta^{13}C_{VPDB}$ ratios around 0±30\textperthousand (e.g., Lyons et al. 2018 and references therein).

Analyses of nitrogen isotopes in Apollo and Luna regolith samples have yielded $\delta^{15}N_{AIR}$[Footnote3] values ranging from *ca.* −200‰ up to around +100‰ (e.g., Becker and Clayton 1975; Kerridge 1975, 1993; Thiemens and Clayton 1980; Frick et al. 1988; Füri et al. 2012; Mortimer et al. 2016). Based on the apparent correlation of increasing $\delta^{15}N_{AIR}$ with decreasing $^{40}Ar/^{36}Ar$ antiquity ratios, it has been suggested that the SW nitrogen isotope composition evolved through time to present-day values around +100‰ (e.g., Becker and Clayton 1975; Kerridge 1975; Thiemens and Clayton 1980). However, based on depth profile analyses of individual mineral grains from soil and regolith samples by SIMS, Hashizume et al. (2000) proposed that the present-day SW is depleted in ^{15}N, with $\delta^{15}N_{AIR}$ values ≤−240‰. A ^{15}N-depleted solar nitrogen was eventually confirmed by analysis of Genesis targets that yielded a best $\delta^{15}N_{AIR}$ estimate of −407±7\textperthousand for SW nitrogen (Marty et al. 2011). Recent high-resolution step-heating analyses of Apollo 12, 14, 15, 16 and 17 surface soil samples and Luna 24 drill core samples are consistent with such a ^{15}N-depleted SW nitrogen (Füri et al. 2012; Mortimer et al. 2016).

Based on depth profile O isotope analyses in metal grains isolated from lunar soils, Hashizume and Chaussidon (2005, 2009) have argued that solar gas is characterised by $\delta^{18}OV_{SMOW}$[Footnote4] < *ca.* −40±15\textperthousand. This is consistent with analysis of Genesis SiC collectors that yielded a $(\delta^{18}O_{VSMOW})$ value of −102±7\textperthousand (McKeegan et al. 2011). However, the fact that such ^{16}O-rich values have not yet been found in lunar samples suggests that either some processes have fractionated the SW oxygen isotope composition during or after its implantation in lunar materials, or that samples with unfractionated SW have not been analysed so far.

The Noble Gases He, Ne and Ar

Heber et al. (2003) carried out high resolution acid etching experiments on ilmenite separates from soil and regolith breccia samples with a wide range of antiquity ages from *ca.* 0.1 to 3.7 Ga (see Sect. 4.3.3 for a discussion of how antiquity ages are estimated). The $(^3He/^4He)_{SW}$ ratios they measured are well correlated with antiquity ages of the samples, suggesting an apparent increase of $(^3He/^4He)_{SW}$ by ~6% per billion years. Similarly, the $(^{20}Ne/^{22}Ne)_{SW}$ ratios measured by Heber et al. (2003) on the samples were also correlated with antiquity ages, with $(^{20}Ne/^{22}Ne)_{SW}$ apparently increasing with time by ~2% per billion years. However, Heber et al. (2003) observed that the apparent temporal changes of $(^3He/^4He)_{SW}$ and $(^{20}Ne/^{22}Ne)_{SW}$ ratios were correlated with each other, mimicking the correlation displayed by single etch steps and indicating progressive etching of deeper sited isotopically heavier solar energetic particles. This led Heber et al. (2003) to conclude that the apparent temporal variations of the $(^3He/^4He)_{SW}$ and $(^{20}Ne/^{22}Ne)_{SW}$ ratios resulted from secondary processes on the lunar surface causing partial removal of surface-correlated SW. The average corrected $(^3He/^4He)_{SW}$ ratio of 4.47 $(\pm 0.26)\times 10^{-4}$ obtained on ilmenite from six samples with antiquity ages from *ca.* 0.1 up to ~3.7 Ga by Heber et al. (2003) is indeed identical to the $(^3He/^4He)_{SW}$ ratio of 4.57 $(\pm 0.16)\times 10^{-4}$ estimated for recent SW implanted in the last 100 Ma on ilmenite grains in lunar soil 71501 (Benkert et al. 1993). The He isotope composition estimated for SW from lunar regolith samples matches very well with the average $(^3He/^4He)_{SW}$ ratio of 4.64 $(\pm 0.18)\times 10^{-4}$ obtained for SW collected for ~2.5 years by the Genesis spacecraft (Heber et al. 2009). The $(^3He/^4He)_{SW}$ ratio derived from lunar soil and Genesis analyses appear to be slightly higher than the average $(^3He/^4He)_{SW}$ ratio of 4.26 $(\pm 0.22)\times 10^{-4}$ for SW collected by the Apollo 11, 12, 14, 15 and 16 collectors (Geiss et al. 2004). However, this average value may be slightly uncertain considering that Apollo 11 and 12 experiments yielded 5.38 $(\pm 0.40)\times 10^{-4}$ and 4.08 $(\pm 0.17)\times 10^{-4}$, respectively, with $(^3He/^4He)_{SW}$ ratios for Apollo 14 $(4.48\pm 0.28\times 10^{-4})$, 15 $(4.33\pm 0.22\times 10^{-4})$ and 16 $(4.42\pm 0.20\times 10^{-4})$ being consistent with the $(^3He/^4He)_{SW}$ ratio derived from lunar soil and Genesis analyses (Benkert et al. 1993; Heber et al. 2003, 2009; Geiss et al. 2004).

As discussed above the apparent increase of $(^{20}Ne/^{22}Ne)_{SW}$ with antiquity measured in Apollo samples by Heber et al. (2009) has been interpreted as an artefact related to secondary processes while samples resided on the lunar surface. The $(^{20}Ne/^{22}Ne)_{SW}$ of 13.80±0.20 measured in ilmenite in sample 71051 (Benkert et al. 1993) is consistent with the $^{20}Ne/^{22}Ne$ isotope ratio of 13.70±0.30 measured on Apollo collectors (Geiss et al. 2004), both measurements being consistent within error with the $(^{20}Ne/^{22}Ne)_{SW}$ ratios of 13.78±0.06 (Heber et al. 2009), 13.97±0.05 (Meshik et al. 2007) and 14.00±0.04 (Pepin et al. 2012) measured in different types of Genesis collectors. Note that the $(^{20}Ne/^{22}Ne)_{SW}$ ratio calculated by Heber et al. (2009) is not consistent with those obtained by Meshik et al. (2007) and Pepin et al. (2012) thought, a discrepancy that is still not yet understood (see Pepin et al. 2012 for further discussion on this). At present, our best estimate for the $(^{20}Ne/^{22}Ne)_{SW}$ ratio is thus *ca.* 13.9±0.1.

Lunar sample studies yielded estimates for the $(^{36}Ar/^{38}Ar)_{SW}$ ratio of *ca.* 5.50 (e.g., Frick et al. 1975; Benkert et al. 1993), which are consistent with both the relatively imprecise $(^{36}Ar/^{38}Ar)_{SW}$ ratio of 5.40±0.30 determined on Apollo collectors (Geiss et al. 2004) and with the much more precise $(^{36}Ar/^{38}Ar)_{SW}$ ratios of 5.50±0.05 obtained on Genesis collectors (Meshik et al. 2007; Heber et al. 2009; Vogel et al. 2011; Pepin et al. 2012). Isotopic analyses of young lunar regolith materials also provided very good estimates for the SW Kr and Xe isotope compositions (Pepin et al. 1995), with Kr and Xe isotope ratios within 1% of current best estimates for SW compositions measured on Genesis samples (e.g., Crowther and Gilmour 2013; Meshik et al. 2014).

Investigating Past Solar Activity and Changes of Galactic Environments

Characterising the evolution of the Sun's activity over the past 4.6 billion years is key to investigating the evolution of inner Solar System planetary surfaces and atmospheres, and by extension the habitability of terrestrial planetary bodies (e.g., Airapetian et al. 2016). Investigation of the meteorite record suggests enhanced solar activity during the early Solar System history (e.g., McKeegan et al. 2000; Aléon et al. 2005; Sossi et al. 2017; Kööp et al. 2018), which is consistent with greater flare activity associated with a shorter rotational period for the young Sun (e.g., Güdel 2007). The lunar regolith can also help characterise the rotational rate of the Sun during its first billion years of evolution since solar flares and associated coronal mass ejection (CME) directly affect the surface of airless bodies and could have driven loss of volatile species such as Na and K from the Moon's surface (e.g., Killen et al. 2012; Saxena et al. 2019). Based on the abundance of Na and K in the lunar crust, and on modelling of the effect of CME on Na and K loss, Saxena et al. (2019) argued that the Sun likely rotated relatively slowly during its first billion years of evolution compared to other young stars.

The lunar regolith also preserves a record of changes of the Solar System galactic environment as a whole (Spudis 1996). Over the past 4.5 billion years the Solar System has orbited the centre of the Milky Way approximately 20 times (e.g., Gies and Helsel 2005) and passed through spiral arms 2 to 6 times (Gillman et al. 2019). While moving around the Galaxy, the lunar regolith could have recorded changes in the flux and types of impactors delivered to the inner Solar System (see Sect. 4: The bombardment history of the inner Solar System). Over this period of time, the Solar System would have also been exposed to changing galactic environment conditions—passing through the Milky Way spiral arms would have enhanced the probability of passing close to supernovae, which could be recorded in the lunar regolith via supernova ejecta implanted in regolith materials and/or as an enhanced galactic cosmic ray (GCR) flux (e.g., Crawford 2017 and references therein). When GCRs interact with geological materials, a wide array of cosmogenic nuclides are produced in response to neutron capture and spallation reactions, including noble gases such as ^{3}He, ^{21}Ne, ^{38}Ar, ^{83}Kr and ^{126}Xe and other isotopes such as ^{10}Be and ^{36}Cl. However, our current Apollo and Luna collections are not ideally suited for investigating possible variations of the GCR flux through time. All these samples were collected at the surface or within the top few meters of the lunar regolith and have undergone constant regolith gardening over the past 3–4 billion years. As a result their exposure history records an integrated GCR flux since they were first exposed, which is generally complicated and poorly constrained (e.g., Levine et al. 2007; Füri et al. 2018). In addition to an increased GCR flux, passing by a supernova could also be recorded by deposition of ejecta enriched in radioactive isotopes such as ^{60}Fe, for which spikes have been identified in ~2–3 and 6–9 Ma-old terrestrial ocean sediments and linked to a supernova origin (see Wallner et al. 2016 and references therein). Spikes of ^{60}Fe have also been measured in Apollo samples and have been linked with deposition of supernova ejecta ~2 Ma ago (Cook et al. 2009; Fimiani et al. 2016). This bears promise for investigating galactic changes through time; however, similarly to examining variations of the GCR flux, detecting older supernova events in the Apollo and Luna collections is complicated by regolith gardening, which would average supernova-related spikes over the period of exposure of studied samples (e.g., Crawford 2017).

Opportunities Offered by Future Sample-Return Missions

Study of soil and regolith samples have provided us with solid estimates for the noble gas isotopic compositions of the SW that were later confirmed by samples returned by the Genesis mission. The composition of the SW for other important elements such as Si, Mg, Fe, S, Cl has not yet been measured in Genesis samples (Burnett et al. 2019), and could be estimated based on targeted analyses on constituents of lunar soil and regolith samples already present in our collections. However, thoroughly tackling questions related to possible changes in solar activity, solar wind composition, and galactic environment through time will require accessing regolith samples that have been exposed to the space environment at a known point in time and for a well constrained duration. Such samples could be returned from palaeoregolith horizons (see Sect. 4.3.3: Secular evolution of types of projectiles hitting the Moon) trapped between datable lava flows or impact melt flows, some of them being potentially accessible to robotic or human explorers through layers exposed in the walls of lava skylights or in layered boulders (Fig. 6) (Crawford et al. 2007, 2010; Fagents et al. 2010; Crawford and Joy 2014).

THE LUNAR MAGNETIC FIELD

State of the Art

The search for evidence of past or present lunar magnetism was one of the objectives of the very first missions to the Moon. The second robotic mission to the Moon, Luna 2 in 1959, carried a magnetometer that found no sign of a detectable global magnetic field (Dolginov et al. 1962). After this was confirmed by Explorer 35 (Sonett et al. 1967), the Moon was generally thought to be magnetically uninteresting, in agreement with the preconception that the Moon was too small to have sustained a dynamo field. However, one of the bigger surprises of the early Apollo missions was the discovery of past lunar magnetism. Examination of Apollo 11 samples revealed strong natural remanent magnetizations (Doell and Grommé 1970) indicative of ancient surface fields with intensities comparable to that of the present-day Earth. Subsequent measurements with the Apollo 12 and subsequent surface magnetometers revealed significant present-day remanent magnetic fields (tens to hundreds of nanotesla—nT) on the surface of the Moon indicating near-surface sources of crustal remanent magnetization (Dyal et al. 1970). Paleomagnetic measurements of lunar samples started right after the return of Apollo 11 and were very active over the whole Apollo sample-return period with about 70 samples studied (see Fuller and Cisowski 1987 for a review). After a period of relative disinterest for more than 20 years, this field of research was reactivated with a series of detailed studies on key samples that benefited from improved analytical techniques and methods for paleomagnetism, as well as better geo- and thermochronology, and an overall more comprehensive understanding of the geological evolution of the Moon (see Weiss and Tikoo 2014 for a review; Tikoo et al. 2017; Lepaulard et al. 2019).

It is now established that the Moon generated a magnetic field through advection in its liquid core for a large portion of its history (Fig. 10). This magnetic field was highest between about 4.25 Ga and 3.5 Ga (during the so called "high-field epoch"), with an average paleointensity of 55 μT, much higher than the maximum value of a few tenths of a μT estimated from dynamo scaling laws (e.g., Laneuville et al. 2014, Evans et al. 2014, 2018). The field then decayed but a dynamo was still operating with surface fields in the several μT range around 1.5–2.0 Ga (Tikoo et al. 2017). Today, lunar surface magnetic fields are mostly below 10 nT, with localised higher values (and a maximum measured surface field of 327 nT at the Apollo 12 landing site; Dyal and Daily 1978). This spatial variability of the lunar surface field reflects lithological heterogeneities and/or different magnetisation histories of the underlying crustal rocks.

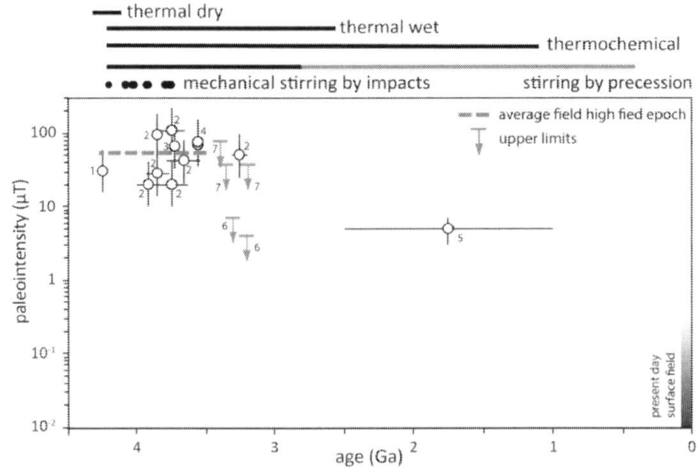

Figure 10. Intensity of lunar surface field versus time. Note the lack of data between 3.0 and 2.0 Ga and the scarcity of data between 2.0 and 1.0 Ga. Bars and dots above the diagram represent the time intervals over which the various dynamo generation mechanisms listed may have operated. References: 1—Gar-

rick-Bethell et al. (2017), 2—Cournède et al. (2012), 3—Shea et al. (2012), 4—Suavet et al. (2013), 5—Tikoo et al. (2017), 6—Tikoo et al. (2012), 7—Tikoo et al. (2014).

Outstanding Questions Requiring Sample-Return Missions (See Summary in Table 4)

Table 4. Science questions and sample requirements for lunar paleomagnetic studies

Sample requirements	Science questions
Regolith breccias younger than 3 Ga	Constrain weak field epoch intensity and duration
Regolith breccias with age between 3 Ga and 2 Ga	Transition between high and low field epochs
Regolith breccias older than 3 Ga	Confirm the high paleointensity during the high-field epoch using thermal paleointensity experiments
One oriented sample from one location	Testing the dipole geometry (if location of the paleopole is assumed), locating the paleopole if a dipolar geometry is assumed
Several oriented samples from one location	Testing the occurrence of reversals, constraining the field geometry, determining the paleopole with implications for true polar wander
Oriented samples from several locations	Testing the dipole geometry and defining the paleopole with implications for true polar wander

Intensity of the Dynamo Field with Time

The paleomagnetic data (Fig. 10) have notable temporal gaps. In particular, the absence of paleomagnetic data between about 3.0 Ga and 1.5 Ga prevents the understanding of the transition between the high and low field epochs. The apparent sharp transition at the precipitous collapse of the dynamo around 3.5 Ga (Tikoo et al. 2014) has been recently re-evaluated with suggestions for paleointensities in the 10–50 μT range between 3.5 Ga and 3.0 Ga (Lepaulard et al. 2019). The extent of the weak field period and the behaviour of the magnetic field during this period are also largely unconstrained with only one published paleointensity estimate over the last 2 Ga of the Moon history (Tikoo et al. 2017). Clearly more paleomagnetic studies of samples with magnetisation ages younger than 3.0 Ga are required.

Regolith breccias are excellent material for lunar paleomagnetism because their glassy matrices contain very fine metal grains that are favourable for recording and maintaining a record of the paleomagnetic field. However, the age at which a regolith breccia was assembled and, therefore, the age of its magnetisation is not easy to determine notably because materials it contains may have been affected by multiple impact events (e.g., Mercer et al. 2015). Basalt samples younger than 3.0 Ga would be a prime target for lunar paleomagnetism. Providing that their formation ages can be accurately determined through high resolution *in situ* geochronological analyses, regolith breccias with formation ages in the interval 3.0–2.0 Ga, and older than 3 Ga, if they exist, are other important targets.

Geometry of the Lunar Surface Field

All lunar paleomagnetic studies were performed on samples collected in the regolith. Therefore, only the paleointensity of the lunar field can be retrieved, but no paleodirectional information. As with terrestrial paleomagnetism, such directional information would be invaluable. It would allow testing the hypothesis that the lunar field was dipolar. Such a geometry is expected by analogy with the dynamo observed today on Earth, Mercury and other bodies. Substantial non-dipolar geometry would imply rapid decay of the field away from the core such that the high surface paleointensities inferred from Apollo samples would indicate even more unrealistic core fields in view of the small size of the lunar core relative to the size of the Moon. Ideally, testing the dipolar hypothesis would require oriented sampling of in-place bedrock samples at different locations on the Moon, preferably at different latitudes. A few outcrops offer the opportunity to sample rocks that have never been transported by impact processes (i.e. they are at the same place than where they formed) at the surface of the Moon. For example, lava pits may offer the possibility to sample such samples through sequences of basaltic lava flows that still have their original orientations (Fig. 6).

A single oriented sample from one location can only test the dipole hypothesis if one assumes the location of the paleo spin pole. However, a sequence of oriented samples collected at a single location would enable testing whether the field reversed. This occurrence, with evidence for two antipodal sets of paleopoles would strengthen the dipole hypothesis and help indentify the location of the paleopole. If a paleopole is found away from the location of the present-day rotation axis of the Moon, this may indicate that the Moon experienced true polar wander, i.e. large differential rotation of the mantle with respect to the core (e.g., Takahashi et al. 2014)

Mechanisms Behind Dynamo Generation

Generating a lunar dynamo requires advection in its liquid core. A number of mechanisms have been proposed that can sustain convection in the Moon: thermal convection, thermochemical convection, and mechanical stirring by mantle precession or impacts. It is noteworthy that these mechanisms are expected to have been restricted to specific, although sometimes overlapping, periods in lunar history (Weiss and Tikoo 2014). For instance, pure thermal convection with a dry mantle is very unlikely to have operated after 4 Ga, impact-generated mechanical stirring is limited to before 3.7 Ga (age of Orientale basin, the youngest impact basin on the Moon) (Suavet et al. 2013), whereas thermochemical and precession generated mechanical stirring could operate almost up to the present-day. Therefore, a better knowledge of the evolution of the field intensity with time will also shed light on which mechanisms lie behind the lunar core convection, with implications for the dynamo theory and modelling community.

Limitation of Current Lunar Rocks

No basalts younger than *ca.* 3.0 Ga are available in the Apollo collection, and regolith breccia samples older than 3.0–3.5 Ga are rare or absent (e.g., Fagan et al. 2014) because these rocks are being modified continuously at the surface of the Moon.

Although the orientation of some samples in the regolith was documented *in situ* or can be determined *a posteriori*, the original orientation of these rocks on the Moon is unknown. The paleohorizontal of some rocks may be assessed using for instance the crystal preferred orientation (e.g., Cournède et al. 2012), or vesicles orientation. Even so, the up and down remain unresolvable, and the constraints on the field geometry remain poor.

Meteorites are another prime source of lunar rocks and, with 378 lunar meteorites for 408.6 kg of material as of August 2019, based on samples classified in the Meteoritical Bulletin Database, they now outweigh the total mass of samples collected by Apollo and Luna missions. Moreover, meteorites sample the lunar surface in a more global way than sample-return missions, and, therefore, are likely to provide more representative sampling of the lunar surface, both spatially and temporally. However, meteorites pose challenges for paleomagnetism. All lunar meteorites are "finds", meteorites found at the Earth surface long after their fall, typically several kyr and up to several hundred kyr for Antarctic meteorites. The magnetic carriers of lunar rocks are Ni-poor iron-nickel alloys that are prone to weathering on Earth. Weathering of magnetic carriers leads to the loss of the primary magnetization, and, worse, possible acquisition of magnetization (chemical remanent magnetisation) in the Earth field. The long stay of these meteorites on Earth also means that they have been in the geomagnetic field for long periods, leading to the acquisition of viscous remanent magnetisation. Although these overprints can be identified and partly erased in some cases, the existence of viscous and chemical remanent magnetisations add to the complexity of lunar paleomagnetism studies. Moreover, 90% of lunar meteorites come from hot deserts, in particular from the Sahara, where they are systematically tested with magnets by private collectors and dealers. Exposure to strong magnets usually totally obliterates the primary magnetisation of the meteorites, which renders them useless for paleomagnetic studies. Among these three magnetic contaminations (viscous by long term exposure to the geomagnetic field, isothermal by exposure to strong fields, and chemical by crystallisation of weathering phases), it is noteworthy that the first two can also affect return samples. Apollo samples are curated in the ambient geomagnetic field and have all been contaminated by viscous remanent magnetisations. Storage in a weak magnetic field would be a simple and efficient solution to implement for future returned samples. Exposure to strong magnetic field can also occur during sampling on the Moon, during return from the Moon, and during sample curation and preparation. The use of non-magnetic tools during curation and the absence of strongly magnetic parts in the sampling and return material are therefore highly recommended for future sample-return missions.

SUMMARY

Despite a large body of research over the past five decades, many new discoveries are being made through continued laboratory investigations of lunar meteorites and samples returned by Apollo and Luna missions. Insights from recent orbital and landed lunar missions have allowed integration of laboratory data acquired on returned samples to address some of the outstanding questions related to the Moon's origin and evolution. At the same time, new fundamental questions have arisen, which can only be addressed by future sample-return missions from areas of the Moon identified as being more representative than was the case for Apollo and Luna missions. As we enter into a new era of sample-return missions from various Solar System bodies, and with an increasing interest in lunar exploration among the worldwide community, new samples returned from the Moon will have immense potential to address key science questions not just related to the Moon but also to the inner Solar System.

NOTES

1. This δD notation corresponds to the per mil deviation of a measured D/H ratio from that of the Vienna Standard Mean Ocean Water (VSMOW –D/H = 1.5576×10^{-4}), given by $[(D/H)_{sample}/(D/H)_{VSMOW} - 1] \times 1000$.
2. This $\delta^{13}C$ notation corresponds to the per mil deviation of a measured 13C/12C ratio from that of the Vienna Pee Dee Belemnite (VPDB–$^{13}C/^{12}C$ = 0.0112), given by $[(^{13}C/^{12}C)_{sample}/(^{13}C/^{12}C)_{VPDB} -1]\times1000$
3. This $\delta^{15}N$ notation corresponds to the per mil deviation of a measured $^{15}N/^{14}N$ ratio from that of the terrestrial air (AIR – $^{15}N/^{14}N$ = 0.003676), given by [(15N/14N)sample/$(^{15}N/^{14}N)_{AIR} - 1] \times 1000$.
4. This $\delta^{18}O$ notation corresponds to the per mil deviation of a measured $^{18}O/^{16}O$ ratio from that of the Vienna Standard Mean Ocean Water (VSMOW – $^{18}O/^{16}O$ = 2.005 × 10–3), given by $[(^{18}O/^{16}O)_{sample}/ (^{18}O/^{16}O)_{VSMOW} - 1] \times$ 1000.

REFERENCES

1. V.S. Airapetian, A. Glocer, G. Gronoff, E. Hébrard, W. Danchi, Prebiotic chemistry and atmospheric warming of early Earth by an active young Sun. Nat. Geosci. 9, 452–455 (2016)
2. J. Aléon, F. Robert, J. Duprat, S. Derenne, Extreme oxygen isotope ratios in the early Solar System. Nature 437, 385–388 (2005)
3. M. Anand, R. Tartèse, J.J. Barnes, Understanding the origin and evolution of water in the Moon through lunar sample studies. Philos. Trans. R. Soc. A 372, 20130254 (2014). https://doi.org/10.1098/rsta.2013.0254
4. ADS E. Anders, N. Grevesse, Abundances of the elements: meteoritic and solar. Geochim. Cosmochim. Acta 53, 197–214 (1989)
5. J.C. Andrews-Hanna, J. Besserer, J.W. Head, C.J.A. Howett, W.S. Kiefer, P.J. Lucey, P.J. McGovern, H.J. Melosh, G.A. Neumann, R.J. Phillips, P.M. Schenk, D.E. Smith, S.C. Solomon, M.T. Zuber, Structure and evolution of the lunar Procellarum region as revealed by GRAIL gravity data. Nature 514, 68–71 (2014)
6. T. Arai, H. Takeda, A. Yamaguchi, M. Ohtake, A new model of lunar crust: asymmetry in crustal composition and evolution. Earth Planets Space 60, 433–444 (2008)
7. T. Arai, B.R. Hawke, T.A. Giguere, K. Misawa, M. Miyamoto, H. Kojima, Antarctic lunar meteorites Yamato-793169, Asuka-881757, MIL 05035, and MET 01210 (YAMM): launch pairing and possible cryptomare origin. Geochim. Cosmochim. Acta 74, 2231–2248 (2010)
8. J.C. Armstrong, Distribution of impact locations and velocities of Earth meteorites on the Moon. Earth Moon Planets 107, 43–54 (2010)
9. ADS MATH J.C. Armstrong, L.E. Wells, G. Gonzalez, Rummaging through Earth's Attic for remains of ancient life. Icarus 160, 183–196 (2002)
10. J.R. Arnold, Ice in the lunar polar regions. J. Geophys. Res. 84, 5659–5668 (1979)
11. G. Avice, B. Marty, The iodine-plutonium-xenon age of the Moon-Earth system revisited. Philos. Trans. R. Soc. A 372, 20130260 (2014). https://doi.org/10.1098/rsta.2013.0260
12. ADS S.J. Barber, P.H. Smith, I.P. Wright, F. Abernethy, M. Anand, K.R. Dewar, M. Hodges, P. Landsberg, M.R. Leese, G.H. Morgan, A.D. Morse, J. Mortimer, H.M. Sargeant, I. Sheard, S. Sheridan, A.

Verchovsky, F. Goesmann, C. Howe, T. Morse, N. Lillywhite, A. Quinn, N. Missaglia, M. Pedrali, P. Reiss, F. Rizzi, A. Rusconi, M. Savoia, A. Zamboni, J.A. Merrifield, E.K. Gibson Jr., J. Carpenter, R. Fisackerly, B. Houdou, ProSPA: the science laboratory for the processing and analysis of lunar polar volatiles within PROSPECT, in *Lunar Planet. Sci. Conf.* (2017), abstract 2171

13. M. Barboni, P. Boehnke, B. Keller, I.E. Kohl, B. Schoene, E.D. Young, K.D. McKeegan, Early formation of the Moon 4.51 billion years ago. Sci. Adv. 3, e1602365 (2017). https://doi.org/10.1126/sciadv.1602365
14. ADS J.J. Barnes, D.A. Kring, R. Tartèse, I.A. Franchi, M. Anand, S.S. Russell, An asteroidal origin for water in the Moon. Nat. Commun. 7, 11684 (2016). https://doi.org/10.1038/ncomms11684
15. ADS J.A. Barrat, B. Zanda, F. Moynier, C. Bollinger, C. Liorzou, G. Bayon, Geochemistry of CI chondrites: major and trace elements, and Cu and Zn isotopes. Geochim. Cosmochim. Acta 83, 79–92 (2012)
16. D.W. Beaty, A.L. Albee, Comparative petrology and possible genetic relations among the Apollo 11 basalts, in *Lunar Planet. Sci. Conf. Proc.* (1978), pp. 359–463
17. R.H. Becker, Evidence for a secular variation in the $^{13}C/^{12}C$ ratio of carbon implanted in lunar soils. Earth Planet. Sci. Lett. 50, 189–196 (1980)
18. R.H. Becker, R.N. Clayton, Nitrogen abundances and isotopic compositions in lunar samples, in *Lunar Sci. Conf. Proc.* (1975), pp. 2131–2149
19. R.H. Becker, R.O. Pepin, Long-term changes in solar wind elemental and isotopic ratios: a comparison of two lunar ilmenites of different antiquities. Geochim. Cosmochim. Acta 53, 1135–1146 (1989)
20. J.J. Bellucci, A.A. Nemchin, M. Grange, K.L. Robinson, G. Collins, M.J. Whitehouse, J.F. Snape, M.D. Norman, D.A. Kring, Terrestrial-like zircon in a clast from an Apollo 14 breccia. Earth Planet. Sci. Lett. 510, 173–185 (2019)
21. J.-P. Benkert, H. Baur, P. Signer, R. Wieler, He, Ne, and Ar from the solar wind and solar energetic particles in lunar ilmenites and pyroxenes. J. Geophys. Res., Planets 98, 13147–13162 (1993)
22. L.E. Borg, C.K. Shearer, Y. Asmerom, J.J. Papike, Prolonged KREEP magmatism on the Moon indicated by the youngest dated lunar igneous rock. Nature 432, 209–211 (2004)

23. L.E. Borg, A.M. Gaffney, C.K. Shearer, D.J. DePaolo, I.D. Hutcheon, T.L. Owens, E. Ramon, G. Brennecka, Mechanisms for incompatible-element enrichment on the Moon deduced from the lunar basaltic meteorite Northwest Africa 032. Geochim. Cosmochim. Acta 73, 3963–3980 (2009)
24. L.E. Borg, J.N. Connelly, M. Boyet, R.W. Carlson, Chronological evidence that the Moon is either young or did not have a global magma ocean. Nature 477, 70–72 (2011)
25. W.F. Bottke, M.D. Norman, The late heavy bombardment. Annu. Rev. Earth Planet. Sci. 45, 619–647 (2017)
26. S.E. Braden, J.D. Stopar, M.S. Robinson, S.J. Lawrence, C.H. van der Bogert, H. Hiesinger, Evidence for basaltic volcanism on the Moon within the past 100 million years. Nat. Geosci. 7, 787–791 (2014)
27. A.D. Brandon, T.J. Lapen, V. Debaille, B.L. Beard, K. Rankenburg, C.R. Neal, Re-evaluating $^{142}Nd/^{144}Nd$ in lunar mare basalts with implications for the early evolution and bulk Sm/Nd of the Moon. Geochim. Cosmochim. Acta 73, 6421–6445 (2009)
28. S.M. Brown, T.L. Grove, Origin of the Apollo 14, 15, and 17 yellow ultramafic glasses by mixing of deep cumulate remelts. Geochim. Cosmochim. Acta 171, 201–215 (2015)
29. D.S. Burnett, A.J.G. Jurewicz, D.S. Woolum, The future of Genesis science. Meteorit. Planet. Sci. 54, 1092–1114 (2019)
30. P.H. Cadogan, Oldest and largest lunar basin? Nature 250, 315–316 (1974)
31. J.T. Cahill, C. Floss, M. Anand, L.A. Taylor, M.A. Nazarov, B.A. Cohen, Petrogenesis of lunar highlands meteorites: Dhofar 025, Dhofar 081, Dar al Gani 262, and Dar al Gani 400. Meteorit. Planet. Sci. 39, 503–529 (2004)
32. D.B. Campbell, B.A. Campbell, L.M. Carter, J.L. Margot, N.J.S. Stacy, No evidence for thick deposits of ice at the lunar South pole. Nature 443, 835–837 (2006)
33. R.W. Carlson, G.W. Lugmair, Sm-Nd constraints on early lunar differentiation and the evolution of KREEP. Earth Planet. Sci. Lett. 45, 123–132 (1979)
34. R.W. Carlson, G.W. Lugmair, The age of ferroan anorthosite 60025: oldest crust on a young Moon? Earth Planet. Sci. Lett. 90, 119–130 (1988)

35. R.W. Carlson, L.E. Borg, A.M. Gaffney, M. Boyet, Rb-Sr, Sm-Nd and Lu-Hf isotope systematics of the lunar Mg-suite: the age of the lunar crust and its relation to the time of Moon formation. Philos. Trans. R. Soc. A 372, 20130246 (2014). https://doi.org/10.1098/rsta.2013.0246
36. ADS J. Carpenter, R. Fisackerly, B. Houdou, Establishing lunar resource viability. Space Policy 37, 52–57 (2016)
37. B. Charlier, T.L. Grove, O. Namur, F. Holtz, Crystallization of the lunar magma ocean and the primordial mantle-crust differentiation of the Moon. Geochim. Cosmochim. Acta 234, 50–69 (2018)
38. R.N. Clark, Detection of adsorbed water and hydroxyl on the Moon. Science 326, 562–564 (2009)
39. R.N. Clayton, M.H. Thiemens, Lunar nitrogen: evidence for secular change in the solar wind, in *Proceedings of the Conference on the Ancient Sun*, ed. by R.O. Pepin, J.A. Eddy, R.B. Merrill (Pergamon, New York, 1980), pp. 463–473
40. B.A. Cohen, T.D. Swindle, D.A. Kring, Support for the lunar cataclysm hypothesis from lunar meteorite impact melt ages. Science 290, 1754–1756 (2000)
41. B.A. Cohen, G.A. Snyder, C.M. Hall, L.A. Taylor, M.A. Nazarov, Argon-40-argon-39 chronology and petrogenesis along the eastern limb of the Moon from Luna 16, 20 and 24 samples. Meteorit. Planet. Sci. 36, 1345–1366 (2001)
42. B.A. Cohen, S.J. Lawrence, N.E. Petro, G.D. Bart, R.N. Clegg-Watkins, B.W. Denevi, R.R. Ghent, R.L. Klima, G.A. Morgan, P.D. Spudis, J.D. Stopar, Identifying and characterizing impact melt outcrops in the Nectaris basin, in *Lunar Planet. Sci. Conf.* (2016), p. abstract 1389
43. A. Colaprete, P. Schultz, J. Heldmann, D. Wooden, M. Shirley, K. Ennico, B. Hermalyn, W. Marshall, A. Ricco, R.C. Elphic, D. Goldstein, D. Summy, G.D. Bart, E. Asphaug, D. Korycansky, D. Landis, L. Sollitt, Detection of water in the LCROSS ejecta plume. Science 330, 463–468 (2010)
44. A. Colaprete, D. Andrews, W. Bluethmann, R.C. Elphic, B. Bussey, J. Trimble, K. Zacny, J.E. Captain, An overview of the Volatiles Investigating Polar Exploration Rover (VIPER) mission, in *American Geophysical Union Fall Meeting* (2019), abstract P34B-03
45. W. Compston, M.J. Vernon, H. Berry, R. Rudowski, The age of the Fra Mauro formation: a radiometric older limit. Earth Planet. Sci. Lett. 12,

55–58 (1971)

46. W. Compston, M.J. Vernon, H. Berry, R. Rudowski, C.M. Gray, N. Ware, Apollo 14 mineral ages and the thermal history of the Fra Mauro formation, in *Lunar Planet. Sci. Conf. Proc.*, ed. by A.E. Metzger, J.I. Trombka, L.E. Peterson, R.C. Reedy, J.R. Arnold (1972), pp. 1487–1501
47. J.N. Connelly, M. Bizzarro, Lead isotope evidence for a young formation age of the Earth–Moon system. Earth Planet. Sci. Lett. 452, 36–43 (2016)
48. J.W. Conrad, F. Nimmo, C.I. Fassett, S. Kamata, Lunar impact history constrained by GRAIL-derived basin relaxation measurements. Icarus 314, 50–63 (2018)
49. D.L. Cook, E. Berger, T. Faestermann, G.F. Herzog, K. Knie, G. Korschinek, M. Poutivtsev, G. Ruge, F. Serefiddin, ^{60}Fe, ^{10}Be, and ^{26}Al in lunar cores 12025/8 and 60006/7: search for a nearby supernova, in *Lunar Planet. Sci. Conf. Proc.* (2009), p. abstract 1129
50. L.M. Corley, P.J. McGovern, G.Y. Kramer, M. Lemelin, D. Trang, J.J. Gillis-Davis, G.J. Taylor, K.E. Powell, W.S. Kiefer, M. Wieczorek, M.T. Zuber, Olivine-bearing lithologies on the Moon: constraints on origins and transport mechanisms from M^3 spectroscopy, radiative transfer modeling, and GRAIL crustal thickness. Icarus 300, 287–304 (2018)
51. C. Cournède, J. Gattacceca, P. Rochette, Magnetic study of large Apollo samples: possible evidence for an ancient centered dipolar field on the Moon. Earth Planet. Sci. Lett. 33, 31–42 (2012)
52. I.A. Crawford, The Moon as a recorder of nearby supernovae, in *Handbook of Supernovae*, ed. by A.W. Alsabti, P. Murdin (Springer, Berlin, 2017), pp. 2507–2522
53. I.A. Crawford, K.H. Joy, Lunar exploration: opening a window into the history and evolution of the inner Solar System. Philos. Trans. R. Soc. A 372, 20130315 (2014). https://doi.org/10.1098/rsta.2013.0315
54. ADS I.A. Crawford, S.A. Fagents, K.H. Joy, Full Moon exploration. Astron. Geophys. 48, 3.18–3.21 (2007)
55. I.A. Crawford, E.C. Baldwin, E.A. Taylor, J.A. Bailey, K. Tsembelis, On the survivability and detectability of terrestrial meteorites on the Moon. Astrobiology 8, 242–252 (2008)
56. I.A. Crawford, K.H. Joy, S.A. Fagents, M.E. Rumpf, Lunar

palaeoregolith deposits as recorders of the galactic environment of the Solar System and implications for astrobiology. Earth Moon Planets 107, 75–85 (2010)

57. S.A. Crowther, J.D. Gilmour, The Genesis solar wind composition and its relationship to planetary xenon signatures. Geochim. Cosmochim. Acta 123, 17–34 (2013)
58. N.M. Curran, K.H. Joy, J.F. Snape, J.F. Pernet-Fisher, J.D. Gilmour, A.A. Nemchin, M.J. Whitehouse, R. Burgess, The early geological history of the Moon inferred from ancient lunar meteorite Miller Range 13317. Meteorit. Planet. Sci. 54, 1401–1430 (2019)
59. E.J. Dasch, C.Y. Shih, B.M. Bansal, H. Wiesmann, L.E. Nyquist, Isotopic analysis of basaltic fragments from lunar breccia 14321: Chronology and petrogenesis of pre-Imbrium mare volcanism. Geochim. Cosmochim. Acta 51, 3241–3254 (1987)
60. J.M.D. Day, R.J. Walker, Highly siderophile element depletion in the Moon. Earth Planet. Sci. Lett. 423, 114–124 (2015)
61. J.M.D. Day, D.G. Pearson, L.A. Taylor, Highly siderophile element constraints on accretion and differentiation of the Earth-Moon system. Science 315, 217–219 (2007)
62. J.W. Delano, Pristine lunar glasses: criteria, data, and implications. J. Geophys. Res. 91, 201–213 (1986)
63. J.W. Delano, K. Livi, Lunar volcanic glasses and their constraints on mare petrogenesis. Geochim. Cosmochim. Acta 45, 2137–2149 (1981)
64. E. Doell, C. Grommé, Survey of magnetic properties of Apollo 11 samples at the Lunar Receiving Laboratory, in *Apollo 11 Lunar Sci. Conf. Proc.* (1970), pp. 2093–2096
65. S.S. Dolginov, E.U. Eroshenko, L.I. Zhuzgov, N.V. Pushkov, A study of the magnetic field of the Moon. The Moon, in *IAU Symposium*, vol. 14 (1962), pp. 45–52
66. K.L. Donaldson Hanna, L.C. Cheek, C.M. Pieters, J.F. Mustard, B.T. Greenhagen, I.R. Thomas, N.E. Bowles, Global assessment of pure crystalline plagioclase across the Moon and implications for the evolution of the primary crust. J. Geophys. Res., Planets 119, 1516–1545 (2014)
67. D.S. Draper, R.L. Klima, S.J. Lawrence, B.W. Denevi, The Inner SOlar system CHRONology (ISOCHRON) Discovery mission: Returning samples of the youngest lunar mare basalts, in *Lunar Planet. Sci. Conf.*

(2019), p. abstract 1110

68. P. Dyal, W.D. Daily, Crustal magnetic scale sizes measured at Apollo landing sites, in *Lunar Planet. Sci. Conf. Proc.* (1978), pp. 273–275
69. P. Dyal, C. Parkin, C. Sonett, Apollo 12 Magnetometer: measurement of a steady magnetic field on the surface of the Moon. Science 169, 762–764 (1970)
70. W.F. Eichelman, W.W. Lauderdale, Apollo scientific experiments data handbook. NASA Technical Memorendum X-58131 (1974)
71. L.T. Elkins-Tanton, E.M. Parmentier, P.C. Hess, Magma ocean fractional crystallization and cumulate overturn in terrestrial planets: implications for Mars. Meteorit. Planet. Sci. 38, 1753–1771 (2003)
72. L.T. Elkins-Tanton, B.H. Hager, T.L. Grove, Magmatic effects of the lunar late heavy bombardment. Earth Planet. Sci. Lett. 222, 17–27 (2004)
73. L.T. Elkins-Tanton, S. Burgess, Q.Z. Yin, The lunar magma ocean: reconciling the solidification process with lunar petrology and geochronology. Earth Planet. Sci. Lett. 304, 326–336 (2011)
74. S. Epstein, H.P. Taylor, $^{18}O/^{16}O$, $^{30}Si/^{28}Si$, D/H, and $^{13}C/^{12}C$ studies of lunar rock and minerals. Science 167, 533–535 (1970)
75. O. Eugster, D. Terribilini, E. Polnau, J. Kramers, The antiquity indicator argon-40/argon-36 for lunar surface samples calibrated by uranium-235-xenon-136 dating. Meteorit. Planet. Sci. 36, 1097–1115 (2001)
76. A.J. Evans, M.T. Zuber, B.P. Weiss, S.M. Tikoo, A wet,heterogeneous lunar interior: lower mantle and core dynamo evolution. J. Geophys. Res., Planets 119, 1061–1077 (2014)
77. A.J. Evans, S.M. Tikoo, J.C. Andrews-Hanna, The case against an early lunar dynamo powered by core convection. Geophys. Res. Lett. 45, 98–107 (2018)
78. A.L. Fagan, K.H. Joy, D.D. Bogard, D.A. Kring, Ages of globally distributed lunar paleoregoliths and soils from 3.9 Ga to the present. Earth Moon Planets 112, 59–71 (2014)
79. S.A. Fagents, M.E. Rumpf, I.A. Crawford, K.H. Joy, Preservation potential of implanted solar wind volatiles in lunar paleoregolith deposits buried by lava flows. Icarus 207, 595–604 (2010)
80. W.M. Farrell, D.M. Hurley, M.I. Zimmerman, Solar wind implantation into lunar regolith: hydrogen retention in a surface with defects. Icarus 255, 116–126 (2015)

81. C.I. Fassett, Analysis of impact crater populations and the geochronology of planetary surfaces in the inner Solar System. J. Geophys. Res., Planets 121, 1900–1926 (2016)
82. C.I. Fassett, J.W. Head, S.J. Kadish, E. Mazarico, G.A. Neumann, D.E. Smith, M.T. Zuber, Lunar impact basins: stratigraphy, sequence and ages from superposed impact crater populations measured from Lunar Orbiter Laser Altimeter (LOLA) data. J. Geophys. Res., Planets 117, E00H06 (2012). https://doi.org/10.1029/2011JE003951
83. ADS W.C. Feldman, S. Maurice, A.B. Binder, B.L. Barraclough, R.C. Elphic, D.J. Lawrence, Fluxes of fast and epithermal neutrons from Lunar Prospector: Evidence for water ice at the lunar poles. Science 281, 1496–1500 (1998)
84. W.C. Feldman, D.J. Lawrence, R.C. Elphic, B.L. Barraclough, S. Maurice, I. Genetay, A.B. Binder, Polar hydrogen deposits on the Moon. J. Geophys. Res. 105, 4175–4195 (2000)
85. V.A. Fernandes, R. Burgess, Volcanism in Mare Fecunditatis and Mare Crisium: Ar-Ar age studies. Geochim. Cosmochim. Acta 69, 4919–4934 (2005)
86. V.A. Fernandes, J. Fritz, B.P. Weiss, I. Garrick-Bethell, D.L. Shuster, The bombardment history of the Moon as recorded by ^{40}Ar-^{39}Ar chronology. Meteorit. Planet. Sci. 48, 241–269 (2013)
87. L. Fimiani, D.L. Cook, T. Faestermann, J.M. Gómez-Guzmán, K. Hain, G. Herzog, K. Knie, G. Korschinek, P. Ludwig, J. Park, R.C. Reedy, G. Rugel, Interstellar ^{60}Fe on the surface of the Moon. Phys. Rev. Lett. 116, 151104 (2016). https://doi.org/10.1103/PhysRevLett.116.151104
88. ADS M. Fischer-Gödde, H. Becker, What is the age of the Nectaris basin? New Re-Os constraints for a pre-4.0 Ga bombardment history of the Moon, in *Lunar Planet. Sci. Conf.* (2011), p. abstract 1414
89. M. Fischer-Gödde, H. Becker, Osmium isotope and highly siderophile element constraints on ages and nature of meteoritic components in ancient lunar impact rocks. Geochim. Cosmochim. Acta 77, 135–156 (2012)
90. E.A. Fisher, P.G. Lucey, M. Lemelin, B.T. Greenhagen, M.A. Siegler, E. Mazarico, O. Aharonson, J.P. Williams, P.O. Hayne, G.A. Neumann, D.A. Paige, D.E. Smith, M.T. Zuber, Evidence for surface water ice in the lunar polar regions using reflectance measurements from the Lunar Orbiter Laser Altimeter and temperature measurements from

the Diviner Lunar Radiometer Experiment. Icarus 292, 74–85 (2017)

91. U. Frick, H. Baur, H. Ducati, H. Funk, D. Phinney, P. Signer, On the origin of helium, neon, and argon isotopes in sieved mineral separates from an Apollo 15 soil, in *Lunar Sci. Conf. Proc.* (1975), pp. 2097–2129
92. U. Frick, R.H. Becker, R.O. Pepin, Solar wind record in the lunar regolith: nitrogen and noble gases, in *Lunar Planet. Sci. Conf. Proc.* (1988), pp. 87–120
93. I. Friedman, J.R. O'Neill, L.H. Adami, J.D. Gleason, K. Hardcastle, Water, hydrogen, deuterium, carbon, carbon-13, and oxygen-18 content of selected lunar material. Science 167, 533–535 (1970)
94. M. Fuller, S.M. Cisowski, Lunar paleomagnetism. Geomagnetism 2, 307–455 (1987)
95. E. Füri, B. Marty, S.S. Assonov, Constraints on the flux of meteoritic and cometary water on the Moon from volatile element (N–Ar) analyses of single lunar soil grains, Luna 24 core. Icarus 218, 220–229 (2012)
96. E. Füri, L. Zimmermann, A.E. Saal, Apollo 15 green glass He-Ne-Ar signatures—In search for indigenous lunar noble gases. Geochem. Perspect. Lett. 8, 1–5 (2018)
97. A.M. Gaffney, L.E. Borg, A young solidification age for the lunar magma ocean. Geochim. Cosmochim. Acta 140, 227–240 (2014)
98. I. Garrick-Bethell, K. Miljković, Age of the lunar South pole-Aitken basin, in *Lunar Planet. Sci. Conf.* (2018), p. abstract 2633
99. I. Garrick-Bethell, M.T. Zuber, Elliptical structure of the lunar South pole-Aitken basin. Icarus 204, 399–408 (2009)
100. I. Garrick-Bethell, J. Wisdom, M.T. Zuber, Evidence for a past high-eccentricity lunar orbit. Science 313, 652–655 (2006)
101. ADS MathSciNet MATH I. Garrick-Bethell, F. Nimmo, M.A. Wieczorek, Structure and formation of the lunar farside highlands. Science 330, 949–951 (2010)
102. I. Garrick-Bethell, B.P. Weiss, D.L. Shuster, S.M. Tikoo, M.M. Tremblay, Further evidence for early lunar magnetism from troctolite 76535. J. Geophys. Res., Planets 122, 76–93 (2017)
103. J. Geiss, P. Bochsler, Nitrogen isotopes in the Solar System. Geochim. Cosmochim. Acta 46, 529–548 (1982)
104. J. Geiss, P. Bochsler, Long time variations in solar wind properties:

possible causes versus observations, in *The Sun in Time*, ed. by C.P. Sonett, M.S. Giampapa, M.S. Matthews (University of Arizona Press, Tucson, 1991), pp. 98–117

105. J. Geiss, F. Bühler, H. Cerutti, P. Eberhardt, C.H. Filleux, J. Meister, P. Signer, The Apollo SWC experiment: results, conclusions, consequences. Space Sci. Rev. 110, 307–335 (2004)
106. J. Geiss, P. Eberhardt, P. Signer, F. Buehler, J. Meister, The Solar-Wind Composition experiment, in *Apollo 11 Preliminary Science Report* (NASA SP-214, Washington, DC, 1969), pp. 183–186
107. E.K. Gibson, G.W. Moore, Volatile-rich lunar soil: evidence of possible cometary impact. Science 179, 69–71 (1973)
108. D.R. Gies, J.W. Helsel, Ice age epochs and the Sun's path through the Galaxy. Astrophys. J. 626, 844–848 (2005)
109. K. Gillet, L. Margerin, M. Calvet, M. Monnereau, Scattering attenuation profile of the Moon: implications for shallow moonquakes and the structure of the megaregolith. Phys. Earth Planet. Inter. 262, 28–40 (2017)
110. M.P. Gillman, H.E. Erenler, P.J. Sutton, Mapping the location of terrestrial impacts and extinctions onto the spiral arm structure of the Milky Way. Int. J. Astrobiol. 18, 323–328 (2019)
111. G.R. Gladstone, D.M. Hurley, K.D. Retherford, P.D. Feldman, W.R. Pryor, J.Y. Chaufray, M. Versteeg, T.K. Greathouse, A.J. Steffl, H. Throop, J.W. Parker, D.E. Kaufmann, A.F. Egan, M.W. Davis, D.C. Slater, J. Mukherjee, P.F. Miles, A.R. Hendrix, A. Colaprete, A. Stern, LRO-LAMP observations of the LCROSS impact plume. Science 330, 472–476 (2010)
112. P. Gleißner, H. Becker, Formation of Apollo 16 impactites and the composition of late accreted material: constraints from Os isotopes, highly siderophile elements and sulfur abundances. Geochim. Cosmochim. Acta 200, 1–24 (2017)
113. R.O. Green, C. Pieters, P. Mouroulis, M. Eastwood, J. Boardman, T. Glavich, P. Isaacson, M. Annadurai, S. Besse, D. Barr, B. Buratti, D. Cate, A. Chatterjee, R. Clark, L. Cheek, J. Combe, D. Dhingra, V. Essandoh, S. Geier, J.N. Goswami, R. Green, V. Haemmerle, J. Head, L. Hovland, S. Hyman, R. Klima, T. Koch, G. Kramer, A.S.K. Kumar, K. Lee, S. Lundeen, E. Malaret, T. McCord, S. McLaughlin, J. Mustard, J. Nettles, N. Petro, K. Plourde, C. Racho, J. Rodriquez, C. Runyon, G.

Sellar, C. Smith, H. Sobel, M. Staid, J. Sunshine, L. Taylor, K. Thaisen, S. Tompkins, H. Tseng, G. Vane, P. Varanasi, M. White, D. Wilson, The Moon Mineralogy Mapper (M^3) imaging spectrometer for lunar science: instrument description, calibration, on-orbit measurements, science data calibration and on-orbit validation. J. Geophys. Res., Planets 116, E00G19 (2011). https://doi.org/10.1029/2011JE003797

114. J. Gross, K.H. Joy, Evolution, lunar: from magma ocean to crust formation, in *Encyclopedia of Lunar Science*, ed. by B. Cudnik (Springer, Berlin, 2016). https://doi.org/10.1007/978-3-319-05546-6_39-1
115. J. Gross, A.H. Treiman, C.N. Mercer, Lunar feldspathic meteorites: constraints on the geology of the lunar highlands, and the origin of the lunar crust. Earth Planet. Sci. Lett. 388, 318–328 (2014)
116. T.L. Grove, M.J. Krawczynski, Lunar mare volcanism: where did the magmas come from? Elements 5, 29–34 (2009)
117. M. Güdel, The Sun in time: activity and environment. Living Rev. Sol. Phys. 4, 3 (2007). https://doi.org/10.12942/lrsp-2007-3
118. ADS S. Guggisberg, P. Eberhardt, J. Geiss, N. Groegler, A. Stettler, G.M. Brown, A. Peckett, Classification of the Apollo-11 mare basalts according to Ar^{39}-Ar^{40} ages and petrological properties, in *Lunar Planet. Sci. Conf. Proc.*, ed. by N.W. Hinners (1979), pp. 1–39
119. A.N. Halliday, D.-C. Lee, Tungsten isotopes and the early development of the Earth and Moon. Geochim. Cosmochim. Acta 63, 4157–4179 (1999)
120. L.J. Hallis, M. Anand, S. Strekopytov, Trace-element modelling of mare basalt parental melts: implications for a heterogeneous lunar mantle. Geochim. Cosmochim. Acta 134, 289–316 (2014)
121. J.K. Harmon, M.A. Slade, Radar mapping of Mercury: full-disk images and polar anomalies. Science 258, 640–643 (1992)
122. W.K. Hartmann, Lunar cratering chronology. Icarus 13, 299–301 (1970)
123. W.K. Hartmann, D.R. Davis, Satellite-sized planetesimals and lunar origin. Icarus 24, 504–515 (1975)
124. W.K. Hartmann, C.A. Wood, Moon: origin and evolution of multi-ring basins. Moon 3, 3–78 (1971)
125. K. Hashizume, M. Chaussidon, A non-terrestrial ^{16}O-rich isotopic composition for the protosolar nebula. Nature 434, 619–622 (2005)

126. K. Hashizume, M. Chaussidon, Two oxygen isotopic components with extra-selenial origins observed among lunar metallic grains—In search for the solar wind component. Geochim. Cosmochim. Acta 73, 3038–3054 (2009)
127. K. Hashizume, M. Chaussidon, B. Marty, F. Robert, Solar wind record on the Moon: deciphering presolar from planetary nitrogen. Science 290, 1142–1145 (2000)
128. K. Hashizume, M. Chaussidon, B. Marty, K. Terada, Protosolar carbon isotopic composition: implications for the origin of meteoritic organics. Astrophys. J. 600, 480–484 (2004)
129. L.A. Haskin, The Imbrium impact event and the thorium distribution at the lunar highlands surface. J. Geophys. Res., Planets 103, 1679–1689 (1998)
130. E.H. Hauri, A.E. Saal, M. Nakajima, M. Anand, M.J. Rutherford, J.A. Van Orman, M. Le Voyer, Origin and evolution of water in the Moon's interior. Annu. Rev. Earth Planet. Sci. 45, 89–111 (2017)
131. P.O. Hayne, B.T. Greenhagen, M.C. Foote, M.A. Siegler, A.R. Vasavada, D.A. Paige, Diviner lunar radiometer observations of the LCROSS impact. Science 330, 477–479 (2010)
132. P.O. Hayne, A. Hendrix, E. Sefton-Nash, M.A. Siegler, P.G. Lucey, K.D. Retherford, J.P. Williams, B.T. Greenhagen, D.A. Paige, Evidence for exposed water ice in the Moon's south polar regions from Lunar Reconnaissance Orbiter ultraviolet albedo and temperature measurements. Icarus 255, 58–69 (2015)
133. J.W. Head III, L. Wilson, Generation, ascent and eruption of magma on the Moon: new insights into source depths, magma supply, intrusions and effusive/explosive eruptions (part 2: observations). Icarus 283, 176–223 (2017)
134. J.W. Head, L. Wilson, Lunar mare volcanism: stratigraphy, eruption conditions, and the evolution of secondary crusts. Geochim. Cosmochim. Acta 56, 2155–2175 (1992)
135. V.S. Heber, H. Baur, R. Wieler, Helium in lunar samples analyzed by high-resolution stepwise etching: implications for the temporal constancy of solar wind isotopic composition. Astrophys. J. 597, 602–614 (2003)
136. V.S. Heber, R. Wieler, H. Baur, C. Olinger, T.A. Friedmann, D.S. Burnett, Noble gas composition of the solar wind as collected by the

Genesis mission. Geochim. Cosmochim. Acta 73, 7414–7432 (2009)

137. G.H. Heiken, D.T. Vaniman, B.M. French, *Lunar Sourcebook* (Cambridge University Press, Cambridge, 1991)
138. P.C. Hess, E.M. Parmentier, A model for the thermal and chemical evolution of the Moon's interior: implications for the onset of mare volcanism. Earth Planet. Sci. Lett. 134, 501–514 (1995)
139. H. Hiesinger, R. Jaumann, G. Neukum, J.W. Head, Ages of mare basalts on the lunar nearside. J. Geophys. Res. 105, 29239–29275 (2000)
140. H. Hiesinger, J.W. Head, U. Wolf, R. Jaumann, G. Neukum, Ages and stratigraphy of mare basalts in Oceanus Procellarum, Mare Nubium, Mare Cognitum, and Mare Insularum. J. Geophys. Res., Planets 108, 5065 (2003). https://doi.org/10.1029/2002JE001985
141. ADS H. Hiesinger, J.W. Head, U. Wolf, R. Jaumann, G. Neukum, Ages and stratigraphy of lunar mare basalts in Mare Frigoris and other nearside maria based on crater size-frequency distribution measurements. J. Geophys. Res., Planets 115, E03003 (2010). https://doi.org/10.1029/2009JE003380
142. ADS H. Hiesinger, J.W. Head, U. Wolf, R. Jaumann, G. Neukum, Ages and stratigraphy of lunar mare basalts: a synthesis, in *Recent Advances and Current Research Issues in Lunar Stratigraphy*, ed. by W.A. Ambrose, D.A. Williams. Geol. Soc. of America Special Paper, vol. 477 (2011), pp. 1–51. https://doi.org/10.1130/2011.2477(01)
143. H. Hiesinger, C.H. van der Bogert, J.H. Pasckert, N. Schmedemann, M.S. Robinson, B. Jolliff, N. Petro, New crater size-frequency distribution measurements of the South pole-Aitken basin, in *Lunar Planet. Sci. Conf.* (2012), p. abstract 2863
144. F. Hörz, R. Grieve, G. Heiken, P. Spudis, A. Binder, Lunar surface processes, in *Lunar Sourcebook*, ed. by G. Heiken, D. Vaniman, B.M. French (Cambridge University Press, Cambridge, 1991), pp. 61–120
145. N. Hosono, S. Karato, J. Makino, T.R. Saitoh, Terrestrial magma ocean origin of the Moon. Nat. Geosci. 12, 418–423 (2019)
146. J. Huang, Z. Xiao, J. Flahaut, M. Martinot, J. Head, X. Xiao, M. Xie, L. Xiao, Geological characteristics of Von Kármán crater, northwestern South pole-Aitken basin: Chang'E-4 landing site region. J. Geophys. Res., Planets 123, 1684–1700 (2018)
147. D. Hurwitz, D.A. Kring, Identifying the geologic context of Apollo 17 impact melt breccias. Earth Planet. Sci. Lett. 436, 64–70 (2016)

148. L. Husain, ^{40}Ar-^{39}Ar chronology and cosmic ray exposure ages of the Apollo 15 samples. J. Geophys. Res. 79, 2588–2606 (1974)

149. G.R. Huss, K. Nagashima, D.S. Burnett, A.J.G. Jurewicz, C.T. Olinger, A new upper limit on the D/H ratio in the solar wind, in *Lunar Planet. Sci. Conf. Proc.* (2012a), abstract 1709

150. G.R. Huss, K. Nagashima, A.J.G. Jurewicz, D.S. Burnett, C.T. Olinger, The isotopic composition and fluence of solar-wind nitrogen in a Genesis B/C array collector. Meteorit. Planet. Sci. 47, 1436–1448 (2012b)

151. R. Jaumann, H. Hiesinger, M. Anand, I.A. Crawford, R. Wagner, F. Sohl, B.L. Jolliff, F. Scholten, M. Knapmeyer, H. Hoffmann, H. Hussmann, M. Grott, S. Hempel, U. Köhler, K. Krohn, N. Schmitz, J. Carpenter, M. Wieczorek, T. Spohn, M.S. Robinson, J. Oberst, Geology, geochemistry, and geophysics of the Moon: status of current understanding. Planet. Space Sci. 74, 15–41 (2012)

152. E.A. Jerde, G.A. Snyder, L.A. Taylor, Y.G. Liu, R.A. Schmitt, The origin and evolution of lunar high-Ti basalts: periodic melting of a single source at Mare Tranquillitatis. Geochim. Cosmochim. Acta 58, 515–527 (1994)

153. B.L. Jolliff, J.J. Gillis, L.A. Haskin, R.L. Korotev, M.A. Wieczorek, Major lunar crustal terranes: surface expressions and crust-mantle origins. J. Geophys. Res. 105, 4197–4216 (2000)

154. B.L. Jolliff, R.C. Watkins, N.E. Petro, D.P. Moriarty, S.J. Lawrence, J.W. Head, C.M. Pieters, J.J. Hagerty, R.L. Fergason, T.M. Hare, L.R. Gaddis, P.O. Hayne, Selecting and certifying a landing site for moonrise in South pole-Aitken basin, in *Lunar Planet. Sci. Conf.* (2017), p. abstract 1326

155. K.H. Joy, Trace elements in lunar plagioclase as indicators of source lithology, in *Lunar Planet. Sci. Conf.* (2013), p. abstract 1033

156. K.H. Joy, T. Arai, Lunar meteorites: new insights into the geological history of the Moon. Astron. Geophys. 54, 4.28–4.32 (2013). https://doi.org/10.1093/astrogeo/att121

157. K.H. Joy, I.A. Crawford, M. Anand, R.C. Greenwood, I.A. Franchi, S.S. Russell, The petrology and geochemistry of Miller range 05035: a new lunar gabbroic meteorite. Geochim. Cosmochim. Acta 72, 3822–3844 (2008)

158. K.H. Joy, R. Burgess, R. Hinton, V.A. Fernandes, I.A. Crawford, A.T.

Kearsley, A.J. Irving, EIMF, Petrogenesis and chronology of lunar meteorite Northwest Africa 4472: a KREEPy regolith breccia from the Moon. Geochim. Cosmochim. Acta 75, 2420–2452 (2011a)

159. K.H. Joy, D.A. Kring, D.D. Bogard, D.S. McKay, M.E. Zolensky, Re-examination of the formation ages of the Apollo 16 regolith breccias. Geochim. Cosmochim. Acta 75, 7208–7225 (2011b)
160. K.H. Joy, I.A. Crawford, N.M. Curran, M. Zolensky, A.F. Fagan, D.A. Kring, The Moon: an archive of small body migration in the Solar System. Earth Moon Planets 118, 133–158 (2016)
161. M. Jutzi, E. Asphaug, Forming the lunar farside highlands by accretion of a companion Moon. Nature 476, 69–72 (2011)
162. R.R. Keays, R. Ganapathy, J.C. Laul, E. Anders, G.F. Herzog, P.M. Jeffery, Trace elements and radioactivity in lunar rocks: implications for meteorite infall, solar-wind flux, and formation conditions of Moon. Science 167, 490–493 (1970)
163. J.F. Kerridge, Solar nitrogen: evidence for a secular increase in the ratio of nitrogen-15 to nitrogen-14. Science 188, 162–164 (1975)
164. J.F. Kerridge, Secular variations in composition of the solar wind: evidence and causes, in *Proc. Conf. Ancient Sun* (1980), pp. 475–489
165. J.F. Kerridge, Long-term compositional variation in solar corpuscular radiation: Evidence from nitrogen isotopes in the lunar regolith. Rev. Geophys. 31, 423–437 (1993)
166. J.F. Kerridge, P. Signer, R. Wieler, R.H. Becker, R.O. Pepin, Long-term changes in composition of solar particles implanted in extraterrestrial materials, in *The Sun in Time*, ed. by C.P. Sonett, M.S. Giampapa, M.S. Matthews (University of Arizona Press, Tucson, 1991), pp. 389–412
167. A. Khan, K. Mosegaard, K.L. Rasmussen, A new seismic velocity model for the Moon from a Monte Carlo inversion of the Apollo lunar seismic data. Geophys. Res. Lett. 27, 1591–1594 (2000)
168. A. Khan, J.A.D. Connolly, A. Pommier, J. Noir, Geophysical evidence for melt in the deep lunar interior and implications for lunar evolution. J. Geophys. Res., Planets 119, 2197–2221 (2014)
169. R.M. Killen, D.M. Hurley, W.M. Farrell, The effect on the lunar exosphere of a coronal mass ejection passage. J. Geophys. Res. 117, E00K02 (2012). https://doi.org/10.1029/2011JE00401
170. ADS T. Kleine, H. Palme, K. Mezger, A.N. Halliday, Hf-W chronometry of lunar metals and the age and early differentiation of the Moon.

Science 310, 1671–1674 (2005)

171. L. Kööp, P.R. Heck, H. Busemann, A.M. Davis, J. Greer, C. Maden, M.M.M. Meier, R. Wieler, High early solar activity inferred from helium and neon excesses in the oldest meteorite inclusions. Nat. Astron. 2, 709–713 (2018)
172. R.L. Korotev, Lunar geochemistry as told by lunar meteorites. Chem. Erde Geochem. 65, 297–346 (2005)
173. R.L. Korotev, B.L. Jolliff, R.A. Zeigler, J.J. Gillis, L.A. Haskin, Feldspathic lunar meteorites and their implications for compositional remote sensing of the lunar surface and the composition of the lunar crust. Geochim. Cosmochim. Acta 67, 4895–4923 (2003)
174. T.S. Kruijer, T. Kleine, Tungsten isotopes and the origin of the Moon. Earth Planet. Sci. Lett. 475, 15–24 (2017)
175. T.S. Kruijer, T. Kleine, M. Fischer-Godde, P. Sprung, Lunar tungsten isotopic evidence for the late veneer. Nature 520, 534–537 (2015)
176. D.R. Lammlein, Lunar seismicity, structure, and tectonic. Philos. Trans. R. Soc. Lond. A 285, 451–461 (1977)
177. M. Laneuville, M.A. Wieczorek, D. Breuer, J. Albert, G. Morard, T. Rückriemen, A long-lived lunar dynamo powered by core crystallization. Earth Planet. Sci. Lett. 40, 251–260 (2014)
178. G. Latham, M. Ewing, J. Dorman, D. Lammlein, F. Press, N. Toksoz, G. Sutton, F. Duennebier, Y. Nakamura, Moonquakes. Science 174, 687–692 (1971)
179. M. Le Feuvre, M.A. Wieczorek, Nonuniform cratering of the Moon and a revised crater chronology of the inner Solar System. Icarus 214, 1–20 (2011)
180. D.C. Lee, A.N. Halliday, G.A. Snyder, L.A. Taylor, Age and origin of the Moon. Science 278, 1098–1103 (1997)
181. M. Lemelin, P.G. Lucey, E. Song, G.J. Taylor, Lunar central peak mineralogy and iron content using the Kaguya Multiband Imager: reassessment of the compositional structure of the lunar crust. J. Geophys. Res., Planets 120, 869–887 (2015)
182. C. Lepaulard, J. Gattacceca, M. Uehara, P. Rochette, Y. Quesnel, R.J. Macke, S.J.W. Kiefer, A survey of the natural remanent magnetization and magnetic susceptibility of Apollo whole rocks. Phys. Earth Planet. Inter. 290, 36–43 (2019)

183. J. Levine, P.R. Renne, R.A. Muller, Solar and cosmogenic argon in dated lunar impact spherules. Geochim. Cosmochim. Acta 71, 1624–1635 (2007)
184. S. Li, R.E. Milliken, Water on the surface of the Moon as seen by the Moon Mineralogy Mapper: distribution, abundance, and origins. Sci. Adv. 3, e1701471 (2017). https://doi.org/10.1126/sciadv.1701471
185. ADS S. Li, P.G. Lucey, R.E. Milliken, P.O. Hayne, E. Fisher, J.P. Williams, D.M. Hurley, R.C. Elphic, Direct evidence of surface exposed water ice in the lunar polar regions. Proc. Natl. Acad. Sci. USA 115, 8907–8912 (2018)
186. C. Li, D. Liu, B. Liu, X. Ren, J. Liu, Z. He, W. Zuo, X. Zeng, R. Xu, X. Tan, X. Zhang, W. Chen, R. Shu, W. Wen, Y. Su, H. Zhang, Z. Ouyang, Chang'E-4 initial spectroscopic identification of lunar far-side mantle-derived materials. Nature 569, 378–382 (2019)
187. Y. Liu, C. Floss, J.M.D. Day, E. Hill, L.A. Taylor, Petrogenesis of lunar mare basalt meteorite Miller Range 05035. Meteorit. Planet. Sci. 44, 261–284 (2009)
188. Y. Liu, Y. Guan, Y. Zhang, G.R. Rossman, J.M. Eiler, L.A. Taylor, Direct measurement of hydroxyl in the lunar regolith and the origin of lunar surface water. Nat. Geosci. 5, 779–782 (2012)
189. J. Liu, M. Sharp, R.D. Ash, D.A. Kring, R.J. Walker, Diverse impactors in Apollo 15 and 16 impact rocks: evidence from osmium isotopes and highly siderophile elements. Geochim. Cosmochim. Acta 155, 122–153 (2015)
190. J. Longhi, Experimental petrology and petrogenesis of mare volcanics. Geochim. Cosmochim. Acta 56, 2235–2251 (1992)
191. D.E. Loper, C.L. Werner, On lunar asymmetries 1. Tilted convection and crustal asymmetry. J. Geophys. Res. 107, 5046 (2002)
192. G.W. Lugmair, R.W. Carlson, The Sm-Nd history of KREEP, in *Lunar Planet. Sci. Conf. Proc.* (1978), pp. 689–704
193. J.R. Lyons, E. Gharib-Nezhad, T.R. Ayres, A light carbon isotope composition for the Sun. Nat. Commun. 9, 908 (2018). https://doi.org/10.1038/s41467-018-03093-3
194. ADS S. Marchi, S. Mottola, G. Cremonese, M. Massironi, E. Martellato, A new chronology for the Moon and Mercury. Astron. J. 137, 4936–4948 (2009)
195. S. Marchi, W.F. Bottke, D.A. Kring, A. Morbidelli, The onset of the

lunar cataclysm as recorded in its ancient crater populations. Earth Planet. Sci. Lett. 325, 27–38 (2012)

196. S. Marchi, W.F. Bottke, B.A. Cohen, K. Wünnemann, D.A. Kring, H.Y. McSween, M.C. De Sanctis, D.P. O'Brien, P. Schenk, C.A. Raymond, C.T. Russell, High-velocity collisions from the lunar cataclysm recorded in asteroidal meteorites. Nat. Geosci. 6, 303–307 (2013)
197. B. Marty, L. Zimmermann, P.G. Burnard, R. Wieler, V.S. Heber, D.L. Burnett, R.C. Wiens, P. Bochsler, Nitrogen isotopes in the recent solar wind from the analysis of Genesis targets: evidence for large scale isotope heterogeneity in the early Solar System. Geochim. Cosmochim. Acta 74, 340–355 (2010)
198. B. Marty, M. Chaussidon, R.C. Wiens, A.J.G. Jurewicz, D.S. Burnett, A ^{15}N-poor isotopic composition for the Solar System as shown by Genesis solar wind samples. Science 332, 1533–1536 (2011)
199. T.B. McCord, J.B. Adams, T.V. Johnson, Asteroid Vesta: spectral reflectivity and compositional implications. Science 168, 1445–1447 (1970)
200. T.J. McCoy, D.W. Middlefehldt, L. Wilson, Asteroid differentiation, in *Meteorites and the Early Solar System II*, ed. by D.S. Lauretta, H.Y. McSween Jr. (University of Arizona Press, Tucson, 2006), pp. 733–745
201. F.M. McCubbin, K.E. Vander Kaaden, R. Tartèse, R.L. Klima, Y. Liu, J. Mortimer, J.J. Barnes, C.K. Shearer, A.H. Treiman, D.J. Lawrence, S.M. Elardo, D.M. Hurley, J.W. Boyce, M. Anand, Magmatic volatiles (H, C, N, F, S, Cl) in the lunar mantle, crust, and regolith: abundances, distributions, processes, and reservoirs. Am. Mineral. 100, 1668–1707 (2015)
202. D.S. McKay, G.A. Heiken, G.M. Taylor, U.S. Clanton, D.A. Morrison, G.H. Ladle, Apollo 14 soils: size distribution and particle types, in *Lunar Sci. Conf. Proc.* (1972), pp. 983–994
203. D.S. McKay, D.D. Bogard, R.V. Morris, R.L. Korotev, P. Johnson, S.J. Wentworth, Apollo 16 regolith breccias: characterization and evidence for early formation in the mega-regolith. J. Geophys. Res. 91, D277–D303 (1986)
204. D.S. McKay, G. Heiken, A. Basu, G. Blanford, S. Simon, R. Reedy, B.M. French, J.J. Papike, The lunar regolith, in *Lunar Sourcebook*, ed. by G. Heiken, D. Vaniman, B.M. French (Cambridge University Press, Cambridge, 1991), pp. 286–356

205. K.D. McKeegan, M. Chaussidon, F. Robert, Incorporation of short-lived ^{10}Be in a calcium–aluminum-rich inclusion from the Allende meteorite. Science 289, 1334–1337 (2000)

206. K.D. McKeegan, A.P.A. Kallio, V.S. Heber, G. Jarzebinski, P.H. Mao, C.D. Coath, T. Kunihiro, R.C. Wiens, J.E. Nordholt, R.W. Moses Jr., D.B. Reisenfeld, A.J.G. Jurewicz, D.S. Burnett, The oxygen isotopic composition of the Sun inferred from captured solar wind. Science 332, 1528–1532 (2011)

207. C.L. McLeod, A.D. Brandon, R.M.G. Armytage, Constraints on the formation age and evolution of the Moon from ^{142}Nd-^{143}Nd systematics of Apollo 12 basalts. Earth Planet. Sci. Lett. 396, 179–189 (2014)

208. H.Y. McSween Jr., R.P. Binzel, M.C. De Sanctis, E. Ammannito, T.H. Prettyman, A.W. Beck, V. Reddy, L. Le Corre, M. Gaffey, T.B. McCord, C.A. Raymond, C.T. Russell, Dawn, the Vesta–HED connection, and the geologic context for eucrites, diogenites, and howardites. Meteorit. Planet. Sci. 48, 2090–2104 (2013)

209. H.J. Melosh, A.M. Freed, B.C. Johnson, D.M. Blair, J.C. Andrews-Hanna, G.A. Neumann, R.J. Phillips, D.E. Smith, S.C. Solomon, M.A. Wieczorek, M.T. Zuber, The origin of lunar Mascon basins. Science 340, 1552–1555 (2013)

210. H.J. Melosh, J. Kendall, B. Horgan, B.C. Johnson, T. Bowling, P.G. Lucey, G.J. Taylor, South pole-Aitken basin ejecta reveal the Moon's upper mantle. Geology 45, 1063–1066 (2017)

211. C.M. Mercer, K.E. Young, J.R. Weirich, K.V. Hodges, B.L. Jolliff, J.A. Wartho, M.C. van Soest, Refining lunar impact chronology through high spatial resolution $^{40}Ar/^{39}Ar$ dating of impact melts. Sci. Adv. 1, e1400050 (2015). https://doi.org/10.1126/sciadv.1400050

212. ADS A. Meshik, J. Mabry, C. Hohenberg, Y. Marrocchi, O. Pravdivtseva, D. Burnett, C. Olinger, R. Wiens, D. Reisenfeld, J. Allton, K. McNamara, E. Stansbery, A.J.G. Jurewicz, Constraints on neon and argon isotopic fractionation in solar wind. Science 318, 433–435 (2007)

213. A. Meshik, C. Hohenberg, O. Pravdivtseva, D. Burnett, Heavy noble gases in solar wind delivered by Genesis mission. Geochim. Cosmochim. Acta 127, 326–347 (2014)

214. A.E. Metzger, E.L. Haines, R.E. Parker, R.G. Radocinski, Thorium concentrations in the lunar surface. I—Regional values and crustal

content, in *Lunar Planet. Sci. Conf. Proc.* (1977), pp. 949–999

215. K. Misawa, M. Tatsumoto, G.B. Dalrymple, K. Yanai, An extremely low U/Pb source in the Moon: U-Th-Pb, Sm-Nd, Rb-Sr, and $^{40}Ar/^{39}Ar$ isotopic systematics and age of lunar meteorite Asuka 881757. Geochim. Cosmochim. Acta 57, 4687–4702 (1993)
216. I.G. Mitrofanov, A.B. Sanin, W.V. Boynton, G. Chin, J.B. Garvin, D. Golovin, L.G. Evans, K. Harshman, A.S. Kozyrev, M.L. Litvak, A. Malakhov, E. Mazarico, T. McClanahan, G. Milikh, M. Mokrousov, G. Nandikotkur, G.A. Neumann, I. Nuzhdin, R. Sagdeev, V. Shevchenko, V. Shvetsov, D.E. Smith, R. Starr, V.I. Tretyakov, J. Trombka, D. Usikov, A. Varenikov, A. Vostrukhin, M.T. Zuber, Hydrogen mapping of the Lunar South Pole using the LRO neutron detector experiment LEND. Science 330, 483–486 (2010)
217. A. Morbidelli, S. Marchi, W.F. Bottke, D.A. Kring, A sawtooth-like timeline for the first billion years of lunar bombardment. Earth Planet. Sci. Lett. 355–356, 144–151 (2012)
218. A. Morbidelli, D. Nesvorny, V. Laurenz, S. Marchi, D.C. Rubie, L. Elkins-Tanton, M. Wieczorek, S. Jacobson, The timeline of the lunar bombardment: revisited. Icarus 305, 262–276 (2018)
219. J.W. Morgan, U. Krähenbühl, R. Ganapathy, E. Anders, Trace elements in Apollo 15 samples: implications for meteorite influx and volatile depletion on the Moon, in *Lunar Sci. Conf. Proc.* (1972), pp. 1361–1376
220. R.V. Morris, Surface exposure indices of lunar soils—A comparative FMR study, in *Lunar Planet. Sci. Conf. Proc.* (1976), pp. 315–335
221. R.V. Morris, The surface exposure (maturity) of lunar soils—Some concepts and I_s/FeO compilation, in *Lunar Planet. Sci. Conf. Proc.* (1978), pp. 2287–2297
222. R.V. Morris, Origins and size distribution of metallic iron particles in the lunar regolith, in *Lunar Planet. Sci. Conf. Proc.* (1980), pp. 1697–1712
223. J. Mortimer, A.B. Verchovsky, M. Anand, Predominantly non-solar origin of nitrogen in lunar soils. Geochim. Cosmochim. Acta 193, 36–53 (2016)
224. V.R. Murthy, M.R. Coscio Jr., Rb-Sr ages and isotopic systematics of some Serenitatis mare basalts, in *Lunar Planet. Sci. Conf. Proc.*, ed. by R.B. Merrill (1976), pp. 1529–1544

225. Y. Nakamura, Shallow moonquakes: how they compare with earthquakes, in *Lunar Planet. Sci. Conf. Proc.* (1980), pp. 1847–1853
226. Y. Nakamura, G.V. Latham, H.J. Dorman, Apollo lunar seismic experiment final summary, in *Lunar Planet. Sci. Conf. Proc.* (1982), pp. A117–A123
227. Y. Nakamura, G.V. Latham, H.J. Dorman, J.E. Harris, Passive Seismic Experiment, Long Period Event Catalog, Final Version (1969 Day 202–1977 Day 273, ALSEP Stations 11, 12, 13, 14, 15, and 16). Univ. Texas Institute Geophys. Tech. Report No. 18 (originally published 19 June 1981, revised 2 October 2008) (2008)
228. National Research Council, The Scientific Context for the Exploration of the Moon (The National Academies Press, Washington, DC, 2007). https://www.nap.edu/catalog/11954/the-scientific-context-for-exploration-of-the-Moon. Accessed 12 June 2019
229. C.R. Neal, Establishing a lunar geophysical network for exploration and Solar System science, in *Annual Meeting Lunar Exploration Analysis Group* (2011), abstract 2052
230. C.R. Neal, G.Y. Kramer, The petrogenesis of the Apollo 14 high-Al mare basalts. Am. Mineral. 91, 1521–1535 (2006)
231. C.R. Neal, L.A. Taylor, Petrogenesis of mare basalts: a record of lunar volcanism. Geochim. Cosmochim. Acta 56, 2177–2212 (1992)
232. C.R. Neal, M.D. Hacker, G.A. Snyder, L.A. Taylor, Y.G. Liu, R.A. Schmitt, Basalt generation at the Apollo 12 site, part 1: new data, classification, and re-evaluation. Meteoritics 29, 334–348 (1994)
233. C.R. Neal, W.B. Banerdt, C. Beghein, P. Chi, D. Currie, S. Del'Agnello, I. Garrick-Bethell, R. Grimm, M. Grott, H. Haviland, S. Kedar, S. Nagihara, M. Panning, N. Petro, N. Schmerr, M. Siegler, R. Weber, M. Wieczorek, K. Zacny, The lunar geophysical network mission, in *Lunar Planet. Sci. Conf.* (2019), abstract 2455
234. A. Nemchin, N. Timms, R. Pidgeon, T. Geisler, S. Reddy, C. Meyer, Timing of crystallization of the lunar magma ocean constrained by the oldest zircon. Nat. Geosci. 2, 133–136 (2009)
235. G. Neukum, B. Ivanov, W.K. Hartmann, Cratering records in the inner Solar System in relation to the lunar reference system. Space Sci. Rev. 96, 55–86 (2001)
236. G.A. Neumann, M.T. Zuber, M.A. Wieczorek, J.W. Head, D.M.H. Baker, S.C. Solomon, D.E. Smith, F.G. Lemoine, E. Mazarico, T.J.

Sabaka, S.J. Goossens, H.J. Melosh, R.J. Phillips, S.W. Asmar, A.S. Konopliv, J.G. Williams, M.M. Sori, J.M. Soderblom, K. Miljković, J.C. Andrews-Hanna, F.W.S. Nimmo, W. Kiefer, Lunar impact basins revealed by gravity recovery and interior laboratory measurements. Sci. Adv. 1, e1500852 (2015). https://doi.org/10.1126/sciadv.1500852

237. ADS M.D. Norman, The lunar cataclysm: reality or "mythconception"? Elements 5, 23–28 (2009)

238. M.D. Norman, R.A. Duncan, J.J. Huard, Identifying impact events within the lunar cataclysm from ^{40}Ar-^{39}Ar ages and compositions of Apollo 16 impact melt rocks. Geochim. Cosmochim. Acta 70, 6032–6049 (2006)

239. S. Nozette, C.L. Lichtenberg, P.D. Spudis, R. Bonner, W. Ort, E. Malaret, M. Robinson, E.M. Shoemaker, The Clementine bistatic radar experiment. Science 274, 1495–1498 (1996)

240. S. Nozette, E.M. Shoemaker, P.D. Spudis, C.L. Lichtenberg, The possibility of ice on the Moon. Science 278, 144–145 (1997)

241. L.E. Nyquist, C.-Y. Shih, The isotopic record of lunar volcanism. Geochim. Cosmochim. Acta 56, 2213–2234 (1992)

242. L.E. Nyquist, B.M. Bansal, H. Wiesmann, Rb-Sr ages and initial ^{87}Sr/^{86}Sr for Apollo 17 basalts and KREEP basalt 15386, in *Lunar Planet. Sci. Conf. Proc.* (1975), pp. 1445–1465

243. L.E. Nyquist, B.M. Bansal, H. Wiesmann, Sr-isotopic constraints on the petrogenesis of Apollo 17 mare basalts, in *Lunar Planet. Sci. Conf. Proc.*, ed. by R.B. Merrill (1976), pp. 1507–1528

244. L.E. Nyquist, H. Wiesmann, B. Bansal, C.Y. Shih, J.E. Keith, C.L. Harper, ^{146}Sm-^{142}Nd formation interval for the lunar mantle. Geochim. Cosmochim. Acta 59, 2817–2837 (1995)

245. L.E. Nyquist, C.Y. Shih, D. Reese, J. Park, D.D. Bogard, D.H. Garrison, A. Yamaguchi, Lunar crystal history recorded in lunar anorthosites, in *Lunar Planet. Sci. Conf.* (2010), p. abstract 1383

246. K.W. Ogilvie, M.A. Coplan, Solar wind composition. Rev. Geophys. 33, 615–622 (1995)

247. M. Ohtake, T. Matsunaga, J. Haruyama, Y. Yokota, T. Morota, C. Honda, Y. Ogawa, M. Torii, H. Miyamoto, T. Arai, N. Hirata, A. Iwasaki, R. Nakamura, T. Hiroi, T. Sugihara, H. Takeda, H. Otake, C.M. Pieters, K. Saiki, K. Kitazato, M. Abe, N. Asada, H. Demura, Y. Yamaguchi, S. Sasaki, S. Kodama, J. Terazono, M. Shirao, A. Yamaji, S. Minami, H.

Akiyama, J.L. Josset, The global distribution of pure anorthosite on the Moon. Nature 461, 236–240 (2009)

248. M. Ohtake, H. Takeda, T. Matsunaga, Y. Yokota, J. Haruyama, T. Morota, S. Yamamoto, Y. Ogawa, T. Hiroi, Y. Karouji, K. Saiki, P.G. Lucey, Asymmetric crustal growth on the Moon indicated by primitive farside high materials. Nat. Geosci. 5, 384–388 (2012)
249. J.B. Paces, C.R. Neal, L.A. Taylor, S. Nakai, A.N. Halliday, D.C. Lee, A strontium and neodymium isotopic study of Apollo 17 high-Ti mare basalts: resolution of ages, evolution of magmas, and origins of source heterogeneities. Geochim. Cosmochim. Acta 55, 2025–2043 (1991)
250. D.A. Paige, M.C. Foote, B.T. Greenhagen, J.T. Schofield, S. Calcutt, A.R. Vasavada, D.J. Preston, F.W. Taylor, C.C. Allen, K.J. Snook, B.M. Jakosky, B.C. Murray, L.A. Soderblom, B. Jau, S. Loring, J. Bulharowski, N.E. Bowles, I.R. Thomas, M.T. Sullivan, C. Avis, E.M. De Jong, W. Hartford, D.J. McCleese, The Lunar Reconnaissance Orbiter Diviner lunar radiometer experiment. Space Sci. Rev. 150, 125–160 (2010a)
251. D.A. Paige, M.A. Siegler, J.A. Zhang, P.O. Hayne, E.J. Foote, K.A. Bennett, A.R. Vasavada, B.T. Greenhagen, J.T. Schofield, D.J. McCleese, M.C. Foote, E. DeJong, B.G. Bills, W. Hartford, B.C. Murray, C.C. Allen, K. Snook, L.A. Soderblom, S. Calcutt, F.W. Taylor, N.E. Bowles, J.L. Bandfield, R. Elphic, R. Ghent, T.D. Glotch, M.B. Wyatt, P.G. Lucey, Diviner Lunar radiometer observations of cold traps in the Moon's South Polar Region. Science 330, 479–482 (2010b)
252. D.A. Papanastassiou, G.J. Wasserburg, Rb-Sr ages of igneous rocks from the Apollo 14 mission and the age of the Fra Mauro formation. Earth Planet. Sci. Lett. 12, 36–48 (1971a)
253. D.A. Papanastassiou, G.J. Wasserburg, Lunar chronology and evolution from Rb-Sr studies of Apollo 11 and 12 samples. Earth Planet. Sci. Lett. 11, 37–62 (1971b)
254. D.A. Papanastassiou, G.J. Wasserburg, Rb-Sr ages and initial strontium in basalts from Apollo 15. Earth Planet. Sci. Lett. 17, 324–337 (1973)
255. J.J. Papike, F.N. Hodges, A.E. Bence, M. Cameron, J.M. Rhodes, Mare basalts: crystal chemistry, mineralogy, and petrology. Rev. Geophys. 14, 475–540 (1976)
256. J.J. Papike, G.W. Fowler, C.K. Shearer, G.D. Layne, Ion microprobe investigation of plagioclase and orthopyroxene from lunar Mg-suite

norites: implications for calculating parental melt REE concentrations and for assessing postcrystallization REE redistribution. Geochim. Cosmochim. Acta 60, 3967–3978 (1996)

257. R.O. Pepin, R.H. Becker, P.E. Rider, Xenon and Krypton in extraterrestrial regolith soil and in the solar wind. Geochim. Cosmochim. Acta 59, 4997–5022 (1995)

258. R.O. Pepin, D.J. Schlutter, R.H. Becker, D.B. Reisenfeld, Helium, neon, and argon composition of the solar wind as recorded in gold and other Genesis collector materials. Geochim. Cosmochim. Acta 89, 62–80 (2012)

259. J.F. Pernet-Fisher, E. Deloule, K.H. Joy, Evidence of chemical heterogeneity within lunar anorthosite parental magmas. Geochim. Cosmochim. Acta (2019a). https://doi.org/10.1016/j.gca.2019.03.033

260. J.F. Pernet-Fisher, F.E. McDonald, R.A. Zeigler, K.H. Joy, 50 years on: legacies of the Apollo programme. Astron. Geophys. 60, 4.22–4.28 (2019b)

261. C.M. Pieters, J. Boardman, B. Buratti, A. Chatterjee, R. Clark, T. Glavich, R. Green, J. Head III, P. Isaacson, E. Malaret, T. McCord, J. Mustard, N. Petro, C. Runyon, M. Staid, J. Sunshine, L. Taylor, S. Tompkins, P. Varanasi, M. White, The Moon Mineralogy Mapper (M3) on Chandrayaan-1. Current. Science 96, 500–505 (2009a)

262. C.M. Pieters, J.N. Goswami, R.N. Clark, M. Annadurai, J. Boardman, B. Buratti, J.P. Combe, M.D. Dyar, R. Green, J.W. Head, C. Hibbitts, M. Hicks, P. Isaacson, R. Klima, G. Kramer, S. Kumar, E. Livo, S. Lundeen, E. Malaret, T. McCord, J. Mustard, J. Nettles, N. Petro, C. Runyon, M. Staid, J. Sunshine, L.A. Taylor, S. Tompkins, P. Varanasi, Character and spatial distribution of OH/H_2O on the surface of the Moon seen by M^3 on Chandrayaan-1. Science 326, 568–572 (2009b)

263. C.M. Pieters, S. Besse, J. Boardman, B. Buratti, L. Cheek, R.N. Clark, J.P. Combe, D. Dhingra, J.N. Goswami, R.O. Green, J.W. Head, P. Isaacson, R. Klima, G. Kramer, S. Lundeen, E. Malaret, T. McCord, J. Mustard, J. Nettles, N. Petro, C. Runyon, M. Staid, J. Sunshine, L.A. Taylor, K. Thaisen, S. Tompkins, J. Whitten, Mg-spinel lithology: a new rock type on the lunar farside. J. Geophys. Res., Planets 116, E00G08 (2011). https://doi.org/10.1029/2010JE003727

264. T.H. Prettyman, J.J. Hagerty, R.C. Elphic, W.C. Feldman, D.J. Lawrence, G.W. McKinney, D.T. Vaniman, Elemental composition of the lunar surface: analysis of gamma ray spectroscopy data from lunar

prospector. J. Geophys. Res., Planets 111, E12007 (2006)

265. I.S. Puchtel, R.J. Walker, O.B. James, D.A. Kring, Osmium isotope and highly siderophile element systematic of lunar impact melt breccias: implications for the late accretion history of the Moon and Earth. Geochim. Cosmochim. Acta 72, 3022–3042 (2008)
266. Y.Q. Qian, L. Xiao, S.Y. Zhao, J.N. Zhao, J. Huang, J. Flahaut, M. Martinot, J.W. Head, H. Hiesinger, G.X. Wang, Geology and scientific significance of the Rümker region in Northern Oceanus Procellarum: China's Chang'E-5 landing region. J. Geophys. Res., Planets 123, 1407–1430 (2018)
267. K. Rankenburg, A.D. Brandon, C.R. Neal, Neodymium isotope evidence for a chondritic composition of the Moon. Science 312, 1369–1372 (2006)
268. J.F. Rapp, D.S. Draper, Fractional crystallization of the lunar magma ocean: updating the dominant paradigm. Meteorit. Planet. Sci. 53, 1432–1455 (2018)
269. J.M. Rhodes, N.J. Hubbard, H. Wiesmann, K.V. Rodgers, J.C. Brannon, B.M. Bansal, Chemistry, classification and petrogenesis of Apollo 17 mare basalts, in *Lunar Planet. Sci. Conf.*, ed. by D.C. Kinsler (1976), pp. 1467–1489
270. J.M. Rhodes, D.P. Blanchard, M.A. Dungan, J.C. Brannon, K.V. Rodgers, Chemistry of Apollo 12 mare basalts: magma types and fractionation processes, in *Lunar Planet. Sci. Conf.*, ed. by R.B. Merril (1977), pp. 1305–1338
271. K. Righter, M.J. Drake, A magma ocean on Vesta: core formation and petrogenesis of eucrites and diogenites. Meteorit. Planet. Sci. 32, 929–944 (1997)
272. K. Righter, C.K. Shearer, Magmatic fractionation of Hf and W: constraints on the timing of core formation and differentiation in the Moon and Mars. Geochim. Cosmochim. Acta 67, 2497–2507 (2003)
273. K. Righter, M.J. Drake, G. Yaxley, Prediction of siderophile element metal-silicate partition coefficients to 20 GPa and 2800°C: the effects of pressure, temperature, oxygen fugacity, and silicate and metallic melt compositions. Phys. Earth Planet. Inter. 100, 115–134 (1997)
274. S.J. Robbins, New crater calibrations for the lunar crater-age chronology. Earth Planet. Sci. Lett. 403, 188–198 (2014)
275. M.S. Robinson, J.W. Ashley, A.K. Boyd, R.V. Wagner, E.J. Speyerer,

B. Ray Hawke, H. Hiesinger, C.H. van der Bogert, Confirmation of sublunarean voids and thin layering in mare deposits. Planet. Space Sci. 69, 18–27 (2012a)

276. K.L. Robinson, A.H. Treiman, K.H. Joy, Basaltic fragments in lunar feldspathic meteorites: connecting sample analyses to orbital remote sensing. Meteorit. Planet. Sci. 47, 387–399 (2012b)

277. S.S. Russell, K.H. Joy, T.E. Jeffries, G.J. Consolmagno, A.T. Kearsley, Heterogeneity in lunar anorthosite meteorites: implications for the lunar magma ocean model. Philos. Trans. R. Soc. A 372, 20130241 (2014). https://doi.org/10.1098/rsta.2013.0241

278. ADS M.J. Rutherford, J.W. Head III, A.E. Saal, E.H. Hauri, L. Wilson, Model for the origin, ascent and eruption of lunar picritic magmas. Am. Mineral. 102, 2045–2053 (2017)

279. G. Ryder, Lunar samples, lunar accretion, and the early bombardment history of the Moon. Eos 71, 322–323 (1990)

280. A.E. Saal, E.H. Hauri, M.L. Cascio, J.A. Van Orman, M.C. Rutherford, R.F. Cooper, Volatile content of lunar volcanic glasses and the presence of water in the Moon's interior. Nature 454, 192–195 (2008)

281. P. Saxena, R.M. Killen, V. Airapetian, N.E. Petro, N.M. Curran, A.M. Mandell, Was the Sun a Slow Rotator? Sodium and Potassium Constraints from the Lunar Regolith. Astrophys. J. Lett. 876, 1 (2019). https://doi.org/10.3847/2041-8213/ab18fb

282. ADS N. Schorghofer, G.J. Taylor, Subsurface migration of H_2O at lunar cold traps. J. Geophys. Res., Planets 112, E02010 (2007). https://doi.org/10.1029/2006JE002779

283. ADS P.H. Schultz, P.D. Spudis, Evidence for ancient mare volcanism, in *Lunar Planet. Sci. Conf.*, ed. by N.W. Hinners (1979), pp. 2899–2918

284. P.H. Schultz, P.D. Spudis, Beginning and end of lunar mare volcanism. Nature 302, 233–236 (1983)

285. P.H. Schultz, B. Hermalyn, A. Colaprete, K. Ennico, M. Shirley, W.S. Marshall, The LCROSS cratering experiment. Science 330, 468–472 (2010)

286. M. Sharp, I. Gerasimenko, L.C. Loudin, J. Liu, O.B. James, I.S. Puchtel, R.J. Walker, Characterization of the dominant impactor signature for Apollo 17 impact melt rocks. Geochim. Cosmochim. Acta 131, 62–80 (2014)

287. E.K. Shea, B.P. Weiss, W.S. Cassata, D.L. Shuster, S.M. Tikoo, J.

Gattacceca, T.L. Grove, M.D. Fuller, A long-lived lunar core dynamo. Science 335, 453–456 (2012)

288. C.K. Shearer, P.C. Hess, M.A. Wieczorek, M.E. Pritchard, E.M. Parmentier, L.E. Borg, J. Longhi, L.T. Elkins-Tanton, C.R. Neal, I. Antonenko, R.M. Canup, A.N. Halliday, T.L. Grove, B.H. Hager, D.C. Lee, U. Wiechert, Thermal and magmatic evolution of the Moon. Rev. Mineral. Geochem. 60, 365–518 (2006)
289. J.W. Shervais, J.J. McGee, Ion and electron microprobe study of troctolites, norite, and anorthosites from Apollo 14: evidence for urKREEP assimilation during petrogenesis of Apollo 14 Mg-suite rocks. Geochim. Cosmochim. Acta 62, 3009–3023 (1998)
290. J.W. Shervais, J.J. McGee, KREEP cumulates in the western lunar highlands: ion and electron microprobe study of alkali-suite anorthosites and norites from Apollo 12 and 14. Am. Mineral. 84, 806–820 (1999)
291. E.M. Shoemaker, R.J. Hackman, Stratigraphic basis for a lunar time scale, in *The Moon*, ed. by Z. Kopal, Z.K. Mikhailov (Academic, San Diego, 1962), pp. 289–300
292. A. Smith, I.A. Crawford, R.A. Gowen, R. Ambrosi, M. Anand, B. Banerdt, N. Bannister, N. Bowles, C. Braithwaite, P. Brown, J. Chela-Flores, T. Cholinser, P. Church, A.J. Coates, T. Colaprete, G. Collins, G. Collinson, T. Cook, R. Elphic, G. Fraser, Y. Gao, E. Gibson, T. Glotch, M. Grande, A. Griffiths, J. Grygorczuk, M. Gudipati, A. Hagermann, J. Heldmann, L.L. Hood, A.P. Jones, K.H. Joy, O.B. Khavroshkin, G. Klingelhoefer, M. Knapmeyer, G. Kramer, D. Lawrence, W. Marczewski, S. McKenna-Lawlor, K. Miljkovic, S. Narendranath, E. Palomba, A. Phipps, W.T. Pike, D. Pullan, J. Rask, D.T. Richard, K. Seweryn, S. Sheridan, M. Sims, M. Sweeting, T. Swindle, D. Talboys, L. Taylor, N. Teanby, V. Tong, S. Ulamec, R. Wawrzaszek, M. Wieczorek, L. Wilson, I. Wright, Lunar Net—a proposal in response to an ESA M3 call in 2010 for a medium sized mission. Exp. Astron. 33, 587–644 (2012)
293. J.F. Snape, A.A. Nemchin, J.J. Bellucci, M.J. Whitehouse, R. Tartèse, J.J. Barnes, M. Anand, I.A. Crawford, K.H. Joy, Lunar basalt chronology, mantle differentiation and implications for determining the age of the Moon. Earth Planet. Sci. Lett. 451, 149–158 (2016a)
294. J.F. Snape, A.A. Nemchin, M.L. Grange, J.J. Bellucci, F. Thiessen, M.J. Whitehouse, Phosphate ages in Apollo 14 breccias: resolving multiple impact events with high precision U–Pb SIMS analyses. Geochim.

Cosmochim. Acta 174, 13–29 (2016b)

295. J.F. Snape, N.M. Curran, M.J. Whitehouse, A.A. Nemchin, K.H. Joy, T. Hopkinson, M. Anand, J.J. Bellucci, G.G. Kenny, Ancient volcanism on the Moon: insights from Pb isotopes in the MIL 13317 and Kalahari 009 lunar meteorites. Earth Planet. Sci. Lett. 502, 84–95 (2018a)
296. J.F. Snape, B. Davids, A.A. Nemchin, M.J. Whitehouse, J.J. Bellucci, Constraining the timing and sources of volcanism at the Apollo 12 landing site using new Pb isotopic compositions and crystallisation ages. Chem. Geol. 482, 101–112 (2018b)
297. J.F. Snape, A.A. Nemchin, M.J. Whitehouse, R.E. Merle, T. Hopkinson, M. Anand, The timing of basaltic volcanism at the Apollo landing sites. Geochim. Cosmochim. Acta (2019). https://doi.org/10.1016/j.gca.2019.07.042
298. G.A. Snyder, L.A. Taylor, C.R. Neal, A chemical model for generating the sources of mare basalts—combined equilibrium and fractional crystallization of the lunar magmasphere. Geochim. Cosmochim. Acta 56, 3809–3823 (1992)
299. G.A. Snyder, D.C. Lee, L.A. Taylor, A.N. Halliday, E.A. Jerde, Evolution of the upper mantle of the Earth's Moon: neodymium and strontium isotopic constraints from high-Ti mare basalts. Geochim. Cosmochim. Acta 58, 4795–4808 (1994)
300. G.A. Snyder, C.R. Neal, L.A. Taylor, A.N. Halliday, Anatexis of lunar cumulate mantle in time and space: clues from trace-element, strontium, and neodymium isotopic chemistry of parental Apollo 12 basalts. Geochim. Cosmochim. Acta 61, 2731–2747 (1997)
301. A.K. Sokol, V.A. Fernandes, T. Schulz, A. Bischoff, R. Burgess, R.N. Clayton, C. Münker, K. Nishiizumi, H. Palme, L. Schultz, G. Weckwerth, K. Mezger, M. Horstmann, Geochemistry, petrology and ages of the lunar meteorites Kalahari 008 and 009: new constraints on early lunar evolution. Geochim. Cosmochim. Acta 72, 4845–4873 (2008)
302. S.C. Solomon, J.W. Head, Lunar Mascon Basins: lava filling, tectonics, and evolution of the lithosphere. Rev. Geophys. 18, 107–141 (1980)
303. C.P. Sonett, D.S. Colburn, R.G. Currie, J.D. Mihalov, The geomagnetic tail topology, reconnection and interaction with the Moon, in *Physics of the Magnetosphere*, ed. by R.L. Carovillano, J.F. McClay, H.F. Radoski (Reidel, Dordrecht, 1967), pp. 461–484

304. P.A. Sossi, F. Moynier, M. Chaussidon, J. Villeneuve, C. Kato, M. Gounelle, Early Solar System irradiation quantified by linked vanadium and beryllium isotope variations in meteorites. Nat. Astron. 1, 0055 (2017). https://doi.org/10.1038/s41550-017-0055

305. ADS T. Spohn, W. Konrad, D. Breuer, R. Ziethe, The longevity of lunar volcanism: implications of thermal evolution calculations with 2D and 3D mantle convection models. Icarus 149, 54–65 (2001)

306. P.D. Spudis, *The Once and Future Moon* (Smithsonian Inst. Press, Washington, 1996)

307. P.D. Spudis, D.E. Wilhelms, M.S. Robinson, The sculptured hills of the taurus highlands: implications for the relative age of serenitatis, basin chronologies and the cratering history of the Moon. J. Geophys. Res., Planets 116, E00H03 (2011). https://doi.org/10.1029/2011JE003903

308. ADS P.D. Spudis, D.B.J. Bussey, S.M. Baloga, J.T.S. Cahill, L.S. Glaze, G.W. Patterson, R.K. Raney, T.W. Thompson, B.J. Thomson, E.A. Ustinov, Evidence for water ice on the Moon: results for anomalous polar craters from the LRO mini-RF imaging radar. J. Geophys. Res., Planets 118, 2016–2029 (2013)

309. A.C. Stadermann, M.R. Zanetti, B.L. Jolliff, H. Hiesinger, C.H. van der Bogert, C.W. Hamilton, The age of lunar mare basalts south of the Aristarchus Plateau and effects of secondary craters formed by the Aristarchus event. Icarus 309, 45–60 (2018)

310. A. Stettler, P. Eberhardt, J. Geiss, N. Grögler, P. Maurer, Ar^{39}-Ar^{40} ages and Ar^{37}-Ar^{38} exposure ages of lunar rocks, in *Lunar Planet. Sci. Conf.* (1973), pp. 1865–1888

311. D. Stöffler, G. Ryder, Stratigraphy and isotope ages of lunar geologic units: chronological standard for the inner Solar System. Space Sci. Rev. 96, 9–54 (2001)

312. D. Stöffler, G. Ryder, B.A. Ivanov, N.A. Artemieva, M.J. Cintala, R.A.F. Grieve, Cratering history and lunar chronology. Rev. Mineral. Geochem. 60, 519–596 (2006)

313. D.E. Stuart-Alexander, D.E. Wilhelms, The Nectarian system, a new lunar time-stratigraphic unit. J. Res. U.S. Geol. Surv. 3, 53–58 (1975)

314. C. Suavet, B.P. Weiss, W.S. Cassata, D.L. Shuster, J. Gattacceca, L. Chan, I. Garrick-Bethell, J.W. Head, T.L. Grove, M.D. Fuller, Persistence and origin of the lunar core dynamo. Proc. Natl. Acad. Sci. USA 110, 8453–8456 (2013)

315. J.M. Sunshine, T.L. Farnham, L.M. Feaga, O. Groussin, F. Merlin, R.E. Milliken, M.F. A'Hearn, Temporal and spatial variability of Lunar hydration as observed by the Deep Impact spacecraft. Science 326, 565–568 (2009)
316. F. Takahashi, H. Tsunakawa, H. Shimizu, H. Shibuya, M. Matsushima, Reorientation of the early lunar pole. Nat. Geosci. 7, 409–412 (2014)
317. M. Tatsumoto, Age of the Moon: an isotopic study of U-Th-Pb systematics of Apollo 11 lunar samples—II, in *Apollo 11 Lunar Sci. Conf. Proc.* (1970), pp. 1595–1612
318. L.A. Taylor, J.W. Shervais, R.H. Hunter, C.-Y. Shih, B.M. Bansal, J. Wooden, L.E. Nyquist, L.C. Laul, Pre-4.2 AE mare-basalt volcanism in the lunar highlands. Earth Planet. Sci. Lett. 66, 33–47 (1983)
319. S.R. Taylor, C.M. Pieters, G.J. MacPherson, Earth-Moon system, planetary science, and lessons learned. Rev. Mineral. Geochem. 60, 657–704 (2006)
320. D.J. Taylor, K.D. McKeegan, T.M. Harrison, Lu-Hf zircon evidence for rapid lunar differentiation. Earth Planet. Sci. Lett. 279, 157–164 (2009)
321. F. Tera, G.J. Wasserburg, U-Th-Pb systematics on lunar rocks and inferences about lunar evolution and the age of the Moon, in *Lunar Sci. Conf. Proc.* (1974), pp. 1571–1599
322. F. Tera, D.A. Papanastassiou, G.J. Wasserburg, Isotopic evidence for a terminal lunar cataclysm. Earth Planet. Sci. Lett. 22, 1–21 (1974)
323. K. Terada, M. Anand, A.K. Sokol, A. Bischoff, Y. Sano, Cryptomare magmatism 4.35 Gyr ago recorded in lunar meteorite Kalahari 009. Nature 450, 849–853 (2007a)
324. K. Terada, Y. Sasaki, M. Anand, K.H. Joy, Y. Sano, Uranium-lead systematics of phosphates in lunar basaltic regolith breccia, Meteorite Hills 01210. Earth Planet. Sci. Lett. 259, 77–84 (2007b)
325. M.H. Thiemens, R.N. Clayton, Ancient solar wind in lunar microbreccias. Earth Planet. Sci. Lett. 47, 34–42 (1980)
326. M.M. Thiemens, P. Sprung, R.O.C. Fonseca, F.P. Leitzke, C. Münker, Early Moon formation inferred from hafnium-tungsten systematics. Nat. Geosci. (2019). https://doi.org/10.1038/s41561-019-0398-3
327. F. Thiessen, A.A. Nemchin, J.F. Snape, M.J. Whitehouse, J.J. Bellucci, Impact history of the Apollo 17 landing site revealed by U-Pb SIMS ages. Meteorit. Planet. Sci. 52, 584–611 (2017)

328. S.M. Tikoo, B.P. Weiss, J. Buz, E.A. Lima, E.K. Shea, G. Melo, T.L. Grove, Magnetic fidelity of lunar samples and implications for an ancient core dynamo. Earth Planet. Sci. Lett. 337–338, 93–103 (2012)
329. S.M. Tikoo, B.P. Weiss, W.S. Cassata, D.L. Shuster, J. Gattacceca, C. Suavet, F. Nimmo, M.D. Fuller, Decline of the lunar core dynamo. Earth Planet. Sci. Lett. 404, 89–97 (2014)
330. S.M. Tikoo, B.P. Weiss, D.L. Shuster, C. Suavet, H. Wang, T.L. Grove, A two-billion-year history for the lunar dynamo. Sci. Adv. 3, e1700207 (2017). https://doi.org/10.1126/sciadv.1700207
331. ADS G.R. Tilton, J.H. Chen, Lead isotope systematics of three Apollo 17 mare basalts, in *Lunar Planet. Sci. Conf. Proc.* (1979), pp. 259–274
332. M. Touboul, I.S. Puchtel, R.J. Walker, Tungsten isotopic evidence for disproportional late accretion to the Earth and Moon. Nature 520, 530–533 (2015)
333. A.H. Treiman, A.K. Maloy, C.K. Shearer Jr., J. Gross, Magnesian anorthositic granulites in lunar meteorites Allan Hills A81005 and Dhofar 309: geochemistry and global significance. Meteorit. Planet. Sci. 45, 163–180 (2010)
334. G. Turner, Argon-40/Argon-39 dating of lunar rock samples. Science 167, 466–468 (1970a)
335. G. Turner, Argon-40/argon-39 dating of lunar rock samples, in *Lunar Sci. Conf. Proc.* (1970b), pp. 1665–1684
336. G. Turner, A Monte Carlo fragmentation model for the production of meteorites: implications for gas retention ages, in *Lunar Planet. Sci. Conf. Proc.* (1979), pp. 1917–1941
337. G. Turner, P.H. Cadogan, C.J. Yonge, Argon selenochronology, in *Lunar Sci. Conf. Proc.* (1973), pp. 1889–1914
338. N. Vogel, V.S. Heber, H. Baur, D.S. Burnett, R. Wieler, Argon, krypton, and xenon in the bulk solar wind as collected by the Genesis mission. Geochim. Cosmochim. Acta 75, 3057–3071 (2011)
339. R.V. Wagner, D.M. Nelson, J.B. Plescia, M.S. Robinson, E.J. Speyerer, E. Mazarico, Coordinates of anthropogenic features on the Moon. Icarus 283, 92–103 (2017)
340. R.J. Walker, Highly siderophile elements in the Earth, Moon and Mars: update and implications for planetary accretion and differentiation. Chem. Erde Geochem. 69, 101–125 (2009)
341. A. Wallner, J. Feige, N. Kinoshita, M. Paul, L.K. Fifield, R. Golser, M.

Honda, U. Linnemann, H. Matsuzaki, S. Merchel, G. Rugel, S.G. Tims, P. Steier, T. Yamagata, S.R. Winkler, Recent near-Earth supernovae probed by global deposition of interstellar radioactive ^{60}Fe. Nature 532, 69–72 (2016)

342. R.D. Warner, G.J. Taylor, G.H. Conrad, H.R. Northrop, S. Barker, K. Keil, M.-S. Ma, R. Schmitt, Apollo 17 high-Ti mare basalts: new bulk compositional data, magma types, and petrogenesis, in *Lunar Planet. Sci. Conf. Proc.*, ed. by N.W. Hinners (1979), pp. 225–247

343. P.H. Warren, J.T. Wasson, The origin of KREEP. Rev. Geophys. 17, 73–88 (1979)

344. K. Watson, B.C. Murray, H. Brown, The behavior of volatiles on the lunar surface. J. Geophys. Res. 66, 3033–3045 (1961)

345. T.R. Watters, R.C. Weber, G.C. Collins, I.J. Howley, N.C. Schmerr, C.L. Johnson, Shallow seismic activity and young thrust faults on the Moon. Nat. Geosci. 12, 411–417 (2019)

346. R.C. Weber, Interior of the Moon, in *Encyclopedia of the Solar System* (2014), pp. 539–554. https://doi.org/10.1016/B978-0-12-415845-0.00024-4. Chap. 24

347. R.C. Weber, P.Y. Lin, E.J. Garnero, Q. Williams, P. Lognonné, Seismic detection of the lunar core. Science 331, 309–312 (2011)

348. B.P. Weiss, S.M. Tikoo, The lunar dynamo. Science 346, 1246753-1–1246753-10 (2014). https://doi.org/10.1126/science.1246753

349. S.C. Werner, Moon, Mars, Mercury: basin formation ages and implications for the maximum surface age and the migration of gaseous planets. Earth Planet. Sci. Lett. 400, 54–65 (2014)

350. J.L. Whitten, J.W. Head III, Lunar cryptomaria: physical characteristics, distribution, and implications for ancient volcanism. Icarus 247, 150–171 (2015a)

351. J.L. Whitten, J.W. Head III, Lunar cryptomaria: mineralogy and composition of ancient volcanic deposits. Planet. Space Sci. 106, 67–81 (2015b)

352. J. Whitten, J.W. Head III, M.I. Staid, C.M. Pieters, J.F. Mustard, R. Clark, J.W. Nettles, R.L. Klima, L.A. Taylor, Lunar mare deposits associated with the Orientale impact basin: New insights into mineralogy, history, mode of emplacement, and relation to Orientale Basin evolution from Moon Mineralogy Mapper (M3) data from Chandrayaan-1. J. Geophys. Res. 116, E00G09 (2011). https://doi.org/10.1029/2010JE003736

353. ADS M.A. Wieczorek, B.L. Jolliff, A. Khan, M.E. Pritchard, B.P. Weiss, J.G. Williams, L.L. Hood, K. Righter, C.R. Neal, C.K. Shearer, I.S. McCallum, S. Tompkins, B.R. Hawke, C. Peterson, J.J. Gillis, B. Bussey, The constitution and structure of the lunar interior. Rev. Mineral. Geochem. 60, 221–364 (2006)
354. M.A. Wieczorek, G.A. Neumann, F. Nimmo, W.S. Kiefer, G.J. Taylor, H.J. Melosh, R.J. Phillips, S.C. Solomon, J.C. Andrews-Hanna, S.W. Asmar, A.S. Konopliv, F.G. Lemoine, D.E. Smith, M.M. Watkins, J.G. Williams, M.T. Zuber, The crust of the Moon as seen by GRAIL. Science 339, 671–675 (2013)
355. R. Wieler, Do lunar and meteoritic archives record temporal variations in the composition of solar wind noble gases and nitrogen? A reassessment in the light of Genesis data. Chem. Erde 76, 463–480 (2016)
356. R. Wieler, V.S. Heber, Noble gas isotopes on the Moon. Space Sci. Rev. 106, 197–210 (2003)
357. R. Wieler, K. Kehm, A.P. Meshik, C.M. Hohenberg, Secular changes in the xenon and krypton abundances in the solar wind recorded in single lunar grains. Nature 384, 46–49 (1996)
358. D.E. Wilhelms, The geologic history of the Moon. USGS Prof. Pap. 1348 (1987)
359. L. Wilson, J.W. Head III, Generation, ascent and eruption of magma on the Moon: new insights into source depths, magma supply, intrusions and effusive/explosive eruptions (part 1: theory). Icarus 283, 146–175 (2017)
360. L. Wilson, J.W. Head III, Controls on lunar basaltic volcanic erupton structure and morphology: gas release patterns in sequential eruption phases. Geophys. Res. Lett. 45, 5852–5859 (2018)
361. S.F. Wolf, M.S. Wang, M.E. Lipschutz, Labile trace elements in basaltic achondrites: can they distinguish between meteorites from the Moon, Mars, and V-type asteroids? Meteorit. Planet. Sci. 44, 891–903 (2009)
362. J.A. Wood, Bombardment as a cause of the lunar asymmetry. Moon 8, 73–103 (1973)
363. J.A. Wood, J.S. Dickey, U.B. Marvin, B.N. Powell, Lunar anorthosites and a geophysical model of the Moon, in *Lunar Sci. Conf. Proc.* (1970), pp. 965–988

364. S. Yamamoto, R. Nakamura, T. Matsunaga, Y. Ogawa, Y. Ishihara, T. Morota, N. Hirata, M. Ohtake, T. Hiroi, Y. Yokota, J. Haruyama, Possible mantle origin of olivine around lunar impact basins detected by SELENE. Nat. Geosci. 3, 533–536 (2010)
365. S. Yamamoto, R. Nakamura, T. Matsunaga, Y. Ogawa, Y. Ishihara, T. Morota, N. Hirata, M. Ohtake, T. Hiroi, Y. Yokota, J. Haruyama, Featureless spectra on the Moon as evidence of residual lunar primordial crust. J. Geophys. Res., Planets 120, 2190–2205 (2015)
366. N.E.B. Zellner, Cataclysm no more: new views on the timing and delivery of lunar impactors. Orig. Life Evol. Biosph. 47, 261–280 (2017)
367. J. Zhao, L. Xiao, L. Qiao, T.D. Glotch, Q. Huang, The Mons Rümker volcanic complex of the Moon: a candidate landing site forthe Chang'e-5 mission. J. Geophys. Res., Planets 122, 1419–1442 (2017)
368. M.H. Zhu, K. Wünnemann, R.W.K. Potter, T. Kleine, A. Morbidelli, Are the Moon's nearside-farside asymmetries the result of a giant impact? J. Geophys. Res., Planets (2019). https://doi.org/10.1029/2018JE005826
369. M.T. Zuber, D.E. Smith, F.G. Lemoine, G.A. Neumann, The shape and internal structure of the Moon from the Clementine mission. Science 266, 1839–1843 (1994)
370. M.T. Zuber, D.E. Smith, M.M. Watkins, S.W. Asmar, A.S. Konopliv, F.G. Lemoine, H.J. Melosh, G.A. Neumann, R.J. Phillips, S.C. Solomon, M.A. Wieczorek, J.G. Williams, S.J. Goossens, G. Kruizinga, E. Mazarico, R.S. Park, D.N. Yuan, Gravity Field of the Moon from the Gravity Recovery and Interior Laboratory (GRAIL) Mission. Science 339, 668–671 (2013)

12

Silica-rich Volcanism in the Early Solar System Dated at 4.565 Ga

Poorna Srinivasan[1,2], Daniel R. Dunlap[3], Carl B. Agee[1,2], Meenakshi Wadhwa[3], Daniel Coleff[4], Karen Ziegler[1,2], Ryan Zeigler[5] & Francis M. McCubbin[5]

[1] Institute of Meteoritics, University of New Mexico, Albuquerque, NM 87131, USA.
[2]Department of Earth and Planetary Sciences, University of New Mexico, Albuquerque, NM 87131, USA.
[3] Center for Meteorite Studies, School of Earth and Space Exploration, Arizona State University, Tempe, AZ 85287, USA.
[4] Jacobs Technology, NASA Johnson Space Center, Mail Code XI3, 2101 NASA Parkway, Houston, TX 77058, USA.
[5]NASA Johnson Space Center, Mail Code XI2, 2101 NASA Parkway, Houston, TX 77058, USA.

Citation: Srinivasan, P., Dunlap, D.R., Agee, C.B. et al. "Silica-rich volcanism in the early solar system dated at 4.565 Ga". Nat Commun 9, 3036 (2018). https://doi.org/10.1038/s41467-018-05501-0

ABSTRACT

The ranges in chemical composition of ancient achondrite meteorites are key to understanding the diversity and geochemical evolution of planetary building blocks. These achondrites record the first episodes of volcanism and crust formation, the majority of which are basaltic. Here we report data on recently discovered volcanic meteorite Northwest Africa (NWA) 11119, which represents the first, and oldest, silica-rich (andesitic to dacitic) porphyritic extrusive crustal rock with an Al–Mg age of 4564.8 ± 0.3 Ma. This unique rock contains mm-sized vesicles/cavities and phenocrysts that are surrounded by quench melt. Additionally, it possesses the highest modal abundance (30 vol%) of free silica (i.e., tridymite) compared to all known meteorites. NWA 11119 substantially widens the range of volcanic rock compositions produced within the first 2.5–3.5 million years of Solar System history, and provides direct evidence that chemically evolved crustal rocks were forming on planetesimals prior to the assembly of the terrestrial planets.

INTRODUCTION

Examination of ancient achondritic meteorites has provided evidence that differentiation was occurring on many planetesimals within the first 10 Ma of Solar System formation. Our record of differentiated meteorites from this time period mainly consists of rocks with relatively low silica content, similar to compositions observed in Earth's oceanic crust. However, newly discovered achondrites have recently provided evidence that crusts on planetesimals early in the Solar System were also comprised of more evolved rocks with higher silica contents than typical achondrites1–3. These evolved lavas, similar to Earth's average continental crust, were long thought to be exclusive to large, planet-sized bodies. Moreover, the origin of Earth's evolved crust arose through the presence of substantial water and plate tectonic processes that allow for tertiary crust formation4, a process that is as of now, thought to be unique to Earth5. Although the presence of evolved granitic clasts in some lunar samples6 and felsic impact spherules in some howardites7 can be attributed to extensive differentiation of mafic parental magmas, it is not clear how well this process can explain the origin of all the evolved achondrite meteorites that have been discovered to date1–3. Furthermore, the geochemistry of these evolved achondrites are not consistent with tertiary crust formation, leaving open the possibility that additional mechanisms exist by which evolved melt compositions can occur in our Solar System. In the present study, we describe a newly discovered evolved achondrite Northwest Africa (NWA) 11119 that is unique among the existing evolved achondrite meteorites in its age, appearance, mineralogy, and bulk composition. NWA 11119 provides important constraints on the early Solar System processes that allow for the formation of evolved melt compositions. The NWA 11119 meteorite was found in Mauritania in December 2016. The recovered specimen is a single stone of 453 g (Fig. 1), and resides at the Maine Mineral and Gem Museum. A 23 g subsample deposited at the Institute of Meteoritics (IOM) at the University of New Mexico in Albuquerque was used for this study. The specimen is highly friable and contains an unusual light-green fusion crust. Broken fragments of the interior show bright green and gray crystals up to 3 mm in size within a medium- to fine-grained matrix. Spherical vesicles and irregularshaped cavities were observed throughout the stone.

RESULTS

Petrology of NWA 11119.

Petrologic examination of NWA 11119 was conducted on the IOM deposit sample (Supplementary Fig. 1; 4.2 × 3.3 × 2.3 cm) as well as a polished slice (0.95 × 0.55 cm) of the deposit sample embedded in a 1-inch epoxy mount. A false-colored Si–Fe–Al X-ray mosaic of the slice is shown in Fig. 2. NWA 11119 exhibits a porphyritic texture composed of millimeter-sized phenocrysts of pyroxene (

Phenocryst compositions

Chemical compositions of the phenocrysts (Table 1) exhibit moderate–minor variations and are represented by the following: clinopyroxene (augite; Fs6.0±2.1

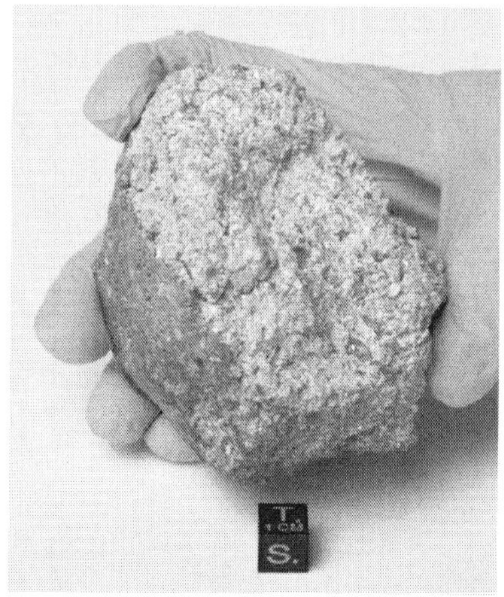

Figure. 1 Photograph of the main mass of NWA 11119 that was found in Mauritania in December 2016. The rock has a light-green fusion crust, and the broken interior shows light-green and gray-colored crystals. Photo credit: © 2017 B. Barrett/Maine Mineral & Gem Museum.

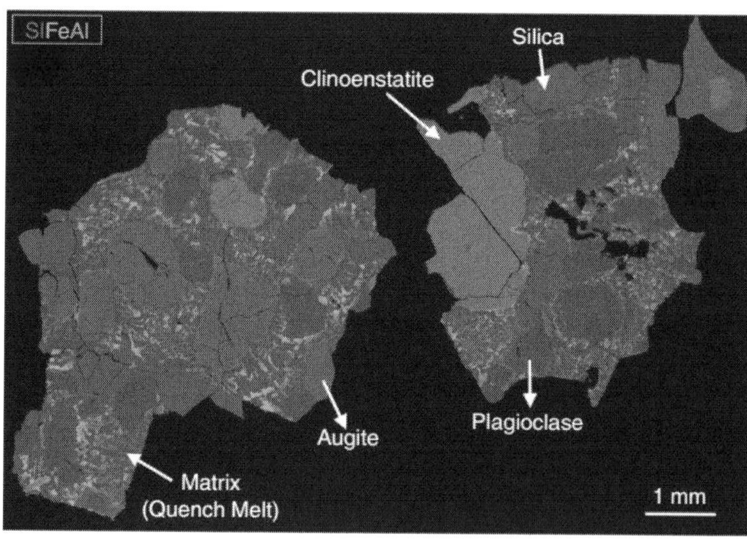

Figure. 2 False-colored X-ray map of NWA 11119 showing mm-sized phenocrysts of silica, clinoenstatite, augite, and plagioclase. Zoning was not observed in silica or pyroxene phenocrysts, but normal zoning appears in plagioclase phenocrysts. The phenocrysts are surrounded by quenched groundmass consisting of highly zoned crystals that range in size from submicrometer to hundreds of micrometers.

$Wo_{38.1\pm1.1}$; Cr_2O_3 = 0.7–1.2 wt.%; Mg# [100*Mg/(Mg + Fe)] = 90), orthopyroxene (enstatite; Fs9.5±0.3Wo4.6±0.1; Cr_2O_3 = 0.8–0.9 wt.%; Mg# = 90) (Supplementary Fig. 3 and Supplementary Data 1), plagioclase (An87.3±7.2Ab12.6±7.1Or0.1±0.2) (Supplementary Fig. 4 and Supplementary Data 1), and silica (~97.2 wt.% SiO_2 and ~0.8 wt.% Al_2O_3) (Supplementary Data 1). The molar Fe/Mn ratio of pyroxene phenocrysts is 4.6–7.8 (Supplementary Fig. 5); with the exception of some winonaites, this range is lower than all

Table 1 Mineral compositions (wt.%) for NWA 11119 determined by EPMA

	Augite							Enstatite						
	All phenocrysts			Core[a]		Rim[a]		All phenocrysts			Core[a]		Rim[a]	
	Average	Min	Max	Average	St. dev.	Average	St. dev.	Average	Min	Max	Average	St. dev.	Average	St. dev.
SiO_2	54.19	52.07	55.63	54.46	0.33	54.00	0.75	56.74	55.73	57.62	56.73	0.36	56.53	0.19
MgO	19.85	16.70	21.51	20.17	0.26	19.63	0.89	32.11	31.32	32.84	32.17	0.31	31.96	0.13
Na_2O	0.08	0.04	0.15	0.08	0.01	0.09	0.02	0.01	0.00	0.03	0.01	0.01	0.01	0.01
Al_2O_3	1.46	1.18	2.28	1.37	0.04	1.74	0.34	0.89	0.79	1.13	0.93	0.10	0.92	0.15
P_2O_5	0.01	0.00	0.05	0.00	0.01	0.01	0.01	0.01	0.00	0.03	0.01	0.01	0.00	0.00
CaO	18.83	16.10	19.34	19.07	0.05	18.41	0.98	2.38	2.31	2.47	2.36	0.03	2.36	0.01
K_2O	0.00	0.00	0.07	0.00	0.01	0.00	0.00	0.00	0.00	0.02	0.00	0.00	0.00	0.00
TiO_2	0.36	0.23	1.08	0.31	0.01	0.48	0.22	0.16	0.13	0.19	0.16	0.02	0.15	0.00
Cr_2O_3	0.79	0.69	1.18	0.74	0.01	0.92	0.15	0.81	0.76	0.85	0.82	0.03	0.80	0.01
NiO	0.01	0.00	0.04	0.01	0.01	0.01	0.01	0.01	0.00	0.04	0.01	0.01	0.01	0.01
FeO	3.79	3.25	9.63	3.36	0.03	4.31	0.98	6.34	6.22	7.61	6.33	0.04	6.33	0.05
MnO	0.72	0.60	1.47	0.66	0.02	0.81	0.14	0.94	0.90	0.99	0.94	0.02	0.94	0.01
Total	100.08	98.18	101.92	100.24	0.59	100.41	0.49	100.39	98.56	101.82	100.48	0.57	100.01	0.32

	Plagioclase							Silica						
	All phenocrysts			Core[a]		Rim[a]		All phenocrysts			Core[a]		Rim[a]	
	Average	Min	Max	Average	St. dev.	Average	St. dev.	Average	Min	Max	Average	St. dev.	Average	St. dev.
SiO_2	47.22	44.69	53.10	46.31	0.09	50.86	2.10	97.36	96.63	98.35	97.00	0.12	97.57	0.54
MgO	0.32	0.11	0.49	0.34	0.01	0.28	0.12	0.00	0.00	0.01	0.00	0.00	0.00	0.00
Na_2O	1.39	0.87	3.68	0.98	0.04	2.76	0.77	0.12	0.08	0.28	0.12	0.01	0.14	0.07
Al_2O_3	34.52	30.14	37.12	35.21	0.32	31.87	1.64	0.74	0.61	1.44	0.71	0.01	0.81	0.29
P_2O_5	0.01	0.00	0.04	0.01	0.01	0.01	0.01	0.00	0.00	0.03	0.00	0.01	0.01	0.01
CaO	17.46	13.11	18.61	18.18	0.08	14.81	1.55	0.09	0.04	0.29	0.09	0.01	0.11	0.08
K_2O	0.02	0.00	0.20	0.01	0.00	0.09	0.06	0.14	0.01	0.17	0.14	0.00	0.13	0.06
TiO_2	0.02	0.00	0.10	0.01	0.01	0.05	0.03	0.10	0.07	0.18	0.09	0.01	0.11	0.03
Cr_2O_3	0.00	0.00	0.02	0.00	0.01	0.00	0.00	0.00	0.00	0.01	0.00	0.00	0.00	0.00
NiO	0.01	0.00	0.04	0.01	0.01	0.01	0.01	0.01	0.00	0.04	0.01	0.02	0.01	0.01
FeO	0.17	0.09	0.50	0.11	0.01	0.36	0.12	0.02	0.00	0.04	0.02	0.01	0.03	0.01
MnO	0.03	0.01	0.06	0.02	0.01	0.04	0.01	0.00	0.00	0.01	0.00	0.01	0.00	0.00
Total	101.17	99.71	101.94	101.19	0.31	101.13	0.26	98.58	98.04	99.78	98.19	0.11	98.92	0.59

[a]Core and rim values were obtained by averaging the first and last three analyses of each line array

achondritic groups8. This Fe/Mn ratio suggests formation in a reduced environment with a low oxygen fugacity below the ironwüstite (IW) buffer and above the Mn–MnO buffer, similar to the reduced silicate mineralogy in winonaites9. The orthopyroxene and clinopyroxene phenocrysts exhibit an OPX–CPX Mg–Fe exchange KD ratio [KD = $(X_{MgO}/X_{FeO})O_{PX}*(X_{FeO}/X_{MgO})$CPX] of 0.88 ± 0.10. This KD value is in the range of equilibrium orthopyroxene and clinopyroxene assemblages in terrestrial rocks such as pyroxene andesites (KD = 0.86) and ultramafic inclusions in basaltic rocks (KD = 0.75–1.51)10. Plagioclase phenocrysts exhibit normal zoning with rims that are more albitic than the cores. X-ray diffraction (XRD) patterns collected from a powdered fragment of NWA 11119 revealed that the silica phase is tridymite. The tridymite phenocrysts appear either as long lathes or large anhedral grains. The pyroxene and plagioclase phenocrysts are subhedral to anhedral.

Matrix material (quench melt)

The matrix consists of grains that are nearly up to ~300 mm in size, are highly zoned, and exhibit substantial variability among matrix grains of an individual phase (Supplementary Data 3). Submicrometer-tomicrometer-sized grains of ulvöspinel, ilmenite, troilite, tranquilityite, zircon, Fe metal, fayalite, and tsangpoite were also observed in the matrix. The texture of the matrix, coupled with the highly variable chemical compositions of matrix minerals, indicates rapid quenching from high temperatures. Consequently, we collected density-corrected quantitative X-ray maps of the groundmass to estimate the composition of the melt interstitial to the phenocryst phases, and this melt composition is represented in Supplementary Data 4. The calculated matrix composition shows low total alkali abundances (<1.5 wt.%) and elevated silica content (~61.28 wt.%). In fact, the bulk composition of the matrix plots near the andesite–dacite boundary on the total alkali versus silica (TAS) diagram (Fig. 3a). To assess the relationship between the matrix and the phenocrysts, we evaluated the Mg–Fe exchange equilibria between pyroxenes and the groundmass. The orthopyroxene and clinopyroxene phenocrysts exhibit meltpyroxene Mg–Fe exchange KD ratios $[K_D = (X_{MgO}/X_{FeO})^{Liquid} * (X_{FeO}/X_{MgO})^{pyroxene}]$ of 0.25 ± 0.05 and 0.22 ± 0.06, respectively. These values overlap with the range of equilibrium values for OPX-melt of 0.2–0.311,12 and CPX-melt of $0.27 \pm 0.03$13, indicating that the phenocrysts could have been in equilibrium with the melt in which they were hosted prior to quench.

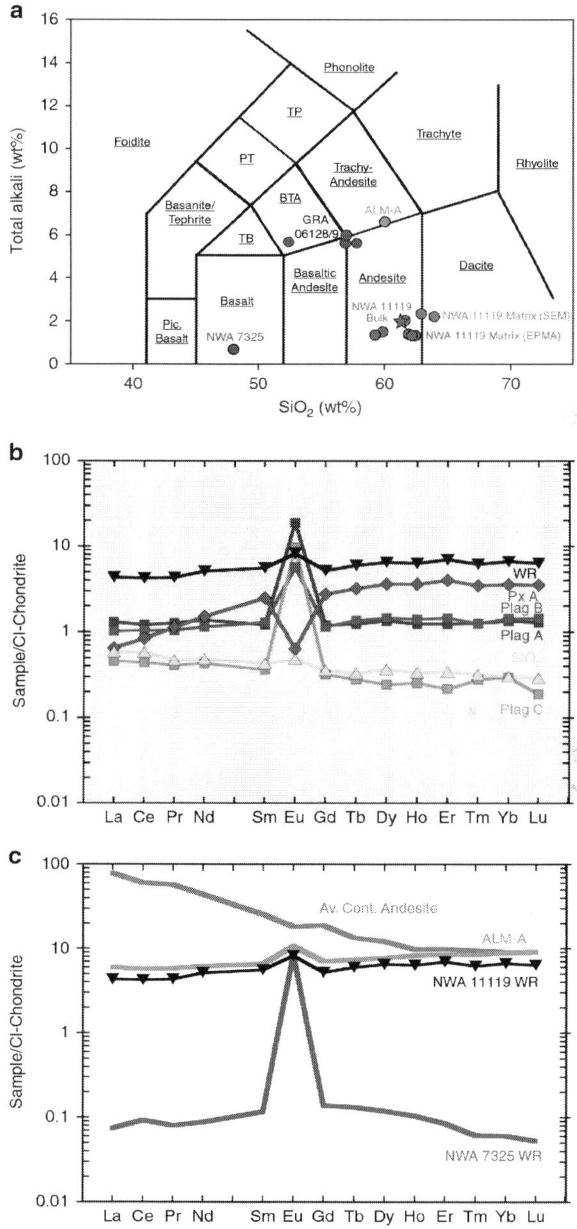

Figure. 3 Chemical composition of NWA 11119. a Plot of total alkalis vs. silica content showing the compositions of the bulk and matrix of NWA 11119. The bulk rock has low alkali content, and measured values for the matrix show an evolved andesitic/dacitic composition. For comparison, bulk compositions for

ALM-A and NWA 7325 are also shown. b CI-normalized rare earth element (REE) abundances in the NWA 11119 whole rock (WR) and mineral separates (Px = pyroxene; Plag = plagioclase; SiO_2 = Silica). c Comparison of CI-normalized REE abundances in whole rocks of NWA 11119 (black inverted diamonds), ALM-A (orange), NWA 7325 (gray), and average continental andesite (green). Compositional data for ALM-A from ref. 3, NWA 7325 from refs. 21,39, average continental andesite from ref.54, and CI chondrites from ref. 55 Errors on our REE data (2 sigma) are typically ±10% and are smaller than the data points.

Major and minor element bulk composition of NWA 11119

The major element bulk composition of NWA 11119 was calculated using the modal abundances of each phase from the XCT data and the phase compositions determined by EPMA (Supplementary Data 4) following the procedures of ref. 14 and densities of ref. 15. The calculated bulk rock composition indicates total alkali abundances (i.e., Na_2O+K_2O) of 0.93 wt.% and a silica content of 61.37 wt.% (Fig. 3a). If the bulk rock composition of NWA 11119 represents that of a melt composition, it is an andesite (Fig. 3a), however, we cannot rule out the possibility that NWA 11119 represents a partial cumulate rock. NWA 11119 has a Mg# of 84, considerably higher than terrestrial orogenic andesites, which average Mg#'s of around 6010. The total alkali content of NWA 11119 is lower than typical terrestrial andesites and dacites, which have values >4 wt.%16. The silica content of the matrix is similar to the bulk rock.

Rare earth element abundances of NWA 11119

Rare earth element (REE) concentrations were determined in a whole-rock (WR) fraction and mineral separates (Px A, Plag A, Plag B, Plag C, and SiO_2) (Fig. 3b, c and Supplementary Data 5). Care was taken to ensure that the mineral separates (purified by hand picking under an optical microscope) were as pure as possible. Nevertheless, small amounts of other phases may be present in these separates as inclusions or minute-attached grains. As such, the REE patterns of the separates are characteristic of their dominant phases and most likely do not represent pure mineral compositions. Specifically, Px A, which shows light-REE depletion (CI-normalized La/Yb=0.2) and a negative Eu anomaly (Eu/Eu*=0.2, where Eu* is the value interpolated between CI-normalized abundances of Sm and Gd), is predominantly pyroxene. The Plag A, B, and C separates show relatively flat REE patterns (CI-normalized La/Yb=0.7–1.6) with a pronounced positive Eu anomaly

(Eu/Eu* = 4.6–28.6). While each of these three is dominated by plagioclase, Plag C is more so than the other two (given the significantly larger Eu anomaly). The tridymite-rich separate has a slight LREE enrichment (CI-normalized La/Yb = 2.0) with a small positive Eu anomaly (Eu/Eu* = 1.2). The NWA 11119 WR has a slight LREE-depletion (CI-normalized La/Yb = 0.7) with La ~4× CI chondrite, and a small positive Eu anomaly (Eu/Eu* = 1.5).

Oxygen isotopic composition of NWA 11119

Oxygen isotope ratios are used to distinguish between planetary bodies and/or planetary reservoirs. The NWA 11119 bulk sample has oxygen isotope ratios of $\delta^{18}O = 8.226, 7.781, 8.121, 7.671$, $\delta^{17}O = 3.364, 3.133, 3.398, 3.046$, and $\Delta^{17}O = -0.979, -0.975, -0.890$, and -1.004 per mil for four bulk rock fragments, respectively (Supplementary Data 6). NWA 11119 plots on the carbonaceous chondrite anhydrous mineral (CCAM) line on an oxygen three-isotope diagram (Fig. 4), which is uncommon for achondrites. Ureilites, which are ultramafic achondrites or fragmental/regolith breccias17·18·19, the trachyandesitic clast ALM-A in the anomalous polymict ureilite Almahata Sitta3, and the basaltic ungrouped achondrite NWA 732520·21·22 are the only other known achondrites that plot at high $\delta^{18}O$ values (>4‰) along the CCAM line. The ungrouped achondrites NWA 313323, NWA 818624·25, and NWA 782226 plot at lower $\delta^{18}O$ (<4‰) along the CCAM line, and show strong links to carbonaceous chondrite meteorite groups.

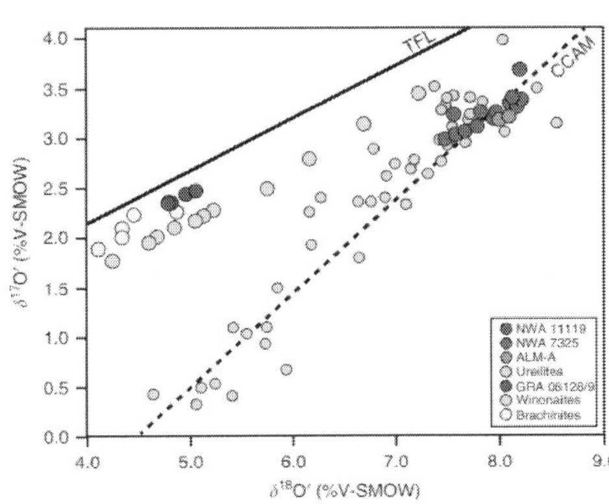

Figure. 4: Oxygen three-isotope plot with various achondritic groups. NWA 11119 plots along the carbonaceous chondrite anhydrous minerals (CCAM) line with ureilites, the trachyandesitic fragment ALM-A from Almahata Sitta, and

ungrouped achondrite NWA 7325. Literature data for all samples besides NWA 11119 are obtained from refs. 1·3·17·20

Age constraints on NWA 11119. The ^{26}Al–^{26}Mg isotope systematics were determined for whole rock (WR) and mineral separates of NWA 11119 (Supplementary Data 7). These define an isochron corresponding to a $^{26}Al/^{27}Al$ ratio at the time of crystallization of this sample of $(1.69 \pm 0.09) \times 10^{-6}$ (Fig. 5). Previously, the highest value reported for an achondrite internal ^{26}Al–^{26}Mg isochron was $(1.28 \pm 0.07) \times 10^{-6}$ (for the ancient cumulate eucrite Asuka 88139427). Relative to the D'Orbigny angrite age anchor28·29, the ^{26}Al–^{26}Mg age of NWA 11119 is 4564.8 ± 0.3 Ma (Fig. 5). It is also possible to calculate a ^{26}Al–^{26}Mg age relative to the first-formed refractory solids in the solar protoplanetary disk that are characterized by a canonical $^{26}Al/^{27}Al$ ratio of 5.2×10^{-5} 30. However, the reported Pb–Pb ages for these refractory solids span ~0.6 Ma30·31·32, leading to a range of ^{26}Al–^{26}Mg ages of 4563.7 ± 0.2–4564.4 ± 0.3 Ma for NWA 11119. Consideration of the full range of these ^{26}Al–^{26}Mg ages (obtained relative to the D'Orbigny angrite or the refractory inclusions) and the Pb–Pb ages of the refractory inclusions30·31·32 indicates that NWA 11119 was formed 2.5–3.5 Ma after the first-formed refractory solids in the solar protoplanetary disk.

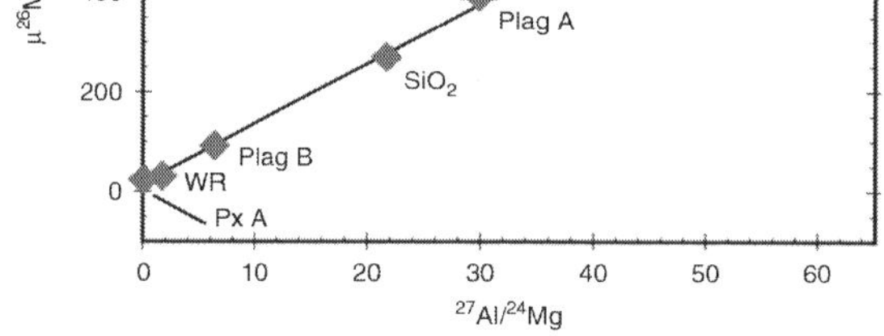

Figure. 5: The Al–Mg internal isochron defined by whole rock (WR) and mineral separates of NWA 11119. This isochron corresponds to a $^{26}Al/^{27}Al$ ratio of $(1.69 \pm 0.09) \times 10^{-6}$. Relative to the D'Orbigny age anchor, NWA 11119 has a ^{26}Al–^{26}Mg age of 4564.8 ± 0.3 Ma. The uncertainties on the $\mu^{26}Mg^*$ values are either the internal 2SE errors (2× the standard error) on repeat analyses of the sample or the 2 SD (2× the standard deviation) external reproducibility based on repeat analyses of standards, whichever is larger; the uncertainties on the $^{27}Al/^{24}Mg$ ratios are ±5%

DISCUSSION

Examples of ancient igneous activity on planetesimals within a few million years of Solar System formation (assumed to coincide with the formation of the first refractory solids in the solar protoplanetary disk[30,31,32]) are dominated by samples that exhibit basaltic–ultramafic bulk rock compositions. For example, internal Pb–Pb and $^{26}Al–^{26}Mg$ isochrons for both the angrite D'Orbigny, which contains vesicles and glass-filled cavities, and the cumulate eucrite Asuka 881394, which has a granulitic texture, indicate that basaltic crusts formed on their distinct differentiated parent bodies well within ~5 Ma of Solar System formation[27,28,33,34]. Additionally, internal Pb–Pb and $^{26}Al–^{26}Mg$ isochrons for the ungrouped achondrite NWA 7325 provide evidence that basaltic crusts were forming on a reduced, differentiated parent body by ~4563 Ma[21,35,36]. Geochronologic data for the recently discovered volcanic and subvolcanic meteorites with rare trachyandesitic compositions are somewhat limited but provide some evidence that their parent bodies also differentiated early. The $^{26}Al–^{26}Mg$ model age for the ungrouped trachyandesitic paired achondrites GRA 06128/91[37,38] suggests that differentiation and plagioclase accumulation on the oxidized, volatile-rich, differentiated body may have occurred as early as ~4565 Ma[38]. Similarly, a $^{26}Al–^{26}Mg$ model age for the trachyandesitic fragment, ALM-A, from the anomalous polymict ureilite Almahata Sitta indicates that evolved crusts may have formed on a volatile-rich, differentiated body by ~4561 Ma[3]. The bulk composition of NWA 11119, along with its early formation age as indicated by its $^{26}Al–^{26}Mg$ internal isochron, demonstrates unambiguously that andesitic–dacitic volcanism occurred on a differentiated parent body within the first 2.5–3.5 Ma of Solar System history, extending both the age and degree of chemical evolution of planetary crustal materials on differentiated bodies in our Solar System.

NWA 11119 shares a number of textural and geochemical similarities with other ancient volcanic achondrites. For example, NWA 11119 has textural similarities to D'Orbigny and other angrites (e.g., vesicles), but it shares isotopic and petrologic links to the achondrites NWA 7325 and ALM-A in regard to oxygen isotopes (Fig. 4) and the presence of Cr-bearing high-Ca pyroxene[3,21]. NWA 11119 also shares similarities in whole-rock REE abundances with the ALM-A fragment (Fig. 3c). Nevertheless, NWA 11119 has other petrologic and geochemical distinctions (e.g., high free silica content and low total alkali content) that make it a unique crustal rock compared to these samples. Figure 3a shows the bulk rock content of NWA 11119 compared to the isotopically similar NWA 7325[39] and

ALM-A3 samples, and the trachyandesitic GRA 06128/91 samples. Although NWA 11119 shows notable geochemical similarities to NWA 7325 and ALM-A, it possesses a distinct bulk rock composition compared to these two samples. Furthermore, the bulk rock composition and pyroxene compositions of NWA 11119 have distinct Fe/Mn ratios compared to NWA 7325 and ALM-A (Supplementary Fig. 5), indicating that NWA 11119 is an andesitic achondrite that may be derived from a previously unsampled parent body.

The formation mechanisms of andesitic crustal materials are highly debated4·40·41. On Earth, ideas range from formation as primary melts from mantle material to secondary products of basaltic partial melts40. Plate tectonics and mantle volatile contents are also thought to play significant roles in the production of andesitic crusts4·41. The formation mechanisms of andesitic materials in asteroidal bodies are more uncertain, and it is still an area of intense research. Melting of chondritic precursors are typically implicated in the formation of achondrites, and the relatively unfractionated REE pattern of the NWA 11119 whole rock (Fig. 3c) is consistent with such a scenario. However, the petrogenetic origin of NWA 11119 cannot be easily linked to known melting processes of chondritic precursor materials. Nevertheless, several experimental studies on partial melting of chondritic materials have reported the formation of evolved melt compositions similar to the SiO_2 content of NWA 1111942·43·44·45. These studies may provide insights into the melting processes that occurred to produce NWA 11119.

Disequilibrium partial melting experiments conducted on an L6 ordinary chondrite produced a basaltic to basaltic–andesite silicate melt composition, but the melt contained significantly higher FeO content (~11–16 wt.%) and a range of total alkali contents (up to ~5 wt.%)43. Experiments on an H chondrite bulk composition produced basaltic to dacitic melt compositions at IW-1, but the FeO abundances (9–14 wt.%) and alkali abundances (3.5–5 wt.%) of the melt were also too high45. Partial melting experiments on an R4 rumuruti chondrite produced a trachyandesitic silicate melt composition, but those experiments produced melts with even higher alkali abundances (~9 wt.%)44. Equilibrium partial melting experiments on an EH4 enstatite chondrite produced an andesitic–dacitic silicate melt with low alkali content42; however, FeO abundances (<0.6 wt.%) were well below those of NWA 11119, owing to the highly reduced nature of the enstatite chondrite parent body.

The EH4, H, and R4 experimental studies highlight the two mechanisms for producing silica-rich melts through equilibrium partial melting of chondritic precursor materials: (1) reducing to highly reducing conditions (akin to the H and EH4 studies) promote SiO_2 enrichment in the lithic fraction by sequestration of Fe in the metallic fraction (as long as the system remains above the oxygen fugacity defined by Si–SiO_2), and (2) even precursor materials with abundant olivine will produce Si-rich melts under low-degree partial melting if melting occurs at a eutectic involving albitic plagioclase (akin to the R4 study44). Both of these mechanisms were used to explain the enigmatic origin of extensive boninitic magmatism on the planet Mercury46. Although mechanism 2 cannot be used to explain the origin of NWA 11119 given its alkali-depleted bulk composition, mechanism 1 may apply to NWA 11119 given that its bulk composition has FeO abundances that are intermediate between the melts produced in the EH4 and H experimental studies. This intermediate FeO abundance indicates that NWA 11119 could have been derived by partial melting of a chondritic precursor at an fO_2 that is intermediate between the fO_2 investigated in the H and EH4 studies (IW-1 and IW-5, respectively). In fact, if the FeO abundance of NWA 11119 is reflective of the FeO abundance of its source, it would imply an fO_2 of approximately IW-4, if its source was in equilibrium with Fe-rich metal (computed using the methods outlined in ref.47). Partial melting studies of chondritic materials between IW-1 and IW-5 are lacking, but future studies aimed at this range in fO_2 may provide important constraints into the origin of NWA 11119, the origin of other evolved and reduced asteroidal samples, and the origin of reduced planetary bodies such as Mercury48. NWA 11119 provides the first unambiguous evidence for evolved volcanism in the earliest stages of Solar System history, prior to the assembly of the terrestrial planets.

METHODS

Sample material

A 4.2 cm × 3.3 cm × 2.3 cm rock sample of NWA 11119 (Fig. 1) was used for this study which is on deposit at the Institute of Meteoritics (IOM) at the University of New Mexico (UNM). This sample weighs 23 g and was removed from the main mass (Supplementary Fig. 1) that was found in Mauritania in December 2016. The IOM deposit sample was internally scanned using XCT at Johnson Space Center (JSC). A ~5 mg fragment was removed from the deposit sample for oxygen isotope studies at UNM. A ~200 mg fragment was removed from the deposit sample for X-ray diffraction studies at UNM. A ~500 mg fragment was removed from the deposit sample for trace element and Al–Mg dating studies at Arizona State University (ASU). A 0.9 cm × 0.55 cm thick section was cut off with a diamond saw and placed into a 1-inch epoxy mount for electron microscopy techniques at UNM and JSC.

X-ray computed tomography

XCT was conducted on the IOM deposit sample using a Nikon XTH 320, enclosed cabinet, micro-focus type CT scanner in the Astromaterials Research and Exploration Science (ARES) E-beam facility at JSC to obtain estimates on mineral modal abundances and vesicle content. The scan was obtained using a 175 kV accelerating voltage and 86 μA current. The deposit sample was enclosed in bubble wrap and placed inside a cylindrical cardboard dispenser to minimize any movement during the ~12 h scan. A 0.5 mm Cu filter was placed in front of the output window of the X-ray source during the scan. The scan consisted of 3141 projections with 16 frames per projection, which ensured a high-resolution and effective pixel size of 23.87 μm. Modal abundances and porosity estimates were calculated using the XCT processing program myVGL 3.0. The vesicles/cavities in the deposit sample can be viewed in Supplementary Movie 1, and the mineralogy can be viewed in Supplementary Movie 2.

Scanning electron microscopy

Scanning electron microscopy (SEM) techniques were conducted on a JEOL 7600F SEM in the ARES/JSC on the slice of NWA 11119. This method was used to identify minerals and obtain images and X-ray maps of the section. The SEM is equipped with an energy dispersive spectrometer (EDS), which allowed for semi-quantitative measurements of phases. Backscattered electron (BSE) images (Supplementary Fig. 2) were acquired with a 15 kV accelerating voltage and 3.89 nA beam current. BSE and X-ray maps were acquired with a 15 kV accelerating voltage and 30 nA beam current. Mineral modal abundances were calculated by pixel counting using the image processing program ImageJ49.

SEM techniques were also conducted on a Tescan Vega3 SEM equipped with a LaB6 gun at the IOM (UNM) on the slice of NWA 11119. This method was used to obtain quantitative data on the average groundmass. Conditions used were a 15 kV accelerating voltage and 1 nA beam current. Data on the matrix were obtained by drawing a free-hand area of interest in the shape of a square, which was then divided into four equal quadrants. The beam was moved in a raster pattern over each quadrant to collect an EDS spectrum, and the acquired data were then averaged for each section. Analyses on diopside and labradorite standards were also taken to test EDS reproducibility, and SiO_2 values were within 3% error of standard values. Density corrections were not applied to this data set. SEM matrix compositions are given in Supplementary Table 3.

Electron probe microanalysis

Electron probe microanalysis (EPMA) techniques were conducted using a JEOL 8530F hyperprobe at ARES/JSC on the slice of NWA 11119. This method was used to obtain quantitative chemical analyses on the phenocrysts and matrix grains. The hyperprobe is equipped with five wavelength-dispersive X-ray spectrometers (WDS) and an EDS system to obtain quantitative chemical analyses on phases. Silicates were analyzed using a 20 nA beam current, 15 kV accelerating voltage, and 1–5 μm spot size. Standards and elements were used as followed: diopside for Si, Ca, rutile for Ti, apatite-wilberforce for P, rhodonite for Mn, orthoclase for K, albite for Na, Sitkin anorthite for Ca, Si, and Al, San Carlos olivine for Mg, Springwater olivine for Si and Fe, chromite for Cr, Rockport fayalite for Fe, and Ni metal for Ni. Count times were 30 s (±15 s background) for each element. Sulfides and oxides were analyzed using a 30 nA beam current, 15 kV accelerating voltage, and 1–5 μm spot size. Standards and elements were used as followed: rutile for Ti, rhodonite for Mn, troilite for Fe, chromite for Cr, ilmenite for Fe and Ti, Ni metal for Ni, Co metal for Co, Zn metal for Zn, V metal for V, Cu metal for Cu, diopside for Ca, zircon for Zr, San Carlos olivine for Mg and Si, and Kakanui hornblende for Al. Count times were 30 s (±15 s background) for each element. Totals below or above 98–102 wt.% were excluded.

EPMA techniques were also conducted using a JEOL 8200 Superprobe at the IOM (UNM) on the slice of NWA 11119. Six sections of the matrix were mapped using quantitative X-ray mapping techniques with the Probe for Windows software. Each map was 300 × 300 pixels in size and acquired in two passes. Conditions used were a 15 kV accelerating voltage, 30 nA beam current, and 1 μm spot size. Standards and elements were used as followed: diopside for Si and Ca, olivine for Mg, albite for Na, spessartine for Mn, alumina for Al, pyrite for S, hematite for Fe, rutile for Ti, and chromite for Cr. Labradorite and ilmenite were included as secondary standards. The quantitative maps were processed using a custom MatLab script that uses element concentrations obtained through the six EPMA maps to identify the minerals in each frame. Totals below or above 90–105 wt.% were excluded. The minerals were then assigned a density to calculate a bulk density-corrected composition. Mineral densities for the matrix phases were assigned based on the average compositions of each phase as follows: plagioclase 2.73 g/cm^3, silica 2.28 g/cm^3, low-Ca pyroxene 3.62 g/cm^3, high-Ca pyroxene 3.44 g/cm^3, and opaques 4.85 g/cm^3. The average density for the matrix material was determined to be 3.12 g/cm^3, which was used to compute the bulk rock composition for NWA 11119. EPMA-calculated matrix compositions with and without density corrections are given in Supplementary Table 3. The XCT results were used to determine the mode of NWA 11119, which was also used to compute the bulk composition of NWA 11119. The XCT results indicated that the lithic portion of NWA 11119 consists of 89% phenocrysts and 11% matrix. The phenocrysts comprise 56% plagioclase, 30% tridymite, and 14% pyroxene. Mineral densities for the phenocrysts were assigned based on the average compositions of each phase as follows: plagioclase 2.74 g/cm^3, silica 2.28 g/cm^3, and pyroxene 3.28 g/cm^3. The density-corrected bulk composition for NWA 11119 is provided in Supplementary Table 3.

X-ray diffraction

X-ray diffraction (XRD) techniques were conducted on a Rigaku Smartlab X-ray diffractometer at UNM to determine the composition of the silica polymorph in NWA 11119. The diffractometer is equipped with a Cu-tube K_α source and a D/teX detector. ~200 mg of the sample was ground and homogenized into a whole-rock powder with a mortar and pestle, and then loaded into the diffractometer. The sample was scanned continuously with a range of 5°–140°, a step size of 0.02°, at a rate of 6°/min. Whole powder peak fitting was completed using MDI Jade software. XRD pattern for the bulk rock is exhibited in Supplementary Figure 5.

Laser fluorination techniques

Oxygen isotopes were obtained from four acid-washed fragments (0.9–1.5 mg) of NWA 11119 in the Center for Stable Isotopes at UNM using a laser fluorination technique. A fraction of NWA 11119 was crushed with a mortar and pestle, and four fractions were separated and washed with dilute HCl (6 M), high-purity alcohol, and distilled water, and then dried in a 60 °C oven. Molecular O_2 was obtained in a BrF_5-atmosphere, and isotope ratios were measured on a Delta PlusXL mass spectrometer, as described in ref.50. Isotopic measurements were obtained for $\delta^{17}O=(\delta^{17}O/\delta^{16}O)_{sample}/(\delta^{17}O/\delta^{16}O)_{SMOW}-1)\times 1000$, $\delta^{18}O=(\delta^{18}O/\delta^{16}O)_{sample}/(\delta^{18}O/\delta^{16}O)_{SMOW}-1)\times 1000$, and $\Delta^{17}O=\delta^{17}O—0.528\times\delta^{18}O$. San Carlos olivine was used for standardization. Oxygen isotope values are listed in Supplementary Table 6.

Inductively coupled plasma mass spectrometry

The Al–Mg isotope analyses were conducted in a clean laboratory in the Isotope Cosmochemistry and Geochronology Lab (ICGL) at ASU. A ~50 mg internal (free of fusion crust) fragment was powdered to obtain a whole-rock (WR) fraction and another internal ~200 mg fragment was processed for mineral separation. After careful crushing, sieving, and density separation, one pyroxene separate (Px A), three plagioclase-rich separates (Plag A, B, and C), and one silica-rich separate (SiO_2) were handpicked. The mineral separates were digested on a hotplate using a 3:1 HNO_3:HF mixture. To ensure complete dissolution, the WR was digested in a high pressure Parr bomb. Aliquots (5%) of the dissolved WR fraction and mineral separates were reserved for determining elemental abundances while the remainder of each solution was processed for Mg purification using methods described in ref.51. Excesses in ^{26}Mg from the decay of ^{26}Al are reported as $\mu^{26}Mg^*$, defined as the non-mass-dependent excesses in the $^{26}Mg/^{24}Mg$ ratio relative to the terrestrial (DSM3) standard in parts per million (ppm) notation. The Al/Mg ratios and Mg isotopes were measured using methods similar to those described in ref.51. We have since improved the precision of our Mg isotope analyses for samples with Mg $\geq$1 µg; specifically, two times the standard deviation on $\mu^{26}Mg^*$ for repeat runs of terrestrial synthetic and rock standards over the course of this study (i.e., 2 SD external reproducibility) was ±6 ppm52. Uncertainties in $\mu^{26}Mg^*$ are reported as either two times the standard error (2SE) from repeat runs of the sample ($n=10$) or the 2 SD external reproducibility (±6 ppm; 54), whichever is larger. The slope and intercept of the ^{26}Al–^{26}Mg internal isochron were calculated using ISOPLOT53. The Al–Mg isotopic data are reported in Supplementary Table 7.

The major and rare earth element (REE) abundances (Supplementary Table 5) were measured in the Keck Laboratory at ASU using a Thermo iCAP-Q ICPMS. A fraction of the reserved 5% aliquot of each mineral separate and WR fraction was diluted in 3% HNO_3 and run in kinetic energy discrimination mode. Internal standards are prepared in house and run along with the samples. Additionally, terrestrial synthetic and rock standards were run after every five samples to monitor the instrument performance. Based on the reproducibility of these standards, we estimate the uncertainties on the major element and REE concentrations to be ~±3% and ~±10%, respectively.

REFERENCES

1. Day, J. M. D. et al. Early formation of evolved asteroidal crust. *Nature* **457**, 179–182 (2009).
2. Hahn, T. M., Lunning, N. G., McSween, H. Y., Bodnar, R. J. & Taylor, L. A. Dacite formation on Vesta: partial melting of the eucritic crust. *Meteorit. Planet. Sci.* **52**, 1173–1196 (2017).
3. Bischoff, A. et al. Trachyandesitic volcanism in the early Solar System. *Proc. Natl Acad. Sci. USA* **111**, 12689–12692 (2014).
4. Campbell, I. H. & Taylor, S. R. No water, no granites – no oceans, no continents. *Geophys. Res. Lett.* **10**, 1061–1064 (1983).
5. Taylor, S. R. & McLennan, S. M. *Planetary Crusts: Their Composition, Origin and Evolution* (Cambridge Univ. Press, New York, 2009).
6. Seddio, S. M., Jolliff, B. L., Korotev, R. L. & Zeigler, R. A. Petrology and geochemistry of lunar granite 12032,366-19 and implications for lunar granite petrogenesis. *Am. Mineral.* **98**, 1697–1713 (2013).
7. Barrat, J. A., Bohn, M., Gillet, Ph & Yamaguchi, A. Evidence for K-rich terranes on Vesta from impact spherules. *Meteorit. Planet. Sci.* **44**, 359–374 (2009).
8. Goodrich, C. A. & Delaney, J. D. Fe/Mg-Fe/Mn relations of meteorites and primary heterogeneity of primitive achondrite parent bodies. *Geochim. Cosmochim. Acta* **64**, 149–160 (2000).
9. Benedix, G. X., McCoy, T. J., Keil, K., Bogard, D. D. & Garrison, D. H. A petrologic and isotopic study of winonaites: evidence for early partial melting, brecciation, and metamorphism. *Geochim. Cosmochim. Acta* **62**, 2535–2553 (1998).
10. Gill, J. B. *Orogenic Andesites and Plate Tectonics* (Springer-Verlag, Berlin, 1981).
11. Bédard, J. H. Trace element partitioning coefficients between silicate melts and orthopyroxene: Parameterizations of D variations. *Chem. Geol.* **244**, 263–303 (2007).
12. Bédard, J. H. Parameterization of the Fe=Mg exhange coefficient (Kd) between clinopyroxene and silicate melts. *Chem. Geol.* **274**, 169–176 (2010).
13. Putirka, K. Clinopyroxene+liquid equilibria to 100 kbar and 2450 K. *Contrib. Mineral. Petrol.* **135**, 151–163 (1999).

14. Santos, A. R. et al. Petrology of igneous clasts in Northwest Africa 7034: Implications for the petrologic diversity of the martian crust. *Geochim. Cosmochim. Acta* **157**, 56–85 (2015).
15. Smyth, J. R. & McCormick, T. C. in *Mineral Physics and Crystallography, A Handbook of Physical Constants* (ed. Ahrens, T. J.) Ch. 1 (American Geophysical Union, Washington, DC, 1995).
16. Blatt, H., Tracy, R. J. & Owens, B. E. *Petrology: Igneous, Sedimentary, and Metamorphic* (W. H. Freeman and Co., New York, 2006).
17. Clayton, R. N. & Mayeda, T. K. Oxygen isotope studies of achondrites. *Geochim. Cosmochim. Acta* **60**, 1999–2017 (1996).
18. Mittlefehldt, D. W., McCoy, T. J., Goodrich, C. A. & Kracher, A. Non-chondritic meteorites from asteroidal bodies. *Rev. Mineral.* **36**, 1–195 (1998).
19. Goodrich, C. A. et al. Origin and history of ureilitic material in the solar system: The view from asteroid 2008 TC_3 and the Almahata Sitta meteorite. *Meteorit. Planet. Sci.* **50**, 782–809 (2015).
20. Irving, A. J. et al. Ungrouped mafic achondrite Northwest Africa 7325: A reduced, iron-poor cumulate olivine gabbro from a differentiated planetary parent body. *Lunar Planet. Sci. Conf.* **44**, 2164 (2013).
21. Goodrich, C. A. et al. Petrogenesis and provenance of ungrouped achondrite Northwest Africa 7325 from petrology, trace elements, oxygen, chromium and titanium isotopes, and mid-IR spectroscopy. *Geochim. Cosmochim. Acta* **203**, 381–403 (2017).
22. Barrat, J. A. et al. Evidence from Tm anomalies for non-CI refractory lithophile element proportions in terrestrial planets and achondrites. *Geochim. Cosmochim. Acta* **176**, 1–17 (2016).
23. Irving, A. J., Bunch, T. E., Rumble, D. III & Larson, T. E. Metachondrites: recrystallized and/or residual mantle rocks from multiple, large chondritic parent bodies. *Meteorit. Soc. Meet.* **68**, 5218 (2005).
24. Agee., C. B. et al. NWA 8186 an ungrouped achondrite from the CK/CV chondrite parent body. *Meteorit. Soc. Meet.* **77**, 5385 (2014).
25. Srinivasan, P. et al. Petrologic and isotopic classifications of ungrouped achondrite NWA 8186: Implications for a CK/CV Asteroidal origin. *Lunar Planet. Sci. Conf.* **46**, 1472 (2015).
26. Ruzicka, A., Grossman, J., Bouvier, A., Herd, C. D. K. & Agee, C. B. The meteoritical bulletin, No. 102. *Meteorit. Soc. Meet.* **50**, 105 (2015).

27. Wadhwa, M. et al. Ancient relative and absolute ages for a basaltic meteorite: implications for timescales of planetesimal accretion and differentiation. *Geochim. Cosmochim. Acta* **73**, 5189–5201 (2009).

28. Brennecka, G. A. & Wadhwa, M. Uranium isotope compositions of the basaltic angrite meteorites and the chonological implications for the early Solar System. *Proc Natl Acad. Sci. USA* **109**, 9299–9303 (2012).

29. Schiller, M. et al. Early accretion of protoplanets inferred from a reduced inner solar system ^{26}Al inventory. *Earth Planet. Sci. Lett.* **420**, 45–54 (2015).

30. Jacobsen, B. et al. ^{26}Al-^{26}Mg and ^{207}Pb-^{206}Pb systematics of Allende CAIs: canonical solar initial $^{26}Al/^{27}Al$ ratio reinstated. *Earth Planet. Sci. Lett.* **272**, 353–364 (2008).

31. Bouvier, A. et al. Absolute chronology of the first solids in the Solar System. In *Workshop on the Formation of the First Solids in the Solar System* (Lunar and Planetary Institute, Kauai, 2011).

32. Connelly, J. N. et al. The absolute chronology and thermal processing of solids in the solar protoplanetary disk. *Science* **338**, 651–655 (2012).

33. Amelin, Y. The U-Pb systematics of angrite Sahara 99555. *Geochim. Cosmochim. Acta* **72**, 4874–4885 (2008).

34. Nyquist, L. E., Reese, Y., Wiesmann, H., Shih, C.-Y. & Takeda, H. Fossil ^{26}Al and ^{53}Mn in the Asuka 881394 eucrite: evidence of the earliest crust on asteroid 4 Vesta. *Earth Planet. Sci. Lett.* **214**, 11–25 (2003).

35. Dunlap, D. R., Wadhwa, M. & Romaneillo, S. R. ^{26}Al-^{26}Mg systematics in the usual ungrouped achondrite NWA 7325 and the eucrite Juvinas. *Lunar Planet. Sci. Conf.* **45**, 2186 (2014).

36. Koefoed, P. et al. U-Pb and Al-Mg systematics of the ungrouped achondrite Northwest Africa 7325. *Geochim. Cosmochim. Acta* **183**, 31–45 (2016).

37. Day, J. M. D. et al. Origin of felsic achondrites Graves Nunataks 06128 and 06129, and ultramafic brachinites and brachinite-like achondrites by partial melting of volatile-rich primitive parent bodies. *Geochim. Cosmochim. Acta* **81**, 94–128 (2012).

38. Shearer, C. K. et al. Non-basaltic asteroidal magmatism during the earliest stages of solar system evolution: A view from Antarctic achondrites Graves Nunatak 06128 and 06129. *Geochim. Cosmochim. Acta* **74**, 1172–1199 (2010).

39. Barrat, J. A. et al. Crustal differentiation in the early solar system: Clues from the unique achondrite Northwest Africa 7325 (NWA 7325). *Geochim. Cosmochim. Acta* **168**, 280–292 (2015).
40. 4Gomez-Tuena, A., Straub, S. M. & Zellmer, G. F. An introduction to orogenic andesites and crustal growth. *Geol. Soc. Lond. Spec. Publ.* **385**, 1–13 (2013).
41. Rudnick, R. L. & Gao, S. Composition of the continental crust. *Treat. Geochem.* **3**, 1–64 (2003).
42. McCoy, T. J., Dickinson, T. L. & Lofgren, G. E. Partial melting of the Indarch (EH4) meteorite: a textural, chemical, and phase relations view of melting and melt migration. *Meteor. Planet. Sci.* **34**, 735–746 (1999).
43. Feldstein, S. N., Jones, R. H. & Papike, J. J. Disequilibrium partial melting experiments on the Leedy L6 chondrite: textural controls on melting processes. *Meteor. Planet. Sci.* **36**, 1421–1441 (2001).
44. Gardner-Vandy, K. G., Lauretta, D. S. & McCoy, T. J. A petrologic, thermodynamic and experimental study of brachinites: partial melt residues of an R chondrite-like precursor. *Geochim. Cosmochim. Acta* **122**, 36–57 (2013).
45. Usui, T., Jones, J. H. & Mittlefehldt, D. W. A partial melting study of an ordinary (H) chondrite composition with application to the unique achondrite Graves Nunataks 06128 and 06129. *Meteor. Planet. Sci.* **50**, 759–781 (2015).
46. Vander Kaaden, K. E. & McCubbin, F. M. The origin of boninites on Mercury: an experimental study of the northern volcanic plains lavas. *Geochim. Cosmochim. Acta* **173**, 246–263 (2016).
47. McCubbin, F. M., Riner, M. A., Vander Kaaden, K. E. & Burkemper, L. K. Is mercury a volatile-rich planet? *Geophys. Res. Lett.* **39**, L09202 (2012).
48. McCubbin, F. M. et al. A low O/Si ratio on the surface of mercury: evidence for silicon smelting? *J. Geophys. Res.* **122**, 2053–2076 (2017).
49. Schneider, C. A., Rasband, W. S. & Eliceiri, K. W. NIH image to ImageJ: 25 years of image analysis. *Nat. Methods* **9**, 671–675 (2012).
50. Sharp, Z. D. A laser-based microanalytical method for the *in situ* determination of oxygen isotope ratios of silicates and oxides. *Geochim. Cosmochim. Acta* **54**, 1353–1357 (1990).

51. Bouvier, A., Spivak-Birndorf, L. J., Brennecka, G. A. & Wadhwa, M. New constraints on early solar system chronology from Al-Mg and U-Pb isotope systematics in the unique basaltic achondrite Northwest Africa 2976. *Geochim. Cosmochim. Acta* **75**, 5310–5323 (2011).
52. Dunlap, D. R., Rai, V. K. & Wadhwa, M. High precision ^{26}Al- ^{26}Mg Systematics of A New Eucrite Northwest Africa 10919 and the Brachinites Northwest Africa 4882 and Brachina. *Lunar Planet. Sci. Conf.* **48**, 2981 (2017).
53. Ludwig, K. R. User's manual for Isoplot 3.75: a geochronological toolkit for Microsoft. *Excel Berkeley Geochron. Cent. Spec. Pub.* **5**, 1–75 (2012).
54. Kelemen, P. B., Hanghj, K. & Greene, A. R. One view of the geochemistry of subduction-related magmatic arcs, with an emphasis on primitive andesite and lower crust. *Treat. Geochem.* **3**, 593–659 (2004).
55. Palme, H., Lodders, K. & Jones, A. Solar system abundances of the elements. *Treat. Geochem.* **2**, 15–36 (2014).

INDEX

N

O

P

Q

R

S

V

W

X